T0201887

SERIES ON SEMICONDUCTOR SCIENCE AND TECHNOLOGY

Series Editors

SERIES ON SEMICONDUCTOR SCIENCE AND TECHNOLOGY

Microcavities

Second Edition

Alexey V. Kavokin

Chair of Nanophysics and Photonics, Physics and Astronomy,
University of Southampton, UK,
Director of Research at the National Research Council, Italy,
and Principal Investigator at the Russian Quantum Center, Moscow, Russia

Jeremy J. Baumberg

Fellow of the Royal Society, Director of NanoPhotonics Centre, Cavendish
Laboratory, University of Cambridge, UK

Guillaume Malpuech

Senior CNRS Researcher and head of the Photon department of the Institut Pascal,
joint Laboratory of Universite Clermont Auvergne and CNRS, France

Fabrice P. Laussy

Principal Lecturer, Director of Studies for Physics,
University of Wolverhampton, UK

OXFORD
UNIVERSITY PRESS

OXFORD

UNIVERSITY PRESS

Great Clarendon Street, Oxford, OX2 6DP,
United Kingdom

Oxford University Press is a department of the University of Oxford.
It furthers the University's objective of excellence in research, scholarship,
and education by publishing worldwide. Oxford is a registered trade mark of
Oxford University Press in the UK and in certain other countries

First Edition published in 2007
First Edition published in paperback 2011
Second Edition published in 2017

Impression: 1

Published in the United States of America by Oxford University Press
198 Madison Avenue, New York, NY 10016, United States of America

British Library Cataloguing in Publication Data
Data available

Library of Congress Control Number: 2017938216

ISBN 978-0-19-878299-5

DOI 10.1093/oso/9780198782995.001

Printed and bound by
CPI Group (UK) Ltd, Croydon, CR0 4YY

PREFACE TO THE 2ND EDITION

Hundreds of research papers reporting several important discoveries have appeared since the publication of the revised edition of our book in 2011. To name just a few: Polariton half-vortices and half-solitons, additional features of superfluidity, polariton lasing at room temperature with electrical injection, concepts of Bosonic cascade lasers, Dirac bosons, non-equilibrium spin-Meissner effect... While the foundations of the field remain the same and keep providing a strong hold onto various arenas of fundamental physics, the field of Polaritonics is continuing its fast-paced expansion and is winning new territories. Clearly, an updated textbook keeping to date with the most important developments is called for. Just as it occured between the first publication of this book (in 2008) and its revision for the paperback release (2011), various reviews and edited volumes devoted to polaritons and cold atoms have appeared.[1] These texts are, however, addressed to the experts and at the edge of the research effort. They offer little help to Master and PhD students, as well as experienced researchers from other fields who wish to contribute to a growing field of research with multi-disciplinary ramifications. These are the main audience for our book and we hope that this new edition will succeed in providing a gentle, but comprehensive overview of a fascinating field.

This second edition comes with the following main additions on the previous versions: all the existing chapters have been thoroughly updated to include the most important discoveries, sometimes consisting in the emergence of new fields altogether. There is now, for instance, a complete and self-contained coverage of superconducting qubits and opto-mechanical resonators. Three new chapters have appeared to provide the most exciting progresses at the level they require. These are:

- Chapter 10: "Quantum fluids", on the topology of polariton condensates, including vortices, half-vortices, solitons, half-solitons, monopoles, skyrmions, etc., and shaping of the condensate by traps, channels, rings, lattices, etc. Other exotic physics such as polariton graphene is discussed here.
- Chapter 11: "Quantum polaritonics", on the quest of harnessing the single-polariton quantum nonlinearity, with polariton blockade, N-photon emitters, or undertaking quantum information processing, with linear polaritonic quantum computing, polariton simulators, polariton cebits and qubits, etc.

---

[1] Important recent reviews and volumes devoted to the central topics of this book are:

* D. Sanvitto and V. Timofeev (Eds.) *Exciton Polaritons in Microcavities*, Springer (2012).
* I. Carusotto and C. Ciuti *Quantum fluids of light*, Review of Modern Physics, **85**, 299 (2013).
* A. Bramati and M. Modugno (Eds.) *Physics of Quantum Fluids*, Springer (2013).
* T. Byrnes, N. Y. Kim and Y. Yamamoto *Exciton–polariton condensates*, Nature Physics, **10**, 803 (2014).
* D. Sanvitto and S. Kéna Cohen *The road towards polaritonic devices*, Nature Materials, **15**, 1061 (2016).

- Chapter 12: "Polariton devices", on the emergence of technological components powered by polaritons, such as polariton transistors and other polariton logical elements, bosonic lasers and bosonic cascades, polariton switches, polariton neurons, etc.

We have also enlarged the historical aspect, with more biographical sketches of the key actors of this scientific journey. Many figures have been updated, redesigned or published in color to enjoy a modern feel and gain in clarity. Finally, the book now comes with an index and more solutions for the exercises.

February 2017,
The authors.

PREFACE TO THE REVISED PAPERBACK EDITION

Since the publication of the first edition of this book in 2007, we have received many comments from our readers, which were of great help when we worked on this paperback version. We thank all those who expressed their positive opinion on the book and encouraged us to improve and update it. We thank also those readers who pointed out misprints and more serious errors in the previous version. We apologise for these imperfections! We have tried to correct all the mistakes we could find in this version.

We are happy that we did not make a mistake in one important aspect: microcavities attract more and more the attention of physicists, material scientists and engineers. This field is enlarging and new findings are extremely interesting. Between 2006 and 2010, the microcavity research was developing at a high pace. Preparing this version of the book, we felt compelled not only to correct the various mistakes found in the previous version, but also to address the most important (in our opinion) effects discovered during these 4 years. In particular, we decided to address the recent works on polariton lasers and diodes, superfluidity of exciton-polaritons, quantised vortices in polariton condensates and strong-coupling with microcavities containing quantum dots. We have incorporated new sections or paragraphs devoted to these aspects in Chapters 5, 6, 8 and 9. We realise that it is virtually impossible to keep the book updated as nowadays, new important papers on the physics of microcavities appear every week.[2] Instead, we see the goal of this book in giving a theoretical background to the complex microcavity physics and in illustrating the basic theoretical concepts by examples from recent experimental works. All of us keep working in this research field, and we keep collaborating on several subjects related to microcavities. We invite the readers of this book to visit with us the wonderland of microcavities. Please do not hesitate to contact us if you have questions, if you noticed mistakes, if you have critical remarks, suggestions, etc.

September 2010,
The authors.

---

[2]Those who wish to know more about the recent developments may find it beneficial to consult:

* H. Deng, H. Haug and Y. Yamamoto, *Exciton-polariton Bose–Einstein condensation*, Review of Modern Physics, **82**, 1489 (2010).
* D. Snoke and P. Littlewood, *Polariton condensates*, Physics Today, **63**, 42 (2010).
* I. Carusotto and D. Sanvitto, *Non-equilibrium Bose–Einstein condensation in a dissipative environment*, in "Quantum Gases: Finite Temperature and Non-Equilibrium Dynamics", ed. by N. Proukakis, S. A. Gardiner, M. J. Davis, M. H. Szymanska and N. Nygaard, Imperial College Press, London (2013).
* I. A. Shelykh, A. V. Kavokin, Y. G. Rubo, T. C. H. Liew and G. Malpuech, *Polariton polarisation sensitive phenomena in planar semiconductor microcavities*, Semiconductor Science and Technology, **25**, 013001 (2010).
* J. Keeling, F. M. Marchetti, M. H. Szymanska and P. B. Littlewood, *Collective coherence in planar semiconductor microcavities*, Semiconductor Science and Technology, **22**(5), R1 (2007).
* E. del Valle, *Microcavity Quantum Electrodynamics*, VDM Verlag, Saarbrücken, Germany (2010).

PREFACE TO THE FIRST EDITION

Effects originating from light–matter coupling have stimulated the development of optics for the last three centuries. Nowadays, the limits of classical optics can be reached in a number of solid-state systems and quantum optics has become an important tool for understanding and interpreting modern optical experiments. Rapid progress of crystal-growth technology in the twentieth century allows the realisation of crystal microstructures that have unusual and extremely interesting optical properties. This book addresses the large variety of optical phenomena taking place in confined solid-state structures: *microcavities*. Microcavities serve as building blocks for many opto-electronic devices, such as light-emitting diodes and lasers. At the edge of research, the microcavity represents a unique laboratory for quantum optics and photonics. The central object of studies in this laboratory is the *exciton-polariton*: a half-light, half-matter quasiparticle exhibiting very specific properties and playing a key role in a number of beautiful effects including parametric scattering, Bose–Einstein condensation, superfluidity, superradiance, entanglement, etc. At present, hundreds of research groups throughout the world work on fabrication, optical spectroscopy, theory and applications of microcavities. The progress in this interdisciplinary field at the interface between optics and solid-state physics is extremely rapid. We expect the appearance of a new generation of opto-electronic devices based on microcavities in the 2010s.

Both rich fundamental physics of microcavities and their intriguing potential applications are addressed in this book, orientated to undergraduate and postgraduate students as well as to physicists and engineers. We describe the essential steps of development of the physics of microcavities in their chronological order. We show how different types of structures combining optical and electronic confinement have come into play and were used to realise first weak and later strong light–matter coupling regimes. We discuss photonic crystals, microspheres, pillars and other types of artificial optical cavities with embedded semiconductor quantum wells, wires and dots. We present the most striking experimental findings of the recent two decades in the optics of semiconductor quantum structures.

The first chapter of this book contains an overview of microcavities. We present the variety of semiconductor, metallic and dielectric structures used to make microcavities of different dimensions, and briefly present a few characteristic optical effects observed in microcavities.

The next two chapters (2 and 3) are devoted to the fundamental principles of optics essential for understanding optical phenomena in microcavities. We provide overviews of both classical and quantum theory of light, discuss the coherence of light, its polarisation, statistics of photons and other quantum characteristics. The reader will find here the basic principles of the transfer matrix technique allowing for easy understanding of linear optical properties of multilayer structures as well as the basics of the second quantisation method. We consider planar, cylindrical and spherical optical cavities, introduce the whispering-gallery modes and Mie resonances.

In Chapters 4 and 5 we give the theoretical background for the most important light–matter coupling effects in microcavities considered from the point of view of classical (Chapter 4) and quantum (Chapter 5) optics. We formulate the semiclassical non-local dielectric response theory and study the dispersion of exciton-polaritons in microcavities. As an important toy model we consider a single exciton state coupled to a single light mode and study many variations of it. We describe important nonlinear effects known in atomic cavities (such as the Mollow triplet).

In Chapter 6 we discuss the physics of *weak coupling*, when interaction of the light field with the exciton acts as a perturbation on its state and energy. We discuss the Purcell effect that symbolises this regime and lasing as its most important application. We also describe nonlinear effects such as bistability.

Chapter 7 addresses the resonant optical effects in the *strong-coupling* regime. We overview the most spectacular experimental discoveries in this area and present the quasimode model of parametric amplification of light. We also discuss the quantum properties of optical parametric oscillators based on microcavities.

Chapters 8 and 9 discuss the future of microcavities. Chapter 8 is devoted to the Bose–Einstein condensation of exciton-polaritons and polariton lasing. At the time of writing, polariton-lasers remain more a theoretical concept than commercial devices, but we believe that in a few years they will become a reality. Thus, for the first time, Bose condensation would be observed at room temperature and used for the creation of a new generation of opto-electronic devices. The path toward this breathtaking perspective and the most serious obstacles in the way that are not yet overcome, are tackled in this chapter.

The subject of Chapter 9 is "spin-optronics": a new subfield of solid-state optics that emerged very recently due to the discoveries made in microcavities and other quantum confined semiconductor structures. How to manipulate the polarisation of light on a nanosecond and micrometre scale? What would be the polarisation properties of polariton-lasers and which mechanisms govern spin-relaxation of exciton-polaritons? These questions are treated in this chapter.

The glossary is addressed to a non-specialist who is searching for the qualitative understanding of the physics of microcavities or to any reader who has no time to go through the entire book, but needs a simple and concise answer to one of the specific questions related to microcavities. In the glossary, a number of important relevant issues are treated without any equations on a simple and accessible level for the general reader. We pay special attention to explanation of terms frequently used in this field of physics, for example, "exciton-polariton", "Rabi splitting", "strong coupling", "Bragg mirror", "VCSEL", "photonic crystal", etc.

The book is intended as a working manual for advanced or graduate students and new researchers in the field. It is written to a high standard of scientific and mathematical accuracy, but to allow an agreeable reading through the essential points unhampered by details, many sophistications or difficulties, as well as side issues or extensions, have been relegated to footnotes. These would be most profitably considered in a second reading.

Exercises are sprinkled throughout the text and are an important part of it. They should be read as a minimum, for otherwise the notions they introduce will be missing for later development of the regular text. Starred exercises are straightforward or systematic, those doubled starred are conceptually challenging or require involved computations, those tripled starred are difficult and almost qualify as research problems. We use the international system of units, while in numerical examples the energies will be given in electron-volts and the distances will be given in micrometres or nanometres. An extensive bibliography used throughout the text appears at the end of the text in Harvard format (identified by first author and date of publication).

Microcavities represent a young and rapidly developing field of physics. Our book covers the state-of-the-art in this field in the first half of 2006 observed from the prism of personal experience of four authors who have actively worked in the physics of microcavities for a large part of their scientific lives. We wanted to give a personal touch to this book and we do not claim to be objective, which at this stage of the field is very difficult. We shall be very grateful for any feedback, comments and critical remarks from our readers!

November 2006,
The authors.

ACKNOWLEDGEMENTS

This book owes much to our collaboration with active researchers in the field. It is our pleasure to express our gratitude to Yuri Rubo, Ivan Shelykh, Kirill Kavokin, Daniele Sanvitto, Pavlos Lagoudakis, Mikhail Glazov, Maurice Skolnick, David Whittaker, Pavlos Savvidis, David Lidzey, Elena del Valle, Carlos Tejedor, Arne Laucht, Norman Hauke, Jonathan Finley, Dario Ballarini, Lorenzo Dominici, Dmitry Solnyshkov, Hugo Flayac, Anton Nalitov, Carlos Sánchez Muñoz, David Colas, Juan-Camilo López Carreño and Amir Rahmani.

Alexey Kavokin
To Sofia Kavokina,

Jeremy Baumberg
To Melissa Murray,

Guillaume Malpuech
To Anne Tournadre,

Fabrice Laussy
To my father, *Raymond Laussy.*
(1953–2006)

CONTENTS

GLOSSARY

This glossary provides a succinct definition of recurrent terms and concepts in microcavity physics, that might otherwise not be found elsewhere in the book.

A

Absorption   of light in a crystal is a measurement that can be made by calorimetres or deduced from the reflectivity and the transmission spectra. It characterises the efficiency of the light–matter coupling.

Acoustic phonon   is a phonon that can be excited by a photon. Interaction of exciton-polaritons with acoustic phonons is one of the most important mechanisms of the polariton energy relaxation in microcavities.

Active layer   In semiconductor lasers, this is the layer of a semiconductor material, e.g., a quantum well, where the inversion of electronic population between the energy levels in the valence and conduction bands is achieved. The stimulated emission of light dominates its absorption in this layer at some frequency.

Anticrossing   is a signature of the strong-coupling regime in a system of two coupled oscillators. Due to the interactions between the oscillators, the eigenfrequencies of the system remain splitted (and this splitting is in fact maximum) when the normal frequencies of the two individual oscillators coincide.

Antinode   (of the light field in a cavity) is the maximum of intensity of the electric field of a standing light mode. Typically an active element (quantum well, wire or dot) is sought to be placed at the antinode of the field as this provides the largest exciton–light coupling strength in the cavity.

B

Bandgap   is the region of forbidden states in the band diagram of a semiconductor. The *bandgap energy* is the energy difference between the conduction and the valence band of a semiconductor, that is, the energy required or released to bring one electron from one to the other.

BCS   is a famous theoretical model explaining low-temperature superconductivity. The term comes from the abbreviation of the surnames of its three authors: Bardeen, Cooper, and Schrieffer. BCS assumes the formation of so-called *Cooper pairs* of electrons, which have bosonic properties in a metal. The Cooper pair's size strongly exceeds the average distance between pairs in contrast with the situation in diluted Bose gases (like exciton gases) where the size of a boson is, in general, much less than the distance between bosons.

Bloch theorem   defines the expression of the electron wavefunction in a periodic-crystal potential and introduces the quasi-wavevector as its quantum number. It also applies to periodic photonic structures (where it gives rise to photonic crystals).

Bloch equations   describe a driven two-level quantum system and can be generalised to describe optical properties of semiconductors (becoming the Maxwell–Bloch equations).

Boltzmann equations   are differential equations that describe the kinetics of the occupation numbers of the eigenstates of an infinite system. They operate with ensemble-averaged populations at a classical level and as such do not describe quantum correlations. Boltzmann equations are a powerful tool for the description of relaxation processes in interacting gases, like an exciton-polariton gas.

Born approximation   consists in decoupling the dynamics of coupled quantum systems such as, e.g., excitons and phonons. This allows typically to factorise the total density matrix into products of density matrices for each subsystem.

Bose–Einstein statistics   describes the energy distribution of quantum particles with integer spin, called *bosons* (e.g., photons). It reduces to the Maxwell–Boltzmann statistics of an ideal classical gas in the high-temperature and high-energy limit. At low temperatures, it predicts the accumulation of bosons in the lowest energy state (Bose–Einstein condensation).

Bosonic cascade   is an ensemble of N localised bosonic condensates that have different energies but overlap in real space. The overlap facilitates transition of bosons between the different condensates by emission or absorption of light, lattice vibration, etc. When energies are equally-spaced, for instance by confining the BEC in an harmonic potential, this realises a bosonic counterpart of the quantum cascade laser.

Bottleneck   The (phonon) bottleneck effect is a slowing down of the rapid polariton relaxation along the lower dispersion branch that is rapid in the exciton-like part due to scattering with acoustic phonons, but then becomes slower in the vicinity of the anticrossing point of the exciton and photon modes because of kinetic blocking of polariton relaxation. The main obstacle for polariton relaxation in the bottleneck region comes from the lack of acoustic phonons that are able to scatter with polaritons of very low effective mass. The bottleneck effect prevents polaritons from relaxing down to their ground state at $\mathbf{k} = \mathbf{0}$, which represents a major problem for the realisation of polariton lasers. The bottleneck effect exists also for bulk or quantum well exciton-polaritons.

Bose–Einstein condensation   (BEC), also simply "Bose condensation", is a phase transition for bosons leading to the formation of a coherent multiparticle quantum state characterised by a wavefunction. The Bose condensate occupies the lowest energy level of the system that coincides with the chemical potential. Strictly speaking, BEC can only take place in infinite systems with dimensionality higher than 2. In finite size and/or low-dimensional systems one can speak of quasi-BEC and Kosterlitz–Thouless phase transition.

Bra   (see *ket* first) In quantum mechanics, the dual state $\langle \psi |$ of a ket $| \psi \rangle$. The product of a bra $\langle \psi |$ with any ket $| \phi \rangle$ gives the *braket* $\langle \psi | \phi \rangle$ that is the inner product of their associated Hilbert space, whence the name.

Bragg mirror   is a mirror formed by alternating layers of semiconductors with different refractive index. Each layer boundary partially reflects the incoming wave and through the effect of constructive interferences, very high reflectivity is achieved. To obtain the strongest interference, the thicknesses of these layers must be chosen equal to a quarter of the wavelength of light in the corresponding material at some frequency referred to as the *Bragg frequency*. The reflection spectrum of a Bragg mirror exhibits a plateau of very high reflectivity centred on the Bragg frequency. This plateau is referred to as a stopband, which represents a one-dimensional photonic bandgap.

Broadening   refers to the fact that a spectral line is never exactly a delta function but always has a width that is due to various mechanisms that blurs the energy definition at which the transition is expected to take place. See "homogeneous" and "inhomogeneous broadenings".

Bright mode   also known as a *superradiant mode*, is a collective state of a few oscillators (atoms, excitons, polaritons, etc.), which has a higher radiative decay rate than any of these oscillators taken in isolation.

Bulk microcavities   are microcavities for which the cavity consists of a bulk semiconductor without any embedded quantum objects, like quantum wells, wires or dots.

C

Coherence is one of the basic characteristics of light. According to the Glauber classification, different orders of the coherence can be defined. The first-order coherence is dependent on temporal correlations of the amplitude of the light field, the second-order coherence is dependent on the intensity correlations, etc. Fully coherent light is fully correlated to all orders. The coherence time and coherence length of light are linked to the first-order coherence.The term is also wildly applied to other concepts, such as quantum coherence or condensate coherence. Often these other meanings themselves split further into more definitions for unrelated concepts.

Coherent state in classical optics is a state of light characterised by a fixed phase. In quantum optics, it is a state that minimises the position–momentum uncertainty in an harmonic potential, spreading equally these uncertainties (otherwise it becomes a *squeezed coherent state*, or "squeezed state" for short. It has a Poisson distribution as its diagonal elements in a density matrix representation (it is a pure state). It contrasts with the thermal state where the vacuum is found with the highest probability.

Collapse The "process" that a quantum state undergoes upon measurement to become the eigenstate $|\omega_{i_0}\rangle$ associated to the eigenvalue ω_{i_0} measured in the course of the experiment. This postulate has been made to match the experimental fact that (immediately) repeated measures of the same observable on a quantum system always return the same result. This assumption is one of the pillars of the Copenhagen interpretation and is also known as the *reduction* of the wavefunction or *quantum jump*. The exact process responsible for it is as yet debated but is described by the theory of decoherence.

Copenhagen interpretation One early *interpretation* of quantum mechanics issued by the joint efforts of Bohr and Heisenberg c. 1927, while collaborating in the capital of Denmark (whence the name).

Coupled cavities are cavities that have a common mirror, usually.

D

Dark exciton also referred to as an *optically inactive exciton* is an exciton that cannot be created by resonant absorption of a photon. Examples include: indirect in real or reciprocal space excitons, excitons with a spin projection on a given axis equal to ± 2, excitons having wavevectors exceeding the wavevector of light in vacuum at their resonance frequency. 384. By analogy, a so-called *bright exciton* is directly coupled to light.

Dark soliton is a solitary dip in a liquid that maintains its shape when propagating with a constant speed. Dark solitons are formed in nonlinear media either due to attractive interaction of quasi-particles with negative effective masses or due to repulsion of quasi-particles with positive effective masses. See also Solitons.

Density matrix is the extension of the concept of the wavefunction into the statistical realm. A quantum state which is described by the wavefunction $|\psi\rangle$ is also described by the density matrix $|\psi\rangle\langle\psi|$, but not all states that are described by a density matrix can also be described by a wavefunction. A state that admits a wavefunction description is called a *pure state*. A state that does not is called a *mixed state*. The latter lacks a full knowledge of the system that is supplemented by classical probabilities.

Detuning refers to the difference in energy between two coupled modes. "Changing the detuning" means bringing the two modes in and out of resonance (equal energy). For instance, the detuning between photon and exciton modes in a microcavity is the difference between the eigenfrequencies of the bare cavity mode and the exciton resonance at zero in-plane wavevector. One speaks of *positive detuning* if the cavity mode is above the exciton resonance and of *negative detuning* if it is below.

Dispersion in optics and in quantum mechanics describes the frequency (energy) dependence on the wavevector. If a wave equation is dispersive, the profile of a wavepacket usually get distorted as it propagates (if combinations of dispersion and nonlinearity concur in maintaining its shape, a *soliton* is formed).

E

Elastic circle is a circle in a two-dimensional reciprocal space centred at $k = 0$ which radius is given by the absolute value of the wavevector of a particle. All states on the elastic circle have the same kinetic energy and are therefore possible final states for the elastic (Rayleigh) scattering of that particle.

Etching is a process to remove unwanted parts of a semiconductor device during its fabrication. Many techniques exist, e.g., *wet etching* using acid chemicals and *dry etching* vaporising the material, *selective* or *anisotropic* etchings allow to shape the structure by etching different parts at different rates.

Exchange interaction is a quantum-mechanical interaction mechanism based on the indistinguishability of quantum particles. For example, if a pair of electrons from quantum states i and j scatters to quantum states m and k, it is impossible to say which one went to the state m and which one to the state k, so that the two scenari (i going to m, j going to k and vice-versa) must be taken into account. If $i = m$ and $j = k$, the first scenario corresponds to the direct interaction while the second one corresponds to the exchange interaction.

Exciton is a Coulomb correlated electron–hole pair in a semiconductor. One can distinguish between Frenkel and Wannier–Mott excitons, the former having much larger binding energies and much smaller Bohr radius than the latter.

Exciton-polariton is a quasiparticle formed by a photon propagating in the crystal and an exciton resonantly excited by light. Exciton-polaritons are true eigenstates of light in crystals in the vicinity of the resonant frequencies of excitonic transitions.

F

Fabry–Pérot resonator is a kind of cavity formed by a dielectric layer sandwiched between two mirrors. Its eigenmodes are standing waves whose wavelength is related to the size of the resonator. If the mirrors are ideal, the integer number of half-wavelengths of an eigenmode should be equal to the thickness of the cavity.

Faraday rotation is a rotation of the polarisation plane of linearly polarised light passing through a media subject to a magnetic field parallel to the light propagation direction. Unlike natural optical activity, the Faraday effect can be accumulated in optical resonators and microcavities due to the multiple round-trips of light.

Fermi–Dirac statistics applies to particles with half-integer spin, called fermions (e.g., electrons and holes in semiconductors). It is based on the Pauli exclusion principle, and introduces a so-called *Fermi energy*, below which all energy levels are occupied and above which all energy levels are empty at zero temperature. At high temperatures, it reduces to the Maxwell–Boltzmann ditribution.

Fock state also referred to as a *number state*, is a quantum state characterised by a fixed number of photons. It has no classical counterpart and is highly sought for quantum information processing.

H

Half soliton in analogy with a half-vortex (see below), this is a topological object in a two-component fluid, with a soliton in one component only co-existing with a homogeneous density distribution in the second component.

Half-vortex is a quantum topological object in a two-component fluid. In contrast to conventional quantised vortices where the phase changes by an integer number of 2π when going around the core, in half-vortices, the overall phase changes by π only. The continuity of a two-component order parameter is ensured by simultaneous rotation of the linear polarisation plane defined by the phase difference between two components of the condensate also by π. A half-vortex is equivalent to a full quantum vortex for one component of the fluid coexisting with no vortex for its other component.

Hanbury Brown–Twiss setup is a photon-counting optical setup that allows one to measure the intensity–intensity correlations in a light beam and extract from them the second-order coherence of light $g^{(2)}$.

Hermitian operator In mathematics, an operator Ω that is self-adjoint, i.e., such that $\Omega^\dagger = \Omega$. As a consequence its eigenvalues are real. Such an operator is typically used to define an *observable*. See the second postulate of quantum mechanics.

Heterostructure The superposition of several thin layers of different (*hetero*) types of semiconductors that together form a *structure* whose bandgap varies with position. A *junction* between two semiconductors is the simplest heterostructure. In the celebrated *double heterostructure*, two semiconductors sandwich a lower-bandgap semiconductor so as to create a potential trap for both electrons and holes. Such a region is the core for semiconductor lasing.

Hilbert space In quantum mechanics, mainly used as a synonym for "the set of quantum states" for a given system. In mathematics, a separable complete vector space that is the foundation for the mathematical formulation of the theory. See the first postulate of quantum mechanics.

Homogeneous broadening is the broadening of a transition due to the lifetime of the particle. It has the shape of a Gaussian.

I

Inhomogeneous broadening is the broadening of a transition due to the potential disorder that randomly shifts the transition energies up or down in different regions of the space. It typically has the shape of a Lorentzian.

J

Jaynes–Cummings model describes the coupling of a single atom with the quantised optical field. It is a rare instance of a fully-integrable fully-quantum Hamiltonian.

Jones vector is a two-component complex vector that describes the polarisation of light. Its components correspond to the amplitudes of two orthogonal linear polarisations.

K

Ket In quantum mechanics, a vector noted $|\psi\rangle$ (by Dirac) part of a Hilbert space that describes the state of a quantum system. See also "Bra".

Kosterlitz–Thouless phase transition is a transition towards a superfluid phase in two-dimensional bosonic systems. It was described for the first time by J.M. Kosterlitz and D.J. Thouless in 1973. In infinite two-dimensional systems, Bose condensation is impossible, while a superfluid can be formed. A superfluid is a collective bosonic state, in which the particles can move throughout space along a phase-coherent, dissipationless path. In infinite-size microcavities, polaritons may undergo Kosterlitz–Thouless-like transitions and form a superfluid if a critical condition linking the concentration of polaritons with temperature is fulfilled. However, the Kosterlitz–Thouless theory cannot be directly applied to exciton-polaritons in microcavities as it ignores the two-component nature of a polariton superfluid coming from the specific spin structure of the exciton-polaritons.

L

Lamb shift is the shift of the emission spectrum due to reabsorption of light.

Landau levels are the quantisation levels for the electron's orbital motion in a magnetic field.

Laser is historically the acronym for *light amplification by stimulated emission of radiation*, but is now a generic term to refer to a device emitting a coherent output with some or all the features of the original laser, even though its specifics can differ (case of the *polariton-laser*, for instance, or the *atom laser* that has nothing to do with light). In its original acceptance the laser generates light (its predecessor the *maser* generates microwaves) from stimulated emission of photons with an inverted population of emitters (atoms, excitons...) with oscillations of the radiation. The oscillation—or positive feedback—is provided by the cavity.

Leaky modes are escape channels in a structure designed to confine. In a planar cavity, they are the light modes with such frequencies and wavevectors that the Bragg mirrors do not confine them in the cavity. In other words, the leaky modes propagate within the transparency ranges of the Bragg mirrors. Exciton-polaritons scattered to the leaky modes easily escape from the cavity.

Light cone is a cone in a 3D space with two axes corresponding to the k_x and k_y projections of the wavevector of light, and the third, perpendicular, axis being the frequency of light ω. The cone is defined by $\sqrt{k_x^2 + k_y^2} = c$ with c the speed of light. In planar structures, optical states situated inside the light cone can be directly probed by light incident from outside. States outside the light cone are frequently referred to as "wave-guiding states"; they cannot be directly accessed from outside and prisms or dielectric gratings are used to populate them.

Liouville equation also known as *Liouville–von Neumann equation* is a linear differential equation that describes the dynamics of a density matrix. It is the counterpart of the Schrödinger equation for a pure state (wavefunction).

Locality is a property of a response function: a local response at a given point in space depends only on the argument of the function at the same point in space.

Longitudinal–transverse splitting, abbreviated as TE–TM splitting, is the splitting between optical modes that have their polarisation vector parallel and perpendicular to the wavevector, respectively. It also applies to exciton-polaritons.

M

Magic angle is the incidence angle of light that allows one to excite the polariton state close to the inflection point of the lowest polariton branch so that the energy and wavevector conservation conditions are fulfilled for the polariton–polariton scattering from this state into the ground state ($\mathbf{k} = \mathbf{0}$) and some higher-energy state belonging to the lower-polariton branch (called "idler"). The resonant polariton–polariton scattering plays a central role in microcavity-based optical parametric amplifiers and oscillators. Excitation at the magic angle allows one to populate quasidirectly the polariton ground state, thus transferring the coherence of the exciting laser pulse to light emitted by the cavity normally to its surface. In typical GaAs-based cavities the magic angle varies between 15 and 20 degrees at detunings close to zero.

Markov approximation consists in the assumption that the time evolution of the quantum state of a system at time t depends on its state and on the external conditions at the same time t only and as such, does not have any memory of its previous dynamics.

Master equation is an equation of motion for a density matrix. It therefore describes the dynamics of the probabilities of the occupation numbers of a quantum system.

Maxwell-Boltzmann distribution describes the probability to find a particle with a given velocity in an ideal classical gas.

Metallic reflection is the reflection of light by materials that have a nonzero imaginary part in their refractive index.

Microsphere, microdisk are semiconductor or dielectric spheres (disks) with a radius comparable to the wavelength of visible light in the media.

Microwires are semiconductor or metallic cylinders that have a diameter in the range of 0.1–10 mm. They can be considered as microcavities, since they usually confine well the visible light due to the contrast of dielectric constants in the wire material and in the surrounding air. The quality factor of whispering gallery modes of light in microwires may be as high as 50 000, while the finesse of these modes is much lower. There have been reports on polariton lasing in ZnO microwires.

Motional narrowing is a quantum effect that consists in the narrowing of a distribution function of a quantum particle propagating in a disordered medium due to averaging of the disorder potential on the size of the wavefunction of a particle. In other words, quantum particles that are never localised at a given point of the space, but always occupy some nonzero volume, have a potential energy that is the average of the potential within this volume. This is why, in a random fluctuation potential, the energy distribution function of a quantum particle is always narrower than the potential distribution function.

N

Nanorods are metal or semiconductor cylinders with a radius ranging from 1 to 100 nm. They provide a strong confinement of carriers and may be used as light emitters with electrically tuneable frequencies (e.g., ZnO nanorods). In contrast to nanowires, nanorods are usually free standing objects, attached to a substrate but not embedded in a crystal matrix. For optical confinement the cylinders of a larger radius termed microwires are better adapted.

Nanowires are one-dimensional metallic or semiconductor objects providing the electronic confinement in two dimensions. They are also referred to as quantum wires. The radius of a nanowire is typically of the order of 1–10 nm.

O

Order parameter of a phase transition is a characteristic of the system that is zero above the critical temperature of the transition and nonzero below. In the case of the Bose–Einstein condensation, the wavefunction of the condensate is an order parameter. Within the second quantisation formalism, the expectation value of the boson creation (annihilation) operator in the condensate plays the same role. From the point of view of experimental observation of the superfluid phase transition in the system of spin-degenerate exciton-polaritons, the spontaneous linear polarisation of the polariton condensate provides an order parameter.

Optical parametric amplifier (OPA) is a process of resonant scattering of two particles (like photons or polaritons) of frequency ω_0 into two particles of frequency $\omega_0 + \omega_1$ and $\omega_0 - \omega_1$ that are called *idler* and *signal*, respectively. In terms of classical optics this is a nonlinear process governed by a χ_3 susceptibility. If a nonlinear media generating the parametric amplification is placed in a resonator, the corresponding device can be referred to as an *optical parametric oscillator* (OPO). In microcavities such a parametric amplification process is extremely efficient if one pumps at the *magic angle*. In this case, the scattering of two pumped polaritons into a ground state (signal) and an excited state (idler) is resonant (conserves both energy and wavevector). The driving force of the scattering is the Coulomb interaction between polaritons. The parametric amplification can be stimulated by a probe pulse that seeds the ground-state, injecting a polariton population larger than one. The process can also be strong enough to be self-stimulated.

Optical phonon is a crystal-lattice vibration mode characterised by the relative motion of cations and anions.

Optical spin Hall effect is the angle-dependent conversion of linear to circular polarised light in microcavities. It is based on the resonant Rayleigh scattering of exciton-polaritons and is governed by their longitudinal-transverse splitting.

Organic materials are made of organic molecules, containing carbon, oxygen and hydrogen atoms, usually.

Oscillator strength characterises the strength of the coupling between light and an oscillating dipole (e.g., an exciton).

P

Pauli principle forbids two fermions to occupy the same quantum state.

Phase diagram shows the functional dependence between the critical parameters (temperature, density, magnetic field, etc.) separating different phases of a system at thermal equilibrium.

Phase transition is a transition between different phases (e.g., the solid and the liquid phase).

Phonon is a quantised mode (longitudinal or transverse) of vibration in a crystal lattice. As such they are the counterpart for sound of what the photon is for light. Phonons can control thermal and electrical conductivities. In particular, long-wavelength phonons transport sound in a solid, whence the name (*voice* in Greek). There are two types of phonons, *acoustic phonons* and *optical phonons*.

Photoluminescence is a powerful method of optical spectroscopy that studies the light emission from a sample illuminated by light of a higher frequency than that that is emitted.

Photonic crystal is a periodic dielectric structure characterised by photonic bands (including allowed bands and gaps). Cavities in photonic crystals allow study of fully localised discrete photonic states.

Pillar microcavity is a pillar etched from a planar microcavity structure. Its diameter is comparable to the wavelength of light at the frequency of the planar cavity mode. It allows one to obtain full (three-dimensional) photonic confinement.

Plasmon is a quantised light mode of propagation on a metal or a highly doped semiconductor. Plasmons can be longitudinal or transverse, localised at the surface or freely propagating in the bulk crystal. In metallic microspheres or other microstructures confined plasmon-polaritons can be formed.

Poincaré sphere is a sphere each point of which surface corresponds to a given polarisation of light. Points of the volume within the sphere describe partially polarised state. The Poincaré sphere can also be used to describe the quantum state of a two-level system, in which case it is known as the *Bloch sphere*.

Polariton is a mixed quasiparticle formed by a photon and a crystal excitation (phonon, magnon, plasmon or exciton). Polaritons can be formed in bulk crystals, at their surfaces, in quantum-confined structures and microcavities. In this book we mostly consider the *exciton-polaritons* (see its entry in the glossary).

Polariton diode is a light-emitting device based on a microcavity in the strong-coupling regime. Light is produced by spontaneous recombination of exciton-polaritons, which are pumped to the system either optically or electronically. In the latter case, the electrons and holes are injected in the microcavity through the contacts, they form excitons, exciton-polaritons and finally recombine emitting photons. Contrary to polariton lasers, in polariton diodes, light is emitted by many quantum states, and an average population of each of these states is less than one.

Polariton laser is a coherent light source based on Bose–Einstein condensation of exciton-polaritons. Contrary to VCSELs, polariton lasers have no threshold linked to the population inversion. Amplification of light is here governed by the ratio between the lifetime of exciton-polaritons and their relaxation time towards the condensate.

Polariton neuron is a bistable polaritonic system confined in a channel. Information propagates in polariton neurons due to the motion of domain walls tha separate "switched on" and "switched off" regions, in analogy with a biological neuron.

Polarisation of light is an important characteristic of a light mode that describes the geometrical orientation and dynamics of the electric field vector.

Pseudospin is a complex vector describing the quantum state of a two-level quantum system in the same manner than the Jones vector describes the polarisation of light.

Purcell effect consists in the modification of the radiative decay rate of an emitter (typically an atom or an exciton) due to the changes in the density of photonic states of the surrounding media. If the density of final state is reduced, emission is inhibited, if it is increased, emission is enhanced. The Purcell effect is the landmark of the weak-coupling regime.

Q

Quantum computation refers to the application of quantum information to process qubits to undergo useful computational procedures (or algorithms) that have been found in some cases to outclass their classical equivalents. For instance, the Shor algorithm factorises large integers in polynomial time and the Grover algorithm speeds up queries in unstructured spaces. The possibility to use microcavities to do quantum computation is in a prehistoric research stage.

quantum cryptography Application of quantum information to communicate a message securely, taking advantage, e.g., of conjugate bases for measurement of a qubit or of EPR correlations. The possibility to use microcavities to do quantum computation is, like quantum computation, in a prehistoric research stage.

quantum dot (QD) is a semiconductor nanocrystal that confines excitation in all three dimensions. It is the ultimate extension of the concept of the reduced dimensionality of a quantum well.

quantum information The formulation of (classical) information theory with quantum systems as the carriers of informations, which proved to be a worthwile extension, yielding as subbranches quantum cryptography and quantum computation.

quantum state A vector in a Hilbert space that fully describes a quantum-mechanical system according to the postulates of quantum mechanics.

quantum well (QW) is a semiconductor heterostructure having a profile of conduction and/or valence band edges in the form of a potential well where the free carriers or excitons can be trapped in one-dimensional sheets and propagate freely in the others two (the so-called plane of the quantum well.)Multiple quantum wells are a system of parallel quantum wells separated by barriers. If the barriers are thin enough to allow for efficient tunnelling between the wells, a system of multiple quantum wells is then called a *superlattice*.

quantum wire is an electrically conducting wire whose dimensions are so small as to impose quantum confinement in the directions normal to the axis. It extends the concept of the reduced dimensionality of a quantum well one step further.

Quasi-Bose–Einstein condensation is a term frequently used to describe the accumulation of a macroscopic quantity of bosons in the same quantum state in a finite-size quantum system, in obvious analogy to Bose–Einstein condensation that strictly speaking is a phase transition for infinite-size systems.

Qubit A quantum two-level system. The term appears in connection with *quantum information* where it is the elementary unit of information carried by a quantum system, and is the support for related effects, like *dense coding*. In microcavities, any two-level system such as the pseudospin of a polariton in principle qualifies as a qubit, provided that the coherence time and control of the state are good enough, which are still open questions. The term qubit has been introduced by Schumacher (1995).

R

Rabi splitting is the splitting of an energy level due to the coupling to a cavity mode. The term came to microcavity physics from atomic physics where an atomic resonance is split in energy. The appearance of Rabi splitting is a signature of the strong-coupling regime in microcavities. In semiconductor microcavities, this term is frequently used instead of exciton-polariton splitting. It can be detected by anticrossing of exciton and cavity-photon resonances in reflection spectra taken at different incidence angles. It should be noted, however, that two dips in reflection can be seen even in the weak-coupling regime, if the exciton inhomogeneous broadening exceeds the cavity mode width. Thus, the dip positions in reflection spectra do not coincide, in general, with the eigenmodes of a microcavity. Typical values of the Rabi splitting are from a few meV in GaAs-based microcavities with a single QW or in bulk microcavities, to more than 100 meV in organic cavities with Frenkel excitons.

Radiative lifetime is the characteristic time of the depopulation of a given quantum state caused by the emission of photons. In semiconductors, radiative recombination follows from the spontaneous recombination of an electron–hole pair.

Rate equations characterise the dynamics of the occupation numbers in a finite quantum system. The system of rate equations for an infinite system is described by the Boltzmann equations.

Reflectivity measures the intensity of light reflected by a sample divided by the intensity of the incident light.

Relaxation is a statistical process of the reduction of a physical parameter (energy, wavevector, etc.)

Resonant excitation in optics, is the generation of a quasiparticle or of an excited electronic state by absorption of photons which energy is equal to the energy of the created quasiparticle (or electronic state).

Resonant Rayleigh scattering of light is an elastic scattering where the wavevector of light changes but not its frequency. It is an important tool of optical spectroscopy of semiconductors.

S

Scattering is a process which changes the wavevector of an incident particle. The elastic scattering conserves energy, while the inelastic scattering does not.

Schrödinger equation is the time evolution equation for a nonrelativistic quantum state. It is the basis for other dynamical equations of quantum systems.

Screening is the attenuation of a force or of some influence due to the presence of a surrounding media.

Second quantisation is a mathematical formalism largely used in quantum field theory and quantum optics where physical processes are described in terms of creation and annihilation operators.

Single-photon emitter is a device that ideally emits a single photon on demand. It has application in quantum cryptography. They are currently usually based on single quantum dots. Due to the Pauli exclusion principle, the dot cannot host more than one electron and one hole at the lowest energy level and in a given spin configuration. Once such an electron–hole pair recombines, emitting a photon, some time is needed to recreate it again in the dot. Therefore, photons are emitted one by one. The same effect can be realised using single atoms, molecules, or defects.

Soliton (also "bright soliton") is a wave-packet that maintains its shape when propagating with a constant velocity. Solitons are formed in nonlinear media either due to attraction of quasiparticles with positive masses or due to repulsion of quasi-particles with negative masses. See also "dark soliton".

Spatial dispersion refers to the dependence of the dielectric constant on the wavevector of light.

Spin is the intrinsic angular momentum of a particle or quasiparticle, as opposed to the orbital angular momentum which is defined as for the case of a classical particle. Spin is a quantum number which can take only specified values and cannot be known simultaneously along all axes. Spin is linked to the statistics of the particles: bosons have integer spin and fermion half-integer spin. The spin of the photon is related to its polarisation.

Spin dynamics studies the evolution of the spin in time.

Spin Meissner effect is the same as full paramagnetic screening in exciton (exciton-polariton) BEC. No Zeeman splitting of the condensate takes place until some critical magnetic field dependent on the occupation number of the condensate and polariton–polariton interaction constants.

Spontaneous symmetry breaking is a signature of any phase transition according to the Landau theory. In microcavities, BEC or the superfluid phase transition of exciton-polaritons require symmetry breaking and the appearance of an order parameter (wavefunction) in the polariton condensate. The signature of spontaneous symmetry breaking in isotropic planar microcavities is the buildup of linear polarisation of light emitted by a polariton condensate. The orientation of the polarisation plane is randomly chosen by the system.

Squeezing refers to the reduction of the quantum uncertainty of an observable by increasing the uncertainty in the conjugate variable, so that the Heisenberg uncertainty relation remains satisfied. It leads to characteristic modifications of the particle-number statistics.

Stimulated scattering is a scattering that is enhanced by the Bose statistics. The probability of scattering becomes proportional to the occupation number (plus one) of the final state to which the bosons are scattered. The term $+1$ is the term independent of the statistics (providing the scattering rate in absence of stimulation). If the final state is macroscopically populated, i.e., forms a Bose condensate, scattering towards such a state is strongly amplified and becomes extremely rapid. Exciton polaritons (which are good bosons) are subject to stimulated scattering that provides the underlying action of polariton lasing.

Stokes shift is a redshift between the absorption and the emission peaks, usually caused by some relaxation or localisation processes.

Stokes vector is a 4-component vector that fully characterises the polarisation of light.

Strong coupling between two systems refers to a regime where the quantum Hamiltonian dynamics predominates over the dissipation of the system. The dynamics cannot be dealt with perturbatively and new quantum states of the system emerge. In a cavity, the strong coupling refers to such a coupling between exciton and light giving rise to polaritons. It manifests itself in the appearance of a splitting between the real parts of the eigenfrequencies of polariton modes, which is maximum at the resonance between bare exciton and photon modes. In this regime, the imaginary parts of two polariton eigenfrequencies coincide at the resonance. The signature of strong coupling is a characteristic anticrossing observed in the reflection (transmission) spectra when the light mode crosses the exciton resonance or vice versa. It requires domination of the exciton–photon coupling strength over different damping factors (acoustic phonon broadening, inhomogeneous broadening etc.) It is the opposite of the *weak-coupling* regime.

Superlattice is a multiple quantum well system where the wells are so close to each other that electrons can easily tunnel between them. As a result, minibands are formed.

Superradiance is the enhancement of the radiative decay rate in a system of coupled oscillators (atoms, excitons, polaritons, etc.)

Superfluidity is a specific property of bosonic liquids at ultralow temperatures. The liquid propagates with zero viscosity and has a linear dependence of the kinetic energy on wavevector. The appearance of superfluidity is a consequence of the repulsive interaction between bosons. According to recent theories, exciton-polaritons in microcavities may become superfluid under certain conditions.

Surface plasmon is a confined light mode formed at the surface of metal. Surface plasmons are characterised by wavevectors outside the light cone (i.e., exceeding the wavevector of light in vacuum at the corresponding frequence). Their polarisation vector must contain a normal to the surface component.

T

Tamm photon is a confined optical mode formed at the boundary of two periodical dielectric structures (Bragg mirrors). The intensity of light in this mode decays when moving from the interface between two mirrors. It may be formed if photonic stopbands of different parity in two mirrors overlap.

Tamm plasmon is a confined optical mode formed at the boundary of a periodical dielectric structure (Bragg mirror) and metal. The intensity of light in this mode decays exponentially both in the metal and in the dielectric structure. Tamm plasmons are characterised by polarisation-dependent parabolic in-plane dispersion. Unlike a surface plasmon a Tamm plasmon can have a zero or small in-plane wave vector.

Terahertz radiation is the electromagnetic radiation in the frequency range of several ThZ (10^{12} s^{-1}), wavelength of the order of fractions of a millimeter and energy of several milli electron Volts.

Thermal state is a the state of a system that has reached thermal equilibrium with its surroundings. It is a mixed state characterised by exponentially decreasing probabilities of the occupation numbers.

Transfer matrix is a mathematical technique that allows to solve the Schrödinger or Maxwell equations in multilayer systems.

Topological insulator is a structure that sustains an electric current (or, generally, a flow of quasiparticles) in one particular direction, while current of the same quasiparticles in the opposite direction cannot be induced due to the symmetry reasons.

Transmission of light is an optical spectroscopy technique that allows to detect the intensity of light passing through a sample.

U

Ultrastrong coupling is the regime of quantum coupling between two modes where the resulting new eigenmodes are split at the anticrossing point by an energy comparable or exceeding the energy of each of the coupled modes.

V

Vacuum Rabi splitting refers to the linear optical regime where the interaction of a single photon with a single atom is implied in the case of atomic cavities. Nonlinear vacuum Rabi splitting has not been evidenced so far so that the distinction has not yet gained importance and "Rabi splitting" is often used to mean "vacuum Rabi splitting".

VCSEL is an acronym for *vertical cavity surface emitting laser*. It is a device based on a microcavity in the weak-coupling regime. Stimulated emission of light by an active element inside the cavity (typically quantum wells, where the inversion of population of electron levels of the conduction and valence bands is achieved due to electrical injection of charge carriers) pumps one of the confined light modes of the cavity. The light emitted by this laser goes out at right angles to the surface of the mirror, contrary to "horizontal" lasers, where the generated light propagates in the plane of the laser cavity.

Vortex is a spinning flow with closed streamlines that circle around a point, the core of the vortex. Classical vortices are easily created in a tea cup with a tea spoon. Quantum vortices form spontaneously in type II superconductors (Abrikosov vortices) and in superfluids. In optics, vortices can be easily created, e.g., by interference of several light beams. In the context of exciton-polariton condensates in microcavities, quantised vortices can be induced by resonant optical excitation or formed spontaneously during the formation of the condensates. One can distinguish between integer vortices, where the phase of the polariton field changes by an integer number of 2π while going around the core of the vortex, and half-integer vortices where the phase changes by π, 3π, etc. In half-integer vortices, the continuity of the polariton field is assured by rotation of the vector polarisation of exciton-polaritons correlated with the phase variation.

W

Weak coupling between two systems refers to the regime opposed to *strong coupling* (see above) where dissipation dominates over the system interaction so that the coupling between the modes can be dealt with pertubatively and both modes retain essentially their uncoupled properties. The weak coupling between exciton and light manifests itself in the appearance of the splitting between the imaginary parts of the eigenfrequencies of exciton-polariton modes at the resonance between bare exciton and photon modes. In this regime the real parts of two polariton eigenfrequencies coincide at the resonance, and two polariton resonances in the reflection or transmission spectra usually coincide (while in the case of a strong imbalance between the widths of the exciton and photon modes the doublet structure in reflection and transmission can be seen even in the weak-coupling regime).

Weak lasing is a phase of a driven-dissipative bosonic system characterised by spontaneously broken real-space symmetry. Weak lasing implies specific phase locking between spatially separated condensates. It is realised due to the interplay between coherent (Josephson) coupling of the condensates and their incoherent, dissipative coupling.

Whispering-gallery modes are standing light modes localised at the equator of a sphere. Contrary to the "breathing modes", the whispering-gallery modes are characterised by high orbital quantum numbers.

Z

Zeeman splitting is the magnetic-field-induced energy splitting of a quantum state into a couple of states characterised by different spin projections onto the magnetic-field direction.

1

OVERVIEW OF MICROCAVITIES

In this chapter we provide an overview of microcavities. We present the variety of semiconductor, metallic and dielectric structures used to make microcavities of different dimensions, and briefly present a few characteristic optical effects observed in microcavities. Many important effects mentioned in this chapter are discussed in greater extent in the following chapters.

Microcavities, Second Edition. Alexey V. Kavokin, Jeremy J. Baumberg, Guillaume Malpuech, Fabrice P. Laussy, Oxford University Press (2017). © Alexey V. Kavokin, Jeremy J. Baumberg, Guillaume Malpuech, Fabrice P. Laussy. DOI 10.1093/oso/9780198782995.001.0001

A microcavity is an optical resonator close to or below the dimension of the wavelength of light. Micrometre- and submicrometre-sized resonators use two different schemes to confine light. In the first, reflection off a single interface is used, for instance from a metallic surface, or from total internal reflection at the boundary between two dielectrics. The second scheme is to use microstructures periodically patterned on the scale of the resonant optical wavelength, for instance a planar multilayer Bragg reflector with high reflectivity, or a photonic crystal (see Fig. 1.1). Since confinement by reflection is sometimes required in all three spatial directions, combinations of these approaches can be used within the same microcavity. In this chapter we will explore a number of basic microcavity designs, and contrast their strengths and weaknesses. For practical purposes in this book, the discussion of microcavities will be limited to cavities in which confining dimensions are below 100 μm.

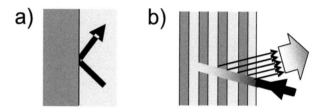

Fig. 1.1: (a) Single interface reflection and (b) interference from multiple interfaces.

1.1 Properties of microcavities

To help survey microcavity designs it is helpful to motivate a comparison of different optical properties of a microcavity. We assume in this section a microcavity with total power reflectivity R, and round-trip optical length L. The resonant optical modes within a microcavity have characteristic lineshapes, wavelength spacings and other properties that control their use. A longitudinal resonant mode has an integral number of half-wavelengths that fit into the microcavity, while transverse modes have different spatial shape. However, in a microcavity this traditional distinction can lose its precision as modes all exist on the same footing.

1.1.1 Q-factor and finesse

The *quality-factor* (or Q-factor) has the same role in an optical cavity as in an LCR electrical circuit, in that it parametrises the frequency width of the resonant enhancement. It is simply defined as the ratio of a resonant cavity frequency, ω_c, to the linewidth (FWHM) of the cavity mode, $\delta\omega_c$:

$$Q = \frac{\omega_c}{\delta\omega_c}. \tag{1.1}$$

The *finesse* of the cavity is defined as the ratio of free spectral range (the frequency separation between successive longitudinal cavity modes) to the linewidth (FWHM) of a cavity mode (see Fig. 1.2):

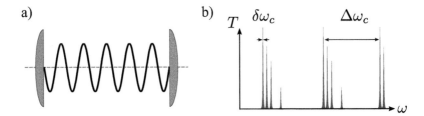

Fig. 1.2: (a) Longitudinal mode has integral wavelengths along the main cavity axis, giving (b) mode spectrum.

$$F = \frac{\Delta\omega_c}{\delta\omega_c} = \frac{\pi\sqrt{R}}{1-R}. \tag{1.2}$$

The Q-factor is a measure of the rate at which optical energy decays from within the cavity (from absorption, scattering or leakage through the imperfect mirrors) and where Q^{-1} is the fraction of energy lost in a single round-trip around the cavity. Equivalently, the exponentially decaying photon number has a lifetime given by $\tau = Q/\omega_c$.

Because the mode frequency separation $\Delta\omega_c = \frac{2\pi c}{L}$ is similar to the cavity mode frequency in a wavelength-scale microcavity, the finesse and the Q-factor are not very different. This is not the situation for a large cavity, in which case the Q-factor becomes much greater than the finesse because of the long round-trip propagation time. Instead, the finesse parametrises the *resolving power* or *spectral resolution* of the cavity.

1.1.2 *Intracavity field enhancement and field distribution*

The on-resonance optical intensity enhancement is given by

$$\frac{I_{\text{intracavity}}}{I_{\text{incident}}} \simeq \frac{1}{1-R} = \frac{F}{\pi\sqrt{R}}, \tag{1.3}$$

assuming the mirror losses dominate the finesse. In a travelling-wave cavity this will be uniformly distributed. However in a standing-wave microcavity, this enhancement is found in the form of spatially localised interference peaks. Hence, it is not always simple to couple an emitter directly to this enhanced optical field. The enhanced optical field inside the microcavity can be usefully harnessed for enhanced nonlinear optical interactions.

1.1.3 *Tuneability and mode separation*

The separation of longitudinal modes in a microcavity, $\Delta\omega_c$, is inversely proportional to the cavity length. However, cavities other than confocal cavities have transverse optical modes at different frequencies and these also scale similarly with cavity length. Hence, microcavities have far fewer optical modes in each region of the spectrum than macroscopic cavities. This can mean that specifically tuning the cavity mode to a particular emission wavelength becomes more important than in large cavities.

Various techniques for spectral tuning of modes have been advanced; however, none is as yet ideal due to the difficulty of modifying the round-trip phase by π without introducing extra loss. The simplest way is to scan the cavity length, although, in cases where this can be altered, it becomes more difficult to then maintain consistently a fixed length once the desired tuning has been reached. For many other monolithic systems, tuning of the cavity modes is extremely difficult and is most advanced for semiconductor lasers (see Section 1.5).

1.1.4   *Angular mode pattern*

Microcavities are typically small in all three spatial directions, with aspect ratios closer to unity than macroscopic cavities. As a result, the angular mode emission patterns, which are Fourier related to the cavity mode spatial distribution, tend to have a wider angular acceptance. This means that microcavities emit into a large solid angle on resonance. On the other hand, emission from a microcavity is still beamed into particular directions. For instance, by embedding an LED active region within a planar microcavity, the light is emitted in a forward-directed cone when the electroluminescence is resonant with the cavity mode.

1.1.5   *Low-threshold lasing*

There are two reasons why microcavities can have lower lasing thresholds: their reduced number of optical modes and their reduced gain volume.

In a microcavity, an embedded emitter has a reduced range of optical states into which it is likely to emit. In free space it can emit into any solid angle and frequency, but the microcavity acts to structure the optical density of states around the emitter. This is a particularly strong effect when the emitter has a large linewidth (e.g. an electron–hole pair in a semiconductor) as the spectral overlap with different cavity modes is reduced by reducing the size of the cavity. Also in a microcavity, the angular acceptance of any particular microcavity mode is much larger. Because the lasing threshold occurs at the point at which a spontaneously emitted photon returns to the emitter and stimulates the next photon emission, reducing the number of cavity modes has the effect of reducing the laser threshold because spontaneously emitted photons are more likely to return to the emitter. This effect is contained within a *spontaneous emission coupling factor*, β, which is defined as the fraction of the total spontaneous emission rate that is emitted into a specific (laser) mode. This is typically below 10^{-5} in bulk lasers, but can be $> 10\%$ in microcavities. The laser threshold is given by

$$P_{\mathrm{thr}} = \frac{\hbar \omega_{\mathrm{c}}^2}{2Q\beta}. \tag{1.4}$$

A small optical cavity means that it contains a smaller volume that is pumped electrically or optically to provide gain. Because such systems have to be pumped so that one of the energy levels is brought into inversion, the total energy needed to reach inversion scales with the volume of active material. As a result of these two effects, microcavities have the smallest known thresholds of any laser, having now reached the state of a single-photon intracavity field threshold level: the first photon emitted turns the laser on.

1.1.6 *Purcell factor and lifetimes*

Embedding an emitter inside a microcavity can lead to additional effects due to the change in the optical density of states. When the emitter linewidth is smaller than that of the cavity mode ($\delta\lambda_c$), the emitter can be considered to couple to an optical continuum and the emission kinetics is given by Fermi's golden rule. This describes the emission lifetime, τ, modified from free space (τ_0), in terms of the detuning between the emitter (λ_e) and cavity:

$$\frac{\tau_0}{\tau} = F_P \frac{2}{3} \frac{|E(\mathbf{r})|^2}{|E_{\max}|^2} \frac{\delta\lambda_c^2}{\delta\lambda_c^2 + 4(\lambda_c - \lambda_e)^2} + f, \qquad (1.5)$$

which is controlled by the Purcell factor given by

$$F_P = \frac{3}{4\pi^2} \frac{\lambda_c^3}{n^3} \frac{Q}{V_{\text{eff}}} \qquad (1.6)$$

where n is the refractive index of the cavity, V_{eff} is the effective volume of the mode, $E(\mathbf{r})$ is the field amplitude in the cavity and $|E_{\max}|^2$ is the maximum of its intensity. The constant f in eqn (1.5) describes the losses into leaky modes (this is further discussed in Chapter 6).

The crucial ratio Q/V_{eff} allows the emitter to emit much faster into the optical field (if both spectral and spatial overlaps are optimised). It also allows decay to be suppressed, although competition with non-radiative recombination generally means that this is accompanied by a decrease in emission efficiency. Typically, the ratio Q/V_{eff} is difficult to enhance arbitrarily since smaller cavities often have restrictions in the maximum Q-factor that is possible. The theory of Purcell effect is presented in Chapter 6.

1.1.7 *Strong vs. weak coupling*

If a resonant absorber is embedded inside a microcavity, then another new regime of optical physics can be reached when the absorption strength is large and narrow-band enough. The absorber in a semiconductor is typically an electron–hole pair, or *exciton*. If the total scattering rates of both the cavity photons and the excited absorber (exciton) are less than the rate at which they couple with each other, new mixed light modes called *polaritons* will result:

$$\psi = \frac{1}{\sqrt{2}} \{\psi_X \pm \psi_C\}, \qquad (1.7)$$

where ψ_X, ψ_C are the exciton and photon wavefunctions, respectively. Spectrally tuning the absorber to the cavity resonance leads to mixing of the photon and absorber, resulting in new polariton states at higher and lower energies (see Fig. 1.3). This effect, known as *strong coupling*, will be dealt with in detail in Chapter 4 and 5. The condition for strong coupling is thus that the light–matter-induced splitting between the new polariton modes (known as the Rabi splitting, Ω) is greater than the linewidths of either cavity photon (γ_c, controlled by the finesse) or the exciton (γ_x, controlled by the inhomogeneous broadening of the excitons in the sample)

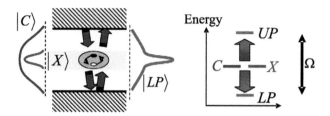

Fig. 1.3: Strong coupling of excitons inside a QW microcavity, with repulsion of initial cavity photon (C) and exciton (X) states producing upper (UP) and lower (LP) polaritons.

1.2 Microcavity realisations

The most common microcavity is the *planar microcavity* in which two flat mirrors are brought into close proximity so that only a few wavelengths of light can fit in between them. To confine light laterally within these layers, a curved mirror or lens can be incorporated to focus the light, or they can be patterned into mesas.

An alternative approach for microcavities uses total internal reflection within a high refractive index convex body, to produce *whispering-gallery modes* that can exist within spheres (3D modes) or disks (2D modes), or more complicated topological structures.

Finally, *photonic crystals* employ periodic patterning in two or three dimensions to confine light to a small volume surrounding a defect of the structure.

The key issues that should be borne in mind when considering microcavities are:

- their optical losses or finesse,
- coupling to incident light,
- optical mode volume,
- fabrication complexity and tolerance,
- incorporation of active emitters,
- practicality of electrical contacting.

1.3 Planar microcavities

The well-known Fabry–Pérot cavity, comprised of two plane mirrors, can perform effectively when the mirror separation, L, is only a few wavelengths of light. The resulting cavity modes are equally spaced in frequency, apart from shifts caused by the variation with wavelength in the phase change on reflection on each of the mirrors. These Fabry–Pérot modes have a characteristic dispersion in their frequency as the angle of incidence, θ, is increased. Essentially, the condition for constructive interference after one round-trip enforces a condition on the wavevector $k_\perp = k \cos \theta$ perpendicular to the mirror surfaces: at higher angles of incidence an additional wavevector parallel to these surfaces means that the total k is larger than $2\pi/\lambda$ and hence the cavity frequency increases, with

$$k_\perp \times 2L = 2m\pi \,, \tag{1.8}$$

hence

$$\omega = \frac{m\pi c/L}{\sqrt{\bar{n}^2 - \sin^2 \theta}} \,, \tag{1.9}$$

Charles **Fabry** (1867–1945) and Alfred **Pérot** (1863–1925).

Fabry developed the theory of multibeam interferences at the heart of the Fabry–Pérot interferometer. He published 15 articles derived from his invention with Pérot in the course of 1896–1902, applying it with great success to spectroscopy, metrology and astronomy. Alone or with others, he derived from the interferometer a system of spectroscopic standards, demonstrated Doppler broadening in the emission line of rare gases and, in 1913, evidenced the ozone layer in Earth's upper atmosphere. When he was elected to a chair at the French "Académie des sciences" in 1927, he gathered 51 votes whereas all other candidates came out with only one, including Paul Langevin. An enthusiastic teacher and populariser of science, some of his lectures were so popular that the doors had to be closed for lack of space half an hour before the beginning. He is quoted as having said *"My whole existence has been devoted to science and to teaching, and these two intense passions have brought me very great joy."*

Pérot did not climb to the same fame as Fabry, beside their joint naming of the interferometer, and the most valued source of information about his life is, in fact, the obituary written for him by Fabry, in "Alfred Pérot", Astrophys. J. **64**, 208 (1926). While his colleague and friend—apart from the theory—would also carry out most of the measurements and calculations, Pérot was mainly involved in the design and construction of the apparatus where he deployed great skills that brought the system immediate fame. He also initiated the project by consulting Fabry on a problem of spark discharges of electrons passing between close metallic surfaces. He later developed an interest in experimental testing of general relativity with some positive outcomes, but a final failure to evidence a gravitational redshift.

Further informations can be found in sources compiled by J. F. Mulligan for the centenary anniversary of the interferometers in "Who were Fabry and Pérot?", Am. J. Phys. **66**, 9 (1998).

where $\bar{n}$ is the average refractive index of the microcavity and c is the speed of light.[3]

Planar microcavities illuminated with plane waves of infinite extent in the plane of the mirrors do not have any additional modes confined in plane. Such a situation is practically unrealistic, and there is always some limit to the lateral extent either from the size of the mirrors, the width of the illuminating beam, or aperturing effects inside the microcavity. The natural basis set for planar microcavities are wavelets of extent in both real and momentum space. One useful variety are Airy modes, because these are anchored around a particular point on the mirror (which we normally impose by

---

[3]The speed of light in vacuum is *exactly* 299 792 458 ms^{-1}, as it is, since 1983, a definition rather than a measurement. In turn, the unit of the SI system, the metre, is defined as the distance light travels in a vacuum in 1/299 792 458 of a second. The second itself is defined as the duration of 9 192 631 770 periods of the radiation in the transition between the two hyperfine levels of the ground state of the caesium-133 atom. Its label c is for "constant". We shall use this notation throughout the text without further mentioning it.

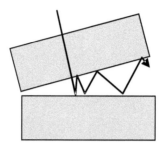

Fig. 1.4: Wedged microcavity, showing walk-off of incident light in multiple reflections.

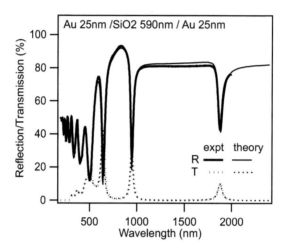

Fig. 1.5: Reflection and transmission of a planar microcavity consisting of a gold-coated 590 nm glass spacer.

illuminating or detecting at a particular position). However, for most purposes plane waves in the transverse direction are used to describe the field distribution.

Microcavities in which the mirrors are not exactly parallel cause incident light to slowly "walk" towards the region of larger cavity length through multiple reflections, as illustrated in Fig. 1.4. The acceleration of the confined light in this direction can be directly tracked in time and space. Thus, this lateral walk-off acts as an extra loss in the cavity modes, reducing the finesse. It is helpful to consider approaches to making these devices through comparison between two sorts of reflectors: metals and distributed Bragg reflector (DBR) stacks.

1.3.1 *Metal microcavities*

The modes of a metal microcavity are limited by the fundamental material parameters of loss and reflectivity in metal films of varying thickness. For wavelengths further into the infrared, this situation improves and the finesse can be high, reaching Q-values of 10^9 for superconducting cavities at microwave frequencies. However, for microcavities around the optical region of the spectrum, the modes have $Q < 500$ (see Fig. 1.5).

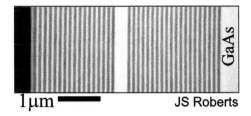

Fig. 1.6: Scanning electron micrograph of GaAs/AlGaAs DBR microcavity on a GaAs substrate, from Savvidis et al. (2000).

The boundary conditions for reflection of light at a metal imply that the optical field is nearly zero at the mirrors, while penetration of light into the metal mirrors is small compared with the wavelength. Note also that the phase change on reflection from the metal varies with wavelength depending on the dielectric constants (see eqn (2.128)).

The ultimate Q-factor for a metal cavity is set by the trade-off between the real and imaginary parts of the dielectric constant, $n = \bar{n} + i\kappa$, which control reflectivity and absorption. On the other hand, in a metal microcavity, the modal extent can be relatively small, with penetration of light into the barriers limited to the exponential decay of the electric field in the metal, $\propto \exp(-\delta z)$ with $\delta = 2\pi\kappa/\lambda$ and z the distance along the normal to the surface.

1.3.2 Dielectric Bragg mirrors

The situation is different when multilayers of many pairs of alternating refractive index are used to make cavity mirrors (see Section 2.6.1). The complete structure has an extra cavity spacer in between the Bragg mirrors and the whole can be considered a 1D photonic crystal cavity with a central defect. A scanning electron micrograph of the cross-section of a typical semiconductor DBR microcavity is shown in Fig. 1.6.

The finesse of this cavity is set by the reflectivity of each mirror that depends on the number of pair repeats and the refractive index contrast between the two materials used. The key condition is that the optical path in each of the layers is a quarter of the desired centre wavelength of reflection. In this case, the resonant field is maximum at the dielectric interfaces and there is significant penetration of light into the surrounding mirror stacks (Fig. 1.7(c)). The reflectivity has a central flat maximum, which drops off in an oscillating fashion either side of the reflection- or stop-band. The spectral bandwidth of the mirror is set by the refractive index difference between the materials (see Section 2.6.1).

The penetration into the mirrors limits the minimum modal length of the DBR microcavity (see eqn 2.119), so the cavity mode volume is larger than for metal Fabry–Pérot microcavities. For instance, the cavity mode and the first mode of the Bragg mirror sidebands have similar extents (Figs. 1.7(c) and (d)).

The finesse of these microcavities is ultimately limited by the number of multilayers that can be conformally deposited, while ensuring that roughness or surface cracking, which creates scattering loss, does not increase unduly. Electron-beam evaporation of dielectrics in general produces the smoothest results for oxides, while Molecular Beam

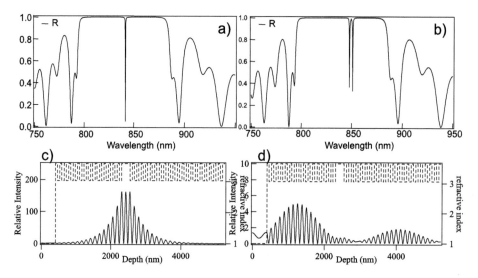

Fig. 1.7: (a,b) Reflection of a planar DBR microcavity consisting of a top and bottom mirror with 15 and 21 repeats of GaAs/AlAs. In (a) a 240 nm thick bulk GaAs cavity is incorporated, while in (b) 3 InGaAs QWs are incorporated in the same region. (c,d) Field distributions corresponding to (a) at wavelengths of (c) 841 nm and (d) 787 nm.

Epitaxy (MBE) of semiconductors produces very high quality epitaxial single-crystal Bragg mirrors. Other techniques have also been investigated, such as controlled etching of layer porosity in Si, polymer multilayers, and chiral liquid-crystalline phases. Finding suitable materials for Bragg mirrors in the UV spectral region is in general more challenging, and the available refractive contrast limits the bandwidth here. In practice, achieving transmissivity ratios of 10^6 on and off the resonance wavelengths is achievable across wide spectral bandwidths.

Some of the earliest active planar microcavities were optically pumped dyes flowing between two DBR mirrors. These demonstrated many of the features resulting from lateral confinement (in this case by the localised optical pump beam) of the microcavity modes in the weak-coupling regime. By controlling the separation between the mirrors, the cavity mode can be tuned into resonance with the dye emission. The far-field pattern shows how the lateral coherence length changes as the stimulated photon emission turns on. By optical pumping two neighbouring positions inside the dye-filled microcavity, the lateral coherence properties could be investigated as a function of their separation.

1.4 Spherical mirror microcavities

In order to fully control the photonic modes in the microcavity, the light has to be confined in the other two spatial directions. The way this is conventionally achieved in macroscopic cavities is to use mirrors with spherical curvature. This can also be achieved in microcavities, although new methods have to be utilised to produce spherical optics with a radius of curvature below 100 μm.

Developing standard routes to spherical mirrors and polishing mirrors to ultra-low roughness can give reflectivity in excess of $R > 0.9999984$. These have been successfully used in 40 μm long cavities for atomic physics experiments, providing Q-factors exceeding 10^8. For such extremely narrow linewidths, active stabilisation of the cavity lengths is mandatory on the level of several picometers using piezoelectric transducers, as discussed by Rempe et al. (1992).

Another route suggested has been to trap micrometre-sized air bubbles in cooling glass and then cut such frozen bubbles in two for subsequent coating. This has been successfully used by Cui et al. (2006) to create microcavities with finesse up to several hundred.

A separate technique has been to develop templating for micrometre-scale mirrors. In this strategy, latex or glass spheres of a selected size are attached to a conducting surface. This is followed by electrochemical growth of reflective metals around them. Subsequent etching of the templating spheres leaves spherical micro-mirrors with smooth surfaces. Such mirrors down to 100 nm radius of curvature have been produced, for instance by Prakash et al. (2004) (see also Fig. 1.9).

A further route developed recently is to use focused ion beams to sculpt a desired 3D shaped micromirror into a substrate, after which it is coated with DBRs to provide a high reflectivity mirror. Extensive optimisation is required to produce arbitrary shapes, ranging from spherical to parabolic, which therefore are able control the transverse mode spectrum of the microcavity.

For all these spherical mirror microcavities, the optical mode spectrum becomes fully discretised. In the paraxial approximation these Laguerre–Gauss modes are given by

$$\omega_{npq} = \frac{c}{L} \left[n + \varepsilon_n + (2p + q) \arctan \sqrt{\frac{L}{R - L}} \right], \tag{1.10}$$

where n, p, q are integers for the longitudinal, axial and azimuthal mode indices, R is the radius of curvature of one mirror (the other mirror is plane) and they are separated by a distance L. The phase shifts from the mirror reflections are $\varepsilon(\lambda)$. Typically in macroscopic-scale cavities, the azimuthally symmetric mode symmetry is broken by slight imperfections in the mirror shape and the astigmatic alignment, and horizontal/vertical TE_{mn} and TM_{mn} modes result.[4] However, for microcavities, the dominant splitting is due to the breakdown of the paraxial approximation, which assumes all rays are nearly parallel to the optic axis so that $\sin \theta \simeq \theta$. In microcavities, the cylindrical mode symmetry is maintained, while mode splittings are more pronounced for fields with different extents in the lateral direction (Fig. 1.9). Enhancing the finesse further requires deposition of DBR mirrors inside the spherical micro-mirror. This becomes progressively more difficult as the radius of the spherical mirror reduces and the mode size shrinks laterally. Hence, there is a minimum limit to the mode volume of spherical microcavities.

---

[4]TE and TM denote modes with transverse electric and magnetic field vectors, respectively.

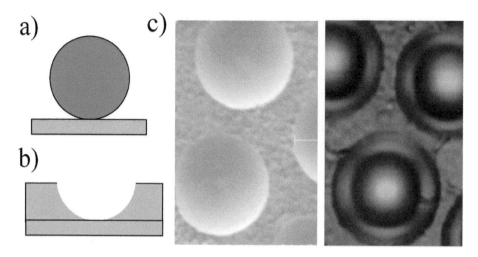

Fig. 1.8: Spherical gold mirrors (a,b) templating process, (c) SEM and optical micrograph, of 5 µm diameter, 5 µm radius of curvature mirrors, from Prakash et al. (2004).

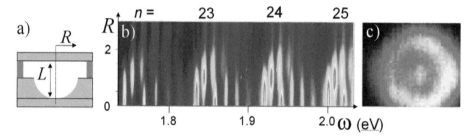

Fig. 1.9: (a) Spherical-planar microcavity, $L = 10$ µm. (b) Cavity mode transmission spectrum vs. radial distance from centre of planar top mirror. (c) Optical near-field image of mode at $\omega = 1.932$ eV, courtesy of B. Pennington.

1.5 Pillar microcavities

Another way to confine the lateral extent of the photonic modes inside planar microcavities is to etch them into discrete mesas (see Fig. 1.10). Total internal reflection is used to confine the light laterally, while the confinement vertically is dependent on reflection from DBR mirrors.

For semiconductors with high refractive indices, like GaAs, the lateral confinement is quite strong. One way to think of these microcavities is as waveguides that have reflectors at each end. Light propagating at any angle less than $73°$ to the external wall surface is totally internally reflected. Once again, the modes have discrete energies, which are further separated in frequency as the pillar area is reduced. From the waveguide point of view, the modes can be labelled as TEM_{pq}. A lowest-order description of the modes in square pillars assumes the electric field vanishes at the lateral surfaces, and hence,

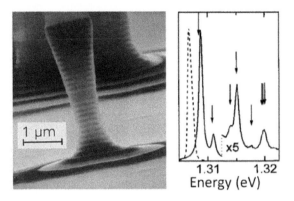

Fig. 1.10: Pillar microcavity from an etched planar DBR semiconductor microcavity (left), with emission mode spectrum (right), from Gérard et al. (1996).

$$\omega_{npq} = \frac{c}{n} \sqrt{k_0^2 + k_x^2 + k_y^2}, \qquad (1.11)$$

where $k_0 = 2\pi n/\lambda_0$ denotes the wavevector of the vertical cavity, and in the lateral directions $k_{x,y} = (m_{p,q} + 1)\pi/D$, for a square pillar of side D with lateral photon quantum numbers $m_{p,q} \in \mathbb{N}$, labelling the transverse modes.

Typically, DBR planar microcavities are used as the basis of the pillar and the etching proceeds to just below the central defect spacer. In such structures, further splitting of degenerate modes is observed, due to the effects of pillar shape (square, elliptical, rectangular), strain (in non-centrosymmetric crystals like III–V semiconductors) and imperfections in the perimeter of the pillar.

The discrete modes of a hard apertured circular microcavity correspond to Airy modes, A_l, in the angular dispersion,

$$A_l(\theta) = \frac{2J_l(kL\cos\theta)}{kL\cos\theta}, \qquad (1.12)$$

where J_l is the first-order Bessel function. The angle-dependent far-field coupling and energies of these modes in larger microcavities falls on top of the in-plane dispersion of the planar microcavity, showing the close connection between mode area and energy splitting. Depending on the geometry of the pillars, the modes may be mixed together, producing orthogonal TE/TM modes, rather than axial symmetric modes as in the spherical microcavities. More details on the optical modes in cylindrical and spherical cavities are given in Chapter 2.

A number of researchers have demonstrated that the growth of thin semiconductor nanowires (such as GaN or ZnO) can produce a similar microcavity effect, with the waveguide modes confined at each end by the refractive index contrast at air and substrate interfaces. In general, however, the smaller the pillar area, the more difficult it is to couple light efficiently into or out of the microcavities. In addition, such pillar microcavities are inherently solid and emitters have to be integrated into the structure.

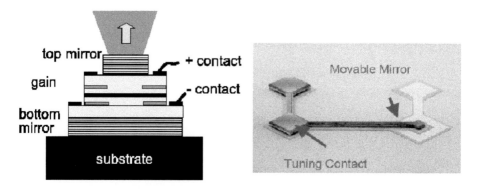

Fig. 1.11: (a) VCSEL design incorporating oxide apertures (b) MEMs VCSEL structure with cantilevered top mirror, from Chang-Hasnain (2000).

Typically for semiconductors, the diffusion length of carriers is micrometres and, thus, once the pillar diameter is reduced below this, non-radiative relaxation on the surfaces of the pillar dominates the emission process reducing the emission intensities.

Typically, pillar DBR microcavities with diameters of $5\,\mu m$ show Q-factors in excess of 10^4 and these can be straightforwardly measured by incorporating an emitting layer, such as InAs quantum dots, in the centre of the cavity stack. These Q-values are limited by imperfections in the DBR mirrors (mostly caused by the buildup of strain within the mirror stacks). The other main loss from the micropillar is the light that can leak out the sides of the pillar. To improve the finesse of such microcavities, they can be coated with metal around their vertical sides.

The most common active pillar-type microcavities are the so-called vertical-cavity surface-emitting lasers (VCSELs), which form the basis of a huge and thriving industry due to their ease of large-scale manufacturing and of in situ testing of devices in wafer form. These VCSELs normally use a combination of pillar etching and further lateral control of the electrical current injection by progressively oxidising an incorporated AlAs layer that thus forms an insulating AlO_x annulus (Fig. 1.11). The oxide also provides extra photon confinement (as the refractive index is much lower than the semiconductor interior), improving the coupling of the optical mode with the electron–hole recombination gain profile.

Tuning of VCSELs has been demonstrated using Movable External Mirror (MEM) technologies, in which the upper mirror is suspended above the active cavity and lower DBR mirror, and can be moved up and down to tune the main cavity mode wavelength. Other cavity tuning schemes use temperature or current through the mirrors to modify the refractive index of the cavity. Wide bandwidth tuning across hundreds of nanometres in the visible and near-infrared has not yet been achieved in integrated structures, although this would be extremely useful for many applications. More information on VCSELs can be found in Chapter 6.

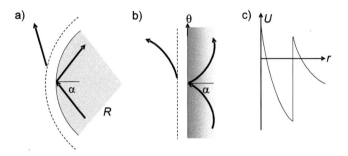

Fig. 1.12: (a) Tunnelling through total internal reflection at curved interface, (b) conformal mapping and (c) optical potential.

1.6 Whispering-gallery modes

When light is incident at a planar interface from a high refractive index n_1 to a low refractive index n_2 medium, it can be completely reflected, provided the angle of incidence exceeds the critical angle, $\theta_c = \arcsin(n_2/n_1)$. This total internal reflection can be used to form extremely efficient reflectors in a microcavity, not dependent on the metal or multilayer properties. Because the light skims around the inside of such a high refractive index bowl, it resembles the whispering-gallery acoustic modes first noticed by Lord Rayleigh in Saint Paul's Cathedral in London. A semiconductor version in the form of a microdisk is shown in Fig. 1.13. The light emission allows direct imaging of the whispering gallery mode.

However, in general, it is not possible to make extremely small microcavities using this principle. The reason is that diffraction plays an increasing role at small scales and any planar surfaces have to be connected by corners that act as leaks for the diffracting light.

Whispering-gallery modes can be classified with integers corresponding to radial, n, and azimuthal, l, mode indices. High-Q whispering-gallery modes generally correspond to large numbers of bounces, l, (thus, at glancing incidence to the walls) with $n \approx l$ localising the mode near the boundary walls (see Section 2.9.2). The limit of these structures is a circle or, in general, a distorted or curved shape. In this regime of curved interfaces, the total internal reflection property is modified, and light can tunnel out beyond a critical distance to escape. A simple way to see this is to map the curved interface into a straight interface by varying the refractive index radially (in a well-defined way, see Fig. 1.12). This leads to an equation for the radial optical potential, which for the perpendicular wavevector, looks like tunnelling through a triangular barrier. Solving this leads to the reflectivity per bounce, R:

$$\ln(1 - R) = -\frac{4\pi\rho}{\lambda} h \left[\ln(h + \sqrt{h^2 - 1}) - \sqrt{1 - h^{-2}} \right], \tag{1.13}$$

where $h = n\cos\alpha$, α is the angle of incidence on the curved boundary and ρ is the radius of curvature.

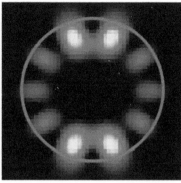

Fig. 1.13: Left: microdisk (lower ring) with upper electrical contact, from Frateschi et al. (1995). Right: microdisk intensity pattern inside disk.

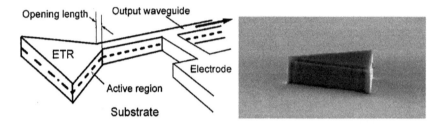

Fig. 1.14: Microdisk triangle laser, with external coupling waveguide, from Lu et al. (2004).

This shows the limits of the total internal reflection even for structures that would give high reflectivity at a planar interface. For instance, for micrometre-sized silica spheres, the reflectivity per round-trip drops below 50% when $l < 5$ bounces.

The variants of whispering-gallery microcavities can be classified according to their geometry, whether the multiple total internal reflections lie in a plane (2D) or circulate also in the orthogonal direction (3D).

1.6.1 *Two-dimensional whispering galleries*

For any sufficiently high refractive index convex shape, light can be totally internally reflected around the boundary. Providing $n > 1/\arcsin 30 = 2$, triangular whispering-gallery modes are the lowest-order number of bounces possible (Fig. 1.14).

In real structures, the third dimension remains important, and confinement in this direction is provided by a waveguide geometry or DBR reflectors. Hence, the geometrical form of 2D whispering galleries is generally a thin disk of high refractive index material on a low effective refractive index substrate. An extreme form of this geometry is provided by the thumb-tack microcavity, in which a semiconductor disk is undercut and supported by a thin central pillar.

The modes of a circular microcavity take the form of spherical Bessel functions inside the disk (Section 2.9). As the microcavity becomes smaller, the photonic mode

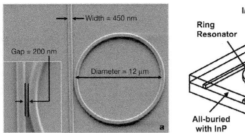
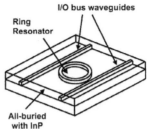

Fig. 1.15: Microring resonator with integrated waveguide coupling, from Xu et al. (2005). On the photo, the gap between the disk and the waveguide is 200 nm, the width of the guide is 450 nm and the diameter of the ring is 12 µm.

Fig. 1.16: Left: bow-tie resonator, from Nöckel (1998). Right: Cog microdisk laser, to pin the azimuthal standing wave field, from Fujita and Baba (2002).

separation increases, and can be estimated from the simple multibounce model based on a ray treatment for constructive interference, $kL = 2\pi m$, after l bounces, hence

$$\omega_{lm} = \frac{m2\pi c}{n2lR\sin(\pi/l)} \simeq \frac{mc}{nR},$$ (1.14)

with the disk radius R and refractive index n.

As most of the field of the whispering-gallery modes is in the outer part of the disk, the inside of the disk can be removed to form microring resonators with very similar characteristics. The lack of emission from the central part when optically pumped reduces the background spontaneous emission, thus reducing the laser threshold as long as scattering losses from the whispering-gallery mode are not also increased.

Since the leakage from a circular microdisk frustrates evanescent coupling to air, it produces a structure that is extremely hard to selectively couple to and from. Favoured versions have included a waveguide in close proximity to provide selective evanescent coupling in particular directions (Fig. 1.15). However, such coupling is exponentially sensitive to the coupling gap between disk and waveguide, giving strict fabrication tolerance on the nanometre scale. To ameliorate this sensitivity, the basic shape of the microdisk can be altered, for instance to a stadium shape that has bow-tie modes in a ray picture. Such shapes have classical chaotic ray orbits and are related to a general class of "quantum billiards" devices for ballistic electron transport. These have more

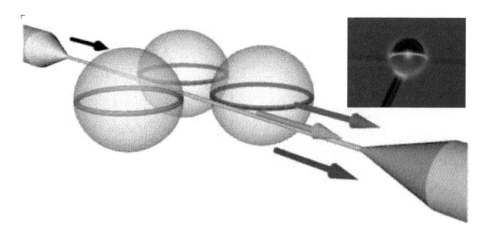

Fig. 1.17: Spherical glass resonator atop optical fibre, with lasing circular whispering-gallery mode, from Vahala (2003).

sharply curved sections, where most of the output coupling occurs and, hence, the microcavity selectively emits in particular directions. A variety of other modified shapes have been produced to induce unidirectional or spatially stabilised output (Fig. 1.16).

Since 2D whispering-gallery structures can be conveniently produced by conventional lithography processes, they are suited to dense integration applications. However, the main issue is that of obtaining efficient outcoupling without destroying the high Q-factor and introducing huge fabrication sensitivity. By using high refractive index contrast GaAs on AlO_x disks, Q-factors $> 10^4$ have been observed in $2\,\mu$m diameter whispering-gallery resonators. Nevertheless, one of the remaining issues is the difficulty of tuning the wavelength of these cavity modes and obtaining access to the internal electric field.

1.6.2   *Three-dimensional whispering-galleries*

Very similar behaviour to the 2D disks is observed for 3D structures. The difference is that the confinement in the third dimension is now provided not by a thin waveguide, but by additional total internal reflection.

The simplest example is the spherical microcavity, which can be simply formed by melting the end of a drawn optical fibre. The resulting sphere produced by surface tension of the glass perches on top of the remaining fibre and for light resonating around the equator forms an excellent high-Q cavity (Fig. 1.17).

Because of the increasing evanescent loss as the curvature increases, the highest Q ($\approx 10^9$) cavities are found in the largest spheres ($> 100\,\mu$m). For micrometre-sized spheres, Q-factors drop to several 1000 and the mode spectral spacing increases to tens of nanometres. Typically, sets of azimuthal modes are seen (with the same l), with different out-of-plane mode indices, m, provided by the extra degree of freedom in 3D (see Section 2.9). These have also been spatially mapped using scanning near-field optical microscopy showing the way the orbits converge around an equatorial orbit.

As in the 2D microdisks, deformations of the sphere provide another control variable for the emission direction and mode symmetry. Investigations of elongated liquid droplets filled with dye have shown the relation between unstable classical ray orbits and diffraction. Once again, the most difficult problem is that of efficiently coupling light into and out of the whispering gallery. As before, smaller spheres show lower Q-factors due to the curvature of the interface that permits light to evanescently escape.

1.7   Photonic-crystal cavities

Photonic crystals arise from multiple photon scattering within periodic dielectrics, and also exist in 2D and 3D versions. The ideal 3D photonic crystal microcavity would be a defect in a perfect 3D photonic lattice with high enough refractive index contrast that there is a bandgap at a particular wavelength in all directions. In principle, this would provide the highest optical intensity enhancement in any microcavity, however, currently no fabrication route has yet demonstrated this. Currently, scattering determines the properties of most 3D photonic crystals.

Instead, 2D photonic crystals etched in thin high refractive index membranes have shown the greatest promise, with the vertical confinement coming from the interfaces of the membrane. These can show Q-factors exceeding 10^5, while producing extremely small mode volumes, which are advantageous for many applications. The main issues for such microcavities are the difficulty of their fabrication, the large surface area in proximity to the active region (which produces non-radiative recombination centres and traps diffusing electron–hole pairs) and the difficulty in tuning their cavity wavelengths. However, they currently offer the best intensity enhancement of any microcavity system. The SEM image of a 2D photonic crystal from Hauke et al. (2010) is shown in Fig. 1.18.

1.7.1   *Random lasers*

Related to regular 3D photonic crystals are media with wavelength-scale strong scatterers that are randomly positioned. Such photonic structures support photonic modes that can be localised, with light bouncing around loops entirely inside the medium. By placing gain materials, such as dyes inside these media, and optically pumping them, so-called "random lasing" can result. While the optical feedback is not engineered in specific orientations in these devices, it is spontaneously formed by the collection of random scatterers.

1.8   Material systems

The initial explorations of solid-state strong-coupling microcavities utilised III–V semiconductor materials because the growth of these layers was highly advanced with atomic precision in the thicknesses. This enabled characterising most of the resulting optical properties discussed in this book, although the exciton localisation from disorder is not completely controlled in these planar quantum wells. This system, predominantly using $In_xGa_{1-x}As$ or GaAs quantum wells in GaAs/AlGaAs DBRs, remains the most reliable realisation of planar microcavities in the strong coupling regime and, thus, remains a workhorse system of the community. However, the major issue is the limited light-matter coupling strength from such III–V emitters, which restricts the energy splitting

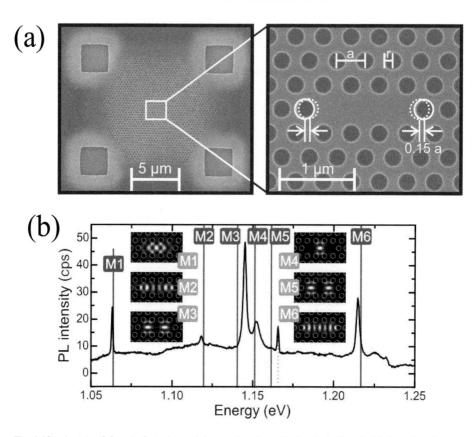

Fig. 1.18: A state-of-the-art photonic crystal resonator, from Hauke et al. (2010). (a) Scanning electron microscopy image with a zoom of the three-hole defect, showing the great control in etching the lattice (with period a) of holes (of radius r), here in silicon material. The accuracy in the hole curvature is ± 2 nm. The outer holes are shifted by 0.15 times the lattice constant to increase the Q factor. (b) Room temperature emission spectra from six modes of the cavity (with $r/a = 0.28$), along with calculated electric field profiles (insets) and their corresponding energy (vertical lines).

arising from the strong coupling and, thus, restricts the maximum temperature at which many polariton effects are observed. To enable room temperature devices, larger strong coupling is required. This is possible for high bandgap semiconductors, such as GaN or ZnO, organic emitters (both polymers and small molecules) and, most recently, transition metal chalcogenides (TMCs). While a number of these show strong coupling at room temperature and pulsed polariton lasing has been observed, none of them yet show the polariton stimulated scattering because of the reduced polariton–polariton nonlinear scattering caused by smaller excitonic Bohr radii and, by localisation, preventing polariton movement. The photonic disorder in these microcavities has also led to highly-localised and highly variable polariton lasing, without the superfluid effects that can be observed in the III–V system.

1.8.1   *GaN microcavities*

Historically, it was realised already over a decade ago that GaN planar microcavities could reach the strong coupling regime. However, it took a number of years before the technology of material growth could enable testing of these structures. The main problem remains how to produce high-quality DBR mirrors from the semiconductor alloy stacks based on the same III–N material systems, without inducing so much strain that cracks appear and photonic disorder limits the cavity Q. The difficulty of lattice matching different alloys, while ensuring they have a sufficient index contrast to give high reflectivity has led to opposite strategies. Using InGaN and AlInN alloys gives a rare lattice matching, but the small index contrast requires many mirror pairs to be grown and produces stopband spectral widths which are 60 nm wide. A different approach grows the inner cavity spacer in the III-N materials (either bulk or quantum wells), but then uses selective chemical etching together with deposition of dielectric DBRs on the top and bottom to form the microcavity. In this case the SiO_2/Ta_2O_5 mirror stack has large index contrast and thus spectral stopband width, but controlling the thickness of the cavity and thus ensuring strong coupling conditions is difficult. A further problem with this mixed approach is that electrical contacting is very difficult since the dielectric stacks are insulating.

However the first successful GaN-based microcavities were achieved in 2007, allowing Christopoulos et al. (2007) to show strong coupling at room temperature, and Baumberg et al. (2008) to produce polariton condensation when pumped with pulsed UV sources. While these used bulk GaN as the cavity spacer with GaN-based DBRs, realisation with quantum wells was later achieved. More recently, an orthogonal approach has been adopted for enabling electrical pumping, with GaN mesas or nanowires grown on doped substrates, and then these side-coated with DBRs so that the microcavity mode emits parallel to the substrate. This allows electrical injection from top contacts, along with room temperature operation.

1.8.2   *ZnO microcavities*

The technology for ZnO planar growth is less advanced than in GaN, and most approaches have used pulsed laser deposition to create multilayers of cavity and DBRs, with Mn or Mg alloys to give index contrast and quantum wells. The large resulting photonic disorder has made observations in these systems less convincing, though lasing has been reported at room temperature. On the other hand, nanowires of ZnO have been more successful and proven to give a version of 1D polaritons, which can also produce polariton lasing under pulsed excitation.

1.8.3   *Organic microcavities*

Several classes of planar organic microcavities in the strong coupling regime have been reported. The most successful ones are based on crystalline layers of conjugated molecules such as porphyrins, in dielectric cavities, which have large collective oscillator strengths and small molecular linewidths with little Stokes shift between absorption and re-emission. Typically, the vibronic coupling that gives phonon sidebands to the absorption lines is also mixed into the strong coupling, developing more congested mixed anticrossing spectra.

The second class of organic microcavities uses small molecules dispersed in a polymeric or gel matrix as the active emitter within the microcavity. A large variety of dye molecules have been successfully used, within both dielectric or metallic mirror cavities. Dye molecules which J-aggregate to produce narrow linewidth high-oscillator strength emitters are also effective. Excitation of these small molecule systems has required the use of co-doping with molecules that can absorb optical pumps at higher energies and then transfer the energy between the molecular species onto the narrow-linewidth dyes, which typically cannot be directly excited at higher energies. This has allowed demonstration of short-pulse excited polariton lasing.

1.8.4 *Transition metal chalcogenides*

Recent advances in fabrication of Transition Metal Chalcogenides (TMCs) have allowed their incorporation into dielectric planar microcavities. At room temperature the 2D confinement of excitons into sheets produces large exciton binding energies, with typical 20 nm linewidths. Although the oscillator strengths are reasonably comparable with molecular layers, it is hard to obtain strong coupling with single TMC layers, such as MoS_2 or $MoSe_2$ without additional photonic confinement, for instance in mesa- or curved-mirror cavities. In addition, the lack of large-area high-quality Chemical Vapour Deposition (CVD) layers of these TMC 2D sheets hampers production of microcavities. It is clear that the major focus on fabrication technologies in the next few years may have a major impact on the availability of such room temperature emitters, and high-quality strong coupling microcavities are an attractive prospect, which may enable observation of nonlinearities and superfluidity at room temperature.

1.8.5 *Plasmonic nanocavities*

Recently, a new class of microcavity has emerged which is based on plasmons localised to ultra-small volumes close to metals. For noble metals such as Au or Ag, a class of localised electromagnetic modes exists at their interface with a dielectric, known as "surface-plasmon polaritons". If the metals are textured on or below the scale of the optical wavelength, these plasmons can be localised in all three directions, producing 0D plasmonic modes. Five implementations are of note: flat metals, metallic voids, spherical metal spheres, coupled metal spheres and nanoparticle-on-mirror, in order of increasingly confined optical fields. They are shown in Fig.1.19. While plasmons bound to flat metal surfaces are free to move along the surface, the plasmons on nanostructures can be tightly localised.

As in any microcavity, these confined modes can be coupled to other excitations, such as excitons in semiconductors, and this has now become of interest as the plasmonic spatial extent can be significantly smaller than the wavelength of light, down to below 1 nm. On the other hand, the presence of absorption from the metal due to plasmon-induced excitation of single electrons places strict limits of the utility of

Fig. 1.19: Plasmonic localisation: on flat noble metals, metallic voids, metal spheres, between metal spheres, and between a metal sphere and mirror ('nanoparticle-on-mirror').

these plasmonic modes. A number of groups have shown that strong coupling is possible for surface plasmons interacting with dyes such as J-aggregates, for instance, in Ag nano-hole arrays. Recent experiments have shown that the Q/V for nanoparticle-on-mirror nanocavities can exceed that of all other microcavities and has enabled the demonstration of strong coupling with single molecules at room temperature in ambient conditions. The development of these plasmonic technologies may thus enable the next generation of quantum emitter devices and nonlinear behaviour at room temperature.

1.9   Microcavity lasers

Microcavities can be used as laser resonators, providing the gain is large enough to make up for the cavity losses. Because of the short round-trip length, the conditions on the reflectivity of the cavity walls are severe. However, these cavities are now the structure of choice (in the form of VCSELs), due to their ease of integratable manufacture and their performance. Small cavity volumes are also advantageous for producing low threshold lasers as the condition for inversion can be reached by pumping fewer electronic states. On the other hand the total power produced from a microcavity is, in general, restricted as eventually the high power density causes problems of thermal loading, extra electronic scattering and saturation. One advantage of a microcavity laser is the reduced number of optical modes into which spontaneous emission is directed, thus increasing the probability of spontaneous emission in a particular mode and thus reducing the lasing threshold. This is discussed in detail in Chapter 6. Polariton lasers are expected to have a lower threshold than VCSELs as they do not require inversion of population (see Chapter 8). They represent one of the currently hottest subjects in the physics of semiconductors. Their characteristics are discussed in detail in the final two chapters.

Another use for microcavities is as nonlinear devices and bistability for optical logic operations has been demonstrated in several different ways. Related to this is the development of trapping potentials for polaritons, either by lithography or by use of carefully sculpted pump profiles, which provide a computer-controlled potential landscape for polariton condensates. These open the way to create arrays of interacting polariton condensates with both on-site and nearest-neighbour Josephson-like interactions, which can produce a variety of unusual states. Along lines suggested by cold atom lattice experiments, the idea is to use such condensate arrays to speed up NP-hard computational problems and since the interactions are very strong in semiconductor polariton systems, such computational solutions could potentially be extremely fast. Such aspects are discussed in more detail in Chapter 11.

The discovery of polariton interferometers, either directly resonantly injected or non-resonantly formed in shaped potentials, offers opportunities for creating sensors for inertial sensing or local magnetic fields. A chip-based version current Sagnac fibre loop technologies would be compact and highly desirable. These are discussed in the final chapter.

Finally, recent observations of the spontaneous magnetisation of condensates that are trapped in optical potentials provide opportunities for true (quantum-based) random-number generators. Streams of random numbers form the basis of cryptography, increasingly in demand for internet-based financial transactions. Quantum-based physical implementations would greatly improve current pseudo-random streams, which can be attacked by modern computation. Condensates that randomly polarise spin-up or spin-down can generate multi-Gb streams of unbiased random numbers on demand and potentially be integrated into large arrays capable of much higher speeds.

1.10   Conclusion

This survey of microcavities shows the wide range of possible designs and blend of optical physics, microfabrication and semiconductor engineering that is required to understand them. These will be taken apart and then put back together in the following chapters.

2

CLASSICAL DESCRIPTION OF LIGHT

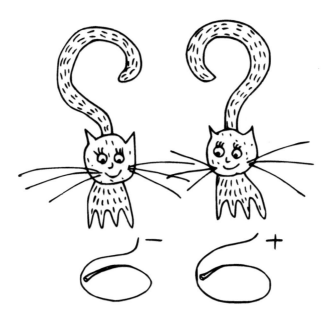

In this chapter we introduce the basic characteristics of light modes in free space and in different kinds of optically-confined structures, including Bragg mirrors, planar microcavities, pillars and spheres. We describe the powerful transfer matrix method that allows for solution of Maxwell's equations in multilayer structures. We discuss the polarisation of light and mention different ways it is modified, including the Faraday and Kerr effects, optical birefringence, dichroism and optical activity.

Microcavities, Second Edition. Alexey V. Kavokin, Jeremy J. Baumberg, Guillaume Malpuech, Fabrice P. Laussy, Oxford University Press (2017). © Alexey V. Kavokin, Jeremy J. Baumberg, Guillaume Malpuech, Fabrice P. Laussy. DOI 10.1093/oso/9780198782995.001.0001

2.1 Free space

2.1.1 *Light-field dynamics in free space*

We start from Maxwell (1865)'s equations in a vacuum (setting the charge density distribution ρ and electric current $\mathbf{J}$ identically to zero):[5]

$$\nabla \cdot \mathbf{E}(\mathbf{r}, t) = 0, \qquad \nabla \times \mathbf{E}(\mathbf{r}, t) = -\partial_t \mathbf{B}(\mathbf{r}, t), \tag{2.1a}$$

$$\nabla \cdot \mathbf{B}(\mathbf{r}, t) = 0, \qquad \nabla \times \mathbf{B}(\mathbf{r}, t) = \frac{1}{c^2} \partial_t \mathbf{E}(\mathbf{r}, t), \tag{2.1b}$$

where $\mathbf{E}$ and $\mathbf{B}$ are the electric and magnetic fields, respectively, both a function of spatial vector $\mathbf{r}$ and time t. However, the equations are much simpler in reciprocal space where fields are expressed as a function of wavevector instead of position, helping to lay down a simpler mathematical structure. The link between the field as we introduced it $\mathbf{E}(\mathbf{r}, t)$ and its weight $\mathscr{E}(\mathbf{k}, t)$ in the plane-wave basis $e^{i\mathbf{k}\cdot\mathbf{r}}$ is assured by Fourier transforms:

$$\mathbf{E}(\mathbf{r}, t) = \frac{1}{(2\pi)^{3/2}} \int \mathscr{E}(\mathbf{k}, t) e^{i\mathbf{k}\cdot\mathbf{r}} \, d\mathbf{k}, \tag{2.2a}$$

$$\mathscr{E}(\mathbf{k}, t) = \frac{1}{(2\pi)^{3/2}} \int \mathbf{E}(\mathbf{r}, t) e^{-i\mathbf{k}\cdot\mathbf{r}} \, d\mathbf{r}. \tag{2.2b}$$

James Clerk **Maxwell** (1831–1879) put together the knowledge describing the basic laws of electricity and magnetism to set up the consistent set of equations (2.1) that form the foundations of electromagnetism.

One of the giants of science, Einstein said of his work it was the "*most profound and the most fruitful that physics has experienced since the time of Newton.*", Planck that "*he achieved greatness unequalled*" and Feynman that "*the most significant event of the 19th century will be judged as Maxwell's discovery of the laws of electrodynamics*". A deeply religious man, Maxwell composed prayers that were later found in his notes. He once said that he was thanking "*God's grace helping me to get rid of myself, partially in science, more completely in society.*"

---

[5] Maxwell's equations in the presence of charge and currents are, in MKS units:

$$\nabla \cdot \mathbf{E}(\mathbf{r}, t) = \frac{1}{\varepsilon_0} \rho(\mathbf{r}, t), \qquad \nabla \times \mathbf{E}(\mathbf{r}, t) = -\partial_t \mathbf{B}(\mathbf{r}, t), \tag{i-a}$$

$$\nabla \cdot \mathbf{B}(\mathbf{r}, t) = 0, \qquad \nabla \times \mathbf{B}(\mathbf{r}, t) = \frac{1}{c^2} \partial_t \mathbf{E}(\mathbf{r}, t) + \frac{1}{\varepsilon_0 c^2} \mathbf{J}(\mathbf{r}, t). \tag{i-b}$$

The properties of differential operators with respect to Fourier transformation turn Maxwell's equations (2.1) into the set of *local equations*:

$$ik \cdot \mathscr{E}(\mathbf{k}, t) = 0, \qquad i\mathbf{k} \times \mathscr{E}(\mathbf{k}, t) = -\partial_t \mathscr{B}(\mathbf{k}, t), \tag{2.3a}$$

$$i\mathbf{k} \cdot \mathscr{B}(\mathbf{k}, t) = 0, \qquad i\mathbf{k} \times \mathscr{B}(\mathbf{k}, t) = \frac{1}{c^2}\partial_t \mathscr{E}(\mathbf{k}, t), \tag{2.3b}$$

with obvious notations: the cursive letter and its straight counterpart are Fourier transform pairs related the one to the other like eqn (2.2). The locality is a first appealing feature, as opposed to real space equations (2.1) where the value of a field at a point depends on other field values in an entire neighbourhood of this point. The picture clarifies further still by separating the transverse $\perp$ and longitudinal $\parallel$ components of the field, $\mathscr{E}(\mathbf{k}, t) = \mathscr{E}_\perp(\mathbf{k}, t) + \mathscr{E}_\parallel(\mathbf{k}, t)$ (with similar definitions for other fields), the longitudinal component at point $\mathbf{k}$ being the projection on the unit vector $\mathbf{e_k} \equiv \mathbf{k}/k$, i.e.,

$$\mathscr{E}_\parallel(\mathbf{k}, t) = \left(\mathbf{e_k} \cdot \mathscr{E}(\mathbf{k}, t) \right) \mathbf{e_k}, \tag{2.4}$$

(with likewise definitions for other fields) and the perpendicular component at point $\mathbf{k}$ being the projection on the plane normal to the unit vector $\mathbf{e_k}$, i.e., $\mathscr{E}_\perp(\mathbf{k}, t) \equiv \mathscr{E}(\mathbf{k}, t) - \mathscr{E}_\parallel(\mathbf{k}, t)$.

Back in real space, $\mathbf{E}_\perp(\mathbf{r}, t)$ and $\mathbf{E}_\parallel(\mathbf{r}, t)$ are obtained, respectively, by Fourier transform of $\mathscr{E}_\parallel(\mathbf{k}, t)$ and $\mathscr{E}_\perp(\mathbf{k}, t)$ as given above, which correspond to the divergence-free and the curl-free components of the field.[6] The field is transverse in vacuum (the magnetic field always is) and these transverse components obey the set of coupled linear equations derived from eqn (2.3):

$$\begin{cases} i\mathbf{k} \times \mathscr{E}_\perp(\mathbf{k}, t) = -\partial_t \mathscr{B}(\mathbf{k}, t) \\ ic^2 \mathbf{k} \times \mathscr{B}(\mathbf{k}, t) = \partial_t \mathscr{E}_\perp(\mathbf{k}, t). \end{cases} \tag{2.5}$$

This linear system is diagonalised by introducing the new mode amplitudes $\mathbf{a}$ and $\mathbf{b}$ as

$$\mathbf{a}(\mathbf{k}, t) \equiv -\frac{i}{2\mathscr{C}(\mathbf{k})}\left(\mathscr{E}_\perp(\mathbf{k}, t) - c\mathbf{e_k} \times \mathscr{B}(\mathbf{k}, t)\right), \tag{2.6a}$$

$$\mathbf{b}(\mathbf{k}, t) \equiv -\frac{i}{2\mathscr{C}(\mathbf{k})}\left(\mathscr{E}_\perp(\mathbf{k}, t) + c\mathbf{e_k} \times \mathscr{B}(\mathbf{k}, t)\right), \tag{2.6b}$$

---

[6]The benefits of decomposing the field into its transverse and longitudinal components are more compelling when charges are taken into account, as is the case in eqn (i-). Then, the dynamics of the field arising from the interplay between its electric and magnetic components (the transverse part) and the dynamics created by the sources (responsible for the static field if they are at rest) are consequently clearly separated. For instance, dotting eqn (2.4) with $\mathbf{k}/k^2$, one gets from the $\mathscr{E}_\parallel(\mathbf{k}, t) = -i\varrho(\mathbf{k})\mathbf{k}/(\varepsilon_0 k^2)$, which, by Fourier transformation of both sides, gives the electric field well-known from electrostatic,

$$\mathbf{E}_\parallel(\mathbf{r}, t) = \frac{1}{4\pi\varepsilon_0} \int \rho(\mathbf{r}', t)\frac{\mathbf{r} - \mathbf{r}'}{|r - r'|^3}\, d\mathbf{r}'. \tag{ii}$$

The fact that eqn (ii) is local in time is a mathematical artifice: only the whole field $\mathbf{E}(\mathbf{r}, t)$ has a physical significance and other instantaneous effects from the transverse field correct those of the longitudinal field.

with $\mathscr{C}$ a real constant as yet undefined (written as such for later convenience).[7] Since $\mathbf{E}$ and $\mathbf{B}$ are real, relation (2.2) implies that $\mathscr{E}(-\mathbf{k}, t)^* = \mathscr{E}(\mathbf{k}, t)$ (the same for $\mathscr{B}$), which allows us to keep only one variable, e.g., $\mathbf{a}$, and write $\mathbf{a}(-\mathbf{k}, t)^*$ instead of $\mathbf{b}(\mathbf{k}, t)$. Inverting, one obtains the expression of the physical fields (in reciprocal space) in terms of its fundamental modes

$$\mathscr{E}_\perp(\mathbf{k}, t) = i\mathscr{C}(\mathbf{k})(\mathbf{a}(\mathbf{k}, t) - \mathbf{a}^*(-\mathbf{k}, t)) \tag{2.7a}$$

$$\mathscr{B}(\mathbf{k}, t) = \frac{i\mathscr{C}(\mathbf{k})}{c}(\mathbf{e}_\mathbf{k} \times \mathbf{a}(\mathbf{k}, t) + \mathbf{e}_\mathbf{k} \times \mathbf{a}^*(-\mathbf{k}, t)). \tag{2.7b}$$

This mathematical formulation has therefore replaced the electric and magnetic fields by a set of complex variables $\mathbf{a}(\mathbf{k}, t)$ (which give the transverse components of the fields) and the phase space variables of sources $(\mathbf{r}_i, \partial_t \mathbf{r}_i/m)$. All quantities of interest that can be expressed in terms of the fields and the dynamical variables can naturally be written with the new set of variables. The energy of the field relevant for our description of the vacuum comes with the Hamiltonian

$$H_\perp = \varepsilon_0 \int \mathscr{C}(\mathbf{k})^2 [\mathbf{a}^*(\mathbf{k}) \cdot \mathbf{a}(\mathbf{k}) + \mathbf{a}(-\mathbf{k}) \cdot \mathbf{a}^*(-\mathbf{k})] \, d\mathbf{k}. \tag{2.8}$$

As a function of this new variable for the field, eqn (2.5) becomes[7]

$$\begin{cases} \partial_t \mathbf{a}(\mathbf{k}, t) = i\omega \mathbf{a}(\mathbf{k}, t) \\ \partial_t \mathbf{a}^*(-\mathbf{k}, t) = -i\omega \mathbf{a}^*(-\mathbf{k}, t) \end{cases}, \tag{2.9}$$

where $\omega = ck$. This is the main result that asserts that *the free electromagnetic field is equivalent to a set of harmonic oscillators*.[8] These oscillators—the complex-valued vectors $\mathbf{a}$ of eqn (2.9)—result from the mathematical manipulations that we have detailed. The physical sense attached to such an oscillator is rather meager in its classical formulation, being at best described as a modal amplitude for the field and therefore chiefly as a mathematical concept. However, this will lead to a straightforward and canonical quantisation scheme.

---

[7] The diagonalisation of eqn (2.5) can be made by evaluating the crossproduct of $\mathbf{k}$ with both sides of the first line, yielding on the l.h.s. $i\mathbf{k} \times (\mathbf{k} \times \mathscr{E}_\perp) = -ik^2\mathscr{E}_\perp$ since $\mathscr{E}_\perp$ is transverse, and on the r.h.s. $-\partial_t(\mathbf{k} \times \mathscr{B})$. Introducing $\omega = ck$, eqn (2.5) then reads

$$\begin{cases} i\omega\mathscr{E}_\perp(\mathbf{k}, t) = \partial_t(c\mathbf{e}_\mathbf{k} \times \mathscr{B}(\mathbf{k}, t)) \\ i\omega c\mathbf{e}_\mathbf{k} \times \mathscr{B}(\mathbf{k}, t) = \partial_t\mathscr{E}_\perp(\mathbf{k}, t). \end{cases} \tag{iii}$$

Summing and subtracting both lines yields

$$\partial_t(\mathscr{E}_\perp(\mathbf{k}, t) \pm c\mathbf{e}_\mathbf{k} \times \mathscr{B}(\mathbf{k}, t)) = \pm i\omega(\mathscr{E}_\perp(\mathbf{k}, t) \pm c\mathbf{e}_\mathbf{k} \times \mathscr{B}(\mathbf{k}, t)) \tag{iv}$$

which, when expressed in terms of $\mathbf{a}(\mathbf{k}, t)$ and $\mathbf{b}(\mathbf{k}, t) = \mathbf{a}(-\mathbf{k}, t)^*$, is the main result (2.9).

[8] Inserting back eqn (2.6) in Maxwell's equations with source terms, eqns (i-), the equations of motion obtained following the same procedure become

$$\partial_t\mathbf{a}(\mathbf{k}, t) = -i\omega\mathbf{a}(\mathbf{k}, t) + \frac{i}{2\varepsilon_0\mathscr{C}(\mathbf{k})}\mathscr{J}_\perp(\mathbf{k}, t). \tag{v}$$

Generally, $\mathscr{J}_\perp(\mathbf{k}, t)$ depends non-locally on $\mathbf{a}$. Interactions with sources therefore couple together the various modes of the fields that are otherwise independent.

2.2 Propagation in crystals

2.2.1 *Plane waves in bulk crystals*

We now describe propagation in dielectric or semiconductor materials. Maxwell's equations get upgraded to the following closely-related expressions:

$$\nabla \cdot \mathbf{D} = \frac{\rho}{\varepsilon_0}, \quad (2.10a) \qquad\qquad \nabla \times \mathbf{E} = -\frac{1}{c}\partial_t \mathbf{B}, \qquad\qquad (2.10c)$$

$$\nabla \cdot \mathbf{B} = 0, \quad (2.10b) \qquad\qquad \nabla \times \mathbf{B} = \frac{1}{\varepsilon_0 c^2}\mathbf{J} + \frac{1}{c^2}\partial_t \mathbf{D}, \qquad (2.10d)$$

where ρ is the *free* electric charge density[9] and $\mathbf{J}$ is the free current density.[10] In the following, we only describe dielectric or semiconductor materials where $\rho = 0$ and $\mathbf{J} = 0$. The *electric displacement field* $\mathbf{D}$ and the magnetic $\mathbf{H}$-field are defined as, respectively,

$$\mathbf{D} = \varepsilon_0 \mathbf{E} + \mathbf{P} = \varepsilon \mathbf{E}, \qquad\qquad (2.11a)$$

$$\mathbf{H} = \frac{1}{\mu_0}\mathbf{B} - \mathbf{M} = \mathbf{B}/\mu, \qquad\qquad (2.11b)$$

where, on the one hand, $\mathbf{P}$ is the *dielectric polarisation* vector and ε is the *dielectric constant* and, on the other hand, $\mathbf{M}$ is the *magnetisation field* and μ is the permeability. In the most general case, ε is a tensor. By a proper choice of the system of coordinates it can be represented as a diagonal matrix

$$\varepsilon = \begin{pmatrix} \varepsilon_1 & 0 & 0 \\ 0 & \varepsilon_2 & 0 \\ 0 & 0 & \varepsilon_3 \end{pmatrix}. \qquad\qquad (2.12)$$

If the diagonal elements of this matrix are not equal to each other, the crystal is optically *anisotropic*. The effect of optical birefringence specific for optically anisotropic media is briefly discussed at the end of this section. Throughout this book, we always assume $\varepsilon_1 = \varepsilon_2 = \varepsilon_3 = \varepsilon$ (unless it is explicitly indicated that we consider an anisotropic case). We largely operate with the quantity $n = \sqrt{\varepsilon/\varepsilon_0}$ known as the *refractive index* of the medium. In crystals having resonant optical transitions, eqns (2.10) have two types of solutions. One of them is given by the condition

$$\nabla \cdot \mathbf{E} = 0. \qquad\qquad (2.13)$$

It corresponds to the transverse waves having electric and magnetic field vectors perpendicular to the wavevector. The transverse waves in media are analogous to the light waves in vacuum. Another type of solution are the longitudinal waves, for which:

$$\varepsilon = 0, \qquad \nabla \cdot \mathbf{E} \neq 0, \qquad \mathbf{B} = \mathbf{0}, \qquad \nabla \times \mathbf{E} = \mathbf{0}. \qquad (2.14)$$

In these modes the electric field is parallel to the wavevector and the magnetic field is equal to zero everywhere. They have no analogy in vacuum (which has $\varepsilon > 0$ at all

---

[9]Free electric charges are those that do not include dipole charges bound in the material.

[10]Free current densities are those that do not include polarisation or magnetisation currents bound in the material.

frequencies), but play an important role in resonant dielectric media where the dielectric constant can vanish at some frequencies. In any case, the solution of eqns (2.10) can be represented as a linear combination of plane waves where the coordinate dependence of the electric field is given by

$$\mathbf{E}(\mathbf{r}, t) = \mathbf{E}_0 \exp\left(i(\mathbf{k} \cdot \mathbf{r} - \omega t)\right). \tag{2.15}$$

$\mathbf{k}$ is the *wavevector* of light and its modulus obeys

$$k = n\frac{\omega}{c}, \tag{2.16}$$

and $\mathbf{E}_0$ is the amplitude of the plane wave. Its vector character is responsible for the polarisation of light. Note that the choice of the sign in the exponential factor of eqn (2.15) is a matter of convention. Note also that the vector can change with time even for a freely propagating plane wave. The longitudinal waves are linearly polarised along the wavevector, by definition. For the transverse waves it is convenient to take the curl of both parts of eqn (2.10c) and substitute the expression for $\nabla \times \mathbf{B}$ from Maxwell's eqn (2.10d). This yields

$$\nabla \times \nabla \times \mathbf{E} = \frac{n^2}{c^2} \partial_t \mathbf{E}, \tag{2.17}$$

therefore, from the identity $\nabla \times \nabla \times \mathbf{A} = \nabla(\nabla \cdot \mathbf{A}) - \nabla^2 \mathbf{A}$ (for any field $\mathbf{A}$) and substituting eqn (2.15), one obtains the wave equation

$$\nabla^2 \mathbf{E} = -k^2 \mathbf{E}. \tag{2.18}$$

The general form of the polarisation for transverse waves can be seen as follows. Consider a plane wave propagating in the z-direction. The vector $\mathbf{E}_0$ can have x- and y-components, in this case,

$$\mathbf{E} = \begin{pmatrix} E_x \\ E_y \end{pmatrix} = \begin{pmatrix} E_{0x} \cos(kz - \omega t) \\ E_{0y} \cos(kz - \omega t + \delta) \end{pmatrix}, \tag{2.19}$$

with[11] $\delta \in [0, 2\pi[$. This vector is the real part of the *Jones vector* of the polarised light introduced by Jones (1941) concisely describing the light polarisation. Since the phase constants can be picked arbitrarily for one component of the vector, we set it to 0 for E_x. After an elementary transformation,

$$\frac{E_y}{E_{y0}} = \cos(kz - \omega t + \delta) = \cos(kz - \omega t) \cos\delta - \sin(kz - \omega t) \sin\delta, \tag{2.20}$$

which yields

---

[11] The notation $[a, b[$ means all the values between a *included* and b *excluded*. It is a common notation in countries such as France or Russia. Another widespread convention uses parenthesis for exclusion, which would read $[a, b)$ in our case. Both are an occasional source for confusion, but are recognised standards of the ISO 31-11 that regulates mathematical signs and symbols for use in physical sciences and technology.

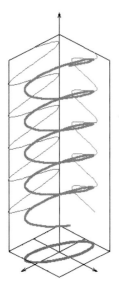

Fig. 2.1: Elliptically polarised light. The curve shows the trajectory followed by the electric-field vector of the propagating wave together with its projections on x- and y-axes and on the plane normal to the direction of motion where the "elliptical" character of the polarisation becomes apparent.

$$\frac{E_x}{E_{x0}}\cos\delta - \frac{E_y}{E_{y0}} = \sin(kz - \omega t)\sin\delta = \sqrt{1 - \left(\frac{E_x}{E_{x0}}\right)^2}\sin\delta\,, \qquad (2.21)$$

and squaring, we finally obtain

$$\left(\frac{E_x}{E_{x0}}\right)^2 + \left(\frac{E_y}{E_{y0}}\right)^2 - 2\frac{E_x}{E_{0x}}\frac{E_y}{E_{0y}}\cos\delta = \sin^2\delta\,, \qquad (2.22)$$

which is the equation for an ellipse in the (E_x, E_y) coordinate system, inclined at an angle ϕ to the x-axis given by

$$\tan(2\phi) = \frac{2E_{0x}E_{0y}\cos\delta}{E_{0x}^2 - E_{0y}^2}\,. \qquad (2.23)$$

The general polarisation is therefore elliptical, as shown in Fig. 2.1. If $\delta = 0$ or π, light is linearly polarised, if $\delta = \frac{\pi}{2}$ or $\frac{3\pi}{2}$ and $E_{0x} = E_{0y}$, it is *circularly polarised*. We call *right circular polarisation* denoted σ^+ for the case where $\delta = \frac{\pi}{2}$, and *left-circular polarisation*, denoted σ^-, for the case where $\delta = \frac{3\pi}{2}$.

The electric-field vector of the circularly polarised light rotates around the wavevector in the clockwise direction or anticlockwise directions for σ^+ and σ^- polarisation, respectively (if one looks in the direction of propagation of the wave).

In reality, light is usually composed of an ensemble of plane waves of the form (2.15) with their phases more or less randomly distributed. As a result, light can be partially polarised or unpolarised. In this case its intensity $I_0 > (|E_{0x}|^2 + |E_{0y}|^2)/2$, while the equality holds for fully polarised light. It is convenient to characterise the partially polarised light by so-called *Stokes parameters* proposed by the English physicist and mathematician Stokes (1852) in 1852. The Stokes parameters $S_{0,1,2,3}$ are defined as a

Henri **Poincaré** (1854–1912) and George Gabriel **Stokes** (1819–1903).

Poincaré was a French mathematician and philosopher. The number of significant contributions he made in numerous fields is overwhelming. Among extraordinary achievements, he discovered the first chaotic deterministic system and understood correctly ahead of his generation the implication of chaos. He formulated the "Poincaré conjecture" one of the most difficult mathematical problems and one of the most important in topology, only recently solved by Grigori Perelman. The Poincaré sphere arose in the context of this conjecture. He was the first to unravel the complete mathematical structure of special relativity—christening Lorentz transformations that he showed form a group—though he missed the physical interpretation. He never acknowledged Einstein's contribution, still referring to it by the end of his life as the "mechanics of Lorentz". He is also famous for the philosophical debates that opposed him to the British philosopher Bertrand Russel. He liked to change the problem he was working on frequently, as he thought the subconcious would still work on the old problem as he would on the new one, a good match to his famous saying "*Thought is only a flash between two long nights, but this flash is everything.*"

Sir George Gabriel Stokes was an Irish mathematician and physicist, also with exceptional productivity in a wide arena of science. He is most renowned for his work on fluid dynamics (especially for the Navier–Stokes equation), mathematical physics (with the Stokes theorem) and optics (with his description of polarisation). In 1852 he published a paper on frequency changes of light in fluorescence, explaining the "Stokes shift". His production is even more remarkable given that he kept unpublished many of his first-rate discoveries, like Raman scattering in the aforementioned work, that Lord Kelvin begged him without success to bring to print, or the spectroscopic techniques (in the form of chemical identification by analysis of the emitted light), which he merely taught to Lord Kelvin (then Sir William Thomson), predating Kirchoff by almost a decade. In a letter he humbly attributed the entire merit to the latter, saying that some of his friends had been over-zealous in his cause. He had a tumultuous mixture of professional and sentimental feelings (not aided by the historical context in Cambridge), and it is reported that his bride-to-be almost called off the wedding upon receiving a 55-page letter of the duties he felt obliged to remind her.

function of the total intensity I_0, the intensity of horizontal (x-linear) polarisation I_1, the intensity of linear polarisation at a $45°$ angle I_2 and the intensity of left-handed circularly polarised light I_3 and then defining

$$S_0 = 2I_0 , \qquad (2.24a)$$

$$S_1 = 4I_1 - 2I_0 , \qquad (2.24b)$$

$$S_2 = 4I_2 - 2I_0 , \qquad (2.24c)$$

$$S_3 = 4I_3 - 2I_0 . \qquad (2.24d)$$

These are often normalised by dividing by S_0. For the fully polarised light, they can also be re-expressed in terms of the electric field as

$$S_0 = E_{0x}^2 + E_{0y}^2 \,, \tag{2.25a}$$

$$S_1 = E_{0x}^2 - E_{0y}^2 \,, \tag{2.25b}$$

$$S_2 = 2E_{0x}E_{0y}\cos\delta \,, \tag{2.25c}$$

$$S_3 = 2E_{0x}E_{0y}\sin\delta \,. \tag{2.25d}$$

In this form, it is clear that a relationship exists connecting these parameters:

$$S_1^2 + S_2^2 + S_3^2 = S_0^2 \,. \tag{2.26}$$

For partially polarised light this equality is not satisfied.

The components of the Stokes vector have a direct analogy with the components of the quantum-mechanical pseudospin introduced for a two-level system, as will be shown in detail in Chapter 9. Condition (2.26) defines a sphere in the (S_1, S_2, S_3) set of coordinates, called the *Poincaré sphere* (Fig. 2.2), introduced by the French mathematician Poincaré (Malykin (1997) gives a good review).

Partially polarised signals can be represented by augmenting the Poincaré sphere with another sphere whose radius is the total signal power, I. The ratio of the radii of the spheres is the degree of polarisation $p = I_p/I$. The difference of the radii is the unpolarised power $I_u = I - I_p$. By normalising the total power I to unity, the inner Poincaré sphere has radius p and will shrink or grow in diameter as the degree of polarisation changes. Modifications in the polarised part of the signal cause the polarisation state to move on the surface of the Poincaré sphere.

2.2.2 Absorption of light

Absorbing media are characterised by a complex refractive index with a positive imaginary part in the convention adopted here, see eqn (2.15):

$$n = \tilde{n} + i\kappa \,. \tag{2.27}$$

An electromagnetic wave propagating in a given direction (say, z-direction) is evanescent in this case and can be represented as

Fig. 2.2: Three-dimensional representation of the *Poincaré sphere*. The Stokes parameters constitute the cartesian coordinates of the polarisation state, which is represented by a point on the surface of the sphere. The radius of the Poincaré sphere corresponds to the total intensity of the polarised part of the signal.

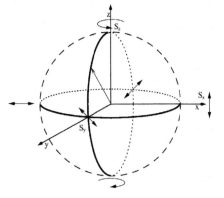

$$\mathbf{E} = \begin{pmatrix} E_{0x}\cos(n\frac{\omega}{c}z - \omega t) \\ E_{0y}\cos(n\frac{\omega}{c}z - \omega t + \delta) \end{pmatrix} \exp(-\kappa\frac{\omega}{c}z). \qquad (2.28)$$

The parameter

$$\alpha = \kappa\frac{\omega}{c} \qquad (2.29)$$

is known as the absorption coefficient. α^{-1} is a typical penetration depth of light in an absorbing medium. It can be as short as a few tens of nanometres in some metals.

2.2.3 Kramers–Kronig relations

The German physicist Ralph Kronig and the Dutch physicist Hendrik Anthony Kramers have established a useful relation between the real and imaginary parts of the dielectric constant, known as the "Kramers–Kronig relations":[12]

$$\text{Re}\{\varepsilon(\omega)\} = \varepsilon_0 + \frac{2}{\pi}\mathcal{P}\int_0^\infty \frac{\Omega\text{Im}\{\varepsilon(\Omega)\}}{\Omega^2 - \omega^2}\,d\Omega, \qquad (2.30\text{a})$$

$$\text{Im}\{\varepsilon(\omega)\} = \frac{2\omega}{\pi}\mathcal{P}\int_0^\infty \frac{\text{Re}\{\varepsilon(\Omega)\} - \varepsilon_0}{\Omega^2 - \omega^2}\,d\Omega, \qquad (2.30\text{b})$$

where the above integrals are to be understood in the sense of Cauchy and $\mathcal{P}$ denotes their Cauchy principal value.[13] In optics, especially nonlinear optics, these relations can be used to calculate the refractive index of a material by the measurement of its absorbance, which is more accessible experimentally. The link between $\tilde{n}$ and α follows as

$$\tilde{n}(\omega) = 1 + \frac{c}{\pi}\mathcal{P}\int_0^\infty \frac{\alpha(\Omega)}{\Omega^2 - \omega^2}\,d\Omega. \qquad (2.31)$$

2.3 Coherence

2.3.1 Statistical properties of light

Physical theories often need to be extended to a statistical description, where lack of knowledge of the system is distributed into probabilities for the ideal or possible situations to arise. In the case of light, the ideal case from Maxwell's equations would be a sinusoidal wave of well-defined frequency ω, amplitude E_0 and phase φ, propagating along some axis x with wavevector $k = \omega/c$. Disregarding the polarisation degree of freedom (for which the argument extends straightforwardly) the solution thus reads

$$E(x, t) = E_0 e^{i(\omega t - kx + \varphi)}. \qquad (2.32)$$

This perfect case is an ideal limit impossible to meet in practice. For instance, in the case where light is generated by an atom, obvious restrictions, such as a finite time of

---

[12] We note $\text{Re}(z)$ and $\text{Im}(z)$ the real and imaginary parts of the complex quantity $z \in \mathbb{C}$, respectively.

[13] The Cauchy principal value of an integral that presents a certain type of singularity is obtained by excluding the singularity from the integration by approaching it asymptotically from both sides (on the real axis).

emission (within the lifetime of the transition), already spoil the monochromatic feature (light of a well-defined frequency): the wave has finite extension in time and therefore a spread in frequency. The uncertainty in the amplitude of the emission results in uncertainty of the phase. Other systems bring further statistics and more complications can, and often must, also be added. An ensemble of atoms can bypass the above shortcoming, as is the case in a laser, where many atoms emit photons made identical by the stimulated process, allowing in this way to overcome the finite lifetime of a single atom and providing a continuously emitting medium, thereby reducing the spread of frequency. Very small linewidths are indeed achieved in this way, but whatever source of light one considers, there is ultimately always a complication that requires a statistical treatment. One recurrent cause calling for statistics is temperature. In the case of the laser, other complications arise from spontaneous emission, that is, the fact that some atoms of the collective ensemble decay independently of the others, bringing noise in the system. The requirement for statistics is ultimately linked to Heisenberg's uncertainty principle, although the argument translates almost verbatim to classical concepts, as illustrated in Fig. 2.3.

2.3.2 *Spatial and temporal coherence*

Coherence measures the amount of perturbation—or noise—in a wave. We shall refer to the later as the "signal" since the notion of coherence in the sense meant in this Section stems from information theory. Namely, coherence refers to the ability of inferring the signal at remote locations (in our case, in space or time) from its knowledge at a given point. In the upper case of Fig. 2.3, there is full coherence, as the value at a given single point determines the state of the field entirely and exactly by a mere translation. In the other cases, this is limited to windows where the field can still be reasonably computed. The exponential decay of the signal in the second case of Fig. 2.3 brings a signal-to-noise type of complication, that is less serious than the randomisation brought by the contribution from multiple sources, leading, respectively, to Lorentzian (slow) and Gaussian (fast) decays of the spectral shape. If the horizontal axis of the left panel of Fig. 2.3 is time, the corresponding interval defines its *coherence time*, if it is space, it defines its *coherence length*. From the symmetry of time (t) and space (x) in eqn (2.32), the two notions share not only similar definition and behaviour, but are linked, the one to the other, by the expression $\tau_c = l_c/c$, with τ_c and l_c the coherence time and length, respectively. A general formula for the coherence time for a wavetrain of spectral width $\Delta\lambda$ centred about λ is

$$\tau_c = \frac{\lambda^2}{c\Delta\lambda}. \tag{2.33}$$

The coherence length of a laser can be as high as hundreds of kilometres. For a discharge lamp it reduces to a few millimetres. It is very small given the speed at which light travels, but it is high enough for experimental measurements by the time of Huygens and Fresnel.

This definition or notion of coherence gives a good vivid picture of the concept, but it is mainly rooted in classical physics. As such, it describes perfectly well coherence

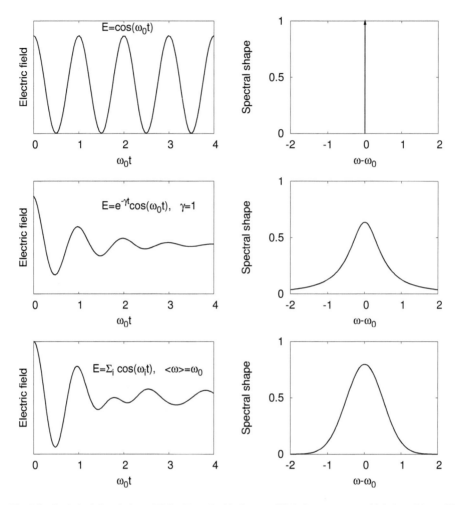

Fig. 2.3: Statistical description of light. Even the ideal case of light in a vacuum, which from Maxwell's equations arise as sinusoidal functions (upper case) of well-defined frequency, need a statistical description. One simple illustration is the finite lifetime of the emitter, resulting in a Lorentzian spread in frequency (central case). Another common case results from superpositions of different wavetrains, producing a Gaussian lineshape (bottom case).

of classical waves like sound or water waves. In a modern understanding of optics, this describes *first-order coherence*, still an important property of light, but definitely not encompassing the whole aspect of the problem.

The principle for measuring (first-order) coherence follows easily from the intuitive description that we have given. The notion that the knowledge of a signal at some point informs about its values at other points is mathematically described by *correlation*, and in this case, since it refers to the signal itself, by *autocorrelation*. The signal itself becomes a *random variable*; here the field E, which means that when prompted for a value (when it is measured, for instance), the "variable" is sampled according to a probability distribution. Physically, when the experimental setup performs a measurement, it draws one possible realisation of the experiment. From our description of coherence, we want to know how much the knowledge of, e.g., the field $E(t_0)$ at a given time t_0, tells us about the values $E(t)$ at other times t (here, E is still a random variable). A way to quantify this is to take the product $E(t_0)E(t)$ and to average over many possible realisations. We denote $\langle E(t_0)E(t)\rangle$ this average. If the field is coherent, that is, if the value at t determines that at t_0, each product will bear a fixed relationship that will survive the average. If, however, the two values are not related to each other, the product is random and averages to zero. In intermediate cases, the degree of correlation defines the degree of coherence. To retain some mathematical properties, the product is taken as hermitian, $\langle E^*(t_0)E(t)\rangle$ and is normalised to make it independent of the field's absolute amplitude, as well as writing $t_0 - t = \tau$ to emphasise the importance of "delay" in "confronting" the two values of the field, and thus, we arrive at the *first-order coherence degree*

$$g^{(1)}(\tau, t) = \frac{\langle E^*(t)E(t+\tau)\rangle}{\langle |E(t)|^2\rangle}. \tag{2.34}$$

This is a complex number in general, satisfying the properties

$$g^{(1)}(0, t) = 1, \qquad g^{(1)}(\tau, t)^* = g^{(1)}(-\tau, t). \tag{2.35}$$

Equation (2.34) depends explicitly on time, as can be the case if the system is not in equilibrium or in a steady-state. Note that such a measure of coherence extends to other degrees of freedom of the fields.[14] Often, however, one considers the time-independent case, which in this context translates as *stationary signals*, i.e., systems that still vary in time "locally" but do not depend on time in an absolute way, as is the case in a pulsed experiment for instance with clear difference before, during and after a pulse, as opposed to continuous pumping, where on average the system does not evolve.

The distinction in terms of the notions we have just presented is that, for a stationary process, the probability distribution is time independent, but the experimental quantity remains the random variable that fluctuates according to its distribution. However, its mean and variance are also constant. An obvious example is when time variations of the fields are limited to fluctuations, although this is only a special case of a stationary process. A periodic signal is also stationary. The *ergodic theorem* asserts that for such processes time and ensemble averages are the same, so that in eqn (2.34), t is taken to assume values high enough so that steady-state is achieved.[15] What kind of average is

---

[14] A more general formula quantifying simultaneously coherence both in time and space reads

$$g^{(1)}(\mathbf{r}_1, t_1; \mathbf{r}_2, t_2) = \langle E^*(\mathbf{r}_1, t_1)E(\mathbf{r}_2, t_2)\rangle / \sqrt{\langle |E(\mathbf{r}_1, t_1)|^2\rangle \langle |E(\mathbf{r}_2, t_2)|^2\rangle}. \tag{2.36}$$

[15] The direct time dependence of $g^{(1)}$ disappears in the steady state. A rigorous definition reads

$$g^{(1)}(\tau) = \lim_{t\to\infty} \frac{\langle E^*(t)E(t+\tau)\rangle}{\langle |E(t)|^2\rangle}. \tag{2.37}$$

Max **Born** (1882–1970) and Emil **Wolf** (b. 1922) emphasised the importance of coherence in optics by developing its statistical theory. They coauthored "Principles of Optics" (1959), which is the classic authoritative text in the field.

Born was awarded the 1954 Nobel prize (half-prize) for *"his fundamental research in quantum mechanics, especially for his statistical interpretation of the wavefunction."* Educated as a mathematician (with Hilbert), he identified the indices of Heisenberg's notations for transitions rates between orbitals as matrix elements. Systematising the approach with Pascual Jordan (then his student), they submitted a paper entitled *"Zur Quantummechanik"* (M. Born and P. Jordan, Zeitschrift für Physik, **34**, 858, (1924)) bearing the first published mention of the term "quantum mechanics."

Wolf is the physicist of optics par excellence, bringing many advances in statistical optics, coherence, diffraction and the theory of direct scattering and inverse scattering. He discovered the "Wolf effect" that is a redshift mechanism to be distinguished from the Doppler effect. As a signature of its eponymous father, the effect follows from partial coherence effects. It was experimentally confirmed the year following its prediction in 1987.

effectively performed is a rather moot point. There is ensemble averaging in most cases because many emitters combine to provide light, all of them independently emitting different "wavetrains" that correspond to an ensemble averaging. On the other hand, the coherence time of light is much too short to be detected directly by a detector that itself performs a time averaging. The importance of statistics and the relevance of correlations especially as defining coherence in optics has been realised in a large measure thanks to the guidance of Born and Wolf (1970). An authoritative reference on optical coherence is the textbook from Mandel and Wolf (1995).

The materialisation of the mathematical procedure outlined above is made in the laboratory with an interferometer, such as the Michelson interferometer or Mach–Zehnder interferometer. In these devices, the field is superimposed with a delayed fraction of itself and the time-averaged intensity of the light is collected at the output. Oscillations in this intensity build up *fringes* with *visibility* defined as

$$\mathcal{V} = \frac{I_{\max} - I_{\min}}{I_{\max} + I_{\min}}, \tag{2.38}$$

with $I_{\min/\max}$ the minimum/maximum intensity of the resulting interference pattern, respectively, and this equates to the modulus of coherence degree

$$\mathcal{V} = |g^{(1)}(\tau)|. \tag{2.39}$$

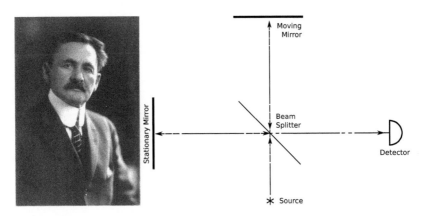

Albert Abraham **Michelson** (1852–1931) received the first American Nobel prize for physics in 1907 "*for his optical precision instruments and the spectroscopic and metrological investigations carried out with their aid.*" He designed the Michelson interferometer (shown on right) that superimposes the light field onto itself with a delay (imparted by the moving mirror), allowing its first order coherence to be measured.

The visibility $\mathcal{V}$ varies with the time delay τ and of course also with position if related to the more general definition of eqn (2.36). It assumes values between zero, for an incoherent field, and one, for a fully coherent one. Intermediate values describe partial first-order coherence.

As a trivial example, the sine wave of eqn (2.32) gives by direct application of the definition of eqn (2.34):

$$g^{(1)}(\tau) = \exp(-i\omega\tau)\,, \tag{2.40}$$

which corresponds to full first-order coherence (as its modulus, eqn (2.39), is one). In comparison, a field that results from two sources, each perfectly coherent (in the sense that they are both a sinusoidal wave of the type of eqn (2.32)) and with a dephasing φ between them, gives as the total scalar field

$$E(x,t)/E_0 = \exp\left(i(k_1z - \omega_1 t)\right) + \exp\left(i(k_2z - \omega_2 t + \varphi)\right). \tag{2.41}$$

We assume they have common amplitude E_0. If φ is kept fixed in the averaging, the same results as for the purely sinusoidal field apply. If, however, φ varies randomly between measurements, the result of Exercise 2.1 is obtained.

Exercise 2.1 $^{(*)}$ *Show that the field given by eqn (2.41) with φ varying randomly yields*

$$\mathcal{V} = |g^{(1)}(\tau)| = |\cos(\frac{1}{2}(\omega_1 - \omega_2)\tau)|\,. \tag{2.42}$$

Another frequent expression for $g^{(1)}$ brings together in eqn (2.45) (derived in Exercise 2.2) the decoherence resulting from two typical atomic dephasing processes: *collision broadening* and *Doppler broadening*.

Collision broadening is associated with the abrupt change of phase due to collisions between emitters, which have probability γ_c per unit time, to see their phase randomised during emission (associated with the chance of colliding with a neighbour).

Doppler broadening results from the Doppler effect that shifts the frequency of emitters as a function of their velocity. From kinetic theory it can be shown that the distribution of frequencies is normal:

$$f(\omega) = \frac{1}{\sqrt{2\pi}\Delta} \exp\left(-\frac{(\omega_0 - \omega)^2}{2\Delta^2}\right), \tag{2.43}$$

centred about the unshifted frequency ω_0 and with root mean square

$$\Delta = \omega_0 \sqrt{k_\mathrm{B} T/(mc^2)}, \tag{2.44}$$

at temperature T for emitters of mass m. This effect is therefore important in a gas at non-vanishing temperatures.

Exercise 2.2 (**) *Model the light field of sources subject to collision and Doppler broadening and show that their first-order coherence degree is given by*

$$g^{(1)}(\tau) = \exp\left(-i\omega_0\tau - \gamma_\mathrm{c}|\tau| - \frac{1}{2}\Delta^2\tau^2\right). \tag{2.45}$$

Observe how in eqn (2.45) the dephasing that ultimately loses completely the phase information results in a decay of $|g^{(1)}|$, rather than in oscillations, as was the case in eqn (2.42), as a result of the finite number of emitters contriving to dephase the system. This will become clear in the next section that links first-order coherence to the emitted spectra.

2.3.3 Wiener–Khinchin theorem

An important relationship was established by Wiener (1930) and Khintchine (1934) between the *spectral density* of a stochastic process and its *autocorrelation function*. Namely, they form a Fourier transform pair.

The spectral density is, physically, the decomposition of a signal (or field) into its components of given frequencies. The relation between frequency ω and energy E,[16]

$$E = \hbar\omega, \tag{2.46}$$

with $\hbar = h/(2\pi)$ the reduced Planck constant,[17] provides the experimental way to record such a spectrum: photons of a given energy are counted over a given time to build up a signal. The name of "*power spectrum*" is also commonly used, since an intensity per unit time is measured. We now discuss what physical quantities it relates to and how to model them mathematically.

---

[16] The relation (2.46) is Planck's hypothesis that energy is emitted by quanta, the quantisation being provided by the box of the blackbody. It was fully developed by Einstein for his explanation of the photoelectric effect.

[17] The reduced Planck constant $\hbar$ is a shortcut for $h/2\pi$, where $h = 4.135\ 667\ 43(35) \times 10^{-15}$ eV s is the Planck constant proper. It is the quantum of action and of angular momentum, or, following uncertainty principle, the "size" of a cell in phase space.

Norbert **Wiener** (1894–1964) and Aleksandr Yakovlevich **Khinchin** (1894–1959)

Wiener was an American mathematician who pioneered an important branch of study of stochastic processes with his generalised harmonic analysis. He is also credited as the founder of "cybernetics". His absent-mindedness and lively nature sparked many famous anecdotes. In his obituary, the Times reports how he "could offend publicly by snoring through a lecture and then ask an awkward question in the discussion". In his essay "the quark and the jaguar", Gell-Mann remembers how Wiener would hinder the circulation in the university by sleeping in the stairs. The first volume of his autobiography is titled "*Ex-Prodigy: My Childhood and Youth*". Working with physicists, he remarked that "one of the chief duties of the mathematician in acting as an adviser to scientists is to discourage them from expecting too much from mathematics." Jerison et al. (1997) edited a fascinating volume celebrating Wiener's life and work in which one can read that Einstein had anticipated the Wiener–Khinchin theorem in 1914, answering a question from a friend on the "power" of meteorological time-series (Kolmogorov also shares some independent paternity in the theorem).

Khinchin was one of the most prominent Russian mathematicians in the field of probabilities, largely dominated at that time by the Soviet school. He unravelled the definition of stationary processes and developed their theoretical foundations. He published "*Mathematical Principles of Statistical Mechanics*" in 1943, which he extended in 1951 to the highly respected text "*Mathematical Foundations of Quantum Statistics*". He aimed at a complete mathematical rigour in his results as characterises well the Wiener–Khinchin theorem, since—applying to signals without a Fourier transform—it is far more involved mathematically than physicists usually appreciate.

Intuitively, given a time-varying signal $E(t)$, one can quantify in the spirit of Fourier how much the harmonic component $e^{i\omega t}$ exists in the signal by overlapping both and averaging to get a number

$$\mathcal{E}(\omega) = \int_{-\infty}^{\infty} E(t)e^{i\omega t}\, dt\,. \tag{2.47}$$

The "weight" amplitude $\mathcal{E}$ is now itself a function of ω. Its modulus square for a given frequency is related to the "strength" of this harmonic in the signal and, thus, would seem to serve as a good spectral density, and sometimes does:

$$\mathcal{S}(\omega) = |\mathcal{E}(\omega)|^2\,. \tag{2.48}$$

However, we already pointed out that an important class of physical signals are stationary, which—being globally time invariant—are not square integrable[18] and consequently do not admit a Fourier transform, which dooms the above approach. Wiener

---

[18] A function f is square integrable if $\int |f|^2$ exists. It is a less stringent condition than integrability that is required for many properties relative to integration to be meaningful.

showed in 1930 that, for a large class of functions z (encompassing all the cases relevant for us), the integral $\Gamma(\tau) = \lim_{T\to\infty} \frac{1}{2T} \int_{-T}^{T} z^*(t)z(t+\tau)\, d\tau$ exists and also its Fourier transform:

$$s(\omega) = \frac{1}{2\pi} \int_{-\infty}^{\infty} \Gamma(\tau) e^{i\omega\tau}\, d\tau. \tag{2.49}$$

$\Gamma(\tau)$ as defined above (and by Wiener) does not refer to an ensemble as there is no averaging. It required Khinchin's inputs on stationary random processes (that we will not detail) and the ergodic theorem to arrive at the *Wiener–Khinchin theorem*, such as it is known today, which is eqn (2.49) together with the definition already discussed $\Gamma(t - t') = \langle E^*(t)E(t')\rangle$ for the autocorrelation function (reverting notation to E for the function that has been z up to now). Hence,

$$s(\omega) = \frac{1}{2\pi} \int_{-\infty}^{\infty} \langle E^*(t)E(t+\tau)\rangle e^{i\omega\tau}\, d\tau. \tag{2.50}$$

Normalising eqn (2.50) gives the *lineshape* of the emission

$$S(\omega) = \frac{1}{2\pi} \int_{-\infty}^{\infty} g^{(1)}(\tau) e^{i\omega\tau}\, d\tau, \tag{2.51}$$

that is, the normalised spectral shape.

Applying formula (2.51) to eqn (2.45) shows that:

- A pure coherent state (without dephasing) emits a delta-function spectrum: there is no spread in frequency, as expected.
- *Homogeneous broadening*, associated to exponential decay of $g^{(1)}$, results in a *Lorentzian lineshape*:

$$S(\omega) = \frac{1}{\pi} \frac{\gamma_c/2}{(\omega - \omega_0)^2 + (\gamma_c/2)^2}. \tag{2.52}$$

- *Inhomogeneous broadening*, associated to Gaussian decay of $g^{(1)}$, results in a *Gaussian lineshape*:

$$S(\omega) = \frac{1}{\Delta\sqrt{2\pi}} \exp\left(-\frac{(\omega - \omega_0)^2}{2\Delta^2}\right). \tag{2.53}$$

By convolution, the general case for $g^{(1)}$ given by eqn (2.45) combining homogeneous and inhomogeneous broadenings yields the following expression of the lineshape (known as the *Voigt lineshape*):

$$S(\omega) = \int_{-\infty}^{\infty} \frac{1}{\Delta\sqrt{2\pi}} \exp\left(-\frac{x^2}{2\Delta^2}\right) \frac{1}{\pi} \frac{\gamma_c/2}{(\omega - x)^2 + (\gamma_c/2)^2}\, dx. \tag{2.54}$$

The Voigt lineshape interpolates between the other two cases, as shown in Fig. 2.4. The quantum version of this Section will be discussed in Section 3.3.6.

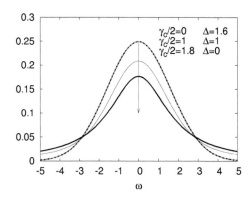

Fig. 2.4: Typical lineshapes (emission spectra) emitted by a source with two mechanisms of broadening resulting in exponential and Gaussian decay of the coherence degree $g^{(1)}$. The limiting cases recover the respective Lorentzian and Gaussian lineshapes (lower and upper lines). The general case in between is known as the *Voigt lineshape*.

2.3.4   *Hanbury Brown–Twiss effect*

Up to now, we have investigated "coherence" as a concept of correlations in the field but in fact, of correlations in the *amplitudes* of the field, as those have been the relative values of E that were compared. Again, this corresponds experimentally to splitting the beam in two and superimposing it onto itself with a delay, as schematised by the Michelson interferometer on page 39. The same approach extends to other quantities deriving from the field. The next step is to quantify correlations in the *intensities*, rather than amplitudes, as has been done by Hanbury Brown et al. (1952), who correlated the signals of two photomultiplier (PM) tubes collecting radio-waves from stars, and formalised by Hanbury Brown and Twiss (1954). The idea sparked when the young Hanbury Brown, working on the development of the radar as part of the British war effort, observed that radio-pulses from two distinct antennas generate signals on the oscilloscope that wiggle similarly to the naked eye. From there on, he conceived the possibility of correlating signals from detectors with very long baselines, possibly by recording the data on tape and post-processing it. Classically, the effect can be readily understood from the constructive interferences of two waves, as shown by Baym (1998). Applying the same idea to a different wavelength, that is, going from radio-waves to light, has been one of those steps that should have been trivial but that presented daunting difficulties, both conceptually and experimentally. It was both demonstrated (by Hanbury Brown and Twiss (1956b)) and discussed (by Hanbury Brown and Twiss (1956c)) that the "HBT effect" (after the initials of the two—not three—authors) also applies in this frequency range. As its most famous application, still by Hanbury Brown and Twiss (1956a), lies the measurement of the radius of Sirius, as sketched on the next page.

The HBT effect evidences positive simultaneous correlations between the two signals, meaning that the detection of a signal on any one branch of the interferometer is matched by detection of a signal on the other branch, more often than if the detections were uncorrelated (which is the case if the signals are not detected simultaneously, for

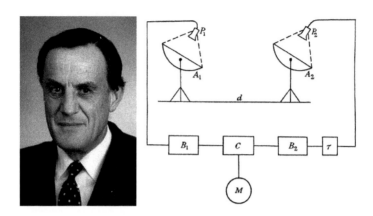

Robert **Hanbury Brown** (1916–2002) invented—with Mathematical support from Richard Twiss—the intensity interferometer. The setup, now used in various "HBT experiments", is schematised by Hanbury Brown and Twiss (1958) for applications in radioastronomy to measure the radius of a star. Two stellar mirrors $A_{1,2}$, spatially separated by a distance d, collect light and focus it on photomultipliers P. B are amplifiers and C operates the coincidence (after delay τ has been imparted on one of the line). M integrates the signal.

The concept of photon interferences was initially met with great opposition, also from the highest authorities of the time such as Richard Feynman. In his autobiography, "BOFFIN : A Personal Story of the Early Days of Radar, Radio Astronomy and Quantum Optics", Hanbury Brown remembers *"As an engineer my education in physics had stopped far short of the quantum theory. Perhaps just as well, otherwise like most physicists I would have come to the conclusion that the thing would not work[...]In fact to a surprising number of people the idea that the arrival of photons at two separated detectors can ever be correlated was not only heretical but patently absurd, and they told us so in no uncertain terms, in person, by letter, in print, and by publishing the results of laboratory experiments, which claimed to show that we were wrong..."*. With an unfailing faith in his effect, even in the face of Twiss' calculations that initially gave credit to their opponent, Hanbury Brown moved to Australia where he got funds to build the Narrabri Stellar Intensity Interferometer. Despite planning to spend only two years there, he dropped his UK tenure and extended his stay for a total of 27 years to have the project complete and flourish.

instance). We shall see in the next chapter that, at the quantum level, this means that photons—the particles actually detected and amplified by the PMs—are lumped together in the light emitted by Sirius, or for that matter, from any thermal light source or even from independent sources (such as two stars from different galaxies). Photons tend to arrive together on the detector. This is known as *bunching*. The HBT effect interpreted at the particle level is counter-intuitive and has been severely criticised when it was first proposed. The correlation, which in the above case is measured as a function of the distance between the detectors, can be made instead as a function of time. With a delay, one measures the likelihood that, given that a first photon is measured at time t, a second one is measured at time $t + \tau$. The HBT setup then consists of a 50/50 beamsplitter directing the light on two photomultipliers (typically avalanche photodiodes, or APD). The time elapsed between the detection of two consecutive photons is measured and the number $n(\tau)$ of photon pairs separated by a time interval τ is counted. This number gives the *probability of joint detection* $P_2(t, t + \tau)$ at times t and $t + \tau$. This probability can be linked to intensity correlations from photodetection theory as has been done by

Mandel et al. (1964).[19] They are proportional:

$$P_2(t, t + \tau) = \alpha^2 \langle I(t) I(t + \tau) \rangle, \tag{2.55}$$

where $I = |E|^2$ is the intensity and α represents the quantum efficiency of the photo-electric detector. Intensity correlations being the straightforward next-order extension of amplitude correlations, eqn (2.34), the notation $g^{(2)}$ is introduced and we call "second-order coherence degree" the quantity

$$g^{(2)}(t, \tau) = \frac{\langle E^*(t) E^*(t + \tau) E(t) E(t + \tau) \rangle}{\langle E^*(t) E(t) \rangle^2}. \tag{2.56}$$

We also introduce the centred, normalised correlation function

$$\lambda(\tau) = \frac{\langle \Delta I(t) \Delta I(t + \tau) \rangle}{\langle I(t) I(t + \tau) \rangle}, \tag{2.57}$$

where $\Delta I = I - \langle I \rangle$. Now eqn (2.55) reads $P_2(t, t+\tau) = \alpha^2 \langle I(t) \rangle \langle I(t+\tau) \rangle [1 + \lambda(\tau)]$ and from the first result of Exercise 2.3, it follows that

$$P_2(t, t) \geq P_2(t, t + \tau). \tag{2.58}$$

Exercise 2.3 (*) *Show that $\lambda(\tau)$ (of eqn (2.57)) satisfies $\lambda(\tau) \leq \lambda(0)$ and also that $\lim_{t \to \infty} \lambda(t) = 0$. Discuss the sign of this quantity.*

Since the detection process is, ultimately, detecting single photons,[20] the HBT experiment is often regarded as the quantum optical measurement par excellence, although it is merely correlating intensities and has been first used by Hanbury Brown and Twiss in a classical context (more of which is investigated in the problem at the end of this chapter). However, it is true that it also characterises light bearing a quantum character, as shall be explained in the next chapter.

2.4 Polarisation-dependent optical effects

We now give a brief account of the main optical effects dealing with the polarisation of light. Most of them have been observed in microcavities. Some of them allow measurement of the most important intrinsic characteristics of microcavities.

2.4.1 Birefringence

Birefringence is the division of light into two components (an *ordinary* and an *extraordinary* ray), found in materials that have different indices of refraction in different directions (n_o and n_e for ordinary and extraordinary rays, respectively). Birefringence is also known as double refraction.

[19]The proportionality between $g^{(2)}$ and the joint detection probability is established in the full quantum case in Exercise 3.21 on page 114.

[20]A single photon is detected in an avalanche photodiode by exciting a single electron–hole pair across the bandgap. The high electric field across the diode accelerates the electron that acquires sufficient energy to excite further electron–hole pairs—the avalanche—and producing a sizable current pulse.

The quantity referred to as *birefringence* is defined as

$$\Delta n = n_{\mathrm{e}} - n_{\mathrm{o}}. \tag{2.59}$$

Crystals possessing birefringence include hexagonal (such as calcite), tetragonal and trigonal crystal classes, and are known as uniaxial. Orthorhombic, monoclinic, triclinic crystal exhibit three indices of refraction. They are known as biaxial. Birefringent prisms include the Nicol prism, Glan–Foucault prism, Glan–Thompson prism and Wollaston prism. They can be used to separate different incident polarisations.

Dichroism is the selective absorption of one component of the electric field of a light wave, resulting in polarisation. *Optical activity* arises when polarised light is passed through a substance containing chiral molecules (or non-chiral molecules arranged asymmetrically), and the direction of polarisation can be changed. This phenomenon is also called *optical rotation*.

The Kerr effect, discovered by the English physicist Kerr (1877), consists in the development of birefringence when an isotropic transparent substance is placed in an electric field **F**. It is used in constructing Kerr cells, which function as variable waveplates with an extremely fast response time, and find use in high-speed camera shutters. Because the effect is quadratic with respect to **F**, it is sometimes known as the *quadratic electro-optical effect*. The amount of birefringence (as characterised by the change in index of refraction) due to the Kerr effect can be parametrised by

$$\Delta n = \lambda_0 K F^2, \tag{2.60}$$

where K is the Kerr constant (ranging between 10^{-12} and $10 \times 10^{-15}\,\mathrm{mV^{-2}}$ for different materials) and λ_0 is the vacuum wavelength. The phase change $\Delta\phi$ introduced in a Kerr cell of thickness d under an applied voltage V is given by

$$\Delta\phi = \frac{2\pi K \lambda V^2}{d^2}. \tag{2.61}$$

Weinberger (2008) gives a modern perspective of the effect rooted in its historical developments.

2.4.2 *Magneto-optical effects*

2.4.2.1 *Faraday effect*   The English physicist Faraday (1846) experimentally discovered diamagnetism and observed what is now called the *Faraday effect*, establishing the first link between light and electromagnetism, that he had been pursuing for over a decade. He demonstrated that, given two rays of circularly polarised light, one with left-hand and the other with right-hand polarisation, the one with the polarisation in the same direction as the electricity of the magnetising current travels with greater velocity. That is why the plane of linearly polarised light is rotated when a magnetic field is applied parallel to the propagation direction (see Fig. 2.5).

The empirical angle of rotation is given by

$$\alpha = VBd, \tag{2.62}$$

where V is the Verdet constant (with units of $\mathrm{rad}/(\mathrm{Tm})$), named after the French physicist Emile Verdet. The Faraday rotation angle is a measure of magnetisation induced in

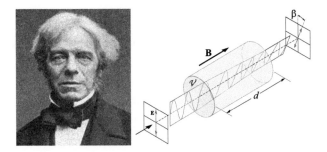

Michael **Faraday** (1791–1867) and an illustration of the *Faraday effect*, similar to optical activity as in both cases the polarisation plane of light rotates as it propagates through the medium. The difference is that the Faraday rotation is independent of the propagation direction and can be accumulated if light makes several round-trips through the medium, while the rotation caused by optical activity changes its sign for light propagating in the opposite direction. Thus, after any round-trip inside an optically-active medium the polarisation of light remains the same.

Faraday was a self-taught genius who developed an interest for science by reading books he had to bind as a poor apprentice working in a bookshop. Through attendance and hard work he gained access to experimental laboratories, where he excelled and developed experimental setups of an unprecedented standard. His evidence that magnetism could affect rays of light proved a relationship between the two, a finding preparing the great first unification in physics of electricity and magnetism as two facets of electromagnetism. To the spectroscopist James Crookes, he once advised "*Work. Finish. Publish.*"

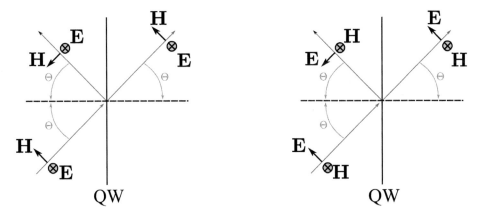

Fig. 2.5: Orientation of electric and magnetic fields in TE- and TM-polarised incident on a planar boundary.

the medium by a magnetic field. In the quantum-mechanical description, it occurs because imposition of a magnetic field B alters the energy levels of atoms or electrons (Zeeman effect). Knudsen (1976) gives a nice historical account (see also Exercise 2.4).

2.4.2.2 *Magneto-optical Kerr effect* This effect has very much in common with the Faraday effect. It also consists in rotation of the polarisation plane of light in the media having a nonzero magnetisation in the direction of light propagation. The difference between the two effects consists in the experimental configuration used to detect the polarisation rotation. For the Faraday effect, the polarisation of the transmitted light is analysed, while for the magneto-optical Kerr effect, the polarisation of the reflected light is compared with polarisation of the incident light. As a large part of reflected signal comes from the surface reflection usually, the Kerr rotation is very sensitive to the surface magnetisation.

Exercise 2.4 $^{(**)}$ *Consider a dielectric slab of thickness d subjected to a magnetic field* **B** *orientated normally to the surface of the slab. Let* t_+ *and* t_- *be amplitude transmission coefficients of the slab for* σ_+ *and* σ_- *polarised light, respectively, and* $|t_+| = |t_-|$. *Find the Verdet constant of the material.*

2.5 Propagation of light in multilayer planar structures

In this section we present the *transfer matrix method*, which solves Maxwell equations in multilayer dielectric structures. We consider the example of a periodical structure (a so-called *Bragg mirror*) and derive general equations for photonic eigenmodes in planar structures. In the beginning, we consider propagation of light normal to the layer planes direction. Then, we generalise to the oblique incidence case. We formulate the transfer matrix approach for Transverse Electric (TE) and Transverse Magnetic (TM) linear polarisations. By definition, TE-polarised (also referred to as *s*-polarised) light has the electric-field vector parallel to the layer planes, TM-polarised light (also referred to as *p*-polarised) has the magnetic-field vector parallel to the planes (see Fig. 2.5).

The behaviour of the electromagnetic field at the planar interface between two dielectric media with different refractive indices is dictated by Maxwell's equations (2.10). They can be solved independently in the two media and then matched for the electric and magnetic fields by the Maxwell boundary conditions at the interface. These conditions require continuity of the tangential components of both fields. They can be microscopically justified for any abrupt interface in the absence of free charges and free currents.

Consider a transverse light wave propagating along the z-direction in a medium characterised by a refractive index n that is homogeneous in the xy-plane, but possibly z dependent. The wave equation (2.18) in this case becomes for the field amplitude

$$\partial_z^2 E = -k_0^2 n^2 E, \tag{2.63}$$

where k_0 is the wavevector of light in vacuum. The general form of the solution of eqn (2.63) reads

$$E = A^+ \exp(ikz) + A^- \exp(-ikz), \tag{2.64}$$

where $k = k_0 n$, A^+ and A^- are coefficients. Using the Maxwell equation (2.10d) one can easily obtain the general form of the magnetic field amplitude:

$$B = nA^+ \exp(ikz) - nA^- \exp(-ikz). \tag{2.65}$$

If we consider the reflection of light incident from the left side of the boundary $(z = 0)$ between two semi-infinite media characterised by refractive indices n_1 (left) and n_2 (right), the matching of the tangential components of electric and magnetic fields gives

$$A_1^+ + A_1^- = A_2^+ , \tag{2.66a}$$

$$(A_1^+ - A_1^-)n_1 = A_2^+ n_2 , \tag{2.66b}$$

where A_1^+, A_1^- and A_2^+ are the amplitudes of incident, reflected and transmitted light, respectively. One can easily obtain the amplitude reflection coefficient

$$r = \frac{A_1^-}{A_1^+} = \frac{n_1 - n_2}{n_1 + n_2} , \tag{2.67}$$

and the amplitude transmission coefficient

$$t = \frac{A_2^+}{A_1^+} = \frac{2n_1}{n_1 + n_2} . \tag{2.68}$$

The *reflectivity*, which is the ratio of reflected to incident energy flux, is given by

$$R = |r|^2 , \tag{2.69}$$

and the *transmittance*, or ratio of transmitted to incident energy flux, is given by

$$T = \frac{n_2}{n_1} |t|^2 . \tag{2.70}$$

In the last formula, the factor n_2/n_1 comes from the ratio of light velocities in the two media.

In multilayer structures, direct application of Maxwell's boundary conditions at each interface requires solving a substantial number of algebraic equations (two per interface). A convenient method allows the number of equations to be solved to be strictly minimised (four in the general case). This is the *transfer matrix method*, which we now briefly describe.

Let us introduce the vector

$$\Phi(z) = \begin{pmatrix} E(z) \\ cB(z) \end{pmatrix} = \begin{pmatrix} E(z) \\ -\frac{i}{k_0} \partial_z E(z) \end{pmatrix} , \tag{2.71}$$

where $E(z)$ and $B(z)$ are the amplitudes of the electric and magnetic field of any light wave propagating in the z-direction in the structure. Note that $\Phi(z)$ is continuous at any point in the structure due to Maxwell's boundary conditions. In particular, it is continuous at all interfaces where n changes abruptly.

By definition, the transfer matrix $\mathbf{T}_a$ across the layer of width a is a matrix that enforces

$$\mathbf{T}_a \Phi|_{z=0} = \Phi|_{z=a} . \tag{2.72}$$

It is easy to verify by substitution of the electric and magnetic amplitudes (2.64) and (2.65) into eqn (2.72) that if n is homogeneous across the layer

$$\mathbf{T}_a = \begin{pmatrix} \cos ka & \frac{i}{n}\sin ka \\ in\sin ka & \cos ka \end{pmatrix}. \tag{2.73}$$

The transfer matrix across a structure composed of m layers is found as

$$\mathbf{T} = \prod_{i=m}^{1} \mathbf{T}_i, \tag{2.74}$$

where $\mathbf{T}_i$ is the transfer matrix across the ith layer. The order of multiplication in eqn (2.74) is essential. The amplitude reflection and transmission coefficients (r_s and t_s) of a structure containing m layers and sandwiched between two semi-infinite media with refractive indices n_{left} and n_{right}, before and after the structure, respectively, can be found from the relation between the fields $\mathbf{\Phi}$ on either side of the structure:

$$\mathbf{T}\begin{pmatrix} 1+r_s \\ n_{\text{left}}(1-r_s) \end{pmatrix} = \begin{pmatrix} t_s \\ n_{\text{right}}t_s \end{pmatrix}. \tag{2.75}$$

One can easily obtain

$$r_s = \frac{n_{\text{right}}t_{11} + n_{\text{left}}n_{\text{right}}t_{12} - t_{21} - n_{\text{left}}t_{22}}{t_{21} - n_{\text{left}}t_{22} - n_{\text{right}}t_{11} + n_{\text{left}}n_{\text{right}}t_{12}}, \tag{2.76a}$$

$$t_s = 2n_{\text{left}}\frac{t_{12}t_{21} - t_{11}t_{22}}{t_{21} - n_{\text{left}}t_{22} - n_{\text{right}}t_{11} + n_{\text{left}}n_{\text{right}}t_{12}}. \tag{2.76b}$$

The intensities of reflected and transmitted light normalised by the intensity of the incident light are given by

$$R = |r_s|^2, \qquad T = |t_s|^2\frac{n_{\text{right}}}{n_{\text{left}}}, \tag{2.77}$$

respectively.

Reciprocally, the transfer matrix across a layer can be expressed via reflection and transmission coefficients of this layer. If the reflection and transmission coefficients for light incident from the right-hand side and left-hand side of the layer are the same, and $n_{\text{left}} = n_{\text{right}} = n$ (the symmetric case of, in particular, a quantum well embedded in a cavity), the Maxwell boundary conditions for light incident from the left and right sides of the structure yield:

$$\mathbf{T}\begin{pmatrix} 1+r_s \\ n(1-r_s) \end{pmatrix} = \begin{pmatrix} t_s \\ nt_s \end{pmatrix} \quad \text{and} \quad \mathbf{T}\begin{pmatrix} t_s \\ -nt_s \end{pmatrix} = \begin{pmatrix} 1+r_s \\ -n(1-r_s) \end{pmatrix}. \tag{2.78}$$

This allows the matrix $\mathbf{T}$ to be expressed as:

$$\mathbf{T} = \frac{1}{2t_s}\begin{pmatrix} t_s^2 - r_s^2 + 1 & -\frac{(1+r_s)^2 - t_s^2}{n} \\ -n((r_s-1)^2 - t_s^2) & t_s^2 - r_s^2 + 1 \end{pmatrix}. \tag{2.79}$$

For a quantum well, as will be shown in Section 2.6, $t_s = 1 + r_s$ and eqn (2.79) becomes

$$\mathbf{T}_{\text{QW}} = \begin{pmatrix} 1 & 0 \\ 2nr_s/t_s & 1 \end{pmatrix}. \tag{2.80}$$

René **Descartes** (1596–1650) and Willebrord **Snellius** (1580–1626).

Descartes is regarded as the father of modern mathematics and natural philosophy. His thinking instigated the revolution in western science that would break from eastern influence and mark the dominance of European thinking. His famous saying "*je pense donc je suis*" (I think therefore I am) in "Discourse on method" ranks among the most influential philosophical statements. His name became synonymous with "rational" and "meaningful", an eponymous tribute whose extent and glory was never equalled after him.

Snellius was a Dutch astronomer and mathematician. He discovered the law of refraction that is now named after him (although in a few countries, especially in France, this law is called Snell–Descartes). Arab astronomers apparently knew it long before Snellius from the work of Ibn Sahl (984). Snellius also worked out a remarkably accurate (for his time) value of the radius of the earth and devised a new method to compute π, the first such improvement since ancient time.

In the oblique incidence case, in the TE-polarisation, one can use the basis $(E_\tau(z), cB_\tau(z))^T$ (T means transposition), where $E_\tau(z)$ and $B_\tau(z)$ are the tangential (in-plane) components of the electric and magnetic fields of the light wave. In this case, the transfer matrix (2.73) keeps its form provided that the following substitutions are made

$$k_z = k \cos\phi, \qquad n \to n \cos\phi, \tag{2.81}$$

where ϕ is the propagation angle in the corresponding medium ($\phi = 0$ at normal incidence).

In the TM-polarisation, following Born and Wolf (1970), we now use the basis $(cB_\tau(z), E_\tau(z))^T$ that still allows the transfer matrix (2.73) to be used, provided that the following substitutions are done:

$$k_z = k \cos\phi, \qquad n \to \frac{\cos\phi}{n}. \tag{2.82}$$

Note that the transfer matrices across the interfaces are still identity matrices, and that eqn (2.74) for the transfer matrix across the entire structure is valid.

In the formulas (2.76)–(2.78) for reflection and transmission coefficients, one should replace, in the TE-polarisation

$$n_{\text{left}} \to n_{\text{left}} \cos\phi_{\text{left}}, \qquad n_{\text{right}} \to n_{\text{right}} \cos\phi_{\text{right}}, \tag{2.83}$$

and in the TM-polarisation

$$n_{\text{left}} \to \frac{\cos\phi_{\text{left}}}{n_{\text{left}}}, \qquad n_{\text{right}} \to \frac{\cos\phi_{\text{right}}}{n_{\text{right}}}, \tag{2.84}$$

where ϕ_{left}, ϕ_{right} are the propagation angles in the first and last media, respectively. The same transformations would be applied to the transfer matrices (2.79) and (2.80). Note that any two propagation angles ϕ_i, ϕ_j in the layers with refractive indices n_i, n_j are linked by the *Snell–Descartes* law:

$$n_i \sin\phi_i = n_j \sin\phi_j, \tag{2.85}$$

which is also valid in the case of complex refractive indices, when the propagation angles formally become complex as well.

2.6 Photonic eigenmodes of planar systems

Consider a multilayer planar structure characterised by a transfer matrix $\mathbf{T}$ being a product of the transfer matrices across all the layers as given by eqn (2.74). Photonic eigenmodes of the structure are the solutions of the Maxwell equations that decay outside the structure and, hence, with the following boundary condition: no light is incident on the structure either from the left ($z \to -\infty$) or from the right side ($n \to +\infty$). This means that the electric field of the eigenmode at $z \to \infty$ can be represented as

$$E_0 \exp(ik_x + ik_y + ik_z), \tag{2.86}$$

with $\mathrm{Re}(k_z) \geq 0$ and $\mathrm{Im}(k_z) \leq 0$, while at $z \to \infty$ the electric field can be represented in the form of eqn (2.86) with $\mathrm{Re}(k_z) \leq 0$ and $\mathrm{Im}(k_z) \geq 0$. In TE-polarisation, let us choose the system of coordinates in such a way that the electric and magnetic fields of the light mode are orientated as

$$E = \begin{pmatrix} 0 \\ E_y \\ 0 \end{pmatrix}, \qquad B = \begin{pmatrix} B_x \\ 0 \\ B_z \end{pmatrix}. \tag{2.87}$$

The transfer matrix $\mathbf{T}_{\text{TE}}$ links the vectors $(E_y, cB_x)^T$ at the left and right boundaries of the structure, so that

$$\mathbf{T}_{\text{TE}} \begin{pmatrix} E_y^{\text{left}} \\ cB_x^{\text{left}} \end{pmatrix} = A \begin{pmatrix} E_y^{\text{right}} \\ cB_x^{\text{right}} \end{pmatrix}, \tag{2.88}$$

where A is a complex coefficient.

Substitution of the electric field (2.64) into the first of Maxwell's equations (2.10a) yields

$$\frac{k_z}{k_0} E_y = cB_x, \tag{2.89}$$

where $k_0 = \omega/c$. This allows us to rewrite eqn (2.88) as

$$\mathbf{T}_{\text{TE}} \begin{pmatrix} 1 \\ \dfrac{k_z^{\text{left}}}{k_0} \end{pmatrix} = A \begin{pmatrix} 1 \\ \dfrac{k_z^{\text{right}}}{k_0} \end{pmatrix}, \tag{2.90}$$

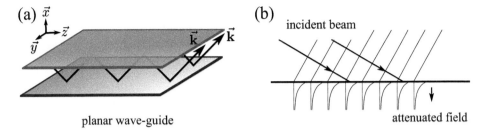

(a) planar wave-guide

(b) incident beam / attenuated field

Fig. 2.6: Propagation of guided modes in planar structures (a) is based on the total internal reflection effect (b).

where k_z^{left} and k_z^{right} are z-components of the wavevector of light on the left and right sides of the structure, respectively. By elimination of A, eqn (2.90) can be easily reduced to a single transcendental equation for the eigenmodes of the structure:

$$t_{11}^{\text{TE}} \frac{k_z^{\text{right}}}{k_0} + t_{12}^{\text{TE}} \frac{k_z^{\text{left}} k_z^{\text{right}}}{k_0^2} - t_{21}^{\text{TE}} - \frac{k_z^{\text{left}}}{k_0} t_{22}^{\text{TE}} = 0 , \qquad (2.91)$$

where t_{ij} are the matrix elements of the transfer matrix $\mathbf{T}_{\text{TE}}$. Solutions of eqn (2.91) are complex frequencies, in general. Only those having a positive real part and negative (or zero) imaginary part have a physical sense. The imaginary part of the eigenfrequency is inversely proportional to the lifetime of the eigenmode, i.e., a characteristic time spent by a photon going back and forth inside the structure before escaping from it to the continuum of free light modes in the surrounding media. So-called *waveguided* or *guided modes* are those that have an infinite lifetime (and consequently, zero imaginary part of the eigenfrequency), see Fig. 2.6.

The equation for eigenfrequencies of TM-polarised modes can be obtained in a similar way. One can choose the system of coordinates in such a way that the electric and magnetic fields of the light mode are orientated as

$$E = \begin{pmatrix} E_x \\ 0 \\ E_z \end{pmatrix} , \qquad B = \begin{pmatrix} 0 \\ B_y \\ 0 \end{pmatrix} . \qquad (2.92)$$

The transfer matrix $\mathbf{T}$ links the vectors $(cB_y, -E_x)^{\text{T}}$ at the left and right boundaries of the structure, so that

$$\mathbf{T}_{\text{TM}} \begin{pmatrix} B_y^{\text{left}} \\ -E_x^{\text{left}} \end{pmatrix} = \begin{pmatrix} B_y^{\text{right}} \\ -E_x^{\text{right}} \end{pmatrix} . \qquad (2.93)$$

The Maxwell equation (2.10a) yields in this case

$$-\frac{k_z}{n^2 k_0} B_y = E_x , \qquad (2.94)$$

where n is the refractive index of the media. This allows us to rewrite eqn (2.93) as

$$\mathbf{T}_{\text{TM}} \begin{pmatrix} 1 \\ \frac{k_z^{\text{left}}}{n_{\text{left}}^2 k_0} \end{pmatrix} = A \begin{pmatrix} 1 \\ \frac{k_z^{\text{right}}}{n_{\text{right}}^2 k_0} \end{pmatrix}, \qquad (2.95)$$

where n_{left} and n_{right} are z-components of the wavevector of light on the left and right sides of the structure, respectively. By elimination of A, eqn (2.95) can be easily reduced to a single transcendental equation for the eigenmodes of the structure:

$$t_{11}^{\text{TM}} \frac{k_z^{\text{right}}}{n_{\text{right}}^2 k_0} + t_{12}^{\text{TM}} \frac{k_z^{\text{left}} k_z^{\text{right}}}{n_{\text{left}}^2 n_{\text{right}}^2 k_0^2} - t_{21}^{\text{TM}} - t_{22}^{\text{TM}} \frac{k_z^{\text{left}}}{n_{\text{left}}^2 k_0} = 0. \qquad (2.96)$$

At normal incidence, equations for the eigenmodes of light in TE- (eqn (2.91)) and TM-polarisations (eqn (2.96)), become formally identical. This is quite reasonable, as at normal incidence there is no difference between TE- and TM-polarisations. One can show using transformations (2.82) and (2.83) that they both reduce to

$$t_{11} n_{\text{right}} - t_{12} n_{\text{right}} n_{\text{left}} - t_{21} + t_{22} n_{\text{left}} = 0, \qquad (2.97)$$

with t_{ij} being elements of the transfer matrix at normal incidence defined by eqns (2.73–2.74). Comparing condition (2.97) and expressions (2.76a) and (2.76b) one can see that reflection and transmission coefficients r_s and t_s become infinite at the complex eigenfrequencies of the system. This result also holds for oblique incidence. It is not unphysical. We note that in optical measurements of reflectivity and transmission we always detect signals at real frequencies and the condition

$$|r_s|^2 + \frac{n_{\text{right}}}{n_{\text{left}}} |t_s|^2 \leq 1. \qquad (2.98)$$

holds (the equality holds in the case of no absorption and scattering).

2.6.1   *Photonic bands of 1D periodic structures*

Consider an infinite structure whose refractive index is homogeneous in the xy-plane and whose dependency on the coordinate z is a periodic function with period d. The shape of this function is not essential, and we shall only assume that a transfer matrix $\mathbf{T}_d$ across the period of the structure can be written as a product of a finite number of matrices of the form of eqn (2.73). Let an electromagnetic wave propagate along the z-direction. For this wave:

$$\mathbf{T}_d \Phi \big|_{z=0} = \Phi \big|_{z=d}, \qquad (2.99)$$

where $\Phi(z)$ is defined by eqn (2.71). According to the Bloch theorem, it can be represented in the form

$$\Phi(z) = e^{iQz} \begin{pmatrix} U_E(z) \\ U_B(z) \end{pmatrix}, \qquad (2.100)$$

where $U_{E,B}(z)$ have the same periodicity as the structure and Q is a complex number in the general case. Note that the factor e^{iQz} is the same for electric (E) and magnetic

(B) fields in a light wave because in the normal incidence case they are linked by the relation

$$B(z) = -\frac{i}{ck_0}\frac{\partial E(z)}{\partial z}.$$ (2.101)

Substitution of eqn (2.100) into eqn (2.99) yields

$$\mathbf{T}_d\Phi\big|_{z=0} = e^{iQd}\Phi\big|_{z=0},$$ (2.102)

thus, e^{iQd} is an eigenvalue of the matrix $\mathbf{T}_d$ and, therefore,

$$\det(\mathbf{T}_d - e^{iQd}\mathbf{1}) = 0.$$ (2.103)

Solving eqn (2.103), we use an important property of the matrix $\mathbf{T}_d$ following from eqns (2.73)–(2.74)

$$\det(\mathbf{T}_d) = 1.$$ (2.104)

Thus, we reduce eqn (2.103) to:

$$1 - (T_{11} + T_{22})e^{iQd} + e^{2iQd} = 0,$$ (2.105)

where T_{ij} are the matrix elements of $\mathbf{T}_d$. Multiplying each term by e^{-iQd}, we finally obtain

$$\cos(Qd) = \frac{T_{11} + T_{22}}{2}.$$ (2.106)

The right-hand side of this equation is frequency dependent. The frequency bands for which

$$\left|\frac{T_{11} + T_{22}}{2}\right| \le 1$$ (2.107)

are allowed photonic bands. In these bands, Q is purely real and the light wave can propagate freely without attenuation. On the contrary, the bands for which:

$$\left|\frac{T_{11} + T_{22}}{2}\right| > 1$$ (2.108)

are usually called *stop-bands* or *optical gaps* (see Fig. 2.7). In these bands, Q has a nonzero imaginary part that determines the decay of propagating light waves. All this is completely analogous to electronic bands in conventional crystals. Equations (2.106)–(2.108) are also valid in the oblique incidence case, while the form of the matrix $\mathbf{T}_d$ is sensitive to the angle of incidence and band boundaries shift as one changes the incidence angle.

We recall that a *Bragg mirror* is a periodic structure composed of pairs of layers of dielectric or semiconductor materials (see Fig. 2.8) characterised by different refractive indices (say n_a and n_b). The thicknesses of the layers (a and b, respectively) are chosen so that

$$n_a a = n_b b = \frac{\bar{\lambda}}{4}.$$ (2.109)

Condition (2.109) is usually called the *Bragg interference condition* due to its similarity to the positive interference condition for X-rays propagating in crystals proposed in

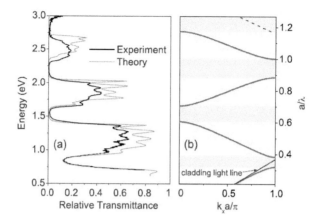

Fig. 2.7: Experimental and theoretical transmittance of a periodic structure composed by Si_3N_4 and SiO_2 dielectric layers (a) compared with the calculated photonic dispersion for this structure (b), from Gerace et al. (2005). The stop-bands are shown in gray.

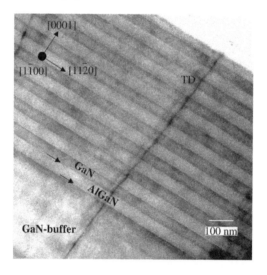

Fig. 2.8: Electronic microscopy image of the GaN/AlGaN Bragg mirror grown by E. Calleja's group in Madrid and reported by Fernández et al. (2001).

1913 by English physicists Sir William Henry Bragg and his son Sir William Lawrence Bragg. The Bragg mirrors are also frequently called *distributed Bragg reflectors* or DBRs. The wavelength of light $\bar{\lambda}$ marks the centre of the stop-band of the mirror. For the wavelengths inside the stop-band the reflectivity of the mirror is close to unity. In the following we assume $n_a < n_b$ (n_a is the refractive index of the first layer from the surface, n_b is the refractive index of the next layer). We describe the optical properties of the mirror within its stop-band using the transfer matrix approach.

William Henry **Bragg** (1862–1942) and William Lawrence **Bragg** (1890–1971).

Father and son, the two Braggs shared the 1915 Nobel Prize in Physics *"for their services in the analysis of crystal structure by means of X-rays."* When a young Bragg, aged 5, broke his arm by falling off his tricycle, he was radiographed by X-rays that his father had recently learned about from Röntgen's experiments. In 1912, aged 22 and a first-year university student, Bragg discussed with his father his ideas of diffraction by crystals that he would develop into Bragg's law. His father developed the spectrometer. Braggs' son became the youngest Nobel prize winner. He is also credited as having played an important role in his support of identifying the DNA double helix, as the then head of the Cavendish laboratory.

At normal incidence, the transfer matrices across the layers that compose the mirror are

$$\mathbf{T}_a = \begin{pmatrix} \cos(k_a a) & \frac{i}{n_a}\sin(k_a a) \\ in_a \sin(k_a a) & \cos(k_a a) \end{pmatrix}, \qquad \mathbf{T}_b = \begin{pmatrix} \cos(k_b b) & \frac{i}{n_b}\sin(k_b b) \\ in_b \sin(k_b b) & \cos(k_b b) \end{pmatrix}, \tag{2.110}$$

where $k_a = \omega n_a/c$ and $k_b = \omega n_b/c$. The transfer matrix $\mathbf{T}$ across the period of the mirror is their product

$$\mathbf{T} = \mathbf{T}_b \mathbf{T}_a. \tag{2.111}$$

An infinite Bragg mirror represents the simplest one-dimensional photonic crystal. Its band structure is given by eqn (2.106). Its solutions with real Q form allowed photonic bands, while solutions with complex Q having a nonzero imaginary part form photonic gaps or stop-bands. At the central frequency of the stop-band, given by

$$\bar{\omega} = 2\pi c/\bar{\lambda}, \tag{2.112}$$

the matrix $\mathbf{T}$ becomes

$$\mathbf{T} = \begin{pmatrix} -\frac{n_a}{n_b} & 0 \\ 0 & -\frac{n_b}{n_a} \end{pmatrix}. \tag{2.113}$$

Its eigenvalues are

$$\exp[iQ(a+b)] = -\frac{n_a}{n_b}, \quad \exp[-iQ(a+b)] = -\frac{n_b}{n_a}. \tag{2.114}$$

The reflection coefficient of a semi-infinite Bragg mirror at $\omega = \bar{\omega}$ can be found from the condition

$$\mathbf{T}\begin{pmatrix} 1+r \\ n_{\text{left}}(1-r) \end{pmatrix} = -\frac{n_a}{n_b}\begin{pmatrix} 1+r \\ n_{\text{left}}(1-r) \end{pmatrix}, \tag{2.115}$$

which readily yields $r = 1$.

In the vicinity of $\bar{\omega}$, one can derive a simple and useful expression for the reflection coefficient, leaving in the matrix $\mathbf{T}$, only terms linear in

$$x = (\omega - \bar{\omega})\frac{\bar{\lambda}}{4c}. \tag{2.116}$$

The matrix is written in this approximation as

$$\mathbf{T} = -\begin{pmatrix} \frac{n_a}{n_b} & i\left(\frac{1}{n_a} + \frac{1}{n_b}\right)x \\ i(n_a + n_b)x & \frac{n_b}{n_a} \end{pmatrix}. \tag{2.117}$$

Equation (2.115) yields in this case

$$r = \frac{n_{\text{left}}\left(\frac{n_a}{n_b} - \frac{n_b}{n_a}\right) - i(n_a + n_b)x}{n_{\text{left}}\left(\frac{n_a}{n_b} - \frac{n_b}{n_a}\right) + i(n_a + n_b)x} \tag{2.118a}$$

$$= \exp\left(i\frac{n_a n_b \bar{\lambda}}{2n_{\text{left}}(n_b - n_a)c}(\omega - \bar{\omega})\right) = e^{i\alpha(\omega - \bar{\omega})}. \tag{2.118b}$$

The coefficient

$$L_{\text{DBR}} = \frac{n_a n_b \bar{\lambda}}{2(n_b - n_a)} = \alpha n_0 c \tag{2.119}$$

is frequently called the *effective length* of a Bragg mirror. Note that it is close, but not exactly equal to the penetration length $\tilde{L}$ of the light field into the mirror at $\omega = \bar{\omega}$. The length $\tilde{L}$ can be easily obtained from the eigenvalues of the matrix (2.113)

$$\tilde{L} = \frac{a+b}{\ln\frac{n_b}{n_a}}. \tag{2.120}$$

One can see from eqn (2.118a) that at $\omega = \bar{\omega}$ the reflection coefficient of the Bragg mirror is equal to 1, which means that the amplitudes of incident and reflected waves have the same sign and absolute value at the surface of the mirror. This is why the maximum (antinode) of the electric field of light is at the surface. We note that this is only true if $n_a < n_b$. In the opposite case, the amplitude reflection coefficient at the centre of a stop-band changes sign and the electric field has a node at the surface.

For a finite-size mirror, the reflection coefficient within the stop-band is different from unity due to the nonzero transmission of light across the mirror. It can be found from the matrix equation

$$\mathbf{T}^N\begin{pmatrix} 1+\bar{r} \\ n_0(1-\bar{r}) \end{pmatrix} = \begin{pmatrix} \bar{t} \\ n_f\bar{t} \end{pmatrix}, \tag{2.121}$$

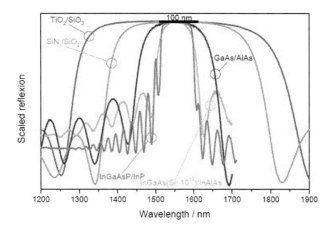

Fig. 2.9: Reflectivity of different Bragg mirror structures. All stop-bands are centred at the same Bragg wavelength of 1550 nm (from the teaching materials of the Institut für Hochfrequenztechnik, Technical University of Darmstadt).

where $\bar{r}$ and $\bar{t}$ are the amplitude reflection and transmission coefficients of the mirror, N is the number of periods in the mirror and n_f is the refractive index behind the mirror. At the centre of the stop-band,

$$\bar{r} = \frac{\left(\dfrac{n_b}{n_a}\right)^{2N} - \dfrac{n_f}{n_0}}{\left(\dfrac{n_b}{n_a}\right)^{2N} + \dfrac{n_f}{n_0}}, \qquad \bar{t} = \frac{\left(-\dfrac{n_b}{n_a}\right)^{N}}{\left(\dfrac{n_b}{n_a}\right)^{2N} + \dfrac{n_f}{n_0}}. \tag{2.122}$$

As follows from these formulas, the higher the contrast between n_a and n_b, the better the reflectivity of the mirror.

A further important characteristic of a Bragg mirror is the width of its stop-band and this can be found from eqn (2.106). The boundaries of the first stop-band are given by the condition

$$\frac{T_{11} + T_{22}}{2} = -1. \tag{2.123}$$

Substituting the matrix elements of the product of matrices (2.110), one easily obtains:

$$\cos^2 \Omega - \frac{1}{2}\left(\frac{n_b}{n_a} + \frac{n_a}{n_b}\right)\sin^2 \Omega = -1, \tag{2.124}$$

where $\Omega = k_a a = k_b b$; therefore

$$\cos \Omega = \pm\frac{n_b - n_a}{n_b + n_a}. \tag{2.125}$$

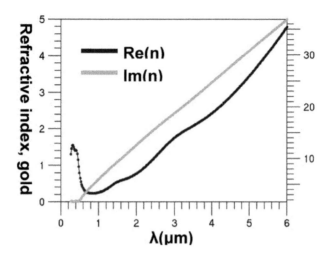

Fig. 2.10: The real and imaginary parts of the refractive index of gold, from Török et al. (1998).

This allows us to obtain the stop-band width (in frequency):

$$\Delta = \frac{8c}{\bar{\lambda}} \left(\frac{\pi}{2} - \arccos \frac{n_b - n_a}{n_b + n_a} \right) \approx \frac{8c}{\bar{\lambda}} \frac{n_b - n_a}{n_b + n_a} . \qquad (2.126)$$

The stop-band width increases with an increase in the contrast between the two refractive indices. Figure 2.9 shows the calculated reflectivity of Bragg mirrors made of different semiconductor and dielectric materials, but all having $\bar{\lambda} = 1550$ nm. One can see that the stop-band width can achieve a few hundred nanometres for high contrast of refractive indices n_a and n_b.

Finally, under oblique incidence the optical thickness of the layers composing a Bragg mirror change. The phase gained by light crossing a layer of thickness a at an angle ϕ_a is given by

$$\theta = \frac{\omega}{c} n_a a \cos \phi_a , \qquad (2.127)$$

where n_a is the refractive index of this layer. It is evident that the frequency that fulfills the Bragg interference condition $\theta = \pi/2$ is higher for oblique angles than for a normal angle. This is why, at oblique angles, stop-bands of any Bragg mirror shift towards higher frequencies. More details on the phases of reflection coefficients of the mirrors at oblique incidence can be found in the textbook by Kavokin and Malpuech (2003).

Metallic reflectivity is usually not so perfect as dielectric reflectivity. Metals reflect light because they have a large imaginary component of the refractive index. Consider an interface between a dielectric having the real refractive index n_1 and a metal having the complex refractive index $n_2 = \tilde{n} + i\kappa$ (we note that the absorption coefficient of the metal is proportional to the imaginary part of its refractive index: $\alpha = \omega\kappa/c$). The reflection coefficient for light incident normally from the dielectric to the metal reads

$$r = \frac{n_1 - \tilde{n} - i\kappa}{n_1 + \tilde{n} + i\kappa} . \qquad (2.128)$$

Clearly, the reflectivity of a metal increases with increase of κ and decrease of $\tilde{n}$. In the limit of $\tilde{n} \to 0$, $\kappa \to \infty$, the reflection coefficient $r \to -1$, which means that the incident and reflected waves compensate each other and the electric-field intensity is close to zero at the surface of the metal.

In everyday life one uses metallic mirrors. A method of backing a plate of flat glass with a thin sheet of reflecting metal came into widespread production in Venice during the sixteenth century; an amalgam of tin and mercury was the metal used. The chemical process of coating a glass surface with metallic silver was discovered by Justus von Liebig in 1835 and this advance inaugurated the modern techniques of mirror making. Present-day mirrors are made by sputtering a thin layer of molten aluminium or silver onto a plate of glass in a vacuum. The metal used determines the reflection characteristics of the mirror; aluminium is cheapest and yields a reflectivity of around 88–92% over the visible wavelength range. More expensive is silver, which has a reflectivity of 95–99% even into the far infrared, but suffers from decreasing reflectivity ($< 90\%$) in the blue and ultraviolet spectral regions. Most expensive is gold, which gives excellent (98–99%) reflectivity throughout the infrared, but limited reflectivity below 550 nm wavelength, resulting in the typical gold colour.

Exercise 2.5 $^{(**)}$ *Find the frequencies of the eigenmodes of an optical cavity composed by a homogeneous layer of width a and refractive index n sandwiched between two mirrors having the amplitude reflection coefficients r.*

Exercise 2.6 $^{(**)}$ *If one of the layers in the infinite Bragg mirror has a different thickness from all other layers, it acts as a single defect or impurity in an ideal crystal. Localised photonic modes appear at such a defect. Find their eigenenergies.*

2.7 Planar microcavities

In microcavities, the cavity layer can be considered as a "defect" layer within a regular Bragg structure (see Exercise 2.6). The Fabry–Pérot confined modes of light appear within the cavity layer under condition (from eqn (2.75))

$$re^{ik_z L_c} = \pm 1, \tag{2.129}$$

where L_c is the cavity width and r is the reflection coefficient of the Bragg mirror. Alternatively, cavities with metallic mirrors can be used. In this case, the reflection coefficient r is given by eqn (2.128). In this section, we only consider dielectric Bragg mirrors characterised by the reflection coefficient $r \approx 1$.

At normal incidence, for the ideal infinite Bragg mirror ($r = 1$) a linear equation for the frequencies of the eigenmodes can be written

$$\alpha(\omega_c - \bar{\omega}) + k_z L_c = j\pi, \quad j \in \mathbb{N}. \tag{2.130}$$

The difference between microcavities and conventional cavities is in the value of L_c. In the case of microcavities, it is of the order of the wavelength of visible light divided by the refractive index of the cavity material (i.e. typically 0.2–0.4 μm). The size of conventional optical cavities is much larger. This is why the index j of the eigenmodes

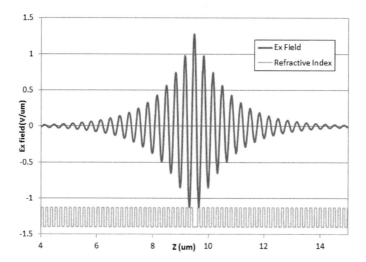

Fig. 2.11: Refractive-index profile and intensity of electric field of the eigenmode of a typical planar microcavity.

of microcavities is low and the spacing between their frequencies is so large that usually each stop-band contains only one microcavity mode. On the contrary, in conventional cavities, the spacing between eigenmodes is small and many of them are present within each stop-band. Usually, the microcavity width is designed to be an integer number of times larger than one of the regular layers in Bragg mirrors, hence $k_z L_c = j\pi$ for $\omega_c = \tilde{\omega}$. The electric-field profile in the eigenmode of a typical planar microcavity is shown in Fig. 2.11.

Transmission spectra of microcavities show peaks at the frequencies corresponding to the eigenmodes. Light is able to penetrate inside the cavity and be transmitted through it at these frequencies. This is an optical interference effect that can be also understood as the resonant tunnelling of photons: the photon from outside excites the eigenmode of the cavity and then jumps out, crossing the mirrors. Figure 2.12 shows the transmission spectrum of a HfO_2/SiO_2 microcavity compared with the reflectivity of the single Bragg mirror.

In the absence of absorption or scattering, the reflectivity R is linked to the transmission T by the simple relation

$$R = 1 - T, \tag{2.131}$$

so that the reflection spectra of dielectric microcavities exhibit dips identical to the peaks of transmission spectra.

One can see that the transmission peak corresponding to the cavity mode is broadened. Broadening is inevitably present because of the finite thickness of the Bragg mirrors and the resulting possibility for light to tunnel through the cavity, even within the stop-bands of the mirrors. The quality factor of the cavity Q is defined as the ratio

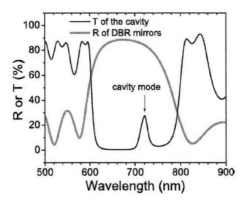

Fig. 2.12: Transmission spectrum of a HfO$_2$/SiO$_2$ microcavity and reflectivity of the single HfO$_2$/SiO$_2$ Bragg mirror containing 7 pairs of quarter-wave layers, from Song et al. (2004).

of the frequency of the cavity mode to the full-width at half-maximum of the peak in transmission corresponding to this mode. The quality factor can be also defined as

$$Q = \operatorname{Re}(\omega_c)\frac{U}{dU/dt}, \qquad (2.132)$$

where U is the electromagnetic energy stored in the cavity, dU/dt is the rate of the energy loss due to the tunnelling of light through the mirrors during a period of time dt, and ω_c is the complex eigenfrequency of the cavity mode given by eqn (2.129). We note that eqn (2.130) is not exact in real structures having finite Bragg mirrors. In particular, it yields purely real solutions, while true eigenfrequencies of the cavity modes are complex. To determine the quality factor of a cavity it is important to know the imaginary part of ω_c, as we show later.

The probability that a photon goes outside is proportional to the number of photons inside the cavity, which yields an exponential dependence of the energy U of the cavity-mode on time:

$$U(t) = U_0 e^{-\operatorname{Re}(\omega_c t/Q)}. \qquad (2.133)$$

This defines $\tau = Q/\operatorname{Re}(\omega_c)$ the lifetime of the cavity mode. Having in mind the link between the energy of an electromagnetic field and its complex amplitude $E(t)$, namely, $U(t) \propto |E(t)|^2$, we obtain

$$E(t) = E_0 \exp(-\operatorname{Re}(\omega_c t/2Q)) \exp(-i\operatorname{Re}(\omega_c)t), \qquad (2.134)$$

where E_0 is the coordinate-dependent amplitude. Standard Fourier transformation gives us the frequency dependence of the field amplitude:

$$E(\omega) = \frac{1}{\sqrt{2\pi}} \int_0^\infty E_0 \exp(-\operatorname{Re}(\omega_c t/2Q)) \exp(-i\operatorname{Re}(\omega_c)t)dt, \qquad (2.135)$$

so that

$$U(\omega) \propto |E(\omega)|^2 \propto \frac{1}{(\omega - \mathrm{Re}(\omega_c))^2 + (\mathrm{Re}(\omega_c)/2Q)^2} \cdot \qquad (2.136)$$

Expression (2.136) determines the transmission spectrum of the cavity. The resonance shape has a full-width at half-maximum equal to $\mathrm{Re}(\omega_c)/Q$, which shows the equality of two definitions of the quality factor we have given. The denominator of expression (2.136) vanishes if

$$\omega = \mathrm{Re}\omega_c + i\mathrm{Re}\omega_c/2Q. \qquad (2.137)$$

Having in mind that the complex eigenfrequency of the cavity is one that corresponds to the infinite transmission (see Section 2.6), we obtain from eqn (2.137):

$$Q = \frac{\mathrm{Re}(\omega_c)}{2\mathrm{Im}(\omega_c)}. \qquad (2.138)$$

For a specific normal mode of the cavity, this quantity is independent of the mode amplitude. The imaginary part of the eigenfrequency of the cavity mode can be easily found from eqn (2.129) as

$$\mathrm{Im}(\omega_c) = -\frac{1}{\alpha}\ln|r|, \qquad (2.139)$$

where the absolute value of the reflection coefficient of the Bragg mirror with a finite number of layers is given by eqn (2.122). In high-quality microcavities, the quality factor can achieve a few thousands.

We note, that the quality factor is different for different eigenmodes of the same cavity, in general. In particular, $Q \to \infty$ (if there is no absorption) for the guided modes that have purely real eigenfrequencies. From eqns (2.133) and (2.138), it follows that the lifetime of the cavity mode is

$$\tau = \frac{1}{2\mathrm{Im}(\omega_c)}. \qquad (2.140)$$

This characterises the average time spent by each photon inside the cavity before going out by tunnelling through the mirrors. The lifetime of guided modes is infinite. Theoretically, light never goes out from the ideal infinite waveguide. In reality, each photonic mode of any structure has a finite lifetime. The photons escape from the eigenmodes due to scattering by defects, interaction with the crystal lattice, absorption, etc. The finesse of the cavity, F (see Chapter 1), is linked to the lifetime of a cavity mode by the relation $F = \Delta\omega_c \tau$, where $\Delta\omega_c$ is the splitting between real parts of the eigenfrequencies of the neighbouring cavity modes.

In the following, we omit the prefix Re when speaking about the real part of the cavity eigenfrequency and will simply denote it as ω_c for brevity. However, we shall remember that it also has an imaginary part, $\mathrm{Im}(\omega_c) = \gamma_c$.

The deviation of the cavity mode frequency from the centre of the stop-band of the surrounding Bragg mirrors $\bar{\omega}$ always takes place in realistic structures, where the thicknesses of all layers change slightly across the sample. The detuning

$$\Delta = \omega_c - \bar{\omega} \qquad (2.141)$$

is an important parameter, which governs the splitting between TE- and TM-polarised cavity modes at oblique incidence.

At $\Delta = 0$ one can find the in-plane dispersion of the cavity Fabry–Pérot mode as the solution to Exercise 2.5:

$$\omega \approx \frac{c\pi j}{n_c L_c} + \frac{ck^2 L_c}{2n_c \pi j}, \qquad (2.142)$$

with n_c being the cavity refractive index, which yields the effective mass of the photonic mode, from $\hbar\omega(k) \approx \hbar\omega_c + \hbar^2 k^2/(2m_{ph})$, where

$$m_{ph} = \frac{\hbar n_c \pi j}{cL_c}. \qquad (2.143)$$

Here, k is the in-plane wavevector.

To take into account the polarisation dependence of the dispersion of microcavity modes, one should take into account the angle dependence of the reflection coefficient of a Bragg mirror. At oblique incidence, one can conveniently define the centre of the stop-band as a frequency $\bar\omega$, for which the phase of the reflection coefficient of the mirror is zero. The transfer matrices are modified in the case of oblique incidence and are different for TE- and TM-polarisations, as described in the previous section (see eqns (2.81) and (2.82)). Condition (2.118a) still holds at oblique incidence. It allows one to obtain the reflection coefficient of the Bragg mirror in the form

$$r_{TE,TM} = \bar{r}_{TE,TM}\exp(i\alpha_{TE,TM}(\omega - \bar\omega_{TE,TM})) \qquad (2.144a)$$

$$= \bar{r}_{TE,TM}\exp\left(i\frac{n_c}{c}L_{DBR}^{TE,TM}\cos\left(\varphi_0(\omega - \bar\omega_{TE,TM})\right)\right), \qquad (2.144b)$$

where for TE-polarisation:

$$\bar{r}_{TE} = \sqrt{1 - 4\frac{n_f \cos\varphi_f}{n_0 \cos\varphi_0}\left(\frac{n_a \cos\varphi_a}{n_b \cos\varphi_b}\right)^{2N}}, \qquad (2.145a)$$

$$\bar\omega_{TE} = \frac{\pi c}{2(a+b)}\frac{n_a \cos\varphi_a + n_b \cos\varphi_b}{n_a n_b \cos\varphi_a \cos\varphi_b}, \qquad (2.145b)$$

$$L_{DBR}^{TE} = \frac{2n_a^2 n_b^2(a+b)\cos^2\varphi_a \cos^2\varphi_b}{n_0^2(n_b^2 - n_a^2)\cos^2\varphi_0}, \qquad (2.145c)$$

where N is the number of periods in the mirror, φ_0 is the incidence angle, $\varphi_{a,b}$ are the propagation angles in layers with refractive indices n_a, n_b, respectively, and φ_f is the propagation angle in the material behind the mirror, which has a refractive index n_f. They are linked by the Snell–Descartes law

$$n_0 \sin\varphi_0 = n_a \sin\varphi_a = n_b \sin\varphi_b = n_f \sin\varphi_f. \qquad (2.146)$$

In TM-polarisation:

$$\bar{r}_{\text{TM}} = \sqrt{1 - 4\frac{n_f}{n_0}\frac{\cos\varphi_0}{\cos\varphi_f}\left(\frac{n_a\cos\varphi_b}{n_b\cos\varphi_a}\right)^{2N}}, \tag{2.147a}$$

$$\bar{\omega}_{\text{TM}} = \frac{\pi c}{2}\frac{n_a\cos\varphi_b + n_b\cos\varphi_a}{n_a n_b(a\cos^2\varphi_a + b\cos^2\varphi_b)}, \tag{2.147b}$$

$$L_{\text{DBR}}^{\text{TM}} = \frac{2n_a^2 n_b^2(a\cos^2\varphi_a + b\cos^2\varphi_b)}{n_0^2(n_b^2\cos^2\varphi_a - n_a^2\cos^2\varphi_b)}. \tag{2.147c}$$

One can see that $\bar{\omega}_{\text{TM}}$ increases faster than $\bar{\omega}_{\text{TE}}$ with an increase of the incidence angle. L_{DBR} increases with the angle in TM-polarisation and decreases in TE-polarisation. Finally, $\bar{r}$ increases with angle in TE-polarisation and decreases in TM-polarisation if $n_0 = n_f$. One can see that the stop-bands move towards higher energies with increase of the in-plane component of the wavevector of light, both in TE and TM polarisations. This is why light from the cavity mode can be resonantly scattered to the so-called leaky modes whose frequencies do not belong to the stop-bands. Via leaky modes, the photons can escape from microcavities, provided that the scattering (e.g., on structural imperfections) is efficient enough. The leaky modes limit quality factors of the planar cavities and contribute to the broadening of the cavity modes.

Substituting into the equation for the cavity eigenmodes (2.129) the renormalised stop-band frequencies (2.134), (2.138) and effective lengths (2.135) and (2.139), one can obtain as shown by Panzarini et al. (1999b), the angle-dependent TE–TM splitting of the cavity modes

$$\omega^{\text{TE}}(\varphi_0) - \omega^{\text{TM}}(\varphi_0) \approx \frac{L_c L_{\text{DBR}}}{(L_c + L_{\text{DBR}})^2}\frac{2\cos\varphi_{\text{eff}}\sin^2\varphi_{\text{eff}}}{1 - 2\sin^2\varphi_{\text{eff}}}\Delta, \tag{2.148}$$

where $\varphi_{\text{eff}} \approx \arcsin\frac{n_0}{n_c}\sin\varphi_0$ and L_{DBR} is given by eqn (2.119). Obviously, the splitting is zero at $\varphi_0 = 0$ as there is no difference between TE- and TM-modes at normal incidence. One can see that the sign of the TE–TM splitting is given by the sign of the detuning between the cavity mode frequency at normal incidence and the centre of the stop-band. Changing the thickness of the cavity one can tune Δ and change the TE–TM splitting within large limits. Usually, the TE–TM splitting is much smaller than the shift shown in eqn (2.131) (note the relation $k = (\omega/c)\sin\varphi_0$).

Finally, note that in addition to the Fabry–Pérot cavity modes described previously, the microcavities possess rich spectra of guided modes. Their spectrum can be found from eqn (2.91) (for TE-polarisation) and eqn (2.96) (for TM-polarisation) at $k > \omega/c$.

In summary to this section, the finite transmittivity of the Bragg mirrors leads to the broadening of the peaks in transmission and dips in reflection, corresponding to the cavity modes. This broadening is related to the imaginary part of the eigenfrequency of the modes and is characterised by a quality factor of the cavity Q. The dispersion of confined light modes in microcavities is parabolic to a good accuracy, while the parabola can have a different curvature in TE- and TM-polarisations. The splitting of TE and TM cavity modes can have a positive or negative sign depending on the difference between the position of the mode at $k = 0$ and the centre of the stop-band of surrounding Bragg mirrors. We conclude with a computation of one of the central quantity of this section, with Exercise 2.7.

Exercise 2.7 [(**)] *Find the quality factor of a GaAs microcavity (refractive index n_c = 3.5, thickness L_c = 244 nm) surrounded by AlAs/Ga$_{0.8}$Al$_{0.2}$As Bragg mirrors containing 10 pairs of layers each (refractive indices of AlAs, n_a = 3.0, of Ga$_{0.8}$Al$_{0.2}$As, n_b = 3.4, layer thicknesses a = 71 nm, b = 63 nm, respectively).*

2.8 Tamm plasmons and photonic Tamm states

Igor Tamm (1895–1971) was a leading Soviet scientist and a theoretical physicist of broad interests, with major contributions to electrodynamics, particle physics and solid-state physics (he is credited as the inventor of the "phonon"). He was awarded the Nobel prize for Physics for the Cherenkov effect in 1958. According to his doctoral student, Ginzburg (2001), this was not his most prized results (he valued more his work on β decay). As a Ph.D adviser, he also mentored Sakharov (with whom he developed the Tokamak) and Keldysh, among others. He was the head of the theory group developing the Soviet hydrogen bomb in the period 1949–1953 in the secret city of Sarov. An avid mountain climber, he introduced Dirac to the practice and his son came to lead the first Soviet Everest expedition.

Tamm states of light are lossless interface modes decaying exponentially in the surrounding media. They can be formed at the boundary between a periodic dielectric structure (Bragg mirror) and a metallic layer, in which case they are referred to as Tamm plasmons. They can also be formed between two periodical dielectric structures, one having a period close to the wavelength of light and another one having a period close to the double of the wavelength. These states are referred to as photonic Tamm states.

In both cases, the dispersion and eigenenergies of Tamm states may be easily obtained by a transfer matrix method. Let us work in the basis of amplitudes of electromagnetic waves propagating in positive and negative directions along x axis:

$$\begin{bmatrix} A_+ \\ A_- \end{bmatrix}. \tag{2.149}$$

Here, x is the growth axis of our structure.

Let us formally introduce a thin vacuum layer between the left and right parts of the structure. A transfer matrix across the vacuum layer of width δL is given by

$$\widehat{T}_{\delta L} = \begin{bmatrix} \exp\left(i\frac{\omega}{c}\delta L\right) & 0 \\ 0 & \exp\left(-i\frac{\omega}{c}\delta L\right) \end{bmatrix}. \tag{2.150}$$

Here, ω is the frequency of light. Requiring that no light is incident from left and right sides to the structure, we readily obtain an equation for the eigenmodes of the structure:

$$\widehat{T}_{\delta L} \begin{bmatrix} r_{left} \\ 1 \end{bmatrix} = \begin{bmatrix} A \\ A r_{right} \end{bmatrix},\qquad (2.151)$$

where r_{left} and r_{right} are reflection coefficients of light incident to the left and right parts of the structure from vacuum, respectively, A is a constant. In the limit $\delta L \to 0$, we have

$$\widehat{T}_0 = \begin{bmatrix} 1 & 0 \\ 0 & 1 \end{bmatrix}.\qquad (2.152)$$

Excluding A, the equation for eigenfrequencies of photonic modes of the structure reduces to

$$r_{left} r_{right} = 1 \qquad (2.153)$$

If the left or right part of the structure is metal, its reflection coefficient can be expressed as

$$r_M = \frac{1 - n_M}{1 + n_M},\qquad (2.154)$$

where, in the Drude model, the refractive index of metal is given by

$$n_M^2 = \varepsilon_B \left(1 - \frac{\omega_p^2}{\omega(\omega + i\gamma)} \right),\qquad (2.155)$$

where ε_B is the background dielectric constant, ω_p is the plasma frequency, γ is the plasma collision rate. For a periodic dielectric structure, the reflection coefficient r_B can be obtained as in Section 2.6.1, as

$$r_B = \pm \exp\left(i\alpha(\omega - \overline{\omega})\right),\qquad (2.156)$$

where $\overline{\omega}$ is the central frequency of the stop-band

$$\alpha = \frac{\pi}{\overline{\omega}} \frac{n_a}{|n_b - n_a|}.\qquad (2.157)$$

n_a and n_b are refractive indices of the layers composing the periodic structure (the level with a refractive index n_a is supposed to be closest to the surface). The negative (positive) sign of r_B corresponds to $n_a > n_b$ ($n_b > n_a$) in the 1st, 3rd, 5th, ..., stop-bands. The opposite is true in the 2nd, 4th, 6th, ..., stop-bands. In order to allow for the appearance of a Tamm state, the signs of r_{left} and r_{right} must be the same (assuming both coefficients to be real). Having in mind that $|\alpha(\omega - \overline{\omega})| < \pi$ within each particular stop-band, this imposes specific conditions on the formation of photonic Tamm states. Typically, such states appear at the boundaries of Bragg mirrors ending by layers of different refractive indices (n_a and n_b) if an even stop-band of one of the mirrors overlaps with an odd subband of another mirror.

It is also important to note that the widths of even stop-bands vanish if the Bragg condition is fulfilled:

$$n_a a = n_b b,\qquad (2.158)$$

where a and b are the widths of the layers having refractive indices n_a and n_b, respectively. Thus, a convenient way to realise a photonic Tamm state would be to match

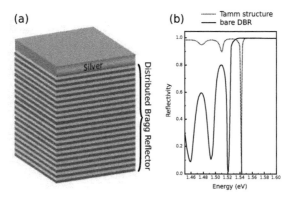

Fig. 2.13: A distributed Bragg reflector covered by a layer of silver: the structure is suitable for the observation of Tamm plasmons (a). The reflectivity of this structure as compared to the reflectivity of a Bragg mirror with no silver on the top (b). One can see a sharp dip attributed to the Tamm plasmon resonance at $1.54\,\mathrm{eV}$. Adapted from Lheureux et al. (2015).

a periodic structure that possesses a nonzero second stopband (this is achieved if the Bragg condition is broken) with a Bragg structure that has a period twice as large. In this case, the Tamm state would be formed within the 2nd stop-band of the first mirror overlapping with the 1st stop-band of the second mirror. For more details on photonic Tamm states, see the discussion by Kavokin et al. (2005b). Tamm plasmons, predicted by Kaliteevski et al. (2007), have been experimentally observed for the first time by Sasin et al. (2008) and then extensively studied by many groups. The advantage of these states compared with conventional surface plasmons is in their accessibility by direct optical excitation. While surface plasmons propagate outside the light cone and need a grating or a prism to be accessed, Tamm plasmons are formed inside the light cone and can be directly optically excited at normal incidence.

Both Tamm plasmons and Tamm photon modes are characterised by parabolic in-plane dispersions similar to dispersions of photon modes in planar microcavities. The typical structure where Tamm plasmons have been studied experimentally, as well as the reflectivity spectrum of this structure showing a narrow dip corresponding to the plasmon state are shown in Fig. 2.13

2.9   Stripes, pillars and spheres: photonic wires and dots

Progress in fabrication of so-called *photonic structures*, i.e., dielectric structures with intentionally modulated refractive indices, has made important the detailed understanding of the spectra, shape and polarisation of the confined light modes in these structures. In general, this is not an easy task as the variety of photonic structures studied till now is huge and the art of designing them ("photonic engineering") is developing rapidly.

The starting idea of the photonic crystal engineering was formulated by the American physicist Yablonovitch (1987) and consisted of creating a bandgap for light using periodic dielectric structures. The interference effects in planar structures can induce formation of the stop-bands or one-dimensional photonic gaps as we have discussed in the previous section in relation to the Bragg mirrors. More complex structures, where

Lord **Rayleigh** (1842–1919) discovered and interpreted what is now known as *Rayleigh scattering* and surface waves, now called solitons.

He is by far much more renowned under his peerage than under his real name, John Strutt. However, he acquired the title by his thirties. Other exceptional achievements include the codiscovery of argon for which he was awarded the Nobel prize in 1904. With the Rayleigh scattering, he was the first to explain why the sky is blue (this was in 1871). A gifted experimentalist, despite tough economy resulting in basic equipment, he pushed the teaching of laboratory courses to undergraduates with fervour. His interests also touched on less mundane topics, such as "insects and the colour of flowers", "the irregular flight of a tennis ball", "the soaring of birds", "the sailing flight of the albatross" and, of course, the problem of the Whispering Gallery. In a presidential British Association address, he said: *"The work may be hard, and the discipline severe; but the interest never fails, and great is the privilege of achievement."*

the refractive index is modulated along three cartesian axes, allow for the creation of three-dimensional photonic gaps. Theoretically, photonic crystals represent ideal non-absorbing mirrors and can be efficiently used for the lossless guiding of light. They have a huge potentiality for applications in future integrated photonic circuits as Fig. 2.14 shows.

In reality, inevitable imperfections in photonic crystals lead to losses because of the Rayleigh scattering of light. A detailed analysis of various photonic crystal systems can be found in the textbooks by Yariv and Yeh (2002) and Joannopoulos et al. (1995). The description of photonic crystals is beyond the scope of this book. We shall mostly address the light–matter coupling in microcavities, i.e., cavities in the photonic structures. Photonic engineering is, indeed, a powerful tool for the control of light–matter coupling strength; the density of states of the photon modes governs efficiently the emission of light by the media (which is referred to as the *Purcell effect*, addressed in detail in Chapter 6). In planar structures, considered in the previous section, the photonic modes have two degrees of freedom linked to the in-plane motion (Fig. 2.15(a)). Additional confinement of light can be achieved in stripes where only the motion along the stripe axis remains free (Fig. 2.15(b)). Stripes, as well as cylinders, can be called *photonic wires*. More radical enhancement of the photonic confinement can be achieved in pillar cavities (Fig. 2.15(c)). Here, in-plane photonic confinement is not perfect, so that the leakage of light from the pillar is possible for a part of the eigenmodes, but the quality factor of the pillar can be high enough to strongly enhance the efficiency of light–matter coupling with respect to the planar cavities. Finally, a realistic object allowing for a three-dimensional photonic confinement is a dielectric (or semiconductor) sphere (Fig. 2.15(d)). Both pillar cavities and spheres can be referred to as *photonic dots*. The light modes confined in such "dots" have a discrete spectrum and quite peculiar polarisation properties. They can be coupled to the optical transitions inside the "dot", which is potentially interesting for observation of the Purcell effect and various nonlinear optical effects. In the rest of this section, we give some basic formulae for the structures having a cylindrical symmetry (cylinders and pillar cavities) and spherical symmetry (spheres).

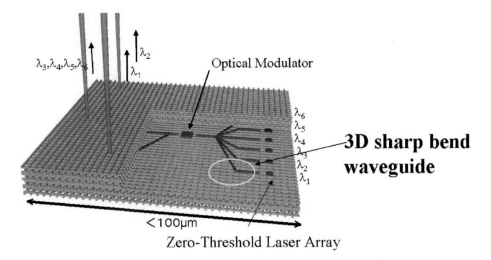

Fig. 2.14: Future concept optical integrated circuits with use of photonic crystals, by Noda et al. (2000). The complete three-dimensional photonic gap would allow lossless propagation of light in bent waveguides.

2.9.1 Cylinders and pillar cavities

Let us solve the wave equation (2.18) in cylindrical coordinates. The Laplacian operator reads in this case

$$\nabla^2 = \frac{1}{r}\frac{\partial}{\partial r}\left(r\frac{\partial}{\partial r}\right) + \frac{1}{r^2}\frac{\partial^2}{\partial\theta^2} + \frac{\partial^2}{\partial z^2}, \tag{2.159}$$

where r, θ, z are cylindrical coordinates (see Fig. 2.16). Let us consider an infinite cylinder of radius a and dielectric constant ε. For simplicity, we only consider the modes with cylindrical symmetry, which means that electric and magnetic fields are independent of θ. Solving the wave equation with the cylindrical Laplacian (2.159) one can represent the z-components of the field amplitudes in this case as

$$F_z(\rho) = J_0(\gamma\rho), \qquad\qquad \rho \leq a, \tag{2.160a}$$
$$F_z(\rho) = AK_0(\beta\rho), \qquad\qquad \rho \geq a, \tag{2.160b}$$

where F_z is either an electric or a magnetic field, J_0 is the zeroth order Bessel function, K_0 is the modified Bessel function, A is a constant that can be determined from Maxwell boundary conditions (see Section 2.5), which, in our case, requires conservation of the z- and θ-components of the fields, $\gamma = ((\omega^2 n^2/c^2) - k_z^2)^{1/2}$, $\beta = (k_z^2 - (\omega^2/c^2))^{1/2}$ and k_z is the wavevector of light along the axis of the cylinder. Other components of the fields can be found using the Maxwell equations (2.10c) and (2.11). Inside the cylinder

$$B_\rho = \frac{ik_z}{\gamma^2}\frac{\partial B_z}{\partial\rho}, \qquad\qquad B_\phi = \frac{in^2 k_z}{c\gamma^2}\frac{\partial E_z}{\partial\rho}, \tag{2.161a}$$

$$E_\rho = \frac{c}{n^2}B_\phi, \qquad\qquad E_\phi = -B_\rho, \tag{2.161b}$$

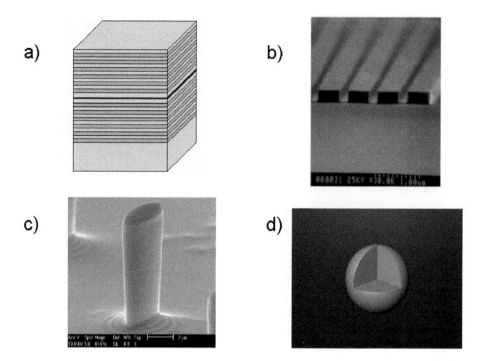

Fig. 2.15: (a) Schematic representation of a *planar microcavity*; (b) a *photonic stripe* as seen by electronic microscopy by Patrini et al. (2002); (c) a *pillar* as seen by electronic microscopy by the group in Sheffield, where the pillars can be grown elliptically to split the polarisation states; (d) schematic representation of a sphere (*photonic dot*).

$k_0 = \omega/c$ and outside the cylinder

$$B_\rho = \frac{-ik_z}{\beta^2}\frac{\partial B_z}{\partial \rho}, \qquad\qquad B_\phi = \frac{-ik_z}{\beta^2 c}\frac{\partial E_z}{\partial \rho}, \qquad (2.162a)$$

$$E_\rho = B_\phi c, \qquad\qquad E_\phi = -B_\rho c. \qquad (2.162b)$$

The triplets (B_z, B_ρ, E_ϕ) (TE-mode) and (E_z, E_ρ, B_ϕ) (TM-mode) are independent of each other. Let us find the spectrum of TE-modes. From eqns (2.161) and (2.162) the field components inside the cylinder can be expressed as

$$B_z = \frac{1}{c}J_0(\gamma\rho), \quad B_\rho = -\frac{ik_z}{c\gamma}J_1(\gamma\rho), \quad E_\phi = \frac{ik_z}{\gamma}J_1(\gamma\rho), \qquad (2.163)$$

and outside the cylinder

$$B_z = \frac{1}{c}AK_0(\beta\rho), \quad B_\rho = -\frac{ik_z A}{c\beta}K_1(\beta\rho), \quad E_\phi = \frac{ik_z A}{\beta}K_1(\beta\rho). \qquad (2.164)$$

Application of Maxwell boundary conditions at $\rho = a$ yields

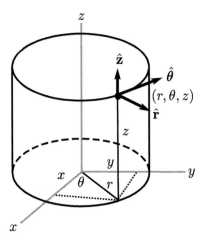

Fig. 2.16: Cylindrical coordinates.

$$-\frac{J_1(\gamma a)}{\gamma J_0(\gamma a)} = \frac{K_1(\beta a)}{\beta K_0(\beta a)}. \tag{2.165}$$

This is a transcendental equation for the eigenfrequencies of the cylinder that determines the dispersion of both guided modes (real eigenfrequency, infinite lifetime and quality factor, $(\omega/c) < k_z \leq n(\omega/c)$) and Fabry–Pérot modes (complex ω having a finite negative imaginary part, finite lifetime and quality factor, $0 \leq k_z \leq (\omega/c)$). In a similar way, one can obtain the spectrum of TM-modes

$$-\frac{n^2 J_1(\gamma a)}{\gamma J_0(\gamma a)} = \frac{K_1(\beta a)}{\beta K_0(\beta a)}. \tag{2.166}$$

In cylindrical waveguides, light can freely propagate along the z-axis. The electric and magnetic fields of the propagating modes can be found by multiplication of the amplitudes found above by an exponential factor $\exp\left(i(k_z z - \omega t)\right)$. In pillar microcavities, light propagation is confined in the z-direction usually by the Bragg mirrors (see Fig. 2.15(c) for an electron microscopy image). Because of the photonic confinement in the z-direction, k_z now takes discrete values approximately given by eqn (2.130), where L_c would be the distance between the two Bragg mirrors. The spectrum of eigenfrequencies of the pillar microcavity is also discrete, which allows it to qualify as a photonic dot. As in the infinite cylinder, the eigenmodes of such a dot can be formally divided in two categories: the "Fabry–Pérot" modes having $0 \leq k_z \leq (\omega/c)$ and the "guided" modes with $(\omega/c) \leq k_z$. The profile of electric and magnetic fields in the z-direction can be obtained as for the planar cavities. If the Bragg mirrors confining the pillar cavity are infinite, the "guided" modes have an infinite lifetime, while "Fabry–Pérot" modes have a finite lifetime and finite quality factor in all cases. In realistic structures, the lifetime of all the modes is finite, while it can become very long in the case of efficient photonic confinement. The quality factor record values, obtained in toroidal microresonators, exceed 10^8, as reported by Armani et al. (2003), which corresponds to a lifetime of the order of 10×10^{-7} s.

Pillar microcavities have attracted attention due to the pioneering experimental observation of the strong coupling of light with individual electron–hole states in semiconductor quantum dots embedded inside cavities. This is further discussed in Chapter 4.

2.9.2   Spheres

To describe the light modes in the dielectric or semiconductor microspheres it is convenient to rewrite the wave equation (2.18) in spherical coordinates. The Laplacian in spherical coordinates reads

$$\nabla^2 = \frac{1}{r^2}\frac{\partial}{\partial r}\left(r^2\frac{\partial}{\partial r}\right) + \frac{1}{r^2\sin^2\phi}\frac{\partial^2}{\partial\theta^2} + \frac{1}{r^2\sin\phi}\frac{\partial}{\partial\phi}\left(\sin\phi\frac{\partial}{\partial\phi}\right), \qquad (2.167)$$

where r, θ, ϕ are the radius, polar and azimuthal angle, respectively (see Fig. 2.17). Let us consider a sphere of radius a and dielectric constant ε surrounded by vacuum. The solution for the amplitudes of the electric and magnetic fields inside the sphere can be represented for a given mode as

$$\begin{pmatrix} F_r^{\text{in}} \\ F_\theta^{\text{in}} \\ F_\phi^{\text{in}} \end{pmatrix} = \begin{pmatrix} a_1 \\ a_2 \\ a_3 \end{pmatrix} j_l(k_{\text{in}}r)P_l^m(\cos\theta)e^{im\phi}, \qquad (2.168)$$

where F_r, F_θ, F_ϕ are the components of either the electric or magnetic field, $j_l(x)$ is the spherical Bessel function of the first kind, $P_l^m(x)$ is the associated Legendre polynomial, l and m are angular and azimuthal mode numbers, a_1, a_2, a_3 are coefficients (different for electric and magnetic fields, of course). An additional radial number of the mode n allows the radial wave number inside the cavity to be linked to the cavity radius a: $k_{\text{in}} \approx (\pi n/a)$ with $n \in \mathbb{N}$.

The fields outside the sphere are given by

$$\begin{pmatrix} F_r^{\text{out}} \\ F_\theta^{\text{out}} \\ F_\phi^{\text{out}} \end{pmatrix} = \begin{pmatrix} b_1 \\ b_2 \\ b_3 \end{pmatrix} h_l(k_{\text{out}}r)P_l^m(\cos\theta)e^{im\phi}, \qquad (2.169)$$

where $h_l(x)$ is the spherical Hankel function of the first kind, k_{out} is the wave number outside the cavity, b_1, b_2, b_3 are coefficients. The links between linear coefficients for the electric and magnetic field are always given by the Maxwell equations (2.10c) and (2.11). In general form, they are rather complex. We address the interested reader to the book by Chew (1995) containing a rigorous derivation of the spectra of the eigenmodes of dielectric spheres. Interestingly, there is no allowed optical modes having a spherical symmetry (i.e. having the angular number $l = 0$). Such a mode would have a diverging magnetic field in the centre of the cavity, which contradicts the Maxwell equation (2.10b). As in pillars, the sets of equations for TE- and TM-modes can be decoupled. TE-modes in this case are defined as those having $E_r^{\text{in}} = E_r^{\text{out}} = 0$ and for TM-modes $B_r^{\text{in}} = B_r^{\text{out}} = 0$.

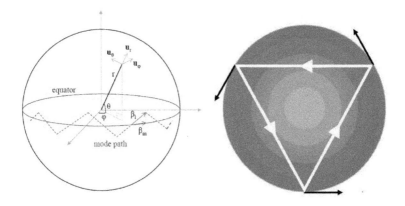

Fig. 2.17: A dielectric sphere and the path of light in the whispering-gallery mode (left); localisation of light in a sphere due to multiple internal reflections (right).

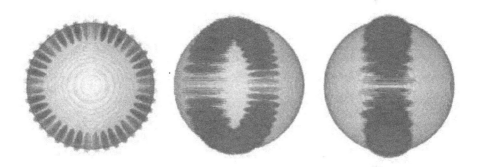

Fig. 2.18: TE whispering-gallery modes with mode numbers $n = 1$, $m = l = 20$ (from lecture notes by Nikka Nitonen).

The spectrum of eigenmodes of the sphere is discrete. It can be obtained by matching of the fields (2.168) and (2.169) by the Maxwell boundary conditions requiring

$$F_\theta^{\text{in}} = F_\theta^{\text{out}}, \quad F_\phi^{\text{in}} = F_\phi^{\text{out}}. \tag{2.170}$$

Substitution of the functions (2.168) and (2.169) into the conditions (2.170) gives the equations for eigenmodes. In TE-polarisation,

$$H_l'(k_{\text{out}}a)J_l(k_{\text{in}}a) = \sqrt{\varepsilon}J_l'(k_{\text{in}}a)H_l(k_{\text{out}}a), \tag{2.171}$$

and, in TM-polarisation,

$$\sqrt{\varepsilon}H_l'(k_{\text{out}}a)J_l(k_{\text{in}}a) = J_l'(k_{\text{in}}a)H_l(k_{\text{out}}a), \tag{2.172}$$

where $J_l(x) = xj_l(x)$ and $H_l(x) = xh_l(x)$, with $'$ meaning derivative.

While an exact spectrum of the light modes in a sphere requires solution of the transcendental equations (2.171) and (2.172), for qualitative understanding of localisation of light in the sphere, the images of ray propagation and arguments of geometrical optics are very helpful. Actually, light can be trapped by total internal reflection near the sphere's surface in a resonant whispering-gallery mode localised around the equator.

In the case of a dielectric sphere the whispering-gallery modes are the eigenmodes having high numbers l and m (high usually means higher than 10 in this context). Figure 2.17 shows schematically how the whispering-gallery modes appear. Light is propagating along the surface of the sphere each time experiencing an almost total internal reflection (not exactly total because of the curvature of the surface). The cyclic boundary conditions determine the energy spectrum of such modes.

These modes have a huge (but finite) quality factor and a long (but finite) lifetime. They can be qualified as "quasiwaveguided" modes. Propagation of whispers in the dome of Saint Paul's cathedral is assured by such "quasiwaveguided" acoustical waves subject to cyclic boundary conditions (see also Exercise 2.8).

Whispering-gallery modes have been studied in micrometre-size liquid droplets and glass spheres from the early days of laser physics. Now, very high quality spheres are obtained by melting a pure silica fibre in vacuum, as done by Collot et al. (1993). The transverse dimensions of the modes can be reduced down to a few micrometres, the sphere's diameter being about 100 μm. The mode is strongly confined. Figure 2.18 shows the calculated field intensity in a TE-polarised whispering-gallery mode in a silica microsphere. The value of the angular number l for such a mode is close to the number of wavelengths of light on the optical length of the equator of the sphere. The value $l - m + 1$ is equal to the number of the field maxima in the polar direction (i.e. perpendicular to the equatorial plane). The radial number n is equal to the number of maxima in the direction along the radius of the sphere and $2l$ is the number of maxima in the azimuthal variation of the field along the equator. The resonant wavelength is determined by the numbers n and l. The modes with lower indices l and m have a lower quality factor and shorter lifetime. They can be referred to as *Fabry–Pérot modes*. These modes are better suited for coupling to the material of the sphere than whispering-gallery modes as they penetrate deeper inside the sphere.

Eli **Yablonovitch**, produced the first artificial *photonic crystals*, the engineered counterparts of such structures as on the right from Yablonovitch (2001): a butterfly wing that—with its incomplete bandgap—produces iridescent colours.

Gustav **Mie** (1869–1957). In background, a computer simulation (with `MiePlot`) for $r = 500\,\mu m$ water drops, superimposed on a photograph of a primary and secondary rainbows.

Mie is known essentially for the solutions he provided to the problem of light described by Maxwell's equations interacting with a spherical particle (commonly but incorrectly called "Mie theory", when instead of a "theory" this is the analytical solution of the equation of an actual theory, namely, electromagnetism). Initially a pure theorist, he indulged in experimental work toward the end of his career.

The German physicist Mie solved in 1908 the problem of scattering of a plane wave by a dielectric sphere and demonstrated the existence of resonances, now known as *Mie resonances*, linked to the eigenmodes of the sphere including the whispering-gallery and Fabry–Pérot modes. Mie theory has allowed, in particular, to describe the scattering of light of the Sun by droplets of water in the clouds. It explains the colour of the sky and the appearance of rainbows and glories. Together with the portrait of Mie are shown the results of calculation of the colour of sky in the presence of two rainbows performed within Mie theory assuming $500\,\mu m$-size water drops randomly distributed in the atmosphere. The simulation result is superimposed with a photograph to demonstrate the accuracy of the model.

Exercise 2.8 (*) *The dome of Aya Sofya Mosque in Istambul has a radius of 31 metres. Find the wavelength of the whispering gallery-mode of this dome having an angular number $l = 31$.*

2.10 Further reading

Many excellent books on light propagation in various photonic structures are available that will usefully supplement the content of this chapter. The interested reader will find further details on the subject of this chapter in Born and Wolf (1970) and Jackson (1975) who give a general picture of classical optics, and the Yariv and Yeh (2002) and Joannopoulos et al. (1995) textbooks, which are devoted to optical properties of dielectric structures including photonic crystals. More details on the transfer matrix method for description of the optical properties of microcavities can be found in the textbook by Kavokin and Malpuech (2003). Rigorous derivation of the spectra of some photonic structures, including spheres is given in the monograph by Chew (1995).

3

QUANTUM DESCRIPTION OF LIGHT

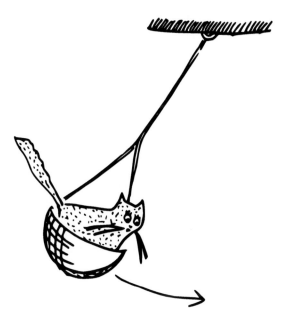

In this chapter we present a selection of important issues, concepts and tools of quantum mechanics, which we investigate up to the level of details required for the rest of the exposition, disregarding at the same time other elementary and basic topics that have less relevance to microcavities. In Chapter 4, we will also need to quantise the material excitation, but for now we limit the discussion to light, which allows us to lay down the general formalism for two special cases—the harmonic oscillator and the two-level system.

Microcavities, Second Edition. Alexey V. Kavokin, Jeremy J. Baumberg, Guillaume Malpuech, Fabrice P. Laussy, Oxford University Press (2017). © Alexey V. Kavokin, Jeremy J. Baumberg, Guillaume Malpuech, Fabrice P. Laussy. DOI 10.1093/oso/9780198782995.001.0001

3.1 Pictures of quantum mechanics

3.1.1 *Historical background*

Historically, quantum mechanics assumed two seemingly different formulations: one by Heisenberg, called *matrix mechanics*, to which we return in Section 3.1.5, and another shortly to follow by Schrödinger, based on *wavefunctions*. Although highly competitive at the start, the two "theories" now framed in modern mathematical notations display clearly their interconnection and unity.[21] As the two *pictures* are useful both for physical intuition and practical use, we study both of them. We start with the Schrödinger picture that offers the best support for the postulates of quantum mechanics in the interpretation of the so-called *Copenhagen* school, which is nowadays the commonly agreed set of working rules to deal with practical issues, although as a worldview this interpretation is now largely discarded.[22]

3.1.2 *Schrödinger picture*

In the Schrödinger picture, one starts with the *Schrödinger equation*:

$$i\hbar\frac{\partial}{\partial t}\left|\psi\right\rangle = H\left|\psi\right\rangle , \tag{3.1}$$

written here with Dirac (1930)'s notation of *bra* and *kets*, an elegant convention capturing the essentials of the mathematical structures, as is discussed below. H is the *quantum Hamiltonian* of the system to be specified for each case under consideration and $\hbar$ is the reduced Planck's constant.

3.1.2.1 *The first postulate of quantum mechanics: the quantum state.* The postulates that govern quantum mechanics, essentially laid down by von Neumann (1932), provide the recipe to use the formalism and relate it to experiments:

I — A quantum system is described by a vector—called the *state* of the system— in a separable, complex Hilbert space $\mathcal{H}$. This vector, in Dirac's notations, is denoted $\left|\psi\right\rangle$, where ψ is the set of variables needed to fully describe the system, but the notation used symbolically affords powerful abstract manipulations.[23]

---

[21]Quantum theory brought about many interesting developments in the history of science for all the controversies among its founding fathers and their personal views that often were the occasion for great drama. Beyond the famous opposition between Bohr and Einstein, there were also even animus feelings between Schrödinger and Heisenberg, and heated opposition amidst political tensions between Heisenberg and Bohr who worked together on the Copenhagen interpretation. A theatrical unravelling on the birth of quantum mechanics based on recently released documents provided the impetus for the play of M. Frayn, *Copenhagen*.

[22]In the field of *interpretation of quantum mechanics*, although there is no consensus, the modern trend favours the theory of *decoherence* and Everett interpretations of consistent realities (or parallel universes). We shall briefly touch upon some of these aspects that intersect with the physics of microcavities, but otherwise will remain oblivious and stick to the conventional Copenhagen interpretation. For further studies, cf., e.g., *Quantum Theory and Measurement*, J. A. Wheeler and W. Zurek (Princeton Series in Physics), 1984.

[23]The main advantage of Dirac's notation is the considerable simplification it brings when handling the *dual* space of $\mathcal{H}$. Whereas a ket is a vector of some given nature, a bra is actually a linear application defined on the ket space. With a little practice, one can almost forget entirely this underlying mathematical structure.

Paul **Dirac** (1902–1984), Werner **Heisenberg** (1901–1976) and Erwin **Schrödinger** (1887–1961) in the Stockholm train station, 1933, before the Nobel prize ceremony to award the 1932 prize to Heisenberg for *the creation of quantum mechanics* and the shared 1933 prize to Schrödinger and Dirac for *the discovery of new productive forms of atomic theory*. The delay in awarding the 1932 Nobel prize was due to the defiance of the Nobel committee towards quantum mechanics.

In this chapter, the quantum system of ultimate interest is light, for which two abstract systems will eventually prove sufficient to describe it fully: a two-level system with associated vector space $\mathcal{H}_2 = \{\alpha\,|0\rangle + \beta\,|1\rangle\,,\ \alpha, \beta \in \mathbb{C}\}$ (which will describe the polarisation state of light) and an harmonic oscillator, which in stark contrast to the simple space $\mathcal{H}_2$ requires a functional space of square modulus integrable functions $\mathcal{H}_a = \{|\psi\rangle\,,\ |\langle\psi|\psi\rangle|^2 < \infty\}$ (and that will describe the oscillations of the normal mode of the light field, cf. Section 2.1.1). These two specific cases will allow us to illustrate in very different cases the mechanism of the theory. We will describe the two-level system

Such shortcuts motivated Bourbaki mathematician Jean Dieudonné to state *"It would appear that today's physicists are only at ease in the vague, the obscure, and the contradictory"*. An interesting discussion of these conflicting approaches is given by Mermin in "What's Wrong with This Elegance?" in the March 2000 issue of Physics Today. We will, of course, make ample use of such simplifications. Such "rules of thumbs" are as follows: the ket $|\psi\rangle$ goes to $\langle\psi|$, coefficients are conjugated, $\alpha\,|\psi\rangle \rightarrow \langle\psi|\,\alpha^*$ and operators are transpose-conjugated, always written in reverse order, so that $\Omega\Lambda\,|\psi\rangle \rightarrow \langle\psi|\,\Lambda^\dagger\Omega^\dagger$. So for instance, the "dual" of Schrödinger equation reads:

$$-i\hbar\frac{\partial}{\partial t}\,\langle\psi| = \langle\psi|\,H\,, \tag{vi}$$

since H is hermitian.

David **Hilbert** (1862–1943) and John **von Neumann** (1903–1957), two pure mathematicians as emblems of the inherent abstract nature of quantum theory. The quantum state is best described as *a vector in a Hilbert space*, as has been axiomatised by von Neumann.

Hilbert's interest in physics started in 1912 and became his main preoccupation. It provided impetus both to quantum mechanics and relativity. He invited Einstein to give lectures on general relativity in Göttingen at which occasion some believe he derived the Einstein field equations. He put forth 23 problems at the International Congress of Mathematicians in Paris in 1900 setting the edge of mathematical knowledge at the new century. The sixth one is "Axiomatise all of physics." It is, as yet, unresolved.

Von Neumann was an extraordinary prodigy. At six, he could mentally divide two eight-digit numbers. He was famous for memorising pages on sight and, as a child, he entertained guests by reciting the phone book. Beside axiomatisation of quantum physics, which he connected to the Hilbert spaces—thereby solving the sixth problem in this particular case—he made crushing contributions—when he did not create the field—to functional analysis, set theory, topology, economics, computer science, numerical analysis, hydrodynamics, statistics, game theory and complexity theory. Many place him as among the greatest geniuses. Fellow mathematician Stanislaw Ulam's biography "Adventures of a Mathematician" is largely a tribute to his mentor with many anecdotes of this peculiar character, famous for his hazardous driving, taste for parties and hypnotisation by women.

in terms of a spin and the harmonic oscillator in terms of a mechanical oscillator, allowing us to recourse to widely used language and intuition. When it is time to return these notions to what we initially planned them for—the quantum description of light—we shall stick to the common practice of keeping the vocabulary of spin and analogies of classical mechanics, so these asides are not completely out of purpose. To later link with the statistical interpretation, we further demand that

$$\langle \psi | \psi \rangle = 1 \,. \tag{3.2}$$

Exercise 3.1 [*] *Show that the normalisation condition, eqn (3.2), remains satisfied at all times through the dynamics of Schrödinger equation (3.1).*

We have noted $|0\rangle$ and $|1\rangle$ two basis vectors of $\mathcal{H}_2$. Mathematically, it is convenient to relate them to the canonical basis, i.e.,

$$|0\rangle = \begin{pmatrix} 1 \\ 0 \end{pmatrix} \quad \text{and} \quad |1\rangle = \begin{pmatrix} 0 \\ 1 \end{pmatrix} \,. \tag{3.3}$$

Physically, we could choose to represent the first state with right circular polarisation of the light mode, $|a\rangle = |0\rangle$, and the second with left circular polarisation, with $|b\rangle = |1\rangle$. We will refer to such states as spin-up and spin-down, respectively (for a true spin $\frac{1}{2}$ particle we might prefer the depiction $|\uparrow\rangle$ and $|\downarrow\rangle$). The first postulate says that the most general state for a two-level system is $|\psi\rangle = (\alpha, \beta)^T/\sqrt{|\alpha|^2 + |\beta|^2}$, with $\alpha, \beta \in \mathbb{C}$ or, if the normalisation has been properly ensured:

$$|\psi\rangle = \alpha|0\rangle + \beta|1\rangle . \tag{3.4}$$

With the description in terms of Jones vectors, such a general state describes an arbitrary polarisation. Basis (3.3) is not always the most convenient. In generic terms, another important basis reads

$$|+\rangle = \frac{|0\rangle + |1\rangle}{\sqrt{2}}, \qquad |-\rangle = \frac{|0\rangle - |1\rangle}{\sqrt{2}} . \tag{3.5}$$

Here, we have chosen the conventional notations for generic two-level systems, today referred to as *qubits* for their fundamental role in *quantum computation*, but of course it transposes immediately with states of polarisations.

Exercise 3.2 (*) *Based on the definitions of the various possible polarisation states, express the states of linear horizontal and vertical polarisation $\{|\leftrightarrow\rangle, |\updownarrow\rangle\}$ and linear diagonal polarisation $\{|\nearrow\rangle, |\nwarrow\rangle\}$ as a function of states $|a\rangle$ and $|b\rangle$. Typical examples would be the right and left circular polarisation of light, given by, respectively,*

$$|a\rangle = \frac{|\leftrightarrow\rangle + i|\updownarrow\rangle}{\sqrt{2}}, \qquad |b\rangle = \frac{|\leftrightarrow\rangle - i|\updownarrow\rangle}{\sqrt{2}} . \tag{3.6}$$

Obtain all other relations between any two bases. Write state (3.4) in each of these bases. How are states of elliptical polarisation described?

Exercise 3.3 (*) *Two bases are said to be conjugate when any vector of the first one has equal projection on all vectors of the second. Study the conjugate character of bases encountered so far.*

The inner product is the vector scalar product, i.e.,

$$\text{if } |\psi\rangle = \begin{pmatrix} \alpha \\ \beta \end{pmatrix} \text{ and } |\phi\rangle = \begin{pmatrix} \gamma \\ \delta \end{pmatrix}, \quad \langle\psi|\phi\rangle = (\alpha^*, \beta^*)\begin{pmatrix} \gamma \\ \delta \end{pmatrix} = \alpha^*\gamma + \beta^*\delta . \tag{3.7}$$

This illustrates the richness of Dirac's notation as the bra vector $\langle\psi|$ is now tentatively associated to $(\alpha, \beta)^*$ and the inner product reads as a product of bra and ket vectors, forming a "braket" (whence the names of each vector in isolation). Being of finite dimension, $\mathcal{H}_2$ is trivially complete and separable.

A choice of basis for $\mathcal{H}_a$ first demands a choice of a space where to project the states of the system. An oscillator could be characterised in real space by its position as it oscillates, or in momentum space by its velocity. A classical oscillator would require specification of both of these at a particular time to be fully specified. In quantum mechanics, as the dynamics is ruled by a first-order differential equation (eqn (3.1)), the state is fully characterised by only one of these pieces of informations. Later we shall see that the simultaneous specification of both is, in fact, impossible.

If the state is projected in real space, a basis could consist of all the states $|x\rangle$ describing an oscillator located at x on a 1D axis (without loss of generality). The first postulate in this case asserts that the most general state for an oscillator is $|\psi\rangle = \int \psi(x) |x\rangle \, dx$. Now the linear superposition requires an integral, as there is a continuous varying set of basis states.

3.1.2.2 *The second and third postulates: observables and measurements.*

II — A physical *observable* is described by an hermitian operator Ω on $\mathcal{H}$. The possible values obtained from an observable are its eigenvalues. If the eigenstates of Ω are found to be $\{|\omega_i\rangle\}$ with associated eigenvalues $\{\omega_i\}$, the result ω_{i_0} is obtained for a system in state $|\psi\rangle$ with *probability* $|\langle\omega_{i_0}|\psi\rangle|^2$.

III — After measurement of an observable Ω that has returned the value ω_{i_0}, the state has *collapsed* to $|\omega_{i_0}\rangle$, so that repeating the observation yields ω_{i_0}, this time with certainty.

An *observable* is a property of the state that one can determine through an appropriate measurement on the system, or vividly "something that can be observed". Such accuracy in defining basic notions has been made compulsory after the counter-intuitive implications of quantum mechanics, of which we shall see a few in the following.

For a quantum system that has variables with classical counterparts, as is the case with a quantum oscillator for which a position and momentum can still be measured, Bohr formulated the prototype of the second postulate in what came to be known as *the correspondence principle*, which asserts that the classical variables x (for position, here in 1D) and p (momentum, also in 1D) are upgraded in quantum theory to hermitian operators X and P defined, in the position basis, as

$$\langle x| X |y\rangle = x\delta(x-y) \quad \text{and} \quad \langle x| P |y\rangle = -i\hbar\delta'(x-y) , \tag{3.8}$$

where δ' is the derivative of the delta function.

Any dynamical variable function of these variables extends to the quantum realm in this way, so for instance the kinetic energy $\frac{1}{2}mv^2$ is written $\frac{1}{2}p^2/m$ and its quantum counterpart reads $\frac{1}{2}P^2/m$. The classical Hamiltonian[24] also extends in this way to a *quantum Hamiltonian*, which appears in the Schrödinger equation (3.1). Therefore, one quantises an harmonic (mechanical) oscillator of mass m and force constant κ in phase space of position–momentum (x,p) starting from its (classical) Hamiltonian $H_c = p^2/(2m) + \kappa x^2/2$ to read, quantum mechanically,

---

[24]The Hamiltonian in classical mechanics is an analytic function that describes the state of a mechanical system in terms of its phase space variables, typically position and momentum variables. It is a reformulation of Newton mechanics that is more suited to shift to quantum mechanics. For most practical use, the Hamiltonian of a system can be understood as the energy of the system written in terms of specified coordinates. For more detailed discussions, see for instance Thornton and Marion (2003).

$$H = \frac{1}{2}\omega(X^2 + P^2),\tag{3.9}$$

through correspondence

$$x \rightarrow (\kappa m)^{-1/4}X \quad \text{and} \quad p \rightarrow (\kappa m)^{1/4}P,\tag{3.10}$$

with $\omega \equiv \sqrt{\kappa/m}$, where along with quantisation of scalars (x, p) we have scaled the operators in terms of a dimensionless variable ω (as we eventually wish to describe oscillations of light modes, without reminiscence of any mechanical embryos).

For a quantum system that has no mechanical counterpart, as is the case with spin,[25] the definition of mathematical observables to describe experimentally measurable properties of a system, is the result of guesswork to adjust with experimental facts. This need not concern us, however, as this procedure was carried out a long time ago for all properties that we will need to describe quantum mechanically in a microcavity.

The second postulate states that in $\mathcal{H}_2$, such an observable is described by a 2×2 hermitian matrix. Any such matrix can be decomposed as a linear superposition, over $\mathbb{C}$, of the identity $\mathbf{1} = \begin{pmatrix} 1 & 0 \\ 0 & 1 \end{pmatrix}$ and the *Pauli matrices*

$$\sigma_x = \begin{pmatrix} 0 & 1 \\ 1 & 0 \end{pmatrix}, \quad \sigma_y = \begin{pmatrix} 0 & -i \\ i & 0 \end{pmatrix}, \quad \sigma_z = \begin{pmatrix} 1 & 0 \\ 0 & -1 \end{pmatrix},\tag{3.11}$$

written here, as will always be the case unless specified otherwise, in the canonical basis (3.3). If we consider, for instance, the observable $S_z = \hbar\sigma_z$, which provides a physical dimension to the result obtained, the second postulate asserts that if S_z is measured on a system in the state (3.4), the possible outputs are $\pm\hbar$ (the eigenvalues of S_z, which are obtained straightforwardly as the operator is diagonal) and $+\hbar$ is obtained with probability $|\alpha|^2$, while $-\hbar$ is obtained with probability $1 - |\alpha|^2$. Postulate III states that after the measurement, $|\psi\rangle$, previously in the superposition (3.4), has collapsed to one of the eigenstates $|0\rangle$ or $|1\rangle$, depending on which eigenvalue has been obtained.

So when a photon is absorbed, it transfers to the detecting material an angular momentum of $\pm\hbar$, depending on which state of circular polarisation it is detected in. The outcome is deterministic if the photon was in one eigenstate $|a\rangle$ or $|b\rangle$. However, according to postulate II and Exercise 3.2, which obtains the possible linear polarisations as a superposition of circular ones, then a linearly polarised photon still impinges one quantum of angular momentum, but now with a given probability. It is only when a beam made up of many photons is considered that distinguishing features of linear polarisation (like zero average angular momentum) appear. Still, all photons in a pure linearly polarised beams are identical.

Statistical interpretation

Quantum mechanics, according to postulate II, is a probabilistic theory: the outcome of a given experiment is in general unknown, the theory can only account for the statistical spread of repeated measurements. In this context, relevant quantities to compute are

---

[25] Although we contend that spin also describes polarisation, which we have seen is a property of classical light as well; this point will be clarified in Section 3.2.3.

average and spread about this average, that is, the value obtained when an experiment is repeated on different systems all in the same quantum state. There should be an ensemble of systems as once a measurement has been made on one of them, it has collapsed on an eigenvalue of the observable, so the next measurement should not be made on the same physical system, but on another system in the same initial quantum state.

Since ω_{i_0} is obtained with probability $|\langle\psi|\omega_{i_0}\rangle|^2$ when Ω is measured on $|\psi\rangle$, the average value of this observable, written $\langle\Omega\rangle$, is the weighting of all possible outcomes (that is, Ω eigenvalues) and so is $\sum_i \omega_i |\langle\psi|\omega_i\rangle|^2$, so that through bra and ket algebra

$$\langle\Omega\rangle = \sum_i \omega_i \langle\psi|\omega_i\rangle\langle\omega_i|\psi\rangle, \tag{3.12a}$$

$$= \langle\psi|\left(\sum_i \omega_i |\omega_i\rangle\langle\omega_i|\right)|\psi\rangle, \tag{3.12b}$$

$$= \langle\psi|\Omega|\psi\rangle, \tag{3.12c}$$

since eqn (3.12b) is Ω spelled out in its eigenstates basis. Note that $\langle\Omega\rangle$ does not specify on which quantum state the average has been taken, which is usually clear from the context. In cases where the specification is important, Dirac's notation once again provide a most convenient alternative, eqn (3.12c).

The dynamics of such an average follows from Schrödinger equation as[26]

$$\frac{\partial}{\partial t}\langle\Omega\rangle = \frac{i}{\hbar}\langle[H,\Omega]\rangle. \tag{3.13}$$

3.1.2.3 *Uncertainty principle.* Coming back to a single experiment on a two-level system, assume that the value $+\hbar$ has been obtained as the result of measuring S_z on state (3.4), so that the system has now collapsed to state $|0\rangle$ according to the third postulate. If another measurement of S_z is performed, the same value $+\hbar$ will be obtained with probability one and $-\hbar$ with probability $|\langle 1|0\rangle|^2 = 0$. So the result is deterministic in this case, but what if the observable associated to, say, $S_x = \hbar\sigma_x$ is now measured? One can check that states (3.5) are the eigenstates of S_x with eigenvalue $\pm\hbar$ so that the system will collapse on one of them as the result of the measurement, with probability $1/2$ (cf. Exercise 3.2). If, after this, one returns to S_z, the result has been randomised completely and the first measurement will yield any possible value $\pm\hbar$ with probability $1/2$. This is a manifestation of the *uncertainty principle* that arises from the second and third postulate from non-commuting operators.

---

[26] Spelling out the derivation of the equation of motion for a quantum average:

$$\frac{\partial}{\partial t}\langle\Omega\rangle = \frac{\partial}{\partial t}\langle\psi|\Omega|\psi\rangle, \tag{vii-a}$$

$$= \left(\frac{\partial}{\partial t}\langle\psi|\right)\Omega|\psi\rangle + \langle\psi|\Omega\left(\frac{\partial}{\partial t}|\psi\rangle\right), \tag{vii-b}$$

$$= \frac{i}{\hbar}\langle\psi|H\Omega - \Omega H|\psi\rangle, \tag{vii-c}$$

which, contracted, yields the result. In eqn (vii-b) we took advantage of the linearity of the differential on Hilbert spaces and their dual remembering that in this case Ω is time independent. In eqn (vii-c) we substituted Schrödinger equation in the ket (3.1) and bra (vi) spaces.

Generally, it is a necessary and sufficient condition for two operators Ω and Λ to share a common basis of eigenstates on the one hand and to commute, $[\Omega, \Lambda] = 0$, on the other hand. The uncertainty principle in its most general form reads as a lower bound for the spread in the distributions of two observables:

$$\text{Var}(\Omega)\text{Var}(\Lambda) \geq \left(\frac{i}{2}\langle[\Omega, \Lambda]\rangle\right)^2, \tag{3.14}$$

where $\text{Var}(\Omega) = (\Omega - \langle\Omega\rangle)^2$.

Exercise 3.4 [**] *Prove eqn (3.14) with Schwarz inequality applied to vectors $(\Omega - \langle\Omega\rangle)\,|\psi\rangle$ and $(\Lambda - \langle\Lambda\rangle)\,|\psi\rangle$.*

From eqn (3.14) it is apparent that the commutation relations between operators are an important ingredient of quantum mechanics. One can check that

$$[X, P] = i\hbar, \tag{3.15}$$

by computing $-i\hbar(x\partial_x - \partial_x x)$ on a generic test function after applying the correspondence principle backward to get back to scalars from the operators. This is, when applied to the general formula, the origin for the most famous form of the uncertainty relation:

$$\Delta x \Delta p \geq \frac{\hbar}{2}, \tag{3.16}$$

3.1.2.4 *Composite systems and symmetry.*

IV — The Hilbert space of a composite system is the Hilbert space tensor product of the state spaces associated with the component systems.

This postulate extends in the expected way the rules of quantum mechanics from a one-dimensional case to many: the dimensionality of the entire system's Hilbert space scales with the number of degrees of freedom to be described quantum mechanically.[27] Observables also inherit this tensor product structure. The additional variable can pertain to the same particle, e.g., be *i*) another spatial dimension, or *ii*) a property of an altogether different character like spin, or *iii*) can be the same variable, but for another particle. Starting with $|\psi_{1D}\rangle$ the state of a particle in $\mathcal{H}_1$ a single-particle Hilbert space, these three cases would lead to, respectively, *i*) $|\psi_{2D}\rangle$ to be projected on $\langle x| \otimes \langle y|$ to give the function of two variables $\psi_{2D}(x, y)$, *ii*) $|\psi_{1D}\rangle \otimes \sigma$ and *iii*) $|\psi_{1D}\rangle \otimes |\psi'_{1D}\rangle$.

The general case of an observable being $\Omega_1 \otimes \Omega_2$, it is customary to drop the explicit tensor sign and abbreviate it into a product, $\Omega_1\Omega_2\,|\psi_1\rangle\,|\psi_2\rangle$ or even and as commonly, simply $|\psi_1, \psi_2\rangle$ or $|\psi_1\psi_2\rangle$ for the state.

Composite systems, however, do not simply transport the quantum "weirdness" of the single-particle case to the higher-dimensional one. They bring one conceptual difficulty of their own rooted in correlations and known as *entanglement*, which is one of the resources for *quantum information processing*.

---

[27] Time is an example of a variable that remains classical in non-relativistic quantum mechanics, i.e., that is not a quantum observable and therefore is not associated to a Hilbert space. This is to be contrasted with special relativity where, by contrast, time is shown to carry equivalent features with space variables.

The counterintuitive physics of entangled systems is best exemplified following Bohm and Aharonov (1957) who consider the singlet spin state of two particles:

$$|\Psi\rangle = \frac{1}{\sqrt{2}} \left(|\uparrow\downarrow\rangle - |\downarrow\uparrow\rangle\right). \qquad (3.17)$$

We repeat that the notation $|\uparrow\downarrow\rangle$ is a shorthand for $|\uparrow\rangle \otimes |\downarrow\rangle$ and that first ket (in this case $|\uparrow\rangle$) refers to one of the particles and the second ket to the other particle. These two particles are separated, although remaining in the state (3.17) (the spatial wavefunction part of the system has not been written, which would change to reflect this spatial separation; the spin state can remain the same for separated particles). When the separation is significantly large, the spin of the "first" (or "left") particle is measured. As a result, the wavefunction collapses on the measured eigenstate. This, however, has the effect of also collapsing simultaneously the state of the other particle. Such an experiment exhibits *nonlocal quantum correlations*, i.e., correlations with no classical counterpart in a sense that we now discuss in greater detail.

There is first the obvious correlation of the measurement that says that if the left branch has measured, say, spin-up, then the other is assured to measure spin-down. This is the correlation part, just as it applies in a classical sense, and that ensures total spin conservation. However, correlations of state (3.17) are also non-classical because they also hold in all other bases, although in these cases the wavefunction does not specify the outcome. Therefore, if one measures the first qubit value in the basis $|\pm\rangle$ of eqn (3.5) (with the same eigenvalues $+1$ and -1) and finds, say, spin-up again, then the other bit is also -1 in the new basis. Although the wavefunction does not specify the values of all components, the correlations always match. These correlations are finally *nonlocal* because this agreement holds even for any separation with the bits possibly measured simultaneously. This has been confirmed experimentally by Weihs et al. (1998).

V — The wavefunction changes or retains its sign upon permutation of two identical particles.

This important postulate[28] can be motivated by the insightful quantum-mechanical property of *indistinguishable identical particles*, which asserts that two particles of the same species bear no absolute or independent role to the wavefunction that describes them both. To make this explicit, consider the wavefunction written as a function of the generalised coordinates $\mathbf{q}_i$ for the ith particle out of N, that is, $\psi(\mathbf{q}_1, \cdots, \mathbf{q}_N)$. The system would remain the same if the jth and kth particles were to be interchanged, provided that they are identical, i.e., refer to particles of the same species that cannot be distinguished experimentally. The quantum state would therefore also remain the same, i.e.,

$$\psi(\mathbf{q}_1, \cdots, \mathbf{q}_j, \cdots, \mathbf{q}_k, \cdots, \mathbf{q}_N) = \alpha\psi(\mathbf{q}_1, \cdots, \mathbf{q}_k, \cdots, \mathbf{q}_j, \cdots, \mathbf{q}_N), \qquad (3.18)$$

---

[28] It is little appreciated that the indistinguishable characters of the quanta is a postulate, motivated by experimental evidence, but that in principle can be violated without undermining quantum mechanics. Messiah and Greenberg (1964) have emphasised this point for elementary particles.

with the phase factor $\alpha = e^{i\theta}$, which does not change the observables (and therefore yields the same quantum state). Doing this twice yields $\alpha^2 = 1$, i.e., $\alpha = \pm 1$. This shows how postulate V derives from invariance of the quantum state upon interchange of identical particles.

Pauli (1940) has shown in the context of relativistic quantum field theory how this property of the wavefunction relates to the spin of the particles thus described and is therefore also an intrinsic property that is always and consistently satisfied. Particles with integer spin are called *bosons* (after Bose) and those of half-integer spin are called *fermions!* (after Fermi). It is shown that wavefunctions for bosons keep the same sign as particles are interchanged, while those for fermions change sign.[29]

This has considerable physical consequences of strikingly different characters depending on the sign, although the mathematical structure assumes a simple unifying form:

$$\psi(\mathbf{q}_{\sigma(1)}, \cdots, \mathbf{q}_{\sigma(N)}) = \zeta^{\xi} \psi(\mathbf{q}_1, \cdots, \mathbf{q}_N), \tag{3.19}$$

with $\zeta = 1$ for bosons and $\zeta = -1$ for fermions,

where $\sigma \in \mathfrak{S}$ is a permutation[30] of $[1, N]$ and ξ the signature of σ.

3.1.3 *Antisymmetry of the wavefunction*

In the case of fermions, eqn (3.19) leads to *Pauli blocking* or *Pauli exclusion*, which asserts that two fermions cannot occupy the same quantum state. This is clear from eqn (3.18) with $\alpha = -1$ since in this case the probability (amplitude) is zero when $\mathbf{q}_j = \mathbf{q}_k$, which makes it impossible to find the system with two particles in the same projections $|\mathbf{q}_i\rangle$. Pauli blocking can also be stated with overall quantum states in which case the entire wavefunction vanishes. If the ith particle out of N is in state $|\phi_i\rangle$, the wavefunction of the whole system reads as a determinant, since this is the mathematical expression to associate sign swapping to function (column) or coordinates (row) permutations:[31]

$$\psi(\mathbf{q}_1, \cdots, \mathbf{q}_N) \propto \det_{1 \leq i,j \leq N} \left(\phi_i(\mathbf{q}_j) \right). \tag{3.20}$$

The constant of normalisation of eqn (3.20) depends on the orthogonality of the states $|\phi_i\rangle$. It diverges as some of the states $|\phi_i\rangle$ overlap to unity, as a manifestation of Pauli exclusion.

---

[29] In addition to possible (small) deviations from Bose and Fermi statistics for elementary particles, which would be of a fundamental character, there are also deviations that result from cooperative or composite effects, or of reduced dimensionalities. Such emerging statistics are typical of solid-state physics and examples are provided in the next chapter with *excitons*.

[30] A *permutation* σ of the set of integers $[1, N] = \{1, 2, \ldots, N\}$ is a one-to-one function from $[1, N]$ unto itself, e.g., $\sigma(1) = 2$, $\sigma(2) = 1$ and $\sigma(3) = 3$ is a permutation of $[1, 3]$. There are $N!$ permutations of $[1, N]$. The set of all permutations is written $\mathfrak{S}$. The signature ξ of a permutation, also known as its parity, is ± 1 according to whether an even or odd number of pairwise swappings is required to bring the sequence $(1, \ldots, N)$ into $(\sigma(1), \ldots, \sigma(N))$.

[31] With only two particles to simplify notations, in respective states $|\phi_i\rangle$, $i = 1, 2$, the total fermion wavefunction reads $\psi(\mathbf{q}_1, \mathbf{q}_2) \propto \phi_1(\mathbf{q}_1)\phi_2(\mathbf{q}_2) - \phi_1(\mathbf{q}_2)\phi_2(\mathbf{q}_1) = \begin{vmatrix} \phi_1(\mathbf{q}_1) & \phi_2(\mathbf{q}_2) \\ \phi_1(\mathbf{q}_2) & \phi_2(\mathbf{q}_1) \end{vmatrix}$.

Exercise 3.5 [(*)] *Show that in the case where the N single-particle states are orthogonal, $\langle \phi_i | \phi_j \rangle = \delta_{i,j}$, the constant of normalisation of the symmetrised or antisymmetrised wavefunction is $1/\sqrt{N!}$.*

3.1.4   Symmetry of the wavefunction

If the sign remains the same in eqn (3.19), there is no cancelling and the probability does not vanish for any given superposition. Rather the opposite tendency holds that the probability is enhanced for two different particles to be found in the same quantum state. The accurate and general formulation states that *the probability that N bosons be found in the same quantum state is $N!$ times the probability that distinguishable particles be found in the same state.*[32]

A more familiar statement is that if N particles are in the same state, the probability for the $(N+1)$th to be found also in this same state is $N+1$ times this probability for distinguishable particles. It is proved with conditional probabilities, if "A" is the statement "the $(N+1)$th particle is in some given state $|\varphi\rangle$", while "B" is the statement "the N other bosons already are in state $|\varphi\rangle$", thus the probability $\mathbb{P}$ (with respect to distinguishable particles) that "A" is realised given that "B" is, is $\mathbb{P}(A \cap B)/\mathbb{P}(B)$, that is $(N+1)!/N!$. This $N+1$ coefficient characterises *bosonic stimulation*.[33]

3.1.4.1   *Solving the Schrödinger equation.*   Now that the postulates and interpretation of quantum mechanics have been laid down, we can carry on with solving equations.

---

[32] The proof is instructive and goes as follows: let $|\Psi\rangle$, the wavefunction of the state, be developed on a basis $|\phi_i\rangle$ of $\mathcal{H}^{\otimes N}$, first assuming distinguishable particles that do not require the symmetry postulate

$$|\Psi\rangle_{\mathrm{C}} = \sum_{i_1,\dots,i_N} \alpha_{i_1,\dots,i_N} |\phi_{i_1}\rangle \cdots |\phi_{i_N}\rangle \,, \tag{viii}$$

(we subscripted the state with C for "classical"), then symmetrising the state to ensure bosonic indistinguishability (3.19)

$$|\Psi\rangle_{\mathrm{B}} = \frac{1}{\sqrt{N!}} \sum_{i_1,\dots,i_N} \sum_{\sigma \in \mathfrak{S}} \alpha_{i_1,\dots,i_N} |\phi_{\sigma(i_1)}\rangle \otimes \cdots \otimes |\phi_{\sigma(i_N)}\rangle \,. \tag{ix}$$

The probability amplitude that all distinguishable particles be in the same quantum state, say $|\phi_1\rangle$, is

$$\langle \phi_1^{\otimes N} | \Psi \rangle_{\mathrm{C}} = \alpha_{1,\cdots,1} \,, \tag{x}$$

while, for indistinguishable particles,

$$\langle \phi_1^{\otimes N} | \Psi \rangle_{\mathrm{B}} = \frac{N!}{\sqrt{N!}} \alpha_{1,\cdots,1} \,. \tag{xi}$$

The ratio of these probabilities is $N!$. Note especially that it is independent of α (which should be nonzero, meaning that one cannot find all particles in the same state if they do not all have a projection in this state). This ratio is therefore independent of any linear combination of the α and thereby of any state of the system, thus showing that the probability to find all particles in the same state if they are indistinguishable bosons is $N!$ the probability for distinguishable particles.

[33] As all results involving conditional probabilities—and in this case further complicated by the quantum interpretation—the bosonic stimulation is more subtle than it appears. However, considered in first order of perturbation theory, it becomes an exact and useful concept in the form of renormalisation scattering rates of emission or in rate equations, as shall be seen in greater detail in later chapters.

Exercise 3.6 (*) *Reduce by separation of variables the Schrödinger equation for a time-independent Hamiltonian to Schrödinger's time-independent equation*

$$H \left| \phi \right\rangle = E \left| \phi \right\rangle , \tag{3.21}$$

with $\left| \phi \right\rangle$ now time independent.

When the Hamiltonian is time independent, a formal solution is obtained as

$$\left| \psi(t) \right\rangle = e^{-iHt/\hbar} \left| \psi(0) \right\rangle . \tag{3.22}$$

All the postulates of quantum mechanics apply on $\left| \psi(t) \right\rangle$, which now just happens to change with time.

Exercise 3.7 (**) *Solve the Schrödinger equation (3.1) for the quadratic potential (3.9). Technically, this requires finding the eigenstates and eigenenergies of H.*[34]

Exercise 3.8 (**) *Using the results of the previous exercise, study the time dynamics of the initial conditions $\langle x | \psi_1 \rangle \propto \exp(-(x - x_0)^2/L^2)$ for various (relevant) sets of parameters (x_0, L).*

3.1.5 Heisenberg picture

At this stage, we have presented the essential facts of quantum theory required for the quantum description of light that we shall soon undertake, of course omitting a lot of material not immediately or crucially needed for that purpose. We now devote further considerations to alternative formulations along with inclusions of other physics, like statistical physics or thermodynamics, because of their importance to microcavity physics.

The formal integration of the Schrödinger equation, eqn (3.22), shows that the time evolution of the state is a rotation in Hilbert space. Taking advantage of this fact, one can use a basis of rotating states and transfer the dynamics from the states to the operators, which were previously fixed, i.e., time independent. Therefore, considering the time-varying average[35] $\langle \Omega \rangle (t)$, which is a quantity physically measurable that should not depend on which formalism is used, and starting from the definition we have given (in the Schrödinger picture)

$$\langle \Omega \rangle (t) = \langle \psi(t) | \, \Omega \, | \psi(t) \rangle , \tag{3.23a}$$

$$= \langle \psi(0) | \, e^{iHt/\hbar} \Omega e^{-iHt/\hbar} \, | \psi(0) \rangle , \tag{3.23b}$$

$$= \langle \psi(0) | \, \tilde{\Omega}(t) \, | \psi(0) \rangle , \tag{3.23c}$$

---

[34]The eigenstates $\left| \phi_n \right\rangle$ of the Schrödinger equation for the harmonic potential are

$$\langle x | \phi_n \rangle = \frac{1}{\sqrt{2^n n!}} \left(\frac{m\omega}{\pi\hbar} \right)^{1/4} \exp\left(-\frac{m\omega x^2}{2\hbar} \right) H_n \left(\sqrt{\frac{m\omega}{\hbar}} x \right) . \tag{xii}$$

The associated energy spectrum is $E_n = (n + 1/2)\hbar\omega$.

[35]Now, the time dependence is shown explicitly everywhere and is accurately attributed to which quantity is time dependent, e.g., $\langle \Omega(t) \rangle$ is very different from $\langle \Omega \rangle (t)$.

we arrive at a time-varying operator $\tilde{\Omega}(t)$

$$\tilde{\Omega} = e^{iHt/\hbar}\Omega e^{-iHt/\hbar}\,, \tag{3.24}$$

which acts on time-independent states (frozen to their initial condition $|\psi(0)\rangle$). This formulation of quantum mechanics, where operators carry the time dynamics and the states are fixed, is called the *Heisenberg picture*. We used the average as an intermediate between the two pictures, but there is an equation of motion for the operators directly, aptly called the *Heisenberg equation*, which reads

$$i\hbar\frac{\partial}{\partial t}\tilde{\Omega}(t) = [\tilde{\Omega}(t), H]\,. \tag{3.25}$$

Exercise 3.9 (*) *Derive the Heisenberg equation (3.25) from the Schrödinger equation (3.1) and also the Schrödinger equation from the Heisenberg equation, thereby proving the complete equivalence of the two formulations. Note that in the commutator of eqn (3.25), $\tilde{\Omega}$ is the time-dependent Heisenberg operator, while H is the time-independent Schrödinger Hamiltonian. In the course of your demonstration, show that for algebraic computation purposes, the Heisenberg equation can be written as*

$$i\hbar\frac{\partial}{\partial t}\tilde{\Omega}(t) = \widetilde{[\Omega, H]}\,, \tag{3.26}$$

where tilde means the operator has been transformed according to eqn (3.24), i.e., $\widetilde{[\Omega, H]} = e^{iHt/\hbar}[\Omega, H]e^{-iHt/\hbar}$. Therefore, the algebra carried out in the commutator is time independent throughout.[36]

Keeping in mind the functional analysis results of the quantum harmonic oscillator, we now consider the problem from an algebraic point of view, therefore closer in spirit

---

[36]From the Schrödinger equation to the Heisenberg equation and back; we compute

$$\frac{\partial}{\partial t}\tilde{\Omega} = \frac{\partial}{\partial t}\left(e^{iHt/\hbar}\Omega e^{-iHt/\hbar}\right)$$

$$= \left(\frac{\partial}{\partial t}e^{iHt/\hbar}\right)\Omega e^{-iHt/\hbar} + e^{iHt/\hbar}\left(\frac{\partial}{\partial t}\Omega\right)e^{-iHt/\hbar} + e^{iHt/\hbar}\Omega\left(\frac{\partial}{\partial t}e^{-iHt/\hbar}\right)$$

$$= \left(\frac{\partial}{\partial t}e^{iHt/\hbar}\right)\Omega e^{-iHt/\hbar} + e^{iHt/\hbar}\Omega\left(\frac{\partial}{\partial t}e^{-iHt/\hbar}\right)\,,$$

since Ω is time independent

$$= \left(\frac{iH}{\hbar}e^{iHt/\hbar}\right)\Omega e^{-iHt/\hbar} + e^{iHt/\hbar}\Omega\left(-\frac{iH}{\hbar}\right)e^{-iHt/\hbar}$$

$$= \frac{i}{\hbar}H\tilde{\Omega} - \frac{i}{\hbar}\tilde{\Omega}H\,,$$

since H commutes with $e^{-iHt/\hbar}$

$$= \frac{i}{\hbar}[H, \tilde{\Omega}]\,.$$

to the Heisenberg approach (this solution is due to Dirac). It pre-figures the analysis we shall make on the more complicated system of the light field in Section 2.1.1.

Let us introduce the *ladder operators*:

$$a = \frac{1}{\sqrt{2\hbar}}(X + iP), \qquad a^\dagger = \frac{1}{\sqrt{2\hbar}}(X - iP), \tag{3.27}$$

where the others quantities have been defined on page 85. It follows straightforwardly from eqn (3.15) and the above definition that

$$[a, a^\dagger] = 1, \tag{3.28}$$

and also that the Hamiltonian (3.9) reads

$$H = \hbar\omega\left(a^\dagger a + \frac{1}{2}\right), \tag{3.29}$$

so that the eigenvalue problem now consists of finding $|\phi\rangle$ such that

$$a^\dagger a |\phi\rangle = \phi |\phi\rangle . \tag{3.30}$$

To do so, dot eqn (3.30) with $\langle\phi|$ (these states are assumed normalised) to get $\phi = \langle\phi| a^\dagger a |\phi\rangle = \|a |\phi\rangle\|^2$ that shows that the eigenvalues ϕ are positive. Using the algebraic relations[37]

$$[a, a^\dagger a] = a, \qquad [a^\dagger, a^\dagger a] = -a^\dagger, \tag{3.31}$$

we obtain the equality $(a^\dagger a)a = a(a^\dagger a - 1)$, which leads to

$$(a^\dagger a)a |\phi\rangle = a(a^\dagger a - 1) |\phi\rangle = a(\phi - 1) |\phi\rangle = (\phi - 1)a |\phi\rangle , \tag{3.32}$$

the two ends showing that $a |\phi\rangle$ is an eigenstate of $a^\dagger a$, with eigenvalue $\phi - 1$, at the possible exception if $a |\phi\rangle = 0$ (the null vector, since it cannot be an eigenstate). As for $a |\phi\rangle$, dotting it with its conjugate yields

$$\|a |\phi\rangle\|^2 = (\langle\phi| a^\dagger)(a |\phi\rangle) = \langle\phi| a^\dagger a |\phi\rangle . \tag{3.33}$$

Iterating eqn (3.32) n times shows that $a^n |\phi\rangle$ is also an eigenvector of $a^\dagger a$, but this time with eigenvalue $\phi - n$, which unless it becomes zero for some value of n (interrupting the iteration), will ultimately become negative, in contradiction to what precedes. Therefore, the zero vector must be hit exactly, i.e., there exists $n \in \mathbb{N}^*$ such that $a^n |\phi\rangle = 0$, but $a^{n-1} |\phi\rangle \neq 0$. Consider the normalised eigenstate $|\phi - n\rangle = a^n |\phi\rangle / \|a^n |\phi\rangle\|$ with eigenvalue $\phi - n$. Applying a on this state yields zero by definition of n, while eqn (3.33) gives its normed squared as $\phi - n$. Equating the two provides

---

[37]Commutation relations between operators derived from the initial definition range from straightforward to intractable with little difference in their shape. Equation (3.31) is easily obtained from eqn (3.28), for example, by brute expansion: $[a, a^\dagger a] = aa^\dagger a - a^\dagger aa = (aa^\dagger - a^\dagger a)a = a$.

the structure of the solutions, $\phi = n$, therefore the eigenstates of $a^\dagger a$ are states $|n\rangle$, where $n \in \mathbb{N}$. The "bottom" state $|0\rangle$ satisfies

$$a\,|0\rangle = 0\,. \tag{3.34}$$

Note that $|0\rangle$ and 0 are two different entities. The latter is the mathematical zero and, in this case, the zero vector of the Hilbert space, while the former is more of a notation for a complicated mathematical object, in this case a Gaussian function once projected in a coordinate space.

Exercise 3.10 (*) *Repeat the previous analysis to prove the counterpart properties for $a^\dagger$, i.e., $a^\dagger\,|\phi\rangle$ is an eigenvector of $a^\dagger a$ and $\|a^\dagger\,|\phi\rangle\| = \sqrt{\phi + 1}$.*

As opposed to a, which eventually reaches the bottom rung $|0\rangle$ past which no other states can be obtained, $a^\dagger$ can increase indefinitely the eigenstate label. These operators are quickly checked to satisfy

$$a\,|n\rangle = \sqrt{n}\,|n-1\rangle\,, \tag{3.35a}$$

$$a^\dagger\,|n\rangle = \sqrt{n+1}\,|n+1\rangle\,, \tag{3.35b}$$

and also, applying them in succession,

$$a^\dagger a\,|n\rangle = n\,|n\rangle\,. \tag{3.36}$$

Higher-order formulas are readily obtained, such as

$$a^{\dagger i} a^j a^{\dagger k}\,|n\rangle = \frac{(n+k)!}{(n+k-j)!}\sqrt{\frac{(n+i+k-j)!}{n!}}\,|n+i+k-j\rangle\,, \tag{3.37}$$

with condition $j \le n + k$, the result being zero otherwise. Equation (3.37) can be used to evaluate expressions often arising in the course of quantum algebra calculations, typically in connection with boson–boson interactions.[38]

---

[38] Applying repeatedly formulas (3.35), written here for convenience with their counterparts in the dual space,

$$a\,|n\rangle = \sqrt{n}\,|n-1\rangle\,, \qquad\qquad \langle n|\,a = \langle n+1|\,\sqrt{n+1}\,, \tag{xiii-a}$$

$$a^\dagger\,|n\rangle = \sqrt{n+1}\,|n+1\rangle\,, \qquad\qquad \langle n|\,a^\dagger = \langle n-1|\,\sqrt{n}\,, \tag{xiii-b}$$

one obtains straightforwardly

$$a^i\,|n\rangle = \sqrt{\frac{n!}{(n-i)!}}\,|n-i\rangle\,, \qquad a^{\dagger i}\,|n\rangle = \sqrt{\frac{(n+i)!}{n!}}\,|n+i\rangle\,, \tag{xiv}$$

and therefore, still iterating:

$$(\text{for } i \le n+j) \qquad a^i a^{\dagger j}\,|n\rangle = \frac{(n+j)!}{\sqrt{n!}\sqrt{(n+j-i)!}}\,|n+j-i\rangle\,, \tag{xv-a}$$

$$(\text{for } i \le n) \qquad a^{\dagger j} a^i\,|n\rangle = \frac{\sqrt{n!}\sqrt{(n+j-i)!}}{(n-i)!}\,|n+j-i\rangle\,. \tag{xv-b}$$

The states $|n\rangle$ are sometimes called *number states* because they are states with a definite number of particles. They are also known, from quantum field theory, as *Fock states* (cf. Section 3.3.1). They are handy for calculations and offer a *de facto* canonical basis, i.e.,

$$|0\rangle = \begin{pmatrix} 1 \\ 0 \\ 0 \\ 0 \\ 0 \\ \vdots \end{pmatrix}, \quad |1\rangle = \begin{pmatrix} 0 \\ 1 \\ 0 \\ 0 \\ 0 \\ \vdots \end{pmatrix}, \quad |2\rangle = \begin{pmatrix} 0 \\ 0 \\ 1 \\ 0 \\ 0 \\ \vdots \end{pmatrix}, \quad \cdots \quad |n\rangle = \begin{pmatrix} 0 \\ \vdots \\ 0 \\ 1 \\ 0 \\ \vdots \end{pmatrix}, \quad \cdots,$$

with 1 as the $(n+1)$th element of the vector. With this basis, the annihilation and creation operators read

$$a = \begin{pmatrix} 0 & 1 & & & & \\ & 0 & \sqrt{2} & & & \\ & & 0 & \sqrt{3} & & \\ & & & \ddots & \ddots & \\ & & & & 0 & \sqrt{n} \\ & & & & & \ddots & \ddots \end{pmatrix}, \quad a^\dagger = \begin{pmatrix} 0 & & & & \\ 1 & 0 & & & \\ & \sqrt{2} & 0 & & \\ & & \sqrt{3} & 0 & \\ & & & \ddots & \ddots \\ & & & & \sqrt{n+1} & 0 \\ & & & & & \ddots & \ddots \end{pmatrix},$$

The result is zero if the condition on the left is not satisfied. Next step gives formula (3.37). With appropriate values for i, j and k, one finds for the following, of frequent use in many problems,

1. $a^2 |n\rangle = \sqrt{n(n-1)} \, |n-2\rangle$

2. $\langle n| \, a^2 = \langle n+2| \, \sqrt{(n+1)(n+2)}$

3. $a^{\dagger 2} |n\rangle = \sqrt{(n+1)(n+2)} \, |n+2\rangle$

4. $\langle n| \, a^{\dagger 2} = \langle n-2| \, \sqrt{n(n-1)}$

5. $a^\dagger a^2 |n\rangle = (n-1)\sqrt{n} \, |n-1\rangle$

6. $\langle n| \, a^\dagger a^2 = \langle n+1| \, n\sqrt{n+1}$

7. $aa^\dagger a |n\rangle = n^{3/2} \, |n-1\rangle$

8. $\langle n| \, aa^\dagger a = \langle n+1| \, (n+1)^{3/2}$

9. $a^2 a^\dagger |n\rangle = \sqrt{n}(n+1) \, |n-1\rangle$

10. $\langle n| \, a^2 a^\dagger = \langle n+1| \, \sqrt{n+1}(n+2)$

11. $aa^{\dagger 2} |n\rangle = \sqrt{n+1}(n+2) \, |n+1\rangle$

12. $\langle n| \, aa^{\dagger 2} = \langle n-1| \, \sqrt{n}(n+1)$

13. $a^\dagger aa^\dagger |n\rangle = (n+1)^{3/2} \, |n+1\rangle$

14. $\langle n| \, a^\dagger aa^\dagger = \langle n-1| \, n^{3/2}$

15. $a^{\dagger 2} a |n\rangle = n\sqrt{n+1} \, |n+1\rangle$

16. $\langle n| \, a^{\dagger 2} a = \langle n-1| \, (n-1)\sqrt{n}$

17. $a^2 a^{\dagger 2} |n\rangle = (n+1)(n+2) \, |n\rangle$

18. $\langle n| \, a^2 a^{\dagger 2} = \langle n| \, (n+1)(n+2)$

19. $a^{\dagger 2} a^2 |n\rangle = (n-1)n \, |n\rangle$

20. $\langle n| \, a^{\dagger 2} a^2 = \langle n| \, (n-1)n.$

Also, note the useful relation obtaining $|n\rangle$ from the vacuum, following from eqn (xv-b),

$$|n\rangle = \frac{a^{\dagger n}}{\sqrt{n!}} \, |0\rangle . \tag{xvi}$$

where the nth column is being shown for the general case. In both cases, the diagonal is zero. All the algebra can be computed with this representation and expressions for operators obtained in this way. The number operator, for instance, is diagonal with values 0, 1, 2, ...

If we return to the Hamiltonian (3.29), the energy spectrum is seen to consist of equally spaced energy levels with energy difference $\hbar\omega$:

$$E_n = \left(n + \frac{1}{2}\right)\hbar\omega. \tag{3.38}$$

The state $|0\rangle$ of lowest energy is the *ground state* of the system and is, by definition, devoid of excitation. It is called the *vacuum state* (or a vacuum state in the case of degeneracy). However, its energy is nonzero since $\langle 0| H |0\rangle = \hbar\omega/2$. This term is associated to quantum fluctuations that arise from the uncertainty principle, of which we shall see more later in a true field-theoretical setting. One more important point is that this structure of equally spaced levels makes compelling the interpretation of $|n\rangle$ as a state with n *quanta* of excitations, each of energy $\hbar\omega$, superimposed on the vacuum fluctuations, which cannot be eliminated.

The effects of the various operators just defined on the state $|n\rangle$ with n excitations entitle them to be called *annihilation*, cf. eqn (3.35a) and *creation operators*, cf. eqn (3.35b). Their net result is indeed to raise or lower the number of excitations.[39] The annihilation operating on the vacuum state, eqn (3.34), results in cancelling the term, which disappears from the process being computed.

The Heisenberg picture has many advantages, including computational or algebraic simplicity, as we shall later appreciate. One further quality is the aid to physical intuition afforded by this formalism as these are the observables that vary in time, in analogy with classical physics.

3.1.6 *Dirac (interaction) picture*

The Dirac picture, also known as the *interaction picture*, is an intermediate case between the Schrödinger picture (with fixed operators and time-varying states) and the Heisenberg picture (with time-varying operators and fixed states) where both operators and states are time dependent. Which fraction of the dynamics is attributed to each of them depends on the decomposition of the Hamiltonian.

First, with the case of a total Hamiltonian H without time dependence, which is separated as

$$H = H_0 + H_I, \tag{3.39}$$

where H_0 is typically a "simple" part of the dynamics, i.e., which can be solved exactly and whose solutions will define the states of the "free" particle, or a "dominant" part that will result in a fast dynamics, and the remaining H_I, typically complicated or of a lesser magnitude, which will be interpreted as an interaction term between the particles defined by H_0.

---

[39] A more accurate terminology would call these operators "*ladder operators*" in the case of a generic quantum harmonic oscillator, comprising a *raising* and *lowering* operator, and to reserve the terms for *annihilation* and *creation* in a field-theoretic context.

The underlying principle of the interaction picture is to concentrate on the complication H_I in the Hamiltonian by embedding H_0 dynamics into the operator by defining

$$\tilde{\Omega} = e^{iH_0t/\hbar}\Omega e^{-iH_0t/\hbar}\,, \tag{3.40}$$

which equation of motion is reminiscent of the Heisenberg equation (3.25), but with some important redefinitions of the quantities involved,

$$i\hbar\frac{\partial}{\partial t}\tilde{\Omega}(t) = [\tilde{\Omega}(t), \tilde{H}_I]\,, \tag{3.41}$$

where tilde means the operator has been transformed to the Heisenberg picture according to eqn (3.40). As opposed to the Heisenberg equation, this transformation applies to what remains of the Hamiltonian as well, $\tilde{H}_I = e^{iH_0t/\hbar}H_Ie^{-iH_0t/\hbar}$, which, however, can remain time independent under this transformation, leading to $\tilde{H}_I = H_I$. In all cases, H_I and not H—like in the Heisenberg or Schrödinger equations—appears in the commutator. The similarities are strong enough, however, for eqn (3.41) to be called the *Heisenberg equation in interaction picture.*

Exercise 3.11 [*] *Derive eqn (3.41).*

If the computation is difficult, one can compute the commutation first, as $[\tilde{\Omega}, \tilde{H}] = \widetilde{[\Omega, H]}$. An efficient approach to solve a general problem is to separate it, as we said, into a "dominant" part, which can be solved exactly or approximated easily, and consider the minor parts left out initially as perturbations. What is meant by dominant depends on the problem at hand, but a typical case would be the time-independent part of a general Hamiltonian $H(t)$, which we could separate as

$$H(t) = H_0 + H_1(t)\,, \tag{3.42}$$

where H_0 is time independent, resulting in a uniform time rotation in Hilbert space $e^{-iH_0t/\hbar}|\psi_0\rangle$, cf. eqn (3.22).

The Heisenberg and Dirac pictures often require a lot of algebra, which means that commutators have to be computed by expansion and simplification, through gathering and cancelling with relations such as eqn (3.28). To allow such an evaluation when faced with a general commutator of the type $[\prod_{i=1}^{n} A_i, \prod_{j=1}^{m} B_j]$, where A_i, B_j are some operators, the most general expansion is a sum over all combinations of commutators $[A_i, B_j]$ with other operators factored outside of the commutator. Their relative position is, of course, important in the most general case where their relative commutation rules are unknown. Their placement is made as follows for the commutator $[A_i, B_j]$: all operators $A_1 \cdots A_{i-1}$ placed before A_i and all operators $B_1 \cdots B_{j-1}$ placed before B_j are placed *before* the commutator $[A_i, B_j]$, in this order, and all operators $A_{i+1} \cdots A_n$ and all operators $B_{j+1} \cdots B_m$ are placed *after* the commutator, in the opposite order of A and B. This is illustrated next, for the case where the operators that remain in the commutator are A_i and B_j:

$$[(A_1 \cdots A_{i-1})A_i(A_{i+1} \cdots A_n), (B_1 \cdots B_{j-1})B_j(B_{j+1} \cdots B_m)]$$

$$(A_1 \cdots A_{i-1})(B_1 \cdots B_{j-1})[A_i, B_j](B_{j+1} \cdots B_m)(A_{i+1} \cdots A_n)$$

Application of this rule on the six arbitrary operators $A, \cdots, F$ written as follows yields

$$[ABC, DEF] = [A, D]EFBC + D[A, E]FBC + DE[A, F]BC \tag{3.43a}$$
$$+ A[B, D]EFC + AD[B, E]FC + ADE[B, F]C \tag{3.43b}$$
$$+ AB[C, D]EF + ABD[C, E]F + ABDE[C, F]. \tag{3.43c}$$

There are in this case nine terms as there are 3×3 combinations for commutators, with operators distributing as illustrated. Such evaluations usually simplify extensively when they are carried over a family of boson operators a_i, obeying the following algebras:

$$[a_i, a_j] = 0, \qquad [a_i, a_j^\dagger] = \delta_{ij}. \tag{3.44}$$

For instance, a useful relation obtained in this way is

$$a^\dagger a^\dagger aa = a^\dagger a(a^\dagger a - 1). \tag{3.45}$$

Exercise 3.12 (*) *Given three Bose operators a_1, a_2 and a_3 obeying commutation relations (3.44), evaluate the following expressions often encountered in the Heisenberg and interaction pictures:* $[a_1, a_1^\dagger a_1]$, $[a_1, a_1^{\dagger 2} a_1^2]$, $[a_1, a_1^\dagger a_1 a_2^\dagger a_2]$, $[a_1, a_1^{\dagger n} a_1^m]$.

3.2 Other formulations

3.2.1 *Density matrix and Liouvillian*

The formulation of the theory so far—in whatever representation—describes so-called *pure states*, which for now we can regard as a synonym for *ket states*, that is, states for which there exists a "wavefunction" $|\Psi\rangle$. A first reason why there would be no such state is if one would attempt to describe only part of a composite system. Namely, if the system of our interest S is in contact with another system R (the notations are for "system" and "reservoir"), the (pure or ket) state to describe the whole system is, in general,

$$|\Psi_{\mathrm{SR}}\rangle = \sum_i \sum_j c_{ij} |\phi_i\rangle |\varphi_j\rangle, \tag{3.46}$$

where the $|\phi_i\rangle$ are basis states for S and $|\varphi_j\rangle$ are basis states for R. Note that in the sense that each state has its associated Hilbert space and set of observables, it is defined independently of the other. If the c_{ij} are such that eqn (3.46) can be written as a (tensor) product

$$|\Psi_{\mathrm{SR}}\rangle = \left(\sum_i c_i^{\mathrm{S}} |\phi_i\rangle \right) \left(\sum_j c_j^{\mathrm{R}} |\varphi_j\rangle \right), \tag{3.47}$$

(that is, if $c_{ij} = c_i^{\mathrm{S}} c_j^{\mathrm{R}}$), then $|\Psi_{\mathrm{SR}}\rangle$ can be likewise decomposed as $|\Phi_{\mathrm{S}}\rangle |\Phi_{\mathrm{R}}\rangle$ and the system S considered in isolation with its quantum state $|\Psi_{\mathrm{S}}\rangle$.

If, however, such a decomposition (3.47) is not possible, in which case the systems S and R are said to be *entangled*, then it is not possible to consider any one of them independently, at least exactly. The density operator is the approximated state that arises when the total wavefunction is averaged over unwanted degrees of freedom of the total system.

Exercise 3.13 (**) *Show that the state that best describes S in isolation is*

$$\rho_S = \mathrm{Tr}_R(|\Psi_{SR}\rangle \langle \Psi_{SR}|), \tag{3.48}$$

where the partial trace[40] is taken over the Hilbert space of R.

Such an operator ρ_S—the so-called *reduced density matrix*—which cannot be described by a ket state, is called a *mixed state*—as opposed to the *pure state*—because of the "averaging" that mixes up quantum states. Every pure state can be written as a density matrix, namely as the projector operator for this state[41]

$$\rho = |\psi\rangle \langle \psi|, \tag{3.50}$$

but only density matrices whose square have trace equal to one can be written as a pure state. For a mixed state, $\mathrm{Tr}(\rho^2) < 1$, and the smaller this trace, the more mixed is the state (for a space of dimension n the minimum is $1/n$).[42]

This notation introduces new techniques to carry out computations, for instance the average of an observable Ω in state (3.50) is now obtained through the formula:

$$\langle \Omega \rangle = \mathrm{Tr}(\rho A), \tag{3.51}$$

as $\langle \Omega \rangle = \langle \psi | \Omega | \psi \rangle = \sum_i \langle \psi | \Omega | \phi_i \rangle \langle \phi_i | \psi \rangle$ by inserting the closure of identity. Now, since $\langle \psi | \Omega | \phi_i \rangle$ and $\langle \phi_i | \psi \rangle$ are two complex numbers, they commute and

$$\langle \Omega \rangle = \sum_i \langle \phi_i | \psi \rangle \langle \psi | \Omega | \phi_i \rangle = \sum_i \langle \phi_i | \rho \Omega | \phi_i \rangle, \tag{3.52}$$

by definition of ρ, which is the result. In this case, it is also easy to turn Schrödinger equation (3.1) into an equation for the density matrix (of a pure state):

---

[40] The *trace* of an operator Ω in some Hilbert space is defined as the sum of its diagonal elements (in any base), i.e., $\mathrm{Tr}(\Omega) = \sum_i \langle \phi_i | \Omega | \phi_i \rangle$, and thus is a number (a real number if the operator is hermitian). The *partial trace*, say over $\mathcal{H}_2$, of an operator Ω acting on a tensor Hilbert space $\mathcal{H}_1 \otimes \mathcal{H}_2$, is the trace over diagonal elements of $\mathcal{H}_2$ in any of its basis, leaving unaffected the projection of Ω on $\mathcal{H}_1$. That is, decomposing Ω as $\Omega = \sum_i \sum_j \omega_{ij} \Omega_i \otimes \Omega_j$,

$$\mathrm{Tr}_{\mathcal{H}_2} \Omega = \sum_i \left(\sum_j \omega_{ij} \sum_k \langle \phi_k | \Omega_j | \phi_k \rangle \right) \Omega_i = \sum_i \tilde{\omega}_i \Omega_i, \tag{3.49}$$

which is an operator on $\mathcal{H}_1$, with $\tilde{\omega}_i$ the term in parentheses. The generalisation to higher dimensions, as well as to any other selection of which spaces to trace out is obvious.

[41] The density operator also makes it clear that the overall phase of a ket state is immaterial, as $|\psi\rangle = \alpha |0\rangle + \beta |1\rangle$ and $e^{i\phi} |\psi\rangle$ both have the same density operator independent of ϕ.

[42] Other mathematical properties of the density matrix are that it is hermitian with trace unity and is positive, i.e., for all states $|\phi\rangle$, $\langle \phi | \rho | \phi \rangle \geq 0$. Reciprocally, any positive operator whose trace equals 1 is eligible to be a density matrix.

Exercise 3.14 (*) *Derive from the Schrödinger equation for* $|\psi\rangle$ *the equation for* $|\psi\rangle\langle\psi|$

$$i\hbar\frac{\partial}{\partial t}\rho = [H, \rho]. \tag{3.53}$$

Another reason to have recourse to a density operator is when statistical mechanics is incorporated into quantum mechanics, i.e., when some indeterminacy is injected in the system as the result of knowledge not available only in practice (in direct analogy to the classical statistical theory and direct opposition to the quantum indeterminacy that is intrinsic to the system). For instance, the state of a radiation mode that can take up possible energies E_i in thermal equilibrium at temperature T requires such a statistical description, as some excitations are randomly poured into or conversely removed from the system by thermal kicks issued by the reservoir. In accordance with thermodynamics, such a state would be described as being in the (pure) quantum state $|E_i\rangle$ of energy E_i with probability $p_i = e^{-E_i/k_b T}/Z$ with $Z = \sum_i e^{-E_i/k_b T}$ the partition function.

The average $\langle\Omega\rangle$ given by eqn (3.51) with the density matrix

$$\rho = \sum_i p_i |E_i\rangle\langle E_i|, \tag{3.54}$$

yields $\langle\Omega\rangle = \sum_i p_i \langle E_i|\Omega|E_i\rangle$, from which it is seen that the average is now the quantum average (3.12c) weighted over the classical probabilities p_i, so eqn (3.54) describes a system where a quantum state is realised with probability p_i.

It is important, although sometimes subtle, to distinguish between the quantum and a classical indeterminacy. For instance, the density matrix

$$\frac{1}{2}\begin{pmatrix} 1 & 1 \\ 1 & 1 \end{pmatrix} = \frac{1}{\sqrt{2}}(|a\rangle + |b\rangle)(\langle a| + \langle b|)\frac{1}{\sqrt{2}}, \tag{3.55}$$

which describes a pure state of linear polarisation (cf. Exercise 3.2), is physically different from the mixed state

$$\frac{1}{2}\begin{pmatrix} 1 & 0 \\ 0 & 1 \end{pmatrix} = \begin{cases} |a\rangle \text{ with probability } 1/2 \\ |b\rangle \text{ with probability } 1/2 \end{cases}, \tag{3.56}$$

which describe a single photon that has either circular polarisation with probability $1/2$ all along. This is different from eqn (3.55), where the circular polarisation becomes left or right only as a result of a measurement in this basis and is the rest of the time otherwise undetermined.

It therefore appears that off-diagonal elements of the density matrix are linked with the pure or mixed character of the state. Indeed, as we shall have many occasions to appreciate in this text, these elements relate to *quantum coherence* and their contribution to the equations of motion discriminates between classical and quantum dynamics.

We shall refer to the Schrödinger equation (3.53) cast with the density operator as *Liouville–von Neumman equation*. It holds for a general density matrix, not merely pure states like eqn (3.50). This results from the fact that ones goes from a pure state to a density matrix by averaging over a statistical ensemble, a procedure that still holds at the level of the equation (3.53).

Exercise 3.15 (**) *Show that the Liouville–von Neumman equation applies for an arbitrary density matrix. Also show that eqn (3.51) is general.*

The generalisation can also be derived easily with a density matrix obtained by partial tracing, like eqn (3.48).

Another and more general way to write eqn (3.53) is in term of a *superoperator*, which is an operator that acts on other operators in much the same way as an operator acts on a state. The generic superoperator is the *Liouvillian* $\mathcal{L}$, which enters in the equation of motion for the density matrix

$$i\hbar\frac{\partial}{\partial t}\rho = \mathcal{L}\rho. \tag{3.57}$$

In the case of the Liouville equation, $\mathcal{L}$ is defined, for all ρ, as

$$\mathcal{L}\rho = H\rho - \rho H. \tag{3.58}$$

The necessity of the Liouvillian arises when the system does not have a purely Hamiltonian dynamics, such as eqn (3.58) above, but includes non-conservative forces, in which case Liouville–von Neumann equation is not enough to describe its dynamics, whereas the Liouvillian can accommodate most of the physics that goes beyond Hamiltonian mechanics, as we will see in Chapter 5 when we include dissipation.

3.2.2 Second quantisation

We have presented in the previous sections the basic concepts of quantum physics from its "mechanical" point of view where the object of quantisation is a mechanical attribute of the particle, like its motion or its spin. When the physical object is a field—as is the case with light—quantum mechanics is upgraded to the status of a *quantum field theory*. There are various possible theoretical formulations, but for the needs of this book we shall be content with a simple and vivid picture, known as *second quantisation*. One conceptual benefit of this reformulation of the theory is the valuable concept it affords of a particle as an excitation of the field, in the terms we are about to present. In the formulation given so far, we have already used repeatedly the term "particle" to describe the object to which to attach the wavefunction or one of its attributes (like the spin). One difficulty, however, arises when the number of particles is not conserved, as is the case in a statistical theory in the grand-canonical ensemble, or if particles are unstable and can decay into other particles (calling for the necessity to remove and add particles in the quantum system). Also, particles are generally considered in a collection, so that the symmetry requirements are to be taken into account.

Second quantisation starts with the *occupation number formalism* that provides an elegant and concise solution to all these desiderata. Let $\mathcal{H}_1$ be a single-particle basis, i.e., a set of states $|\phi_i\rangle$ that are orthonormal and—assuming a discrete basis—such that any possible quantum state $|\psi\rangle$ can be written as $|\psi\rangle = \sum_i \alpha_i |\phi_i\rangle$ for a suitable choice of α_i (which is also unique by orthogonality of the basis). Now consider a state $|\Psi\rangle$ of the system with n_i particles in state $|\phi_i\rangle$. Without any symmetry requirement, such a state would read

$$|\Psi\rangle = |\phi_1\rangle^{\otimes n_1} \otimes \cdots \otimes |\phi_i\rangle^{\otimes n_i} \otimes \cdots \qquad (3.59)$$

A more convenient way to represent the state (3.59) is to specify directly the numbers of particles in each of the (previously agreed) states $|\phi_i\rangle$, i.e.,

$$|\Psi\rangle = |n_1, \ldots, n_i, \ldots\rangle \qquad (3.60)$$

using once more the elegant Dirac's notation. By the fourth postulate, the wavefunction for a collection of particles needs to be symmetrised. The advantage of the occupation number formalism then becomes compelling, as the details of the symmetrisation can be hidden in the abstract notation of eqn (3.60). For fermions, all n_i equal at most 1 and the exchange of two particles result in a change of the sign of the wavefunction. For bosons, there are no limits in the number of particles and exchange of two particles leaves the wavefunction unchanged.

We have introduced with eqn (3.27) the so-called ladder operators a and $a^\dagger$ that annihilate and create, respectively, excitations of the quantum harmonic oscillator. Because the energy levels are equally spaced, cf. eqn (3.38), we can understand such a state, say the nth one with energy $(n + 1/2)\hbar\omega$, as consisting of n excitations each with energy $\hbar\omega$, with a remainder of $\hbar\omega/2$ in energy when $n = 0$, which is therefore identified as the energy of the vacuum. The field-theoretic version of the ladder operators are *creation* and *annihilation* operators, which bring a state with n particles to one with $n - 1$ and $n + 1$ particles, respectively. If there is no particle to annihilate, application of the annihilation operator yields zero, that is, the process is cancelled or does not appear in the calculation.

Let us give an example to make clear the overall picture. Let us assume three possible quantum states of bosonic particles, that we call "s", "p" and "i", with associated Bose operators $a_{\mathrm{s,p,i}}$, respectively. The basis is afforded by states $|n_{\mathrm{s}}, n_{\mathrm{p}}, n_{\mathrm{i}}\rangle$ with $n_{\mathrm{s,p,i}} \in \mathbb{N}$. Each operator acts on its relevant part of the total state, e.g.,

$$a_{\mathrm{s}} |n_{\mathrm{s}}, n_{\mathrm{p}}, n_{\mathrm{i}}\rangle = \sqrt{n_{\mathrm{s}} - 1} \, |n_{\mathrm{s}} - 1, n_{\mathrm{p}}, n_{\mathrm{i}}\rangle \,, \qquad (3.61)$$

if $n_{\mathrm{s}} \geq 1$, 0 otherwise. A process that scatter-off two particles from state p to redistribute them in states s and i would read, in the second quantisation formalism

$$a_{\mathrm{s}}^\dagger a_{\mathrm{i}}^\dagger a_{\mathrm{p}}^2 \,. \qquad (3.62)$$

Such a process would bring an initial state $|0, 2, 0\rangle$ into the final state $|1, 0, 1\rangle$. The reverse process that brings back the particle into the state p is the hermitian conjugate of eqn (3.62). The state $|1, 1, 0\rangle$, for instance, vanishes when eqn (3.62) is applied on it, so it decouples from the dynamics and from all the calculations (like transition rates) for an Hamiltonian based on eqn (3.62). Physically, such a process corresponds to parametric scattering where a state that is pumped (p) scatters off particles into a signal (s) and a so-called "idler" state (i). Such a dynamic will be amply discussed in the chapters to follow. Of course, the procedure extends to spaces of infinite dimensions. A two-particle interaction in free space, such as the Coulomb interaction, reads in second quantisation

$$V = \frac{1}{2} \sum_{\substack{\mathbf{k}_1, \mathbf{k}_2 \\ \mathbf{k}_3, \mathbf{k}_4}} \langle \mathbf{k}_3 \mathbf{k}_4 | \, V \, | \mathbf{k}_1 \mathbf{k}_2 \rangle \, a_{\mathbf{k}_3}^\dagger a_{\mathbf{k}_4}^\dagger a_{\mathbf{k}_2} a_{\mathbf{k}_1} \, . \tag{3.63}$$

To such a process one can attach a *Feynman diagram*, which displays graphically one term of the Born expansion of an interaction, as shown in Fig. 3.1. We have given an example for the case of bosons, but the same would apply for fermions using σ operators instead, that follow Fermi's algebra.

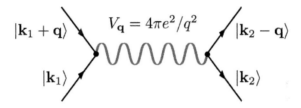

Fig. 3.1: A Feynman diagram for Coulomb interaction, whereby two particles of momenta $\mathbf{k}_1$ and $\mathbf{k}_2$ in the initial state scatter to final states $\mathbf{k}_1 + \mathbf{q}$ and $\mathbf{k}_2 - \mathbf{q}$ by exchanging momentum $\mathbf{q}$.

Second quantisation also sheds much light on the dilemma of wave–particle duality and the so-called *complementarity*[43] that is naively perceived as a photon behaving sometimes like a particle, sometimes like a wave. A particle is a quantum state of the field that reads $|0, \dots, 0, 1, 0 \dots \rangle$, the position of the 1 corresponding to which quantum state the particle is in. Taken in isolation, this always behave as a particle in the classical sense (of one discrete lump of matter or energy). With repeated measurements over single-particle states, wave-like behaviour starts to emerge. For example if one tracks the position on the screen where a single electron is detected after crossing a double slit, one always detect a single spot corresponding to a single particle. Repeating the measurement many times—each time tracking a single electron—results in a wave-like interference patterns for all the collected positions. The behaviour of a collection of particles gives rise to the notion of many-particle quantum states. We discuss more states in the next sections. For now, we complete the second quantisation of light by promoting the electromagnetic field modal decomposition of the previous chapter to matrix algebra.

---

[43] It is interesting to recall Anderson who comments in Nature, **437**, 625, (2005) that *"Niels Bohr's 'complementarity principle'—that there are two incompatible but equally correct ways of looking at things—was merely a way of using his prestige to promulgate a dubious philosophical view that would keep physicists working with the wonderful apparatus of quantum theory. Albert Einstein comes off a little better because he at least saw that what Bohr had to say was philosophically nonsense. But Einstein's greatest mistake was that he assumed that Bohr was right—that there is no alternative to complementarity and therefore that quantum mechanics must be wrong."*

3.2.3 *Quantisation of the light field*

The quantisation of the electromagnetic spectrum is known as *quantum electrodynamics* (QED). For the spectral window of light that corresponds to high frequencies, but still moderate quanta energies, QED can be framed in a suitable way to make the most of the various energy and time scales, which is known as *quantum optics* and is the topic of this section.

We tackle the quantisation of the field in the vacuum, leaving the complicated problem of the interaction with matter to the next chapter that will bring many different techniques to tackle this issue in various approximations. We therefore proceed from the classical equations obtained in Section 2.1.1. We quantise the field following Dirac (1927) and Fermi (1932) by canonical quantisation of the variables a related to the field through eqn (2.6a) and that is known from eqn (2.9) to undergo harmonic oscillation. To settle notations, we regard free space as a large cubic box of size L, with boundary conditions for the electromagnetic field of running waves[44] $e^{i(\omega_{\mathbf{k}}t - \mathbf{k} \cdot \mathbf{r})}$, as opposed to standing waves. A wavevector of these solutions is defined by

$$\mathbf{k} = \left(\frac{2\pi n_x}{L}, \frac{2\pi n_y}{L}, \frac{2\pi n_z}{L} \right), \tag{3.64}$$

where $n_{x,y,z} \in \mathbb{N}$ and L is taken as high as necessary for the sought precision. The field amplitude in this box is

$$\mathcal{E}_{\mathbf{k}} = \sqrt{\frac{\hbar \omega_{\mathbf{k}}}{2\epsilon_0 V}}. \tag{3.65}$$

Completing the modal expansion with quantised operators, the expression of the field $\mathbf{E}$ (now also an operator) reads

$$\mathbf{E} = \sum_{\mathbf{k}} \mathcal{E}_{\mathbf{k}} a_{\mathbf{k}} e^{i(\mathbf{k} \cdot \mathbf{r} - \omega t)} + \text{h.c.} \tag{3.66}$$

The vector nature of a is associated to polarisation. We split these two aspects—field amplitude and polarisation—to deal with them separately, which allows us to apply the results already presented regarding the formal quantum systems of the harmonic oscillator and the two-level system. We assume the vector variable a to be transverse and we choose an orthogonal 2D basis in the transverse $\mathbf{k}$ plane:

$$\mathbf{a}(\mathbf{k}, t) = a_\uparrow(\mathbf{k}, t) \mathbf{e}_\uparrow(\mathbf{k}) + a_\downarrow(\mathbf{k}, t) \mathbf{e}_\downarrow(\mathbf{k}), \tag{3.67}$$

where $(\mathbf{e}_\uparrow(\mathbf{k}), \mathbf{e}_\downarrow(\mathbf{k}), \mathbf{k})$ form an orthogonal basis of unit vectors. When need arises we will use the full vector expression, but without limitation we now focus on one projection only, say $a_\downarrow$, which for brevity we shall denote a, thereby coming back to the elementary one-dimensional quantum mechanics, which was our starting point.

---

[44]Boundary conditions of a running wave are such that once the wavefront reaches the end of the free space within the box and touches the border, it goes through and reappears instantaneously on the other side of the box. In this sense, this boundary condition is not really physically relevant and should be viewed as a mathematical trick to model the infinite universe in a more intuitive way, with labels L to track our wavevectors.

Vladimir **Fock** (1898–1974) gave his name to the Fock space and Fock state after building the Hilbert space for Dirac's theory of Radiation, Zs. f. Phys., **49** (1928) 339.

Many other results and methods due to him such as the Fock proper time method, the Hartree–Fock method, the Fock symmetry of the hydrogen atom and still others make him one of the most popular names in quantum field theories.

3.3   Quantum states

We now discuss some commonly encountered quantum states in the light of the formalism of second quantisation.

3.3.1   *Fock states*

An important and intuitive building block is the *Fock state*, which is the canonical basis state of the corresponding *Fock space*. It is the state with a definite number of particles (in the case of the electromagnetic field, a definite number of photons) and is, in this respect, a physical state as well as a mathematical pillar of second quantisation. The vacuum $|0\rangle$ and the single particle $|1\rangle$ are two typical, important and "relatively" easy states to prepare in the laboratory.[45] The arbitrary case $|n\rangle$ with $n \in \mathbb{N}$ becomes increasingly difficult with large values of n. Small values have, however, been reported in the literature and cavities are precious tools to this end, as demonstrated, for instance, by Bertet et al. (2002).

From the mathematical point of view, Fock states are useful for many computational needs. When quantum electrodynamical calculations are conducted pertubatively, the Fock state plays a major role as each successive order of the approximation describes processes that increase or decrease the number of photons.

3.3.2   *Coherent states*

The *coherent state* was initially introduced by Schrödinger as the quantum state of the harmonic oscillator that minimises the uncertainty relation (3.14) for the observables P and X. From the derivation of the inequality of the generalised Heisenberg uncertainty relation of Exercise 3.4, we know that eqn (3.14) is the Schwarz inequality in disguise applied on vectors

$$(\Omega - \langle \Omega \rangle)\,|\alpha\rangle \quad \text{and} \quad (\Lambda - \langle \Lambda \rangle)\,|\alpha\rangle \,, \tag{3.68}$$

---

[45]To prepare a single photon state it suffices to dispose a thin absorbing media in front of a light with thickness increasing until an avalanche photodiode registers separate detections. Each detection corresponds to a quantum. This is assured to work provided that one is disposed to wait the necessary time. To provide a single-photon source *on demand* is an altogether more difficult problem, subject to active research because of its application in quantum cryptography (see the problem at the end of this chapter).

with now $\Omega = P$ and $\Lambda = X$, and with uncertainty assumed to be equally distributed[46] in both position P and impulsion X. The inequality is optimised to its minimum when these vectors are aligned, which yields

$$(P - \langle P \rangle) |\alpha\rangle = i(X - \langle X \rangle) |\alpha\rangle \,, \qquad (3.69)$$

or, written back in terms of a, $a^\dagger$,

$$a |\alpha\rangle = \frac{1}{\sqrt{2\hbar}} (\langle X \rangle + i \langle P \rangle) |\alpha\rangle \,, \qquad (3.70)$$

so that the *coherent state* that minimises the uncertainty relation appears as the eigenstate of the annihilation operator (we come back later to the more general case where the uncertainty is not equally distributed).

It was expected that α could be complex since a is not hermitian. From eqn (3.70) it is seen that it actually spans the whole complex space. The phase associated with this complex number (in the sense of a polar angle in the complex plane) maps to the physical notion of phase since it arises as a complex relationship between the phase-space variables. The physical meaning of the (unsqueezed[46]) coherent state is that of the most classical state allowed by quantum physics, since it has the lowest uncertainty allowable in its conjugate variables and can be located in the complex plane as the position in the phase space of the quantum oscillator, with position on the real axis and momentum on the imaginary axis. The Hamiltonian being time independent, the evolution of $|\alpha\rangle$ is obtained straightforwardly from the propagator:

$$\begin{aligned} |\alpha(t)\rangle &= e^{-i\omega(a^\dagger a + 1/2)(t-t_0)} |\alpha(t_0)\rangle \\ &= e^{-i\omega(t-t_0)/2} \left| e^{-i\omega(t-t_0)} \alpha(t_0) \right\rangle \,, \end{aligned} \qquad (3.71)$$

so that the free propagation of the coherent state is rotation in the complex space, or harmonic oscillations in real space. Definitely, it is a state of well-defined phase.

In terms of Fock states, it reads

$$|\alpha\rangle = \exp(-|\alpha|^2/2) \sum_{n=0}^{\infty} \frac{\alpha^n}{\sqrt{n!}} |n\rangle \,, \qquad (3.72)$$

with $\alpha \in \mathbb{C}$.

Coherent states are normalised though not orthogonal:[47]

---

[46] When the uncertainty of a coherent state is not equally shared, i.e., $\Delta X \neq \Delta P$, although the product remains $\hbar/2$, the state is said to be *squeezed* in the variable whose root mean square goes below $\hbar/2$. In the dynamical picture provided by Exercise 3.8, squeezed states correspond to the ground state of the harmonic oscillator, i.e., a Gaussian wavefunction, whose width is mismatched with the harmonic potential characteristic length, so that as the state bounces back and forth in the trap, its wavepacket spreads or narrows periodically (it is said to "breathe").

[47] The non-orthogonality of coherent states close in phase space was one compelling support in favour of Glauber (1963)'s argument that the light emitted by the newly discovered maser called for a quantum explanation over the classical model advocated by Mandel and Wolf (1961).

Roy **Glauber** (b. 1925) advocated the importance of quantum theory in optics and its implications for the notion of *coherence*. As such he is widely regarded as the father of quantum optics and was awarded the 2005 Nobel prize in physics (half of the prize) "for his contribution to the quantum theory of optical coherence".

Before being awarded the Nobel prize, Glauber was an active and long-time supporter of its parody—the "Ig Nobel" prize—where he traditionally swept off the stage the paper planes sent on it by the audience. He made a testimonial for the 1998 Ig Nobel Physics prize awarded for "unique interpretation of quantum physics as it applies to life, liberty, and the pursuit of economic happiness".

$$\langle \beta | \alpha \rangle = \exp \left(-\frac{1}{2}(|\alpha|^2 + |\beta|^2 - 2\alpha\beta^*) \right), \tag{3.73}$$

with indeed $\langle \alpha | \alpha \rangle = 1$, but $\langle \beta | \alpha \rangle \neq \delta(\alpha - \beta)$, being smaller the farther apart α, β in $\mathbb{C}$.

3.3.3 *Glauber–Sudarshan representation*

Sudarshan (1963) promoted coherent states as a basis for decomposition of the density matrix ρ, replacing it by a scalar function P of the complex argument α of the coherent state $|\alpha\rangle$:

$$\rho(t) \equiv \int P(\alpha, \alpha^*, t) |\alpha\rangle \langle\alpha| d^2\alpha, \tag{3.74}$$

which allows one to carry out quantum computations with tools of functional analysis rather than operator algebra. Sudarshan incorrectly deduced that this proved the equivalence of the quantum formulation and classical one. Glauber, who also had in his first publication the insight of this decomposition, showed him to be wrong.[48]

We shall see many applications and usage of the Glauber–Sudarshan (or simply Glauber) representation later on. Now we proceed to give its expression for some states of interest.[49] For the coherent state, it is straightforward by identification to obtain for representation of $|\alpha_0\rangle$ that

$$P(\alpha, \alpha^*) = \delta(\alpha - \alpha_0). \tag{3.75}$$

[48] Upon award of the Nobel prize in 2005 to Glauber, Sudarshan sparked a controversy, questioning this decision and writing to the Nobel committee and the Times (who did not publish his letter) where he claimed priority on the representation. An extract of the text of Sudarshan to the Nobel committee reads "[. . .]*While the distinction of introducing coherent states as basic entities to describe optical fields certainly goes to Glauber, the possibility of using them to describe 'all' optical fields (of all intensities) through the diagonal representation is certainly due to Sudarshan. Thus there is no need to 'extract' the classical limit [as stated in the Nobel citation]. Sudarshan's work is not merely a mathematical formalism. It is the basic theory underlying all optical fields. All the quantum features are brought out in his diagonal representation[. . .]*". He concludes "*Give unto Glauber only what is his.*"

[49] The P representation for Fock states is highly singular, involving generalised functions of a much higher complexity than for the coherent state, which somehow restrain its applicability. This can be linked to the non-classical character of such a field.

3.3.4 Thermal states

The incoherent superposition of many uncorrelated sources generates chaotic or a so-called *thermal state*, as is the case of the light emitted from a lightbulb, where each atom emits independently of its neighbour. The convolution rule and the *central limit theorem* combine to provide the P function of a such a state: *a chaotic state without phase or amplitude correlations has a Glauber distribution that is Gaussian in the complex plane*:

$$P(\alpha, \alpha^*) = \frac{1}{\pi\bar{n}} \exp(-|\alpha|^2/\bar{n}), \tag{3.76}$$

where $\bar{n} = \langle a^\dagger a \rangle$ is the average number of particles in the mode. From this expression the Fock-state representation can be obtained:

Exercise 3.16 (*) *Show that the density matrix of the thermal state in the Fock states basis is*

$$\rho = \sum_{n=0}^{\infty} \frac{\bar{n}^n}{(\bar{n}+1)^{n+1}} |n\rangle \langle n| . \tag{3.77}$$

Observe that the density matrix eqn (3.77) is diagonal (all the terms $|n\rangle \langle m|$ with $n \neq m$ are zero). This means that there is no quantum coherence and the superposition is classical. This is the fully mixed state, where the uncertainty is entirely statistical. The result can be alternatively obtained directly from the statistical argument: in the canonical ensemble, the density matrix for a system with Hamiltonian H at temperature T is given by[50]

$$\rho = \frac{\exp(-H/k_B T)}{\text{Tr}(\exp(-H/k_B T))}, \tag{3.78}$$

which can be evaluated exactly when H is the Hamiltonian for an harmonic oscillator, given by eqn (3.29):

Exercise 3.17 (*) *Show that eqn (3.78) evaluates to*

$$\rho = \left[1 - \exp\left(-\frac{\hbar\omega}{k_B T}\right)\right] \sum_{n=0}^{\infty} \exp\left(-\frac{n\hbar\omega}{k_B T}\right) |n\rangle \langle n| . \tag{3.79}$$

for the harmonic oscillator Hamiltonian. Deduce the following relation between the average occupancy of the mode and temperature:

$$\bar{n} = \langle a^\dagger a \rangle = \frac{1}{\exp\left(-\dfrac{\hbar\omega}{k_B T}\right) - 1}. \tag{3.80}$$

---

[50]This follows from the fundamental postulate for the canonical ensemble, which states that if a system is in equilibrium at temperature T, the probability that it is found with energy E_n is $(1/Q) \exp(-E_n/(k_B T))$, where $Q = \sum_n \exp(-E_n/(k_B T))$ (known as the *partition function*) and E_n is the energy of any of the states in which the system can be found. Therefore, taking $|i\rangle$ as the state with energy E_i and Ω an operator, from the definition of quantum average that we have given, cf. eqn (3.51), it follows that $\langle \Omega \rangle = (1/Q) \sum_{|i\rangle} \langle i| \Omega |i\rangle \exp(-E_i/(k_B T))$.

The important formula eqn (3.80)—the thermal distribution of bosons at equilibrium (here for a single mode)—is named the *Bose–Einstein distribution*.[51]

3.3.5 Mixture states

An important property of P functions is that the superposition of uncorrelated fields each described by its P function amounts to a field whose own P function is obtained by the convolution of that of the constituting fields. So, for instance, if we superpose two non-correlated fields described, respectively, by functions P_1 and P_2, the total field is given by $P = P_1 * P_2$, or

$$P(\alpha, \alpha^*) = \int P_1(\beta, \beta^*) P_2(\alpha - \beta, \alpha^* - \beta^*) d\beta d\beta^* . \tag{3.81}$$

This has application for an important class of states, which interpolate between the coherent state eqn (3.75) and the thermal state eqn (3.76), it being rare in practice that a state is realistically completely coherent.[52] When there is a large number of particles, the physically relevant case is that of an essentially coherent state, with, say, n_c particles, to which is superimposed a fraction of a thermal state, with n_t particles. Such a state is obtained in the optical field by interfering ideal coherent radiation with that emitted by a blackbody. In the resulting field, we will conveniently refer to such particles as coherent and incoherent respectively, though of course once the two fields are merged, a particle no longer belongs to any one part of this decomposition but is indistinguishable from any other. This is just a vivid picture to describe a collective state that has some phase and amplitude spreading. Only $n_c + n_t = \langle n \rangle$ is well defined.

The P state that results from the convolution of a Gaussian centred about $\alpha_{\text{coh}} \in \mathbb{C}$ and a delta function centred about α_{th} is a Gaussian centred about $\alpha_{\text{coh}} - \alpha_{\text{th}}$, so that the P state of the mixed state is a Gaussian centred about α_0, as depicted in Fig. 3.2. Its analytical expression reads

$$P_{\text{m}}(\alpha, \alpha^*) = \frac{1}{\pi n_t} e^{-|\alpha - n_c e^{i\varphi}|^2 / n_t} , \tag{3.82}$$

where φ is the mean phase of the state, inherited from the phase of the coherent state.

We conclude by illustrating the power of the P function as a mathematical tool by providing a few techniques that we shall use later. With proper notation, eqn (3.70) reads

$$a \left| \alpha \right\rangle = \alpha \left| \alpha \right\rangle , \tag{3.83}$$

which states that the coherent state is an eigenstate for the annihilation operator (reciprocal equation is $\langle \alpha | a^\dagger = \langle \alpha | \alpha^*$). Such properties make the evaluation of many states

---

[51] The general formula for the Bose–Einstein distribution, with all modes (and possible degeneracy such as that imparted by polarisation) and with a chemical potential is readily obtained along the same lines as Exercise 3.17 and will be a central ingredient of the physics of Chapter 8.

[52] From photodetection theory one can show that the inefficiency of the detector results in a broadening of the counting statistics of a coherent state of the kind of the mixed states for the density matrix of a Bose single mode.

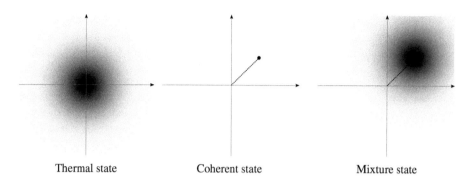

| Thermal state | Coherent state | Mixture state |

Fig. 3.2: Schematic representation of P representations for a thermal state (Gaussian), a coherent state (δ function) and a superposition, or mixture, of the two, in which case the Gaussian gets displaced onto the δ location.

straightforward, translating the operator as its eigenvalue. It also results in many simplifications of the mathematical analysis of equations, as operator algebra gets mapped to complex calculus. However, during this translation from operators to c-numbers, the need for the following can (and does) arise:

$$a^\dagger \left|\alpha\right\rangle , \tag{3.84}$$

(or the reciprocal $\left\langle\alpha\right| a$). This is made technically easy with a few tricks that can be devised from the Fock representation of the coherent state, eqn (3.72), as in this case:

Exercise 3.18 $^{(*)}$ *By going back and forth to the Fock and coherent basis, derive the following rule of thumb:*

$$a^\dagger \left|\alpha\right\rangle = \left(\frac{\partial}{\partial\alpha} + \frac{\alpha^*}{2}\right)\left|\alpha\right\rangle , \tag{3.85}$$

and following the same principle, compute the exhaustive set of possible combinations that arise in the conversion of the master equation with operators into a so-called Fokker–Planck equation for c-numbers functions:

$$a\left|\alpha\right\rangle\left\langle\alpha\right| a^\dagger , \qquad a^\dagger a\left|\alpha\right\rangle\left\langle\alpha\right| , \qquad \left|\alpha\right\rangle\left\langle\alpha\right| a^\dagger a , \tag{3.86a}$$

$$\left|\alpha\right\rangle\left\langle\alpha\right| aa^\dagger , \qquad a^\dagger\left|\alpha\right\rangle\left\langle\alpha\right| a . \tag{3.86b}$$

The master equation for a matrix can in this sense be directly rewritten as a Fokker–Planck equation, which is an equation of diffusion and drift for a probability distribution. Many insights into the quantum picture can be gained through this approach as we will see when investigating the quantum interaction of light with matter in Chapter 5.

3.3.6 *Power spectrum*

The power spectrum, introduced classically in Section 2.3.3, can be understood in quantum mechanics as the mean number of photons with frequency ω, i.e., by definition of the number operator:

$$s(\omega) = \langle a^\dagger(\omega)a(\omega)\rangle,\qquad(3.87)$$

where the average is over the quantum state rather than over a statistical ensemble. This is proportional to the intensity of photons emitted by the mode a, say a cavity, at this frequency and so, up to some numerical factor reflecting the collection efficiency and geometry of the setup, corresponds to the photoluminescence spectrum of the cavity mode. It will be convenient, throughout, to deal with normalised spectra:

$$S(\omega) = s(\omega)\Big/\int_0^\infty \langle a^\dagger a\rangle(t)dt\,,\qquad(3.88)$$

so that eqn (3.88) is now the density of probability that a photon emitted by the system has frequency ω. The annihilation operator $a(\omega)$ is related to the familiar quantity in time, $a(t)$, by a Fourier transform. This brings us, however, to a two-time correlator.

Exercise 3.19 $^{(*)}$ *Show that, going through the Fourier transform,*

$$s(\omega) = \frac{1}{2\pi}\iint_0^\infty \langle a^\dagger(t_1)a(t_2)\rangle e^{i\omega(t_2-t_1)}dt_1 dt_2\,,\qquad(3.89)$$

the power spectrum can be expressed in terms of the first-order time autocorrelator eqn (3.102) for positive time delay $\tau = t_2-t_1 \geq 0$, to yield the Fourier-pair relationship between the power spectrum and the autocorrelation function:

$$S(\omega) = \frac{1}{\pi\int_0^\infty\langle a^\dagger a\rangle(t)dt}\Re\int_0^\infty\int_0^\infty G^{(1)}(t,\tau)e^{i\omega\tau}\,d\tau dt\,.\qquad(3.90)$$

Equation (3.90) holds as such in the spontaneous emission case. In the steady state case, care must be taken with cancellation of infinities brought by the ever-increasing time t, since both the numerator and the denominator are infinite quantities. Their ratio, however, produces a finite quantity, which recovers the Wiener–Khinchin theorem, eqn (2.51). This is seen setting apart the transient (TR) part of the population and the autocorrelator, from their steady-state (SS) limit:

$$\langle a^\dagger a\rangle(t) = \langle a^\dagger a\rangle^{\mathrm{TR}}(t) + \lim_{t\to\infty}\langle a^\dagger a\rangle(t)\,,\qquad(3.91a)$$

$$G^{(1)}(t,\tau) = \langle a^\dagger(t)a(t+\tau)\rangle^{\mathrm{TR}} + \lim_{t\to\infty}\langle a^\dagger(t)a(t+\tau)\rangle\,.\qquad(3.91b)$$

We rewrite eqn (3.90) as the time integration of the Fourier transform until time T, that is left to increase without bounds:

$$S(\omega) = \lim_{T\to\infty}\frac{1}{\pi\int_0^T\langle a^\dagger a\rangle(t)dt}\Re\int_{t=0}^T\int_{\tau=0}^\infty G^{(1)}(t,\tau)e^{i\omega\tau}\,d\tau dt\,.\qquad(3.92)$$

Substituting eqns (3.91) in this expression, we can keep track of the terms that cancel:

$$S^{\mathrm{SS}}(\omega) = \frac{1}{\pi}\lim_{T\to\infty}\frac{1}{N^{\mathrm{TR}}+Tn_a^{\mathrm{SS}}}\Re\int_{\tau=0}^\infty\Big[M^{\mathrm{TR}}(\tau)+T\lim_{t\to\infty}\langle a^\dagger(t)a(t+\tau)\rangle\Big]e^{i\omega\tau}d\tau\,.\qquad(3.93)$$

The quantities $N^{\mathrm{TR}} \equiv \int_0^\infty\langle a^\dagger a\rangle^{\mathrm{TR}}(t)dt$ and $M^{\mathrm{TR}}(\tau) = \int_0^\infty\langle a^\dagger(t)a(t+\tau)\rangle^{\mathrm{TR}}e^{i\omega\tau}\,dt$ for all $\tau > 0$, are bounded, as can be understood from their definition and physical sense or checked explicitly when applying the formula to a particular case. Since the norm of the Fourier transform of $M^{\mathrm{TR}}(\tau)$ is also bounded (that, again, can be checked from explicit results), the limit in T yields

$$S^{\mathrm{SS}}(\omega) = \frac{1}{\pi} \frac{1}{n_a^{\mathrm{SS}}} \lim_{t\to\infty} \Re \int_0^\infty \langle a^\dagger(t)a(t+\tau)\rangle e^{i\omega\tau}\,d\tau\,, \tag{3.94}$$

which is the quantum counterpart of the Wiener–Khinchin theorem, eqn (2.50).

Eberly and Wódkiewicz (1977) took another approach to compute the spectrum, rooting their derivation not in mathematical properties of stationary random processes à la Wiener–Khinchin but in an actual physical process of photon detection. They obtain a time dependent "physical" spectrum, computed as

$$S(\omega,t) = \Gamma_f^2 \int_{-\infty}^t \int_{-\infty}^t e^{-(\Gamma_f - i\omega)(t-t_1)} e^{-(\Gamma_f + i\omega)(t-t_2)} \langle a(t_1)^\dagger a(t_2)\rangle dt_1 dt_2\,, \tag{3.95}$$

with Γ_f the detector resolution (the passband width of the filter, in their derivation that tackles all the quantities on physical grounds, ω is the "setting" frequency of the filter).

To compute the spectra, it is needed in any case to compute two-time dynamics of the quantum correlations. We have seen at length how to compute single time dynamics, in various pictures of quantum mechanics. It would seem that the Heisenberg picture is the most suited, since it can provide the dynamics of the operator $a(t)$. However, in practical terms, it is usually difficult to compute and, when at all possible, the expressions are heavy, as we will have occasion to see in Chapter 5.

A more efficient machinery can be used, known as the *quantum regression theorem*, that provides the equations of motion for the two-time dynamics in terms of those for the single-time dynamics. Formally, the quantum regression theorem states that, given a Liouvillian $\mathcal{L}$ and a set of operators $C_{\{\eta\}}$ that satisfy

$$\mathrm{Tr}(C_{\{\eta\}}\mathcal{L}\Omega) = \sum_{\{\lambda\}} M_{\{\eta\lambda\}}\mathrm{Tr}(C_{\{\lambda\}}\Omega) \tag{3.96}$$

for any operator Ω, the equations of motion for the two-time correlators read

$$\partial_\tau \langle \Omega(t)C_{\{\eta\}}(t+\tau)\rangle = \sum_{\{\lambda\}} M_{\{\eta\lambda\}}\langle \Omega(t)C_{\{\lambda\}}(t+\tau)\rangle\,. \tag{3.97}$$

To attack a given problem, the first task is to identify the set $C_{\{\eta\}}$, where we denote $\{\eta\}$ the set of degrees of freedoms required to describe the dynamics, that is typically a tuple of integers. Then one must compute the M tensor that, for any operator, satisfy eqn (3.96). Once this is done, one has the equation of motions for the two times correlators, given by eqn (3.97), that can be solved analytically, semi-analytically or, in any case, numerically. Del Valle (2010b) works out many cases in details. We shall also have occasion in Chapter 5 to explicit this abstract procedure in a few particular important problems. As a one-particle observable, however, the power spectrum is typically identical in both the classical and quantum formulation of a problem.

3.3.7 $g^{(2)}$ and other Glauber correlators

To go beyond classical physics, one must turn to two-particle (or more) observables. The zero-delay second-order coherence degree, $g^{(2)}$, presents strong experimental and theoretical assets in this regard. The $\tau = 0$ coincidence case embeds a lot of information into a single quantity, in particular as an unambiguous criterion for the non-classical character of the field. It can be computed easily in any of the quantum-mechanical pictures we have presented earlier in this chapter. We now discuss this pillar of quantum optics in more details, along with its higher-order counterparts.

We emphasised in Section 2.3 how a realistic description of the optical field needs to take into account its statistical character, at the classical level as has been seen in Chapter 2, but also at the quantum level with additional features of a specific quantum nature. The importance of statistics in optics has been realised by Mandel and Wolf (1963), but they missed the importance of quantum mechanics and it was for Glauber (1963) to formalise the theory of quantum coherence. Fluctuations of light are responsible for the optical coherence. Since there are always at least the quantum fluctuations of the field, a fundamental definition of coherence is required at the quantum level. Glauber provided such a definition by emphasising the role of correlations. One of his main contributions in this respect was to separate the notion of optical coherence from that of monochromaticity and the notion of interference from that of phase.

A generic property of the light field can depend on arbitrary high-order correlation functions. On the other hand, a statistical classical model—that for all relevant details would be modelled after a stationary Gaussian stochastic process—has all its information contained already in its frequency spectrum (first order in the correlators). The first such physical property to generally rely on a higher-order correlation than $g^{(1)}$ was the bunching of photons in the Hanbury Brown–Twiss experiment presented in Chapter 2 and that motivated the work of Glauber even though in the original configuration of HBT, this was within reach of classical fields only. At a time when the Hanbury Brown–Twiss claim was still highly controversial, Purcell (1956) stood in its favour and pointed out the quantum character of the effect from Bose statistics, even predicting antibunching for fermions. In the abstract of his paper published in Nature, he comments:

> *Brannen and Ferguson (preceding abstract) have suggested that the correlation between photons in coherent beams observed by Brown and T. (cf. above), if true, would require a revision of quantum mechanics. It is shown that this correlation is to be expected from quantum mech. considerations and is due to a clumping of photons. If a similar experiment were performed with electrons a neg. cross-correlation would be expected.*

His semiclassical insight is based on quantum statistics of bosons along the lines or arguments used in Section 3.1.4 on the symmetric states of bosonic particles. Let us consider two particles, a and b that can be detected by any one of two detectors, 1 and 2. In the case where particles are classical (distinguishable), the probability P_c of detecting one particle on each detector is the sum of the probabilities of all possibilities, namely detecting a in 1 and b in 2 or vice versa, with respective probabilities $|\langle a|1\rangle\langle b|2\rangle|^2$ and $|\langle a|2\rangle\langle b|1\rangle|^2$ so that

$$P_c = 2|ab|^2 . \tag{3.98}$$

If particles are bosons, however, they are indistinguishable and there is quantum interference of their trajectories at detection, and the amplitudes sum, rather than the probabilities (which is one tenet of how quantum mechanics extends classical mechanics). The probability remains the modulus square of the total amplitude, so that

$$P_Q = |\langle a|1\rangle\langle b|2\rangle + \langle a|2\rangle\langle b|1\rangle|^2 , \tag{3.99a}$$
$$= |2ab|^2 = 2P_c , \tag{3.99b}$$

which shows how the probability of joint detection increases for thermal bosons as compared with classical particles. As the latter display no correlation, this increase translates as a bunching of particles, that is, a tendency to cluster and arrive together at the detector.

Exercise 3.20 $^{(**)}$ *Spell out the procedure outlined above for plane wave modes for the photons (with $|a\rangle$, $|b\rangle$ dotting with momentum $\langle \mathbf{k}|$ and position $\langle \mathbf{r}|$) and by using annihilation operators to model the detection. Follow the routes of the particles and how they interfere destructively. Recover the result obtained previously in a classical picture.*

We now systematise this idea to more general quantum states, a necessity following from an insight by Glauber (1963), who inaugurated a series of publications to account for the Hanbury Brown–Twiss correlations in a fully quantum picture, in competition with the classical approach of Mandel et al. (1964). First, Glauber derives the quantum analogue of eqn (2.55) with quantum fields and operators and obtains eqn (3.100).

Exercise 3.21 $^{(**)}$ *The photodetection can be modelled by the ionisation of atoms of the active medium of the detector, these being at positions $\mathbf{r}_1$ and $\mathbf{r}_2$, respectively. Calling w_i the constant transition probability for an atom excited by the beam show (summing over final electron states) that the probability of coincidence detection $w(t_1, t_2)$ at t_1 and t_2 is given by*

$$w(t_1, t_2) = w_1 w_2 \frac{\mathrm{Tr}\big(\rho E^{(-)}(\mathbf{r}_1, t_1) E^{(-)}(\mathbf{r}_2, t_2) E^{(+)}(\mathbf{r}_1, t_1) E^{(+)}(\mathbf{r}_2, t_2)\big)}{\mathrm{Tr}\big(\rho E^{(-)}(\mathbf{r}_1, t_1) E^{(+)}(\mathbf{r}_1, t_1)\big) \mathrm{Tr}\big(\rho E^{(-)}(\mathbf{r}_2, t_2) E^{(+)}(\mathbf{r}_2, t_2)\big)} . \tag{3.100}$$

A more direct route is to agree on some canonical quantisation of the classical notions of coherence developed in Section 2.3, indeed still largely valid in the quantum regime. This requires only the field $\mathbf{E}$—which was a c-number quantity in the previous chapter—to now be promoted to its operator form, eqn (3.66). If we consider a single mode (noting a the associated annihilation operator), we find the expressions of $g^{(n)}$ for quantum fields with $n = 1, 2$ given by:[53]

---

[53] Higher-order correlations, with $n \geq 3$, are defined in the same way, but they find little practical value even in fields where these quantities have been mastered for a long time. Theoretically, however, Glauber crowns the definition of coherence by stating that a field is *all order coherent* if $g^{(n)}(t, t + \tau) = 1$ for all $n \in \mathbb{N}$ and all $\tau \in \mathbb{R}$. That is, the correlations all factorise into products of single-operator averages. This is the case for a monochromatic wave, for instance, or in terms of states, for a coherent state $|\alpha\rangle$.

$$g^{(1)}(\tau, t) = \frac{\langle a^\dagger(t) a(t+\tau) \rangle}{\langle a^\dagger(t) a(t) \rangle}, \tag{3.101a}$$

$$g^{(2)}(\tau, t) = \frac{\langle a^\dagger(t) a^\dagger(t+\tau) a(t+\tau) a(t) \rangle}{\langle a^\dagger(t) a(t) \rangle^2}, \tag{3.101b}$$

which matches with eqn (3.100). It is often convenient to deal with unnormalised version of the above, e.g.,

$$G^{(1)}(t, \tau) = \langle a^\dagger(t) a(t+\tau) \rangle. \tag{3.102}$$

Once they are evaluated on the quantum state of a given system, the correlators eqn (3.101) become c-number functions that can be processed in the same way as before, e.g., the $g^{(1)}$ fed into eqn (2.51) provides the emitted spectra of the system, as in the classical case, but this time taking into account the quantum dynamics of the system and quantum fluctuations of the state. At zero time delay, the second-order correlator becomes

$$g^{(2)}(0) = \frac{\langle (a^\dagger)^2 a^2 \rangle}{\langle a^\dagger a \rangle^2}. \tag{3.103}$$

Observe how this quantity is sensitive to the quantum state considered, while the spontaneous-emission power spectrum would be identical:

Exercise 3.22 (*) *Show that the following values of $g^{(2)}(0)$ are obtained for the associated quantum states:*

$g^{(2)}(0)$	State
$1 - \dfrac{1}{n}$	Fock state $|n\rangle$
1	Coherent state $|\alpha\rangle$ (cf. eqn (3.72))
2	Thermal state (cf. eqn (3.79)).

The most noteworthy result of Exercise (3.22) is $g^{(2)}(0)$ for the Fock state that is zero for $n = 1$ and is smaller than 1 for all $n \in \mathbb{N}$. This contradicts eqn (2.58), if one remembers that the two-time coincidence probability P_2 is proportional to $g^{(2)}(\tau)$ and that $g^{(2)}(\infty) = 1$, which translates as the constraint:

$$g^{(2)}(0) \geq 1 \quad \text{(from classical model)}. \tag{3.104}$$

This is because there is no classical state that can correctly describe a Fock state, which thus poses itself as a pure *quantum state* without any classical analogue. On the other hand, coherent and thermal states, and all the mixed states interpolating between them do have such classical counterparts and the effects they display (like bunching or no correlations) can be explained classically or semiclassically. However, *antibunching*, typically—that is the decrease of probability of a second detection once one has been registered—cannot be reproduced by any classical model. How quantisation of the field

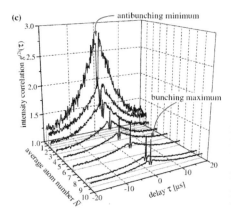

(c)

Fig. 3.3: Bunching (front) and antibunching (back) of light as observed by Hennrich et al. (2005) by varying the average number of atoms in a cavity. At zero delay, $g^{(2)}$ has a peak when light arrives in lumps and a drop when photons avoid each other. The curves are symmetric in time and tend asymptotically (though slowly) towards 1 for long delays.

bears on this problem can be illustrated as follows: If instead of the light of a star or from a light-bulb, a single photon is fed to the HBT setup, maximum anticorrelation results from the annihilation of this one-quantum, which results in an abrupt annihilation of the whole signal. Indeed, the branch that detects the photon destroys it and the probability to detect it on the other branch becomes exactly zero (whereas if collecting from a thermal source, photons are emitted chaotically and their bosonic properties increase their chance on arriving by groups and thus feeding both branches of the interferometer).

How to say whether the "quantum character" of a field is relevant can be decided from the Glauber P representation: any state whose P representation has a singularity stronger than a Dirac delta function, is a quantum state without any classical counterpart. Indeed, one can show that the P distribution for the Fock state $|n\rangle$ is a $2n$ partial derivative of the delta function.[54] For other non-classical states, it can be that P becomes negative. Physically, whenever P cannot be interpreted as a well-behaved probability distribution, the field is quantum. Otherwise, a classical description with statistical weight of plane waves weighted by the P function will provide an adequate substitute. The first report of such a field without a classical counterpart—by Kimble et al. (1977) who observed photon antibunching in resonance fluorescence—still ranks among as one of the strongest pieces of direct evidence of quantisation of the optical field.

Typical experimental results for $g^{(2)}(\tau)$ are shown in Fig. 3.3 where light emitted from an atomic ensemble coupled to a single cavity mode offers the nice feature of changing from bunching to antibunching with the number of atoms, as reported by Hennrich et al. (2005). There is a peak at zero delay in the bunching case (in front of the plot) and a dip in the antibunching case (in back). Observe, however, that despite the dip, $g^{(2)}(0)$ is still higher than 1 and eqn (3.104) is not violated in this case, but $g^{(2)}(0) \geq g^{(2)}(\tau)$ is and this also is a signature of a quantum field, namely, it is a violation of Cauchy–Schwarz inequality. This is one reason why it is interesting to

---

[54]Explicitly, the expression reads

$$P_{\text{Fock}}(\alpha, \alpha^*) = \frac{\exp(|\alpha|^2)}{n!} \partial^{2n}_{\alpha^n, \alpha^{*n}} \delta(\alpha) \,. \qquad \text{(xvii)}$$

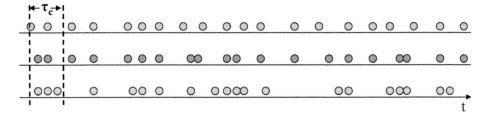

Fig. 3.4: Computer-generated time series for detection of photons for:
i) antibunched light such as emitted by resonance fluorescence of an atom (with $g^{(2)} < 1$),
ii) coherent light such as emitted by a laser or other coherent source ($g^{(2)} = 1$),
iii) bunched light such as emitted by a candle, the Sun or any such incoherent source ($g^{(2)} > 1$).

know $g^{(2)}(\tau)$ also at nonzero values of τ, although quantum characteristics are usually more marked at zero delay and a decay towards uncorrelated values of the field are, of course, eventually obtained at long times (the timescale of this decay is another good reason to measure or compute at nonzero delay). While the temporal accuracy is often a problem in such measurements, Wiersig et al. (2009) pioneered a method of photon counting with a streak camera, allowing ultrafast measurements and access to still higher orders of correlations.

A graphical representation of a distribution of photons as a function of time (say on their way toward a detector) is shown in Fig. 3.4 for the three main quantum states discussed previously, namely a Fock state (antibunching), a coherent state (no correlation) and a thermal state (bunching). Observe how in the upper sequence—associated to Fock states and therefore to a field of a highly quantum nature—the stream consists of well-spaced photons reaching the target one by one, as if emitted by a "photon gun". Antibunched states are more easily obtained with fermions, for instance they are easily formed by electrons passing through a large resistance due to Coulomb interaction. In this sense, antibunched light is a way to have photons (that are bosons) behave as fermions. Such emitters are highly prized for quantum information processing. If an additional degree of freedom like polarisation can be controlled for each emission, it is possible with such a device to set up a completely secure (unbreakable) cryptographic system, as is studied in the problem at the end of this chapter.

The *statistics of the state* is the probability of a given state to be found with a given occupancy number n (to be found in the state $|n\rangle$). These probabilities are obtained from the diagonal elements of the density matrix in the basis of Fock states. Therefore, this statistics is defined as $p(n) = \langle n| \rho |n\rangle$ with ρ the density matrix. From the definition (eqn (3.101b)) and the boson algebra, it is straightforward to obtain the formula that links $g^{(2)}$ to $p(n)$ in the Fock-state basis:

$$g^{(2)}(0) = \frac{\displaystyle\sum_{n=0}^{\infty} n(n-1)p(n)}{\left(\displaystyle\sum_{n=0}^{\infty} np(n)\right)^2}, \tag{3.105}$$

It is easy to compute the statistics of most of the states considered so far. The results are compiled in Table 3.1.

Table 3.1 Probability distribution and moments of basic quantum states.
i) First row is the probability to have n excitations
ii) Second row is the mean number of excitations
iii) Third row is the variance
iv) Fourth row is the profile for averages $\langle a^\dagger a \rangle = 1, 5$ and 10.

	Fock state $\lvert n_0 \rangle$	Coherent state $\lvert \alpha \rangle$ cf. eqn (3.72)	Thermal state cf. eqn (3.77)
Distribution $p(n)$	δ_{n,n_0}	$e^{-\lvert\alpha\rvert^2}\dfrac{\lvert\alpha\rvert^{2n}}{n!}$	$\dfrac{1}{\bar{n}+1}\left(\dfrac{\bar{n}}{\bar{n}+1}\right)^n$
$\langle a^\dagger a \rangle = \sum_n n p(n)$	n_0	$\lvert\alpha\rvert^2$	$\bar{n}$
$\langle (a^\dagger a)^2 \rangle - \langle a^\dagger a \rangle^2$	0	$\lvert\alpha\rvert^2$	$\bar{n}^2 + \bar{n}$

Other states present more sophisticated distributions. This is the case of the mixture of a coherent with a thermal state, for instance, which despite being merely an offset Gaussian in the Glauber representation, (cf. eqn (3.82)), has a complex $p(n)$.

Exercise 3.23 $^{(**)}$ *Consider the state that is the mixture of a thermal state with intensity n_t and a coherent field with intensity n_c, the P representation of which is given by eqn (3.82). Show that the distribution of this state is*

$$p_m(n) = \exp\left(-\frac{\bar{n}\chi}{1+\bar{n}(1-\chi)}\right)\frac{(\bar{n}(1-\chi))^n}{(1+\bar{n}(1-\chi))^{n+1}}L_n\left(-\frac{\chi}{(1-\chi)(1+\bar{n}(1-\chi))}\right),$$
(3.106)

where L_n is the nth Laguerre polynomial[55] and χ is the coherent ratio (percentage of "coherent particles")

---

[55]The Laguerre polynomials are a canonical basis of solutions for the differential equation $xy'' + (1-x)y' + ny = 0$. They can be written as $L_n(x) = \frac{e^x}{n!}\frac{d^n}{dx^n}\left(e^{-x}x^n\right)$ and computed by carrying out the derivation. The first Laguerre polynomials are for $n \geq 0$, $L_n(x) = 1$ (constant), $-x + 1$, $\frac{1}{2}(x^2 - 4x + 2)$, ... See E. W. Weisstein, "Laguerre Polynomial." From MathWorld at http://mathworld.wolfram.com/LaguerrePolynomial.html.

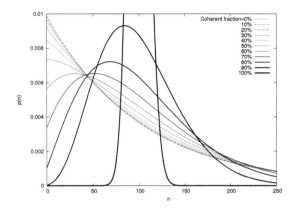

Fig. 3.5: Statistics of mixtures of thermal and coherent states in proportions ranging from 0 (thermal state) to 100% (coherent state) coherence, for an average number of particle $\bar{n} = 100$. At 90% coherence, the deviation is already very significant. The pure coherent state that has been truncated extends four times higher than is displayed, see Fig. 3.6 where this state is fully displayed. On the other hand, the curves with 10% coherence and the thermal state are practically superimposed.

$$\chi = \frac{n_c}{n_c + n_t} \, . \tag{3.107}$$

The distribution (eqn (3.106)) of mixture states is plotted in Fig. 3.5 for values of χ ranging from 0 to 100% in steps of 10%, also with the two limiting cases of the pure coherent state ($\chi = 1$) and the pure thermal state ($\chi = 0$). The mathematical expression is heavy, but as a function it is well behaved and brings no problem for a numerical treatment. As this distribution flattens very quickly with small incoherent fractions, it can profitably be replaced by a Gaussian that is a much better approximation than to consider pure coherent states, as seen in Fig. 3.6. The first two moments of P_m can be retained exactly, namely the mean $\bar{n}$ and the variance computed as

$$\sigma^2 = \bar{n} + n_t^2 + 2n_c n_t \, , \tag{3.108}$$

which, moreover, links χ and $g^{(2)}$ through

$$g^{(2)}(0) = 2 - \chi^2 \, . \tag{3.109}$$

3.3.8 *Polarisation*

The polarisation state of a photon and of a beam of photons is described by direct transposition of the Jones vector into a quantum superposition of basis states. This can be described using a second-quantised formalism as well, as is done now in the most general case of elliptical polarisation.

A photon with circular polarisation degree given by $P \equiv \cos^2 \theta - \sin^2 \theta$ is the coherent superposition of a spin-up photon with probability $\cos^2 \theta$ and of a spin-down photon with probability $\sin^2 \theta$, therefore, its quantum state can be created from the

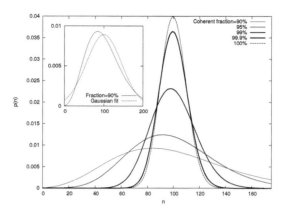

Fig. 3.6: Statistics of mixtures with high fractions of coherence, in respective proportions of 90% (also displayed on Fig. 3.5), 95%, 99%, 99.9% and 100% (coherent state) coherence. For high values of coherence, the exact expression (3.106) can be approximated by a simple Gaussian, as illustrated in the inset. The approximation becomes much better at higher ratios of coherence.

vacuum $|0, 0\rangle$ (zero spin-up and zero spin-down photon) by application of the following operator:

$$|1, \theta, \phi\rangle \equiv (\cos\theta a_{\uparrow}^{\dagger} + e^{i\phi} \sin\theta a_{\downarrow}^{\dagger}) |0, 0\rangle \,, \qquad (3.110)$$

where the angle ϕ is the in-plane orientation of the axis of the polarisation ellipse. This defines $a_{\theta,\phi}^{\dagger}$ the creation operator for an elliptically polarised photon as

$$a_{\theta,\phi}^{\dagger} \equiv \cos\theta a_{\uparrow}^{\dagger} + e^{i\phi} \sin\theta a_{\downarrow}^{\dagger} \,. \qquad (3.111)$$

The superposition of n such correlated photons is obtained by recursive application of the creation operator:

$$|n, \theta, \phi\rangle = a_{\theta,\phi}^{\dagger n} |0\rangle = \frac{1}{\sqrt{n!}} (\cos\theta a_{\uparrow}^{\dagger} + e^{i\phi} \sin\theta a_{\downarrow}^{\dagger})^n |0, 0\rangle \,, \qquad (3.112)$$

which we have normalised (here $|0\rangle$ is the vacuum in the basis of elliptically polarised states).

Exercise 3.24 (***) *The BB84 protocol of single photon quantum cryptography.*

One desirable goal of cryptography is the secure communication of a key, which is a binary digit list of arbitrary length that two parties A and B should be able to communicate at will. The exact information carried by the key is not important as long as it is known by the two parties and them only.

Assume that they use a quantum channel over which they can send single photons that reach B in the polarisation state that has been encoded by A. They also use a classical channel over which they can communicate any information they need, given, however, that this information is thus also made available to potential code-breakers.

By using the concept of conjugate bases, cf. Exercise 3.3, design a process by which A is able to generate and communicate to B a key that they both know for sure has not been observed by any third party.

3.4   Outlook on quantum mechanics for microcavities

Quantum physics at large is a significantly active and important area of research today. It is at the outset of numerous series of topics that diverge from each other as they get more specialised. One of these routes leads to microcavities.

What makes this topic especially attractive is the depth and extent that it affords, a precious and rather uncommon quality in today's research, where specialisation reduces the physics to its most intricate details. This point can be illustrated from the photograph at the end of the chapter, taken on the occasion of the fifth Solvay conference. The first edition in 1911 was also the first international conference (the series is still continued to this date, it is held every 3 years) on the topic of *Radiation and the Quanta*. Einstein attended as the youngest participant (at 25 years old). We have put in bold the names of the scientists whose work is central to the physics that makes the topic of this book and, without whose knowledge, one cannot pursue useful research. As one can clearly see, microcavities physics essentially brings again to the fore the fundamental physics of the fathers of modern science.

A. Piccard, E. Henriot, P. Ehrenfest, Ed. Herzen, Th. De Donder, **E. Schrödinger**, E. Verschaffelt, **W. Pauli**, **W. Heisenberg**, R. H. Fowler, **L. Brillouin**, (upper row); *P. Debye, M. Knudsen, W. L. Bragg, H. A. Kramers, P. A. M. Dirac, A. H. Compton, L. de Broglie, M. Born, N. Bohr, (middle);* I. Langmuir, **M. Planck**, M. Curie, **H. A. Lorentz**, **A. Einstein**, **P. Langevin**, Ch. E. Guye, C. T. R. Wilson and O. W. Richardson (lower row) posing for the 1927 Solvay conference.

3.5 Further reading

This chapter has reviewed some of the basic aspects of quantum mechanics and their relevance to the optical field to cover our needs for the more specific treatment in what follows, and in a form and context such as they would typically be encountered in the physics of microcavities. However, dealing with such general issues, this chapter is also the most remote from the object of study of this book and, therefore, requires much supplementary reading to appreciate the subject in some depth. General quantum mechanics can be obtained from countless sources, e.g., Merzbacher (1998). For quantum field theory, attention should be directed towards texts written with statistical physics or condensed-matter in mind, as there is little to be gained from the more popular relativistic formulation that has different concerns. Renowned classics are the textbooks by Abrikosov et al. (1963), Negele and Orland (1998) and Fetter and Walecka (2003). For a study with special attention paid to the light field, while still maintaining a strong field-theoretic approach, one can refer to the textbook by Cohen-Tannoudji et al. (2001). For quantum optics proper, Mandel and Wolf (1995) provide the best reference, accompanied by a very large literature with such textbooks as those by Loudon (2000) or Scully and Zubairy (2002) among the most significant.

4

SEMICLASSICAL DESCRIPTION OF LIGHT–MATTER COUPLING

In this chapter we consider light coupling to elementary semiconductor crystal excitations—excitons—and discuss the optical properties of mixed light–matter quasiparticles named exciton-polaritons, which play a decisive role in optical spectra of microcavities. Our considerations are based on the classical Maxwell equations coupled to the material relation accounting for the quantum properties of excitons.

Microcavities, Second Edition. Alexey V. Kavokin, Jeremy J. Baumberg, Guillaume Malpuech, Fabrice P. Laussy, Oxford University Press (2017). © Alexey V. Kavokin, Jeremy J. Baumberg, Guillaume Malpuech, Fabrice P. Laussy. DOI 10.1093/oso/9780198782995.001.0001

4.1 Light–matter interaction

4.1.1 *Classical limit*

Although this chapter refers to a *semiclassical* treatment of light–matter interaction, there naturally exists a "pure classical treatment" where matter is put on an equal "classical" footing with light (cf. Chapter 2). This description is, moreover, a very powerful one and one that will lay down an important concept: the *oscillator strength*. It is, also, simple enough to serve as an introduction to the so-called semiclassical treatment, where parts of the quantum concepts are involved in a rather vague and intuitive way into the material excitation. This helpful and short description in full classical terms is what we address now.

Lorentz proposed a fully classical picture of light–matter interactions where light is modelled by Maxwell's equations and the atom by a mechanical system of two masses—the nucleus and an electron—bound together by a spring. ω_0 is therefore the natural frequency of oscillation. This purely mechanical oscillator will carry along many concepts into the quantum picture. The spring is set into motion when light irradiates the atom.

Hendrik A. **Lorentz** (1853-1928), who received the 1902 Nobel Prize in physics (with Zeeman) for his work on electromagnetic radiation, provided an insightful classical description of light–matter interactions.

His doctoral thesis in 1875 developed Maxwell's theory of 1865 to explain reflection and refraction of light. He is also noted for the *Lorentz transformation* of space and time dilations and contractions that would be at the heart of Einstein's special theory of relativity.

The gist of the physical implications of this assumption is retained in the simplest case where the atom is fixed and the electron a distance $x(t)$ away, moving under the influence of an applied electric field $E(t)$, with an equation of motion

$$m_0\ddot{x} + m_0 2\gamma\dot{x} + m_0\omega_0^2 x = -eE(t)\,, \qquad (4.1)$$

where m_0 is the mass of the electron, $-e$ its charge and $m_0\omega_0^2$ the harmonic potential binding the electron (the spring). Excluding γ, this is merely Newton's equation $F = m_0 a$ of a dipole whose acceleration a is driven by a force F. The loss term arises in this model from the fact that an oscillating dipole radiates energy.[56] With this understanding, its expression can be related to fundamental constants by computing the rate of energy loss of a dipole.

[56]That a moving charge radiates energy is one of the problems with the classical picture of atoms where the electron is thought of as orbiting (hence, moving around) the nucleus, therefore doomed to spiral into it as it loses energy. Bohr postulated that the electron cannot move smoothly towards its nucleus because of interferences along certain orbits, giving birth to the original quantum theory.

Exercise 4.1 $^{(**)}$ *Show that* $\gamma = e^2\omega_0^2/(3m_0c^3)$.

In our case where the oscillator models an atom, it is a good approximation to model the exciting field as an harmonic function of time—the vacuum solution of Maxwell's equations—with $E(t) = E_0\cos(\omega t)$. One can solve eqn (4.1) and obtain the steady-state of the system, where the electron also oscillates harmonically with the frequency of the external force, but with a different amplitude and a different phase:

$$x(t \to \infty) = \mathcal{A}\cos(\omega t - \phi). \qquad (4.2)$$

One can see the intrinsic simplicity of the Lorentz oscillator. We are going to see now its considerable richness. When we say "different amplitude and phase" of oscillation, we mean essentially the change in the response as a function of the exciting frequency ω (see Fig. 4.1):

$$\mathcal{A}(\omega) = \frac{-eE_0}{m_0} \frac{1}{\sqrt{(\omega^2 - \omega_0^2)^2 + (2\gamma\omega)^2}}, \qquad (4.3a)$$

$$\phi(\omega) = \arctan\left(\frac{2\gamma\omega}{\omega_0^2 - \omega^2}\right). \qquad (4.3b)$$

The external frequency that maximises the amplitude $\mathcal{A}$ for a given set of parameters, ω_0 and γ, is the resonant frequency. We find it by taking the derivative of the amplitude function and setting it to zero:

$$\omega_{\text{res}} = \sqrt{\omega_0^2 - 2\gamma^2}. \qquad (4.4)$$

At this frequency, the field is transferring energy most efficiently to the atom's electron. In the inset of Fig. 4.1 one can see how the resonant frequency $\omega_{\text{res}}(\gamma)$ varies with the damping. There is only a genuine resonance for cases with $\gamma < \omega_0/\sqrt{2}$, otherwise the system is overdamped and does not oscillate. On the other hand, if $\gamma = 0$ the resonant frequency is simply the natural frequency of the system. In the limiting case where $\gamma \ll \omega_0/\sqrt{2}$ (for frequencies close to ω_0), the energy–transfer distribution is a Lorentzian:

$$\frac{\gamma}{(\omega - \omega_0)^2 + \gamma^2}. \qquad (4.5)$$

A more general method to solve eqn (4.1), is to find the "output" or "answer" $x(t)$ of the system to the "input" field $E(t) = \exp(i\omega t)$ through its *transfer function* $\mathcal{H}$:

$$x(t) = \mathcal{H}(\omega)E(t). \qquad (4.6)$$

The linearity of eqn (4.1) allows one to compute $\mathcal{H}$ directly (by direct substitution) as

$$\mathcal{H}(\omega) = \frac{e/m_0}{\omega^2 - \omega_0^2 - 2i\omega\gamma} \qquad (4.7)$$

Then, the time evolution of the oscillator for an arbitrary excitation $E(t)$—other than $\exp(i\omega t)$ for which eqn (4.6) applies—can be obtained directly as

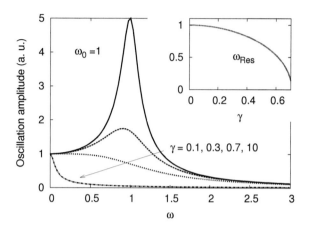

Fig. 4.1: The amplitude of the oscillations of the electron, driven by a sinusoidal external field of frequency ω, shows a resonance when $\omega = \omega_{\text{res}}$. In the inset, we can see how the resonant frequency ω_{res} varies with the damping coefficient γ, becoming ω_0 for the undamped case. The natural frequency of the system is set to 1.

$$x(t) = \int_{-\infty}^{\infty} h(t - \tau)E(\tau)\,d\tau\,, \qquad (4.8)$$

where h is the Fourier transform of $\mathcal{H}$. The imaginary part of $\mathcal{H}$ is called the *absorption signal* as it provides similar information as the amplitude $\mathcal{A}$ we have analysed in our particular case.

4.1.2 *Einstein coefficients*

A fundamental problem of light–matter interaction that falls out of the scope of classical physics, as noted by the beginning of the twentieth century, is that of blackbody radiation.[57] So fundamental was this problem that any mathematical trick to derive the solution was providing a physical insight into a worldview going much beyond that afforded by the then existing models, which would flourish into quantum mechanics. The attempt by Planck culminated in his law for blackbody radiation:

$$I(\omega) = \frac{2\hbar\omega^3}{\pi c^3} \frac{1}{e^{\hbar\omega/(k_{\text{B}}T)} - 1}\,, \qquad (4.9)$$

where I is the spectral energy density (with dimensions of Joule per cubic meter per second^{-1}), ω the frequency,[58] k_{B} Boltzmann's constant[59] and T the temperature. Its

---

[57] A *black body* is an object that radiates energy originating from intrinsic emission by the object and is not the result of reflection or transmission from external radiation. Therefore, all these radiations from outside the object and that impinge on it are absorbed by it (and later re-radiated, but only as a result of how the black body stores energy to maintain its thermal equilibrium). This is an ideal limiting case (since all objects reflect or transmit light to some extent) to investigate the thermodynamic of the electromagnetic field.

[58] Strictly speaking, $\omega/(2\pi)$ is the frequency and ω is the *pulsation*. We shall follow the standard usage that is to call ω the frequency as well.

[59] Boltzmann's constant is $1.380\,650\,4 \times 10^{-23}\,\text{J K}^{-1}$ and $8.617\,343 \times 10^5\,\text{eV K}^{-1}$. It connects many macroscopic quantities from the microscopic realm, its most fundamental connection being between

main merit is the perfect accord with experimental data, previously afforded only when separating short and long wavelengths. Expression (4.9) gave impetus to the concept of the photon.[60] The attempt by Bose led to the new statistics of bosons and in 1916, the attempt of Einstein himself—already a key actor in the two previous approaches—led to the fundamental processes of semiclassical light–matter interaction. These are the topics of this section.

Einstein proposed the existence of three fundamental processes in the interaction of light with matter, i.e., in the view of the time, in the interaction of a photon with an atom. These are (see Fig. 4.2):

1. Absorption.
2. Spontaneous emission.
3. Stimulated emission.

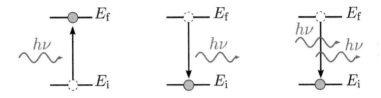

Fig. 4.2: The three fundamental processes of light–matter interaction in the semiclassical paradigm: *Absorption, spontaneous emission* and *stimulated emission*.

The first two processes present no difficulty once the postulate of quantised energy levels is accepted: a photon of energy $E_f - E_i$ can be absorbed by an atom, thus exciting it from energy level E_i to energy level E_f (energy is conserved and—this is the novelty arising from the semiclassical/old quantum theory—exchanged in discrete amount defined by the atom structure).

Spontaneous emission is the reverse process where energy is released by the atom and a photon created that carries it away. It is called *spontaneous* because an excited atom will decay by itself into a lower energy level (until it reaches the ground state, where it stays until it gets excited again). The average time for this transition is, like the possible energies of the atom, a property of the atom.[61] It will later be for quantum mechanics to calculate it; at this stage this is a given constant τ_{sp}.

---

the entropy and the logarithm of the number of microscopic states available to the system. Although this relation is engraved on Boltzmann's tombstone in the Zentralfriedhof in Vienna, the constant, its name k and its numerical value were first provided by Planck some 23 years after the link envisioned by Boltzmann.

[60] Einstein interpreted Planck's hypothesis in terms of "quanta" of light—one reason for which he is credited as a founding father of quantum mechanics—which he used for his explanation of the photoelectric effect. The name "photon" itself was coined in by the chemist Gilbert Lewis as a support of a theory that was soon refuted and abandoned.

[61] Spontaneous emission is not an *intrinsic* property of the atom as it can be modified by changing the optical environment of the atom. This important point is discussed in detail in Chapter 6.

To account for the Planck distribution, however, Einstein requires the introduction of the third process, *stimulated emission* (see Fig. 4.2).[62] Because of this process, an excited atom—in addition to the decay by spontaneous emission (first process)—can also decay if it is interacting with a photon, by emitting a clone of this photon. This is not very intuitive but we now show that if it is accepted, the blackbody radiation spectrum is readily derived by mere rate equations and detailed balance arguments. Each of these processes is associated to a probability of occurrence: the probability per unit time of spontaneous emission is called A, i.e., if the density (number per unit volume) of atoms in the excited (resp. ground) state is n_2 (resp. n_1), there is a transfer from the excited to the ground-state populations following

$$\frac{dn_1}{dt} = An_2 . \tag{4.10a}$$

Equation (4.10a) implies that $A = \tau_{\rm sp}^{-1}$.

Absorption is ruled by the Einstein coefficient B_{12} that gives the probability per unit time and per unit energy density of the radiation field that the atom initially in state 1 absorbs a photon and jumps to state 2 causing a change in the number density of atoms in the ground state of

$$\frac{dn_2}{dt} = B_{12}n_1 I(\omega) , \tag{4.10b}$$

with $I(\omega)$ the spectral intensity of the radiation field at the frequency of radiation $\omega = (E_2 - E_1)/\hbar$. This equation merely spells out the definition of B given above, with $I(\omega)$ quantifying the number of photons available to excite the atom.

Stimulated emission, as noted above, is induced by a photon bringing an atom in the excited state to its ground-state. It is therefore the same process as eqn (4.10b) only reversed, with an excited atom as the starting state and finishing with one more ground-state atom. Calling the corresponding probability of this event B_{21},

$$\frac{dn_1}{dt} = B_{21}n_2 I(\omega) . \tag{4.10c}$$

At equilibrium, all these three processes concur to establish the steady-state conditions $dn_i/dt = 0$. In this regime, the average change in the populations of ground and excited states is zero, being balanced by the losses and gains. The principle of "*detailed balance*" postulates that such an equilibrium in population exchange is reached by pairwise compensations of the type $dn_1/dt = -dn_2/dt$, which leads to

$$n_2 A - n_1 B_{12} I(\omega) + n_2 B_{21} I(\omega) = 0 , \tag{4.11}$$

which provides the energy density as a function of other parameters as

$$I(\omega) = \frac{A}{(n_1/n_2)B_{12} - B_{21}} . \tag{4.12}$$

---

[62]Historically, one was speaking of "induced", rather than "stimulated" emission, but the new term that is now prevailing is more common and vivid for other similar manifestations, such as "stimulated scattering".

We are investigating the energy distribution of the radiation field. That of the atoms (which form a classical system like a gas or a solid, and nothing like a Bose condensate where their statistics could play a role) was well known from the earlier work of Boltzmann and Maxwell. The kinetic theory of gases that they developed gives the populations of atomic states after the ratio of their energy to the temperature:

$$n_i = g_i \exp\left(-\frac{E_i}{k_B T}\right) \tag{4.13}$$

(with g_i the degeneracy of the state), so that finally

$$I(\omega) = \frac{A}{(g_1/g_2)\exp(\hbar\omega/k_B T)B_{12} - B_{21}}. \tag{4.14}$$

Identifying this result with eqn (4.9) provides the following expressions for Einstein coefficients:

$$A = \frac{\hbar\omega^3}{\pi^2 c^3} B_{21}, \tag{4.15a}$$

$$g_1 B_{12} = g_2 B_{21}. \tag{4.15b}$$

4.2 Optical transitions in semiconductors

An arena of choice for microcavity physics is that of semiconductor physics. A semiconductor is a solid whose electrical conductivity has behaviour and magnitudes in between metals and insulators. This comes from the energy levels of such systems that form *bands* separated by *gaps* of forbidden energies (or states). Consequently, semiconductors afford a great control of electronic excitations. We note here the most essential features of the structure of optical transitions in semiconductors.[63]

Table 4.1 Classification of solids.

	Fermi level	Energy gap width	Conductivity ($\Omega^{-1}m^{-1}$)
metals	inside the band	any	Up to 6.3×10^7 (silver)
semiconductors	inside the gap	< 4 eV	Varies in large limits
dielectric	inside the gap	4 eV	Can be as low as 10^{-10}

The discrete electronic levels of individual atoms form *bands* in crystals where thousands of atoms are assembled in a periodic structure. There are also gaps between the allowed bands where no electronic states exist in an ideal infinite crystal. Those crystals that have a Fermi level[64] inside one of the allowed bands are *metals*, while the crystals having a Fermi level inside the gap are *semiconductors* or *dielectrics* (see Table 4.1).

---

[63] Much more information on this subject can be found in Charles Kittel, Introduction to Solid State Physics (Wiley: New York, 1996) and Neil W. Ashcroft and N. David Mermin, Solid State Physics (Harcourt: Orlando, 1976).

[64] The Fermi energy is the energy below which, at zero temperature, all the electronic states are occupied and above which all the states are empty. The Fermi level is the set of states with Fermi energy.

The difference between semiconductors and dielectrics is quantitative: the materials where the bandgap containing the Fermi level is narrower than about $4\,eV$ are usually called semiconductors, the materials with wider bandgaps are dielectrics. In this chapter we consider only semiconductor crystals.

The eigenfunctions of electrons inside the bands have a form of so-called *Bloch waves*. The concept of the Bloch waves was developed by the Swiss physicist Felix Bloch in 1928, to describe the conduction of electrons in crystalline solids. The *Bloch theorem* states that a wavefunction of an electronic eigenstate in an infinite periodic crystal potential $V(\mathbf{r})$ can be written in the form (see Section 2.6.1)

$$\Psi_{\mathbf{k},n} = U_{\mathbf{k},n}(\mathbf{r})e^{i\mathbf{k}\cdot\mathbf{r}}, \tag{4.16}$$

where $U_{\mathbf{k},n}$ (called the *Bloch amplitude*) has the same periodicity as the crystal potential, $\mathbf{k}$ is a so-called *pseudowavevector* of an electron (hereafter we shall omit "pseudo" while speaking about this quantity) and n is the index of the band.

Substitution of the wavefunction (4.16) into the Schrödinger equation for an electron propagating in a crystal,

$$-\frac{\hbar^2}{2m_0}\nabla^2\Psi_{\mathbf{k},n} + V(\mathbf{r})\Psi_{\mathbf{k},n} = E_{\mathbf{k},n}\Psi_{\mathbf{k},n}, \tag{4.17}$$

with m_0 being the free-electron mass, one obtains for the Bloch amplitude the equation

$$-\frac{\hbar^2}{2m_0}\Delta U_{\mathbf{k},n} + \left(V(\mathbf{r}) + \frac{\hbar^2 k^2}{2m_0} + \frac{\hbar}{m_0}(\mathbf{k}\cdot\mathbf{p})\right)U_{\mathbf{k},n} = E_{\mathbf{k},n}U_{\mathbf{k},n}, \tag{4.18}$$

where $\mathbf{p} = \frac{\hbar}{i}\nabla$. Consideration of the operators in the parentheses as a perturbation constitutes the method of the $\mathbf{k}\cdot\mathbf{p}$ *perturbation theory*, which readily enables solving the shape of the electronic dispersion in the vicinity of $\mathbf{k} = 0$ points of all bands and that appears to be strongly different from the free-electron dispersion in vacuum. The approximation

$$E_{\mathbf{k},n} \approx E_{0,n} + \frac{\hbar^2 k^2}{2m_n*} \tag{4.19}$$

is called the *effective mass approximation* with m_n* the electron effective mass in the nth band:[65]

$$\frac{1}{m_n*} = \frac{1}{m_0} + \frac{2}{m_0^2}\sum_{l\neq n}\frac{|\langle U_{0,l}|\mathbf{p}|U_{0,n}\rangle|^2}{E_{0,l} - E_{0,n}}. \tag{4.20}$$

The frequencies and polarisation of the optical transitions in direct gap semiconductors are governed by the energies and dispersion of the two bands closest to the Fermi level[66], referred to as the conduction band (first above the Fermi level) and the valence band (first below the Fermi level; often several close bands are important).

---

[65] In general, the effective mass is a tensor. It reduces to a scalar in crystals having a cubic symmetry.

[66] In semiconductors, the Fermi level is situated in the gap. The width of this gap, E_g, governs the optical absorption edge.

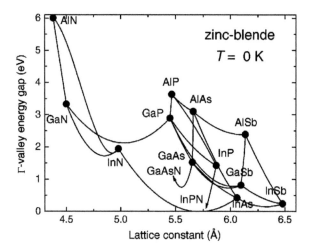

Fig. 4.3: Energy gaps and lattice constants in some direct-bandgap semiconductors and their alloys, from Vurgaftman et al. (2001).

Semiconductors can be divided into those with direct bandgaps and those with indirect bandgaps. In indirect gap semiconductors (like Si and Ge) the electron and hole occupying the lowest energy states in conduction and valence bands cannot directly recombine emitting a photon due to the wavevector–conservation requirement. While a weak emission of light by these semiconductors due to phonon-assisted transitions is possible, they can hardly be used for fabrication of light-emitting devices and studies of light–matter coupling effects in microcavities. In the following, we shall only consider the direct gap semiconductor materials like GaAs, CdTe, GaN, ZnO, etc. (see Fig. 4.3). Most of them have either a zincblende or a wurtzite crystal lattice[67] (see Fig. 4.4). In zincblende semiconductors, the valence band splits into three sub-bands referred to as the heavy-hole, light-hole and spin-off bands (see Fig. 4.5). At $k = 0$ the heavy- and light-hole bands are degenerate in bulk crystals, while this degeneracy can be lifted by strain or external fields. In the wurzite semiconductors, the valence band is split into three non-degenerate sub-bands referred to as A, B and C bands.

Dispersion of the light and heavy holes in zincblende semiconductors can be conveniently described by the *Luttinger Hamiltonian* proposed in a famous paper by Luttinger and Kohn (1955):

$$H = \frac{\hbar^2}{2m_0} \left[\left(\gamma_1 + \frac{5}{2}\gamma_2 \right) \Delta - 2\gamma_3 (\nabla \mathbf{J})^2 + 2(\gamma_3 - \gamma_2) \left(J_x^2 \frac{\partial^2}{\partial x^2} + J_y^2 \frac{\partial^2}{\partial y^2} + J_z^2 \frac{\partial^2}{\partial z^2} \right) \right],$$
(4.21)

where $J_\alpha = \pm\frac{1}{2}, \pm\frac{3}{2}$, $\alpha = x, y, z$ and γ_1, γ_2, γ_3 are Luttinger band parameters dependent on the material.

---

[67] A cubic phase is somewhat more exotic. It is found for GaN, for example.

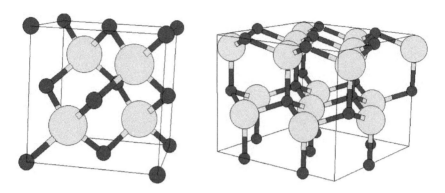

Fig. 4.4: Zincblende (left) and wurtzite (right) crystal lattices.

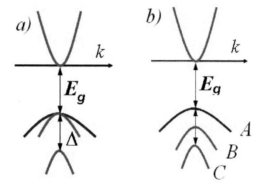

Fig. 4.5: Schematic band structure of a zincblende semiconductor (a) with a conduction band (on the top), degenerated heavy- and light-hole bands (in the middle), and the spin-off band (at the bottom) and a wurtzite semiconductor (b) with A, B and C valence subbands.

In the bulk zincblende samples, at $\mathbf{k} = 0$ the degenerate light- and heavy-hole states mean that the probability of an allowed optical transition from a heavy hole to an electron state is three times higher than the probability of a transition from a light-hole state (see Fig. 4.6). This is why illumination of a semiconductor crystal by circularly polarised light leads to preferential creation of electrons with a given spin projection. This effect referred to as *optical orientation* will be discussed in more detail in Chapter 9.

Optical absorption spectra in semiconductors are governed by the density of electronic states in the valence and conduction bands, $g(E) = \frac{\partial n}{\partial E}$, where n is the number of quantum states per unit area. In bulk crystals, inside the bands the density of states behaves as $\sqrt{E}$, which results in the corresponding shape of the interband absorption spectra. Besides this, at low temperatures, the absorption spectra of semiconductors exhibit sharp peaks below the edge of interband absorption (i.e. at frequencies $\omega < E_{\mathrm{g}}/\hbar$,

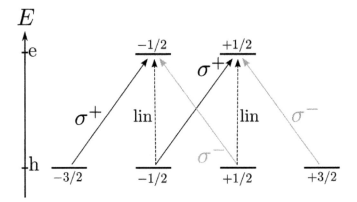

Fig. 4.6: Polarisation of the interband optical transitions in zincblende semiconductor crystals. Black, gray and dashed lines show σ^+, σ^- and linearly polarised transitions, respectively.

where E_g is the bandgap energy). These peaks manifest the resonant light–matter coupling in semiconductors. They are caused by the excitonic transitions that will remain the focus of our attention throughout this book.

4.3 Excitons in semiconductors

4.3.1 *Frenkel and Wannier–Mott excitons*

In the late 1920s, narrow photoemission lines were observed in the spectra of organic molecular crystals by Kronenberger and Pringsheim (1926) as well as by I. Obreimov and W. de Haas (see Zakharchenya (1994) for an historical account). These data were interpreted by the Russian theorist Frenkel (1931) who introduced the concept of excitation waves in crystals and, later, Frenkel (1936) coined the term *exciton*.

By definition, the exciton is a Coulomb-correlated electron–hole pair. Frenkel treated the crystal potential as a perturbation to the Coulomb interaction between an electron and a hole belonging to the same crystal cell. This scenario is most appropriate in organic molecular crystals. The binding energy of Frenkel excitons (i.e. the energy of its ionisation to a non-correlated electron–hole pair) is typically of the order of 100–300 meV. Frenkel excitons have been searched for and observed in alkali halides by Apker and Taft (1950). At present they are widely studied in organic materials where they dominate the optical absorption and emission spectra.

At the end of the 1930s, Wannier (1937) and Mott (1938) developed the concept of excitons in semiconductor crystals where the rate of electron and hole hopping between different crystal cells much exceeds the strength of their Coulomb coupling with each other. Unlike Frenkel excitons, Wannier–Mott excitons have a typical size of the order of tens of lattices constants and a relatively small binding energy (typically, a few meV, see Fig. 4.7).

Besides Frenkel and Wannier–Mott excitons, there are a few other types of excitons. The *charge–transfer excitons* are spatially separated Coulomb-bound electron–hole pairs having a spatial extension of the order of the crystal lattice constant. The

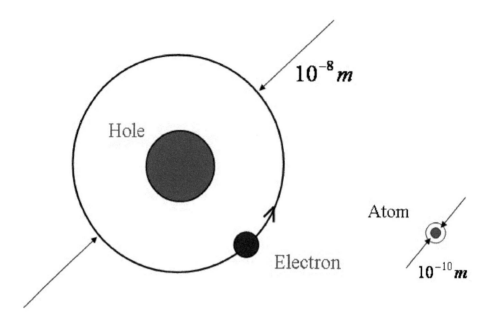

Fig. 4.7: A Wannier–Mott exciton is the solid-state analogy of a hydrogen atom, while they have very different sizes and binding energies. Unlike atoms, the excitons have a finite lifetime.

lowest energy charge–transfer exciton usually extends over two nearest-neighbour molecules in a molecular crystal and creates a so-called donor–acceptor complex. See, for instance, Silinish (1980) for a more thorough treatment.

Another example of a few-particle exciton complex in a quantum-confined semiconductor system is the so-called *anyon exciton* appearing in the regime of the quantum Hall effect as described by Rashba and Portnoi (1993). The energy of anyon excitonic transitions lies in the far-infrared range, thus these quasiparticles cannot be coupled to light in optical cavities.

Agranovich et al. (1997) have proposed a concept of hybrid Frenkel–Wannier–Mott excitons that can be formed in mixed organic–inorganic structures. Such excitons would combine a huge binding energy and relatively large size. Extensive information on the Frenkel or hybrid excitons and their coupling with light in organic microcavities can be found in the volume edited by Agranovich and Bassani (2003).

In the present section we only discuss the Wannier–Mott excitons in semiconductors. Such excitons can be conveniently described within the effective mass approximation that allows the periodic crystal potential to be neglected, and describes electrons and holes as free particles having a parabolic dispersion and characterised by effective masses dependent on the crystal material. Usually, the effective masses of carriers are smaller than the free-electron mass in vacuum m_0. For example, in GaAs the electron effective mass is $m_e = 0.067 m_0$, the heavy-hole mass is $m_{hh} = 0.45 m_0$.

Consider an electron–hole pair bound by the Coulomb interaction in a crystal having a dielectric constant ε. The wavefunction of relative electron–hole motion $f(r)$ can be

Yakov Il'ich **Frenkel** (1894–1952), Sir Nevill Francis **Mott** (1905–1996) and Grégory **Wannier** (1911–1983) gave their names to the two main categories of excitons.

Frenkel was a versatile physicist who made his main contributions in solid-state physics. He wrote the first paper devoted to the quantum theory of metals. He is now remembered for the exciton bearing his name and the *Frenkel defect*. A very prolific writer, Dirac was using the "Frenkel" as a unit of book writing's speed and Kapitza reportedly once told him *"You would be a genius if you published ten times less than you do"* (his most noted work is the textbook "Kinetic Theory of Liquids"). A good overview of his life and work is given by Lopatnikov and Cheng (2005).

Mott, the Nobel prize-winner for Physics in 1977 for "fundamental theoretical investigations of the electronic structure of magnetic and disordered systems", is most renowned for his mechanism explaining why material predicted to be conductors by band theory are in reality (so-called Mott) insulators and for describing the transition of substances from metallic to non-metallic states (Mott transition).

Wannier authored a series of important papers on the properties of crystals. His main achievement is a complete set of orthogonal functions, known as "Wannier functions", which provide an alternative representation of localised orbitals to the usual Bloch orbitals. According to his graduate student Hofstadter (1984), *"it is not what [he] would want to be known for primarily. He was so involved in so many areas of physics, and his breadth was so refreshing, compared to the narrow range of most physicists today, that I think he would wish to be remembered for that breadth and for his style, a style that stressed beauty and purity and fundamentality"*.

found from the Schrödinger equation analogous to one describing the electron state in a hydrogen atom:

$$-\frac{\hbar^2}{2\mu}\nabla^2 f(r) - \frac{e^2}{4\pi\varepsilon\varepsilon_0 r}f(r) = Ef(r) \,, \qquad (4.22)$$

with $\mu = m_e m_h/(m_e + m_h)$ the reduced mass, $r = \sqrt{x^2 + y^2 + z^2}$ the distance between electron and hole. The solutions of eqn (4.22) are well known as they correspond to the states of the hydrogen atom with the renormalisations:[68]

$$m_0 \to \mu \,, \qquad e^2 \to e^2/\varepsilon \,, \qquad (4.23)$$

For example, the wavefunction of the 1s state of exciton reads

---

[68]In the hydrogen-atom problem the reduced mass is equal, in good approximation, to the electron mass m_0 because of the very large mass of the nucleus.

$$f_{1s} = \frac{1}{\sqrt{\pi a_{\mathrm{B}}^3}} e^{-r/a_{\mathrm{B}}} , \qquad (4.24)$$

with the Bohr radius a_{B} given as

$$a_{\mathrm{B}} = \frac{4\pi\hbar^2 \varepsilon\varepsilon_0}{\mu e^2} . \qquad (4.25)$$

The binding energy of the ground exciton state is

$$E_{\mathrm{B}} = \frac{\mu e^4}{(4\pi)^2 2\hbar^2 \varepsilon\varepsilon^2} = \frac{\hbar^2}{2\mu a_{\mathrm{B}}^2} . \qquad (4.26)$$

Given the difference between the reduced mass μ and the free-electron mass, and taking into account the dielectric constant in the denominator, one can estimate that the exciton binding energy is about three orders of magnitude less than the Rydberg constant. Table 4.2 shows the binding energies and Bohr radii for Wannier–Mott excitons in different semiconductor materials.

Table 4.2 Exciton characteristics in the most popular semiconductors.

Semiconductor crystal	E_g (eV)	m_e/m_0	E_B (eV)	a_B (A)
PbTe*	0.17	0.024/0.26	0.01	17 000
InSb	0.237	0.014	0.5	860
Cd$_{0.3}$Hg$_{0.7}$Te	0.257	0.022	0.7	640**
Ge	0.89	0.038	1.4	360
GaAs	1.519	0.066	4.1	150
InP	1.423	0.078	5.0	140
CdTe	1.606	0.089	10.6	80
ZnSe	2.82	0.13	20.4	60
GaN***	3.51	0.13	22.7	40
Cu$_2$O	2.172	0.96	97.2	38****
SnO$_2$	3.596	0.33	32.3	86****

* Strongly anisotropic conduction and valence bands, direct transitions far from the centre of the Brillouin zone.

** In the presence of a magnetic field of 5 T.

*** An exciton in hexagonal GaN.

**** The ground state corresponds to an optically forbidden transition, data for the $n = 2$ state.

The exciton excited states form a number of hydrogen-like series. Observation of such a series of excitonic transitions in the photoluminescence spectra of Cu$_2$O in 1951 was the first experimental evidence for Wannier–Mott excitons (see the illustration below the photograph on the facing page). This discovery was made by the Russian spectroscopist Evgeniy Gross who worked in the same institution—the Ioffe Physico-Technical Institute in Leningrad—as Ya. I. Frenkel at that time.

Evgenii Fedorovich **Gross** (1897–1972) and the experimental discovery of the exciton: the hydrogen-like "yellow" series in emission of Cu_2O observed by Gross et al. (1956), with its numerical fit. Besides discovering experimentally the exciton, he is also noted for pioneering in 1930 the experimental observation of Rayleigh scattering fine structure due to Brillouin–Mandelstam light scattering on acoustic waves. A short account of his scientific achievements can be found in "Evgeni Fedorovich Gross", B. P. Zakharchenya and A. A. Kaplyanski, Sov. Phys. Uspekhi, **11**, 141 (1968).

$$E_n = 2{,}17244 - \frac{0{,}0972}{n_0^2}; \quad n_0 = 2{,}3{,}....$$

4.3.2 *Excitons in confined systems*

Since the beginning of the 1980s, progress in the growth technology of semiconductor heterostructures encouraged study of Wannier–Mott excitons in confined systems including quantum wells, quantum wires and quantum dots. The main idea behind the development of heterostructures was to artificially create potential wells and barriers for electrons and holes, combining different semiconductor materials. The shape of the potential in conduction and valence bands is determined in these structures by the positions of the corresponding band edges in the materials used as well as by the geometry of the structure. The *band engineering* in semiconductor structures by means of high-precision growth methods has allowed the creation of a number of electronic and opto-electronic devices including transistors, diodes and lasers. It has also permitted discovery of important fundamental effects including the integer and fractional quantum Hall effects, Coulomb blockade, light-induced ferromagnetism, etc.

The large size of Wannier–Mott excitons makes them strongly sensitive to nanometre-scale variations of the band-edge positions that can be easily obtained in modern semiconductor nanostructures. The energy spectrum and wavefunctions of quantum-confined excitons can be strongly different from those of bulk excitons. Here, we consider by means of an approximate but efficient variational method the excitons in quantum wells, wires and dots (see Fig. 4.8). We will use the effective-mass approximation. When we refer to wavefunctions we always mean the envelope functions, neglecting the Bloch amplitudes of electrons and holes. Note that in these examples we neglect the complexity of the valence-band structure and consequent anisotropy of the hole effective mass that sometimes strongly affects the excitonic spectrum in real semiconductor systems. More information on excitons in confined systems can be found in the books by Bastard (1988), Ivchenko and Pikus (1997) and Ivchenko (2005).

4.3.3 *Quantum wells*

The Schrödinger equation for an exciton in a quantum well (QW) reads

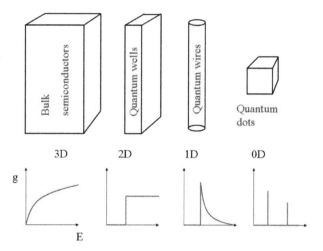

Fig. 4.8: Reduction of the dimensionality of a semiconductor system from 3D to 0D from a bulk semiconductor to a quantum dot. The electronic density of states $g(E) = dN/dE$—with dN the number of electron quantum states within the energy interval dE—changes drastically between systems of different dimensionalities as is shown schematically in the figure. This variation of the density of states is very important for light-emitting semiconductor devices.

$$\left(-\frac{\hbar^2}{2m_e}\nabla_e^2 - \frac{\hbar^2}{2m_h}\nabla_h^2 + V_e(z_e) + V_h(z_h) - \frac{e^2}{4\pi\varepsilon\varepsilon_0|\mathbf{r}_e - \mathbf{r}_h|}\right)\Psi = E\Psi, \quad (4.27)$$

with $V_{e,h}(z_{e,h})$ the confining potential for electron, hole on the z-axis, which is the growth axis of the structure. Solving exactly eqn (4.27) is not an easy task. We approach the problem variationally over a class of trial functions having the form

$$\Psi(\mathbf{r}_e, \mathbf{r}_h) = F(\mathbf{R})f(\boldsymbol{\rho})U_e(z_e)U_h(z_h), \quad (4.28)$$

where

$$\mathbf{R} = \frac{m_e\mathbf{r}_e + m_h\mathbf{r}_h}{m_e + m_h} \quad (4.29)$$

is the exciton centre of mass coordinate and

$$\boldsymbol{\rho} = \boldsymbol{\rho}_e - \boldsymbol{\rho}_h \quad (4.30)$$

is the *in-plane* radius-vector of electron and hole relative motion, $\mathbf{r} = (\boldsymbol{\rho}, z)$. Four components of the trial function (4.28) describe the exciton centre of mass motion, the relative electron–hole motion in the plane of the QW, and electron and hole motion normal to the plane direction. The factorisation of the exciton wavefunction makes sense when the QW width is less than or comparable with the exciton Bohr diameter in the bulk semiconductor. In this case, the electron and hole are quantised independently of each other. On the other hand, in larger QWs, one can assume that the exciton is confined as a whole particle and keeps the internal structure of a 3D hydrogen atom. Here and later, we shall consider narrow QWs where eqn (4.28) represents a good approximation. The four terms that compose the exciton wavefunction are normalised to unity:

$$\int |U_e(z_e)|^2 dz_e = 1, \qquad \int_0^\infty |f(\rho)|^2 2\pi\rho d\rho = 1, \qquad (4.31a)$$

$$\int |U_h(z_h)|^2 dz_h = 1, \qquad \int_0^\infty |F(R)|^2 2\pi R dR = 1. \qquad (4.31b)$$

After substitution of the trial function (4.28) and integration over R, eqn (4.27) becomes

$$\left\{ -\frac{\hbar^2}{2m_e}\frac{\partial^2}{\partial z_e^2} - \frac{\hbar^2}{2m_h}\frac{\partial^2}{\partial z_h^2} - \frac{1}{\rho}\frac{\partial}{\partial\rho}\left(\frac{\hbar^2}{2\mu}\rho\frac{\partial}{\partial\rho}\right) + V_e(z_e) + V_h(z_h) \right.$$

$$\left. - \frac{e^2}{4\pi\varepsilon\varepsilon_0\sqrt{\rho^2+(z_e-z_h)^2}} - \frac{P_{\text{exc}}^2}{2(m_e+m_h)} - E \right\} f(\rho)U_e(z_e)U_h(z_h) = 0, \quad (4.32)$$

where P_{exc} is the excitonic momentum, $P = 0$ for the ground state. Equation (4.32) can be transformed into a system of three coupled differential equations, each defining one of the components of our trial function. The equation for $f(\rho)$ is obtained by multiplication of both parts of eqn (4.32) by $U_e^*(z_e)U_h^*(z_h)$ and integrating over z_e and z_h. This yields

$$\left\{ -\frac{1}{\rho}\frac{\partial}{\partial\rho}\left(\frac{\hbar^2}{2\mu}\rho\frac{\partial}{\partial\rho}\right) - \frac{e^2}{4\pi\varepsilon\varepsilon_0}\iint \frac{|U_e(z_e)|^2|U_h(z_h)|^2}{\sqrt{\rho^2+(z_e-z_h)^2}} dz_e dz_h \right\} f(\rho) = -E_B^{QW} f(\rho),$$

$$(4.33)$$

where E_B^{QW} is the exciton binding energy. The electron and hole confinement energies E_e and E_h, and wavefunctions $U_{e,h}(z_{e,h})$, can be obtained by multiplying eqn (4.32) by $f^*(\rho)U_{h,e}^*(z_{h,e})$ and integrating over $z_{e,h}$ and ρ:

$$\left\{ -\frac{\hbar^2}{2m_{e,h}}\nabla_{e,h}^2 + V_{e,h} - \frac{e^2}{4\pi\varepsilon\varepsilon_0}\iint \frac{|f(\rho)|^2|U_{h,e}(z_{h,e})|^2}{\sqrt{\rho^2+(z_e-z_h)^2}} 2\pi\rho d\rho dz_{h,e} \right\} U_{e,h}(z_{e,h})$$

$$= E_{e,h}U_{e,h}(z_{e,h}). \quad (4.34)$$

In the ideal 2D case, $|U_{e,h}(z_{e,h})|^2 = \delta(z_{e,h})$ and eqn (4.33) transforms into

$$\left\{ -\frac{\hbar^2}{2\mu}\frac{1}{\rho}\frac{\partial}{\partial\rho}\left(\rho\frac{\partial}{\partial\rho}\right) - \frac{e^2}{\varepsilon\rho} \right\} f(\rho) = E_B^{2D} f(\rho), \qquad (4.35)$$

which is an exactly solvable 2D hydrogen atom problem. For the ground state,

$$f_{1S}(\rho) = \sqrt{\frac{2}{\pi}}\frac{1}{a_B^{2D}}\exp(-\rho/a_B^{2D}), \qquad (4.36)$$

with

$$a_B^{2D} = a_B/2, \qquad (4.37)$$

and a_B the Bohr radius of the three-dimensional exciton given by eqn (4.25). The binding energy of the two-dimensional exciton exceeds by a factor of 4 the bulk exciton binding energy:

$$E_\mathrm{B}^{2\mathrm{D}} = 4E_\mathrm{B} \,, \tag{4.38}$$

For realistic QWs, eqns (4.32) and (4.33) still can be decoupled if the Coulomb term in eqn (4.33) is neglected. This allows the functions $|U_{e,h}(z_{e,h})|$ to be found independently from each other, as well as $f(\rho)$. Solving eqn (4.32) with a trial function

$$f(\rho) = \sqrt{\frac{2}{\pi}} \frac{1}{a} \exp(-\rho/a) \,, \tag{4.39}$$

where a is a variational parameter, one can express the binding energy as

$$E_\mathrm{B}^{QW}(a) = -\frac{\hbar^2}{2\mu a^2} + \frac{e^2}{4\pi\varepsilon\varepsilon_0} \iiint \frac{|f(\rho)|^2 |U_e(z_e)|^2 |U_h(z_h)|^2}{\sqrt{\rho^2 + (z_e - z_h)^2}} 2\pi\rho d\rho dz_e dz_h \,. \tag{4.40}$$

Maximisation of $E_\mathrm{B}^{QW}(a)$ finally yields the exciton binding energy in a QW, which ranges from E_B to $E_\mathrm{B}^{2\mathrm{D}}$, and depends on the QW width and barrier heights for electrons and holes. The binding energy increases if the exciton confinement strengthens. This is why the dependence of the binding energy on the QW width is non-monotonic: for wide wells the confinement increases with the decrease of the QW width, while for ultranarrow wells the tendency is inverted due to tunnelling of electron and hole wavefunctions into the barriers (Fig. 4.9).

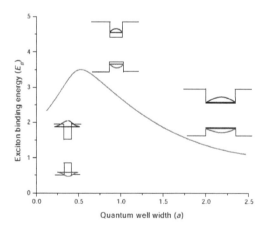

Fig. 4.9: Exciton binding energy as a function of the QW width (schema). The insets show the QW potential and wavefunctions of electron (convex shape) and hole (concave shape) for different QW widths.

4.3.4 Quantum wires and dots

Variational calculation of the ground exciton state energy and wavefunction in quantum wires or dots can be done using the same method of separation of variables and decoupling of equations as for QWs. There exist a number of important peculiarities of wires and dots with respect to wells, however.

For a wire, the Schrödinger equation for the wavefunction of electron–hole relative motion $f(z)$, with z the axis of the wire, reads

$$\left\{-\frac{\hbar^2}{2\mu}\nabla_z^2 - \frac{e^2}{4\pi\varepsilon\varepsilon_0}\iint \frac{|U_e(\rho_e)|^2|U_h(\rho_h)|^2}{\sqrt{z^2 + (\rho_e - \rho_h)^2}}d\rho_e d\rho_h\right\}f(z) = -E_B^{QWW}f(z),$$

$$(4.41)$$

with the kinetic energy for relative motion being along the axis of the wire. $U_{e,h}(\rho_{e,h})$ is the electron, hole wavefunction in the plane normal to the wire axis and E_B^{QWW} is the exciton binding energy in the wire. Despite visible similarity to eqn (4.33) for electron–hole relative motion wavefunction in a QW, eqn (4.41) has a different spectrum and different eigenfunctions. As a quantum particle in a 1D Coulomb potential has no ground-state with a finite energy, the exciton binding energy in a quantum wire is drastically dependent on spreading of the functions $U_{e,h}(\rho_{e,h})$, and can, theoretically, have any value between E_B and infinity. The trial function cannot be exponential (as it would have a discontinuous first derivative at $z = 0$ in this case). The Gaussian function is a better choice in this case. Usually, realistic quantum wires do not have a cylindrical symmetry (most popular are "T-shape" and "V-shape" wires, see Fig. 4.10), which makes computation of $U_{e,h}(\rho_{e,h})$ a separate, difficult task. Moreover, the realistic wires have a finite extension in the z-direction that is comparable with the exciton Bohr diameter in many cases. Even if the wire is designed to be much longer than the exciton dimension, inevitable potential fluctuations in the z-direction lead to the exciton localisation. This makes realistic wires similar to elongated quantum dots (QDs).

An exciton is fully confined in a QD, and if this confinement is strong enough its wavefunction can be represented as a product of electron and hole wavefunctions:

$$\Psi = U_e(\mathbf{r}_e)U_h(\mathbf{r}_h),$$ $$(4.42)$$

where the single-particle wavefunctions $U_{e,h}(\mathbf{r}_{e,h})$ are given by coupled Schrödinger equations

$$\left\{-\frac{\hbar^2}{2m_{e,h}}\nabla_{e,h}^2 + V_{e,h} - \frac{e^2}{4\pi\varepsilon\varepsilon_0}\int \frac{|U_{h,e}(\mathbf{r}_{h,e})|^2}{|\mathbf{r}_e - \mathbf{r}_h|}d\mathbf{r}_{h,e}\right\}U_{e,h}(\mathbf{r}_{e,h}) = E_{e,h}U_{e,h}(\mathbf{r}_{e,h}),$$

$$(4.43)$$

with $V_{e,h}$ is the QD potential for an electron, hole. In this case, the exciton binding energy is defined as

$$E_B^{QD} = E_e^0 + E_h^0 - E_e - E_h,$$ $$(4.44)$$

where E_e^0 and E_h^0 are energies of the non-interacting electron and hole, respectively, i.e., the eigenenergies of the Hamiltonian (4.43) without the Coulomb term.

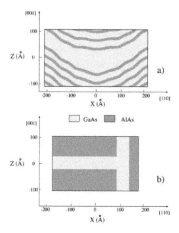

Fig. 4.10: Cross-sections of V-shape (a) and T-shape (b) quantum wires, from Di Carlo et al. (1998).

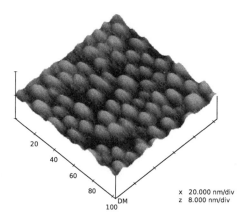

Fig. 4.11: Transmission electron microscopy image from Widmann et al. (1997) of the self-assembled QDs of GaN grown on AlN

In small QDs, Coulomb interaction can be considered as a perturbation to the quantum-confinement potential for electrons and holes. The exciton binding energy can be estimated using perturbation theory as

$$E_B \approx \frac{e^2}{4\pi\varepsilon\varepsilon_0} \iint \frac{|U_e(\mathbf{r}_e)U_h(\mathbf{r}_h)|^2}{|\mathbf{r}_e - \mathbf{r}_h|} d\mathbf{r}_e d\mathbf{r}_h . \qquad (4.45)$$

As in the wire, the exciton binding energy in the dot is strongly dependent on the spatial extension of the electron and hole wavefunctions, and can range from the bulk exciton binding energy to infinity, theoretically. In realistic wires and dots, the binding energy rarely exceeds $4E_B$, however. At present, small QDs are mostly fabricated by the so-called Stransky–Krastanov method of molecular beam epitaxy and have either pyramidal or ellipsoidal shape (see Fig. 4.11). In large quantum dots ("large" meaning "of

Solomon I. **Pekar** (1917–1985), John J. **Hopfield** (b. 1933) and Vladimir M. **Agranovich** (b. 1929).

Pekar was the leading theorist of the Physical Institute in Kiev. He created the theory of adiabatic polarons in 1946 (the term polaron comes from him) and with his work on the so-called *additional light wave*—stating that under excitation by monochromatic light, the dielectric constant's spatial dispersion near an excitonic resonance gives rise to an additional propagating polariton wave—he originated the theory of exciton-polaritons.

Hopfield, after important contributions to physics, turned his interest to biology, where he made his most significant contribution to science with his associative neural network now know as the *Hopfield network*. At the time of writing, he is professor in the department of Molecular Biology at Princeton University and the President of the American Physical Society.

Agranovich is the head of the Theoretical Department of the Institute of Spectroscopy of the Russian Academy of Sciences. He made seminal contributions to the theory of excitons especially in organic crystals and is one of the founders of the theory of polaritons. A prolific author, he fathered among other important work, "Crystal Optics with Spatial Dispersion, and Excitons" with Nobel laureate Ginsburg, a "Theory of Excitons" and recently the monograph "Electronic Excitations in Organic Based Nanostructures". He currently holds the special position of "Pioneer of Nano-Science" at the University of Texas at Dallas.

a size exceeding the exciton Bohr diameter") excitons are confined as whole particles and their binding energy is equal to the bulk exciton binding energy. Good examples of large quantum dots are spherical microcrystals that may serve also as photonic dots.

Exercise 4.2 [*] *Find the binding energies of the first excited states of 2D and 1D excitons.*

4.4 Exciton–photon coupling

It has been clear since the very beginning of experimental studies of excitons that the easiest way to create these quasiparticles is by optical excitation. In the mid-1950s, theorists understood that coupling to light strongly influences the physical properties of excitons and their energy spectrum. The Ukrainian physicist Pekar (1957) was the first to describe these changes of the exciton energy spectrum due to coupling to light in terms of *additional waves* appearing in the crystal. Almost simultaneously, the term *polariton* appeared in the works of Agranovich (1957) (Russia) and Hopfield (1958) (USA), devoted to the description of photon-exciton coupling.[69]

---

[69]The *polaritons* in their most generally accepted terminology refer to mixed light–matter states in crystals. They do not necessarily imply excitons, but can also be formed, typically, by optical phonons. In this text we use the term "polariton" to mean *exciton-polaritons* only. The reader can find a recent starting point on phonon-polaritons in, e.g., the text of Stoyanov et al. (2002).

We begin the description of exciton-polaritons from equations proposed by Hopfield (1958), adapting his notations to our exposition.

The first equation reads:

$$\frac{\varepsilon_B}{c^2}\frac{\partial^2}{\partial t^2}\mathbf{E}(\mathbf{r},t) + \nabla \times \nabla \times \mathbf{E}(\mathbf{r},t) = -\frac{1}{c^2}\frac{\partial^2}{\partial t^2}\mathbf{P}(\mathbf{r},t), \tag{4.46}$$

which directly follows from the wave equation (2.17) and material relation eqn (2.11). ε_B is the normalised background dielectric constant[70] that does not contain the excitonic contribution. The link between the polarisation and the electric field is given by the second Hopfield equation:

$$\left[\frac{\partial^2}{\partial t^2} + 2\gamma\frac{\partial}{\partial t} + \omega_0^2 - \frac{\hbar\omega_0}{M_x}\nabla^2\right]\mathbf{P}(\mathbf{r},t) = \varepsilon_B\omega_p^2\mathbf{E}(\mathbf{r},t), \tag{4.47}$$

where ω_p is the so-called *polariton Rabi frequency*, ω_0 is the exciton transition energy that is dependent on the difference of energies between an electron and a hole composing the exciton, and on the exciton binding energy, and $M_x = m_e + m_h$ is the exciton translation mass. The second Hopfield equation is a direct consequence of the Lorentz dipole oscillator model. Formally, it can be derived from eqn 4.1.

Equation (4.47) is derived assuming a linear optical response of the system and considering each exciton as an harmonic oscillator having its eigenfrequency corresponding to the energy of the excitonic transition, with damping caused by exciton interaction with acoustic phonons. The polarisation created by excitons is taken to be proportional to the amplitude of the harmonic oscillator, which constitutes the so-called *dipole approximation*. A derivation in full details appears in the textbook by Haug and Koch (1990). A double Fourier transform of eqn (4.47) yields:

$$\mathbf{P}(\omega,\mathbf{k}) = \frac{\varepsilon_B\omega_p^2\mathbf{E}(\omega,\mathbf{k})}{\omega_0^2 - \omega^2 - 2i\omega\gamma + \hbar\omega_0 k^2/M_x}. \tag{4.48}$$

In the vicinity of the resonant frequency, one can express the normalised dielectric function from eqn (4.48) as:

$$\varepsilon(\omega,k) = \varepsilon_B + \frac{\varepsilon_B\omega_{LT}}{\omega_0 - \omega + \hbar k^2/(2M_x) - i\gamma}, \tag{4.49}$$

where $\omega_{LT} = \omega_p^2/(2\omega_0)$ is the so-called longitudinal-transverse splitting. For $\gamma = 0$ and $M_x \to \infty$, it is equal to the splitting between the frequencies at which the dielectric constant goes to infinity (ω_T) and to zero ($\omega_L = \omega_T + \omega_{LT}$). This splitting is a direct measure of the coupling strength between the exciton and light, and is proportional to the exciton *oscillator strength* f:

$$f = \frac{4\pi\varepsilon_0\sqrt{\varepsilon_B}}{\pi}\frac{m_0 c}{e^2}\omega_{LT}. \tag{4.50}$$

For the ground exciton state in GaAs, $\hbar\omega_{LT} = 0.08\,\text{meV}$, while in wide-bandgap materials (GaN, ZnO) it is an order of magnitude larger.

---

[70] ε_B is normalised so that it is equal to unity in vacuum.

The dependence of the tensor of the dielectric susceptibility of a crystal on the wavevector of light is referred to in crystal optics as *spatial dispersion*. Equation (4.49) describes the spatial dispersion induced by excitons in semiconductors.

The relation between frequency and wavevector of the transverse polariton modes is given by

$$k^2 = \varepsilon(\omega, k)\frac{\omega^2}{c^2},$$

(4.51)

which is a biquadratic equation with solutions

$$k_{1,2}^2 = -\frac{M_x}{\hbar}\left(\omega_0 - \omega - i\gamma - \frac{\hbar\varepsilon_B}{2M_x}\frac{\omega^2}{c^2}\right)$$

$$\pm\frac{M_x}{\hbar}\sqrt{\left(\omega_0 - \omega - i\gamma - \frac{\hbar\varepsilon_B}{2M_x}\frac{\omega^2}{c^2}\right)^2 + \frac{2\hbar}{M_x}\varepsilon_B\omega_{LT}\frac{\omega^2}{c^2}}.$$

(4.52)

For the longitudinal modes, the condition $\varepsilon(\omega, k) = 0$ yields

$$k_L^2 = \frac{2M_x}{\hbar}(\omega - \omega_0 - \omega_{LT} + i\gamma).$$

(4.53)

One can see that, at each frequency, two transverse and one longitudinal polariton modes with different wavevectors can propagate in the same direction. The appearance of additional light modes in crystals at the exciton resonance frequency as a result of spatial dispersion was theoretically predicted by Pekar (1957) and confirmed experimentally by Kiselev et al. (1973). Description of additional light modes in finite-size crystal slabs requires additional boundary conditions (ABC) that have to be imposed on the dielectric polarisation in spatially dispersive media. Pekar proposed the ABC with

$$P = 0$$

(4.54)

at the surface of the crystal. The condition comes from the physical argument that the excitons that are responsible for the appearance of the polarisation P exist only inside the crystal. Physically, the Pekar conditions assume that the exciton wavefunction is confined within the crystal slab, which acts on the exciton as a potential well with infinitely high barriers. The choice (4.54) of ABC is not the only possible one. In a number of studies the concept of a so-called "dead-layer" is used, assuming that the exciton centre of mass cannot approach the interface closer than some critical length of the order of the exciton Bohr radius. In general, Neumann-type conditions on the polarisation and its derivative may be formulated. Though Pekar ABC have proven to provide a good agreement between theoretical and experimental spectra of exciton-polaritons, the debate on the exact form of ABC still continues.

Here, we use the Pekar ABC to obtain the eigenfrequencies of exciton-polariton modes in a crystal slab of thickness L. Assuming zero in-plane wavevector of all light

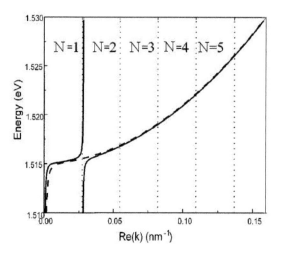

Fig. 4.12: Dispersion of the transverse (solid) and longitudinal (dashed) exciton-polariton modes in GaAs as calculated by Vladimirova et al. (1996). Vertical dotted lines show positions of quantum confined polariton modes in a 1148 Å thick film of GaAs.

modes (normal incidence geometry), one can readily obtain from eqn (4.54) the conditions on wavevectors k_j, $(j = 1, 2, L)$:

$$k_j = \frac{N\pi}{L}, \qquad N \in \mathbb{N}. \tag{4.55}$$

Figure 4.12 shows the dispersion of transverse (solid lines) and longitudinal (dashed line) polariton modes in GaAs calculated with eqns (4.52) and (4.53), respectively. The vertical dotted lines show those values of the wavevectors that satisfy the condition (4.55) for a given value of L. Crossing points of the dotted lines and the dispersion curves yield the discrete spectrum of eigenfrequencies of exciton-polaritons in the thin film. These frequencies correspond to resonances in reflection or transmission spectra. The splitting between neighbouring eigenfrequencies increases with decrease of the exciton mass and the thickness L. The fit of optical spectra of exciton-polaritons in thin films allows one to obtain the exciton mass with a good accuracy.

4.4.1 Surface polaritons

Surface polaritons result from exciton coupling with light modes having a component of wavevector in the plane of the surface, $k_x > \omega/c$, i.e., outside of the light cone (we assume that TM-polarised light propagates along the x-axis, in the plane, so that the magnetic field vector of the light-wave is parallel to the y-axis also in the plane). In this case, the light wave propagating along the surface of the crystal decays in vacuum. If we consider the right surface, the electric field vector of such a mode in vacuum behaves as

$$\mathbf{E}^+ = \mathbf{E}_0^+ \exp\left(ik_x x - \sqrt{k_x^2 - \frac{\omega^2}{c^2}}\, z \right), \tag{4.56}$$

where the z-axis is normal to the surface. The electric field vector in the crystal reads

$$\mathbf{E}^- = \mathbf{E}_0^- \exp\left(ik_x x + \sqrt{k_x^2 - \frac{\omega^2}{c^2}\varepsilon(\omega)}z \right), \tag{4.57}$$

where $\varepsilon(\omega)$ is the same as in eqn (4.49). Dependent on the sign and value of $\varepsilon(\omega)$, the electric field may decay or may not inside the crystal. From eqns (4.56) and (4.57), one can easily obtain the ratio of the x- and z-components of the field as

$$\frac{E_z^+}{E_x^+} = \frac{k_x}{i\sqrt{k_x^2 - \frac{\omega^2}{c^2}}}, \qquad \frac{E_z^-}{E_x^-} = -\frac{k_x}{i\sqrt{k_x^2 - \varepsilon\frac{\omega^2}{c^2}}}. \tag{4.58}$$

The dispersion of surface polariton modes can be obtained from the Maxwell boundary conditions, which require

$$E_x^- = E_x^+, \qquad \varepsilon(\omega)E_z^- = E_z^+. \tag{4.59}$$

The second condition comes from the continuity of the magnetic field at the surface. From (4.58) and (4.59) we obtain

$$\varepsilon(\omega)\sqrt{k_x^2 - \frac{\omega^2}{c^2}} = -\sqrt{k_x^2 - \varepsilon(\omega)\frac{\omega^2}{c^2}}, \tag{4.60}$$

thus

$$\omega = ck_x\sqrt{\frac{1 + \varepsilon(\omega)}{\varepsilon(\omega)}}, \qquad \varepsilon < 0. \tag{4.61}$$

To analyse the dispersion equation (4.61), let us consider the limit $\gamma \to 0$ and $M_x \to \infty$. $\varepsilon(\omega)$ is a real function schematically displayed in Fig. 4.13. Equation (4.61) yields the real exciton-polariton eigenfrequencies if $\varepsilon(\omega) < -1$. One can see that this condition can only be satisfied within the frequency range

$$\omega_0 \leq \omega < \omega_0 + \omega_{LT} - \delta, \tag{4.62}$$

where δ can be found from the condition

$$\varepsilon(\omega_0 + \omega_{LT} - \delta) = -1. \tag{4.63}$$

The dispersion of surface polaritons can now be easily understood: it starts at $\omega = \omega_0$, $k_x = \omega/c$ and goes to $\omega \to \omega_0 + \omega_{LT} - \delta$, $k_x \to \infty$ as shown on Fig. 4.14.

Exercise 4.3 $^{(**)}$ *A short pulse of light centred at the exciton resonance frequency is reflected from a semi-infinite semiconductor crystal. Calculate the time dependence of the intensity of reflected light neglecting the spatial dispersion of exciton-polaritons.*

Exercise 4.4 $^{(***)}$ *A short pulse of light centred at the exciton resonance frequency is transmitted through a semiconductor crystal slab of thickness d. Calculate the time dependence of the intensity of transmitted light neglecting the spatial dispersion of exciton polaritons.*

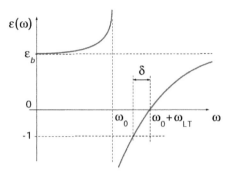

Fig. 4.13: Dielectric constant in the vicinity of the exciton resonance (scheme). The broadening and special dispersion are neglected.

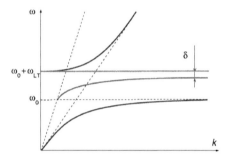

Fig. 4.14: Schematic dispersions of the surface polariton (solid), the transverse bulk polaritons (dashed) and the longitudinal bulk polariton (dotted). Spatial dispersion and broadening of the exciton resonance are neglected.

4.4.2   Exciton–photon coupling in quantum wells

In quantum confined structures the second Hopfield equation (4.47) cannot be directly used as the exciton is no more a free moving particle and its wavevector in the confinement direction is not defined. On the other hand, the theory of spatial dispersion in optical media can still be applied to describe the dielectric response of quantum structures containing excitons, where the exciton wavefunction plays the role of a correlation function. Indeed, once an exciton is created, the dielectric polarisation changes in all points where its wavefunction spreads.

The first theoretical description of exciton-polaritons in 2D structures was given by Agranovich and Dubovskii (1966). In this Section we will follow the so-called nonlocal dielectric response theory, developed in the beginning of the 1990s by Andreani et al. (1991) and Ivchenko (1992) to describe the optical response of excitons in QWs. We consider the simplest case of reflection or transmission of light through a QW in the vicinity of the exciton resonance at normal incidence. We neglect the difference

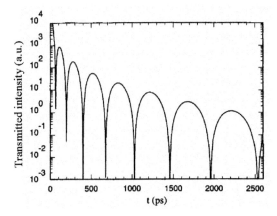

Fig. 4.15: Transmission intensity through a slab of Cu_2O of thickness $d = 0.91$mm as calculated by Panzarini and Andreani (1997).

between background dielectric constants of the well and barrier materials (which is usually small), and only take into account the exciton-induced resonant reflection.

The non-local dielectric response theory is based on the assumption that the exciton-induced dielectric polarisation can be written in the form

$$P_{\text{exc}}(z) = \int_{-\infty}^{\infty} \chi(z, z')E(z')dz' \,, \tag{4.64}$$

where

$$\chi(z, z') = \tilde{\chi}(\omega)\Phi(z)\Phi(z') \,, \tag{4.65}$$

with

$$\tilde{\chi}(\omega) = \frac{Q}{\omega_0 - \omega - i\gamma}, \qquad Q = \varepsilon_B \omega_{\text{LT}} \pi a_B^3 \,. \tag{4.66}$$

Here, $\Phi(z) = \Psi_{\text{exc}}(R = 0, \rho = 0, z_e = z_h = z)$ is the exciton wavefunction (4.28) taken with equal electron and hole coordinates, ω is the frequency of the incident light, γ is the *homogeneous* broadening of the exciton resonance, same as in the Hopfield equations, ω_{LT} and a_B are the longitudinal-transverse splitting and Bohr radius of exciton in the bulk material. Once the polarisation (4.64) is introduced, eqn (4.46) becomes an integro-differential equation and can be solved exactly using the Green's function method. In this method the solution of eqn (4.46) is represented in the form

$$E(z) = E_0 \exp(ikz) + k_0^2 \int P_{\text{exc}}(z')G(z - z')dz' \,, \tag{4.67}$$

where E_0 is the amplitude of the incident light, $k_0 = \omega/c$ and the Green's function G satisfies the equation

$$\left(\frac{\partial^2}{\partial z^2} + k^2\right) G(z) = -\delta(z), \qquad k = \sqrt{\varepsilon_B}k_0 \,. \tag{4.68}$$

Bearing in mind that $\int_{-\infty}^{\infty} f(z')\delta(z-z')dz = f(z')$, one easily checks that G is given by

$$G(z) = \frac{i\exp(ik|z|)}{2k}. \tag{4.69}$$

Equation (4.67) can be solved with respect to $E(z)$. In order to do it, multiply the left and right parts of eqn (4.67) by $\Phi(z)$ and integrate over z. This yields

$$\int E\Phi(z)dz = E_0 \int \Phi(z)\exp(ikz)dz+$$

$$+ k_0^2\tilde{\chi} \iint \Phi(z)\Phi(z')G(z-z')dzdz' \int E\Phi(z'')dz'', \tag{4.70}$$

which means that

$$\int E\Phi(z)dz = \frac{E_0 \int \Phi(z)\exp(ikz)dz}{1 - k_0^2\tilde{\chi} \iint \Phi(z)\Phi(z')G(z-z')dzdz'}. \tag{4.71}$$

We now return to eqn (4.67) and substitute eqn (4.71) into its right-hand side:

$$E = E_0\exp(ikz) + k_0^2\tilde{\chi} \int \Phi(z')G(z-z')dz' \int E(z'')\Phi(z'')dz''$$

$$= E_0 \left[e^{ikz} + \frac{k_0^2\tilde{\chi} \int \Phi(z')G(z-z')dz' \int e^{ikz}\Phi(z'')dz''}{1 - k_0^2\tilde{\chi} \iint \Phi(z)\Phi(z')G(z-z')dzdz'} \right]. \tag{4.72}$$

Using eqn (4.69), we finally obtain

$$E(z) = E_0 e^{ikz} + \frac{\frac{ik_0}{2\sqrt{\varepsilon_B}}QE_0 \int \Phi(z'')e^{ikz''}\,dz'' \int \Phi(z')e^{ik|z-z'|}dz'}{\omega_0 - \omega - i\gamma - Q\frac{ik_0}{2\sqrt{\varepsilon_B}} \iint e^{ik|z'-z''|}\Phi(z')\Phi(z'')dz'dz''}. \tag{4.73}$$

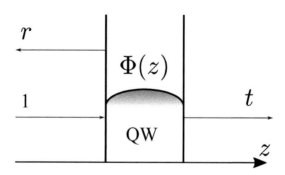

Fig. 4.16: Schema of multiple reflections of light within a crystal slab.

The amplitude reflection (r) and transmission (t) coefficients of the QW can then be obtained as

$$r = \frac{E(z) - E_0(z)e^{ikz}}{E_0(z)e^{-ikz}}\bigg|_{z\to\infty} \quad \text{and} \quad t = \frac{E(z)}{E_0 e^{ikz}}\bigg|_{z\to\infty}. \tag{4.74}$$

If we consider a ground exciton state in a QW, $\Phi(z)$ is an even function and the integrals on the right-hand side of eqn (4.73) can be easily simplified.

In the case $z \to \infty$,

$$\int \Phi(z')G(z - z')dz' = \frac{i}{2k}\int \Phi(z')e^{ik(z-z')}dz' = \frac{ie^{ikz}}{2k}\int \cos(kz')\Phi(z')dz', \tag{4.75}$$

and in the case $z \to -\infty$,

$$\int \Phi(z')G(z - z')dz' = \frac{i}{2k}\int \Phi(z')e^{-ik(z-z')}dz' = \frac{ie^{-ikz}}{2k}\int \cos(kz')\Phi(z')dz', \tag{4.76}$$

while

$$\iint G(z - z')\Phi(z)\Phi(z')dzdz' = \frac{i}{2k}\left[\int \Phi(z)\cos(kz)dz\right]^2 - \frac{1}{2k}\iint \Phi(z)\Phi(z')\sin(k|z - z'|)dzdz'. \tag{4.77}$$

This allows us to obtain the reflection and transmission coefficients of the QW in a simple and elegant form:

$$r(\omega) = \frac{i\Gamma_0}{\tilde{\omega}_0 - \omega - i(\Gamma_0 + \gamma)}, \tag{4.78a}$$

$$t(\omega) = 1 + r(\omega), \tag{4.78b}$$

where

$$\Gamma_0 = \frac{Qk_0}{2\sqrt{\varepsilon_B}}\left[\int \Phi(z)\cos(kz)dz\right]^2, \tag{4.79}$$

is an important characteristic further referred to as the *exciton radiative broadening* and

$$\tilde{\omega}_0 = \omega_0 + \frac{Qk_0}{2\sqrt{\varepsilon_B}}\iint \Phi(z)\Phi(z')\sin(k|z - z'|)dzdz' \tag{4.80}$$

is the renormalisation of the exciton resonance frequency due to the polariton effect.

The radiative broadening Γ_0 is connected to the exciton radiative lifetime τ by the relation

$$\tau = \frac{1}{2\Gamma_0}, \tag{4.81}$$

which follows from the time-dependence of the intensity of light reflected by a QW excited by an infinitely short pulse of light

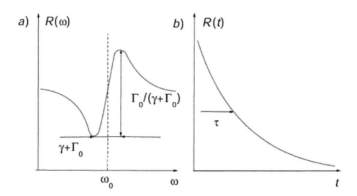

Fig. 4.17: Typical frequency- (a) and time-resolved (b) reflection spectra of a structure containing quantum wells (schema). In frequency-resolved reflectivity, the exciton resonance usually induces a characteristic modulation on the top of strong background reflectivity dependent on the refractive index of the barrier materials.

$$R(t) = \left| \frac{1}{2\pi} \int_{-\infty}^{\infty} r(\omega) e^{-i\omega t} d\omega \right|^2 = \Gamma_0^2 e^{-2\Gamma_0 t} . \tag{4.82}$$

A finite exciton radiative lifetime is a peculiarity of confined semiconductor systems. In an infinite bulk crystal, an exciton-polariton can freely propagate in any direction and its lifetime is limited only by non-radiative processes such as scattering with acoustic phonons. On the contrary, in a QW the exciton-polariton can disappear by giving its energy to a photon which escapes the QW plane (see Fig. 4.17). The polariton effect (sometimes referred to as the *retardation effect*) consists, in this case, in the possibility for the emitted photon to be reabsorbed once again by the same exciton. The chain of virtual emission-absorption processes leads to a finite value of τ and is also responsible for renormalisation of the exciton frequency (4.80). This renormalisation does not exceed a few µeV in realistic QWs, although it becomes more important in QDs. It does not play an essential role in microcavities and we shall neglect it hereafter. The radiative lifetime τ is about 10 ps in typical GaAs-based QWs. Although it is extremely hard to observe free-excitons in the photoluminescence as often photoluminescence is governed mainly by excitons localised at imperfections of a QW, a lifetime of 10 ± 4 ps for a free-exciton has been measured experimentally by Deveaud et al. (1991) in a record-quality (for that time) 100 Å-thick GaAs/AlGaAs QW (see Fig. 4.18).

4.4.3 Exciton–photon coupling in quantum wires and dots

Quantum wires and dots scatter light. In the vicinity of the exciton resonance, this scattering has a resonant character and polariton effects take place. If a wave of light is incident on an array of quantum wires or dots, the interference of waves scattered by different individual wires or dots results in enhanced signals in reflection and transmission directions. A regular array of identical wires or dots would diffract light in certain directions similarly to a crystal lattice diffracting X-rays. In realistic semiconductor structures, the inevitable potential disorder leads to random fluctuations of the

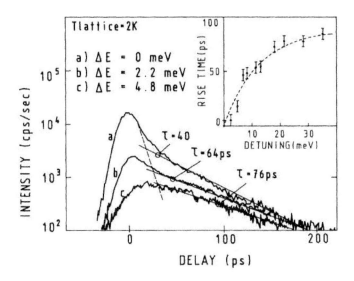

Fig. 4.18: Time-resolved photoluminescence spectra of a GaAs/AlGaAs QW measured by Deveaud et al. (1991) at different detunings between the excitation energy and the exciton resonance. These data allow an exciton radiative lifetime of about 10 ps to be extracted.

exciton resonance frequency from wire to wire and from dot to dot. This leads to the *inhomogeneous broadening* of exciton resonances in reflection or transmission spectra. The stronger the inhomogeneous broadening, the larger the ratio of scattered to reflected light intensity.

Polariton effects in systems of quantum wires and dots manifest themselves in finite exciton radiative lifetime, appearance of collective exciton-polariton states due to optical coupling of different wires or dots and appearance of a specific polarisation-dependent fine-structure of exciton resonances. All these phenomena can be conveniently described within the nonlocal dielectric response theory in a similar way to that done for quantum wells (Fig. 4.17).

The dielectric polarisation induced by an exciton in a plane array of N quantum wires or dots reads

$$\mathbf{P}_{\text{exc}} = \sum_{n=1}^{N} T_n \Phi_n(\mathbf{r} - \mathbf{R}_n) \int \mathbf{E}(\mathbf{r}') \Phi_n(\mathbf{r}' - \mathbf{R}_n) d\mathbf{r}' \qquad (4.83)$$

with

$$T_n = \frac{\varepsilon_B \omega_{\text{LT}} \pi a_B^3}{\omega_n - \omega - i\gamma}, \qquad (4.84)$$

where k_0 is the wavevector of the incident light in a vacuum, ω_{LT} is the longitudinal-transverse splitting in bulk, ε_B is the background dielectric constant, a_B is the exciton Bohr radius in bulk, γ is the homogeneous broadening and $\Phi_n(\mathbf{r})$ is the exciton wavefunction in the nth wire or dot taken with equal electron and hole coordinate $\mathbf{r}$. Solving the wave equation using the Green's function method, one represents the electric field as

$$\mathbf{E}(\omega, \mathbf{r}) = \mathbf{E}_0 \exp(i\mathbf{k} \cdot \mathbf{r})$$

$$+ k_0^2 \sum_{n=1}^{N} T_n \int \Phi_n(\mathbf{r}' - \mathbf{R}_n) G_0(\mathbf{r} - \mathbf{r}') d\mathbf{r}' \int \mathbf{E}(\mathbf{r}'') \Phi_n(\mathbf{r}'' - \mathbf{R}_n) d\mathbf{r}'', \quad (4.85)$$

where $k = k_0 \sqrt{\varepsilon_B}$ is the wavevector of the incident light in the medium. In the case of infinite quantum wires orientated along the y-axis

$$G_0(x, y, z) = \frac{i}{4} H_0^{(1)}(k_x x + k_z z), \quad (4.86)$$

where $H_0^{(1)}$ is the Hankel function. G_0 is a Green's function that satisfies the equation

$$-\left(\frac{\partial^2}{\partial x^2} + \frac{\partial^2}{\partial y^2} + k^2 \right) G_0 = \delta(x)\delta(z). \quad (4.87)$$

In the case of QDs, eqn (4.85) is the Green's function for a zero-dimensional system. For realistic systems, the "quantum dot" model is more relevant, as the excitons in quantum wires are inevitably localised in the y-direction due to potential fluctuations. This makes the wires similar to elongated dots. In the following, we therefore consider an array of QDs.

The integral equation (4.85) can be treated analytically, see, for instance, the discussion by Parascandolo and Savona (2005), but the calculation becomes heavy if the number of dots is large. A compact analytical expression for the electric field can be obtained if we assume that the wavefunctions of excitons in all QDs are identical and that $E(\omega, \mathbf{r} + \mathbf{R}_n) \approx e^{i\mathbf{r} \cdot \mathbf{R}_n} E(\omega, \mathbf{r})$. This would be true for a regular grating of identical QDs characterised by the exciton wavefunction $\Phi(\mathbf{r})$. In all other cases, this is a more or less accurate approximation depending on the degree of disorder in the system. Multiplying eqn (4.85) by $\Phi(\mathbf{r})$ and integrating over $\mathbf{r}$ we obtain in this case

$$\mathbf{E}(\omega, \mathbf{r}) = \mathbf{E}_0 \exp(i\mathbf{k} \cdot \mathbf{r})$$

$$+ k_0^2 \sum_{n=1}^{N} T_n e^{i\mathbf{k} \cdot \mathbf{R}_n} \int \Phi(\mathbf{r}' - \mathbf{R}_n) G_0(\mathbf{r} - \mathbf{r}') d\mathbf{r}' \frac{\int \mathbf{E}_0 e^{i\mathbf{k} \cdot \mathbf{r}} \Phi(\mathbf{r}) d\mathbf{r}}{1 - \sum_{n=1}^{N} \Theta_n}, \quad (4.88)$$

where $\Theta_n = k_0^2 T_n e^{i\mathbf{k} \cdot \mathbf{R}_n} \iint G_0(\mathbf{r} - \mathbf{r}') \Phi(\mathbf{r}' - \mathbf{R}_n) \Phi(\mathbf{r}) d\mathbf{r}' d\mathbf{r}$.

The Fourier transform of the electric field (4.88) yields its directional dependence, which can be represented in the form

$$\mathbf{E}_d(\omega, \mathbf{k}_s) = \mathbf{E}_0 \delta_{\mathbf{k}, \mathbf{k}_s} + \mathbf{E}_0 \sum_{m=1}^{N} \frac{i\Gamma_0^{QD}}{\omega_m - \omega - i\gamma} \frac{1}{1 - \sum_{n=1}^{N} \Theta_n} \exp(i\mathbf{k}_s \cdot \mathbf{R}_m), \quad (4.89)$$

where

$$\Gamma_0^{QD} = \frac{1}{6} \omega_{LT} k_0^3 a_B^3 \left[\int \cos(\mathbf{k} \cdot \mathbf{r}) \Phi(\mathbf{r}) d\mathbf{r} \right]^2, \quad (4.90)$$

and $\delta_{\mathbf{k}, \mathbf{k}_s} = 1$ if $\mathbf{k} = \mathbf{k}_s$, and is zero otherwise.

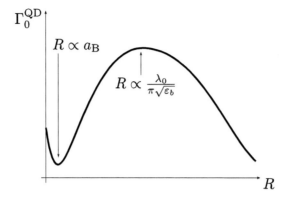

Fig. 4.19: Exciton radiative decay rate as a function of the radius of a spherical quantum dot R (schema).

The quantity Γ_0^{QD} is the exciton radiative decay rate in a single QD. This is an important characteristic of light–matter coupling in QDs. It changes non-monotonically with the dot size, as Fig. 4.19 shows. If the size of the dot R is much smaller than the wavelength of light at the exciton resonance frequency λ_0, one can neglect the cosine in the right part of eqn (4.90). In this case, the radiative damping rate is directly proportional to the volume occupied by the exciton wavefunction. This dependence has been obtained theoretically for the first time by Rashba and Gurgenishvili (1962) for impurity-bound excitons in semiconductors. The minimum of Γ_0^{QD} corresponds to the strongest exciton confinement at $R \approx a_{\text{B}}$. At very small R the exciton is less confined due to penetration of the electron and hole wavefunctions into the barriers. At larger R the exciton is confined as a whole particle inside the dot and the volume occupied by its wavefunction increases proportionally to the size of the dot. When the size of the dot approaches the wavelength of light, the cosine in the right part of eqn (4.90) can no longer be neglected, as was shown by Gil and Kavokin (2002). Γ_0^{QD} has its maximum about $R \approx \lambda_0/(\pi\sqrt{\varepsilon_{\text{B}}})$ and then decreases.

Equation (4.89) describes all kinds of coherent optical experiments (reflection, transmission and Rayleigh scattering) within the same semiclassical formalism, and takes properly into account the polariton effect. In the specular reflection direction, one can neglect the small portion of scattered light. Substituting summation by integration in eqn (4.89) we obtain the reflection and transmission coefficients of an array of QDs, r_{QD} and t_{QD}, as

$$r_{\text{QD}} = \frac{\beta}{1-\beta} \quad \text{and} \quad t_{\text{QD}} = \frac{1}{1-\beta}, \tag{4.91}$$

where

$$\beta = i\Gamma_r \int \frac{f(\nu)}{\nu - \omega - i\gamma} d\nu, \tag{4.92a}$$

$$\Gamma_r = \frac{k}{2d^2} \omega_{\text{LT}} \pi a_{\text{B}}^3 \int \left[\Phi(\mathbf{r}) \cos(\mathbf{k} \cdot \mathbf{r}) d\mathbf{r}\right]^2, \tag{4.92b}$$

Fig. 4.20: An electron microscopy image of a GaAs microcavity with GaAlAs/AlAs Bragg mirrors and the calculated profile of electric field of the cavity mode.

where d is the average distance between the dots, $f(\nu)$ is the distribution function of exciton resonance frequencies in the dots ($\int_{-\infty}^{\infty} f(\nu)d\nu = 1$). The ratio of reflected and average scattered intensities can be estimated from eqns. (4.89)–(4.90), assuming that the light waves scattered by QDs lying within a Γ_r vicinity interfere positively in the reflection direction:

$$\frac{I_r}{I_s} \approx N \left(\frac{\Gamma_r}{\Delta}\right)^2, \tag{4.93}$$

where N is given by the number of QDs within the spot of light that illuminates the sample. In real systems, $\Gamma_r \gg \Gamma_0^{\mathrm{QD}}$, which means that optically coupled QDs emit light much faster than single QDs. The radiative lifetime of an exciton in a single QD varies in large limits as a function of the QD size, typically in the range 10^{-10}–10×10^{-8} s.

Exercise 4.5 [**] *Find the reflection coefficient of a system of two identical quantum wells parallel to each other and separated by a distance d.*

4.4.4 *Dispersion of polaritons in planar microcavities*

4.4.4.1 *Bulk microcavities:* In Chapter 2, we considered the eigenmodes of planar microcavities in the absence of exciton–photon coupling. Imagine now that the cavity is made from a material having an excitonic transition at the frequency close to the eigenfrequency of the photonic mode of the cavity. Neglecting the spatial dispersion of exciton-polaritons in the cavity layer, one can solve the eigenvalue problem using the transfer matrix technique as we did in Chapter 2. Imposing the boundary condition of no light incident from left and right sides on the cavity, one can readily obtain the matrix equation for the eigenfrequencies:

$$\begin{pmatrix} \cos(kL_c) & \frac{i}{\sqrt{\varepsilon(\omega)}}\sin(kL_c) \\ i\sqrt{\varepsilon(\omega)}\sin(kL_c) & \cos(kL_c) \end{pmatrix} \begin{pmatrix} 1+r_B \\ \sqrt{\varepsilon(\omega)}(r_B-1) \end{pmatrix} = A\begin{pmatrix} 1+r_B \\ \sqrt{\varepsilon(\omega)}(r_B-1) \end{pmatrix}$$

(4.94)

where L_c is the cavity width and

$$\varepsilon(\omega) \approx \varepsilon_B + \frac{\varepsilon_B \omega_{LT}}{\omega_0 - \omega - i\gamma}, \qquad (4.95)$$

as follows from eqn (4.49) if the wavevector-dependent term in the denominator is neglected. Eliminating A we obtain

$$r_B e^{ikL_c} = \pm 1. \qquad (4.96)$$

Near the frequency $\bar{\omega}$, which is the centre of the stop-band of the mirrors, r_B can be approximated by

$$r_B = \sqrt{R}\exp\left[i\frac{n_c L_{DBR}}{c}(\omega - \bar{\omega})\right], \qquad (4.97)$$

where L_{DBR} is the effective length of the mirror (see Section 2.5). Assuming $\omega_0 = \bar{\omega}$, $\omega_{LT} \ll \omega_0$, $\Gamma \ll \omega_0$ and $1 - R \ll 1$ one can reduce eqn (4.96) to the two-coupled oscillator problem:

$$(\omega_0 - \omega - i\gamma)(\omega_c - \omega - i\gamma_c) = V^2, \qquad (4.98)$$

where $\omega_c - i\gamma_c$ is the complex eigenfrequency of the cavity mode in the absence of exciton–photon coupling and

$$V = \sqrt{\frac{2\omega_0 \omega_{LT} d}{L_{DBR} + d}}. \qquad (4.99)$$

Equation (4.98) has two complex solutions:

$$\omega_{1,2} = \frac{\omega_0 + \omega_c}{2} - \frac{i}{2}(\gamma + \gamma_c) \qquad (4.100a)$$

$$\pm \sqrt{\left(\frac{\omega_0 - \omega_c}{2}\right)^2 + V^2 - \left(\frac{\gamma - \gamma_c}{2}\right)^2 + \frac{i}{2}(\omega_0 - \omega_c)(\gamma_c - \gamma)}. \qquad (4.100b)$$

The parameter V has the sense of the coupling strength between the cavity photon mode and the exciton. If $\omega_0 = \omega_c$, the splitting between the two values is given by $\sqrt{4V^2 - (\gamma - \gamma_c)^2}$. If

$$V > \left|\frac{\gamma - \gamma_c}{2}\right|, \qquad (4.101)$$

the anticrossing takes place between the exciton and photon modes, which is characteristic of the *strong-coupling regime*. In this regime, two distinct exciton-polariton branches manifest themselves as two optical resonances in the reflection or transmission spectra. The splitting between these two resonances is referred to as the *vacuum-field*

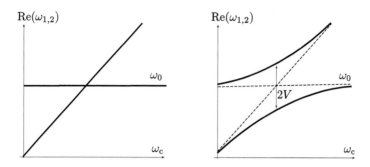

Fig. 4.21: Real parts of the eigenfrequencies of the exciton-polariton modes in the weak-coupling regime (left) and strong coupling regime (right).

Rabi splitting. It reaches 4–15 meV in current GaAs-based microcavities, up to 30 meV in CdTe-based microcavities and is found to be as large as 50 meV in GaN cavities. If

$$V < \left| \frac{\gamma - \gamma_c}{2} \right|,\qquad(4.102)$$

the weak-coupling regime holds, which is characterised by crossing of the exciton and photon modes, and an increase of the exciton decay rate at the resonance point (see Fig. 4.21). This regime is typically used in vertical-cavity surface-emitting lasers (VCSELs).

The spatial dispersion of exciton-polaritons leads to the appearance of additional resonances in reflection and transmission and additional eigenmodes of the microcavity. In order to take into account the spatial dispersion one should use the expression (4.49) for the dielectric function, which links it not only with the frequency, but also with the wavevector. The wavevector is no longer a unitary function of frequency, so that the simple transfer matrix method described in Chapter 2 and used to obtain eqn (4.98) fails. In order to calculate the optical spectra of semiconductor films containing exciton resonances we apply the generalised transfer matrix method. As was discussed in Section 4.4, the usual Maxwell boundary conditions are not sufficient in this case. We use Pekar's additional boundary conditions

$$P|_{z=\pm L_c/2} = 0,\qquad(4.103)$$

where P is the exciton-induced dielectric polarisation and $z = \pm L_c/2$ corresponds to the boundaries of the cavity. Equation (4.103) yields a series of exciton-polariton eigenmodes (see Fig. 4.12). Their energies depend on L_c and the exciton mass. The modes can be characterised by a quantum number N equal to the number of nodes in $F(R)$ for the corresponding state plus one. These are polariton modes uncoupled to the cavity photon mode. To introduce the coupling one should substitute the amplitudes of all existing modes of electromagnetic field into the Maxwell boundary conditions. For simplicity, we consider only the normal incidence geometry (in-plane wavevector equal to 0) and only the transverse light modes. In order to write down the transfer matrix taking into account spatial dispersion of exciton-polaritons, one should describe propagation of four waves with amplitudes $E_1^\pm$, $E_2^\pm$ at the beginning and the end of the layer, respectively, where "+" denotes a wave propagating in the positive direction and "−" denotes a wave propagating in the negative direction.

The matrix Q connecting $E_1^\pm$ and $E_2^\pm$ so that

$$\begin{pmatrix} E_2^+ \\ E_2^- \end{pmatrix} = Q \begin{pmatrix} E_1^+ \\ E_1^- \end{pmatrix},$$ (4.104)

can be written as

$$Q = -\frac{\varepsilon_1}{\varepsilon_2} \begin{pmatrix} \lambda_2^+ & \lambda_2 \\ \lambda_2 & \lambda_2^+ \end{pmatrix}^{-1} \begin{pmatrix} \lambda_1^+ & \lambda_1 \\ \lambda_1 & \lambda_1^+ \end{pmatrix},$$ (4.105)

with $\varepsilon_j = c^2 k_j^2/\omega^2 - \varepsilon_B$ and

$$\lambda_j = \exp\left(ik_j \frac{d}{2}\right), \qquad \lambda_j^+ = \exp\left(-ik_j \frac{d}{2}\right),$$ (4.106)

k_j, $(j = 1, 2)$ are wavevectors of the transverse polariton modes given by eqn (4.52). The transfer matrix across the cavity layer, i.e., the matrix that connects the in-plane components of the electric and magnetic fields at the boundaries, can be written as:

$$T = \left[\begin{pmatrix} \lambda_1^+ & \lambda_1 \\ n_1\lambda_1^+ & -n_1\lambda_1 \end{pmatrix} + \begin{pmatrix} \lambda_2^+ & \lambda_2 \\ n_2\lambda_2^+ & -n_2\lambda_2 \end{pmatrix} Q \right]$$
$$\times \left[\begin{pmatrix} \lambda_1 & \lambda_1^+ \\ n_1\lambda_1 & -n_1\lambda_1^+ \end{pmatrix} + \begin{pmatrix} \lambda_2 & \lambda_2^+ \\ n_2\lambda_2 & -n_2\lambda_2^+ \end{pmatrix} Q \right]^{-1},$$ (4.107)

where $n_j = ck_j/\omega$. Once its elements are known, the eigenmodes of the cavity can be found using the standard procedure described in Section 2.5 from the equation:

$$T \begin{pmatrix} 1 + r_B \\ \sqrt{\varepsilon(\omega)}(r_B - 1) \end{pmatrix} = A \begin{pmatrix} 1 + r_B \\ \sqrt{\varepsilon(\omega)}(1 - r_B) \end{pmatrix}.$$ (4.108)

Generalisation of this equation into the oblique incidence case requires substantial modifications of the transfer matrix (4.107). In the case of TM-polarised light, the longitudinal polariton modes can be excited and their amplitudes should also be substituted into the Maxwell boundary conditions. We address the interested reader to the paper by Vladimirova et al. (1996) and show here the dispersion curves of exciton-polaritons in a model bulk GaAs microcavity calculated in that study (Fig. 4.22). Remarkably, only the modes with even N are coupled to light: this is because the exciton wavefunction should be of the same parity as the cavity mode to be coupled.

One can see that the light mode of the cavity goes through different polariton resonances in the layer of GaAs forming a series of anticrossings. The splitting (Rabi splitting) at the lowest of them ($N = 2$) is given with a good accuracy by $2V$, where V is defined by eqn (4.99).

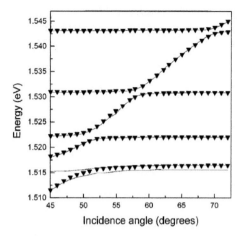

Fig. 4.22: Calculated exciton-polariton eigenenergies as a function of the incidence angle of light in TE- (triangles) and TM- (solid line) polarisations, from Vladimirova et al. (1996).

The Rabi splitting for higher exciton states (Ω_N) is related to the splitting for the lowest state approximately as the ratio of squared overlap integrals of the excitonic polarisation and electric field of the light mode

$$\frac{\Omega_N}{\Omega_2} = \left(\frac{P_N}{P_2}\right)^2, \quad N = 2, 4, 6 \ldots, \tag{4.109}$$

where

$$P_N \approx \int_{-\pi/2}^{\pi/2} \sin(Nx)\sin x\,dx = \frac{2}{N^2 - 1}. \tag{4.110}$$

This result easily follows from the quantum model as the matrix element of light-exciton coupling is proportional to P_N. In real bulk microcavities, the absorption of free e–h pairs above E_g produces a large imaginary component to ε_B and washes out the higher anticrossings.

4.4.4.2 *Microcavities containing quantum wells:* Microcavities containing quantum wells (QWs) are the one most commonly used in practice. They allow for observation of the strong coupling of a single exciton resonance and the cavity mode. The use of multiple QWs also helps to enhance the Rabi splitting and make the cavity polaritons more stable. In order to obtain the dispersion equation of the polaritons in structures containing QWs it is convenient to use the transfer matrices written in the basis of amplitudes of electromagnetic waves propagating in positive and negative directions along the axis of the cavity.

Let us represent the electric field at the point z of the structure as

$$E(z) = E^+(z) + E^-(z), \tag{4.111}$$

where $E^{\pm}(z)$ is the complex amplitude of a light wave propagating in the positive/negative direction. One can define the transfer matrix M_a by its property

$$M_a \begin{pmatrix} E^+(0) \\ E^-(0) \end{pmatrix} = \begin{pmatrix} E^+(a) \\ E^-(a) \end{pmatrix}.$$ (4.112)

Consider a few particular cases. If the refractive index n is constant across the layer a, the transfer matrix has the simple form:

$$M_a = \begin{pmatrix} e^{ika} & 0 \\ 0 & e^{-ika} \end{pmatrix}.$$ (4.113)

The transfer matrix across the interface between a medium with refractive index n_1 and a medium with refractive index n_2 is

$$M_a = \frac{1}{2n_2} \begin{pmatrix} n_1 + n_2 & n_2 - n_1 \\ n_2 - n_1 & n_1 + n_2 \end{pmatrix}.$$ (4.114)

It can be obtained using the condition (4.112) applied to the light waves incident from the left and right side of the interface, keeping in mind the well-known expressions for the reflection and transmission coefficients of interfaces

$$r = \frac{n_1 - n_2}{n_1 + n_2}, \qquad t = \frac{2n_1}{n_1 + n_2}.$$ (4.115)

A transfer matrix across a structure containing m layers has the form

$$M = \prod_{j=1}^{2m+1} M_{2m+2-j},$$ (4.116)

where $j = 1, \ldots, 2m + 1$, labels all the layers and interfaces of the structure from its left to right side. The amplitude reflection and transmission coefficients (r_s and t_s) of a structure containing m layers, and sandwiched between two semi-infinite media with refractive indices n_{left}, n_{right} before and after the structure, respectively, can be found from the relation

$$M \begin{pmatrix} 1 \\ r_s \end{pmatrix} = \begin{pmatrix} t_s \\ 0 \end{pmatrix},$$ (4.117)

as

$$r_s = \frac{m_{21}}{m_{11}} \quad \text{and} \quad t_s = \frac{1}{m_{11}}.$$ (4.118)

Note that here the refractive indices n_{left}, n_{right} do not appear explicitly as they are contained in the transfer matrices across both surfaces of the structure.

If the reflection and transmission coefficients for light incident from the right- and left-hand sides of the layer are the same and $n_{\text{left}} = n_{\text{right}} = n$, Maxwell's boundary conditions for light incident from the left and right sides of the structure yield, in addition to eqn (4.117), the equation

$$M \begin{pmatrix} 0 \\ t_s \end{pmatrix} = \begin{pmatrix} r_s \\ 1 \end{pmatrix}. \tag{4.119}$$

In this case, the transfer matrix across a symmetric object (a QW embedded in the cavity, for example) can be written as

$$M = \frac{1}{t_s} \begin{pmatrix} t_s^2 - r_s^2 & r_s \\ -r_s & 1 \end{pmatrix}. \tag{4.120}$$

Consider a symmetric microcavity with a single QW embedded in the centre. In the basis of amplitudes of light waves propagating in positive and negative directions along the z-axis, the transfer matrix across the QW has the form

$$T_{\text{QW}} = \frac{1}{t} \begin{pmatrix} t^2 - r^2 & r \\ -r & 1 \end{pmatrix}, \tag{4.121}$$

where r and t are the angle- and polarisation-dependent amplitude reflection and transmission coefficients of the QW derived previously. The transfer matrix across the cavity from one Bragg mirror to the other one is the product

$$T_c = \begin{pmatrix} e^{ikL_c/2} & 0 \\ 0 & e^{-ikL_c/2} \end{pmatrix} \frac{1}{t} \begin{pmatrix} t^2 - r^2 & r \\ -r & 1 \end{pmatrix} \begin{pmatrix} e^{ikL_c/2} & 0 \\ 0 & e^{-ikL_c/2} \end{pmatrix}, \tag{4.122}$$

where L_c is the cavity width. The matrix elements read

$$T_{11}^c = \frac{t^2 - r^2}{t} e^{ikL_c}, \qquad\qquad T_{12}^c = \frac{r}{t}, \tag{4.123a}$$

$$T_{21}^c = -\frac{r}{t}, \qquad\qquad T_{22}^c = \frac{1}{t} e^{-ikL_c}. \tag{4.123b}$$

To find the eigenfrequencies of the exciton-polariton modes of the microcavity, one should search for non-trivial solutions of Maxwell's equations under the requirement of no light incident on the cavity from outside. This yields

$$T_c \begin{pmatrix} r_{\text{B}} \\ 1 \end{pmatrix} = A \begin{pmatrix} 1 \\ r_{\text{B}} \end{pmatrix}, \tag{4.124}$$

where r_{B} is the angle-dependent reflection coefficient of the Bragg mirrors for light incident from inside the cavity, introduced in Section 2.6.

Eliminating the coefficient A from eqn (4.124), one obtains the equation for polariton eigenmodes as

$$\frac{T_{21}^c r_B + T_{22}^c r_B}{T_{12}^c r_B + T_{11}^c r_B} = r_B \,. \tag{4.125}$$

This is already a dispersion equation because the coefficients of the transfer matrix and r_B are dependent on the in-plane wavevector of light. Substituting the coefficients (4.123a) into eqn (4.125), one can represent the dispersion equation in the form

$$\left(r_B(2r+1)e^{ikL_c} - 1\right)\left(r_B e^{ikL_c} + 1\right) = 0 \,. \tag{4.126}$$

Here, we have used the relation $t - r = 1$, cf. eqn (4.78b). Solutions of eqn (4.126), coming from zeros of the second bracket on the left-hand side, coincide with pure *odd* optical modes of the cavity. These modes have a node at the centre of the cavity where the QW is situated. Therefore, they are not coupled with the ground exciton state having an *even* wavefunction. The first bracket on the left-hand side of eqn (4.126) contains the reflection coefficient of the QW, which is dependent on excitonic parameters. The zeros of this bracket describe the eigenstates of exciton-polaritons resulting from coupling of even optical modes with the exciton ground state. From now on we shall consider only these states, neglecting excited exciton states that may be coupled to odd cavity modes.

For the even modes and normal incidence, if we take

$$r_B = \bar{r}\exp(i\alpha(\omega - \omega_c)) \approx \bar{r}(1 + iL_{DBR}\frac{n_c}{n}(\omega - \omega_c)) \,, \tag{4.127}$$

where $\bar{r}$ is close to 1 (see Section 2.5) and assume $e^{ikL_c} \approx 1 + i(\omega - \omega_c)n_c L_c/c$, we obtain, using the explicit form for the reflection coefficient r,

$$\bar{r}(1+i(L_{DBR}+L_c)\frac{n_c}{c}(\omega-\omega_c))(\omega_0-\omega-i(\gamma-\Gamma_0)) = \omega_0 - \omega - i(\gamma+\Gamma_0)\,, \tag{4.128}$$

which finally yields, after trivial transformations,

$$(\tilde{\omega}_0 - \omega - i\gamma)(\omega_c - \omega - i\gamma_c) = V^2 \,, \tag{4.129}$$

where

$$\gamma_c = \frac{1 - \bar{r}}{\bar{r}\frac{n_c}{c}(L_{DBR} + L_c)} \,, \tag{4.130}$$

$$V^2 = \frac{1 + \bar{r}}{\bar{r}} \frac{\Gamma_0 c}{n_c(L_{DBR} + L_c)} \,. \tag{4.131}$$

Here, quadratic terms in $(\omega_c - \omega)$ have been omitted. In all further calculations we assume for simplicity $\tilde{\omega}_0 = \omega_0$. Equation (4.129) is an equation for eigenstates of a system of two coupled harmonic oscillators, namely, the exciton resonance and the cavity mode. In this form eqn (4.129) was published for the first time by Savona et al. (1995) while its general form (4.125) was later obtained by Kavokin and Kaliteevski (1995). Its solutions have the form (4.100). The weak-to-strong coupling threshold is

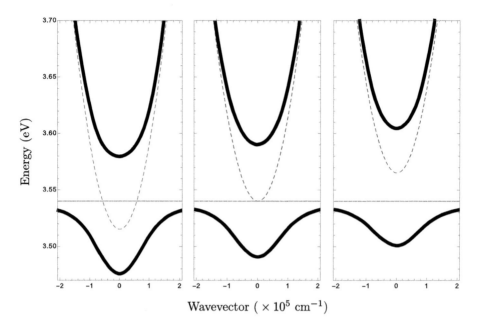

Fig. 4.23: Energies of exciton-polaritons at (a) negative, (b) zero and (c) positive detuning between the bare photon and exciton modes. Solid lines show the in-plane dispersion of exciton-polariton modes. Dashed lines show the dispersion of the uncoupled exciton and photon modes. The calculations have been made with parameters typical of a GaN microcavity, from Kavokin and Gil (1998), with a photon mass of $0.5 \times 10^{-4} m_0$ and a Rabi splitting of 50 meV.

defined in the same way as for the bulk cavities (cf. eqns. (4.101) and (4.102)). Note that all the above theory neglects the effect of disorder on the exciton resonance. Taking into account the inevitable inhomogeneous broadening of the exciton resonance and Rayleigh scattering of exciton-polaritons, one should also modify the criterion for the weak-to-strong coupling threshold.

The detuning of bare photon and exciton modes in a microcavity is an important parameter that strongly affects the shape of polariton dispersion curves in the strong coupling regime, as Fig. 4.23 shows.

4.4.4.3   *Oblique incidence case:*   Equation (4.129) can be easily generalised for the oblique incidence case by introduction of the dependence of the cavity and exciton eigenmode frequencies ω_c and ω_0 on the in-plane wavevector k_{xy}:

$$\omega_c = \frac{\hbar k_{xy}^2}{2m_{\text{ph}}}, \qquad \omega_0 = \frac{\hbar k_{xy}^2}{2M_{\text{exc}}}, \qquad (4.132)$$

where M_{exc} is the sum of the electron and hole effective masses in the QW plane and m_{ph} is the photon effective mass. In an ideal λ-microcavity, the normal-to-plane component of the wavevector of the eigenmode is given by $k_z = 2\pi/L_c$. The energy of the mode is

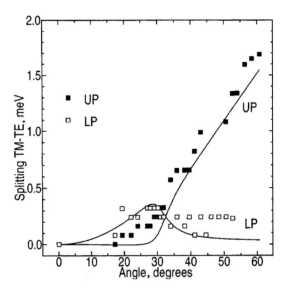

Fig. 4.24: Theoretical (lines) and experimental (squares) values of longitudinal-transverse splitting of the upper (UP) and lower (LP) polariton branches in a GaAs-based microcavity with embedded QWs, by Panzarini et al. (1999a).

$$\omega_c = \frac{c}{n_c}\sqrt{k_{xy}^2 + k_z^2} \approx \frac{c}{n_c}k_z\left(1 + \frac{k_{xy}^2}{2k_z^2}\right) = \frac{2\pi c}{n_c L_c} + \frac{\hbar k_{xy}^2}{2m_{\mathrm{ph}}}, \qquad (4.133)$$

and thus $m_{\mathrm{ph}} = hn_c/(cL_c)$. This mass is extremely light in comparison to the exciton mass as it usually amounts to 10^{-5}–$10^{-4}m_0$, where m_0 is the free-exciton mass. Note also that the in-plane wavevector k_{xy} is related to the angle of incidence of light illuminating the structure, φ, by the relation

$$k_{xy} = \frac{\omega}{c}\sin\varphi. \qquad (4.134)$$

By measuring the angle dependence of the resonances in the reflection or transmission spectra of microcavities, one can restore the true dispersion curves of exciton-polaritons.

The coupling constant V is renormalised in the case of oblique incidence and it becomes polarisation dependent. In TE-polarisation, $\Gamma_0^s = \Gamma_0/\cos\varphi_c$, where φ_c is the propagation angle within the cavity. In TM-polarisation, $\Gamma_0^p = \Gamma_0\cos\varphi_c$. The effective length of the Bragg mirrors, L_{DBR}, is also angle and polarisation dependent: it decreases with angle in TE-polarisation, but increases in TM-polarisation. Also, the coefficient $\bar{r}$ slightly depends on the angle (see eqns. (2.145)–(2.147)). All these factors make the coupling constant increase with angle in TE-polarisation, while in TM-polarisation the opposite tendency occurs. This is why the strong coupling regime can be lost in TM-polarisation at some critical angle. Note also that in TM- and TE-polarisations the eigenfrequencies of pure cavity modes $\omega_c^{\mathrm{TE,TM}}$ are split (see eqn (2.148)). Figure 4.24

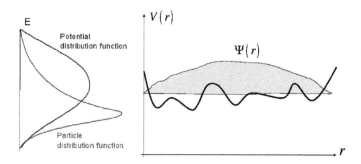

Fig. 4.25: Due to their finite de Broglie wavelength the quantum particles see the averaged potential. The averaging is done on the scale of the particle's wavefunction. This is why the distribution function of quantum particles localised within some fluctuation potential is usually narrower than the potential distribution function.

by Panzarini et al. (1999b) shows the longitudinal-transverse splitting of cavity polariton modes calculated and experimentally measured in a GaAs microcavity containing InGaAs QWs.

The longitudinal-transverse splitting becomes essential at large in-plane wavevectors and it has an impact on the spin-relaxation of exciton-polaritons, as we shall discuss in detail in Chapter 9. The splitting is dependent on two main factors: the TE–TM splitting of the bare cavity mode and the detuning between cavity and exciton modes. Therefore, it can be efficiently controlled and tuned within a wide range by changing the cavity width.

4.4.5 *Motional narrowing of cavity polaritons*

Motional narrowing is the narrowing of a distribution function of a quantum particle propagating in a disordered medium due to averaging of the disorder potential over the size of the wavefunction of a particle. In other words, a quantum particle, which is never localised at a given point in space but always occupies some nonzero volume, has a potential energy that is the average of the potential within this volume. This is why, in a random fluctuation potential, the energy distribution function of a particle is always narrower than the potential distribution function (see Fig. 4.25).

Motional narrowing of exciton-polaritons in microcavities was the subject of scientific polemics at the end of the 1990s. The debate was initiated by an experimental finding of the Sheffield University group reported by Whittaker et al. (1996). Measuring the sum of full-widths at half-minimum (FWHM) for two exciton-polariton resonances in reflection spectra of microcavities as a function of incidence angle, the experimentalists found a minimum of this function at the anticrossing of exciton and cavity modes (see Fig. 4.26). This result contradicts what one could expect from a simple model of two coupled oscillators. Actually, if the dispersion of microcavity polaritons is given by eqn (4.129) then the sum of the imaginary parts of the two solutions of this equation is always $-(\gamma + \gamma_c)$, independently of the detuning $\delta = \omega_0 - \omega_c$. This follows from a more general property of any system of coupled harmonic oscillators to keep constant the sum of eigenfrequencies, independently of the coupling strength.

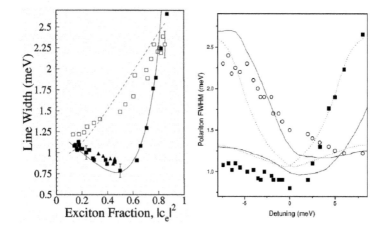

Fig. 4.26: (Left) Linewidth of lower (solid) and upper (open) polaritons in an InGaAs microcavity as a function of exciton fraction (or detuning), demonstrating linewidth narrowing on resonance, reported by Whittaker et al. (1996). (Right) Full-width at half-minimum of two polariton resonances in reflectivity spectra of a GaAs-based microcavity with an embedded QW, from Whittaker et al. (1996), as a function of the detuning $\delta = \omega_0 - \omega_c$ (circles correspond to the upper branch, squares correspond to the lower branch) in comparison with a theoretical calculation accounting for (solid lines) or neglecting (dashed lines) asymmetry in the exciton frequency distribution from Kavokin (1997).

Clearly, eqn (4.129) is no longer valid if, instead of a single free-exciton transition, one has an infinite number of resonances distributed in energy. This is what happens in realistic QWs, where in-plane potential fluctuations caused by the QW width and alloy fluctuations induce the so-called *inhomogeneous broadening* of an exciton resonance. An idea has been proposed that exciton-polaritons, having a smaller effective mass than bare excitons, are less sensitive to the disorder potential, which is a manifestation of their motional narrowing. Thus, the inhomogeneous broadening of exciton-polariton modes is less than that of a pure exciton state, which is a consequence of the polariton motional narrowing effect. At the anticrossing point this effect is especially strong, since at this point both upper and lower-polaritons are half-excitons half-photons.

Further analyses by Whittaker et al. (1996) and Ell et al. (1998) have shown, however, that experimentally observed narrowing of polariton lines at the anticrossing point is, indeed, caused by exciton inhomogeneous broadening, but not by the motional narrowing effect. On the other hand, the motional narrowing may manifest itself in resonant Rayleigh scattering or even photoluminescence.

In order to understand the inhomogeneous broadening effect on the widths of polariton resonances in microcavities, let us first consider its influence on the optical spectra of QWs.

We suppose that the in-plane wavevector of any exciton interacting with the incident light is the same as the in-plane wavevector of light $\mathbf{q}$, while the frequency of exciton resonance ω_0 is distributed with some function. This is a particular case of a microscopic model considering all exciton states as quantum-dot like. It is well adapted for the description of reflection or transmission, i.e., experiments that conserve the in-plane component of the wavevector of light. Note that most scattered light does not contribute to reflection and transmission spectra, while a small part of the scattered light can re-obtain the initial value of q after a second, third, etc., scattering act. The main reason why motional narrowing has almost no influence on reflection spectra is that it is an effect that originates from the finite in-plane size of the exciton-polariton wavefunction, or, in other words, implies scattering of exciton-polaritons in the plane of the structure. The impact of scattering is, however, negligibly small in reflection and transmission experimental geometries.

As in Section 4.4.2, we shall operate with the dielectric susceptibility in eqn (4.65), while also taking into account that the exciton resonance frequency is distributed with some function $f(\omega - \omega_0)$. We assume that in this case the dielectric susceptibility of a QW can be written in the form

$$\tilde{\chi}(\omega) = \int \chi(\omega - \nu) f(\nu - \omega_0)\, d\nu. \tag{4.135}$$

If f is a Gaussian function, this integral can be computed analytically,

$$\tilde{\chi}(\omega) = \frac{1}{\sqrt{\pi}\Delta} \int \chi(\omega - \nu) \exp\left[-\left(\frac{\nu - \omega_0}{\Delta}\right)^2\right] d\nu = \frac{i\sqrt{\pi}\Theta}{\Delta} \exp(-z^2)\mathrm{erfc}(-iz), \tag{4.136}$$

with

$$\chi(\omega - \omega_0) = \frac{\varepsilon_\infty \omega_{LT}\pi a_B^3 \omega_0^2/c^2}{\omega_0 - \omega - i\gamma}, \quad \Theta = \varepsilon_\infty \omega_{LT}\pi \omega_0^2 a_B^3/c^2 \tag{4.137a}$$

$$\text{and} \quad z = \frac{\omega - \omega_0 + i\gamma}{\Delta} \tag{4.137b}$$

where erfc is the complementary error function, ε_∞ is the high frequency limit of the normalised dielectric constant and Δ is a width parameter of the Gaussian distribution that describes exciton inhomogeneous broadening. We assume $\Delta, \gamma > 0$ and consider a normal incidence case for simplicity. Substituting the susceptibility (4.135) into eqn (4.64) and carrying out the same transformations as in Section 4.4.2, we obtain finally the amplitude reflection and transmission coefficients of a QW in the form

$$r = \frac{i\alpha\tilde{\chi}}{1 - i\alpha\tilde{\chi}}, \quad t = 1 + r, \tag{4.138}$$

where $\alpha = \Gamma_0/\Theta$ and Γ_0 is the radiative damping rate of the exciton in the case of no inhomogeneous broadening. This yields, using eqn (4.136),

$$r = -\frac{\sqrt{\pi}\Gamma_0 \exp(-z^2)\mathrm{erfc}(-iz)}{\Delta + \sqrt{\pi}(\Gamma_0 + i(\tilde{\omega}_0 - \omega_0))\exp(-z^2)\mathrm{erfc}(-iz)}, \quad t = 1 + r. \tag{4.139}$$

We shall neglect renormalisation of the exciton resonance frequency due to the polariton effect, since it is much less than Γ_0, and also assume $\tilde{\omega}_0 - \omega_0 \approx 0$. In the limit of small inhomogeneous broadening, the complementary error function becomes

$$\lim_{|z| \to \infty} \exp(-z^2) \operatorname{erfc}(-iz) = \frac{i}{\sqrt{\pi} z} \,, \tag{4.140}$$

which allows one to reduce eqn (4.139) to the "homogeneous" formula (4.78) of Andreani et al. (1998).

In realistic narrow QWs, the exciton resonance frequency may have a more complex non-Gaussian distribution. Quite often it is asymmetric due to the so-called excitonic motional narrowing effect (to be distinguished from the motional narrowing of exciton-polaritons). This effect comes from the blueshift of the lower-energy wing of the excitonic distribution due to the lateral quantum confinement of localised excitons.

Exciton inhomogeneous broadening of any shape necessarily modifies the equation for exciton-polariton eigenmodes in a microcavity. Equation (4.129), obtained for the model of two coupled oscillators, is no longer valid and one should use the general formula (4.126) instead. For the coupled exciton and photon modes, it reduces to

$$r_{\mathrm{B}}(2r + 1)e^{ikL_{\mathrm{c}}} = 1 \,. \tag{4.141}$$

Using the same representation for the reflection coefficient of the Bragg mirror as in eqn (4.127) and eqn (4.138) for r with $\tilde{\chi}$ given by eqn (4.136), we obtain after transformations analogous to those in section 4.4.2:

$$\omega_{\mathrm{c}} - \omega - i\gamma_{\mathrm{c}} = V^2 \int_{-\infty}^{\infty} \frac{f(\nu - \omega_0)}{\nu - \omega - i\gamma} \, d\nu \,. \tag{4.142}$$

In the limit of a small inhomogeneous broadening, it reduces to eqn (4.129). As for eqn (4.142), it has between zero and two complex solutions depending on the shape of the distribution function $f(\nu - \omega_0)$. In the general case, the sum of the imaginary parts of its eigenfrequencies varies as a function of detuning, which is not the case for the pure two coupled oscillator problem. Figure 4.27 shows the dependencies of the FWHM of two polariton resonances on detuning $\delta = \omega_0 - \omega_{\mathrm{c}}$, in calculated reflection spectra of a GaAs-based microcavity with an embedded QW, in comparison with experimental data. One can see that the broadenings of the two polariton modes coincide near the zero-detuning point. At this point, both in calculation and experiment, the sum of the two FWHM has a pronounced minimum.

This minimum is a specific feature of an inhomogeneously broadened exciton state coupled to the cavity mode. It can be interpreted in the following way. The coupling to light has a different strength for excitons from the centre and from the tails of an inhomogeneous distribution. As the density of states of excitons has a maximum at $\omega = \omega_0$, these excitons have the highest radiative recombination rate and the strongest coupling to light. Now, the strong-coupling regime holds only for excitons situated in the vicinity of ω_0, while the tails remain in the weak-coupling regime. The central part of the excitonic distribution is, of course, less broadened than the entire distribution. Therefore, the two polariton modes that arise due to its coupling with the cavity photon are also narrower than one would expect for the case of all excitons equally coupled to light. At zero detuning, both polariton modes are far enough from the bare exciton energy, so that the tails of the exciton resonance give no contribution to the FWHM of polariton

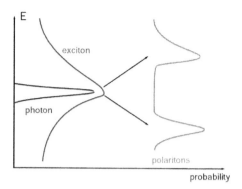

Fig. 4.27: Schematic energy distributions functions of bare excitons, bare photons and exciton-polaritons in microcavities. The sum of the broadenings of two polariton modes at the anticrossing point is less than the sum of exciton and photon mode broadenings because of the preferential photon coupling with the exciton states close to the centre of the inhomogeneous distribution of excitons.

resonances. On the contrary, for strong negative or positive detunings, one of two polariton states almost coincides in energy with a bare exciton state, so that the lineshape of the corresponding spectral resonance is necessarily affected by the tails of the excitonic distribution.

Note that this interpretation does not involve any motional narrowing. On the other hand, specific effects of motional narrowing play a role in resonant Rayleigh scattering of the cavity polaritons and may also provide narrowing of the photoluminescence lines. Only wavevector-conserving optical spectroscopies, such as reflection and transmission, are not sensitive to the motional narrowing.

Another remark concerns the asymmetric behaviour of the broadenings of the two polariton peaks in Fig. 4.26. This comes from the asymmetry of the excitonic distribution. It is a manifestation of the *exciton* motional narrowing that we have described above. Actually, the lower-polariton branch is the result of mixing between the photon mode and the lower part of the excitonic distribution, which is sharper than the upper part. This is why the lower branch has a narrower linewidth at the anticrossing.

4.4.6 *Microcavities with quantum wires or dots*

Planar microcavities with embedded quantum wires or quantum dots can exhibit the weak or strong exciton–photon coupling regime similarly to bulk cavities or microcavities with embedded quantum wells. An essential difference comes from the enhanced resonant scattering of light by excitons in these structures. If the scattering is strong, exciton states in quantum dots or wires are coupled to the whole ensemble of the cavity photon modes with different in-plane wavevectors. In this case, the polariton eigenstate can hardly be represented as a linear combination of plane waves and calculation of the spectrum of exciton-polaritons becomes a non-trivial task. However, for the cavities with embedded arrays of quantum wires or dots, scattering of light can be neglected if the spacing between neighbouring wires (dots) is less than the wavelength of light and the variation of the exciton resonance frequency from wire to wire (from dot to dot)

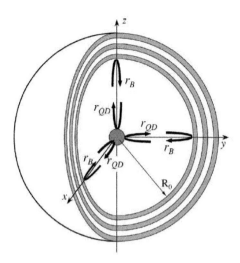

Fig. 4.28: Schematic diagram by Kaliteevski et al. (2001) of a spherical microcavity with an embedded quantum dot. The central core of radius r_{QD} is surrounded by a spherical Bragg reflector constructed from alternating layers with refractive indices n_a and n_b.

is less than the Rabi splitting. In this case, one can still use eqn (4.126) for polariton eigenfrequencies, replacing the QW reflection coefficient r by the reflection coefficient of the array of wires or dots in eqn (4.91).

Microcavities with a three-dimensional photonic confinement can exhibit strong coupling between their confined photon modes and an exciton resonance in a single QD. Experimental evidence for strong coupling of a single QD exciton to a cavity mode has been reported by Reithmaier et al. (2004), Yoshie et al. (2004) and Peter et al. (2005) (see Section 6.1). Here, we consider the simplest model of a spherical microcavity of radius r_{QD} with a spherical QD embedded in its centre (Fig. 4.29).

The eigenfrequencies of exciton-polariton modes can be found by the transfer matrix method generalised for spherical waves. Using the Green's function technique to resolve the Maxwell equations for a spherical wave incident on the QD in a similar way as we used above to describe scattering of a plane wave by a QD, one can obtain the reflection coefficient

$$r_{QD} = 1 + \frac{2i\Gamma_{sp}^{QD}}{\omega_0 - \omega - i(\gamma + \Gamma_{sp}^{QD})}, \qquad (4.143)$$

as described in more detail by Kaliteevski et al. (2001), with

$$\Gamma_{sp}^{QD} = \frac{2}{3}\pi k^4 V_0^2 \omega_{LT} a_B^3, \qquad (4.144)$$

$V_0 = \int \Phi(r)j_0(kr)dr$ and $k = n_c\omega/c$, (j_0 is the zero-order spherical Bessel function and Φ the exciton wavefunction in the QD taken with equal electron and hole coordinates and assumed to have a spherical symmetry). The term 1 in the right part of eqn (4.143) comes from the fact that, for a spherical wave incident on a centre, the

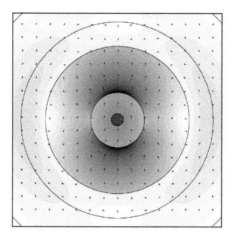

Fig. 4.29: Schematic distribution by Kaliteevski et al. (2001) of the magnitude of the electric-field intensity (grey scale) for the $l = 1$, $m = 1$ TM mode in a spherical microcavity with an embedded QD. The arrows show directions of the electric-field vector of the mode.

transmission contributes to reflectivity. The generalised transfer matrix method yields an equation for the eigenfrequencies of the cavity polaritons, being a spherical analogue of eqn (4.126):

$$h_1^{(2)}(kR_0) = r_B r_{QD} h_1^{(1)}(kR_0)\,, \qquad (4.145)$$

where the functions h are related to the spherical Hankel functions H by $h_l^{(1,2)}(x) = \sqrt{\pi/2x} H_{l+1/2}^{(1,2)}(x)$ and r_B is the reflection coefficient of the Bragg mirror for the spherical wave incident from inside the cavity.

Equation (4.145) yields the dispersion of all existing cavity modes including those coupled to the QD exciton. Each mode can be characterised by an orbital quantum number l and magnetic number m and also TE- or TM-polarisation (corresponding to the modes where electric or magnetic fields have no radial component, respectively). A detailed classification of spherical polariton modes has been given by Ajiki et al. (2002). Note that there exist no allowed optical mode having a perfect spherical symmetry ($l = 0$) and the photon mode with $l = 1$ has the lowest allowed energy (see Fig. 4.29). In particular cases, eqn (4.145) can be reduced to the problem of two coupled harmonic oscillators familiar in planar microcavities. Assuming zero broadening of both exciton and photon modes and approximating r_B by

$$r_B \approx \exp(i\beta\frac{\omega - \omega_B}{\omega_B})\,, \qquad (4.146)$$

with $\beta = \pi n_a n_b / n_c (n_b - n_a)$, one can rewrite eqn (4.145) as

$$\beta\frac{\omega - \omega_b}{\omega_b} - \frac{2\Gamma_{sp}^{QD}}{\omega - \omega_{ex}} + 2kR_0 = 2\pi(N + 1)\,, \qquad N \in \mathbb{N}. \qquad (4.147)$$

In particular, if the uncoupled cavity mode has a frequency equal to ω_B, the eigen-frequencies of exciton-polariton modes are given by

$$\omega_{1,2} = \omega_b \pm \sqrt{\frac{2\Gamma_{sp}^{QD}\omega_b}{\beta + 2\frac{n_c\omega_b}{c}R_0}} . \tag{4.148}$$

If the splitting exceeds the line broadening of exciton and photon modes, the strong coupling regime can be observed.

Exercise 4.6 $^{(***)}$ *Find the frequencies of exciton-polariton eigenmodes in a planar microcavity with an embedded periodical grating of infinite quantum wires (QWWs). Consider diffraction effects to the first order.*

5

QUANTUM DESCRIPTION OF LIGHT–MATTER COUPLING

In this chapter we study with the tools developed in Chapter 3 the basic models that are the foundations of light–matter interaction. We start with Rabi dynamics, then consider the optical Bloch equations that add phenomenologically the lifetime of the populations. As decay and pumping are often important, we cover the Lindblad form, a correct, simple and powerful way to describe various dissipation mechanisms. Then we go to a full quantum picture, quantising also the optical field. We first investigate the simpler coupling of bosons and then culminate with the Jaynes–Cummings model and its solution to the quantum interaction of a two-level system with a cavity mode. Finally, we investigate a broader family of models where the material excitation operators differ from the ideal limits of a Bose and a Fermi field.

Microcavities, Second Edition. Alexey V. Kavokin, Jeremy J. Baumberg, Guillaume Malpuech, Fabrice P. Laussy, Oxford University Press (2017). © Alexey V. Kavokin, Jeremy J. Baumberg, Guillaume Malpuech, Fabrice P. Laussy. DOI 10.1093/oso/9780198782995.001.0001

5.1 Historical background

Many important concepts of light–matter interaction and especially their terminology are rooted in the physics of nuclear magnetic resonance (NMR), where the effects were first observed and the models first developed. During World War II there was a strong interest in radars and researchers investigated the radio-frequency range of the electromagnetic field with great scrutiny, especially the means of creation and the efficient detection of such waves. Purcell, then at the Massachusetts Institute of Technology, noted that magnetic nuclei—such as ^{1}H, ^{13}C or ^{31}P—could absorb radiowaves when placed in a magnetic field of specific strength. The perturbation of this state allows accurate measurements that are the basis of NMR and derived techniques. As the effect was understood to be linked to the intrinsic magnetic properties of the nucleus, the dynamics of its spin under the action of strong electromagnetic fields was studied. The physics of spin–radiowave interaction shares many similarities to that of atom–light interaction: the two levels of the spin-up/spin-down configurations become the ground and excited states of an atomic resonance, and the electromagnetic field merely changes in frequency, so that many results and concepts obtained in the former context reappear in light–matter interaction. We start our investigation of the physics of light–matter interaction with Rabi's approach to the problem of nuclear induction. He had the simplest model, minimally quantised. His model cannot be further simplified without reducing to the Lorentz oscillator. He neglected lifetime, pumping and decoherence, and eliminated all "complications", even those he could have taken into account easily. For this reason, he unravelled the most important features of the problem and had his name pinned in the physics of light–matter coupling. All subsequent approaches rely on his simple result, as we shall see in this chapter.

5.2 Rabi dynamics

Rabi investigated the coupling of a quantum two-level system driven by a sinusoidal wave, modelling a classical optical field interacting with a spin. We shall use in the following the terminology of atoms. In the approximations of Rabi, we write the atomic wavefunction as

$$|\psi(t)\rangle = C_{\mathrm{g}}(t)\,|\mathrm{g}\rangle + C_{\mathrm{e}}(t)\,|\mathrm{e}\rangle\,, \tag{5.1}$$

the dynamics of which is entirely contained in the two complex coefficients C_{g} and C_{e}, and the Hamiltonian

$$H = E_{\mathrm{g}}\,|\mathrm{g}\rangle\langle\mathrm{g}| + E_{\mathrm{e}}\,|\mathrm{e}\rangle\langle\mathrm{e}| \tag{5.2a}$$

$$+\big(V_{\mathrm{ge}}\,|\mathrm{g}\rangle\langle\mathrm{e}| + V_{\mathrm{eg}}\,|\mathrm{e}\rangle\langle\mathrm{g}|\,\big)E(t)\,. \tag{5.2b}$$

Throughout, subscripts "g" and "e" refer to "ground" and "excited" (level, energy...),[71] V_{eg} is the dipole moment of the transition (discussed more later) and V_{ge} its complex conjugate. Only the atom is treated quantum mechanically so the light energy does

---

[71]Later on, we shall use the qubit notation of $|0\rangle$ for "ground" and $|1\rangle$ for "excited", which is more convenient when algebra is involved.

not enter into the Hamiltonian, it only acts as a time-dependent interaction through the c-function E. In the dipole approximation, the electric field is

$$E(t) = E_0 \cos \omega_a t, \tag{5.3}$$

where the amplitude E_0 and the frequency ω_a are constant. To emphasise the temporal dynamics, we introduce the frequency associated to the levels energies:

$$E_g = \hbar \omega_g, \qquad E_e = \hbar \omega_e. \tag{5.4}$$

Exercise 5.1 (*) *Show that for the system (5.1)–(5.4), Schrödinger equation (3.1) becomes*

$$\begin{cases} \dot{C}_g = -i\omega_g C_g - 2i\Omega_R e^{i\phi} \cos(\omega_a t) C_e \\ \dot{C}_e = -i\omega_e C_e - 2i\Omega_R e^{-i\phi} \cos(\omega_a t) C_g \end{cases} \tag{5.5}$$

with ϕ the complex phase of $V_{eg} = |V_{eg}| e^{i\phi}$ and in terms of the Rabi frequency

$$\Omega_R = \frac{|V_{eg} E_0|}{2\hbar}. \tag{5.6}$$

The quantity given by formula (5.6) is an important parameter for the quantum dynamics of a two level system. It is often referred to as an energy, $\hbar \Omega_R$ (the *Rabi energy*), and more often still as twice this quantity under the name of *Rabi splitting*. We have already encountered this in the semiclassical treatment of Chapter 4 in eqn (4.47) as a term quantifying the magnitude of the splitting (hence the name) of two resonances in the polarisation. This was obtained without reference to quantum dynamics, as this is a result that in most cases can also be derived from a classical perspective, but the origin of this popular term[72] is in the quantum treatment, eqns (5.5).

In the interaction picture (or in the language more suited to this problem, in "rotating frames"), the interesting dynamics is in the slow evolution due to the coupling Ω_R not in the rapid and trivial one imparted by the optical frequency, so that we redefine

$$c_g = C_g e^{i\omega_g t} \quad \text{and} \quad c_e = C_e e^{i\omega_e t}, \tag{5.7}$$

which lead to

$$\begin{cases} \dot{c}_g = -i\Omega_R e^{i\phi} (e^{i\Delta t} + e^{-i(\omega_a + \omega_\sigma)t}) c_e \\ \dot{c}_e = -i\Omega_R e^{-i\phi} (e^{-i\Delta t} + e^{i(\omega_a + \omega_\sigma)t}) c_g \end{cases}, \tag{5.8}$$

where we have introduced the detuning Δ:

$$\Delta = \omega_a - \omega_\sigma, \tag{5.9}$$

between the cavity frequency ω_a and the atomic transition $\omega_\sigma = \omega_e - \omega_g$. We also wrote the cosine as $(e^{i\omega_a t} + e^{-i\omega_a t})/2$ to deal with complex exponentials only. Despite their

---

[72]Numerous other accommodations of the adjective apply, such as "Rabi flop" or "Rabi oscillation" to denominate the dynamics of Fig. 5.1)

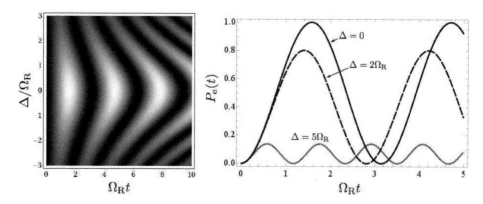

Fig. 5.1: Rabi oscillations in the probability P_e (greyscale coding, light colour corresponds to high probability) as a function of time and detuning (in units of Δ/Ω_R and $\Omega_R t$). On the right, slices for the given values of detunings. Departing from resonance increases the transition rate frequency but spoils its efficiency.

simple appearance, these equations are not easy to solve exactly because of the many terms that appear by combinations of the two terms on the right-hand sides. They become straightforward, however, if a single term is retained. Now let us remind ourselves that the purpose of eqns (5.7) is precisely to separate the slow dynamics of c coefficients from the rapid oscillation at rate $\omega_a + \omega_\sigma$, so that on a small time interval over which eqns (5.8) are integrated, the temporal evolution of $\exp\big(\mp i(\omega_a + \omega_\sigma)t\big)c_{e/g}$ is essentially given by the exponential that is oscillating so quickly that it averages to zero even over a small interval where c changes by a little amount. This is called the *rotating wave approximation* (RWA) because of the way frequencies contrive to stabilise or cancel the time evolution depending on whether they oscillate with or against the frame rotating with the field. In the RWA, the solution follows

$$c_g(t) = e^{i\frac{\Delta}{2}t}\Big[\big(\cos(\Omega t) - i\frac{\Delta}{2\Omega}\sin(\Omega t)\big)c_g(t) - i\frac{\Omega_R}{\Omega}\sin(\Omega t)c_e(t)\Big], \quad (5.10a)$$

$$c_e(t) = e^{-i\frac{\Delta}{2}t}\Big[\big(\cos(\Omega t) + i\frac{\Delta}{2\Omega}\sin(\Omega t)\big)c_e(t) - i\frac{\Omega_R}{\Omega}\sin(\Omega t)c_g(t)\Big], \quad (5.10b)$$

in terms of a newly introduced and important quantity:

$$\Omega = \sqrt{\Omega_R^2 + \left(\frac{\Delta}{2}\right)^2}. \quad (5.11)$$

From the interpretation of the wavefunction in quantum mechanics, $|c_g|^2$ and $|c_e|^2$ are probabilities to find the atom in its ground or excited state, respectively. If the atom is initially in its ground state (at $t = 0$), the probability to find it in its excited state at time t is, from eqns (5.10),

$$P_e(t) = |c_e|^2 = \left(\frac{\Omega_R}{\Omega}\right)^2 \sin^2(\Omega t). \quad (5.12)$$

Edward **Purcell** (1912–1997) and Felix **Bloch** (1905–1983), the 1952 Nobel prize-winners in physics for *"their development of new methods for nuclear magnetic precision measurements"*.

Purcell worked during World War II at the MIT Radiation Laboratory on the development of the microwave radar under the supervision of Rabi. Back in Harvard, he discovered nuclear magnetic resonance (NMR) with Pound and Torrey, published in Phys. Rev, **69**, 37 (1946). Other similar and important contributions include spin-echo relaxation and negative spin temperature. In another field, but still revolving about the radio spectrum, he made (with Ewen) the first detection of the famous "21cm line", due to the hyperfine splitting of hydrogen. This seeded the field of radioastronomy and allowed a breakthrough in the study of galactic structure. Most importantly for the cavity community, he gave his name to the enhancement or inhibition of dipole radiation by boundary conditions (see Chapter 6).

Bloch's education—like that of Purcell—was initially that of an engineer, but he soon turned to physics which gave him the opportunity to study with Schrödinger, Weyl and Debye, among others. When Schrödinger left Zurich, Bloch worked with Heisenberg instead and immediately after with Pauli, Bohr, Fermi and Kramers as parts of fellowships he earned. In 1928, he formulated in his doctoral dissertation the theorem bearing his name (that we presented on page 130) to describe the conduction of electrons in crystalline solids. He met with Purcell in 1945 at a meeting of the American Physical Society, where they realised the unity of their work on nuclear resonance. They agreed to share its experimental investigations, in crystals for Purcell's group, in liquids for Bloch.

The Rabi frequency, which at resonance is proportional to the amplitude of the light field, E_0, and to the matrix element V_{eg}, becomes renormalised with detuning at the same time as the transition efficiency gets spoiled, as seen in Fig. 5.1. The probability oscillates in time, so that continuously exciting a system is not the best way to get it into the excited state: once the excitation is created, further continuous excitation now works towards bringing back the system to its ground state. Note that no mechanism of decay or relaxation has been included in this simplest of pictures and that if the external excitation is shut off, the system stays where it is forever. In effect, this tendency of an external excitation to induce the atom to de-excite is the *stimulated emission* foreseen by Einstein and introduced in Chapter 3, for which we have just provided a microscopic derivation.

5.3   Bloch equations

The model developed by Rabi contains the key elements of the dynamics, associated to the Rabi frequency Ω_R. However, it is not realistic in many respects, if only because it lacks any form of dissipation. Losing excitation from the system or coupling it to some exterior reservoir will result in dephasing of the state,[73] which is clearly not the case in the previous model where the quantum state is a wavefunction. Dissipation will induce a loss of quantum coherence in the sense of losing the quantum correlations between states that are embedded in the off-diagonal elements of the density matrix. A way to modify this limitation in the above approach is therefore to rewrite the equation with a density matrix instead of with a wavefunction and add phenomenological decay terms that reduce the elements of the density matrix. First, rewriting the equation in terms of

$$\rho = |\psi\rangle\langle\psi| \, , \tag{5.13}$$

gives, in the simple case of eqn (5.1),

$$\rho = \begin{pmatrix} |C_g(t)|^2 & C_g(t)C_e(t)^* \\ C_g(t)^*C_e(t) & |C_e(t)|^2 \end{pmatrix} , \tag{5.14}$$

and subsequently, its time equation of motion:

Exercise 5.2 [*] *Derive the following dynamics from eqn (3.53) applied to eqns (5.13) and (5.1):*

$$\begin{cases} \dot{\rho}_{gg} = \dfrac{i}{\hbar}\left(V_{eg}E\rho_{eg} - \text{c.c.}\right), \\[2mm] \dot{\rho}_{ee} = -\dfrac{i}{\hbar}\left(V_{eg}^*E\rho_{ge} - \text{c.c.}\right), \\[2mm] \dot{\rho}_{ge} = -i\omega_a\rho_{ge} - \dfrac{i}{\hbar}V_{eg}E(\rho_{ee} - \rho_{gg}), \\[2mm] \dot{\rho}_{eg} = i\omega_a\rho_{eg} + \dfrac{i}{\hbar}V_{eg}^*E(\rho_{ee} - \rho_{gg}). \end{cases} \tag{5.15}$$

This is, up to now, strictly equivalent to eqns (5.5) as all elements of the density matrix can be reconstructed from the knowledge of the two amplitudes $C_{g/e}$. Also observe the symmetries among the matrix elements, namely, $\dot{\rho}_{gg} = -\dot{\rho}_{ee}$ and $\dot{\rho}_{ge} = \dot{\rho}_{eg}^*$. The latter follows from hermiticity of the density matrix. The former is an expression of the conservation of probabilities: if the atom gets excited, the ground state becomes depopulated. Because these matrix elements are real numbers, the complex conjugation is not necessary (in the equations it is corrected by the off-diagonal element).

Lifetime and dephasing are now added phenomenologically to eqn (5.15), by the introduction of two characteristic times, known as T_1 and T_2.[74] They account for lifetime (decay in population) and dephasing (decay in phase), respectively. Two typical physical mechanisms responsible for these terms are spontaneous emission and atom–atom

---

[73]The interaction of the system with reservoirs that dephase it implies that even in the case where the initial state of the system is well known (pure state), it evolves with time into a mixture of states where only probabilistic information remains.

[74]In NMR, T_1 and T_2 are called the longitudinal and transverse relaxation times, respectively, because they cause the decay of orthogonal projections of the magnetic spin along its precession axis.

collision, respectively. The spontaneous decay rate also dephases the system so that a typical expression for T_2 could be

$$\frac{1}{T_2} = \frac{1}{2T_1} + \frac{1}{T_2^*}, \tag{5.16}$$

where T_2^* is the mean time between atomic collisions that reset the phases of the emitters. A microscopic derivation of these quantities is possible, but of course, more involved mathematically. In the former case it requires quantisation of the optical field (which here is still a c-function E), and in the latter introduction of a bath of other carriers and a statistical treatment. On the other hand, the phenomenological treatment is clear enough (we shall see more about their microscopic origin in what follows). With these additions, eqns (5.15) read

$$\begin{cases} \dot{\rho}_{ee} = -\dfrac{i}{\hbar}\left(V_{eg}^* E \rho_{ge} - \text{c.c.}\right) - \dfrac{\rho_{ee}}{T_1}, \\[2mm] \dot{\rho}_{ge} = -i\omega_a \rho_{eg} - \dfrac{i}{\hbar} V_{eg} E (\rho_{ee} - \rho_{gg}) - \dfrac{\rho_{ge}}{T_2}, \end{cases} \tag{5.17}$$

where we have also limited the discussion to independent terms only and performed the rotating wave approximation.[75]

Equations (5.17) are called the *optical Bloch equations*, after Bloch who derived them for a spin in an oscillatory electric field in a form suitable to be mapped onto the Bloch sphere (or Poincaré sphere). They do not admit analytical solution in the general case, but limiting cases of practical interest exist and are presented now.

Exercise 5.3 $^{(**)}$ *Consider trial solutions of the form* $\tilde{\rho}_{ij}(t) = \tilde{\rho}_{ij}(0)\exp(\lambda t)$. *Show that at zero detuning and for* $V_{eg} \geq (1/2T_1)$, *the linearisation thus afforded admits solutions*

$$\lambda_1 = 0, \quad \lambda_2 = -\frac{1}{T_1}, \quad \lambda_3 = -\frac{3}{2}\frac{1}{T_1} + i\lambda \quad \text{and} \quad \lambda_4 = -\frac{3}{2}\frac{1}{T_1} - i\lambda, \tag{5.18}$$

where

$$\lambda = \sqrt{|V_{eg}|^2 - 1/(2T_1)^2}. \tag{5.19}$$

Consider now the case $V_{eg} \ll 1/T_1$ *where the same method yields*

$$\lambda_1 = 0, \quad \lambda_2 = -\frac{2}{T_1}, \quad \lambda_3 = -\frac{1}{T_1} + i\Delta \quad \text{and} \quad \lambda_4 = -\frac{1}{T_1} - i\Delta. \tag{5.20}$$

5.4 Full quantum picture

We now turn to a complete quantum treatment where the optical field gets quantised as well. Successful descriptions of light–matter interactions in full quantum-mechanical terms is, of course, a complicated problem. It culminates in the theory of quantum

---

[75] See the discussion on page 178.

electrodynamics (QED), which is, to date, the most successful, accurate and inspiring physical theory, among other points having served as a blueprint for a vast family of sister theories—collectively known as the "standard model"—to describe the force experimentally observed in nature (QED being the quantisation of the electromagnetic force). Only gravity is resisting the axiomatisation and interpretation of QED.

5.4.1   Light–matter interaction Hamiltonian

Back to our concern of light–matter interaction, one starting point[76] is to acknowledge that matter—being essentially neutral—couples with light through the electric field component arising from *fluctuations* in the electric charges (since the total charge cancels on the average). In the multipolar expansion of the field, the first nonzero fluctuations are the *dipolar* fluctuations for most cases of interest, including the simplest and generic case of an atom made up from a positive nucleus and an orbiting negative electron. This case naturally also depicts the exciton, cf. Fig. 4.7. The potential energy U of a classical dipole[77] $\mathbf{d}$ in an electric field $\mathbf{E}$ is easily derived from Newton's law and electrostatic energy, to yield

$$U = -\mathbf{d} \cdot \mathbf{E}. \tag{5.21}$$

We have already quantised the field $\mathbf{E}$, cf. eqn (3.66). We now consider the atom dipolar moment $\mathbf{d}$. We will simplify the above interaction by restraining to only two atomic levels and a single mode of the electromagnetic field. This is a bold approximation that leaves out a lot of the physics described by eqn (5.21) that, even for the simplest atom—the electron–proton pair bound as hydrogen—describes the coupling of two infinite sets of energy levels. The considerable simplification that we shall make proves, however, to be sound, especially for microcavities whose merit is to filter out modes of the electromagnetic field by imposing boundary conditions. For the atom, it suffices to select two levels whose energy difference matches the energy singled out by the optical mode to make all other transitions negligible in comparison. In the absence of a cavity this can be approached by using laser light that mimics a single mode thanks to small energy fluctuations.

We recall our notations for the two atomic modes $|e\rangle$ and $|g\rangle$ with energies E_e and E_g for "excited" and "ground", respectively. At this stage, and with such nomenclature, we have virtually already completely abstracted the "real" or "physical" atom away. Let us keep in mind, however, that $|g\rangle$ can equally well be the lower of two excited states of the physical atom. In the spirit of Dirac, $|g\rangle$ is a simple notation for a potentially complicated wavefunction. To illustrate the point we shall consider the $|\Psi_{100}\rangle$ state of the hydrogenoid atom as the ground $|g\rangle$ state and $|\Psi_{211}\rangle$ as the excited $|e\rangle$ state.

---

[76] Another, even more popular quantisation scheme, is that of the Hamiltonian $H = V(r) + (\mathbf{p} - q\mathbf{A}(\mathbf{r}, t))^2/2m$ of the vector potential $\mathbf{A}$. This Hamiltonian, or that derived from eqn (5.21), are essentially equivalent and are exactly so at resonance. The so-called Ap Hamiltonian (of this footnote) is better adapted to delocalised systems like electrons and holes in a bandstructure. Localised systems, on the contrary, like atoms or quantum dots, find a better starting point with the dipole Hamiltonian.

[77] A *dipole* is most simply visualised as the limiting case of two point-like particles of opposite charges $+q$ and $-q$, and separation δ, being $\mathbf{d} = q\delta\mathbf{u}_d$, where $\mathbf{u}_d$ is the unit vector pointing from the plus to the minus charge. Physical dipoles are two actual opposite charges separated by some distance small in comparison to the distance at which the dipole is observed.

The expressions for these states follow from the Schrödinger equation with Coulomb potential as

$$\Psi_{100}(r) = \frac{1}{\sqrt{\pi a_B^3}} \exp\left(-\frac{r}{a_B}\right),\tag{5.22a}$$

$$\Psi_{211}(\mathbf{r}) = \sqrt{\frac{1}{64\pi a_B^3}} \left(\frac{r}{a_B}\right) \exp\left(-\frac{r}{2a_B}\right) \sin\theta e^{i\phi},\tag{5.22b}$$

with $\mathbf{r} = (r, \theta, \phi)$ in eqn (5.22b).

Exercise 5.4 (**) *Show that the dipole element operator[78] for states (5.22) is*

$$\mathbf{d} = |e\rangle\langle g| \mathbf{d}_{eg} + |g\rangle\langle e| (\mathbf{d}_{eg})^*,\tag{5.23}$$

where

$$\mathbf{d}_{eg} = e \int \Psi_{211}^*(\mathbf{r})\mathbf{r}\Psi_{100}(\mathbf{r})\,d\mathbf{r},\tag{5.24a}$$

$$= -\frac{2^7}{3^5} e a_B (\mathbf{x} - i\mathbf{y}).\tag{5.24b}$$

For the following, it is therefore mathematically advantageous to consider the complicated "object" $\mathbf{d}$ as a mere two-by-two matrix on a Hilbert space spanned by

$$|e\rangle = \begin{pmatrix} 1 \\ 0 \end{pmatrix}, \qquad\qquad \langle e| = (1,0),$$

$$|g\rangle = \begin{pmatrix} 0 \\ 1 \end{pmatrix}, \qquad\qquad \langle g| = (0,1).$$

In this space, $\mathbf{d}$ can be expressed as a function of $\sigma = (\sigma_x + i\sigma_y)/2$ and $\sigma^\dagger$ (see eqn (3.11)) to become

$$\mathbf{d} = d_{eg}\mathbf{x}(\sigma^\dagger + \sigma),\tag{5.25}$$

where, in this case, the orientation of the dipole has been arbitrarily taken along the x-axis so as to cancel the imaginary part (orientation along the latter would result in a minus sign in eqn (5.25) and a complex matrix element). So finally, the light–matter interaction reads $\hbar g(\sigma^\dagger + \sigma)(a - a^\dagger)$, where

$$\hbar g = \sqrt{\frac{\hbar \omega_a}{2\epsilon_0 V}}\, d_{eg},\tag{5.26}$$

with V a volume of quantisation. Equation (5.26) is a "typical" result for the cases we have been considering. It specifically depends on the system at hand and how it

---

[78] A note of caution on notations: in eqn (5.23), $\mathbf{d}$ is an operator, while $\mathbf{d}_{eg}$ is a cartesian vector. Some authors use a hat to denote explicitly an operator, i.e., they would write $\hat{\mathbf{d}} = |e\rangle\langle g| \mathbf{d}_{eg} + $ h.c. As a matter of fact, $\mathbf{d}$ is also a vector, as is seen in eqn (5.24b), whose components are operators. We do not follow this practice because it is often clear enough from the context what the mathematical nature of a variable is, and it is helpful to retain the scope of an equation in both classical and quantum domains.

is modelled. In this case we have even left the dipole element in its general form. If considering the complete, multimode light field, eqn (3.66) should be used instead, as will be the case in the next chapter where the continuum of modes is needed to compute the radiative lifetime of a dipole in the vacuum.

With the free energy of the optical mode $\hbar\omega_a$ and of the atom $\hbar\omega_\sigma$ added, the "generic" complete Hamiltonian is

$$H = \hbar\omega_a a^\dagger a + \frac{1}{2}\hbar\omega_\sigma \sigma^\dagger \sigma + \hbar g(\sigma^\dagger + \sigma)(a - a^\dagger).\tag{5.27}$$

This system is the basis for what follows. To start with, we investigate quantum coupling in simpler cases than eqn (5.27), which despite its apparent simplicity is highly non-trivial. We shall, in all cases, reduce the problem in this chapter to single modes. This could be the case when large splitting of energies separate higher states from those considered, typically one cavity mode near resonance with an excitonic transition. More elaborate multimode couplings are treated in subsequent chapters.

For now, neglecting off-resonant terms[79] like $a^\dagger \sigma^\dagger$, the Hamiltonian becomes

$$H = \hbar\omega_a a^\dagger a + \hbar\omega_\sigma \sigma^\dagger \sigma + \hbar g(a\sigma^\dagger + a^\dagger \sigma).\tag{5.28}$$

One mode, a, which describes light, will be in all cases a pure Bose operator with commutation relation (3.28). The operator σ, that describes the material excitation, is in the derivation above the two-level system operator. In microcavities, other situations are possible, and depending on the model of the matter field, one could range from another Bose operator following the same commutation relations as a (in which case we would note it b), to a Fermi operator with anticommutation relations, passing through more complicated expressions to take into account spin, excited states, etc.[80] These two limiting cases—coupling a single cavity mode to a bosonic or fermionic field—are also of tremendous importance and many experimental configurations refer to them in some approximations. Their simple mathematical forms allow exact solutions to be obtained and, therefore, many insights to be gained. We consider them in turn in what follows, starting with the coupling of two Bose fields.

5.4.2 Dressed bosons

The most elementary problem of the kind of eqn (5.28) is the case where b is also a Bose operator, namely

$$[b, b^\dagger] = 1\tag{5.29}$$

(cf. eqn (3.28)). This is the coupling of two harmonic oscillators (hence, the problem is referred to as *linear* and this linearity will transpire in all that follows). We investigate the solutions in the various pictures of quantum mechanics introduced in Chapter 3.

The analysis of eqn (5.28) can be made directly in the basis $|i, j\rangle$ with i excitations in the matter field and j in the photonic field, $i, j \in \mathbb{N}$. These are called *bare states* in

---

[79]Neglecting terms that do not conserve energy like $a^\dagger \sigma^\dagger$ or $a\sigma$—since they simultaneously create or annihilate two excitations, respectively—is the second quantised version of the rotating wave approximation.

[80]A particular case that interpolates between Bose and Fermi statistics will be constructed in Section 5.7.3.

contrast to the *dressed states* that we consider hereafter. The value of this approach is that the excitation, loss and dephasing processes generally pertain to the bare particles. For instance, matter excitations are usually created by an external source (pumping) and light excitations can be lost by transmission through the cavity mirror. This physics is best expressed in the bare-states basis.

As the system is linear, the integration is straightforward in the bare-states basis. In the Heisenberg picture:

Exercise 5.5 (*) *Show that the time evolution of the operators a and b under the dynamics of Hamiltonian (5.28) is given by*

$$a(t) = \exp(-i\bar{\omega}t)\left(a(0)\left[\cos(\Omega t) - i\frac{\Delta}{2\Omega}\sin(\Omega t)\right] - ib(0)\frac{g}{\Omega}\sin(\Omega t)\right), \qquad (5.30a)$$

$$b(t) = \exp(-i\bar{\omega}t)\left(-ia(0)\frac{g}{\Omega}\sin(\Omega t) + b(0)\left[\cos(\Omega t) + i\frac{\Delta}{2\Omega}\sin(\Omega t)\right]\right), \qquad (5.30b)$$

where $\bar{\omega} = (\omega_a + \omega_b)/2$ and $\Omega = \sqrt{g^2 + (\Delta/2)^2}$.

Note that the commutation relation (3.28) remains well behaved at all times, thanks to the intermingling of a and b operators. This can be illustrated in the case $\Delta = 0$, where the expressions (5.30) simplify considerably to

$$a(t) = e^{-i\bar{\omega}t}[a(0)\cos(gt) - ib(0)\sin(gt)], \qquad (5.31a)$$
$$b(t) = e^{-i\bar{\omega}t}[b(0)\cos(gt) - ia(0)\sin(gt)], \qquad (5.31b)$$

in which case one gets

$$[a(t), a^\dagger(t)] = [a(0), a^\dagger(0)]\cos^2(gt) + [b(0), b^\dagger(0)]\sin^2(gt) = 1. \qquad (5.32)$$

We shall see in the Schrödinger representation how this result manifests in complicated correlations between the two states. This result should also make clear that equal-time commutations hold for Bose operators. Observe how $[a(0), a^\dagger(t)]$ oscillates between 0 and 1 with time.

Observables can be obtained directly in the Heisenberg picture from the solutions (5.30). For instance, the population is obtained in Exercise 5.6.

Exercise 5.6 (*) *Show that the population operators for the coupling of two oscillators with initial condition $|\Psi\rangle = \alpha|1,0\rangle + \beta|0,1\rangle$ are given by*

$$\langle a^\dagger a\rangle(t) = |\alpha|^2\cos^2(\Omega t) + \frac{\left|\frac{\Delta\alpha}{2} + \beta g\right|^2}{\Omega^2}\sin^2(\Omega t)$$

$$- \Im(\alpha\beta^*)\frac{2g}{\Omega}\cos(\Omega t)\sin(\Omega t), \qquad (5.33)$$

and use this result to recover the Rabi oscillations of Fig. 5.1.

On the other hand, eqn (5.28) assumes a straightforward expression in the basis of so-called *dressed states* that diagonalise the Hamiltonian, as we now show explicitly going back to the Schrödinger picture. The most general substitution that can be attempted is[81]

$$p = \alpha a + \beta b, \tag{5.34a}$$

$$q = \gamma a + \delta b, \tag{5.34b}$$

with $\alpha, \beta, \gamma, \delta \in \mathbb{C}$. If we require that p and q remain boson operators, i.e.,

$$[p, p^\dagger] = [q, q^\dagger] = 1, \tag{5.35}$$

then

$$|\alpha|^2 + |\beta|^2 = |\gamma|^2 + |\delta|^2 = 1. \tag{5.36}$$

We also require $[p, q] = [p, q^\dagger] = 0$. The second one implies $\alpha\gamma^* + \beta\delta^* = 0$, while the first is automatically satisfied. Relations (5.34) reversed read

$$a = \frac{\delta p - \beta q}{\alpha\delta - \beta\gamma}, \tag{5.37a}$$

$$b = \frac{-\gamma p + \alpha q}{\alpha\delta - \beta\gamma}. \tag{5.37b}$$

Their substitution in eqn (5.28) yields

$$(\alpha\delta - \beta\gamma)^2 H = p^\dagger p\{\hbar\omega(|\delta|^2 + |\gamma|^2) - \hbar g\Im(\delta^*\gamma)\} \tag{5.38a}$$

$$+ q^\dagger q\{\hbar\omega(|\beta|^2 + |\alpha|^2) - \hbar g\Im(\beta^*\alpha)\} \tag{5.38b}$$

$$+ p^\dagger q\{\hbar\omega(-\beta\delta^* - \gamma^*\alpha) + \hbar g(\alpha\delta^* + \beta\gamma^*)\} + \text{h.c.} \tag{5.38c}$$

We require that line (5.38c) be zero, which reduces to $\alpha\delta^* + \beta\gamma^* = 0$. Fitting the above conditions, we are led to $\alpha = \cos\theta$, $\beta = \sin\theta$, $\gamma = -\sin\theta$ and $\delta = \cos\theta$, ensuring that $\alpha\delta - \beta\gamma = 1$ (canonical unitary transformation):

$$p = \cos\theta a + \sin\theta b, \tag{5.39a}$$

$$q = -\sin\theta a + \cos\theta b, \tag{5.39b}$$

where

$$\cos\theta = \frac{1}{\sqrt{2}} \frac{\Delta/2 + \Omega}{\sqrt{(\Delta/2)^2 + g^2 + \Omega\Delta/2}}, \tag{5.40}$$

θ is known as the *mixing angle*. Then H reads

$$H = \hbar\omega_p p^\dagger p + \hbar\omega_q q^\dagger q, \tag{5.41}$$

where

$$\omega_{p/q} = \bar{\omega} \pm \Omega. \tag{5.42}$$

The eigenfrequencies $\omega_{p/q}$ given by eqns (5.42) are plotted in Fig. 5.2.

---

[81]We shall later see how this approach is the archetype of so-called *Bogoliubov* transformations, first considered in connection with high-density, weakly interacting Bose condensates.

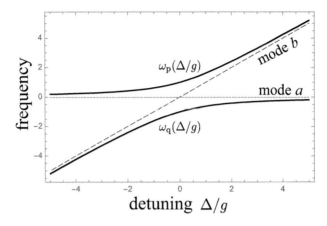

Fig. 5.2: Thick lines: eigenfrequencies of the system eqn (5.28) as a function of the detuning Δ (in units of the interaction strength g). At zero detuning, an anticrossing is observed. At large detunings, the bare modes (thin lines) are recovered.

For clarity we shall note $|i, j\rangle$ the dressed states, i.e., the eigenstates of eqn (5.41) with i dressed particles of energy $\hbar\omega_p$ and j of energy $\hbar\omega_q$. We call a *manifold* of excitation the set of states with a fixed number of excitations. In the dressed-states basis, it reads for the case of n excitations,

$$\mathcal{H}_N = \{|n, m\rangle \; ; \; n, m \in \mathbb{N} \text{ with } n + m = N\} . \tag{5.43}$$

Its energy diagram is computed as

$$H|n, m\rangle = E_{n,m}|n, m\rangle , \tag{5.44}$$

with

$$E_{n,m} = \hbar(n\omega_p + m\omega_q) . \tag{5.45}$$

It appears in Fig. 5.3 for manifolds with zero (vacuum), one, two, three and four excitations. When an excitation escapes the system while in manifold $\mathcal{H}_N$, a transition is made to the neighbouring manifold $\mathcal{H}_{N-1}$ and the energy difference is carried away, either by the leaking out of a cavity photon, or through exciton emission into a radiative mode other than that of the cavity, or a non-radiative process. The detailed analysis of such processes requires a dynamical study, but in some cases the knowledge of only the energy-level diagrams gives a good first idea of the radiation spectra. The important feature of this dissipation is that, although such processes involve a or b (rather than p or q), they nevertheless still result in removing one excitation out of one of the oscillators. Hence, only transitions from $|n, m\rangle$ to $|n - 1, m\rangle$ or $|n, m - 1\rangle$ are allowed, bringing away, respectively, $\hbar\omega_p$ and $\hbar\omega_q$ of energy, accounting for the so-called Rabi doublet (provided the initial n and m are nonzero in which case only one transition is allowed). Physically, it comes from the fact that, as in the classical case, the coupled system acts as two independent oscillators vibrating with frequencies $\omega_{p/q}$.

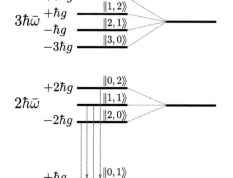

Fig. 5.3: The dressed bosons ladder. The coupling of degenerate quantum modes leads to a splitting in energy between the new states, known as *dressed states*. In the case case of two Bose fields, $N + 1$ dressed states arise from the possible ways to share the N excitations. Each set of such states form a manifold of excitation. Transitions between manifolds, in the linear case, proceed by exchange of one dressed excitation, therefore only transitions from $|n, m\rangle$ to $|n - 1, m\rangle$ or $|n, m - 1\rangle$ are allowed, as depicted for the case of the first and second manifold. Each manifold is very far apart from its neighbours as compared with the splitting between dressed states, i.e., $\hbar\bar{\omega} \gg \hbar g$, which is sketched by the symbol $\approx$. Otherwise the system is in the so-called ultrastrong coupling regime, which does not allow to diagonalise the Hamiltonian into manifolds of excitations.

In the case of vacuum-field Rabi splitting, a single excitation is shared between the two fields, and so the manifold $\mathcal{H}_1$ is connected to the single line of the vacuum manifold. In this case, there is obviously no possibility beyond a doublet. It is straightforward to compute the transition amplitudes between the two manifolds by mean of the bare-states annihilation operators. The rates, which are proportional to the four components of the dressed states, are

$$\mathcal{M}_1 = \langle 0, 0| \, a \, |1, 0\rangle = \alpha(\Delta/g) \,, \tag{5.46a}$$
$$\mathcal{M}_2 = \langle 0, 0| \, a \, |0, 1\rangle = \gamma = -\alpha(-\Delta/g) \,, \tag{5.46b}$$
$$\mathcal{M}_3 = \langle 0, 0| \, b \, |1, 0\rangle = \beta = \alpha(-\Delta/g) \,, \tag{5.46c}$$
$$\mathcal{M}_4 = \langle 0, 0| \, b \, |0, 1\rangle = \delta = \alpha(\Delta/g) \,. \tag{5.46d}$$

The amplitudes of transitions $\mathcal{M}_i$ given by eqns (5.46) are displayed in Fig. 5.4. Their square is the physical quantity one is interested in:

$$\mathcal{I}_1 = |\alpha(\Delta/g)|^2 \quad \text{and} \quad \mathcal{I}_2 = |\alpha(-\Delta/g)|^2 \,. \tag{5.47}$$

The physical sense of these results is that when the mode annihilates an excitation in one of the dressed states, it only "sees" it through its weight in the total (or dressed) state, so the strength comes out as proportional to these amplitudes. Intuitively, if the "polariton" becomes more "exciton-like" because of detuning, the cavity emission disappears. The intensities degenerate into two lines out of four for the amplitudes because

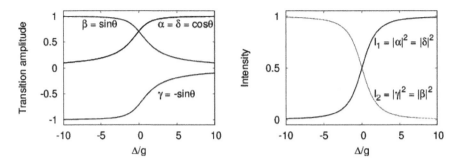

Fig. 5.4: Amplitudes $\mathcal{M}_i$ (left) and corresponding intensities $|\mathcal{M}_i|^2$ (right) of transitions between the manifolds of one excitation and the vacuum, as a function of the detuning Δ in units of the interaction strength g.

of the symmetry of the two modes (here we have two harmonic oscillators, in our case the two modes are different). The transition rates are antisymmetric with Δ and, as one goes from 0 (from $-\infty$) to 1 (from $+\infty$), the other does the reverse. The emission of one given mode is therefore two lines, which degenerate at zero detuning (where the rates match), and with the detuning, one vanishes while the other increases.

The dressed state $\|n, 0\rangle$ in terms of bare states reads

$$\|n, 0\rangle = \frac{1}{\sqrt{n!}} p^{\dagger n} |0, 0\rangle = \sum_{k=0}^{n} \sqrt{\binom{n}{k}} \cos^k \theta \sin^{n-k} \theta \, |k, n - k\rangle . \tag{5.48}$$

The general case is computed along these lines, $\|n, m\rangle = p^{\dagger n} q^{\dagger m} |0, 0\rangle / \sqrt{n! m!}$, as

$$\|n, m\rangle = \sum_{\nu=0}^{n+m} \sum_{l=0}^{\nu} (-1)^l \binom{n}{\nu - l} \binom{m}{l} \sqrt{\frac{\binom{n+m}{n}}{\binom{n+m}{\nu}}}$$
$$\times \cos^{\nu-l} \theta \sin^l \theta \sin^{n-\nu+l} \theta \cos^{m-l} \theta \, |\nu, n + m - \nu\rangle . \tag{5.49}$$

Exercise 5.7 [*] *Derive eqn (5.49) above.*

This result gives the probability of measuring the bare state $|\nu, n + m - \nu\rangle$ when the system is prepared in the dressed state $\|n, m\rangle$:

$$p(\nu) = \left| \sum_{l=0}^{\nu} (-1)^l \binom{n}{\nu - l} \binom{m}{l} \sqrt{\frac{\binom{n+m}{n}}{\binom{n+m}{\nu}}} \cos^{m+\nu-2l} \theta \sin^{n-\nu+2l} \theta \right|^2 . \tag{5.50}$$

We conclude this section with the basis of coherent states and the P representation of the oscillators (cf. eqn (3.75)). If we call α the variable relating to oscillator a and β that relating to b, the density matrix ρ of the coupled system (5.28) decomposes as $\rho = \iint P(\alpha, \alpha^*, \beta, \beta^*, t) |\alpha\beta\rangle \langle\beta\alpha| \, d\alpha d\beta$. Following the procedure detailed in Section 3.3.3, the operator equation (5.28) provides its c-number counterpart:

$$\dot{P} = i\bar{\omega}(\alpha\partial_\alpha - \alpha^*\partial_{\alpha^*} + \beta\partial_\beta - \beta^*\partial_{\beta^*})P$$
$$+ ig(\alpha\partial_\beta - \alpha^*\partial_{\beta^*} + \beta\partial_\alpha - \beta^*\partial_{\alpha^*})P, \tag{5.51}$$

which can be integrated to yield the exact solution

$$P(\alpha, \alpha^*, \beta, \beta^*, t) = F\left(\frac{i}{2}(\alpha^* + \beta^*)e^{-i(\bar{\omega}+g)t}, \frac{i}{2}(-\alpha^* + \beta^*)e^{-i(\bar{\omega}-g)t},\right.$$
$$\left.\frac{i}{2}(\alpha - \beta)e^{-i(-\bar{\omega}+g)t}, \frac{i}{2}(-\alpha - \beta)e^{i(\bar{\omega}+g)t}\right), \quad (5.52)$$

where F is a differentiable function.

Exercise 5.8 (*) *Check that the partial differential equation* $axu_x + byu_y + czu_z + dwu_w = 0$ *has solutions of the type* $u(x, y, z, w) = F(y^a/x^b, z^a/x^c, w^a/x^d)$, *where* F *is a differentiable function. By diagonalisation of eqn (5.51), obtain eqn (5.52).*

By matching eqn (5.52) with the initial conditions, one can obtain the time evolution of the complete wavefunction. For instance, the case of two states that are mixtures of coherent and thermal states (in the sense of Section 3.3.5), with respective intensities $|\alpha_0|^2$ and n for one oscillator, and $|\beta_0|^2$ and m for the other, leads to

$$P(\alpha, \alpha^*, \beta, \beta^*, t) = \frac{1}{\pi m} \exp\left(-|\alpha|^2 \left[\frac{\cos^2(gt)}{m} + \frac{\sin^2(gt)}{n}\right]\right.$$
$$-\alpha \left[\frac{-\alpha_0^* e^{i\bar{\omega}t}\cos(gt)}{m} + \frac{-i\beta_0^* e^{i\bar{\omega}t}\sin(gt)}{n}\right]$$
$$-\alpha^* \left[\frac{-\alpha_0 e^{-i\bar{\omega}t}\cos(gt)}{m} + \frac{i\beta_0 e^{-i\bar{\omega}t}\sin(gt)}{n}\right]$$
$$\left.-|\alpha_0|^2 \left[\frac{1}{m}\right]\right)$$
$$\times \frac{1}{\pi n} \exp\left(-|\beta|^2 \left[\frac{\sin^2(gt)}{m} + \frac{\cos^2(gt)}{n}\right]\right.$$
$$-\beta \left[\frac{-i\alpha_0^* e^{i\bar{\omega}t}\sin(gt)}{m} + \frac{-\beta_0^* e^{i\bar{\omega}t}\cos(gt)}{n}\right]$$
$$-\beta^* \left[\frac{i\alpha_0 e^{-i\bar{\omega}t}\sin(gt)}{m} + \frac{-\beta_0 e^{-i\bar{\omega}t}\cos(gt)}{n}\right]$$
$$\left.-|\beta_0|^2 \left[\frac{1}{n}\right]\right)$$
$$\times \exp\left(-\alpha^*\beta \left[\frac{i\cos(gt)\sin(gt)}{m} + \frac{-i\sin(gt)\cos(gt)}{n}\right]\right.$$
$$\left.-\alpha\beta^* \left[\frac{-i\cos(gt)\sin(gt)}{m} + \frac{i\sin(gt)\cos(gt)}{n}\right]\right). $$
$$(5.53)$$

This result shows how, due to the coupling, even though it is linear and of the simplest kind that can affect two modes, complex *correlations* build up between the two

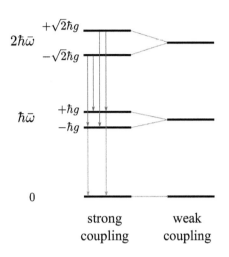

$3\hbar\bar\omega$

$+\sqrt{3}\hbar g$

$-\sqrt{3}\hbar g$

$2\hbar\bar\omega$

$+\sqrt{2}\hbar g$

$-\sqrt{2}\hbar g$

$\hbar\bar\omega$

$+\hbar g$

$-\hbar g$

0

strong
coupling

weak
coupling

Fig. 5.5: The Jaynes–Cummings ladder. As compared to the case of two Bose fields, Fig. 5.3, with which it shares only the first two manifolds, the coupling of a Bose field with a two-level system leads to manifolds (with n excitations) of always two dressed states (instead of $n+1$), splitted by $2\sqrt{n}\hbar g$ (rather than constant to $2\hbar g$). All transitions between two neighbouring manifolds are allowed, leading to anharmonic resonant frequencies. Very high in the ladder, two of the four transitions have similar frequencies, leading to the Mollow triplet. Here also, each manifold is very far apart from its neighbours as compared with the splitting between dressed states, i.e., $\hbar\bar\omega \gg \hbar g$, sketched by the symbol $\approx$. Otherwise the system is in ultrastrong coupling regime, but now with a two-level system.

modes. The main behaviour of the coupled system remains that of two mixture states oscillating in their populations, which are represented by the first two exponentials of the above product, with variables α and β factoring out, but a third exponential inextricably links them.

5.4.3 *Josephson coupling*

The Josephson effect is rooted in superconductor physics, where it was discovered by a young Ph.D aspirant in Cambridge (Brian Josephson) who was attending Anderson's lectures (then on sabbatical leave from Bell Labs), at which occasion a preprint of Cohen et al. (1962) on the tunnelling between a superconductor and a normal metal was discussed. There was stiff opposition from Bardeen (1961) himself regarding the possibility for Cooper pairs to tunnel through a barrier and the configuration of tunnelling between two superconductors, that was contemplated by Cohen et al., was not further investigated by leading theorists. It was left to Josephson (1962) to predict that a supercurrent flows between two superconductors in absence of an applied voltage, driven by their phase difference (superconductors have a complex order parameter). In the presence of an applied voltage, the supercurrent oscillates, giving rise to the AC–Josephson version. The effect became of paramount importance on several accounts and earned a Nobel prize to the youngest recipient to date. Let alone technological applications through the SQUID (superconducting quantum interference device), it brought a striking manifestation of symmetry breaking, Gauge invariance and quantum coherence (some of the themes of Anderson's lectures in Cambridge).

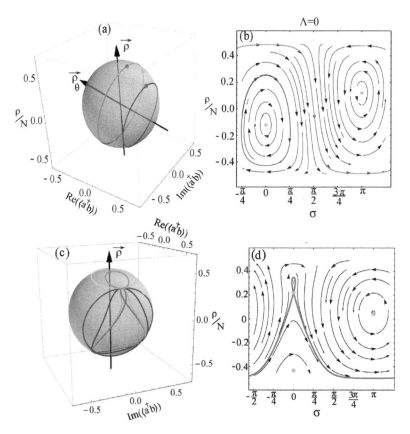

Fig. 5.6: **Bosonic Josephson oscillations**, from Rahmani and Laussy (2016). Polaritons as coupled-condensates (of excitons and photons) cast their dynamics in the framework of Josephson physics. In absence of interactions, there are two stable fixed points (the lower and upper polaritons) shown in panel (b) and the dynamics is simply a circle on the Bloch sphere, normal to the $\vec{\rho}_\theta$ that joins the polaritons, panel (a). The relative-phase is running or oscillating, respectively, depending on whether this trajectory encircles or not the $\vec{\rho}$ axis of observables (exciton-photon). In the presence of interactions, new fixed points appear, one unstable, when $\Lambda > \Lambda_c$. This allows to distinguish unambiguously the Rabi and Josephson regimes. In panel (c), the blue, purple and green lines are in the Rabi regime, while the red, cyan and orange are in the Josephson regime, with increasing values of Λ. Panel (d) shows the phase-space with the four fixed points (in green) and the red trajectory slightly above the threshold.

Despite its ties to superconductivity, the Josephson effect is really an attribute of coupled condensates (the BCS state serving as the condensate in the superconducting version), and as such is also relevant to superfluids and BECs. To distinguish the more fundamental case that involves pure bosons from that of Cooper pairs, which remains of great importance, one speaks of "Bosonic Josephson Junctions". Gati and Oberthaler (2007) give an excellent review. The best contextualisation of the Josephson phenomenon was provided by Leggett (1975), who defines it as "*the dynamics of N bosons restricted to occupy the same two-dimensional single particle Hilbert space*".

Leggett (1999) introduced three regimes for such systems depending on the relationship between tunelling and interactions, namely the Rabi (non-interacting), Josephson (weakly-interacting) and Fock (strongly-interacting) regimes. The Fock regime leads to genuine quantum phases (with no classical counterparts) while the Rabi and Josephson coupling are well described by classical fields in a mean-field approximation. These are the two regimes of interest in this Section as they are tightly related to Rabi oscillations as described previously in this chapter. The dynamics of the Bosonic Josephson effect has been considered extensively by Raghavan et al. (1999). It describes the coupling between two weakly-interacting Bose fields, a and b with free energies $\epsilon_{a,b}$, ruled by the Hamiltonian $H = H_0 + V$ where

$$H_0 = \epsilon_a a^\dagger a + \epsilon_b b^\dagger b + g(a^\dagger b + b^\dagger a) \quad \text{and} \quad V = v_b(b^\dagger b^\dagger bb) + v_a(a^\dagger a^\dagger aa). \quad (5.54)$$

Here, a and b would typically describe left and right bosons in two spatially separated traps. Leggett (1999) also pointed out that an "internal" Josephson effect can take place between internal degrees of freedom of a BEC, giving the example of different hyperfine Zeeman states of alkali gases (the Josephson oscillation in a single junction of BECs involving two condensates in space was first observed by Albiez et al. (2005)). Voronova et al. (2015) have observed that polaritons are therefore predisposed for Josephson physics from the very nature of their light-matter composition, exhibiting innately this internal type of such Josephson dynamics where the exchange is not between two spatially separated condensates but between the two internal degrees of freedom that make up the polariton, namely, its exciton and photon components. This is an adequate picture, since Dominici et al. (2014) had demonstrated that condensates of polaritons are condensates of photons and excitons. Consider the case of lower polaritons. According to eqn (3.72),

$$|\alpha, 0\rangle = e^{-|\alpha|^2/2} \sum_{n=0}^{\infty} \frac{\alpha^n}{\sqrt{n!}} |n, 0\rangle , \quad (5.55)$$

where $\alpha \in \mathbb{C}$ refers to a coherent state and $n \in \mathbb{N}$ to a Fock state. One can check (Exercise 5.9) that the entanglement $\alpha |0, 1\rangle + \beta |1, 0\rangle$ cancels out when weighted by the Poisson weights of the coherent superposition, to yield

$$|\alpha, 0\rangle = |\alpha \cos\theta, \alpha \sin\theta\rangle , \quad (5.56)$$

i.e., a factorised (non-entangled) product of coherent states of photons and excitons (keeping in mind that $\|L, U\rangle\rangle$—with double strokes—refers to Lower and Upper polaritons and $|C, X\rangle$—with single strokes—refers to Cavity photons and Exciton, each in a coherent state if using Greek letters or a Fock state if using a Latin index. As such, eqn (5.56) indeed states that a lower-polariton condensate (coherent state) is a product of exciton and photon condensates. In the linear regime, $v_a = v_b = 0$, the values of α and β are given by $\langle a \rangle$ and $\langle b \rangle$ from eqns (5.30) and only the interpretation changes, from vacuum Rabi coupling to coupled condensates. The dynamics is otherwise the same. Such connections are carried further in Section 11.1.1, where the classical/quantum relationship of polaritons is discussed in more details.

Exercise 5.9 [(*)] *Derive eqn (5.56) and express similarly $\|\alpha, \beta\rangle$ in the exciton-photon basis.*

The (Bosonic) Josephson viewpoint of the light-matter coupling of polaritons from Voronova et al. (2015) is therefore already relevant at the level of their very fabric. There has been some interest for the polariton Josephson physics before that, but in the framework of external oscillations between two spatially separated polariton condensates, in most cases with no need of an internal structure, and focusing instead on spin and/or out-of-equilibrium features. There have been several theoretical proposals, e.g., from Sarchi et al. (2008a), Wouters (2008) and Shelykh et al. (2008) followed by experimental observations reported in both the linear (Rabi) regime, by Lagoudakis et al. (2010), and nonlinear (self-trapping) regime, by Abbarchi et al. (2013) regime. The polariton implementation of Josephson effects is increasingly investigated with notable works from Pavlovic et al. (2013), Khripkov et al. (2013), Racine and Eastham (2014), Gavrilov et al. (2014), Zhang and Zhang (2014), Ma et al. (2015), Rayanov et al. (2015), which do not exhaust the list.

In Voronova et al. (2015)'s picture, the Rabi coupling acts as the Josephson tunnelling and interactions are then for the excitonic component only, bringing a variation on the atomic counterpart in space. Detuning of the free energies between the modes act as the external potential, so the analogy is essentially complete. The Josephson dynamics is typically described in terms of i) the population imbalance $\rho \equiv (\langle a^\dagger a \rangle - \langle b^\dagger b \rangle)/2$ between the two modes and ii) their relative phase $\sigma = \arg \langle a^\dagger b \rangle$. The importance of the phase has been prominent since the original Josephson (1962) report of its driving the supercurrent with no need of an applied voltage. Anderson (1966) became an ardent proponent of the role of the phase in superflow phenomena (for instance associating dissipation to phase-slippage). The first transposition of its physics to the case of BECs was considering non-interacting particles Javanainen (1986) and the role of the phase difference as the source of the superflow was then the focus of attention. The question of the phase of macroscopically degenerate quantum states remained anchored in the phenomenon but also took a separate route of its own with (sometimes conflicting) discussions from, e.g., Leggett and Sols (1991), Castin and Dalibard (1997) and Molmer (1997), that is still investigated to this day, e.g., by Javanainen and Rajapakse (2015) or, in the context of polaritons, by Antón et al. (2014). The phase was also prominently featured in the analysis of Voronova et al. (2015), who observed that detuning could bring the system from a regime of oscillating phase (typically associated to the Rabi regime) to one of running phase (typically associated to the Josephson regime). It was shown by Rahmani and Laussy (2016) that this follows from chosing a basis of observables (typically that of dressed states) that is not the eigenbasis of polaritons. The pure Rabi regime of non-interacting particles, $v_a = v_b = 0$, has a simple dynamic of circles on the Bloch sphere, as shown in Fig. 5.6(a), with equation

$$|\langle a_\theta^\dagger b_\theta \rangle|^2 + \rho_\theta^2 = (N(t)/2)^2 \,, \tag{5.57}$$

where $a_\theta = \cos(\theta)a + \sin(\theta)b$ and $b_\theta = -\sin(\theta)a + \cos(\theta)b$ are the annihilation operators for the polaritons, $\rho_\theta = \langle a_\theta^\dagger a_\theta \rangle - \langle b_\theta^\dagger b_\theta \rangle$ and $N_\theta = \langle a_\theta^\dagger a_\theta \rangle + \langle b_\theta^\dagger b_\theta \rangle$ are the relative and total number of particles, respectively. The equation still holds with dynamically evolving variables in presence of pumping and decay. Detuning thus merely change the observer's viewpoint for a dynamic that is otherwise the same, with no role of the relative phase as a driving agent and no special meaning to its drifting or oscillating behaviour. Rahmani and Laussy (2016) propose as a criterion to identify the Josephson regime to consider the number and stability of the fixed points, as they generalise the conventional case of resonant and equal-interaction condensates to the case ruled by effective detuning ΔE and interactions Λ:

$$\Delta E \equiv \left[-(\epsilon_a - \epsilon_b) + N(v_b - v_a)\right]/g\,, \tag{5.58a}$$
$$\Lambda \equiv (v_a + v_b)N/g\,. \tag{5.58b}$$

There is threshold Λ_c such that $\Lambda \geq \Lambda_c$ results in four (instead of two) fixed points, one of them unstable (instead of all stable). The value of Λ_c is

$$\Lambda_c = \sqrt{4 + \Delta E^2 + \frac{6(2\Delta E^2)^{2/3}}{\Xi^{1/3}} + 3(2\Delta E^2)^{1/3}\Xi^{1/3}}\,, \tag{5.59}$$

where $\Xi \equiv 4 + \Delta E^2 + |4 - \Delta E^2|$. The typical configuration where $\Delta E = 0$ reduces to the usual Josephson threshold $\Lambda_c = 2$. The behaviour of the phase, however, does not reflect the regime, as shown in Fig. 5.6(c) for the dynamics of two coupled BECs in presence of interactions, with three cases such that $\Lambda < \Lambda_c$ (blue, purple and green lines) and three others such that $\Lambda \geq \Lambda_c$ (green, cyan, orange). Strong interactions also lead to circle-on-the-sphere trajectories that are, however, far from the Rabi dynamics. This analysis shows that the nature of light-matter coupling can be subtle and fall within a wider range of theoretical descriptions all leading to essentially the same observable behaviours, but with different underlying interpretations. The presence of pumping and decay further complicate the picture. Rahmani and Laussy (2016) provide a detailed classification of fixed points in the several combinations that this simple problem offers.

5.4.4 Jaynes–Cummings model

All the previous material has been leading us towards the full quantum treatment of the two-level system interacting with a light mode, where both the atom (or exciton) *and* the photon are quantised, according to eqn (5.28). The cases we have been dealing with so far, where the material excitation is modelled as a harmonic quantum oscillator, do not encompass the fermionic limit, which is the more important in the case of material excitations.

Different physics occurs when such excitations are described by fermionic rather than bosonic statistics only in the nonlinear regime, where more than one excitation resides in the system so that higher energy states are probed. The ground and first excited states are otherwise identical in the Bose and Fermi cases. With the advent of lasers, such cases are, however, not difficult to realise experimentally. In the case of cavity QED, the matter part is usually a beam of atoms passing through the cavity, and a single excitation is the independent response of the atoms to the light-field excitation. The simplest situation is that of a dilute atomic beam where a single atom (driven at

Edwin Thompson **Jaynes** (1922–1998) and Frederick **Cummings**

Jaynes challenged the mainstream of physics, most notably with his proposition to use Bayesian probabilities to formulate a reinterpretation of statistical physics as inferences due to incomplete information. In quantum optics, he rejected the Copenhagen interpretation that he qualified as *"a considerable body of folklore"*. He wrote of Oppenheimer (whom he called Oppy and felt would be an unsuitable Ph.D. advisor to carry out independent research):

Oppy would never countenance any retreat from the Copenhagen position, of the kind advocated by Schrödinger and Einstein. He derived some great emotional satisfaction from just those elements of mysticism that Schrödinger and Einstein had deplored, and always wanted to make the world still more mystical, and less rational. This desire was expressed strongly in his 1955 BBC Reith lectures (of which I still have some cherished tape recordings which recall his style of delivery at its best). Some have seen this as a fine humanist trait. I saw it increasingly as an anomaly—a basically anti-scientific attitude in a person posing as a scientist—that explains so much of the contradictions in his character.

In quantum optics, he questioned the need of full quantisation of the optical field, e.g., to explain effects such as blackbody radiation, spontaneous emission or the Lamb shift, the latter two he claimed could be obtained in the realm of his so-called neoclassical theory, whose fields offer the additional advantage to be *"conspicuously free from many of the divergence problems of quantum electrodynamics."* Ironically, he is today most remembered in quantum optics for the Jaynes and Cummings (1963) model the main merit of which is to be an integrable fully-quantised system. This "drosophila" of quantum mechanics was initially derived in a form and with approximations allowing for comparison with a classical description in favour of which Jaynes was ready to bet, like for the origin of the Lamb shift (the outcome of his bet with Lamb was left undecided).

Frederick Cummings was Jaynes' Ph. D. student who, in his (Cummings (1950)) reminiscences imagined *"that if another of Ed's students (e.g. Joe Eberly) had come into his office fifteen minutes before me, the 'Jaynes–Cummings' (JC) model would now perhaps be called otherwise, e.g., the 'JE' model"*. He passed on the problem to his own student, Tavis and Cummings (1968), to tackle the case of N emitters. He is now a professor Emeritus at the University of California, Riverside. His interest turned to biophysics in the mid-1980s.

resonance so that it appears as a two-level system) is coupled to a Fock state of light. This case has been described theoretically by Jaynes and Cummings (1963) to yield the model that now bears their name.

The model is similar to the Bloch equations except that the classical field is now upgraded to a quantum field, and for our single mode, to the Bose annihilation operator a. We rewrite the Hamiltonian (5.28) in the interaction picture and with the approximation of Jaynes and Cummings,

$$H = \hbar g(e^{i\Delta t}\sigma a^{\dagger} + e^{-i\Delta t}\sigma^{\dagger}a), \tag{5.60}$$

where we have again denoted Δ the detuning in frequency between the cavity mode (with boson operator a) and the atom (with fermion operator σ). Observe that at resonance, $\Delta = 0$, the Hamiltonian is time independent, which allows for straightforward solutions. Out of resonance, more mathematics are involved, but in the limit of very high detunings, conservation of the number of photons and excitation in the atom is recovered, and another limiting case is worthy of interest.

To compute with σ and $\sigma^{\dagger}$, one can go to its matrix representation

$$\sigma^{\dagger} = \begin{pmatrix} 0 & 1 \\ 0 & 0 \end{pmatrix}, \tag{5.61}$$

with $|0\rangle = (0, 1)^{T}$ and $|1\rangle = (1, 0)^{T}$. The direct application of these definitions, i.e., $\sigma|1\rangle = |0\rangle$ and $\sigma|0\rangle = 0$, show that one can subtract one excitation from the excited state but not from the vacuum, and the reciprocal $\sigma^{\dagger}|0\rangle = |1\rangle$ and $\sigma^{\dagger}|1\rangle = 0$, that one can excite the ground state but, from Pauli exclusion, not the one already excited. All the needed expressions follow from the definition above, such as $\sigma\sigma^{\dagger}\sigma = \sigma$. A general result for evaluating products in normal order reads

$$\sigma^{\dagger m}\sigma^{n}|\mu\rangle = \delta_{m0}(\delta_{n0}\delta_{\mu0} + \delta_{n1}\delta_{\mu1})|0\rangle + (\delta_{m0}\delta_{n0}\delta_{\mu1} + \delta_{m1}\delta_{n0}\delta_{\mu0} + \delta_{m1}\delta_{n1}\delta_{\mu1})|1\rangle. \tag{5.62}$$

Commutation relations are also useful, especially when computing in the Heisenberg picture or evaluating averages.[82] Coupled to the results of Section 3.1.5, these results allow to obtain the equations of motion for the Jaynes–Cummings model. In this case, because the excitation can only be transferred from the atom to the cavity or vice versa, but none is ever lost or created under eqn (5.28), all the dynamics of a fixed number n of total excitations is contained within the manifold

$$\mathcal{H}_{n} = \{|0, n\rangle, |1, n-1\rangle\}, \tag{5.63}$$

provided that $n \geq 1$. The associated energy diagram appears in Fig. 5.5, with two states in each manifold (in our conventions $|0, n\rangle$ refers to the bare states with the atom in the ground state and n photons, while $|1, n-1\rangle$ has the atom in the excited state and $n-1$ photons). The total wavefunction $|\Psi(t)\rangle$ is a superposition of these states. If the number of excitations is not fixed, one can still decouple independent dynamics of $|\Psi\rangle$ by projecting the state onto $\mathcal{H}_{n}$, which can always be solved independently following

---

[82] A general result of particular help in many algebraic computations is

$$[\sigma^{\dagger\mu}\sigma^{\nu}, \sigma^{\dagger\eta}\sigma^{\theta}] = \nu\theta(\eta - \mu)\sigma + \mu\eta(\nu - \theta)\sigma^{\dagger} + [(1-\mu)\nu\eta(1-\theta) - \mu(1-\nu)(1-\eta)\theta](1 - 2\sigma^{\dagger}\sigma).$$

Other cases with more operators can be obtained by iterating the results above.

the procedure below, and put back together again at the end. The formal solution of Schrödinger equation reads

$$|\Psi(t)\rangle = \exp\left(-\frac{i}{\hbar}Ht\right)|\Psi(0)\rangle \qquad (5.64a)$$

$$= \exp\left(-ig(\sigma a^\dagger + \sigma^\dagger a)t\right)|\Psi(0)\rangle \ . \qquad (5.64b)$$

To compute the exponential in eqn (5.64b), since $\exp(-\omega\Omega) = \sum_{n\geq 0}(-\omega)^n\Omega^n/n!$ with $\omega \in \mathbb{C}$ and $\Omega \in \otimes_{n\geq 0}\mathcal{H}_n$, we decompose it into its real and imaginary parts,

$$\exp\left(-ig(\sigma a^\dagger + \sigma^\dagger a)t\right) = c - is, \qquad (5.65)$$

where

$$c = \sum_{n=0}^\infty \frac{(-1)^n(gt)^{2n}}{(2n)!}(\sigma a^\dagger + \sigma^\dagger a)^{2n}, \qquad (5.66a)$$

$$s = \sum_{n=0}^\infty \frac{(-1)^n(gt)^{2n+1}}{(2n+1)!}(\sigma a^\dagger + \sigma^\dagger a)^{2n}(\sigma a^\dagger + \sigma^\dagger a), \qquad (5.66b)$$

and we are reduced to algebraic computation of the type $(\sigma a^\dagger + \sigma^\dagger a)^{2n}$.

Exercise 5.10 $^{(**)}$ *Taking advantage of the algebra of σ, especially of such properties as $\sigma^2 = \sigma^{\dagger 2} = 0$, show that*

$$(\sigma a^\dagger + \sigma^\dagger a)^{2n} = \begin{pmatrix} (aa^\dagger)^n & 0 \\ 0 & (a^\dagger a)^n \end{pmatrix} \qquad (5.67)$$

in the basis of bare states and consequently that c and s as defined by eqns (5.66) have the following matrix representations:

$$c = \begin{pmatrix} \cos(gt\sqrt{aa^\dagger}) & 0 \\ 0 & \cos(gt\sqrt{a^\dagger a}) \end{pmatrix} \quad and \quad s = \begin{pmatrix} 0 & \frac{gt\sqrt{aa^\dagger}}{\sqrt{aa^\dagger}}a \\ \frac{gt\sqrt{a^\dagger a}}{\sqrt{a^\dagger a}}a^\dagger & 0 \end{pmatrix}, \qquad (5.68)$$

where we noted, e.g., $(\sqrt{a^\dagger a})^{2n} = (a^\dagger a)^n$, so as to carry out the series summation exactly with the new operator defined unambiguously (and as one can check, correctly) on the Fock-state basis.

Observe that the time propagator has off-diagonal elements. They correspond to virtual processes where an odd number of excitations is exchanged between the two fields. The knowledge of eqns (5.68) provides the dynamics, including such effects, by direct computation. Let us consider the case of fixed n and the atom in a quantum superposition of ground and excited states:

$$|\Psi(0)\rangle = (\chi_g\,|g\rangle + \chi_e\,|e\rangle) \otimes |n\rangle\,, \qquad (5.69)$$

which is separable because there is no summation to entangle with other configurations. The label n does not intervene directly and one can write

$$|\Psi(0)\rangle = \chi_g\,|g,n\rangle + \chi_e\,|e,n\rangle\,, \qquad (5.70)$$

which, from eqn (5.64b), leads to

$$
\begin{align}
|\Psi(t)\rangle &= (c-is)(\chi_g\,|g,n\rangle + \chi_e\,|e,n\rangle) \tag{5.71a}\\
&= \chi_g[\cos(\sqrt{n}gt)\,|e,n\rangle - i\sin(\sqrt{n}gt)\,|g,n-1\rangle] \tag{5.71b}\\
&\quad + \chi_e[\cos(\sqrt{n+1}gt)\,|e,n\rangle - i\sin(\sqrt{n+1}gt)\,|g,n+1\rangle]\,. \tag{5.71c}
\end{align}
$$

This dynamics bears much resemblance to the Rabi oscillations of Section 5.2, but the more pronounced quantum character results in a more complicated dynamics.

If the dynamics is constrained from the initial condition to hold in a manifold, e.g., if the initial state is

$$|\Psi(0)\rangle = |e,n-1\rangle\,, \qquad (5.72)$$

then the only other state available to the dynamics under eqn (5.28) is $|g,n\rangle$ as one can check from the closure relations

$$
\begin{align}
\sigma a^\dagger\,|e,n-1\rangle = |g,n\rangle\,, \qquad && \sigma^\dagger a\,|e,n-1\rangle = 0\,, \tag{5.73a}\\
\sigma a^\dagger\,|g,n\rangle = 0\,, \qquad && \sigma^\dagger a\,|g,n\rangle = |e,n-1\rangle\,. \tag{5.73b}
\end{align}
$$

The dynamics is therefore closed in $\mathcal{H}_n$, where the Hamiltonian can be diagonalised exactly. For the resonant condition, the dressed states for this manifold are split by an energy $2\sqrt{n}\hbar g$. In the general case, all four transitions between the states in manifolds $\mathcal{H}_n$ and $\mathcal{H}_{n-1}$ are possible, and this would result in a quadruplet in the emitted spectrum. It is difficult to resolve this quadruplet, but it has been done in a Fourier transform of time-resolved experiments by Brune et al. (1996). It is simpler to consider photoluminescence directly under continuous excitation at high intensity (where the fluctuations of particles number have little effect). In this case, with $n \gg 1$, the two intermediate energies are almost degenerate and a triplet is obtained with its central peak being about twice as high as the two satellites. This is the Mollow (1969) triplet of resonance fluorescence. We investigate this problem in more detail in Section 5.6.3.

Exercise 5.11 $^{(*)}$ *Consider now the out-of-resonance case where the two-level transition does not match the cavity photon energy, i.e., $\Delta \neq 0$ in eqn (5.60). The Hamiltonian is now time dependent and the formal integration eqn (5.64a) is no longer possible. Taking advantage of the closure relations (5.73), consider the ansatz*

$$|\Psi(t)\rangle = \sum_{n=0}^{\infty}[\psi_{g,n}(t)\,|g,n\rangle + \psi_{e,n-1}(t)\,|e,n-1\rangle]\,, \qquad (5.74)$$

where the time dependence is in the coefficients $\psi_{g,n}$, $\psi_{e,n-1}$ only. Show, therefore, that the Schrödinger equation with Hamiltonian (5.60) applied on state (5.74) yields

$$\begin{cases} \dot{\psi}_{g,n} = -ig\sqrt{n}e^{i\Delta t}\psi_{e,n-1} \\ \dot{\psi}_{e,n-1} = -ig\sqrt{n}e^{-i\Delta t}\psi_{g,n} \,. \end{cases} \tag{5.75}$$

Evoking as they do the flopping between two states, those equations are also reminiscent of Rabi dynamics.

Exercise 5.12 $^{(**)}$ *Use a technique of your choice to solve eqns (5.74). Show that the result reads*

$$\psi_{g,n}(t) = e^{i\Delta t/2}\left[\left(\cos\left(\Omega_n t\right) - i\frac{\Delta}{2\Omega_n}\sin\left(\Omega_n t\right)\right)\psi_{g,n}(0)\right. \tag{5.76a}$$
$$\left. - i\frac{g\sqrt{n}}{\Omega_n}\sin\left(\Omega_n t\right)\psi_{e,n-1}(0)\right],$$

$$\psi_{e,n-1}(t) = e^{-i\Delta t/2}\left[\left(\cos\left(\Omega_n t\right) + i\frac{\Delta}{2\Omega_n}\sin\left(\Omega_n t\right)\right)\psi_{e,n-1}(0)\right. \tag{5.76b}$$
$$\left. - i\frac{g\sqrt{n}}{\Omega_n}\sin\left(\Omega_n t\right)\psi_{g,n}(0)\right],$$

in terms of the generalised complex Rabi frequencies

$$\Omega_n = \sqrt{ng^2 + (\Delta/2)^2}\,. \tag{5.77}$$

This result in the full-quantum picture recovers the semiclassical eqns (5.10) with $\Omega_n \to \Omega$ and $g\sqrt{n} \to \Omega_R$. Equations (5.76) and (5.77) represent one of the most important results of quantum optics. They are the crowning achievement towards which all the results converge, either from simplified models, such as Bloch optical equations or Rabi flopping between two states, or from more refined theories, such as non-rotating wave Hamiltonians. This general solution connects directly to the resonant case treated previously from operator algebra. In the detuned case, where $2\sqrt{n}g \ll |\Delta|$, a serial expansion of $\psi_{e,n-1}(t)$ shows that the atom is inhibited in its transition and remains in the same state up to some phase fluctuations induced by the perturbation of the interaction.

One can describe decay to some extent by providing an imaginary part to the bare energies, that become $\omega_{a,\sigma} - i\gamma_{a,\sigma}/2$ (here, we use the notation of decay rates, γ, rather than lifetimes, T). The diagonalisation in this case provides the complex dressed states with energy

$$E^n_\pm/\hbar = n\omega_a - \frac{\Delta}{2} \pm R_n - i\frac{(2n-1)\gamma_a + \gamma_\sigma}{4}\,, \tag{5.78}$$

in terms of the nth-*manifold (half) Rabi splitting*

$$R_n = \sqrt{ng^2 - \left(\gamma_- + i\frac{\Delta}{2}\right)^2}\,. \tag{5.79}$$

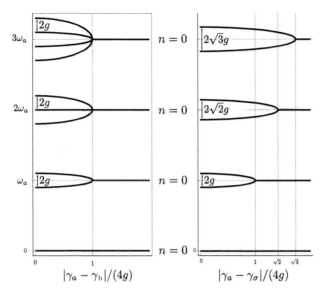

Fig. 5.7: Dressed bosons and Jaynes–Cummings ladder with dissipation, adapted from del Valle et al. (2009a). This offers the complete picture of strong and weak coupling of two Bose fields, left (cf. Fig. 5.3), and of a Bose field with a two-level system, right (cf. Fig. 5.5). Whereas n excitations split into as many dressed states in the linear case, leading to an essentially invariant picture of strong-coupling along the ladder, the saturation of the two-level system leads to renormalisation of the light–matter interaction. This makes strong-coupling a richer and more complex phenomenon, with direct quantisation of the field and sensibility at the one-quantum level, and relevant units as square roots of integers. Observe, for instance, how coherent interactions can be triggered at any value of the dissipation by climbing sufficiently high in the ladder.

In the expression above we have introduced the shortcut γ_- from a general notation that we shall use later:

$$\gamma_\pm = \frac{\gamma_a \pm \gamma_\sigma}{4}. \tag{5.80}$$

The most popular case is the vacuum Rabi splitting that provides the condition for strong-coupling with one excitation:

$$g > |\gamma_a - \gamma_\sigma|/4. \tag{5.81}$$

The same can also be done for coupled bosons. This allows to see the dressed bosons ladder, Fig. 5.3, and the Jaynes–Cummings ladder, Fig. 5.5, in a new light, namely, with dissipation. This has been done in Fig. 5.7, which is an important representation since it brings together in a natural way the quantum, or strong, coupling with the so-called weak-coupling. The real part of $E_\pm^n$ provides the energy of the dressed-state and the imaginary part its decay rate. This dissipative version is much richer than the conventional pictures. It shows, in particular, how dressed states open at different values of dissipation in the Jaynes–Cummings case, a natural consequence of their different splitting. This makes a general definition of strong-coupling in such systems a delicate problem (of which the coupled bosons do not suffer, even in the nonlinear regime). In

Section 5.6.3, we will see that this expedient is fully justified when rigorously derived from within an accurate dissipative master equation description.

5.4.5 *Dicke model*

Robert **Dicke** (1916–1997) was a versatile experimentalist, having contributed to the fields of radar physics, atomic physics, quantum optics, gravity, astrophysics, and cosmology. His contribution to superradiance displays his aptitudes in the theoretical field as well. He wrote his autobiography, dated 1975, but it has not been published (it is now possessed by NAS).

He once wrote: "*I have long believed that an experimentalist should not be unduly inhibited by theoretical untidiness. If he insists on having every last theoretical T crossed before he starts his research the chances are that he will never do a significant experiment. And the more significant and fundamental the experiment the more theoretical uncertainty may be tolerated.*"

Closely related to the linear coupling of the previous section lies the Dicke (1954) model that yields qualitatively similar results at low densities. In this model, the matter excitation is upgraded to creation operator J_+ for an excitation of the "matter field" that distributes the excitation throughout an assembly of N identical two-level systems described by fermion operators $\sigma_i^\dagger$, so that the material excitation operator $\sigma^\dagger$ in eqn (5.28) maps to J_+ with

$$J_+ = \sum_{i=1}^{N} \sigma_i^\dagger . \tag{5.82}$$

One checks readily that J_+, and $J_- = J_+^\dagger$ thus defined, obey an angular momentum algebra with magnitude $N(N+1)$ (and maximum z projection of J_z equal to N). In this case, the Rabi doublet arises in the limit where the total number of excitations μ (shared between the light and the matter field) is much less than the number of atoms, $\mu \ll N$, in which case the usual commutation relation $[J_-, J_+] = -2J_z$ becomes $[J_-/\sqrt{N}, J_+/\sqrt{N}] \approx 1$, which is the commutation for a bosonic field. This comes from the expression of a Dicke state with μ excitations shared by N atoms given as the angular momentum state $|-N/2 + \mu\rangle$. Therefore, the annihilation/creation operators J_-, J_+ for one excitation shared by N atoms appear in this limit like renormalised Bose operators $\sqrt{N}a$, $\sqrt{N}a^\dagger$, resulting in a Rabi doublet of splitting $2\hbar g\sqrt{N}$. Such a situation corresponds, e.g., to an array of small QDs inside a microcavity such that in each dot electron and hole are quantised separately, while our model describes a single QD that can accommodate several excitons. The corresponding emission spectra are close to those obtained here below the saturation limit $\mu \ll N$, while the nonlinear regime $N \gg 1$, $\mu \gg 1$ has peculiar behaviour, featuring non-Lorentzian emission lineshapes and a non-trivial multiplet structure, like the "Dicke fork" obtained by Laussy et al. (2005).

5.5 Lindblad dissipation

Göran **Lindblad** with his work "on the generators of quantum mechanical semigroups", Comm. Math. Phys. **48** (1976), 119, attached his name to terms that turn the Liouville–von-Neuman equation into a dissipative system.

Lindblad is a mathematical physicist and Emeritus Professor of the KTH ("Kungliga Tekniska högskolan", the Royal Institute of Technology) in Stokholm, from which he retired in 2005.

We have been able to introduce dissipation in the Rabi picture (dealing with a pure state) and in the Bloch picture (dealing with a density matrix). In these cases, however, the decay was phenomenological, adding a memoryless decay term to the matrix element to reproduce a sought effect. We now describe a popular formalism to introduce dissipation in a quantum system on a firm theoretical basis, which also gives many insights into the origin and nature of dissipation in a quantum system. We shall illustrate our point on the simple system of an harmonic oscillator, cf. eqn (3.29), whose dynamics has already been investigated from many different viewpoints.

The postulates of quantum mechanics do not accommodate well indirect attempts to "reproduce" dissipation. For instance, canonical quantisation of the equations of motion of the damped oscillator eqn (4.1) with the methods of Chapter 3 yields[83]

$$[X(t), P(t)] = e^{-\gamma t}[X(0), P(0)], \tag{5.83}$$

and the commutation relation and its derived algebra—which define the quantum field and such properties as its statistics—are lost in time. The dissipation introduced in this way washes away the quantum character of the system, as well as its dynamics. This is not a very satisfying picture: an atom that relaxes to its fundamental state should still remain an atom.

In the previous section where we have investigated the dynamics of two coupled quantum oscillators, we have seen how the excitation of one was transferred to the other. This is a dissipation insofar as the first system is considered over this interval of time. Then, the Hamiltonian dynamics brings back the excitation of the second oscillator to the first one, and the process continues back and forth cyclically. Here lies the key idea of a correct model for dissipation in quantum mechanics. It is not a fundamental characteristic of a system that needs to be quantised, but it is a feature of its dynamics: the quantum system gives its energy away and takes it back when it couples to another state. Imagine now that the system is coupled not to one, but to numerous other modes, which together form a *reservoir*, so that as the system's energy is exchanged with the

---

[83] The Hamiltonian equations with dissipation being $\dot{X} = P/m_0$ and $\dot{P} = -\gamma P - m_0\omega^2 X$, the correspondence principle eqn (3.8) makes $d[X, P]/dt = \dot{X}P + P\dot{X} - \dot{X}P - X\dot{P} = -\gamma[X, P]$ which integrates to the result eqn (5.83) from eqn (3.15).

reservoir, each mode has little probability to be the recipient, but the system has a high probability of losing its excitation. Once enough time has elapsed so that energy is with high probability in the reservoir, it will continue being exchanged in an oscillatory way, as follows from Hamiltonian dynamics that demands that all configurations of the systems be visited with the same weight. However, the system is so insignificant, compared with the reservoir, that the probability of a full return of its initial expenditure is vanishingly small with increasing size of the reservoir. In effect, this accounts for dissipation in quantum mechanics in the same way that, in classical physics, the Newtonian dynamics that violates thermodynamics (being reversible) is suppressed by the law of large numbers.

We detail the mathematical form in an explicit, simple case before turning to the general expression. Consider a single harmonic oscillator a coupled linearly to a set of oscillators b_i that all merge together to form the bath. All oscillators a and b_i follow the algebra of eqn (3.27). The coupling Hamiltonian reads

$$H = \hbar\omega_a a^\dagger a + \sum_i \hbar\omega_i b_i^\dagger b_i + \sum_i \hbar(g_i^* a b_j^\dagger + g_i a^\dagger b_j). \tag{5.84}$$

We have made the same approximations as in the previous section. The dynamics of this simple system is clear physically: the system a loses its excitation that is transferred to one of the b_i modes through the process $ab_i^\dagger$ weighted by the strength of the coupling g_i. The reverse process $a^\dagger b_i$, where the excitation is destroyed in the mode b_i and put back in the system exists. It follows, as we have already emphasised, from the Hamiltonian character of this dynamics. We now proceed to make the approximations that are in order when the set of b_i is treated as a reservoir. Namely, first of all we neglect the dynamics of the reservoir (being in equilibrium) and its direct correlations to the system. We write the total density matrix ϱ for the combined system described by eqn (5.84) as

$$\varrho(t) = \rho(t) \otimes R_0, \tag{5.85}$$

where R_0 is the time-independent density matrix of bosons at equilibrium:

$$R_0 = \bigotimes_i \left(1 - \exp\left(-\frac{\hbar\omega_i}{k_B T}\right)\right) \exp\left(-\frac{\hbar\omega_i b_i^\dagger b_i}{k_B T}\right). \tag{5.86}$$

Note that we have assumed the separability of the density matrix ϱ into a product of the density matrix of the system and of the reservoir. This is called the "Born approximation". It is physically reasonable for a reservoir, as each mode quickly washes away the coherence that it develops with the system through its interaction. In fact, note that R_0 in eqn (5.86) is also decorrelated for all modes. The important changes retained are the dephasing and loss of energy of the system alone. The latter now, moreover, accounts for the entire time dependency. To make it more sensible still, we shift to the interaction picture so that the rapid and trivial oscillation due to the optical frequency is removed. From now on, therefore,

$$a(t) = a e^{-i\omega_a t} \quad \text{and} \quad a^\dagger(t) = a^\dagger e^{i\omega_a t}, \tag{5.87}$$

as detailed in Section 3.1.6 ($a(0)$ and a of the Schrödinger picture being equal). The interacting Hamiltonian becomes

$$H(t) = \hbar\left[a(t)B^\dagger(t) + a(t)^\dagger B(t)\right],\tag{5.88}$$

where we have introduced the "reservoir operators"

$$B(t) = \sum_i g_i b_i e^{-i\omega_i t} \quad\text{and}\quad B^\dagger(t) = \sum_i g_i^* b_i^\dagger e^{i\omega_i t}.\tag{5.89}$$

It remains to eliminate the complicated structure of operators (5.89) and focus on the dynamics of ρ alone. This can be done by averaging over the degrees of freedom of the reservoir. Let us first obtain the equation of motion for ρ. If we formally integrate the Liouville–von Neumann equation, cf. eqn (3.53) for ϱ, we get

$$i\hbar\varrho(t) = i\hbar\varrho(-\infty) + \int_{-\infty}^{t} [H(t), \rho(t)]\, dt,\tag{5.90}$$

where, for a while, we write everywhere the time dependency. We have assumed the initial condition is at $t \to -\infty$.

Exercise 5.13 $^{(*)}$ *By inserting the exact result eqn (5.90) back into eqn (3.53) (for ϱ), show that the following integro-differential is obtained:*

$$\dot\varrho(t) = -\frac{1}{\hbar^2}\int_{-\infty}^{\tau} [H(t), [H(\tau), \varrho(\tau)]]\, d\tau.\tag{5.91}$$

It is enough in this case to limit to order two in the commutator, which yields the dynamics of the populations while neglecting that of the fluctuations. Later in this book, e.g., when we investigate the dynamics of the laser field in weak-coupling microcavities, we shall pursue this kind of expansion of eqn (5.90) further. They are known as *Born expansions*.

Exercise 5.14 $^{(**)}$ *Carry out the algebra of $[H(t), [H(\tau), \varrho(\tau)]]$ with definitions given by eqns (5.88) and (5.89) and show that under the approximation of separability given by eqn (5.85):*

$$\dot\rho = -\frac{1}{\hbar^2}\int_0^t \Big[(a^2\rho(\tau) - a\rho(\tau)a)e^{-i\omega_a(t+\tau)}\langle B^\dagger(t)B^\dagger(\tau)\rangle + h.c.\tag{5.92a}$$

$$+ (a^{\dagger 2}\rho(\tau) - a^\dagger\rho(\tau)a^\dagger)e^{i\omega_a(t+\tau)}\langle B(t)B(\tau)\rangle + h.c.\tag{5.92b}$$

$$+ (aa^\dagger\rho(\tau) - a^\dagger\rho(\tau)a)e^{-i\omega_a(t-\tau)}\langle B^\dagger(t)B(\tau)\rangle + h.c.\tag{5.92c}$$

$$+ (a^\dagger a\rho(\tau) - a\rho(\tau)a^\dagger)e^{i\omega_a(t-\tau)}\langle B(t)B^\dagger(\tau)\rangle + h.c.\Big]\, d\tau,\tag{5.92d}$$

with, for the reservoir correlators, $\langle B^\dagger(t)B^\dagger(\tau)\rangle = \sum_{i,j} g_i^* g_j^* e^{i(\omega_i t + \omega_j \tau)}\mathrm{Tr}(R_0 b_i^\dagger b_j^\dagger)$ *and* $\langle B(t)B(\tau)\rangle = \sum_{i,j} g_i g_j e^{-i(\omega_i t + \omega_j \tau)}\mathrm{Tr}(R_0 b_i b_j)$, *and more importantly*

$$\langle B^\dagger(t)B(\tau)\rangle = \sum_i |g_i|^2 e^{i\omega_i(t-\tau)}\bar{n}_T(\omega_i)\,, \tag{5.93a}$$

$$\langle B(t)B^\dagger(\tau)\rangle = \sum_i |g_i|^2 e^{-i\omega_i(t-\tau)}(\bar{n}_T(\omega_i)+1)\,, \tag{5.93b}$$

where $\bar{n}_T$ is given by the Bose–Einstein distribution, cf. eqns (3.80) and (5.86).

Lines (5.92)a and d cancel exactly because of repeated annihilation on the diagonal density operator eqn (5.86). Evaluation of eqns (5.93) is more involved. The physical idea that we wish to emphasise over direct attempts towards evaluating the mathematical expressions, is that given a time evolution for $\rho(\tau)$ that is much smaller than that of $e^{\pm i\omega_a(t-\tau)}$ and $\langle B^\dagger(t)B(\tau)\rangle$, $\langle B(t)B^\dagger(\tau)\rangle$ also with rapid dependence of the type $e^{\pm i\omega_i(t-\tau)}$, it is a good approximation to write

$$\rho(\tau) \approx \rho(t) \tag{5.94}$$

in eqn (5.92), which is called the *Markov approximation*, since the density matrix $\dot{\rho}(t)$ at time t on the left-hand side of eqn (5.92) loses the dependency on its value at a previous time $\tau < t$, i.e., it has no memory of its past. It is especially clear in the form in Exercise 5.15:

Exercise 5.15 [*] *Combining the arguments given above and using the representation* $\int e^{i(\omega_i-\omega_a)(t-\tau)}\,d\tau = \pi\delta(\omega_i - \omega_a)$, *reduce eqn (5.92) to the following first-order differential, Markovian equation:*

$$\dot{\rho} = A(2a^\dagger\rho a - aa^\dagger\rho - \rho aa^\dagger) \tag{5.95a}$$
$$+ B(2a\rho a^\dagger - a^\dagger a\rho - \rho a^\dagger a) \tag{5.95b}$$

where

$$A = \pi\sum_i |g_i|^2 \bar{n}_T(\omega_i)\delta(\omega_a - \omega_i)\,, \tag{5.96a}$$

$$B = \pi\sum_i |g_i|^2 (\bar{n}_T(\omega_i) + 1)\delta(\omega_a - \omega_i)\,. \tag{5.96b}$$

Coefficients (5.96) are those that would be obtained by Fermi's golden rule. They exhibit a delta-singularity because of the Markov approximation and the long-time average of the interaction between the system and the reservoir that requires conservation of the energy. By broadening the modes of the reservoir, they provide two rates that we associate in the following way to the master equation

$$\dot{\rho} = \bar{n}_T\frac{\kappa_a}{2}(2a^\dagger\rho a - aa^\dagger\rho - \rho aa^\dagger) \tag{5.97a}$$

$$+ (\bar{n}_T + 1)\frac{\kappa_a}{2}(2a\rho a^\dagger - a^\dagger a\rho - \rho a^\dagger a)\,. \tag{5.97b}$$

where $\kappa_a = 2\pi\sigma(\omega_a)|g(\omega_a)|^2$, $\sigma(\omega)$ being the density of states of the oscillators in the reservoir at frequency ω_a (σ is a smooth function of κ_a that is the continuous limit

Isidor Isaac **Rabi** (1898–1988) and Benjamin **Mollow**

Rabi studied in the 1930s the hydrogen atom and the nature of the force binding the proton to the atomic nuclei, out of which investigations he conceived molecular-beam magnetic-resonance, a new and very accurate detection method for which he was awarded the exclusive Nobel Prize for Physics in 1944, "for his resonance method for recording the magnetic properties of atomic nuclei". He wanted to be a theorist and had accepted only because the invitation was from the prestigious Stern. In his biography he is quoted as saying *"Whenever one of my students came to me with a scientific project, I asked only one question, 'Will it bring you nearer to God?'"*

Mollow is currently a Physics Professor at the University of Massachusetts Boston where he studies quantum optics.

of ω_i). Expression (5.97) is a popular one in quantum optics. Note that in the interaction picture, the present form eqn (5.97) is completely devoid of Hamiltonian dynamics. As we expect and as was the aim of the construction, it precisely describes dissipation, as is investigated further below. Before concentrating on this form, however, let us give the result of the most general possible form for a master equation under the assumption that the evolution is Markovian (as is the case for many models). It is known as the *Lindblad master equation* and reads

$$\dot{\rho} = -\frac{i}{\hbar}[H, \rho] - \frac{1}{\hbar} \sum_{n,m} h_{n,m} \left(\rho L_m L_n + L_m L_n \rho - 2 L_n \rho L_m \right) + \text{h.c.}, \qquad (5.98)$$

where the part from the Schrödinger equation $-\frac{i}{\hbar}[H, \rho]$, has been put back, and the terms on the right, L_m, are operators being, along with the constants $h_{n,m}$, defined by the system. In the case of the harmonic oscillator coupled to the bath, $L_0 = a$ and $L_1 = a^\dagger$, all others being zero. In the thermal case, coefficients A and B of eqn (5.95) are interrelated (see eqn (5.97)), but there is no reason to constrain them a priori.

Equation (5.98)—which we have derived from the Schrödinger equation in the thermal case after a physical model of coupling of a small system to a big reservoir—correctly reproduces features of dissipation, as can be seen by computing the equation of motions of the observables of interest. Consider, for example, the average number of excitations in an oscillator part of a reservoir, $n_a = \text{Tr}\left(\rho a^\dagger a \right)$. Its equation of motion is given by (we assume we are now in the Schrödinger picture to keep a time independent but the case of the interaction picture follows straightforwardly)

$$\frac{\partial n_a}{\partial t} = \text{Tr}\left[\left(\frac{\partial \rho}{\partial t}\right) a^\dagger a\right] \tag{5.99a}$$

$$= \bar{n}_T \frac{\kappa_a}{2} \text{Tr}\left[\left(2a^\dagger \rho a - aa^\dagger \rho - \rho aa^\dagger\right) a^\dagger a\right] \tag{5.99b}$$

$$+ (\bar{n}_T + 1)\frac{\kappa_a}{2} \text{Tr}\left[\left(2a\rho a^\dagger - a^\dagger a\rho - \rho a^\dagger a\right) a^\dagger a\right] . \tag{5.99c}$$

The approach of computing equations of motion of operators from the master equation is outlined above. It only remains to simplify the expression (5.99) using algebraic relations presented in Chapter 3. Let us carry out explicitly the case of line (5.99b): by cyclic permutation of the trace,[84] the density matrix can be factored out to give $(\bar{n}_T \kappa_a/2)\text{Tr}\left(\rho(2aa^\dagger aa^\dagger - a^\dagger aaa^\dagger - aa^\dagger a^\dagger a)\right)$ that give rise to new operators that simplifies further still in terms of quantities already known, for instance by noting that

$$2aa^\dagger aa^\dagger - a^\dagger aaa^\dagger - aa^\dagger a^\dagger a = aa^\dagger(aa^\dagger - a^\dagger a) + (aa^\dagger - a^\dagger a)aa^\dagger, \tag{5.100a}$$

$$= aa^\dagger[a, a^\dagger] + [a, a^\dagger]aa^\dagger, \tag{5.100b}$$

$$= 2aa^\dagger, \tag{5.100c}$$

using expressions like eqn (3.44). In this way, the total expression finally reduces to $\bar{n}_T \kappa_a \langle a^\dagger a\rangle$ and if the same is carried out for line (5.99c), one gets

$$\frac{\partial n_a}{\partial t} = -\kappa_a(n_a - \bar{n}_T), \tag{5.101}$$

i.e., the population relaxes to the Bose distribution $\bar{n}_T$. This integrates immediately to

$$n_a(t) = n_a(0)e^{-\kappa_a t} + \bar{n}_T(1 - e^{-\kappa_a t}), \tag{5.102}$$

which is the exact behaviour that one would expect for dissipation of an harmonic oscillator: a decay from its initial population $n_a(0)$ towards the mean value of the reservoir $\bar{n}_T$, on a timescale κ_a^{-1}.

Note how line (5.97b) is linked to the decay in the sense that it empties the mode, whereas line (5.97a) has the effect of a pump that can populate the mode. The Lindblad terms can be used to that effect in a large number of systems investigated in the Schrödinger picture, as we shall see in next Section.[85]

---

[84]Operators can be permuted cyclically in the trace, i.e., for arbitrary operators A, B and C,

$$\text{Tr}(ABC) = \text{Tr}(BCA) = \text{Tr}(CAB), \tag{xviii}$$

whereas it is not true, in general, for other permutations, e.g., $\text{Tr}(ABC) \neq \text{Tr}(BAC)$.

[85]In the Heisenberg picture, the average of the reservoir operator serves as a fluctuating force that, by the action of the "fluctuation–dissipation" theorem, also results in a decay of the averages. There is less emphasis on the quantum aspect of the decay in this case since there is always the possibility or temptation to understand the dynamical averages as semiclassical and the whole formalism becomes one that favours classical interpretations or analogies.

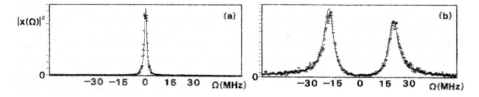

Fig. 5.8: Transmission spectra (around the 3P–3S transition at 589 nm) through a $L = 1.7$ nm cavity with finesse of 20 000 that is (a) empty or (b) contains sodium atoms, from Raizen et al. (1989).

Exercise 5.16 $^{(**)}$ *Show that the master equation (5.97) written for the Glauber P function becomes (cf. relations (3.86))*

$$\frac{\partial P}{\partial t} = \left[\frac{\kappa_a}{2} \left(\frac{\partial}{\partial \alpha} \alpha + \frac{\partial}{\partial \alpha^*} \alpha^* \right) + \kappa_a \bar{n}_T \frac{\partial^2}{\partial \alpha \partial \alpha^*} \right] P . \qquad (5.103)$$

Exercise 5.17 $^{(*)}$ *Check that the following expression for P satisfies eqn (5.103) with initial condition $\delta(\alpha - \alpha_0)$ in P space ($|\alpha_0\rangle$ with Dirac notation)*

$$P(\alpha, \alpha^*, t) = \frac{1}{\pi \bar{n}_T (1 - \exp(-\kappa_a t))} \exp \left[-\frac{|\alpha - \alpha_0 \exp(-(\kappa_a/2)t)|^2}{\bar{n}_T (1 - \exp(-\kappa_a t))} \right] . \qquad (5.104)$$

5.6 Quantum dynamics with decay and pumping

After considering the pure quantum dynamics of a coupled light–matter system free of any dissipation, and then in presence of decay, it remains to add pumping to complete our description.[86] In the time domain, pumping can be pulsed or continuous. In the energy domain, pumping can be coherent or incoherent. All combinations are possible. We will address particularly the case of *incoherent* and *continuous* pumping. This regime probes the intrinsic dynamics of the system, and is, in many respects, an extension of the spontaneous emission case, which is the most fundamental. With coherent excitation, new phenomena arise thanks to the coherent properties of the driving source. Another name for this regime is "resonant" excitation. This will be studied in further details in Chapter 7. The pulsed regime also opens new horizons in the quantum dynamics of light–matter coupling, which we will not, however, further consider in this text. Coherent pumping enters the Hamiltonian directly, while incoherent pumping can be described thanks to the Lindblad formalism outlined in the previous Section.

The first observations of the vacuum-Rabi splitting were made with atomic systems, by Raizen et al. (1989), who showed that the atomic decay rate was halved on resonance, because the atoms spent half their time in the de-excited state with the energy in the cavity photon field. This is shown in Fig. 5.8. Because of the small optical cross section for atoms, much improved cavity finesse was needed before. McKeever et al. (2003) demonstrated emission and lasing from a single atom in the strong-coupling regime.

Although the exciton-polariton in planar microcavities as first observed by Weisbuch et al. (1992) had been initially thought to provide the solid-state realisation of

---

[86]Except some contrived particular cases, pumping requires decay as otherwise the system would diverge.

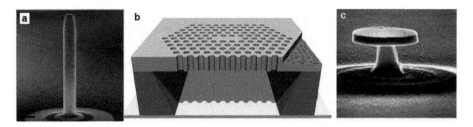

Fig. 5.9: The three seminal systems in which have been reported the first observations of strong coupling between a quantum dot and a single-mode microcavity. (a) A pillar microcavity containing elongated quantum dots from Reithmaier et al. (2004), (b) a photonic crystal microcavity from Yoshie et al. (2004) and (c) a microdisk microcavity from Peter et al. (2005). The result of the first experiment is shown in Fig. 5.15.

quantum (or strong) coupling as realised in atomic cavity QED, the 2D polaritons represent in fact a different system with little in common with the seminal paradigm set by atoms. The semiconductor systems that most closely describes the fundamental physics of quantum coupling of light and matter are those where this interaction is confined to zero-dimension, that is, quantum dots in single-mode microcavities. Strong-coupling in such systems has been reported for the first time by Reithmaier et al. (2004) and Yoshie et al. (2004), in two consecutive letters to *Nature*. Shortly later the claim was repeated by Peter et al. (2005) in *Physical Review Letters*. Their respective systems are shown in Fig. 5.9. One might think that the delay to achieve this regime, decades after its observation with atoms, was merely a technical problem, and that the theory was already laid down in 20 years of literature. However, there is now much evidence that the semiconductor case brings forward many specificities of its own. We shall discuss some of them in the following. An important asset playing in favour of the solid-state implementation is that one has a much better control of the system. A quantum dot stuck in the cavity can be kept immobile, while in the atomic cavity case, the excitations are beams of atoms with much difficulty to single out one atom or to deal with it for prolonged periods of time. This also restricts how much the cavity can be shrunk. In fact, strong-coupling with one and the same atom has been reported the same year as the semiconductor case, by Boca et al. (2004). The advantages of solid state implementation is obvious for technological purposes.[87] In Fig. 5.10 is shown the sketch of a recent design and successful fabrication of a microcavity QED setup, by Laucht et al. (2009b), enabling one to control electrically weak and strong coupling, both quickly and easily, with a straightforward applied voltage. One can speculate on the basis of such achievements how close is the operation at room temperature of more elaborate devices, which will bring to our daily life the wonders of quantum technology.

5.6.1 *Single-time dynamics of coupled Bose fields*

We revisit the content of previous Sections in order and start with the coupling of two Bose fields, or, in a zero-dimensional system, of two-harmonic oscillators. The master

---

[87]In the words of Weisbuch et al. (1992): "*Besides its relying on a much simpler implementation—the solid-state system is monolithic—the effect should lead to useful applications.*"

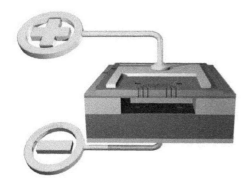

Fig. 5.10: On-chip microcavity QED device, as designed, fabricated and characterised by Laucht et al. (2009b). Tuning of the dot and cavity resonances is here performed (through the confined Stark effect) by simply applying a bias voltage. This allows to quickly and reversibly control the quantum light–matter coupling. Such devices open the way to electrically tunable quantum technology.

equation reads

$$\frac{d\rho}{dt} = i[\rho, H_a + H_b + H_{ab}] \tag{5.105a}$$

$$+ \frac{\gamma_a}{2}(2a\rho a^\dagger - a^\dagger a\rho - \rho a^\dagger a) + \frac{\gamma_b}{2}(2b\rho b^\dagger - b^\dagger b\rho - \rho b^\dagger b) \tag{5.105b}$$

$$+ \frac{P_a}{2}(2a^\dagger \rho a - aa^\dagger \rho - \rho aa^\dagger) + \frac{P_b}{2}(2b^\dagger \rho b - bb^\dagger \rho - \rho bb^\dagger), \tag{5.105c}$$

where $H_c = \hbar\omega_c c^\dagger c$, with $c = a, b$ and $H_{ab} = \hbar g(a^\dagger b + ab^\dagger)$, as before. Line (5.105a) contains the Hamiltonian (coherent, or quantum) dynamics, line (5.105b) the decay of both modes and line (5.105c) their pumping. The novelty in this Section is to be found in this third line. Note that, in addition to the excitation of the quantum dot itself, P_b, which is the natural source of excitation since electron–hole pairs are created in the system, we have also included a cavity pumping term, P_a. There has been an ample literature devoted to the identification of this cavity feeding mechanism, which populates the system directly with cavity photons. Its existence was betrayed as it is still effective even when the dot is far detuned from the cavity mode, an oddity particularly pointed out in experimental works such as those of Hennessy et al. (2007) or Kaniber et al. (2008). Underlying mechanisms have been investigated by Suffczyński et al. (2009), who identify dephasing as responsible for this cavity feeding, and Hohenester et al. (2009) and Hohenester (2010), who linked it to the phonon bath. Investigations by Winger et al. (2009) have also shown the role of the continuum of multiexciton states in the cavity feeding at larger detunings, a mechanism further identified by time-resolved experiments from Laucht et al. (2010), showing temporal correlations between the cavity emission and multiexciton states.

A more detailed description of eqns (5.105) at the quantum optical level is also possible, to take into account spin, other excited states, different charges configurations, etc. Yamaguchi et al. (2009) thus modelled an exciton complex and Ritter et al. (2010) provided a description of the microscopic structure by analysing Lindblad terms of the type

$$\sum_{i,j} \frac{\gamma_{ij}}{2} \left(2a_i^\dagger a_j \rho a_j^\dagger a_i - a_j^\dagger a_i a_i^\dagger a_j \rho - \rho a_j^\dagger a_i a_i^\dagger a_j \right), \tag{5.106}$$

that describe incoherent transitions between levels $|i\rangle$ and $|j\rangle$ at the rate γ_{ji}, using, in their case, four levels to model two confined quantum dot states for holes ($|1\rangle$, $|2\rangle$) and electrons ($|3\rangle$, $|4\rangle$). The "inverse relaxation" rate γ_{41} describes external pumping (excitation of electrons and holes) and γ_{34}, γ_{12} the ones intrinsic to this microscopic configuration. This allows the description of a much richer picture and takes into account many phenomena relevant to the solid-state system.

The theoretical models to describe at various degrees of accuracy and approximations the mechanisms above may differ in their details from the basic description afforded by eqns (5.105), which retains the concept of cavity feeding at its most fundamental level. Other approaches, such as that of Kaer et al. (2010), go beyond Markov approximations. In the following, we will not enter into such lines of active research and consider only Lindblad terms of the form of eqns (5.105), that will unravel the most important features of the problem. As compared with the previous Section, we do not assume any relationship between the pumping parameters P_c and the decay rate γ_c, although such a relationship can exist and may be established by some underlying microscopic derivations of eqns (5.105). For instance, assuming that cavity pumping P_a is due to other dots in weak-coupling will provide such a nontrivial relationship, as shown by del Valle and Laussy (2010). In any case, one can of course trivially adapt this general theory to any particular case which is of the Lindblad form above, by mere substitution of *ad hoc* expressions for the parameters. For instance, by comparing with the derivation of eqn (5.97), we can obtain the parameters for a thermal reservoir for a bosonic mode a (with frequency ω_a and $H_a = \omega_a a^\dagger a$) at temperature T. The effective rate of excitation reads

$$P_a = \kappa_a \bar{n}_T \,, \tag{5.107}$$

with $\bar{n}_T$ given by the reservoir Bose–Einstein distribution. It vanishes at $T = 0$. In thermal equilibrium, the system is losing excitations at a larger rate of

$$\gamma_a = \kappa_a(1 + \bar{n}_T) = \kappa_a + P_a \,. \tag{5.108}$$

The parameter κ_a is the *spontaneous emission* (SE) rate at $T = 0$. The steady state thermal equilibrium reads

$$n_a = \langle a^\dagger a \rangle = \frac{P_a}{\gamma_a - P_a} = \frac{P_a}{\kappa_a} = \bar{n}_T \,, \tag{5.109}$$

as anticipated in eqn (5.102). Substituting expressions (5.107)–(5.108) in eqns (5.105) and, therefore, in all the results that follow from it, will provide the particular case of a thermal cavity pumping. Given the triviality of such a substitution if the need arises, there is no reason to restrict the master equation (5.105) to a given type of reservoir (such as a thermal bath). A microcavity QED system is a complicated solid state system, open to many sources of excitations. In the most general case, one can include gain effects. Granted all together, the microscopic coefficients are terms of the form

$$\gamma_a = \kappa_a(1 + \bar{n}_T) + G_a \bar{m}_{-T'} \,, \tag{5.110a}$$
$$P_a = \kappa_a \bar{n}_T + G_a(1 + \bar{m}_{-T'}) \,, \tag{5.110b}$$

that is, including loss media and gain media. Net losses in the case of a cavity mode comes, among other reasons, from the fact that the photons can escape the cavity through the imperfect mirrors. Net gain may come from surrounding off-resonance or weakly coupled QDs, high energy QD levels or the wetting layer. Gardiner and Zoller (2000) studied a bath of uncorrelated two-level emitters that are kept on average in the excited state, providing such a gain.

Because of the incoherent pumping, that brings excitations at a random time, there is no deterministic expression for the operators. A stochastic noise term can be included in some cases. It is usually more convenient to consider averages directly. The single-time mean values of interest for this problem are obtained by solving the equation of motion of the coupled system, using the techniques of Chapter 3 and of the present chapter. We introduce the effective broadening

$$\Gamma_a = \gamma_a - P_a \quad \text{and} \quad \Gamma_b = \gamma_b - P_b, \tag{5.111}$$

where $\Gamma_\pm$ is defined as

$$\Gamma_\pm = \frac{\Gamma_a \pm \Gamma_b}{4}. \tag{5.112}$$

The equations for the population n_a and n_b are closed with the equation for n_{ab}, which together form the set of linearly coupled equations

$$\partial_t \begin{pmatrix} n_a \\ n_b \\ n_{ba} \\ n_{ab} \end{pmatrix} = \begin{pmatrix} P_a \\ P_b \\ 0 \\ 0 \end{pmatrix} + \begin{pmatrix} -\Gamma_a & 0 & ig & -ig \\ 0 & -\Gamma_b & -ig & ig \\ ig & -ig & -i\Delta - 2\Gamma_+ & 0 \\ -ig & ig & 0 & i\Delta - 2\Gamma_+ \end{pmatrix} \begin{pmatrix} n_a \\ n_b \\ n_{ba} \\ n_{ab} \end{pmatrix}, \tag{5.113}$$

where, as usual, $\Delta = \omega_a - \omega_b$ is the detuning between the bare modes. This covers both the case of spontaneous emission, by setting $P_{a,b}$ to zero and choosing the initial conditions, and the steady state, by setting the left-hand side of eqn (5.113) and solving the linear equation. This procedure introduces the *complex (half) Rabi frequency*

$$R = \sqrt{g^2 - \left(\Gamma_- + i\frac{\Delta}{2}\right)^2}, \tag{5.114}$$

that arises as a direct extension of the dissipationless case, eqn (5.11). This also recovers, rigorously, the case $n = 1$ of expression (5.79), which was obtained with complex energies in the Hamiltonian.

Exercise 5.18 (**) *Solve eqn (5.113) for the most general initial conditions*

$$n_a^0 \equiv n_a(0), \quad n_b^0 \equiv n_b(0), \quad \text{and} \quad n_{ab}^0 \equiv n_{ab}(0), \tag{5.115}$$

and with arbitrary (positive) pumping P_a and P_b. Check that your solution recovers the particular case below of the cavity population when pumping terms are both zero:

$$
\begin{aligned}
n_a(t) = e^{-2\gamma_+ t} \Bigg\{ & \left[\cos\left(R_r t\right) + \cosh\left(R_i t\right) \right] \frac{n_a^0}{2} \\
& - \left[\cos\left(R_r t\right) - \cosh\left(R_i t\right) \right] \frac{(\frac{\Delta^2}{4} + \gamma_-^2)n_a^0 + g^2 n_b^0 + g\Delta\Re n_{ab}^0 - 2g\gamma_- \Im n_{ab}^0}{2|R|^2} \\
& + \left[\frac{\sin\left(R_r t\right)}{R_r} + \frac{\sinh\left(R_i t\right)}{R_i} \right] \left(g\Im n_{ab}^0 - \gamma_- n_a^0 \right) \\
& + \left[\frac{\sin\left(R_r t\right)}{R_r} - \frac{\sinh\left(R_i t\right)}{R_i} \right] \times \\
& \quad \frac{\gamma_-(\frac{\Delta^2}{4} + \gamma_-^2 - g^2)n_a^0 + g\Delta\gamma_- \Re n_{ab}^0 + g(\frac{\Delta^2}{4} - \gamma_-^2 + g^2)\Im n_{ab}^0}{|R|^2} \Bigg\},
\end{aligned}
$$

$$(5.116)$$

where we decompose the complex Rabi frequency as

$$
R = R_r + iR_i . \tag{5.117}
$$

Check also that your solution converges to the steady state solutions

$$
n_a^{SS} = \frac{g^2 \Gamma_+ (P_a + P_b) + P_a \Gamma_b (\Gamma_+^2 + (\frac{\Delta}{2})^2)}{4g^2 \Gamma_+^2 + \Gamma_a \Gamma_b (\Gamma_+^2 + (\frac{\Delta}{2})^2)}, \tag{5.118a}
$$

$$
n_{ab}^{SS} = \frac{\frac{g}{2}(\gamma_a P_b - \gamma_b P_a)(i\Gamma_+ - \frac{\Delta}{2})}{4g^2 \Gamma_+^2 + \Gamma_a \Gamma_b (\Gamma_+^2 + (\frac{\Delta}{2})^2)}. \tag{5.118b}
$$

From the above, can you get the expression for $n_b(t)$ easily? The crossed mean value $n_{ab}(t)$ that reflects the coherent coupling may require more work.

The mean value $n_a(t)$ is plotted in Fig. 5.11(a) for three initial conditions: an exciton, a photon and a polariton. In the steady state, both photonic and excitonic reduced density matrices are diagonal. They correspond to thermal distributions of particles with the above mean numbers, as discussed by Alicki (1989):

$$
\rho_{n,p}^a = \sum_m \rho_{n,m;p,m} = \delta_{n,p} \frac{(n_a^{SS})^n}{(1 + n_a^{SS})^{n+1}}, \tag{5.119a}
$$

$$
\rho_{m,q}^b = \sum_n \rho_{n,m;n,q} = \delta_{m,q} \frac{(n_b^{SS})^m}{(1 + n_b^{SS})^{m+1}}. \tag{5.119b}
$$

Behind their forbidding appearance, eqns (5.118) enjoy a transparent physical meaning, that they inherit from the semi-classical—and, therefore, intuitive—picture of rate equations. When the coupling strength between the two modes, g, vanishes, the solutions are those of a source and sink problem for bosons, eqn (5.109):

$$
n_a^{SS}(g = 0) = \frac{P_a}{\gamma_a - P_a}, \tag{5.120}
$$

(the same for b throughout by interchanging a and b indexes), i.e., they are solutions of $\partial_t n_a = -\gamma_a n_a + P_a(n_a + 1)$, featuring the famous Bose stimulation effect, whereby

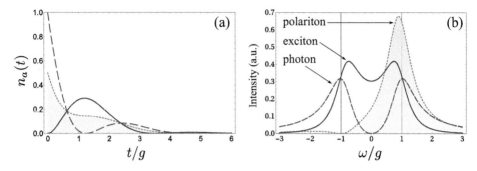

Fig. 5.11: (a) Cavity population as a function of time $n_a(t)$, eqn (5.116), for three initial conditions: an exciton (solid line), a photon (dashed) and the upper polariton (dotted), from Laussy et al. (2009). The starting point modulates the dynamics, that is otherwise same, in particular, the Rabi frequency is identical (note that all lines intersect at common points). (b) The corresponding power spectra for these three cases, with corresponding drawing style. Vertical lines are Rabi resonances. The initial condition affects considerably the spectral shape. The Rabi doublet is better resolved for the case of a photon than for an exciton. The difference becomes qualitative for the case of a polariton, where only a single line, rather than a doublet, is observed.

the probability of relaxation towards the final state is increased by its population. In the general case, where $g \neq 0$, the mean numbers can also be written in the form of eqn (5.120):

$$n_a^{\mathrm{SS}} = \frac{P_a^{\mathrm{eff}}}{\gamma_a^{\mathrm{eff}} - P_a^{\mathrm{eff}}},\tag{5.121}$$

(the same for $a \leftrightarrow b$), in terms of effective pump and decay rates

$$P_a^{\mathrm{eff}} = P_a + \frac{Q_a}{\Gamma_a + \Gamma_b}(P_a + P_b),\tag{5.122a}$$

$$\gamma_a^{\mathrm{eff}} = \gamma_a + \frac{Q_a}{\Gamma_a + \Gamma_b}(\gamma_a + \gamma_b),\tag{5.122b}$$

with Q_a the rate at which mode a exchanges particles with mode b,

$$Q_a = \frac{4(g^{\mathrm{eff}})^2}{\Gamma_b},\tag{5.123}$$

in terms of the effective coupling strength at nonzero detuning

$$g^{\mathrm{eff}} = \frac{g}{\sqrt{1 + \left(\frac{\Delta/2}{\Gamma_+}\right)^2}}.\tag{5.124}$$

Q_a is a generalisation of the *Purcell rate* $\gamma_a^{\mathrm{P}} = 4g^2/\gamma_b$, which is the rate at which the population $n_a(t)$, [cf. eqn (5.116)], decays in weak coupling when $\gamma_b, \gamma_a^{\mathrm{P}} \gg \gamma_a$. From the point of view of mode a, the coupling with mode b is both adding particles, contributing to P_a^{eff}, and removing them, contributing to γ_a^{eff}. The total effective decay is

$$\Gamma_a^{\mathrm{eff}} = \gamma_a^{\mathrm{eff}} - P_a^{\mathrm{eff}} = \Gamma_a + Q_a.\tag{5.125}$$

Note that the generalised Purcell rate Q_a appears in the same way in both effective parameters in eqns (5.122), due to the symmetry with which the coupling both brings in and removes excitations. The mean value of the coherence can also be expressed in terms of these quantities:

$$n_{ab}^{SS} = \frac{2g^{eff}}{\Gamma_a^{eff} + \Gamma_b^{eff}} \frac{\gamma_a P_b - \gamma_b P_a}{\Gamma_a \Gamma_b} e^{i\phi}, \tag{5.126}$$

where $\phi = \arctan(2\Gamma_+/\Delta)$.

The quantities defined in eqns (5.122) and eqn (5.125) are all positive when $\Gamma_b > 0$ ($Q_a > 0$) and all negative when $\Gamma_b < 0$ (if there exists a solution for the steady state). The conditions for the pumping terms P_a, P_b to yield a physical state (a steady state), are, therefore, those for which the mean values $n_{a,b}^{SS}$ are positive and finite, implying

$$\Gamma_+ > 0, \tag{5.127a}$$

$$4(g^{eff})^2 > -\Gamma_a \Gamma_b. \tag{5.127b}$$

The first condition requires that pumps P_a, P_b are not *simultaneously* larger than their respective decay rates γ_a, γ_b. The second condition only represents a restriction when one of the effective parameters, either Γ_a or Γ_b, is negative. Then, it reads explicitly $4(g^{eff})^2 > |\Gamma_a \Gamma_b|$. Note that, out of resonance, the pumping rates appear both in g^{eff} and Γ_a, Γ_b, and therefore the explicit range of physical values for them needs to be found self-consistently.

5.6.2   Two-time dynamics of coupled Bose fields

The same analysis as above can be undertaken for the two-time dynamics, i.e., rather than tracking a quantity such as $n_a(t) = \langle a^\dagger(t)a(t)\rangle$, one can track that of $G^{(1)}(t, \tau) = \langle a^\dagger(t)a(t+\tau)\rangle$, eqn (3.102), which we have related to the power spectrum in Chapter 3. In the steady state, we will shortly see that as it embeds the coherence of the system, it evidences whether the system is in strong or weak coupling better than the single-time dynamics. Using the quantum regression technique exposed in Section 3.3.6, with basis of operators $a^m b^n$ (i.e., with $\{\eta\} = (m, n) \in \mathbb{N}$), the tensor M is a matrix (of infinite size) defined by

$$M_{mn \atop mn} = -i(m\omega_a + n\omega_b) - m\frac{\Gamma_a}{2} - n\frac{\Gamma_b}{2}, \tag{5.128a}$$

$$M_{\substack{mn \\ m+1,n-1}} = M_{\substack{nm \\ n-1,m+1}} = -ign, \tag{5.128b}$$

and zero everywhere else. With this matrix, one can obtain the equation of motion of any two-time correlator. For the particular case of interest of $G^{(1)}(t, \tau)$, the equation is closed together with that of $\langle a^\dagger(t)b(t+\tau)\rangle$:

$$\partial_\tau \begin{pmatrix} \langle a^\dagger(t)a(t+\tau)\rangle \\ \langle a^\dagger(t)b(t+\tau)\rangle \end{pmatrix} = \begin{pmatrix} -i\omega_a - \frac{\Gamma_a}{2} & -ig \\ -ig & -i\omega_b - \frac{\Gamma_b}{2} \end{pmatrix} \begin{pmatrix} \langle a^\dagger(t)a(t+\tau)\rangle \\ \langle a^\dagger(t)b(t+\tau)\rangle \end{pmatrix}, \tag{5.129}$$

and can thus be readily solved. The solution reads (at positive τ)

$$\langle a^\dagger(t)a(t+\tau)\rangle = \frac{1}{2R}e^{-\Gamma_+\tau}e^{-i(\omega_a-\frac{\Delta}{2})\tau}$$
$$\times \left\{ e^{iR\tau}[(R+i\Gamma_- -\Delta/2)n_a(t) - g\,n_{ab}(t)] \right.$$
$$\left. + e^{-iR\tau}[(R-i\Gamma_- +\Delta/2)n_a(t) + g\,n_{ab}(t)] \right\}. \quad (5.130)$$

Density plots of eqn (5.130) are shown in Fig. 5.12 for the case of spontaneous emission, so that both the real time dynamics (in t) and the auto-correlation time dynamics (in τ) can be compared. Panels (a) and (b) show the dynamics of a system in strong-coupling, and (c) in weak-coupling. Panel (a) and (c) correspond to the case of an exciton as an initial condition, and panel (b) to a polariton. Only panel (a) exhibits clear oscillations in both the t and τ dynamics. Oscillations in real time are smaller the stronger the coupling for the case of a polariton, and vanish completely in a perfect system. Oscillations in τ, however, are clearer when strong coupling is better, since they reflect coherent dynamics proper of this regime. On the other hand, in a system in weak coupling, an exchange of population can be observed in real time, while autocorrelation is monotonously decreasing. The oscillations in τ, rather than in t, are thus the mark of strong coupling.

From the Wiener–Khinchin formula, the Fourier transform of $G^{(1)}$ gives the power spectrum of mode a, that corresponds, from an experimental point of view, to the photoluminescence spectrum of the cavity. It is easy to compute it in the case of two-coupled Bose fields from the analytical expression (5.130). The formal structure of the emission spectrum follows as

$$S(\omega) = \frac{1}{2}(\mathcal{L}^1 + \mathcal{L}^2) - \frac{1}{2}\Im\{W\}(\mathcal{L}^1 - \mathcal{L}^2) - \frac{1}{2}\Re\{W\}(\mathcal{A}^1 - \mathcal{A}^2), \quad (5.131)$$

where $\mathcal{L}(\omega)$ is a Lorentzian function—characteristic of the emission of a single mode—and $\mathcal{A}(\omega)$ a dispersive function—characteristic of some interference from another mode or from an external drive:

$$\mathcal{L}^{1,2}(\omega) = \frac{1}{\pi}\frac{\Gamma_+ \pm R_i}{(\Gamma_+ \pm R_i)^2 + (\omega - (\omega_a - \frac{\Delta}{2} \mp R_r))^2}, \quad (5.132a)$$

$$\mathcal{A}^{1,2}(\omega) = \frac{1}{\pi}\frac{\omega - (\omega_a - \frac{\Delta}{2} \mp R_r)}{(\Gamma_+ \pm R_i)^2 + (\omega - (\omega_a - \frac{\Delta}{2} \mp R_r))^2}. \quad (5.132b)$$

Their features (position and broadening) are entirely specified by

- the complex Rabi frequency, eqn (5.114), decomposed in its real and imaginary parts as $R = R_r + iR_i$,
- the effective broadening Γ_+, eqn (5.112),
- the detuning Δ.

The structure of the spectral shape is clear. The Lorentzians correspond to the dressed states. Their position is given by the real part of the complex Rabi and their broadening by its imaginary part. When the splitting-to-broadening ratio is large, and the dressed

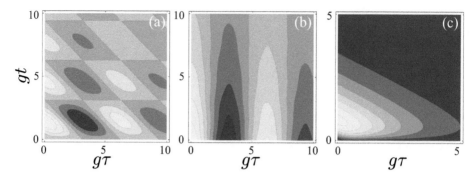

Fig. 5.12: Quantum dynamics of the cavity field $\langle a^\dagger(t)a(t+\tau)\rangle$, eqn (5.130), in real and autocorrelation times. Lighter shades correspond to higher values, black being zero, but the pattern of oscillation is the point of interest here. Fig. (a) and (b) correspond to the spontaneous emission of an exciton and of an upper polariton, respectively, both in strong coupling (with $\gamma_a = 2\gamma_b = 0.1g$) and Fig. (c) corresponds to the spontaneous emission of an exciton in weak coupling (with $\gamma_a = 100\gamma_b = 0.1g$). In (a) the system oscillates both in real and autocorrelation time. In (b), the Rabi oscillations in the population, that is, in the t dynamics, are vanishing since the system is in an eigenstate of the Hamiltonian. Because it is not an exact eigenstate in presence of decay, there is a "wobbling" of the contour lines. In (c), oscillations are absent at all times in the τ dynamics, in weak-coupling, which, on the other hand, displays one oscillation in the t dynamics. The oscillations in τ, rather than in t, are the mark of strong coupling. Adapted from Laussy et al. (2009).

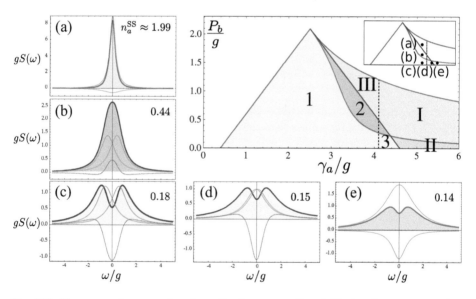

Fig. 5.13: Phase space of strong and weak coupling between two Bose fields, from Laussy et al. (2009). Light shades of grey/latin numerals correspond to regions of strong coupling and dark shades of grey/roman numerals to weak coupling, as defined by the oscillating or damped nature of the complex Rabi frequency, respectively. Zone 1 is the conventional strong coupling, zone 2 is strong coupling that is not resolved in the photoluminescence emission of the cavity, because of too large broadening-to-dressed-modes-splitting ratio, zone 3 is in strong coupling including pumping when it would be in weak coupling in the spontaneous emission case. Zone I is the conventional weak coupling, zone II is in weak coupling but features two peaks at resonances from an interference effect, zone III is in weak coupling because of pump-induced decoherence when it would be in strong-coupling neglecting pumping. In inset, five points are selected in these regions of interest, whose optical spectra $S(\omega)$ are displayed in panels (a)–(e), showing, in solid, the photoluminescence spectral line, in dotted, the Lorentzian emission (the dressed-states when in strong coupling) and in dash-dotted, the dispersive correction due to strong coupling. Line-splitting at resonance in the photoluminescence spectrum neither implies nor is implied by strong coupling.

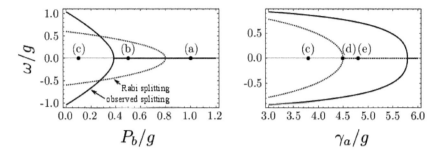

Fig. 5.14: Comparison between the observed (solid) and Rabi (dotted) splittings. The former is measured in the cavity photoluminescence spectrum and the latter is the intrinsic splitting between the dressed states, regardless of the dynamics or channel of detection. The points are those in Fig. 5.13. Starting from point (c), on the left panel, one loses PL splitting before strong coupling (increasing pumping) and on the right panel, one loses strong coupling before PL splitting (increasing cavity decay rate).

states (polaritons) are dominant, this Lorentzian contribution is maximum. The polaritons are well defined and the peaks in the photoluminescence spectrum are a faithful mapping to them. When they overlap, some interference gives rise to the dispersive correction $\mathcal{A}$. Such a term has a strong influence in cases of not-very-strong coupling, as we shall discuss shortly. This fundamental structure is weighted in various ways to reflect the dynamics. For instance, in the case of the spontaneous emission of a polariton, one Lorentzian should cancel the other polariton. This is taken into account by the coefficient W.

Exercise 5.19 $^{(**)}$ *By going through all the required algebra, show that W is given by* $[\Gamma_- + i(\frac{\Delta}{2} + gD)]/R$, *where* $D = \int_0^\infty \langle a^\dagger b\rangle(t)\, dt / \int_0^\infty \langle a^\dagger a\rangle(t)\, dt$. *This is valid for both spontaneous emission and a steady state.*

Examples of such spectra are given in Fig. 5.11(b), in the spontaneous emission of an exciton, a photon or a polariton, and in Fig. 5.13 (along with the underlying Lorentzian and dispersive parts), in the steady state established under various values of P_b and γ_a.[88]

In the spontaneous emission case, although the dynamics of the population is fairly identical in all three cases, at least qualitatively, it results in markedly different line-shapes, in particular, the Rabi doublet from the decay of a bare state becomes a single line from the decay of a polariton. The difference is less profound between an exciton or a photon as an initial state, but still quite important. Note how much better resolved is the splitting in the latter case. These differences are due to the interfering factors $\mathcal{A}$. The splitting is slightly different in both cases. Remembering that the real part of the complex Rabi gives the position of the dressed states (the Lorentzian peaks), it is tempting to approximate the Rabi splitting by the observed splitting, and this has been done many times. We now show that, in general, such a practice is inaccurate.

At resonance, the observed splitting, in both strong and weak coupling, can be computed from eqn (5.131). It is given by

---

[88] An applet to display these lineshapes is available at http://goo.gl/FDfJkQ.

Leonid Veniaminovich **Keldysh** (b. 1931) gave several key contributions to solid state physics, including the theory of electric field effect on the absorption edge in semiconductors, known as the Franz–Keldysh effect, and proposals for Bose–Einstein condensation of excitons (1968) and BCS-BEC transition in semimetals. Keldysh spent 10 years doing his PhD under the supervision of V.L. Ginzburg. His thesis presented a new diagram technique (Keldysh diagrams) which is now famous and widely used in the physics of non-equilibrium systems. Instead of a PhD degree, Keldysh was directly awarded the highest Russian academic degree (Doctor of Science) for this outstanding work. Later on, he was awarded the Lenin and Lomonosov prizes and the Prize of European Physical Society. Keldysh directed the Lebedev Physical institute during the difficult years 1988–1994. He now works in Texas A&M university.

$$\Delta\omega_O = 2g\sqrt{\sqrt{\left(1+\frac{P_b}{P_a}\right)^2 - 4\frac{\Gamma_+}{g}\left(\frac{P_b}{P_a}\frac{\Gamma_-}{g} - \frac{\Gamma_b}{2g}\right)} - \frac{P_b}{P_a} - \left(\frac{\Gamma_b}{2g}\right)^2}, \qquad (5.133)$$

when this quantity is real (and is zero otherwise). This expression is general. In the spontaneous emission case, the limit of vanishing pumping, $P_a \to 0$ and $P_b \to 0$, should be taken. The ratio P_a/P_b (or P_b/P_a if P_b is exactly zero), however, affects strongly the result. This is the quantity that one reads from the observed splitting, rather than the Rabi splitting, that is given by twice the real part of eqn (5.114). The strong difference between these two expressions is shown even better in Fig. 5.14, where possible contradictory conclusions are obtained: on the one hand, the system is in strong coupling (with a positive Rabi splitting), although there is only one peak in the photoluminescence spectrum—this is observed on point (b) in the left panel—and on the other hand, the system is in weak coupling (no Rabi splitting), although two peaks are observed in the photoluminescence spectrum—this is observed on point (e) on right panel. The reasons for these discrepancies are understood in Fig. 5.13, where the Lorentzian parts—which manifest the polaritons—and the $\mathcal{A}$ correction (dot-dashed lines)—to be added to them to provide the observed lineshape (solid lines)—have been plotted (with dotted lines) together with the spectra. It is clear that when the splitting-to-broadening ratio is small, as in the case of panel (b), one cannot discriminate the dressed states. In such cases of not-very-strong coupling, interferences are completely blurring the picture and little confidence should be entrusted to the observed splitting. Note, for instance, that in the case of panel (d), although in strong coupling, but with a very small splitting, the observed spectrum still displays a neat doublet, because of the interference. A slightly worse system (lower Q of the cavity) as shown in (e) goes into weak-coupling. The Rabi splitting cancels completely, however the photoluminescence spectrum has changed very little, the doublet being now due to an interference, which carves a hole in the lineshape. This latter interference effect has been analysed from a different perspective by Keldysh et al. (2006). The conclusion from this picture is that, although a strong qualitative change occurs in the system, with coherent oscillations in the τ dynamics

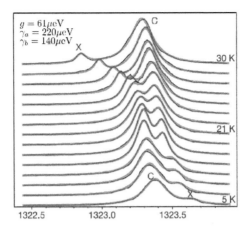

Fig. 5.15: The seminal observation of strong coupling with a quantum dot in a microcavity by Reithmaier et al. (2004) with, superimposed, the first quantitative description of this physics by Laussy et al. (2008), using eqn (5.131). The exciton X is brought in resonance with the cavity C by changing the temperature. At resonance, around 21 K, the polariton modes are identified in the cavity emission by the splitting that exhibit anticrossing. Theoretical analysis shows that this is thank to the photon-like effective quantum state that is realised in the system.

and emergence of dressed states, these are not so dramatically echoed in the observed spectra, which on the opposite vary smoothly to distort their lineshape with little regard for the underlying structure. This does not imply that the dressed states are abstract objects. They can still be detected, for instance by measuring the direct exciton emission, or by altering the effective quantum state realised in the system by the interplay of the two types of pumping (exciton and cavity pumping). It is, however, important to have the above picture in mind when claiming strong coupling, which is typically done in an anticrossing experiment, namely, bringing one mode on top of the other and observing whether two peaks are observed throughout (at resonance). The results above show that a weakly-coupled system can exhibit an apparent anticrossing, or that, on the contrary, a strongly-coupled system can exhibit an apparent crossing. As discussed by González-Tudela et al. (2010b), the closed-form expression eqn (5.133) can be obtained only at resonance. When analysing an anticrossing experiment, it is therefore, counter-intuitively, more convenient to study the lineshape (for which there is a closed-form expression for all detunings, eqns (5.131–5.132)), rather than the maxima of the peaks. Such an analysis is shown in Fig. 5.15, where Laussy et al. (2008) have fitted with the above formalism the seminal observation of strong-coupling of a quantum dot in a micropillar, by Reithmaier et al. (2004). The fitting keeps the coupling strength g and the bare more decay rates γ_a and γ_b globally constant for all curves. Only the detuning is varied, as should be, along with pumping rates (the detuning is varied experimentally by tuning temperature, which affects the excitation scheme). Laucht et al. (2009a) and Münch et al. (2009) have undertaken similar global fitting analysis of their data, as a function of temperature and pumping power. In the case of increasing pumping, either some mechanism should keep the number of excitation small enough so that the differences between the Bose and the Fermi models are negligible (for instance, dephasing), or the dot should indeed satisfy Bose–Einstein statistics (for instance, because it is a large dot, made so to get a large oscillator strength and able to accommodate many excitons). Otherwise, one should turn to the Jaynes–Cummings model, which we address in Section 5.6.3.

5.6.3   *The two-level system coupled to a Bose field*

When the quantum dot is a two-level system, one can still describe the light–matter coupling with the result of Section 5.6.2, as long as the probability to have two excitations in the system is negligible. As we have discussed before, if the second manifold of excitation is not probed, the results are the same up to the first order correlations, which include spectral shapes. It is therefore interesting—so as to pinpoint specificities of the two-level systems—to reach these higher manifolds of excitation, or, in a pictorial description, to climb the Jaynes–Cummings ladder. Most results so far have been obtained in systems other than semiconductors, which still lack sufficient figures of merits to clearly exhibit quantum nonlinearities. With superconducting qubits, which have exceedingly good coupling strengths (they are leading the race towards ultrastrong coupling), stunning demonstrations of the Jaynes–Cummings structure in the nonlinear regime have been brought forward by Fink et al. (2008) and Bishop et al. (2009). Clear signatures have also been reported for atoms in cavities, for instance by Schuster et al. (2008). With quantum dots in microcavities, indirect evidence has been produced with coherent excitation, for instance by Faraon et al. (2008), who found production of quantum light in bunching and antibunching consistent with the Jaynes–Cummings structure. Kasprzak et al. (2010) have obtained promising results towards a more explicit characterisation of the ladder with four-wave mixing. In the wake of the previous discussion, we shall, however, focus on incoherent pumping.

The Jaynes–Cummings Hamiltonian, eqn (5.28), is not exactly solvable in presence of incoherent pumping and decay. The main complication is that dissipation connects together the various rings of the Jaynes–Cummings ladder. In the limit of vanishing excitation, the system recovers the limiting case of the coupled Bose fields. With non-vanishing excitations, the system must be solved numerically in most cases, although some analytical or semi-analytical features can sometimes be obtained. To mark the difference with the previous case of a bosonic matter field, with operator b, we will revert to the notation σ for the two-level system with Fermi statistics. It is convenient, here as well, to introduce the fermionic effective broadening

$$\Gamma_\sigma = \gamma_\sigma + P_\sigma, \tag{5.134}$$

with $\Gamma_\pm = (\Gamma_a \pm \Gamma_\sigma)/4$ defined as in eqn (5.112), but with $b \leftrightarrow \sigma$ (it should be obvious from the context whether $\Gamma_\pm$ refers to b or σ). Note that compared with the Bose case, eqn (5.111), the emitter effective decay rate, eqn (5.134), comes with a plus sign.

In the boson case, it was enough to know the average photon (n_a) and exciton (n_b) numbers together with the off-diagonal element $n_{ab} = \langle a^\dagger b \rangle$. The counting statistics was trivially defined by its thermal value, from the very nature of pumping. In the most

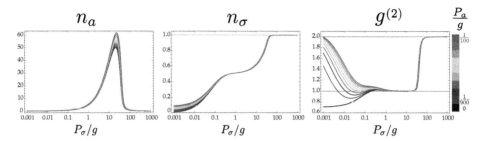

Fig. 5.16: Cavity population n_a, quantum dot population n_σ and two-photon counting statistics $g^{(2)}(0)$ as a function of incoherent pumping only (solid) or in addition to cavity pumping $P_a/g = 1/100$ (dotted). Cavity pumping affects essentially the statistics at low pumping, where it spoils the quantum character by providing thermal-like fluctuations in the system. The system goes from the quantum regime to lasing and is ultimately quenched by the incoherent pumping. Adapted from del Valle and Laussy (2011).

general case of the fermion system, one deals with a countably infinite number of independent parameters. The new order of complexity brought by the fermion system is illustrated for even the simplest observable. Instead of a closed relationship that provides, e.g., the populations in terms of the system parameters and pumping rates, only relations between observables can be obtained in the general case. For instance, for the populations:

$$\Gamma_a n_a + \Gamma_\sigma n_\sigma = P_a + P_\sigma. \qquad (5.135)$$

This expression is formally the same as for the coupling of two bosonic modes. The differences are in the effective dissipation parameter $\Gamma_\sigma = \gamma_\sigma + P_\sigma$ (instead of the bosonic one, $\gamma_b - P_b$) and the constrain on the exciton population, $0 \le n_\sigma \le 1$. Numerical results are shown in Fig. 5.16.

As for the power spectra, they too require a countably infinite number of two-time correlators, e.g., the cavity emission is ruled by the equation of motion for

$$\partial_\tau \mathbf{v}_a(t, t+\tau) = \mathbf{M}\mathbf{v}_a(t, t+\tau) \qquad (5.136)$$

where $\mathbf{M}$ is defined in Exercise 5.20 and

$$\mathbf{v}_a(t, t+\tau) = \begin{pmatrix} \langle a^\dagger(t)a(t+\tau)\rangle \\ \langle a^\dagger(t)\sigma(t+\tau)\rangle \\ \langle a^\dagger(t)(a^\dagger a^2)(t+\tau)\rangle \\ \langle a^\dagger(t)(a^\dagger a\sigma)(t+\tau)\rangle \\ \langle a^\dagger(t)(a^2\sigma^\dagger)(t+\tau)\rangle \\ \vdots \end{pmatrix}, \qquad (5.137)$$

is infinite in size. One therefore needs to adopt a truncation scheme, the most direct one being to limit to correlators that assume values above a threshold ϵ, which is small enough so that results do not appreciably vary when repeating the procedure for a smaller ϵ.

Exercise 5.20 (**) *Show that for the Jaynes–Cummings Hamiltonian, the* **M** *tensor reads*

$$M_{\substack{mn\mu\nu \\ mn\mu\nu}} = i\omega_a(m-n) + i\omega_\sigma(\mu-\nu) - \frac{\gamma_a - P_a}{2}(m+n) - \frac{\gamma_\sigma + P_\sigma}{2}(\mu+\nu),$$

$$\tag{5.138a}$$

$$M_{\substack{mn\mu\nu \\ m-1,n-1,\mu\nu}} = P_a mn, \quad M_{\substack{mn\mu\nu \\ mn,1-\mu,1-\nu}} = P_\sigma \mu\nu, \tag{5.138b}$$

$$M_{\substack{mn\mu\nu \\ m-1,n,1-\mu,\nu}} = M^*_{\substack{nm\nu\mu \\ n,m-1,\nu,1-\mu}} = igm(1-\mu), \tag{5.138c}$$

$$M_{\substack{mn\mu\nu \\ m,n+1,\mu,1-\nu}} = M^*_{\substack{nm\nu\mu \\ n+1,m,1-\nu,\mu}} = -ig\nu, \tag{5.138d}$$

$$M_{\substack{mn\mu\nu \\ m,n+1,1-\mu,\nu}} = M^*_{\substack{nm\nu\mu \\ n+1,m,\nu,1-\mu}} = 2ig\nu(1-\mu), \tag{5.138e}$$

and is zero everywhere else. Compare with eqns (5.128).

Repeating what was done in the boson case, a mathematically straightforward and physically transparent way to solve eqn (5.136) is by diagonalisation of the linear system, thereby constructing the solution in the basis of the eigenstates (or dressed states) of the system. To do so, we introduce the matrix **E** of normalised eigenvectors of **M** and $-\mathbf{D}$ the diagonal matrix of eigenvalues:

$$-\mathbf{D} = \mathbf{E}^{-1}\mathbf{M}\mathbf{E}. \tag{5.139}$$

The formal solution is then $\mathbf{v}_c(t, t+\tau) = \mathbf{E}e^{-\mathbf{D}\tau}\mathbf{E}^{-1}\mathbf{v}_c(t,t)$. The spectral part is obtained exactly as $\int e^{(-\mathbf{D}+i\omega)\tau}d\tau$ and application of the Wiener–Khinchin theorem to the appropriate row of $\mathbf{v}_a$ provide the emission spectra of the cavity in a form reminiscent of eqns (5.131)–(5.132), but now with an infinite sum:

$$S_c(\omega) = \frac{1}{\pi}\sum_{p=1}^{\infty}\left(L_p\frac{\gamma_p}{(\omega-\omega_p)^2+\gamma_p^2} - K_p\frac{\omega-\omega_p}{(\omega-\omega_p)^2+\gamma_p^2}\right). \tag{5.140}$$

Various quantities have been introduced, that one can group in two main classes. The first one, ω_p and γ_p, embed the intrinsic spectral properties of the Jaynes–Cummings model and form the skeleton of the system, as they are independent from all the details of a particular configuration. Such details are taken into account by the second class of parameters, namely, the coefficients L_p and K_p. This is, again, a generalisation of the linear case: the dressed states emit Lorentzians and these possibly interfere with each other. The coefficients ω_p and γ_p indeed correspond to the resonant frequencies, ω_p, and their broadenings, γ_p. They arise from the formalism as the imaginary and real parts of D_p, respectively:

$$\gamma_p + i\omega_p = D_p. \tag{5.141}$$

These are shown in Fig. 5.17 for the case of vanishing pumping, in which case they can be obtained exactly. There are two resonances for the first manifold, given by

$$D_{\frac{1}{2}} = \Gamma_+ + i\left(\omega_a - \frac{\Delta}{2} \mp \sqrt{g^2 - \left(\Gamma_- + i\frac{\Delta}{2}\right)^2}\right), \tag{5.142}$$

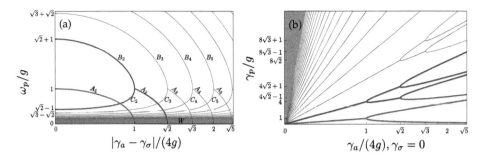

Fig. 5.17: (a) Transitions and (b) their broadening in the Jaynes–Cummings model. Only positive values are shown for (a), which is symmetric with respect to the horizontal axis. In thick black, the results are identical to the coupled bosons case. The splitting closes when the system enters weak-coupling. In thick grey are shown transitions from the second manifold of excitation to the first, cf. Fig. 5.7. Until 1, there are four transitions possible (two positives, as shown here). When the first manifold closes, only two transitions remain, until $\sqrt{2}$ where the second manifold closes as well, resulting in emission at the cavity mode, placed at the origin. The opposite happens with γ_p. Broadening increases very sharply for high manifolds of excitations. Both of these structures are intrinsic to the system and are independent of the dynamics.

and four for each manifold of higher order $k > 1$, given by, for $4k - 5 \le p \le 4k - 2$,[89]

$$D_p = \Gamma_k + i\left(\omega_a + \text{sgn}\big(p - (8k - 7)/2\big)R_k + (-1)^p R_{k-1}^*\right), \qquad (5.143)$$

with R_k as defined by eqns (5.79) and Γ_k by the the kth-*manifold (half) broadening*:

$$\Gamma_k = (2k - 3)\gamma_- + (2k - 1)\gamma_+ = (k - 1)\gamma_a + \frac{\gamma_\sigma}{2}. \qquad (5.144)$$

The D_p have a natural interpretation in terms of transitions between the manifolds of the dissipative Jaynes–Cummings ladder, Fig. 5.7. They correspond to transitions between the dressed states, with complex energy given by eqns (5.78):

$$D_{4k-5} = i[E_-^k - (E_+^{k-1})^*], \qquad D_{4k-4} = i[E_-^k - (E_-^{k-1})^*], \qquad (5.145a)$$
$$D_{4k-3} = i[E_+^k - (E_+^{k-1})^*], \qquad D_{4k-2} = i[E_+^k - (E_-^{k-1})^*]. \qquad (5.145b)$$

Consider the case $k = 1$ for illustration. Only the two peaks common with the linear regime arise, $D_{1,2} = iE_\mp^1$, given respectively by eqns (5.145a) and (5.145b) with $E^0 = 0$. The fact that the D_p correspond to $i[E^k - (E^{k-1})^*]$ shows that, although the positions of the lines are given by a difference, their broadenings are given by a sum (because of complex conjugation). Physically, the uncertainties in the initial and final states add up in the uncertainty of the transition energy. This means that in a system dominated by the coherent (strong) coupling, with small decay rates, one can expect a faithful reconstruction of the Jaynes–Cummings ladder in the emitted spectrum. This is shown in Fig. 5.18, in a system of exceedingly good quality, where lifetimes are very long and consequently, resonances very sharp and the transitions between them well

---

[89] In the expression for D_p, $\text{sgn}(x)$ is defined as 0 for $x = 0$ and $x/|x|$ otherwise.

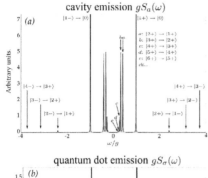

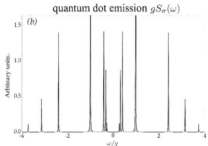

Fig. 5.18: Transitions in the Jaynes–Cummings ladder in a system in very strong coupling $((\gamma_a, \gamma_\sigma)/g = (10^{-3}, 0))$, as seen directly in the cavity emission (a) and in the direct exciton emission (b). Pumping rate is $P_\sigma/g = 10^{-3}$. This case shows explicit transitions between rungs of the Jaynes–Cummings ladder and how these, in particular the outer transitions, are better resolved through the direct quantum-dot emission rather than through the cavity.

defined. This allows to justify the Jaynes–Cummings ladder in presence of dissipation, which was introduced with eqn (5.78) on the assumption of a decay introduced with complex energies in the Hamiltonian. Alternatively, as was initially done by del Valle et al. (2009a), the dissipative ladder can be reconstructed—at last in this case—from the rigorous derivation above. This is for the case of decay only. With pumping, the situation becomes more complicated and such a "reconstruction" of the dressed states from their emission pattern is not so straightforward. When the pump parameters are comparable to, or higher than, the decay parameters, the manifold picture in terms of Hamiltonian eigenenergies breaks, in a way rather similar than it does when the RWA does not hold and the system enters ultrastrong coupling. The underlying spectral structure must be computed numerically and exhibit a complex behaviour, as discussed by del Valle et al. (2009a). New channels of coherence flow are opened by the excitation, that render almost useless the notion of dressed state, or even result in a proliferation of them. This unexpected aspect of light–matter coupling has been studied in detail in an exactly solvable toy model by del Valle (2010a).

Rather than describe in much detail what is happening with the dressed states in such cases of high pumping, we shall directly look at the final result, in terms of the emission spectra in some cases of interest. A thorough picture is given by del Valle (2010b).

When broadening is comparable to the coupling strength, the transitions overlap and interfere, and there is, as a result, a poor phenomenology in terms of qualitative manifestation of the underlying Jaynes–Cummings physics. Increasing excitation (in an attempt to probe higher manifolds of excitations) typically collapses the Rabi doublet to a single peak. The case of a system good enough to display quantitative features is

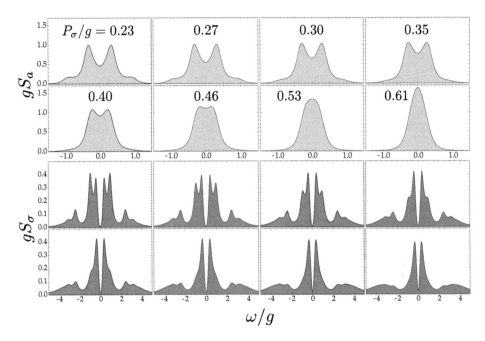

Fig. 5.19: Spectra of emission for the cavity and the dot, in a system slightly better than the state of the art, as exciton pumping is increased, from del Valle et al. (2009a). Two upper rows show the cavity emission, with elbows as the Rabi doublet, dominated by inner transitions in the Jaynes–Cummings ladder. These collapse as the system enters lasing, with emission in a single, narrowing, line. Two bottom rows show the exciton emission, that display a much neater structure, with transition high in the Jaynes–Cummings ladder emerging into a Mollow triplet. The central peak still has a strong depletion due to coherent scattering of the excitation into the cavity mode.

shown in Fig. 5.19, as pumping of the emitter is increased. The figure shows both the cavity emission (two upper rows) and the direct exciton emission (two lower rows). The dominant doublet that is observed in the cavity emission arises from transitions between excited manifolds, all overlapping to unite in common peaks (as opposed to the situation in Fig. 5.18, where the fine-structure splitting was clearly resolved). This is thus a manifestation of quantum nonlinearities. When the Rabi doublet—still observed as elbows for values of $P_\sigma/g < 1/4$—clearly coexists with the inner peaks, the resulting Jaynes–Cummings fork provides a sufficiently clear proof of Jaynes–Cummings physics at work. This structure has not, however, yet been reported experimentally, so far probably only for lack of good enough semiconductor systems. More interesting than the cavity emission is, however, the direct exciton emission. It is different in character than the optical mode emission. This was already apparent in a system in very strong coupling, Fig. 5.18. Quantum features are generally better observed when probing the quantum emitter, rather than the cavity, whose close connections with the classical oscillator tend to surface rapidly and dominate strongly. This can be understood at a fundamental level by the way emission is distributed preferentially in the cavity mode, through the intensity

$$I_a^{(\pm \to \mp)} \propto |\langle n-1, \mp| a |n, \pm \rangle|^2 = |\sqrt{n} - \sqrt{n-1}|^2/4 \approx 0, \qquad (5.146a)$$

$$I_a^{(\pm \to \pm)} \propto |\langle n-1, \pm| a |n, \pm \rangle|^2 = |\sqrt{n} + \sqrt{n-1}|^2/4 \approx n. \qquad (5.146b)$$

This shows the predominance of the inner peaks versus the outer ones. The transitions formed by the inner peaks are therefore enhanced and clearly identifiable in an experiment while the outer peaks are suppressed. On the other hand, in the direct exciton emission, the counterparts of eqns (5.146) are manifold-independent and equal for both the inner and outer peaks:

$$I_\sigma^{(\pm \to \mp)} \propto |\langle n-1, \mp| \sigma |n, \pm \rangle|^2 = 1/4, \qquad (5.147a)$$

$$I_\sigma^{(\pm \to \pm)} \propto |\langle n-1, \pm| \sigma |n, \pm \rangle|^2 = 1/4. \qquad (5.147b)$$

In this case, therefore, one can expect similar strength of transitions for both the inner and outer peaks with a richer multiplet structure for the direct exciton emission.

The bottom row of Fig. 5.19 shows in the direct QD emission a striking transition from a complex structure made up of a multitude of lines to a structure of much reduced complexity, namely, a smooth triplet. This transition is shown in greater details over a wider range of pumping in Fig. 5.20, from the Rabi doublet at low pumping to a triplet reminiscent of the famous Mollow (1969) structure at high pumping. The relationship between this Mollow triplet formed with a quantised light-field and the conventional one with a classical c-number field has been described in details by del Valle and Laussy (2010). The main difference is that instead of the coherent state driving the system, the microcavity QED counterpart arises under incoherent excitation and from the coherence built up self-consistently in the system. This gives rise to many interesting features, like a sublinear (instead of linear) splitting of the satellite peaks, a different broadening and, remarkably, a Rayleigh peak (the δ function part of the spectrum) that can be both negative (coherent absorption) or positive (coherent scattering), as opposed to the conventional case where it is only positive. This is possibly the simplest cross over from the quantum to the classical regime. Its observation would be a giant progress in the nonlinear spectroscopy of such systems.

We conclude this Section by a rapid overview of some of the further phenomenology that has been predicted, and possibly observed, in this system, as shown in Fig. 5.21. Quantum nonlinearities remain to be explicitly observed. The indirect approach of Kasprzak et al. (2010) to probe the higher rungs remains the most direct evidence so far. Instead of neat anharmonic nonlinearities producing a Jaynes–Cummings fork, as seen on the first column, various reports have been made of mysterious spectral triplets, as shown on the second column. These have been observed by Hennessy et al. (2007) and more recently in a configuration more related to Fig. 5.21, by Ota et al. (2009). A triplet can be obtained theoretically by adding pure dephasing to the model, that is, a term that decays the off-diagonal element in the density matrix without affecting the population. Such a term originates typically from high excitation powers or high temperatures. Such dependencies have been studied in details by Laucht et al. (2009a), who could quantify it by fitting their data with a model of strong-coupling including dephasing along the lines exposed in Section 5.6.2. Theoretically, this is described by adding a term $\mathcal{L}_{\gamma_\sigma^\phi} \rho =$

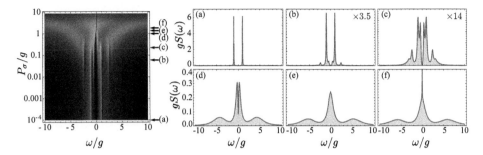

Fig. 5.20: A quantum-to-classical transition with a quantum-dot (two-level system) in a cavity, as observed through the QD emission. (a) At vanishing pumping P_σ, the system exhibits the vacuum Rabi doublet of strong-coupling, identical to that of two-coupled oscillators. (b) As pumping is increased, nonlinear quantum features from higher-manifold transitions appar in the spectrum. (c) Increasing pumping even more, the quantised structure of peaks from transition between manifolds melts into another, smoother, spectral shape. (d) At high enough pumping, the transition to a Mollow triplet is complete, with only three peaks instead of the myriad of peaks when transitions can still be resolved. Until the cavity-field has fully formed a coherent state, interestingly, a depletion is observed in the central peak, that is the Rayleigh peak, but with an absorbing character instead of scattering. (e) At this pumping, the cavity has achieved its coherence and the Mollow triplet appears naked with no Rayleigh peak. (f) Increasing pumping even more, just as in the conventional Mollow case where light is an external coherent field, the Rayleigh peak is now scattering and takes the familiar form of a δ function sitting on top of the central peak.

$\gamma_\sigma^\phi(S_z\rho S_z - \rho)$ to the density matrix master equation, where $S_z = \frac{1}{2}[\sigma^\dagger, \sigma]$, as discussed by Agarwal and Dutta Gupta (1990). The impact of this dephasing on the resonances ω_p in the nonlinear regime has been studied by González-Tudela et al. (2010a), who find that, without losing strong-coupling (as testified by the outer peaks from the Jaynes–Cummings transitions), dephasing brings together the closely-spaced resonances that unite in a common peak, emitting at the cavity mode. Other models have been suggested to explain this spectral triplet: by their fathers themselves, Hennessy et al. (2007) and Ota et al. (2009), who propose a loss of strong-coupling, due to, for instance, a charge that intermittently brings the dot out of resonance with the cavity, thereby producing a triplet as a result of overlapping a Rabi doublet with the cavity mode. Hughes and Yao (2009) show that an interference of the fields can also result in such a triplet and Yamaguchi et al. (2009) find one thanks to the quantum anti-Zeno effect. There is, so far, no evidence of which mechanism is responsible for this striking experimental observation. The striking transition of Fig. 5.20 that we have described in details is less impressive (unless system parameters are very good) when observed through the cavity emission, which is however the natural experimental configuration. Instead of a melting of the quantised nonlinear quantum features from the Jaynes–Cummings Hamiltonian into a Mollow triplet, one observes a simpler transition from a Rabi doublet to a single lasing line, as shown in the third column of Fig. 5.21. This "lasing in strong-coupling" has been claimed by Nomura et al. (2010). Finally, coming back to the Mollow triplet observed in the QD emission, as shown in the last column of Fig. 5.21, it must be emphasised that it is not strictly equal to the conventional Mollow triplet due to some thermal component of the cavity that drives the two-level system. This structure remains to be observed in any system.

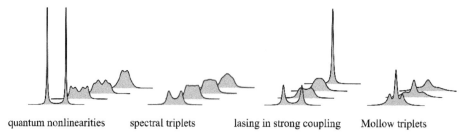

quantum nonlinearities   spectral triplets   lasing in strong coupling   Mollow triplets

Fig. 5.21: An overview of some of the spectral shapes expected (or observed) in the Jaynes–Cummings physics under incoherent pumping, such as is realised with quantum dots in microcavities. First set, in good systems—where the coupling strength neatly dominates the decay rates—quantum nonlinearities are observed as peaks sitting at anharmonic frequencies, resulting in a Jaynes–Cummings fork (four peaks) and a transition from the Rabi doublet to a doublet with splitting $\sqrt{\langle n \rangle + 1} - \sqrt{\langle n \rangle}$. Second set, in systems with pure dephasing, the Jaynes–Cummings fork is replaced by a spectral triplet instead, the central lines having collapsed onto a common transition, without, however, losing strong-coupling. Third set, in systems not good enough to resolve quantum nonlinearities, the overlap of the nonlinear transitions results in lasing in strong-coupling, a trend recently observed by Nomura et al. (2010). Fourth set, high in the Jaynes–Cummings ladder, a counterpart of the Mollow triplet is observed, where the coherence acquired by the cavity substitutes the laser of resonance fluorescence.

5.7   Excitons in semiconductors

In the following chapters, we pursue the investigation of light–matter coupling in both the semiclassical and quantum regimes, putting more emphasis on specificities of microcavities. To bring forward these results we give now more elements on the material excitations of semiconductors that parallel the exposition of the previous chapter, but from a quantum-mechanical perspective. At this stage, we shall change notations for the fields to follow popular customs, so that, for instance, a that was previously a cavity mode (annihilating a photon) will now typically refer to the polariton. The new notations will be introduced as we proceed.

5.7.1   *Quantisation of the exciton field*

The second-quantised Hamiltonian of a semiconductor at the fermionic level reads, in real space,

$$H = \int \Psi(\mathbf{r})^\dagger \left(-\frac{\hbar^2}{2m} \nabla^2 + V(\mathbf{r}) \right) \Psi(\mathbf{r}) \, d\mathbf{r} \tag{5.148a}$$

$$+ \frac{1}{2} \int \int \Psi(\mathbf{r})^\dagger \Psi(\mathbf{r}')^\dagger \frac{e^2}{|\mathbf{r} - \mathbf{r}'|} \Psi(\mathbf{r}') \Psi(\mathbf{r}) \, d\mathbf{r} \, d\mathbf{r}' \,, \tag{5.148b}$$

where $\Psi(\mathbf{r})$ is the electron annihilation field operator and V the Coulomb. We expand Ψ in terms of $\varphi_{i,\mathbf{k}}(\mathbf{r}) = \langle \mathbf{r} | i, \mathbf{k} \rangle$ the single-particle wavefunction labelled by the quantum number $\mathbf{k}$ in the ith semiconductor band, in terms of the *electron annihilation operator* $e_{i,\mathbf{k}}$:

$$\Psi(\mathbf{r}) = \sum_{i \in \{c,v\}} \sum_{\mathbf{k}} \varphi_{i,\mathbf{k}}(\mathbf{r}) e_{i,\mathbf{k}} \,. \tag{5.149}$$

Because of interactions and correlations, the determination of $\varphi_{i,\mathbf{k}}(\mathbf{r})$ is a difficult task, typically solved numerically. The full many-body problem, eqn (5.148), can be approximated to an effective single-body problem through the so-called *Hartree–Fock approximation*, which introduces an effective potential V_{eff}. The resulting Schrödinger equation with Hamiltonian $H_{\text{HF}} = -\hbar^2\nabla^2/2m + V_{\text{eff}}$ is nonlinear since the potential depends on the wavefunction φ and so the problem remains one of considerable difficulty. Bloch's theorem, however, allows a statement of general validity:

$$\varphi_{i,\mathbf{k}}(\mathbf{r}) \propto e^{i\mathbf{k}\cdot\mathbf{r}}u_{i,\mathbf{k}}(\mathbf{r}), \tag{5.150}$$

with u having the same translational symmetry as the crystal.

Once all the algebra has been gone through, the semiconductor Hamiltonian (5.148) becomes, in reciprocal space,

$$H = \sum_{i\in\{c,v\}}\sum_{\mathbf{k}} E_i(\mathbf{k})e_{i,\mathbf{k}}^\dagger e_{i,\mathbf{k}} \tag{5.151a}$$

$$+ \frac{1}{2}\sum_{i\in\{c,v\}}\sum_{\mathbf{k},\mathbf{p},\mathbf{q}\neq 0} V(\mathbf{q})e_{i,\mathbf{k}+\mathbf{q}}^\dagger e_{i,\mathbf{p}-\mathbf{q}}^\dagger e_{i,\mathbf{p}} e_{i,\mathbf{k}} \tag{5.151b}$$

$$+ \sum_{\mathbf{k},\mathbf{p},\mathbf{q}\neq 0} V(\mathbf{q})e_{c,\mathbf{k}+\mathbf{q}}^\dagger e_{v,\mathbf{p}-\mathbf{q}}^\dagger e_{v,\mathbf{p}} e_{c,\mathbf{k}}, \tag{5.151c}$$

with E_i the dispersion relation for the ith band and $V(\mathbf{q})$ the Fourier transform of the Coulomb interaction.

The limit of very low density (in fact, in the limit where the ground state is devoid of conduction-band electrons) allows some analytical solutions to be obtained after performing approximations. One simplification, both conceptual and from the point of view of the formalism, is the introduction of the hole (fermionic) operator h as

$$h_{\mathbf{k}} = e_{v,-\mathbf{k}}^\dagger, \tag{5.152}$$

(spin is also reversed if granted). This allows elimination of the negative effective mass of valence electrons and enables us to deal with an excitation as an "addition" of a particle, rather than annihilation (in terms of valence electrons, the ground state is full of electrons and gets depleted by excitations). Conceptually, it replaces a sea of valence electrons by a single particle, making it easier to conceive the exciton as a bound state. In terms of electrons $e_{\mathbf{k}}$ and holes $h_{\mathbf{k}}$, eqn (5.151) now reads

$$H = \sum_{\mathbf{k}}[E_e(\mathbf{k})e_{\mathbf{k}}^\dagger e_{\mathbf{k}} + E_h(\mathbf{k})h_{\mathbf{k}}^\dagger h_{\mathbf{k}}] \tag{5.153a}$$

$$+ \frac{1}{2}\sum_{\mathbf{k},\mathbf{p},\mathbf{q}\neq 0} V(\mathbf{q})[e_{\mathbf{k}+\mathbf{q}}^\dagger e_{\mathbf{p}-\mathbf{q}}^\dagger e_{\mathbf{p}} e_{\mathbf{k}} + h_{\mathbf{k}+\mathbf{q}}^\dagger h_{\mathbf{p}-\mathbf{q}}^\dagger h_{\mathbf{p}} h_{\mathbf{k}}] \tag{5.153b}$$

$$- \sum_{\mathbf{k},\mathbf{p},\mathbf{q}\neq 0} V(\mathbf{q})e_{\mathbf{k}+\mathbf{q}}^\dagger h_{\mathbf{p}-\mathbf{q}}^\dagger h_{\mathbf{p}} e_{\mathbf{k}}, \tag{5.153c}$$

with explicit expression for electron and hole dispersion (as a function of their effective mass):

$$E_e(\mathbf{k}) = E_{\text{gap}} + \frac{(\hbar \mathbf{k})^2}{2m_e^*},$$ (5.154a)

$$E_h(\mathbf{k}) = \frac{(\hbar \mathbf{k})^2}{2m_h^*}.$$ (5.154b)

In the low-density limit, if one neglects line (5.153b) in the Hamiltonian, it can be diagonalised by introducing the *exciton operator*

$$X_\nu(\mathbf{k}) \equiv \sum_{\mathbf{p}} \varphi_\nu(\mathbf{p}) h_{\mathbf{k}/2-\mathbf{p}} e_{\mathbf{k}/2+\mathbf{p}},$$ (5.155)

with $\varphi_\nu(\mathbf{p})$ the Fourier transform of Wannier equation eigenstates (with ν the quantum numbers, as for the hydrogen atom; the spectrum of energy we call E_ν).

The exciton Hamiltonian becomes

$$H = \sum_{\nu,\mathbf{k}} E_X^\nu(\mathbf{k}) X_\nu^\dagger(\mathbf{k}) X_\nu(\mathbf{k}),$$ (5.156)

with

$$E_X^\nu(\mathbf{k}) = E_\nu + E_e(\mathbf{k}) + E_h(\mathbf{k}).$$ (5.157)

5.7.2 *Excitons as bosons*

Excitons behave as true bosons when the commutator of the field operators satisfies the relation

$$[X_\nu(\mathbf{k}), X_\mu^\dagger(\mathbf{q})] = \delta_{\nu,\mu} \delta_{\mathbf{k},\mathbf{q}}.$$ (5.158)

Direct evaluation of the commutator with explicit expression (5.155) yields

$$[X_\nu(\mathbf{k}), X_\mu^\dagger(\mathbf{q})] = \delta_{\nu,\mu} \delta_{\mathbf{k},\mathbf{q}} - \sum_{\mathbf{p}} |\varphi_{1s}(\mathbf{p})|^2 (c_{\mathbf{k}}^\dagger c_{\mathbf{q}} + h_{-\mathbf{k}}^\dagger h_{-\mathbf{q}}),$$ (5.159)

so that the diagonalisation is legitimate at low densities. In particular, $\langle [X_0, X_0^\dagger] \rangle = 1 - O(N a_B^2)$, where N is the density of excitons and a_B is the Bohr radius associated with φ_{1s}. One can therefore treat excitons as bosons with confidence, in the limit $N a_0^2 \ll 1$.

5.7.3 *Excitons in quantum dots*

We now show one possible route of extending the results of the previous sections. We have investigated the limiting cases where the material excitation that couples to the field is either a boson, or a fermion. In actual systems, composite particles are neither exactly one nor the other. The importance of this distinction can become important in a quantum dot (QD), where the excitations are located in a tiny region of real space, so that their wavefunctions overlap appreciably. If the confining potential of the dot is much stronger than the Coulomb interaction, electrons and holes, which are elementary excitations of the system, will be quantised separately, whereas if Coulomb interactions dominate over the confinement, one electron–hole pair will bind as an exciton and therefore behave rather like a boson. Here, we investigate a model of interactions of light with excitons in QDs of varying size, where their boson or fermion character is tunable.

It is more relevant to carry out the analysis in real space since the QD states are localised. We note

$$\varphi_{n_e}^e(\mathbf{r}_e) = \langle \mathbf{r}_e | \varphi_{n_e}^e \rangle \quad \text{and} \quad \varphi_{n_h}^h(\mathbf{r}_h) = \langle \mathbf{r}_h | \varphi_{n_h}^h \rangle \tag{5.160}$$

the set of their basis wavefunctions with $\mathbf{r}_e$ and $\mathbf{r}_h$ the positions of the electron and hole, respectively. Subscripts n_e and n_h are multi-indices enumerating all quantum numbers of electrons and holes. The specifics of the three-dimensional confinement manifests itself in the discrete character of n_e and n_h components. We restrict our considerations to direct-bandgap semiconductors with non-degenerate valence bands. Such a situation can be experimentally achieved in QDs formed in conventional III–V or II–VI semiconductors, where the light-hole levels lie far below, in energy, the heavy-hole ones due to the effects of strain and size quantisation along the growth axis. Therefore, only electron/heavy–hole excitons need to be considered. Moreover, we will neglect the spin degree of freedom of the electron–hole pair and assume all carriers to be spin polarised. To carry out the same formalism as presented in the previous sections, we need to build the second quantised operator for the QD. We define it as

$$X^\dagger = \sum_{n_e, n_h} C_{n_e, n_h} e_{n_e}^\dagger \varsigma_{n_h}^\dagger , \tag{5.161}$$

where e_{n_e} and h_{n_h} are fermion creation operators for an electron and a hole in state $\left| \varphi_{n_e}^e \right\rangle$ and $\left| \varphi_{n_h}^h \right\rangle$, respectively:

$$e_{n_e}^\dagger |0\rangle = \left| \varphi_{n_e}^e \right\rangle , \quad h_{n_h}^\dagger |0\rangle = \left| \varphi_{n_h}^h \right\rangle , \tag{5.162}$$

with $|0\rangle$ denoting both the electron and hole vacuum fields. The (single) exciton wavefunction $|\varphi\rangle$ results from the application of $X^\dagger$ on the vacuum. In real-space coordinates,

$$\langle \mathbf{r}_e, \mathbf{r}_h | \varphi \rangle = \varphi(\mathbf{r}_e, \mathbf{r}_h) = \sum_{n_e, n_h} C_{n_e, n_h} \varphi_{n_e}^e(\mathbf{r}_e) \varphi_{n_h}^h(\mathbf{r}_h). \tag{5.163}$$

At this stage, we do not specify the wavefunction (that is, the set of coefficients C_{n_e, n_h}), which depends on various factors such as the dot geometry, electron and hole effective masses and dielectric constant. Rather, we consider the n-excitons state that results from successive excitation of the system through $X^\dagger$:

$$|\Psi_n\rangle = (X^\dagger)^n |0\rangle . \tag{5.164}$$

The associated normalised wavefunction $|n\rangle$ reads

$$|n\rangle = \frac{1}{N_n} |\Psi_n\rangle , \tag{5.165}$$

where, by definition of the normalisation constant,

$$\mathcal{N}_n = \sqrt{\langle \Psi_n | \Psi_n \rangle}. \tag{5.166}$$

The creation operator $X^\dagger$ can now be obtained explicitly. We define α_n the nonzero matrix element that lies below the diagonal in the exciton representation:

$$\alpha_n = \langle n | X^\dagger | n-1 \rangle, \tag{5.167}$$

which, by comparing eqns (5.164) and (5.167) turns out to be

$$\alpha_n = \frac{\mathcal{N}_n}{\mathcal{N}_{n-1}}. \tag{5.168}$$

The coefficients α_n can be linked to the coefficients C_{n_e, n_h} (the latter assuming a specific value when the system itself is known):

Exercise 5.21 $^{(***)}$ *Show that the normalisation coefficients $\mathcal{N}$ necessary to compute the matrix elements α_n (through eqn (5.168)), can be computed by the following recurrent relation:*

$$\mathcal{N}_n^2 = \frac{1}{n} \sum_{m=1}^{n} (-1)^{m+1} \beta_m \mathcal{N}_{n-m}^2 \prod_{j=0}^{m-1} (n-j)^2, \tag{5.169}$$

with $\mathcal{N}_0 = 1$ and β_m the irreducible m-excitons overlap integrals, $1 \leq m \leq n$:

$$\beta_m = \int \left(\prod_{i=1}^{m-1} \varphi^*(\mathbf{r}_{e_i}, \mathbf{r}_{h_i}) \varphi(\mathbf{r}_{e_i}, \mathbf{r}_{h_{i+1}}) \right) \varphi^*(\mathbf{r}_{e_m}, \mathbf{r}_{h_m}) \varphi(\mathbf{r}_{e_m}, \mathbf{r}_{h_1})$$

$$d\mathbf{r}_{e_1} \dots d\mathbf{r}_{e_m} d\mathbf{r}_{h_1} \dots d\mathbf{r}_{h_m}. \tag{5.170}$$

The procedure to calculate the matrix elements of the creation operator is as follows:[90] One starts from the envelope function $\varphi(\mathbf{r}_e, \mathbf{r}_h)$ for a single exciton. Then, one calculates all overlap integrals β_m as given by eqn (5.170), for $1 \leq m \leq n$ where n is the highest manifold to be accessed. Then the norms can be computed with eqn (5.169). Finally the matrix elements α_n are obtained as the successive norms ratio, cf. eqn (5.168).

The limiting cases of Bose–Einstein and Fermi–Dirac statistics are recovered in the limits of large and shallow dots, respectively. This is made most clear through consideration of the explicit case of two excitons ($n = 2$). Then the wavefunction reads

$$\Psi_2(\mathbf{r}_{e_1}, \mathbf{r}_{e_2}, \mathbf{r}_{h_1}, \mathbf{r}_{h_2}) = \varphi(\mathbf{r}_{e_1}, \mathbf{r}_{h_1}) \varphi(\mathbf{r}_{e_2}, \mathbf{r}_{h_2}) - \varphi(\mathbf{r}_{e_1}, \mathbf{r}_{h_2}) \varphi(\mathbf{r}_{e_2}, \mathbf{r}_{h_1}), \tag{5.171}$$

with its normalisation constant (5.166) readily obtained as

$$\mathcal{N}_2^2 = \int |\Psi_2(\mathbf{r}_{e_1}, \mathbf{r}_{e_2}, \mathbf{r}_{h_1}, \mathbf{r}_{h_2})|^2 d\mathbf{r}_{e_1} \dots d\mathbf{r}_{h_2} = 2 - 2\beta_2, \tag{5.172}$$

---

[90]The numerical computation of the β_m and α_n values needs to be carried out with great care. The cancellation of the large numbers of terms involved in eqn (5.169) requires a high-precision computing of β_m.

where β_2, the two-exciton overlap integral, reads explicitly

$$\beta_2 = \int \varphi(\mathbf{r}_{e_1}, \mathbf{r}_{h_1}) \varphi(\mathbf{r}_{e_2}, \mathbf{r}_{h_2}) \varphi(\mathbf{r}_{e_1}, \mathbf{r}_{h_2}) \varphi(\mathbf{r}_{e_2}, \mathbf{r}_{h_1}) d\mathbf{r}_{e_1} \dots d\mathbf{r}_{h_2}. \qquad (5.173)$$

This integral is the signature of the composite nature of the exciton. The minus sign in eqn (5.172) results from the Pauli principle: two fermions (electrons and holes) cannot occupy the same state. Assuming $\varphi(\mathbf{r}_e, \mathbf{r}_h)$ is normalised, $\mathcal{N}_1 = 1$, so according to eqn (5.168),

$$\alpha_2 = \sqrt{2 - 2\beta_2}. \qquad (5.174)$$

Since $0 \le \beta_2 \le 1$ this is smaller than or equal to $\sqrt{2}$, the corresponding matrix element of a true boson creation operator. This result has a transparent physical meaning: since two identical fermions from two excitons cannot be in the same quantum state, it is "harder" to create two real excitons, where the underlying structure is probed, than two ideal bosons. We note that if L is the QD lateral dimension, $\beta_2 \sim (a_B/L)^2 \ll 1$ when $L \gg a_B$. Thus, in large QDs the overlap of excitonic wavefunctions is small, so $\alpha_2 \approx \sqrt{2}$ and the bosonic limit is recovered. On the other hand, in a small QD, where Coulomb interaction is unimportant compared to the dot potential confining the carriers, the electron and hole can be regarded as quantised separately:

$$\varphi(\mathbf{r}_e, \mathbf{r}_h) = \varphi^e(\mathbf{r}_e)\varphi^h(\mathbf{r}_h). \qquad (5.175)$$

In this case, all $\beta_m = 1$ and subsequently all $\alpha_m = 0$ with the exception of $\alpha_1 = 1$. This is the fermionic limit where $X^\dagger$ maps to the Pauli matrix σ_+.

5.7.3.1 *Gaussian toy model*

We now turn to the general case of arbitrary-sized QDs, interpolating between the (small) fermionic and (large) bosonic limits. We do not attempt for this conceptual presentation to go through the lengthy and complicated task of the numerical calculation of the exciton creation operator matrix elements for a realistic QD. Rather, we consider a model wavefunction that can be integrated analytically and illustrates some expected typical behaviours.

We consider a QD strongly confined in one direction (along the z-axis) and having a symmetrical shape in the xy-plane with, possibly, larger dimensions. This corresponds to realistic self-assembled semiconductor QDs Widmann et al. (1997). We assume a Gaussian form for the wavefunction that allows evaluation analytically of all the required quantities. This follows from a harmonic confining potential, as has been considered for instance by Que (1992). As numerical accuracy is not the chief goal of this approach, we further assume in-plane coordinates x and y to be uncorrelated to ease the computations. The wavefunction reads

$$\varphi(\mathbf{r}_e, \mathbf{r}_h) = \mathcal{C} \exp(-\gamma_e \mathbf{r}_e^2 - \gamma_h \mathbf{r}_h^2 - \gamma_{eh} \mathbf{r}_e \cdot \mathbf{r}_h), \qquad (5.176)$$

properly normalised with

$$\mathcal{C} = \frac{\sqrt{4\gamma_e\gamma_h - \gamma_{eh}^2}}{\pi}, \qquad (5.177)$$

provided that $\gamma_{eh} \in [-2\sqrt{\gamma_e\gamma_h}, 0]$ with $\gamma_e, \gamma_h \ge 0$. The γ parameters allow interpolation between the large and small dot limits within the same wavefunction (cf. Section 4.3.3). To connect these parameters γ_e, γ_h and γ_{eh} to physical quantities, eqn (5.176) is regarded as a trial wavefunction that is to minimise the Hamiltonian H_{QD}, confining the electron and hole in a quadratic potential where they interact through Coulomb interaction

$$H_{QD} = \sum_{i=e,h} \left(\frac{\mathbf{p}_i^2}{2m_i} + \frac{1}{2}m_i\omega^2\mathbf{r}_i^2 \right) - \frac{e^2}{4\pi\epsilon\epsilon_0|\mathbf{r}_e - \mathbf{r}_h|} . \tag{5.178}$$

Here, $\mathbf{p}_i$ is the momentum operator for the electron and hole, $i = $ e, h, respectively,

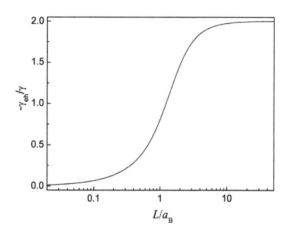

Fig. 5.22: Ratio of parameters $-\gamma_{eh}$ and γ (with $\gamma = \gamma_e = \gamma_h$) as a function of L/a_B. For large dots, where $L \gg a_B$, $-\gamma_{eh}/\gamma \approx 2$, which corresponds to the bosonic limit where the electron and hole are strongly correlated. For shallow dots, where $L \ll a_B$, $-\gamma_{eh}/\gamma \approx 0$, with electron and hole quantised separately. The transition is shown as the result of a variational procedure, with an abrupt transition when the dot size becomes comparable with the Bohr radius. From Laussy et al. (2006a).

m_e, m_h the electron and hole masses, ω the frequency that characterises the strength of the confining potential, e the charge of the electron and ϵ the background dielectric constant screening the Coulomb interaction. This Hamiltonian defines the two length scales of our problem, the 2D Bohr radius a_B and the dot size L:

$$a_B = \frac{4\pi\epsilon\epsilon_0\hbar^2}{2\mu e^2} , \tag{5.179a}$$

$$L = \sqrt{\frac{\hbar}{\mu\omega}} , \tag{5.179b}$$

where $\mu = m_e m_h/(m_e+m_h)$ is the reduced mass of the electron–hole pair. To simplify the following discussion we assume that $m_e = m_h$, resulting in $\gamma_e = \gamma_h = \gamma$. The trial wavefunction (5.176) separates as $\varphi(\mathbf{r}_e, \mathbf{r}_h) = C\Phi(\mathbf{R})\phi(\mathbf{r})$, where $\mathbf{r} = \mathbf{r}_e - \mathbf{r}_h$ is the radius-vector of relative motion and $\mathbf{R} = (\mathbf{r}_e + \mathbf{r}_h)/2$ is the centre-of-mass position:

$$\Phi(\mathbf{R}) = \frac{\sqrt{2(2\gamma + \gamma_{eh})}}{\sqrt{\pi}} \exp\left(-\mathbf{R}^2[2\gamma + \gamma_{eh}]\right), \tag{5.180a}$$

$$\phi(\mathbf{r}) = \frac{\sqrt{2\gamma - \gamma_{eh}}}{\sqrt{2\pi}} \exp\left(-\mathbf{r}^2 \left[\frac{2\gamma - \gamma_{eh}}{4}\right]\right), \tag{5.180b}$$

Equation (5.180a) is an eigenstate of the centre-of-mass energy operator and equating its parameters with those of the exact solution yields the relationship $2\gamma + \gamma_{eh} = 2/L^2$. This constraint allows us to minimise eqn (5.180b) with respect to a single parameter, $a = -\gamma_{eh}/2 + 1/(2L^2)$, which eventually amounts to minimising $4a_B/a^2 + a_B a^2/L^4 - 2\sqrt{\pi}/a$. On doing so, we obtain the ratio $-\gamma_{eh}/\gamma$ as a function of L/a_B, displayed in Fig. 5.22. The cross over from the bosonic to the fermionic regime is seen to occur sharply when the dot size becomes commensurate with the Bohr radius. For large dots, i.e., for large values of L/a_B, the ratio is well approximated by the expression

$$-\gamma_{eh}/\gamma = 2 - (a_B/L)^2, \tag{5.181}$$

so that in the limit of big dots where $a_B/L \to 0$, eqn (5.176) becomes $\varphi(\mathbf{r}_e, \mathbf{r}_h) \propto \exp(-(\sqrt{\gamma_e}\mathbf{r}_e - \sqrt{\gamma_h}\mathbf{r}_h)^2)$ with vanishing normalisation constant. This mimics a free-exciton in an infinite quantum well. It corresponds to the bosonic case. On the other hand, if L is small compared with the Bohr radius, with $\gamma_{eh} \to 0$, the limit (5.175) is recovered with $\varphi \propto \exp(-\gamma_e \mathbf{r}_e^2) \exp(-\gamma_h \mathbf{r}_h^2)$. This corresponds to the fermionic case. The trial wavefunction is of course not exact,[91] but the exciton operator that it yields is exact, as are all the intermediates.[92] Together with eqns (5.177), (xxii) and

[91] One can check that eqn (5.176) gives, in the case $\gamma_{eh} \to -2\sqrt{\gamma_e\gamma_h}$, an exciton binding energy that is smaller by only 20% than that calculated with a hydrogenic wavefunction, which shows that the Gaussian approximation should be tolerable for qualitative and semiquantitative results.

[92] It can be seen, for instance, that the overlap integrals (5.170) take a simple form in terms of multivariate Gaussians as a function of a matrix A defined as

$$\beta_m = \mathcal{C}^{2m} \int \exp(-\mathbf{x}^T A \mathbf{x}) \, d\mathbf{x} \int \exp(-\mathbf{y}^T A \mathbf{y}) \, d\mathbf{y}, \tag{xix}$$

where

$$\mathbf{x}^T = (x_{e_1}, x_{e_2}, \ldots, x_{e_m}, x_{h_1}, x_{h_2}, \ldots, x_{h_m}), \tag{xx-a}$$

$$\mathbf{y}^T = (y_{e_1}, y_{e_2}, \ldots, y_{e_m}, y_{h_1}, y_{h_2}, \ldots, y_{h_m}), \tag{xx-b}$$

are the $2m$-dimensional vectors that encapsulate all the degrees of freedom of the m excitons-complex and A is a positive-definite symmetric matrix that equates eqn (5.170) and (xix), i.e., which satisfies

$$\mathbf{x}^T A \mathbf{x} = 2\gamma_e \sum_{i=1}^{m} x_i^2 + 2\gamma_h \sum_{i=m+1}^{2m} x_i^2 + \gamma_{eh} x_m x_{m+1}$$

$$+ \gamma_{eh} \sum_{i=1}^{m} x_i x_{m+i} + \gamma_{eh} \sum_{i=1}^{m-1} x_i x_{m+i+1}, \tag{xxi}$$

and likewise for $\mathbf{y}$ (to simplify notation we have not written an index m on $\mathbf{x}$, $\mathbf{y}$ and A, but these naturally scale with β_m). The identity for $2m$-fold Gaussian integrals is

$$\int \exp(-\mathbf{x}^T A \mathbf{x}) \, d\mathbf{x} = \frac{\pi^m}{\sqrt{\det A}}. \tag{xxii}$$

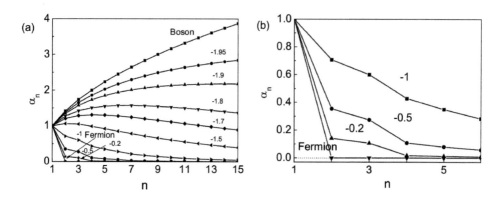

Fig. 5.23: (a) Matrix elements α_n of the exciton creation operator $X^\dagger$ calculated for $n \leq 15$ for various Gaussian trial wavefunctions corresponding to various sizes of the dot. The top curve shows the limit of true bosons, where $\alpha_n = \sqrt{n}$, and the bottom curve the limit of true fermions, where $\alpha_n = \delta_{n,1}$. Intermediate cases are obtained for values of γ_{eh} from $-1.95\sqrt{\gamma_e\gamma_h}$ down to $-0.2\sqrt{\gamma_e\gamma_h}$, interpolating between the boson and fermion limits. (b) Magnified region close to the fermion limit. Values displayed are everywhere given in units of $\sqrt{\gamma_e\gamma_h}$. From Laussy et al. (2006a).

(xxiii), expression (xix) in footnote 92, provides the β_m in the Gaussian approximation. One can see the considerable complexity of the expressions despite the simplicity of the model wavefunction. Even a numerical treatment meets with difficulties owing to manipulations of a series of large quantities that sum to small values. α_n are obtained by exact algebraic computations to free coefficients from numerical artifacts.

Figure 5.23 shows the behaviour of α_n for different values of γ_{eh} interpolating from the bosonic case ($\gamma_{eh} = -2\sqrt{\gamma_e\gamma_h}$) to the fermionic case ($\gamma_{eh} = 0$). The cross over from the bosonic to the fermionic limit can be clearly seen: for γ_{eh} close to $-2\sqrt{\gamma_e\gamma_h}$, the curve behaves like $\sqrt{n}$, the deviations from this exact bosonic result becoming more

The problem is now reduced to the determinant of A, which, being a sparse matrix, also admits an analytical solution, although this time a rather cumbersome one. The determinant of the matrix A can be computed as

$$\det A = \gamma_{eh}^{2m} \sum_{k=0}^{m} \sum_{l=0}^{m-k} (-1)^{\lfloor m/2 \rfloor + k} \mathcal{A}_m(k,l) \left(\frac{\gamma_e\gamma_h}{\gamma_{eh}^2} \right)^k . \tag{xxiii}$$

Here, we introduced a quantity

$$\mathcal{A}_m(k,l) = \mathcal{A}'_m(k,l) + \sum_{i=1}^{m} \left(\mathcal{A}'_{m-i}(k,l-i) - \mathcal{A}'_{m-i-1}(k,l-i) \right) , \tag{xxiv}$$

and

$$\mathcal{A}'_m(k,l) = \sum_{\eta=1}^{p(l)} \frac{(\sum_i \nu_\eta^l(i))!}{\prod_i \nu_\eta^l(i)!} \times \binom{m-l}{\sum_i \nu_\eta^l(i)} \binom{m-l-\sum_i \nu_\eta^l(i)}{k - \sum_i \nu_\eta^l(i)} , \tag{xxv}$$

with $k \in]0, m]$, $l \in [0, m]$ and $p(l)$ and $\nu_\eta(i)$ already introduced as the partition function of l and the number of occurrences of i in its ηth partition. For the case $k = 0$ the finite size of the matrix implies a special rule that reads $\mathcal{A}_m(0,l) = 4\delta_{m,l}\delta_{m\equiv2,0}$.

pronounced with increasing n. For γ_{eh} close to $-2\sqrt{\gamma_e\gamma_h}$, the curve initially behaves like $\sqrt{n}$, the deviations from this exact bosonic result becoming more pronounced with increasing n. The curve is ultimately decreasing beyond a number of excitations that is smaller the greater the departure of γ_{eh} from $-2\sqrt{\gamma_e\gamma_h}$. After the initial rise, as the overlap between electron and hole wavefunctions is small and bosonic behaviour is found, the decrease follows as the density becomes so large that Pauli exclusion becomes significant. Then, excitons cannot be considered as structureless particles, and fermionic characteristics emerge. With γ_{eh} going to 0, this behaviour is replaced by a monotonically decreasing α_n, which means that it is "harder and harder" to add excitons in the same state in the QD; the fermionic nature of excitons becomes more and more important.

5.8   Exciton–photon coupling

The polaritons discussed in Section 4.4.4.2 can be seen in a simplified, but very accurate, model as the new eigenstates that arise from the coupling of two oscillators, i.e., the photon and the exciton. Vividly, the polariton is then seen as the chain process where the exciton annihilates, emitting a photon with the same energy E and momentum $\mathbf{k}$, which is later reabsorbed by the medium, creating a new exciton with the same $(E, \mathbf{k})$, and so on until the excitation finds its way out of the cavity (resulting in the annihilation of the polariton), or the electron or hole is scattered.

In this part, we neglect for brevity spins and states other than 1s for the exciton. The dipole moment $-e\mathbf{r}$ of the electron–hole pair couples to the light field $\mathbf{E}$ and adds the following coupling Hamiltonian to (5.148):

$$H_{CX} = \int \Psi^\dagger(\mathbf{r})[-e\mathbf{r}\cdot\mathbf{E}(\mathbf{r})]\Psi(\mathbf{r})\,d\mathbf{r}. \tag{5.182}$$

The same procedure to obtain eqn (5.153) from eqn (5.148) including eqn (5.182) leads to the following exciton–photon coupling Hamiltonian:

$$H = \sum_\mathbf{k} E_\mathrm{C}(\mathbf{k})B_\mathbf{k}^\dagger B_\mathbf{k} + \sum_\mathbf{k} E_X(\mathbf{k})X_\mathbf{k}^\dagger X_\mathbf{k} + \sum_\mathbf{k} \hbar g(\mathbf{k})(B_\mathbf{k}^\dagger X_\mathbf{k} + X_\mathbf{k}B_\mathbf{k}^\dagger), \tag{5.183}$$

with $\hbar g(\mathbf{k}) = \mu_{cv}\varphi_{1s}(0)\sqrt{E_\mathrm{C}(\mathbf{k})/(2\epsilon_0\epsilon)}$, μ_{cv} being the dipole matrix element dotted with electron and hole, B the photon annihilation operator and X its exciton counterpart. Hamiltonian (5.183) can be diagonalised provided that X operators obey the bosonic algebra of eqn (5.158), following exactly the same procedure as those employed previously for two linearly-coupled harmonic oscillators, starting with eqns (5.34). In such an approximation, the Hamiltonian (5.183), bilinear in bosonic operators, is diagonalised with the Hopfield transformation

$$a_\mathbf{k}^\mathrm{L} \equiv \mathcal{X}_\mathbf{k}X_\mathbf{k} - \mathcal{C}_\mathbf{k}B_\mathbf{k}, \tag{5.184a}$$

$$a_\mathbf{k}^\mathrm{U} \equiv \mathcal{C}_\mathbf{k}X_\mathbf{k} + \mathcal{X}_\mathbf{k}B_\mathbf{k}, \tag{5.184b}$$

where the so-called *Hopfield coefficients* $\mathcal{C}_\mathbf{k}$ and $\mathcal{X}_\mathbf{k}$ satisfy $\mathcal{C}_\mathbf{k}^2 + \mathcal{X}_\mathbf{k}^2 = 1$, so that the transformation is canonical and a operators follow the bosonic algebra as well. As previously with excitons, Hamiltonian (5.183) reduces to free-propagation terms only:

$$H = \sum_{\mathbf{k}} E_U(\mathbf{k}) a_{\mathbf{k}}^{U\dagger} a_{\mathbf{k}}^U + \sum_{\mathbf{k}} E_L(\mathbf{k}) a_{\mathbf{k}}^{L\dagger} a_{\mathbf{k}}^L, \tag{5.185}$$

for *upper* and *lower-polariton branches*, with second quantised annihilation opera-tors a^U and a^L, respectively. The dispersion relations for these branches are

$$E_{\substack{U\\L}}(\mathbf{k}) = \frac{1}{2}(E_X(\mathbf{k}) + E_C(\mathbf{k})) \pm \sqrt{\hbar^2 g(\mathbf{k})^2 + (\Delta_{\mathbf{k}}/2)^2}, \tag{5.186}$$

(the U subscript is associated with the plus sign, L with minus), where E_C is given by expression (4.133) and E_X by eqn (5.157), and $\Delta_{\mathbf{k}}$ is the energy mismatch, i.e., detuning, between the cavity and exciton modes:

$$\Delta_{\mathbf{k}} \equiv E_C(\mathbf{k}) - E_X(\mathbf{k}). \tag{5.187}$$

The Hopfield coefficients used to diagonalise this Hamiltonian are related to the mixing angle, cf. eqn (5.40) and Fig. 5.4. Their squares correspond to the photon (resp. exciton) fraction of the polariton, that is, to the probability for the mixed exciton–photon particle to be found in either one of these states. These probabilities are conveniently given as a function of the dispersion relations:

$$|\mathcal{C}_{\mathbf{k}}|^2 = \frac{E_U(\mathbf{k})E_X(\mathbf{k}) - E_L(\mathbf{k})E_C(\mathbf{k})}{(E_C(\mathbf{k}) + E_X(\mathbf{k}))\sqrt{\Delta_{\mathbf{k}}^2 + 4\hbar^2 g(\mathbf{k})^2}}, \tag{5.188a}$$

$$|\mathcal{X}_{\mathbf{k}}|^2 = \frac{E_U(\mathbf{k})E_C(\mathbf{k}) - E_L(\mathbf{k})E_X(\mathbf{k})}{(E_C(\mathbf{k}) + E_X(\mathbf{k}))\sqrt{\Delta_{\mathbf{k}}^2 + 4\hbar^2 g(\mathbf{k})^2}}. \tag{5.188b}$$

Again, because of their connection with the mixing angle, a plot of eqns (5.188) appears on Fig. 5.4b.

5.8.1   *Polariton splitting*

Equation (5.186) is one of the major results of microcavity polariton physics for the various consequences this relation bears on many key issues that we are going to address in the next chapter. It is plotted in solid lines in Fig. 4.23 where are also plotted in dashed lines the dispersions for the exciton, eqn (5.157), and the photon, eqn (4.133). The first and third of these figures display negative (where bare dispersions cross each other) and positive (where they do not) detunings, respectively, while the central figure displays zero detuning (resonance at $\mathbf{k} = 0$). The splitting at normal incidence is the counterpart for polaritons in quantum wells of the resonant coupling of quantum dots in microcavities. It is the most important characterisation of strong-coupling in such systems.

Experimentally, the polariton splitting can be measured in reflectivity (R), transmis-sion (T), absorption (A) and photoluminescence (PL) and is in general different in all these cases, as well as being different from the exciton–photon coupling constant g as discussed earlier for the case of quantum dots. Yet it has been shown by Savona et al. (1998) that the general following relation holds:

$$\Delta E_R \geq \Delta E_T \geq \Delta E_A , \tag{5.189}$$

and also $\hbar\Omega_R \geq \Delta E_A$. Only the splitting in absorption unambiguously proves strong coupling, while the splitting in transmission or reflectivity is a necessary but not sufficient condition. It might therefore be the most important experimental expression, which is given by Savona et al. (1998) as[93]

$$\Delta E_A = \sqrt{\hbar^2\Omega_R^2 - \frac{(\gamma_c^2 + \gamma_x^2)}{2}} . \tag{5.190}$$

5.8.2 The polariton Hamiltonian

Now that the free-polariton Hamiltonian has been obtained, one can proceed with deriving next-order or additional processes that will account for the dynamics of polaritons. Such additions include polariton–polariton interactions (from the underlying exciton–exciton interaction coming from Coulomb interaction), polariton–phonon interaction and possibly such terms as polariton–electron interaction if there is residual doping, polariton coupling to the external electromagnetic field giving them a chance to escape the cavity, a term that is ultimately responsible for decay, and its counterpart that injects particles, behaving as a pump. Below we provide the expression for these terms without detailing their derivations. We come back to more detailed discussion of these terms below; see also the discussion by Savona et al. (1998).

In terms of the annihilation and creation operators a_k, $a_k^\dagger$ for polaritons and b_k, $b_k^\dagger$ for phonons (where k is a two-dimensional wavevector in the plane of the microcavity), obeying the usual bosonic algebra, our model Hamiltonian in the interaction representation reads

$$H = H_{\text{pump}} + H_{\text{lifetime}} + H_{\text{pol-phon}} + H_{\text{pol-el}} + H_{\text{pol-pol}} . \tag{5.191}$$

The usual models considered for these various terms are as follow:

$$H_{\text{pump}} = \sum_{k} g(\mathbf{k})(K_{\text{pump}}a_k^\dagger + K_{\text{pump}}^* a_k) . \tag{5.192}$$

This term describes the pumping of the system by a classical light field of amplitude K_{pump} (a scalar as opposed to a_k, which is an operator). $g(\mathbf{k})$ is the wavevector-dependent coupling strength between the two fields. This Hamiltonian is adapted to describe the resonant pumping or the non-resonant pumping case depending on the approximations performed:

$$H_{\text{lifetime}} = \sum_{k} \gamma(\mathbf{k})(\alpha_k a_k^\dagger + \alpha_k^\dagger a_k) . \tag{5.193}$$

H_{lifetime} describes the linear coupling between the polariton field and an empty external light field responsible for the polariton lifetime. What happens in reality is that photons escape the discrete cavity mode into a continuum from where their probability

---

[93] Adding dissipation brings no new results as compared to the semiclassical result, eqn (4.100).

of return is zero. This translates as a decay. $\alpha_{\mathbf{k}}$, $\alpha_{\mathbf{k}}^{\dagger}$ are the annihilation-creation operators of the external light field. $\gamma(\mathbf{k})$ is the wavevector-dependent coupling strength between the two fields.

$$H_{\text{pol}-\text{phon}} = \sum_{\mathbf{k}\neq 0, \mathbf{q}\neq 0} U(\mathbf{k}, \mathbf{q}) e^{\frac{i}{\hbar}(E(\mathbf{k})+\omega_{\mathbf{q}}-E(\mathbf{k}+\mathbf{q}))t} a_{\mathbf{k}}^{\dagger} b_{\mathbf{q}} a_{\mathbf{k}+\mathbf{q}} + \text{h.c.} \qquad (5.194)$$

describes the coupling between the polariton a and the phonon b fields. U is the Fourier transform of the interaction potential for polariton–phonon scattering. $\omega_{\mathbf{q}}$ is the phonon dispersion, $E(\mathbf{k})$ is the dispersion of the lower-polariton branch.

$$H_{\text{pol}-\text{el}} = \frac{1}{2} \sum_{\substack{\mathbf{k}\neq 0, \mathbf{p}\neq 0 \\ \mathbf{q}\neq 0}} U^{\text{el}}(\mathbf{k}, \mathbf{p}, \mathbf{q}) e^{\frac{i}{\hbar}\left(E(\mathbf{k}+\mathbf{q})-E(\mathbf{k})+\frac{\hbar^2}{2m_e}(|\mathbf{p}-\mathbf{q}|^2-|\mathbf{p}|^2)\right)t} a_{\mathbf{k}+\mathbf{q}}^{\dagger} a_{\mathbf{k}} e_{\mathbf{p}-\mathbf{q}}^{\dagger} e_{\mathbf{p}}$$

$$(5.195)$$

describes the coupling between the polariton a and the electron e fields. $e_{\mathbf{k}}$, $e_{\mathbf{k}}^{\dagger}$ are the annihilation-creation operators of the electron field, m_e is the electron mass, U^{el} is the Fourier transform of the interaction potential for polariton-electron scattering.

Polariton-free carrier scattering may be important if polaritons are created by electrical injection of electrons and holes, and in doped microcavities. Note also that some residual doping is inevitably present in semiconductor samples. The term describing polariton-hole interactions has the same form as eqn (5.195). Finally,

$$H_{\text{pol}-\text{pol}} = \frac{1}{2} \sum_{\substack{\mathbf{k}\neq 0, \mathbf{p}\neq 0 \\ \mathbf{q}\neq 0}} V(\mathbf{k}, \mathbf{p}, \mathbf{q}) e^{\frac{i}{\hbar}\left(E(\mathbf{k}+\mathbf{q})+E(\mathbf{p}-\mathbf{q})-E(\mathbf{k})-E(\mathbf{p})\right)t} a_{\mathbf{k}+\mathbf{q}}^{\dagger} a_{\mathbf{p}-\mathbf{q}}^{\dagger} a_{\mathbf{k}} a_{\mathbf{p}},$$

$$(5.196)$$

describes the polariton–polariton interaction. V is the Fourier transform of the interaction potential for this scattering.

This Hamiltonian will be a starting point for Chapters 7 and 8 where various approximations are made to single out the resonant dynamics where polariton–polariton interactions dominate, or out-of-resonance where relaxations bring the system in quasi-equilibrium with reservoirs as a function of pump and decay. In Chapter 9, an extension of eqn (5.196) will be carried out to include the spin degree of freedom.

6

WEAK-COUPLING MICROCAVITIES

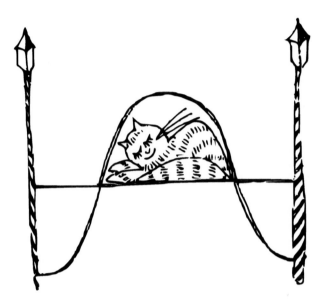

In this chapter we address the optical properties of microcavities in the weak-coupling regime and review the emission of light from microcavities in the linear regime. We present a derivation of the Purcell effect and stimulated emission of radiation by microcavities and consider how this develops towards lasing. Finally, we briefly consider nonlinear properties of weakly coupled semiconductor microcavities. The functionality of vertical-cavity surface-emitting lasers (VCSELs) is also described.

Microcavities, Second Edition. Alexey V. Kavokin, Jeremy J. Baumberg, Guillaume Malpuech, Fabrice P. Laussy, Oxford University Press (2017). © Alexey V. Kavokin, Jeremy J. Baumberg, Guillaume Malpuech, Fabrice P. Laussy. DOI 10.1093/oso/9780198782995.001.0001

6.1 Purcell effect

6.1.1 *The physics of weak coupling*

We are interested in this chapter in the so-called *weak coupling* of light with matter in the sense that the effect of the radiation field can be dealt with as a perturbation on the dynamics of the system. The dynamics we have in mind is typically the spontaneous emission of the system initially in its excited state. As emphasised by Kleppner (1981) in the atomic case, the atom releases its energy because of its interaction with the vacuum of the optical field, so that if the interaction could be "switched off", the atom would remain forever in its excited state. This idea is an extension of one reported much earlier by Purcell (1946) who was seeking the opposite effect, namely, to increase the interaction so as to speed up the release of the excitation.[94] Intuitively, if the dipole is resonant with the cavity mode, the photon density of states seen by the dipole is increased with respect to the vacuum density of states. The spontaneous emission rate is therefore enhanced: the dipole decays radiatively faster than in vacuum and the photons are emitted in the cavity mode. On the other hand, if the dipole is placed out of resonance, namely in a photonic gap, the photon density of state seen by the dipole is smaller than in vacuum and the spontaneous emission rate is reduced. The Purcell effect therefore perfectly illustrates the role played by an optical cavity that is to locally modify the photon density of states. The control of spontaneous emission through the Purcell effect is a way to reduce the threshold of lasers and the effect has been actively looked for with atoms placed in cavities, for instance by Goy et al. (1983), and more recently with quantum dots placed in micropillars, microdisks or photonic crystals, for instance in the work of Gérard et al. (1998). We review in more details the experimental realisations later.

The above description of the emission neglects reabsorption. For dipoles in free space, one can easily believe intuitively that this effect is weak. In fact, it is responsible for the energy shift known as the *Lamb shift* that is, indeed, orders of magnitude smaller than the radiative broadening. In quantum electrodynamics, this shift is interpreted as the influence of virtual photons emitted and reabsorbed by the dipole. The situation changes dramatically when the dipole is placed in a cavity. Photons emitted are then reflected by the mirrors and remain inside the cavity. This increases the probability of reabsorption of the photons by the dipole. If the confinement is so good that the probability of re-absorption of a photon by the dipole is larger than its probability of escaping the cavity, the perturbative weak-coupling regime breaks and instead the so-called *strong coupling* takes place. This means that the eigenmodes of the coupled dipole-cavity system are no longer bare modes, but mixed light–dipole modes. Their energies are strongly modified with respect to the bare modes. The strong-coupling regime is addressed in detail in Chapters 7 and 8, but in discussing weak coupling, one should

---

[94]Purcell was motivated by a practical goal in nuclear magnetic resonance: to bring spins into thermal equilibrium at radio frequencies, with the relaxation time for the nuclear spin in vacuum of the order of 10^{21} s. He calculated that the presence of small metallic particles would, thanks to the Purcell enhancement, be reduced to an order of minutes. The much-quoted reference that records this landmark of QED—E. M. Purcell, Phys. Rev., **69**, 681, (1946)—is actually a short proceedings abstract.

bear in mind that strong coupling is regarded as implying richer physics from the fundamental point of view[95] and that it is typically harder to obtain than the weak coupling that "historically" precedes it in a given system until enough control of the system and its interactions are reached to give preponderance to the quantum Hamiltonian dynamics over the dissipations.

6.1.2 Spontaneous emission

We now give a mathematical derivation of the above ideas, relating spontaneous emission with weak (or perturbative) coupling to the optical field. Our starting point is eqn (3.66) for the electric-field operator that we write here in the dipolar approximation and for a given state of polarisation

$$\mathbf{E} = \sum_{\mathbf{k}} u_{\mathbf{k}} \sqrt{\frac{\hbar \omega_{\mathbf{k}}}{2\epsilon_0 L^3}} \mathbf{e}_{\mathbf{k}} (a_{\mathbf{k}}^{\dagger} + a_{\mathbf{k}}). \tag{6.1}$$

We recall that photon modes are resulting from the quantisation of the electromagnetic field in a box of size L. From there, we go to the Jaynes–Cummings model (see Section 5.4.4) for the multimode field operator eqn (6.1), which turns the coupling term eqn (5.60) into, written directly in the interaction picture:

$$V = \sum_{\mathbf{k}} \sqrt{\frac{\hbar \omega_{\mathbf{k}}}{2\epsilon_0 L^3}} (\mathbf{e}_{\mathbf{k}} \cdot \mathbf{d}) e^{-i(\omega_0 - \omega_{\mathbf{k}})t} (a_{\mathbf{k}}^{\dagger} \sigma + \sigma^{\dagger} a_{\mathbf{k}}). \tag{6.2}$$

We consider as the initial state the atom in an excited state and all photon modes empty, $|e, 0_{\mathbf{k}_1}, 0_{\mathbf{k}_2}, \cdots, 0_{\mathbf{k}_n}, \cdots\rangle$. The final states are states with the atom in the ground state and one photon in one of the final states, $|g, 0_{\mathbf{k}_1}, 0_{\mathbf{k}_2}, \cdots, 1_{\mathbf{k}_m}, \cdots\rangle$. We also take into account only the term $a_{\mathbf{k}}^{\dagger} \sigma$ of the Hamiltonian, which destroys the atomic excitation and creates a photon. The reverse process $a_{\mathbf{k}} \sigma^{\dagger}$ is neglected. This means that we assume that the photon is escaping quickly far away from the atom and cannot be reabsorbed. The matrix element between the initial state and one of the final state therefore reads

$$M_{\mathbf{k}_n} = \langle e, 0_{\mathbf{k}_1}, 0_{\mathbf{k}_2}, \cdots, 0_{\mathbf{k}_n}, \cdots | H | g, 0_{\mathbf{k}_1}, 0_{\mathbf{k}_2}, \cdots, 1_{\mathbf{k}_n}, \cdots\rangle = \sqrt{\frac{\hbar \omega_{\mathbf{k}}}{2\epsilon_0 L^3}} (\mathbf{e}_{\mathbf{k}} \cdot \mathbf{d}).$$
$$\tag{6.3}$$

One can then apply the Fermi golden rule that stands as (we denote "at" for atom)

$$\Gamma_0^{\text{at}} = \frac{2\pi}{\hbar} \sum_{\mathbf{k}} |M_{\mathbf{k}}|^2 \delta(E_0 - E_{\mathbf{k}}), \tag{6.4}$$

where the sum is on the set of final states, which becomes a continuum as $L \to \infty$. We now pass to the thermodynamic limit, making the size of the system go to infinity. The sum is replaced by an integral using the rule

---

[95] It is not so obvious from an applied point of view that strong coupling is intrinsically better than weak coupling; all the physics of lasers that is addressed later in this chapter pertains to the weak coupling.

$$\sum_{\mathbf{k}} \rightarrow \frac{2}{\left(\frac{2\pi}{L}\right)^3} \int_0^{2\pi} d\phi \int_0^{\pi} \sin\theta d\theta \int_0^{\infty} k^2 dk \,. \tag{6.5}$$

The factor 2 in the numerator stands for the two different polarisations, $\mathbf{e_k} \cdot \mathbf{d} = d\cos\theta$ with θ the angle between the dipole axis and the electric field. This gives

$$\Gamma_0^{\text{at}} = \frac{c}{4\pi^2\epsilon_0} d^2 \int_0^{2\pi} d\phi \int_0^{\pi} \sin\theta\cos^2\theta d\theta \int_0^{\infty} \delta(E_0 - E_{\mathbf{k}})k^3 dk \tag{6.6a}$$

$$= \frac{c}{3\pi\epsilon_0} d^2 \int_0^{\infty} \delta(E_0 - E_{\mathbf{k}})k^3 dk \,. \tag{6.6b}$$

We then use $k = E/(\hbar c)$, which yields

$$\Gamma_0^{\text{at}} = \frac{1}{3\pi c^2 \hbar^3 \epsilon_0} d^2 \int_0^{\infty} \delta(E_0 - E_{\mathbf{k}})E^3 dE \,, \tag{6.7}$$

which finally gives the usual formula for the emission rate of an atom in the vacuum:

$$\Gamma_0^{\text{at}} = \frac{\omega_0^3}{3\pi\hbar c^3\epsilon_0} d^2 \,. \tag{6.8}$$

An approach involving the dynamics—also based on the Jaynes–Cummings Hamiltonian and known as the *Wigner–Weisskopf theory*—also assumes that at time $t = 0$ the atom is in the excited state and the field modes are in the vacuum state. The state vector therefore reads

$$|\psi(t)\rangle = c_{\text{e}}(t)\,|e, 0_{\mathbf{k}}\rangle + \sum_{\mathbf{k}} c_{\text{g},\mathbf{k}}(t)\,|g, 1_{\mathbf{k}}\rangle \,, \tag{6.9}$$

with the initial time amplitude of probabilities given by $c_{\text{e}}(0) = 1$ and $c_{\text{g},\mathbf{k}}(0) = 0$.

We now determine the state of the atom and the state of the light field at some later time, when the atom starts to emit. We therefore write the Schrödinger equation $(\partial/\partial t)\,|\psi(t)\rangle = -(i/\hbar)V\,|\psi(t)\rangle$ with V given by eqn (6.2). Projecting this equation on the different basis vectors, one gets the equations of motion for the probability amplitudes:

$$\dot{c}_{\text{e}}(t) = -i\sum_{\mathbf{k}} M_{\mathbf{k}} e^{i(\omega_0-\omega_{\mathbf{k}})t} c_{\text{b},\mathbf{k}}(t) \,, \tag{6.10a}$$

$$\dot{c}_{\text{b},\mathbf{k}}(t) = -i M_{\mathbf{k}} e^{i(\omega_0-\omega_{\mathbf{k}})t} c_{\text{e}}(t) \,. \tag{6.10b}$$

We formally integrate eqns (6.10b) and substitute the result in eqn (6.10a) yielding

$$\dot{c}_{\text{e}}(t) = -\sum_{\mathbf{k}} |M_{\mathbf{k}}|^2 \int_0^t e^{i(\omega_0-\omega_{\mathbf{k}})(t-t')} c_{\text{e}}(t')dt' \,. \tag{6.11}$$

This expression is still exact. We now perform the so-called "Wigner–Weisskopf" approximation, which amounts to the following substitution:

$$\int_0^t e^{i(\omega_0 - \omega_{\mathbf{k}})(t-t')} c_e(t') dt' \approx c_e(t) \int_0^\infty e^{i(\omega_0 - \omega_{\mathbf{k}})t'} dt' . \qquad (6.12)$$

This approximation is also known more generally as a *Markov approximation*, which we shall use repeatedly in subsequent chapters to derive kinetic equations. This approximation consists physically in neglecting memory effects, that is, to make the time evolution of a quantity depend on its value at the present time and not on its history (its values in the past). Mathematically, it consists of replacing $c_e(t')$ by $c_e(t)$ in the integral of eqn (6.12). The second approximation that is performed is to replace the upper bound of the integral by infinity. This approximation is justified if $\omega_0 t \gg 1$. We now use the Cauchy formula, which is

$$\frac{1}{2\pi} \int_0^\infty e^{i\omega t} dt = \frac{1}{2} \left(\delta(\omega) - \frac{1}{i\pi} P\left(\frac{1}{\omega}\right) \right) , \qquad (6.13)$$

where P stands for the principal value of the integral.[96] Equation (6.12) becomes

$$\dot{c}_e(t) = \left(-\frac{1}{2}\Gamma_0 + i\Delta\omega \right) c_e(t) , \qquad (6.14)$$

where $\Gamma_0 = (2\pi/\hbar) \sum_{\mathbf{k}} |M_{\mathbf{k}}|^2 \delta(E_0 - E_{\mathbf{k}})$ has the same expression as in eqn 6.7 and

$$\Delta\omega = -\frac{1}{\hbar} \sum_{\mathbf{k}} |M_{\mathbf{k}}|^2 P\left(\frac{1}{E_0 - E_{\mathbf{k}}} \right) . \qquad (6.15)$$

This last quantity is the Lamb shift. It is, as already discussed, the renormalisation of the frequency of emission of the atom induced by the reabsorption of the light by the atom after its initial emission. This frequency shift can be calculated using the Wigner–Weisskopf approach, whereas it is naturally absent from derivations based on the Fermi golden rule. This shift is, however, usually small.

6.1.3 *Quantum dots, 2D excitons and 2D electron–hole pairs*

This aspect has been treated at length in Chapter 4 and we merely discuss here how the main results compare with respect to the case of atoms. The QD Hamiltonian and the procedure that can be used to calculate the decay of a QD excitation is exactly similar to the one we have just detailed. The only difference comes from the shape of the matrix element of coupling.

The situation is slightly different for QWs excitons for which the decay has been found in Chapter 4 as a solution of Maxwell's equations. The operators describing excitons are bosonic and not fermionic as for atoms. This has, however, no impact on the result since we are dealing with occupation numbers smaller than one. Another important difference is that a QW exciton with a given in-plane wavevector is coupled to a continuum of states in a single direction of the reciprocal space, instead of three for

---

[96] See footnote 13 on page 34.

atoms or QDs. In the framework of the Fermi golden rule, the decay rates of a QW exciton having a null wavevector in the plane reads

$$\Gamma_0^{\mathrm{QW}} = \frac{2\pi}{\hbar} \sum_{k_z} |M_{k_z}^{\mathrm{QW}}|^2 \delta(E_0 - E_{k_z}), \qquad (6.16)$$

with the matrix element of interaction between photons and the QW exciton now being

$$M_{k_z}^{\mathrm{QW}} = \frac{\mu_{\mathrm{cv}}}{2\pi a_{\mathrm{B}}^{\mathrm{2D}}} \sqrt{\frac{E_{k_z}}{\epsilon\epsilon_0 L}}. \qquad (6.17)$$

where L is the QW width. Going to the thermodynamic limit as before, this gives

$$\Gamma_0^{\mathrm{QW}} = \frac{2nL}{h^2 c} |M_{E_0}^{\mathrm{QW}}|^2 = \frac{n\mu_{\mathrm{cv}}^2 E_0}{4\pi^2 h^2 c\epsilon\epsilon_0 (a_{\mathrm{B}}^{\mathrm{2D}})^2}. \qquad (6.18)$$

For electron–hole pairs in a quantum well the matrix element of interaction simply reads

$$M_{k_z}^{\mathrm{eh}} = \mu_{\mathrm{cv}} \sqrt{\frac{E_{k_z}}{2\epsilon\epsilon_0 L^3}}, \qquad (6.19)$$

which, with the same procedure as above, yields

$$\Gamma_0^{\mathrm{eh}} = \frac{n\mu_{\mathrm{cv}}^2 E_0}{h^2 c\epsilon\epsilon_0 L^2}. \qquad (6.20)$$

6.1.4 *Fermi's golden rule*

Both inhibition and enhancement of the spontaneous emission are essentially related to Fermi's golden rule.[97] For the case of an electric dipole $\mathbf{d}$ interacting at point $\mathbf{r}$ and time t with the light field $\mathbf{E}(\mathbf{r}, t)$, the spontaneous emission rate for an emitter with energy $\hbar\omega_{\mathrm{e}}$, reads

$$\frac{1}{\tau} = \frac{2\pi}{\hbar^2} |\mathbf{d} \cdot \mathbf{E}(\mathbf{r}, t)|^2 \rho(\omega_{\mathrm{e}}), \qquad (6.21)$$

with $\rho(\omega_{\mathrm{e}})$ the photon density of states at the energy $\hbar\omega_{\mathrm{e}}$ of the emitter. In the vacuum, it is given by

$$\rho_{\mathrm{v}}(\omega) = \frac{\omega^2 V n^3}{\pi^2 c^3}. \qquad (6.22)$$

---

[97]Fermi's golden rule states that if $|i\rangle$, $|f\rangle$ are eigenstates of a Hamiltonian H_0 subject to a perturbation $H(t)$, the probability of transition from an initial state $|i\rangle$ to a continuum of final states $|f\rangle$ is given by the formula

$$\frac{1}{\tau} = \frac{2\pi}{\hbar^2} |\langle f| H' |i\rangle|^2 \rho_f, \qquad (\text{xxvi})$$

with ρ_f the density of final states. If H' is time independent, ρ_f becomes an energy-conserving δ function.

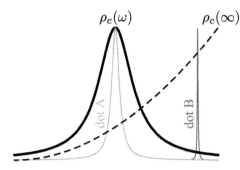

Fig. 6.1: Density of states of the vacuum, ρ_v (thick dashed line), and of a cavity single mode (thick solid), ρ_c, as a function of the frequency ω. The lines of two emitters, e.g., quantum dots, are sketched in a configuration where the emitter is in resonance with the cavity (A) with an enhanced probability of emission into the cavity mode or strongly detuned with no final state to decay, resulting in an increase of its lifetime.

In a single-mode cavity, however, with energy ω_c and quality factor Q, the density of states becomes a Lorentzian

$$\rho_c(\omega) = \frac{2}{\pi} \frac{\Delta\omega_c}{4(\omega - \omega_c)^2 + \Delta\omega_c^2} . \tag{6.23}$$

Both densities are sketched in Fig. 6.1. The localisation of modes available for the final state (decay) of the emitter around the cavity mode allows us to enhance (respectively inhibit) spontaneous emission by tuning (respectively detuning) the emitter with the cavity mode. As a result of its reduced (respectively increased) lifetime, the line gets correspondingly broadened (respectively sharpened). When the characteristic emission time given by eqn (6.21) for the vacuum and cavity case are compared, one gets

$$\frac{\Gamma_c}{\Gamma_0} = \frac{3Q(\lambda_c/n)^3}{4\pi^2 V_{\text{eff}}} \frac{\Delta\omega_c^2}{4(\omega_e - \omega_c)^2 + \Delta\omega_c^2} \frac{|\mathbf{E}(\mathbf{r})|^2}{|\mathbf{E}_{\text{max}}|^2} \left(\frac{\mathbf{d} \cdot \mathbf{E}(\mathbf{r})}{dE} \right)^2 . \tag{6.24}$$

Equation (6.24) is the central equation behind the Purcell effect. It neatly puts together all the modifications that the cavity forces on the lifetime of an enclosed emitter:

- The term $3Q(\lambda_c/n)^3/(4\pi^2 V_{\text{eff}})$ depends only on parameters of the cavity, its quality factor Q, wavelength λ_c, refractive index n and effective volume V_{eff}. As such, it is a *figure of merit* of the cavity, which quantifies the efficiency of Purcell enhancement for an ideal emitter coupled in an ideal way to the cavity. It is this quantity that appears in Purcell et al. (1946)'s seminal paper, in honour of whom it is now called the *Purcell factor*:

$$F_P = \frac{3Q(\lambda_c/n)^3}{4\pi^2 V_{\text{eff}}} . \tag{6.25}$$

- The term $\Delta\omega_c^2/\left[4(\omega_e - \omega_c)^2 + \Delta\omega_c^2\right]$ comes from the density of states of a single mode in Fermi's golden rule formula. It shows the effect of the detuning on the efficiency of the coupling, on which it acts through a phase-space-filling effect. As this quantity is smaller than one, it contributes always towards inhibition of emission. In this approach, fully focused on the Purcell effect, it means that the lifetime of the dot can be made arbitrarily long by detuning it further from the

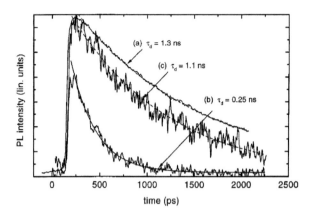

Fig. 6.2: Time-resolved photoluminescence of a quantum dot when placed, (a) in the bulk of the material (GaAs matrix), and (b, c) in a pillar microcavity of the same material, as observed by Gérard et al. (1998). In case (b) the dot is in resonance with the single mode of the cavity and decays about five times quicker than when it is coupled to a continuum of modes, case (a). In case (c) the dot still inside the cavity is detuned with the mode and, as a result, displays only a small enhancement of its lifetime as compared with the bulk case.

cavity. In a real situation, there are other modes to which the emitter couples as well, especially leaky modes that act as a dissipation. For instance, emission through the sides of a pillar put a limit to emission inhibition, as observed by Bayer et al. (2001), who could partially enhance the situation by coating the sides. Finally, there is always a channel of non-radiative decay, which adds a constant term to eqn (6.24) (as is the case in eqn (1.5) where it is appears as $+f$).

- The term $|\mathbf{E}(\mathbf{r})|^2/|\mathbf{E}_{\max}|^2$ underlines the importance of fluctuations in the systems. In the solid-state case, a given sample is more stable than its atomic counterpart as the dot stays fixed at the same position. However, the possible configurations change from sample to sample, and it can require many trials until a good sample is found where the location, oscillator strength and other properties of the dot are suitable to display nontrivial physics (this is especially true when trying to achieve strong coupling, as we shall see in the subsequent chapters). The theoretical fits in Fig. 6.2, for instance, were made by averaging over eqn (6.24).

- The last term $(\mathbf{d}\cdot\mathbf{E})^2/(dE)^2$, which in some cases or for some authors would fall in the same category as above, as adding an element of randomness weakening the optimal Purcell enhancement as quantified by the Purcell factor. However, recent results such as those obtained by Unitt et al. (2005) indicate that a deterministic pining of the dipole along crystallographic axis allows polarisation-selective Purcell enhancement.

The Purcell enhancement of spontaneous emission is neatly demonstrated in Fig. 6.2 where emission is released according to curve (b) in resonance or (c) out of resonance, displaying a much quicker decay in the former case as opposed to the latter that is essentially the same without mode coupling (curve (a)).

An important issue is dimensionality. An atom and its semiconductor counterpart, the quantum dot, is naturally emitting spherical light waves. In a plane-wave representation of light, this means that an atom is intrinsically coupled to all possible plane-wave directions. The decay of such a system will be well described by the application of Fermi's golden rule. On the other hand, the achievement of the Purcell effect or of strong coupling will require light to be confined in the three directions of space, namely to use a three-dimensional optical cavity. The case of bulk excitons considered at length in Chapter 4 is radically different. In the bulk, an exciton is a plane wave characterised by a well-defined wavevector. This exciton is coupled to a single photon state. The strong coupling is achieved and eigenstates of the system are exciton-polaritons that only decay because the photon ultimately escapes through the edges of the sample. In QWs the situation becomes similar again to the atom/dot case. Indeed, the transitional invariance is broken in one direction and kept in the other two. In these latter two directions, the one-to-one coupling holds, whereas in the third direction, the QW exciton is coupled to a continuum of photonic modes. QW excitons therefore radiatively decay in this direction. The achievement of the Purcell effect or of strong coupling in this latter case requires the use of a planar microcavity that confines the light in the direction perpendicular to the QW.

6.1.5 *Dynamics of the Purcell effect*

We now investigate a toy model illustrating the Purcell effect from a quantum dynamical point of view. We consider the case of an atom placed within a three-dimensional optical cavity of volume $V_c = L_c^3$. The fundamental mode of this cavity has resonance energy $\hbar\omega_c$ and is characterised by a Q factor $Q = \omega_c/\gamma_c$, where γ_c is the width of the cavity mode. We recall that a picturesque definition of the quality factor Q is that it numbers how many round-trips light makes in the cavity before leaking out. Formally, the atom is now coupled to a single decaying mode. The Fermi golden rule can no longer be used. We describe the coherent coupling of the atom with a single cavity mode that is dissipatively coupled to a bath of external photon modes. The system we consider is sketched in Fig. 6.3.

We define the atom–light states involved in coherent/dissipative couplings of the first manifold as

$$|1\rangle = |e, 0\rangle, \qquad |2\rangle = |g, 1\rangle, \quad \text{and} \quad |3\rangle = |g, 0\rangle. \qquad (6.26)$$

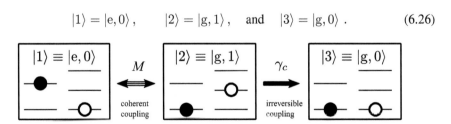

Fig. 6.3: Sketch of the physics involved in the coupling of one atom in a leaky cavity mode. There is coherent coupling between states $|e, 0\rangle$ and $|g, 1\rangle$, through Jaynes–Cummings dynamics, and dissipative coupling to a reservoir, bringing the system to the vacuum $|g, 0\rangle$. These three states form a basis of states called $|1\rangle$, $|2\rangle$ and $|3\rangle$, respectively.

With these notations, the density matrix of the system reads

$$\rho = \sum_{i=1}^{3} \sum_{j=1}^{3} \rho_{i,j} \, |i\rangle\langle j| \, . \tag{6.27}$$

The equation of motion for the density matrix taking into account Lindblad dissipation of Section 5.5 (now for the atom case) gives, in this special case where the Hilbert space is spanned by the basis $\{|1\rangle, |2\rangle, |3\rangle\}$,

$$\frac{\partial}{\partial t}\rho = -\frac{i}{\hbar}[H, \rho] + L\rho \, , \tag{6.28}$$

with

$$H = \begin{pmatrix} \hbar\omega_0 & M & 0 \\ M & \hbar\omega_c & 0 \\ 0 & 0 & 0 \end{pmatrix} \quad \text{and} \quad L = \begin{pmatrix} 0 & -\frac{\omega_c}{2Q} & 0 \\ -\frac{\omega_c}{2Q} & -\frac{\omega_c}{Q} & 0 \\ 0 & 0 & \frac{\omega_c}{Q} \end{pmatrix} \, . \tag{6.29}$$

It follows, for the matrix elements,

$$\dot{\rho}_{11} = i\frac{M}{\hbar}(\rho_{12} - \rho_{21}) \, , \tag{6.30a}$$

$$\dot{\rho}_{22} = -i\frac{M}{\hbar}(\rho_{12} - \rho_{21}) - \frac{\omega_c}{Q}\rho_{22} \, , \tag{6.30b}$$

$$\dot{\rho}_{12} - \dot{\rho}_{21} = i\frac{M}{\hbar}(\rho_{11} - \rho_{22}) + \left[i(\omega_0 - \omega_c) - \frac{\omega_c}{2Q}\right](\rho_{12} - \rho_{21}) \, , \tag{6.30c}$$

$$\dot{\rho}_{33} = \frac{\omega_c}{Q}\rho_{22} \, . \tag{6.30d}$$

The Fourier transform of this linear system gives the equations determining the eigenfrequencies, reminiscent of the renormalisations already encountered several times throughout this book. They read

$$\omega_{\pm} = \frac{1}{2}\left(\omega_0 + \omega_c + i\frac{\omega_c}{Q} \pm \sqrt{\left(\omega_0 - \omega_c - i\frac{\omega_c}{Q}\right)^2 + 4\left(\frac{M}{\hbar}\right)^2}\right) \, . \tag{6.31}$$

One can see that in the general case, the real and imaginary parts of the modes are both renormalised with respect to the case of the bare states. We now analyse some particular cases.

First, we consider the resonance between the dipole and the cavity mode: $\omega_0 = \omega_c$. In this case, two different regimes take place, depending on the sign of the quantity that is below the radical. If $\omega_C/Q > 2M/\hbar$, the square root in eqn (6.31) is imaginary. The two eigenmodes have the same real part, but different imaginary parts, one larger and one smaller than $\omega_c/(2Q)$. This regime is the weak coupling. If $\omega_c/Q \gg 2M/\hbar$, the expansion of the square root gives

$$\omega_+ \approx \omega_c + i\frac{Q}{\omega_c}\left(\frac{M}{\hbar}\right)^2 = \omega_c + i\gamma_c \, , \tag{6.32a}$$

$$\omega_- \approx \omega_c(1 + i/Q) \, . \tag{6.32b}$$

One mode has the energy and decay of the bare cavity mode. The other mode has the same energy as the bare mode, but the decay constant is enhanced by the presence of the cavity, which is the manifestation of the Purcell effect. Using the explicit expression for M, the decay of an atom is usually given as

$$\gamma_c = \frac{Q}{\omega_c}\left(\frac{M}{\hbar}\right)^2 = F_P\Gamma_0\,, \tag{6.33}$$

where Γ_0 is the free atom decay and F_P is the Purcell factor given by eqn (6.25).

One can see that F_P—the Purcell enhancement factor—is proportional to the Q factor of the cavity and is maximal for small cavity volumes that maximise the overlap between the confined photon mode and the atom. Physically, this means that the atom will emit light only in the cavity mode and will decay much faster than in the vacuum.

On the other hand, if $\omega_c/2 < 2M/\hbar$, the radical of eqn (6.31) is real. In this case, the two eigenmodes have the same imaginary part, but two different real parts. This regime is the strong coupling. A photon emitted by the atom does not tunnel out, but is virtually reabsorbed and re-emitted several times by the atom before it leaks out. The eigenmodes of the system are, therefore, mixed light–matter modes. In the limiting case, where $\omega_c/Q \ll 2M/\hbar$, the splitting between the two eigenfrequencies is given by $2M/\hbar$. The strong coupling regime is presented in detail in the subsequent chapters.

In the strong off-resonance case, when $|\omega_0 - \omega_c| \gg M/\hbar$ and also $|\omega_0 - \omega_c| \gg \omega_c/Q$, the square root of eqn (6.31) can be developed keeping only terms contributing to the imaginary part:

$$\frac{\omega_c}{Q} \pm \sqrt{\left(\omega_0 - \omega_c - i\frac{\omega_c}{Q}\right)^2 + 4\left(\frac{M}{\hbar}\right)^2} \tag{6.34}$$

$$\approx (\omega_0 - \omega_c)\left(1 - i\left(\frac{\omega_c}{Q(\omega_0 - \omega_c)^2} + \frac{(4(M/\hbar)^2 - (\omega_c/Q)^2)(\omega_c/Q)}{2(\omega_0 - \omega_c)^3}\right)\right).$$

In contrast with the other case, the photonic mode remains unperturbed, whereas the emission from the atom is strongly inhibited.

The formalism presented above to calculate the Purcell enhancement factor is valid for any single mode emitter coupled to a single optical mode. The value achieved by the enhancement factor only depends on the dimensionality. Hence, the enhancement factor of a QD in a 3D cavity is the same as that calculated above for an atom. The case of QWs excitons and electron–hole pairs is, however, different. We use the formula and we replace the optical quantisation length L of the equation by the cavity thickness L_c, and we find

$$F_P^{2D} = \frac{1}{4\pi}\frac{\lambda_c}{nL_c}Q\,. \tag{6.35}$$

This formula has the same form as its zero-dimensional counterpart. In normal microcavities, the cavity length is of the order of the wavelength, which makes an enhancement factor of the order of $Q/(4\pi)$.

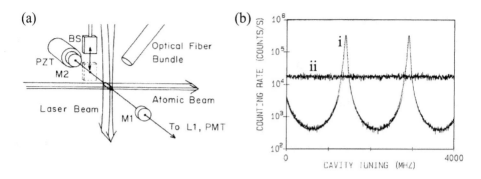

Fig. 6.4: (a) Atomic cavity experiment from Heinzen et al. (1987) for Purcell modification of (b) spontaneous emission with cavity open (i) or blocked (ii).

Our previous analyses showed that the effect depends on the value achieved by the matrix element of interaction that itself is composed of an intrinsic part (used in the Hamiltonian) and of a part induced by the non-ideal overlapping between the cavity mode and the dipole mode. It is easy and instructive to consider as an example the impact of the position of a QW in a planar microcavity. For a cavity surrounded by perfect mirrors, the amplitude of the electric field along z goes like $E(z) = E_0 \cos(2\pi z/L)$. This is valid if the well is placed at the maximum of the electric field. Otherwise, one should go back to the formula replacing M by $M \cos(2\pi z/L)$. The Purcell enhancement factor therefore becomes

$$F_P^{2D}(z) = \frac{1}{4\pi} \frac{\lambda_c}{nL_c} Q \cos^2\left(\frac{2\pi z}{L}\right). \tag{6.36}$$

One can see that an inappropriate placement of the well inside the cavity can lead to a strong decrease of the Purcell factor and even to a complete suppression of emission. In practice, this constraint is not that strong in planar microcavities, where a very good control of the QWs position can be achieved. This constraint is, however, much stronger for QDs in 3D cavities. Indeed, the position of QDs is weakly controlled by crystal-growth techniques and the quality of overlapping between an optical mode and a QD is often governed by chance, although site-controlled techniques are developing and are increasingly being used.

6.1.6   Experimental realisations

Experimental realisations of the Purcell effect have been held back because typical strong dipoles emit at high rates (lifetimes of order 1 ns), which implies that high Q-factors and small cavities (cavity lifetimes $\simeq Qc/L$) are needed to perturb the photon density of states sufficiently.

For atoms in an optical microcavity, this required the development of cold atomic beams and suitable continuous-wave tuneable lasers. The first realisation of the Purcell experiment in a larger macrocavity ($L = 5\,\text{cm}$) by Heinzen et al. (1987) showed a few per cent change in the emission rates, primarily due to the low solid angle in the confocal cavity mode (Fig. 6.4). In the same year, experiments by Jhe et al. (1987) in

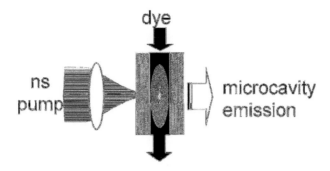

Fig. 6.5: Optically pumped dielectric DBR planar microcavity laser with flowing dye.

the near-infrared on plane–plane Au-coated microcavities separated by $1.1\,\mu m$ showed that the μs decay times of Cs atoms could be increased by 60%. Also in the same year, de Martini et al. (1987) demonstrated the same effect for dye molecules flowing in a solution between two plane–plane dielectric mirrors whose tuneable separation (down to below $100\,nm$) produced photoluminescence lifetime enhancements of up to 300% (see Fig. 6.5).

Similar effects for semiconductor quantum dots embedded in microcavities have been demonstrated by Bayer et al. (2001). In $5\,\mu m$ GaAs microdisks, the embedded InAs quantum dot lifetimes could be 300% longer and corresponding effects have been seen with quantum dots in 1–$20\,\mu m$ diameter micropillars. The difficulty of greatly increasing the spontaneous lifetimes in semiconductors is the extra contribution from other non-radiative recombination processes, which in general also increase when patterning devices into optical microcavities. As a result, such experiments have been carried out at low lattice temperatures, $T \approx 4\,K$, reducing the excited phonon mode occupations of the solid.

Subsequent improvements in this type of experiment have allowed emission experiments on single ions or single semiconductor quantum dots inside microcavities. However, Purcell factors are difficult to greatly improve, as it remains difficult to control precisely the spatial position of the emitter within the microcavity. In order to have the slowest emission, the emitter should be in a field minimum. At the same time, good spectral matching should be achieved. A pioneering effort has been made by Hennessy et al. (2004) and Badolato et al. (2005) to achieve such a deterministic coupling, by positioning the dot (within $25\,nm$ accuracy) and growing a spectrally-matched cavity around it. Such a feat of engineering at the quantum level is shown in Fig. 6.6, where a single dot has been put at the maximum of the optical field, computed numerically. Deterministic techniques to bring light and matter together have since been continued, for instance by Dousse et al. (2008), Gallo et al. (2008), Schneider et al. (2008), Thon et al. (2009), Yang et al. (2015), Kuruma et al. (2016), Yang et al. (2016), etc. Photonic crystal defect cavities with semiconductor quantum dots have shown a great potential for realising the Purcell effect. This is mainly because the mode volume

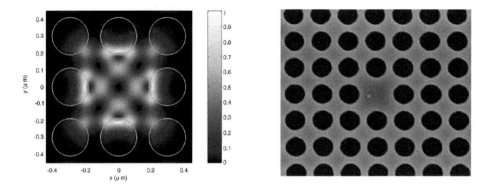

Fig. 6.6: Deterministic coupling of a quantum dot in a microcavity, by Hennessy et al. (2004). On the left, the electric field intensity is computed for this particular photonic crystal configuration of a single hole defect. The light shade reflects intense optical field. By ingenious combinations of growth techniques, gold-markers and micro-photoluminescence, a dot has been successfully placed at one of these bright zones. The dot is actually neatly visible on the scanning electron microscopic image of the microcavity, on the right.

is so small that the cavity linewidth is much smaller than the linewidth of the electronic transition at room temperature, implying that the number of photon round-trips before phase scattering inside the solid is what controls the emission characteristics. Increasing the coupling strength between light and matter sufficiently, one enters strong-coupling, where quantum dynamics dominates the system. This regime has already been discussed in Chapter 5, since theoretically, strong-coupling is the starting point of light–matter interaction (when granted at its simplest and most fundamental level, i.e., that of the Hamiltonian). This should be extended to take into account dissipation, which is a complication unavoidable in a realistic system. Experimentally, one approaches the problem from the other direction: dissipation dominates the dynamics, and only with extraordinary efforts—or much luck—in the design and control of these microscopic laboratories, can one arrive to give predominance to the quantum interaction.

6.2 Lasers

The laser is a direct application of Einstein's theory of light–matter interaction based on the A and B coefficients that have been presented in Section 4.1.2. Prokhorov (1964) discusses in his Nobel lecture why this "obvious" application—the principle of which had been realised by many—took so much time for its realisation, and through the skills of the radiowave community, rather than from optics. The main principle of generating gain in an inverted population by overcoming absorption with emission thanks to stimulation was, however, clear. Moreover, the wide range of possible types of lasers, including semiconductors, was realised early on. In fact, in the Nobel lectures accompanying that of Prokhorov the emphasis is made simultaneously and equally both for atoms by Townes (1964) and for semiconductors by Basov (1964). These main principles having to do with the interplay of gains and losses are presented now. Later, we will turn more specifically to the semiconductor laser, which was rapidly demonstrated as

Basov envisioned it, showing that the main ideas were sound, but quickly requiring ingenious manufacturing to operate efficiently. The description of these elaborations and specificities are discussed in Section 6.2.1.

Charles **Townes** (b. 1915), Nikolai **Basov** (1922–2001) and Aleksandr **Prokhorov** (1916–2002), the 1954 Nobel Prize in Physics for *"fundamental work in the field of quantum electronics, which has led to the construction of oscillators and amplifiers based on the maser-laser principle"*.

Townes gained expertise with microwave techniques from his design of radar bombing systems during World War II. From spectroscopy he gradually went on to develop the concept but also the physical realisation of the maser (he also gave the device its name). In the late 1950s, he discussed with Gordon Gould, then a Ph.D student (this was never completed), about the *optical maser*, nowadays known under the term that was used for the first time by Gould: the *laser*. Gould referred to his knowledge of optical pumping to bring the maser into the optical window, while Townes and his brother-in-law, Schawlow, were undertaking important founding work on the topic. This sparked one of the most famous patent fights in history that lasted over 30 years and finally saw Gould victorious over Townes' design, deemed by the court as not eventually working. Gould also won his court battles against laser manufacturers and became a multimillionaire from royalties of his patents. Controversy, especially sparked by Townes and Schawlow, still surrounds Gould as the inventor of the laser.

Prokhorov was Basov's Ph.D advisor. They both served in the Red Army during the war (Prokhorov being wounded twice) and earned many distinctions from the Soviet union. Prokhorov was chief editor of the *Great Soviet Encyclopedia* from 1969. They made the breakthrough in the maser effect simultaneously and independently from Townes, using a cavity reflecting light at both ends to amplify a microwave beam. Their work displays an harmonious mastery of both experimental and theoretical treatments.

6.2.1   *The physics of lasers*

Einstein's A and B coefficients are fundamental parameters of the system. The equilibrium considerations in Section 4.1.2 served as a useful particular case to investigate them, but they still apply out of equilibrium when the system is excited or driven in some way. In the following, for simplicity of notations, we consider non-degenerate cases so that $B_{12} = B_{21}$, which we will denote simply as B.

If an excited atom can be "induced" to emit a photon by another photon, there is the possibility of starting a chain reaction in a population of inverted atoms with each additional photon stimulating another one, in turn stimulating other photons. Quantitatively, in a population of $n = n_1 + n_2$ atoms (per unit volume), n_1 of which are in the ground state and n_2 in the excited state, one has $n_1 BI$ photons absorbed by the atoms (with I the photon energy density) and $n_2 BI$ emitted by stimulation, hence $\delta n BI$ photons number of photons gained per second and per unit volume, with $\delta n = n_2 - n_1$.

The number of photons per unit area exchanged (emitted or absorbed) with those already present and accounting for a flux ϕ in the small distance Δz is $\delta n B I \Delta z$. Going to the limit of infinitesimal distance (and flux) yields $d\phi = \delta n B I dz$, which integrates as a function of z as

$$\phi(z) = \phi(0) \exp(\delta n I B z). \tag{6.37}$$

According to the sign of δn, eqn (6.37) will display either exponential attenuation or amplification. The condition $\delta n > 0$ is called *population inversion* and is realised when there are more atoms in the excited than in the ground state.

The amplification by stimulation results from the interplay of emission and absorption, the former having spontaneous emission in addition to the stimulated channel that in other respects is similar to absorption. To quantify this, we define I_s the radiation energy density that equates spontaneous emission and stimulated emission:

$$A = I_s B. \tag{6.38}$$

As we discussed in Section 4.1.2, the detailed balance of the rate equation $dn_1/dt = -dn_2/dt = n_2 A + (n_2 - n_1) I B$ becomes, at equilibrium and in terms of I_s,

$$n_2 I_s + (n_2 - n_1) I = 0, \tag{6.39}$$

solving for atomic populations

$$n_1 = \frac{I_s + I}{I_s + 2I} n \quad \text{and} \quad n_2 = \frac{I}{I_s + 2I} n, \tag{6.40}$$

with $n = n_1 + n_2$. These equations show that $n_2 < (1/2) < n_1$ for all values of I: it is impossible to have more atoms in the excited state than in the ground state by direct pumping as the result of the intrinsic nature of light–matter interactions that have a channel of *stimulated* decay in addition to the intrinsic or spontaneous one. The more intense the radiating field to excite atoms, on the one hand, the more it stimulates decay of the atoms already excited, on the other.

This shows that a *two-level system*, that is, one with only two populations of atomic states, cannot lase: arbitrary high excitation will only approach the equal population configuration (and displays *optical transparency* as a beam will cross the material without being absorbed or emitted). It is, however, possible to obtain inversion of population if there are other transient states. For instance, if the final state 1 of laser radiation has zero lifetime and is always empty, decaying toward the ground state 0, however small the population is of an excited state 2 that decays into 1, it will realise an inversion of population. Amplification is therefore obtained by turning to at least three-level systems, as we show now.

Consider a population of n atoms with possible energy states 0, 1 and 2 with energies E_0, E_1 and E_2, respectively, such that $E_0 < E_1 < E_2$. The population of state i is denoted n_i.

The transition of interest remains the lasing transition, shown in Fig. 6.7 with energy $E_2 - E_1$. Level 2 decays into mode 1 with associated lifetime τ_{21} and into the

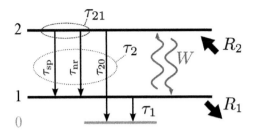

Fig. 6.7: Sketch of transition rates of a three-level system suitable for amplification by stimulated emission. The lasing transition is 2–1. Level 2 has decay channels into level 1 labelled sp for spontaneous emission and nr for all other non-radiative transitions, and a decay channel into the ground state 0 labelled 20. They all contribute to a total lifetime τ_2 for this level. Other such transitions are labelled in the same way. Also mentioned are the pumping rates R_2 populating the excited state and R_1 depopulating it. Finally, when the radiation field starts to become important, stimulated emission and absorption W also enter the picture.

ground-state with lifetime τ_{20}. If there are many channels of decay for a given transition, typically one decay by spontaneous emission with characteristic time $\tau_{\rm sp}$ and a non-radiative decay with time $\tau_{\rm nr}$, then the total lifetime builds up as follows:

$$\tau_{21}^{-1} = \tau_{\rm sp}^{-1} + \tau_{\rm nr}^{-1} . \tag{6.41}$$

Without pumping, the system quickly decays into its equilibrium state with all atoms in the ground state. Efficient and typical pumping schemes involve pumping the excited state of the laser transition at a rate R_1 and depopulating it at a rate R_2. The rate equations taking into account these transitions only, are, by definition of the quantities involved:

$$\frac{dn_2}{dt} = R_2 - \frac{n_2}{\tau_2} , \tag{6.42a}$$

$$\frac{dn_1}{dt} = -R_1 - \frac{n_1}{\tau_1} + \frac{n_2}{\tau_{21}} . \tag{6.42b}$$

The steady-state solution is readily found in this case and the population difference $\delta n = n_2 - n_1$ reads

$$\delta n_{\rm ss} = R_2 \tau_2 \left(1 - \frac{\tau_1}{\tau_{21}} \right) + R_1 \tau_1 , \tag{6.43}$$

where subscript "ss" means "steady state". Large values of $\delta n_{\rm ss}$ (and therefore high values of optical gain) are obtained when the pumping rates R_i are high and when τ_2 is large so as to build up a high population of excited states relative to level 1, and with a small value of τ_1 if $R_1 < (\tau_2/\tau_{21})R_2$. Otherwise, the population of level 1 becomes detrimental and it is better if the level is quickly depopulated thanks to the short lifetime to compensate an inefficient R_1.

When the radiation field builds up it triggers transitions between levels 1 and 2 by absorption (depopulating 1 for 2) and stimulated emission (inducing the opposite transition from 2 to 1). The equations now become

$$\frac{dn_2}{dt} = R_2 - \frac{n_2}{\tau_2} - n_2 W + n_1 W \,, \tag{6.44a}$$

$$\frac{dn_1}{dt} = -R_1 - \frac{n_1}{\tau_1} + \frac{n_2}{\tau_{21}} + n_2 W - n_1 W. \tag{6.44b}$$

Observe that the new term $\delta n W$ cancels in the sum of eqns (6.44). The population difference for this case is also obtained readily in the steady state:

$$\Delta n_{ss} = \frac{\delta n_{ss}}{1 + \tau_s W} \,, \tag{6.45}$$

where δn_{ss} is the population difference in the absence of the radiation field, eqn (6.43); and τ_s is the so-called *saturation time constant*:

$$\tau_s = \tau_2 + \tau_1 \left(1 - \frac{\tau_2}{\tau_{21}} \right). \tag{6.46}$$

Of course δn_{ss} and Δn_{ss} coincide when $W \to 0$.

Exercise 6.1 (*) *Show that τ_s is a well-behaved time that is always positive. As a result show that*

$$\Delta n_{ss} \leq \delta n_{ss} \,. \tag{6.47}$$

Equation (6.46) shows a very important aspect of light–matter interactions as the result of Einstein processes that is displayed in Fig. 6.8. Stimulated emission is a desirable effect for amplification (and its coherent properties as the result of cloning the stimulated emitted photon with the stimulating one), but it is detrimental to the population inversion. When the radiation field becomes important enough, stimulated emission and absorption dominate, with equal weights as they have equal probabilities. In this case, the dilemma of the two-level system springs up again and strong radiation tends towards equalisation of populations (this time from above, though).

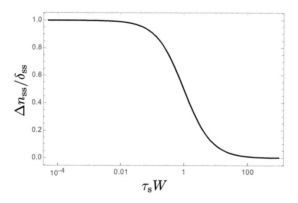

Fig. 6.8: Normalised population difference in the steady state $\Delta n_{ss}/\delta_{ss}$ as a function of the radiation field intensity W. When $W = \tau_s^{-1}$, Δn_{ss} has been halved from its optimum value δn_{ss}, the population inversion in the absence of radiation.

6.2.2   *Semiconductors in laser physics*

Semiconductors are important materials in light–matter physics thanks to the radiative recombination of electrons and holes, although not at equilibrium where their densities are too small to produce a detectable quantity of light even with high doping.[98] It is, however, easy to operate a semiconductor out of equilibrium by applying an electric voltage to it and generating large populations of carriers. Indeed, a forward-biased p-n gallium arsenide junction generates strong light in the infrared as reported in the early 1960s by Hall et al. (1962), Nathan et al. (1962) and Quist et al. (1962).[99] Holonyak and Bevacqua (1962) could obtain emission in the visible window[100] by using the GaAsP compound. By increasing the pumping of the structure to the point where electrons and holes undergo an inversion of population, the diode reaches the stage where gain by stimulated emission overcomes losses and an input signal on the active region is amplified. It remains to engineer the device so that this input is levied from its output to trigger the laser oscillations. The cavity in this case is provided by the semiconductor crystal itself whose facets have been cleaved, i.e., terminated along the crystal axis to create a perfectly flat endface, perpendicular to the axis of the junction[101]. When the light generated by recombination of electrons and holes gets to this surface, it is partially reflected back by internal reflection. The reflectivity is consequently quite low for such lasers, about 30% (the facets can be coated for better reflection).

These preliminary diode sources were not efficient lasers as the active region where electron and hole recombine is spread out across the junction with great losses and requiring significant threshold currents to compensate. Only the short-pulse regime is possible before melting the device. A solution was envisioned on how to constrain carriers effectively, theoretically by Kroemer (1963) and later realised in the Ioffe institute by Alferov: the double heterostructure (DH). It consists of a thin region of semiconductor with a small energy gap sandwiched between two oppositely doped semiconductors with a wider bandgap. When forward biased, carriers flow into the active region and recombine more efficiently because the potential barriers of the heterostructure confine the carriers to the active region. Practical and soon efficient operation was achieved and the device became one key element in the computer and information era, with maybe its most significant impact in the data storage with optical reading of CD and DVD types

---

[98]The first connection of light to semiconductors was made in the beginning of twentieth century with cat's whisker detectors by a collaborator of Marconi, H. J., Round, who reported his finding of a green glow from SiC in "A note on carborundum" in Elect. World **19**, 309 (1907).

[99]Biard and Pittman could prevail on the observation of radiation by a junction while working on GaAs diodes, dating it to 1961. Not expecting light emission, they noticed it using an infrared microscope. Their contribution has been acknowledged for the record by patent issues (for which they received $1 each) although their oldest publication on this topic under the title "GaAs Infrared Source" by Biard, Bonin, Carr and Pittman, is also dated 1962 (in the PGED Electron Device Conference.)

[100]Bringing radiation emission from semiconductors *"in the visible spectrum where the human eye sees"*, to quote Holonyak, made his paper the most quoted of *Applied Physics Letters*.

[101]The idea to use the crystal itself as the cavity comes from Hall who polished the surfaces. He shared it with Holonyak who was planning to use an actual external cavity. Holonyak had the further idea to cleave the crystal. It proved too difficult, however, and Hall came first with the semiconductor laser.

Zhores **Alferov** (b. 1930) and Herbert **Kroemer** (b. 1928)

The 2003 Nobel prize-winners (with Kilby), enabled semiconductors the means to revolutionise laser physics and shape the era of telecommunications.

Kroemer, now a Professor at University of California, Santa Barbara, proposed the concept of the double heterostructure in 1963. A major publication on this topic was rejected for Applied Physics Letters and published in Proc. IEEE instead. His favourite saying—as claimed in his Nobel lecture—is *"if in discussing a semiconductor problem, you cannot draw an energy band diagram, then you don't know what you are talking about."*

Alferov's name is an icon for the Ioffe institute in Saint Petersburg, where he worked from 1953, and as its director in the period between 1987 and 2004. A gifted administrator, he managed to save the institute infrastructures from disaster in the 1990s. In 1971 he received the USA Franklin's institute gold medal for his pioneering works on semiconductor heterolasers. In its history, the Franklin medal has been awarded to four Russian physicists: Kapitsa, Bogoliubov, Sakharov and Alferov, all but Bogoliubov having later received the Nobel prize.

of optical disks (still widely used today for these applications.)[102] These structures that are now called *classical heterostructures* rely on the profile of the energy bands for providing potential traps for the carriers. Their size varies in the range between a few hundred μm and a few mm. The idea was pushed forward by reducing further still the area of localisation to the point where size-quantisation plays a role, opening the way to *quantum heterostructures*, quantum wells, quantum wires and quantum dots.

These various schemes of lasing with semiconductors are sketched in Fig. 6.9. Lasing with a simple junction is a brute-force approach that requires high threshold currents.[103] Consequently, the device can only be operated in pulsed mode with much loss in the conversion. By confining the carrier in the active region with the heterostructure potential trap (simulating the action of a *quasi-electric field*—a term coined by Kroemer—a feast that no genuine external field can achieve), the lasing could be operated for threshold currents J_{th} reduced by two orders of magnitude, a trend that has continued by following the road towards evermore quantisation, as is illustrated

---

[102]Other applications of semiconductor lasers for public use include fibre-optics communication, laser printers, laser surgery, barcode readers, laser pointers... They are also useful for research and the military.

[103]Holonyak and Bevacqua (1962) report 11×10^3 A cm^{-2} as a threshold to superlinear emission and linewidth narrowing.

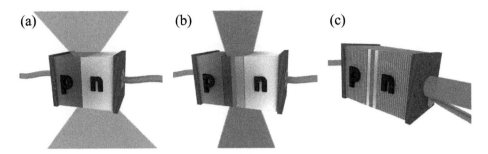

Fig. 6.9: Sketch of semiconductor lasers. *a*) the *p-n* junction where electron–hole recombinations at the interface serve as the active population; this scheme is more viable for LED operation, *b*) the edge-emitting laser where the active region is confined by a heterostructure and *c*) the VCSEL where localisation is pushed to the quantum limit and emission made from the surface. Artwork by Guillermo Guirales.

in Fig. 6.10 where a further two orders of magnitude gain on the current threshold is attained with QDs as the most recent realisation.

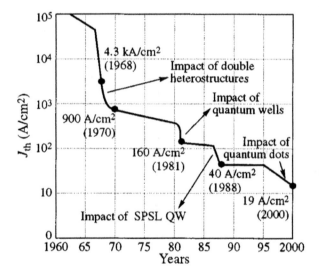

Fig. 6.10: Evolution of the threshold current J_{th} of semiconductor lasers, from Alferov (2001).

The main advantage of edge-emitting lasers is that the size of the active region allows one to store significant amounts of energy compared with the microscopic or smaller volumes involved in diodes where the active region lies in a quantum heterostructure. For this reason, the light–matter interaction must be compensated by constraining the photons inside the active region. This is achieved by shifting the optical axis from the plane of the structure (where photons travel a few times over millimetres) to the growth axis (where photons travel back and forth thousands of times over a micrometre). The mirror that, in the case of the edge-emitting laser is provided by index contrast, must now have a reflectivity of the order of 99.9%. This is possible only with dielectric mirrors, namely Bragg mirrors, that form a microcavity that confine the active region. Such structures are known as VCSELs (for vertical-cavity surface-emitting laser).

6.2.3 *Vertical-cavity surface-emitting lasers*

Until the late 1970s, semiconductor lasers exclusively used the stripe geometry with cavity lengths longer than $100\,\mu$m. However, with the production of high-quality integrated DBR mirrors, it became possible to rotate the orientation of the emission so that it emerged normal to the growth-layer planes. The principal advantage of this surface-emission geometry of vertical-cavity surface-emitting lasers (VCSELs) is the ability to make and test large numbers of lasers on a single wafer without having to cleave the wafer into individual lasers with facets, with a large reduction in the cost of quality control, manufacture and ease of packaging devices. Another advantage is the high spatial quality and non-astigmatic laser emission, which makes it easy to also match to optical fibres. Subsequently, it was realised that very fast modulation could be obtained in VCSELs with low power consumption.

However, in order to get these devices to work, the hundred-fold reduction in cavity length (and, hence, round-trip gain) has to be recovered in decreased round-trip loss (hence the need for very high mirror reflectivities). This became possible with the introduction of lattice-matched DBR designs producing 99% to 99.9% intensity reflection coefficients. The cavity lifetime typically can reach $\tau_c = 1$ ps (i.e. Q values up to 1000), which is similar to that in conventional stripe lasers in which the photons reflect off the cavity mirrors hundreds of times less often.

In order to efficiently pump the device and retain single-mode operation, the current transport and optical emission within the planar microcavities have to be laterally confined. Typically, this is achieved in two ways: either by etching a circular mesa with an annular top contact injecting holes into the upper p-type material, or by using an Al-rich layer within the active region that is oxidised from an outside trench to produce an insulating annulus that forces current through the centre of the device. However, other fabrication procedures can give the same effect including growing in buried holes and proton bombardment. The oxide technology has some useful advantages in that as well as confining the injection current, it also confines photons (since the central core has a higher refractive index than aluminium oxide annulus), which results in better overlap of the light and electron–hole pairs.

Typically, to get maximum gain, quantum wells or quantum dots are used as the active material and placed within the structure at the peak of the antinodes of the intracavity electric field. To overcome the round-trip loss, the gain needs to be as high as possible and thus the carrier density in the active region is universally above the ionisation threshold for excitons. Hence, these devices operate without excitonic contributions to their gain spectrum, distinguishing them from the polariton-lasers discussed later.

One of the advantages of VCSELs is that their active volume can be made very small. Typical small mode areas can be 1–$10\,\mu$m^2 and active cavity material volumes can be as low as $V = 0.05\,\mu$m^3, which is about 100 times smaller than conventional

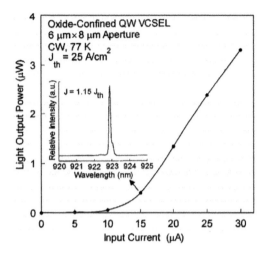

Fig. 6.11: Light output curve for a quantum dot VCSEL, as reported by Zou et al. (2000).

stripe lasers. Hence, the threshold in a VCSEL is correspondingly small since the threshold current is approximately $I_{th} = eVN_t/\tau_c$ when the cavity loss rate is small, controlled by the need for carrier densities above the transparency value $N_t \simeq 10^{18} \text{cm}^{-3}$. Thresholds below $10\,\mu\text{A}$ have recently been reported for quantum dot VCSELs by Zou et al. (2000) (Fig. 6.11).

The growth of such multilayers is now at a highly sophisticated level, with special designs being able to simultaneously control the electrical transport through the DBR layers (using specific doping profiles that provide useful band bending), to control the photon confinement (through the oxide apertures), to control the thermal management (by matching the thermal impedance mismatches through the DBR stacks) and to control the electrical carrier spatial distributions to maximise the gain. The resulting overlaps with the optical fields can be of the order of 5% in the vertical direction (overlap with QWs at the centre of cavity) and up to 80% in the lateral direction, leading to confinement factors of 4% or so.

The power conversion (or wallplug) efficiency of VCSELs is also very high when pumped above threshold, because little voltage drop exists across the p-n junction except at the active region, leading to $\eta > 50\%$. However, as the current increases, the transverse mode of the VCSEL has a tendency to switch, due to spatial hole burning and current spreading. To prevent this, structures can be designed that are antiguiding and operate well at higher power. In addition, another switching can appear at higher power due to the birefringence of the devices, originating either from shape anisotropies or intrinsic strain in the layers. The polarisation output of these devices is always linear, but can switch axes between two near degenerate cavity modes as the temperature of the device rises. Full polarisation dynamics of VCSELs have been discussed by Gahl et al. (1999).

We now provide a more thorough discussion of the Boltzmann equations (i.e. semiclassical rate equations) following an analysis by Björk and Yamamoto (1991) and

Björk et al. (1993), which is modified to provide a clear comparison with the inversion-less polariton-lasers discussed in Chapter 8. We define generally a population density N_x pumped at a rate P, producing laser emission from state N_0 through the steady-state rate equations

$$V\frac{dN_x}{dt} = \frac{P}{\hbar\omega} - \Gamma_{\mathrm{nr}}VN_x - N_0R(N_x) - S(N_x) = 0\,, \qquad (6.48a)$$

$$\frac{dN_0}{dt} = -\Gamma_0 N_0 + N_0 R(N_x) + S(N_x) = 0\,, \qquad (6.48b)$$

where N_0 is the number of photons in a conventional laser and the number of polaritons in a matter-wave laser. For a conventional semiconductor laser of active volume V, the spontaneous S, and stimulated R, scattering rates are

$$S = \beta\Gamma_s VN_x\,, \qquad (6.49a)$$

$$R = \beta\Gamma_s V(N_x - N_t)\,, \qquad (6.49b)$$

with non-radiative decay rate $\Gamma_{\mathrm{nr}} = 1/\tau_{\mathrm{nr}} + (1-\beta)\Gamma_s$, where Γ_0 is the photon cavity escape rate, Γ_s is the spontaneous emission rate and τ_{nr} is the non-radiative lifetime. The fraction of photons that are spontaneously emitted into the lasing mode is β, which is a key parameter for enhancing the emission from microcavities.

These equations can be simply solved to yield the output power as the pump rate is increased, giving a lasing threshold

$$P_{\mathrm{th}} = \frac{\hbar\omega\Gamma_0}{\beta}(1 + \Gamma_{\mathrm{nr}}/\Gamma_s)\left(1 + \frac{\beta\Gamma_s VN_t}{\Gamma_0}\right)\,. \qquad (6.50)$$

Björk and Yamamoto (1991) have shown how this threshold behaves in a vertical microcavity laser in the weak-coupling regime, and emphasised that the values of both β and Γ_{nr} are critical to low threshold action. Typically, in a semiconductor Fabry–Pérot laser, thresholds are around $10\,\mathrm{mW}$ in a $300\,\mu\mathrm{m}$ long device. This contrasts with VCSELs with optimised $\beta \approx ^{-2}$, allowing thresholds below $1\,\mathrm{mW}$ (Fig. 6.12).

The form of these equations is the same for most lasers and is underpinned by the deep relationship between spontaneous and stimulated emission through the Einstein coefficients of a transition. The result is that population inversion is a necessary condition for lasing and in a semiconductor laser this requires a sufficient carrier density N_t to bleach the absorption. At these densities above the Mott density, screening by the Coulomb interaction and phase space filling is sufficiently strong that the electron–hole plasma screens out the exciton binding and thus exciton lasing is impossible. The situation is completely altered when the polariton pair scattering discussed in Chapter 7 is used to feed energy into a lasing transition that is in the strong-coupling regime.

One further important characteristic of the VCSEL is the possibility of turning it on and off at high frequencies. Typically, this rate is controlled by the relaxation oscillation frequency, f_r, the natural oscillation rate between cavity photons and electronic excitations produced by the coupled equations (6.48):

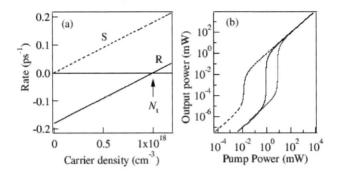

Fig. 6.12: VCSEL characteristics: (a) spontaneous (S) and stimulated (R) scattering rates from excitons to photons as the carrier density increases. $\tau_{nr} = 50\,ps$, $\Gamma_0^{-1} = 30\,ps$, $\Gamma_s^{-1} = 3\,ps$, $\beta = 3.6 \times 10^{-3}$, $V = 0.15\,\mu m^3$. (b) Output vs. input power for the conditions in (a) (solid), for $\tau_{nr} = 50\,ns$ (dashed) and for a Fabry–Pérot laser with $\tau_{nr} > 10\,ns$ $\Gamma_0^{-1} = 3\,ps$, $\Gamma_s^{-1} = 3\,ns$, $\beta = 3 \times 10^{-5}$, $V = 60\,\mu m^3$ (dotted).

$$f_r = \frac{\Gamma_s}{2\pi}\sqrt{\frac{1}{\Gamma_s \tau_c}\left(\frac{I}{I_{th}} - 1\right)}. \tag{6.51}$$

For typical parameters this frequency exceeds 10 GHz in VCSELs compared with factors of 10 smaller for conventional lasers.

We also note here that there has been some discussion about the extent to which coherence in the electronic system can play a role in semiconductor lasers and the coherence in the system remains difficult to calculate due to the many-body nature of the problem.

6.2.4 Resonant-cavity LEDs

Recently, a popular alternative to VCSELs for strong light emission that is incoherent has been the resonant-cavity light-emitting diode (RCLED). Because the gain of a wavelength-thick semiconductor layer is typically small, high-quality mirrors are required to produce effective lasers and these require sophisticated fabrication of many precise semiconductor layers. On the other hand, for many applications, the coherence of the light emission from the device is not so important, and it is the efficiency and directionality that is key. Typical applications are in display technologies, in xerography, in biophotonics and in general lighting applications. In an RCLED, the semiconductor emitting layer is clad with low-finesse mirrors (unbalanced so they have smaller reflectivity on the top side), which modify the angular emission pattern and suppress emission into in-plane waveguide modes that do not escape efficiently from the sample. With these modifications, the LED efficiencies can exceed 50% and make such emitters competitive even with incandescent lighting. The rise of such devices has tracked that of the GaN-based technologies so that UV/blue emitters can now be tuned across the visible spectrum using either phosphor-based light conversion or In doping of the emitting GaN to produce ranges of colours.

Typically, the design of RCLEDs is similar to VCSELs. To enhance the emission, the active p-n junction is placed at the antinodes of a microcavity whose length is close

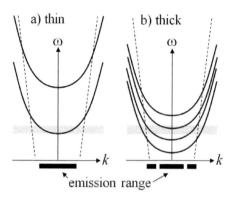

Fig. 6.13: (a) Cavity dispersion (curves) and LED active emission spectrum (shaded box) for (a) wavelength-scale and (b) thick microcavity, showing different emission patterns inside the light cone (dashed).

to a few optical wavelengths. This ensures that the linewidth of the cavity mode set by the Q-factor is as large as possible, while the angular emission is beamed as much as possible in the surface normal direction. In thick cavities of refractive index n, the fraction of resonances that exist within the light cone (from which light can escape) is given by $1/2n^2$ and this emission is shared between the number of possible microcavity resonances (see Fig. 6.13). This also shows intuitively that the enhancement will be better for active LEDs with a narrower spectral width.

The simplest design for a RCLED uses a buried conducting DBR and a metal mirror as the top contact, with the light extracted through the substrate (which therefore has to be transparent). In such a structure the extraction efficiency can exceed 20%. The remainder is emitted into guided modes (which are absent as they are below the cutoff in half-wavelength thick cavities) or remains trapped in the substrate.

6.2.5   *Quantum theory of the laser*

In the above sections we have developed theories of the laser dealing with average values both of populations of atomic (or carrier) populations and of the radiation field intensity (number of photons).

A full quantum treatment of the laser requires some approximations that one can consider in some depth in the classical or semiclassical theories, such as multimode operation, spatial inhomogeneities, temporal drifts, inhomogeneous broadenings and so on. The mathematical complications brought by dealing with operators in the quantum counterpart makes it awkward to draw a clear parallel. All that pertain to approximations that concern the average populations can be stripped from a quantum perspective as they appear as classical averages anyway. The simplest Hamiltonian of interaction is sought and a posteriori investigations show that for most purposes it is enough to consider

$$H = \hbar g \left\{ a\sigma^\dagger u(\mathbf{r}) e^{i(\omega_0 - \omega)(t - t_0)} + a\sigma u(\mathbf{r}) e^{-i(\omega_0 + \omega)(t - t_0)} + \text{h.c.} \right\} \tag{6.52}$$

in the interaction picture, where a is the photon annihilation operator, σ the atom annihilation operator, $u(\mathbf{r})$ the normalised cavity mode function and g the interaction strength derived in Chapter 4. We have already considered a system close to this one to derive the Bloch equations, but an important approximation has been retained here that hints at the difference in the laser case, namely the rotating wave approximation of the second term.

The density matrix of the system is the combined atom (carrier)–photon field system. At initial time in the absence of correlations between them, it reads

$$\rho(t_0) = \rho_A(t_0) \otimes \rho_F(t_0), \tag{6.53}$$

with ρ_A (respectively ρ_F) the density matrix for the atom (respectively photon field). The state of ρ_F is obtained by tracing over the atomic variables. If we write the Liouville–von Neumann equation of motion as its Born expansion to infinite order before doing so, we get

$$\rho_F(t) = \rho_F(t_0) + \mathrm{Tr}_A \left(\sum_{n=1}^{\infty} \frac{1}{(i\hbar)^r} \int_{t_0}^{t} \int_{t_0}^{t_1} \cdots \right.$$
$$\left. \int_{t_0}^{t_r-1} [H(t_1), [H(t_2), [\ldots [H(t_r), \rho(t_0)]\ldots]]] dt_1 dt_2 \cdots dt_r \right). \tag{6.54}$$

To successfully describe laser action—even in the simplest setting—with eqns (6.52) and (6.54), one must carry out the algebra to high order in the commutators. One gets

$$[H, \rho] = \hbar g[\sigma a^\dagger \rho_F u^*(\mathbf{r}) - \sigma^\dagger \rho_F a u(\mathbf{r})], \tag{6.55}$$

and successively, up to the fourth-order commutator that yields

$$[H, [H, [H, [H, \rho]]]] = (\hbar g)^4 |u(\mathbf{r})|^4 [\sigma^\dagger \sigma(aa^\dagger aa^\dagger \rho_F + 3aa^\dagger \rho_F aa^\dagger + \text{h.c.})$$
$$- 4\sigma\sigma^\dagger(a^\dagger aaa^\dagger \rho_F a + \text{h.c.})]. \tag{6.56}$$

To proceed one considers the evolution from the initial atomic position in both ground and excited states. For instance, in the excited state, eqn (6.56) becomes, tracing over atomic variables

$$\mathrm{Tr}_A([H, [H, [H, [H, \rho]]]]) = (\hbar g)^4 |u(\mathbf{r})|^4 (aa^\dagger aa^\dagger \rho_F$$
$$+ 3aa^\dagger \rho_F aa^\dagger - 4a^\dagger aaa^\dagger \rho_F a + \text{h.c.}). \tag{6.57}$$

Up to now, we have dealt with a single atom interacting with a single mode of the cavity mode. The latter approximation is reasonable but the single atom does not describe a conventional laser,[104] which properly involves an assembly of atoms as its active media. Rigorous but heavy methods have been developed, for instance by Lamb Jr. (1964) or Scully and Lamb Jr. (1967). Even textbooks specialising in this topic find

---

[104]The necessity to consider an assembly of atoms in the model of a laser is made more stringent by the fact that there exist single-atom lasers that would dispense with these requirements, demanding in exchange a more thorough consideration of its quantum-mechanical features.

it difficult to attain such feats of meticulousness. Mandel and Wolf (1995) propose a shortcut—which a posteriori proves to be essentially equivalent—to consider the effects of an assembly of atoms by considering a coarse-grained lifetime average with probability distribution $P(\Delta t) = e^{-\Delta t/T_2}/T_2$, where T_2 is the lifetime of the excited state (level 2) and multiplied by R_2 the pumping rate of this level. Also, the equation is averaged spatially over the active medium. An equation of motion is obtained for the gain mechanism (since the atom was in its excited state). The same procedure can be started again for the atom initially in its ground state, yielding another master equation. The sum of which (since there is no coherence between the atoms) provides the final master equation for the photon field, which reads

$$\frac{\partial \rho_F}{dt} = -\frac{1}{2} A [aa^\dagger \rho_F - a^\dagger \rho_F a + \text{h.c.}] - \frac{1}{2} C [a^\dagger a \rho_F - a \rho_F a^\dagger + \text{h.c.}]$$

$$+ \frac{1}{8} B [aa^\dagger aa^\dagger \rho_F + 3 aa^\dagger \rho_F aa^\dagger - 4 a^\dagger aa^\dagger \rho_F a + \text{h.c.}], \quad (6.58)$$

where the above-mentioned derivation (introducing quantities such as $\eta(\mathbf{r})$ the density of active atoms and η_1 that of loss atoms) provides coefficients

$$A = 2(R_2/N)(gT_2)^2 \int \eta(\mathbf{r}) |u(\mathbf{r})|^2 \, d\mathbf{r}, \quad (6.59a)$$

$$B = 8(R_2/N)(gT_2)^4 \int \eta(\mathbf{r}) |u(\mathbf{r})|^4 \, d\mathbf{r}, \quad (6.59b)$$

$$C = 2(R_1/N)(gT_2)^2 \int \eta_1(\mathbf{r}) |u(\mathbf{r})|^2 \, d\mathbf{r}. \quad (6.59c)$$

Coefficients (6.59) characterise gain, nonlinearity and losses of the laser, respectively. Observe that B is of the order of the square of A and C.

Equation (6.58) is a typical single-mode laser master equation. Coefficients would vary for other systems derived under other approximations (or more rigorously derived), but the main principles remain with nonlinear terms displaying such asymmetric repartition about the density operator. This results in coupling the diagonal elements to off-diagonal elements and plays a role in the coherence of the field. As for the diagonal elements, their equation of motion is readily obtained by dotting the master equation to get the equation of motion of $p(n,t) = \langle n| \rho_F |n \rangle$ as

$$\frac{\partial p(n,t)}{\partial t} = -A(n+1)\left(1 - \frac{B}{A}(n+1)\right)p(n,t) + An\left(1 - \frac{B}{A}n\right)p(n-1,t)$$

$$+ C(n+1)p(n+1,t) - Cnp(n,t), \quad (6.60a)$$

$$\approx -\frac{A(n+1)}{1 + (B/A)(n+1)}p(n,t) + \frac{An}{1 + (B/A)n}p(n-1,t)$$

$$+ C(n+1)p(n+1,t) - Cnp(n,t). \quad (6.60b)$$

Equation (6.60) is a rate equation similar to those already encountered in the classical theory of lasers, but this time describing the flow in probability space, rather than for averages (the latter can, of course, be obtained by summing the weighted probabilities). In the same way, detailed balance can be applied to obtain the steady-state solution of eqn (6.60) from the compensation of configurations differing by one photon. This is seen clearly in eqns (6.60) where a term involving A cancels with the term involving C of the opposing sign. The other two terms also cancel in this way; they are in fact equivalent substituting n for $n + 1$:

$$\frac{A(n+1)}{1 + (B/A)(n+1)}p(n,t) = -C(n+1)p(n+1,t)\,, \tag{6.61a}$$

$$\frac{An}{1 + (B/A)n}p(n-1,t) = -Cnp(n,t)\,. \tag{6.61b}$$

The corresponding transitions are sketched in Fig. 6.14.

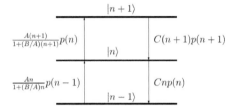

Fig. 6.14: Flow of probabilities between configurations with $n - 1$, n and $n + 1$ photons. At equilibrium, steady state is established by detailed balancing of the neighbouring terms that equate to each other through substitutions $n \to n + 1$.

Equation (6.61) can be solved by recurrence, yielding $p(n)$ knowing $p(n - 1)$ through

$$p(n) = \frac{A/C}{1 + (B/A)n}p(n-1)\,, \tag{6.62}$$

which repeated application yields

$$p(n) = p(0)\prod_{i=1}^{n}\frac{A/C}{1 + i(B/A)}\,, \tag{6.63}$$

with $p(0)$, the starting point, being determined by normalisation condition

$$\sum_{n=0}^{\infty}p(n) = 1\,. \tag{6.64}$$

A solution in Fig. 6.15 is compared to the distribution of an ideal coherent state.

Exercise 6.2 (*) *Show that the polynomial expansion for the equation of motion of* $\langle n \rangle = \sum_n np(n)$ *derived from eqn (6.60b) is of the type*

$$\frac{d\langle n \rangle}{dt} = \alpha\langle n \rangle - \beta\langle n^2 \rangle + \gamma\,. \tag{6.65}$$

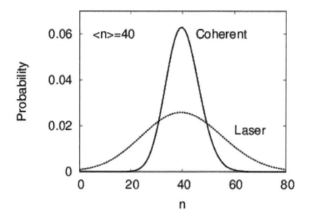

Fig. 6.15: Statistics $p(n)$ of a laser given by eqn (6.63) as compared to the Poisson distribution of a coherent state. Coefficients $A/C = 1.2$ and $B/C = 0.05$ result in an average number of photons $\langle n \rangle = 40$. Even above threshold a laser still has large deviation from the ideal coherent case.

Analyse this equation providing physical meanings of the parameters α, β and γ, and link them to microscopic parameters.

In the semiclassical theory, the linewidth is obtained from the Fourier transform of $\langle E(t) \rangle$, where E is the photon field operator so that, in the Schrödinger picture,

$$\langle E \rangle (t) = \sqrt{\hbar\omega/2eV}\,\sin(kz)\mathrm{Tr}\big(\rho(t)a^\dagger\big)e^{i\nu t} \tag{6.66a}$$

$$\propto \sin(kz)\sum_{n=0}^{\infty}\sqrt{n+1}\rho_{n,n+1}(t)e^{i\nu t}, \tag{6.66b}$$

where $\rho_{n,n+1} = \langle n|\,\rho\,|n+1\rangle$ is the upper diagonal element of the density matrix. Conversely to diagonal elements, the off-diagonal elements do not form a closed set of equations and couple to all other elements of the density matrix (including diagonal elements), showing that the dynamics of coherence that is of a quantum character is more complicated than the dynamics of population that is of a classical character.

Dotting eqn (6.58) with $|n\rangle$ and $|n+1\rangle$, one gets for the equation of motion

$$\dot{\rho}_{n,n+1} = -\left[\left(A - B(n+\tfrac{3}{2})\right)(n+\tfrac{3}{2}) + \tfrac{1}{8}B + C(n+\tfrac{1}{2})\right]\rho_{n,n+1}$$
$$+ \left(A - B(n+\tfrac{1}{2})\right)\sqrt{n(n+1)}\rho_{n-1,n} \tag{6.67}$$
$$+ C\sqrt{(n+1)(n+2)}\rho_{n+1,n+2}.$$

High above the threshold where n assumes high values over which ρ varies smoothly, $\rho_{n,n+1}$ can be approximated by its neighbour value $\rho_{n,n}$, for which recursive closed relations are known, cf. eqn (6.63). This turns eqn (6.67) into

$$\rho_{n-1,n} \approx C\sqrt{(A-Bn)(A-B(n+1))}\rho_{n,n+1} \tag{6.68a}$$

$$\rho_{n+1,n+2} = C\sqrt{(A-B(n+1))(A-B(n+2))}\rho_{n,n+1}. \tag{6.68b}$$

Injecting this expression back into eqn (6.67) gives

$$\dot{\rho}_{n,n+1} = -\frac{1}{2}D\rho_{n,n+1}, \tag{6.69}$$

with

$$D \approx \frac{1}{2}\frac{A}{\langle n \rangle}. \tag{6.70}$$

From the Fourier transform of eqn (6.66), the lineshape of the laser turns out to be

$$S(\omega) = \frac{|\langle E(0)\rangle|^2}{(\omega-\nu)^2 + (D/2)^2}, \tag{6.71}$$

that is, it is a Lorentzian centred on the laser transition with width D. The notable feature is that D varies inversely with the photon field intensity: the laser has a very narrow line as a result of the photon compression in phase space, an effect first realised by Schawlow and Townes. This is, however, more of a theoretical limit as other factors broaden the line much beyond the value given by eqn (6.70).

6.3 Nonlinear optical properties of weak-coupling microcavities

By placing a material inside a microcavity, its nonlinear optical properties are enhanced. The first enhancement arises simply from the enhancement in internal optical intensity due to the finesse, which thus reduces the external threshold light intensity to get a certain nonlinear optical response. The advantage of using a microcavity is that the buildup time for the optical field is short, as well as the transit time, and hence the device operation remains nearly as fast as that of the intrinsic nonlinear material. Another advantage is that the refractive part of the nonlinear response is converted into a transmission nonlinearity due to the optically-induced spectral shifting of the cavity modes (Fig. 6.16).

The nonlinear optical process may arise directly from occupation of the upper state of two-level systems (in atoms, or in semiconductors) that depends on the fermionic statistics of electrons. It may also arise from Coulomb interactions between optically excited states (such as the exchange interaction) and is also typically divided into "real" and "virtual" processes, which correspond to whether switching optical energy ultimately ends up absorbed inside the medium (the former), or the energy remains within the optical field (the latter). In effect, the "virtual" process is a transient state inside the medium that re-emits the optical energy (for instance, if an atom is strongly excited off-resonantly), but transiently produces nonlinear responses as above. Such a process will be intrinsically faster than a "real" process, in which the absorbed energy must be lost (e.g through recombination or spontaneous emission) before the next switching event can take place.

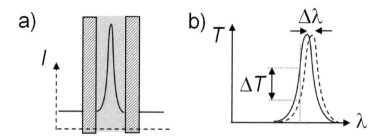

Fig. 6.16: (a) Microcavity with enhanced optical intensity within a nonlinear medium. (b) Optically induced spectral shift of cavity modes from refractive-index change produces a change in transmission at a near-resonant wavelength.

The use of microcavities in these applications has been studied since the 1980s, typically in semiconductor Fabry–Pérot interferometers, for possible ultrafast optical switching elements. A typical example is a gold-coated semiconductor slab containing a quantum well that is optically excited to the long-wavelength side of the exciton resonance. The creation of a virtual population of excitons blueshifts the exciton resonance, thus changing the refractive index at the cavity resonance and producing an enhanced ultrafast response. In general, the problem with such devices is that the nonlinear response from a small-volume microcavity is limited.

6.3.1   Bistability

New effects occur when the cavity mode can be spectrally shifted by more than the cavity linewidth. In this case, optical bistability can occur in which there are conditions for which two stable states of the cavity transmission exist, "high" and "low". The idea is to set up the cavity response and the pump-laser tuning in such a way that an increase in incident optical power spectrally shifts the cavity closer into resonance with the excitation laser. This further increases the power fed into the cavity and thus provides a positive feedback that clamps the transmission to maximum. The reverse situation occurs as the incident power is reduced, in that the internal optical field is sufficiently strong so that the cavity resonance remains closer to the incident laser wavelength than expected from the incident optical power alone, until a critical minimum power at which the whole effect switches off. Two regimes are possible for bistability in microcavities, with the nonlinear response primarily either absorptive or dispersive, as shown by Gibbs (1985). Bistability is also observed in the atom-filled microcavity system in the strong-coupling regime, in which case measurements of the transmission at higher power become distorted and shifted, as seen in Fig. 6.17(c) by Gripp et al. (1997).

A bistable optical response can also be seen in polarisation switching in VCSELs. In such lasing microcavities, only one mode lases at any time; however, perturbations (for instance, in current injection, or incident light) can switch the lasing between two orthogonally polarised nearly degenerate lasing wavelengths. The balance between these states is controlled by the spin relaxation of the excitons inside the active quantum well regions, as well as strain within the fabricated pillar microcavities.

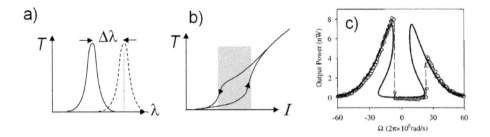

Fig. 6.17: (a) Optically-induced spectral shift of cavity modes locks cavity to input laser wavelength at high power. (b) Transmitted intensity response vs. input intensity showing the region of bistability (shaded). (c) Transmission through strongly coupled atom–cavity showing hysterisis as the incident light is tuned, from Gripp et al. (1997).

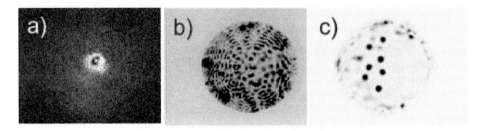

Fig. 6.18: Pattern formation in electrically contacted microcavities: (a) spatial soliton in liquid-crystal microcavity as observed by Hoogland et al. (2002) and (b) pattern formation and (c) seven stable spatial solitons in semiconductor VCSELs reported by Hachair et al. (2004).

For any microcavity system that is bistable, there remain the transverse degrees of freedom that allow optical pattern formation within the microcavity. At one extreme, this can lead to the formation of spatial solitons, in which the light within a region of the microcavity that is switched "high" suffers nonlinear diffraction in such a way that the lateral shape of the resonant optical mode within the region is preserved.

In other cases, 2D grating patterns can emerge either statically or in a constantly changing dynamic pattern evolution. The exact response depends critically on the illumination conditions, the cavity length and mode spectrum, and the boundary conditions, as discussed by Hachair et al. (2004).

A number of realisations of pattern formation within microcavities have been demonstrated including (a) atoms on resonance (though this is not in a microcavity but extended over cm lengths), (b) liquid crystals within planar microcavities and (c) semiconductor quantum wells in large area VCSELs (Fig. 6.18). In general, pattern formation is a sensitive phenomena and is thus perturbed strongly by imperfections in the microcavity properties. Use of this phenomenon, for instance for switching of pixels, is thus problematic.

6.3.2 *Phase matching*

One further use for nonlinear microcavities has been to act as optimised optical frequency-doubling devices. By carefully controlling the Bragg reflector mirror stack, it is possible to produce a microcavity that is resonant at both ω and 2ω, with a selectable phase difference between the two per round-trip due to different DBR penetrations. This can thus act as a phase-matching device, when non-critical phase matching is difficult.

An equivalent use of microcavities has been as a pulsed photodiode to measure ultra-short optical pulses, using two-photon absorption to generate a dc electrical current even at small input intensities. By surrounding the active region of a two-photon photodiode with a microcavity tuned to the input wavelength, the electrical current measured, which depends on the peak field of the pulses, is amplified by 10^4, while the short cavity length ensures minimal broadening of the temporal response.

6.4 Conclusion

In this chapter, a basic overview of emission from microcavities in the weak-coupling regime shows a number of their benefits including lower thresholds, fast response, and controllable emission characteristics. In Chapter 7, we show how these are modified in the strong-coupling regime.

STRONG COUPLING: RESONANT EFFECTS

This chapter presents experimental studies performed on planar semiconductor microcavities in the strong-coupling regime. The first section reviews linear experiments performed in the 1990s that evidence the linear optical properties of cavity exciton–polaritons. The chapter is then focused on experimental and theoretical studies of resonantly excited microcavity emission. We mainly describe experimental configurations in which stimulated scattering was observed due to formation of a dynamical condensate of polaritons. Pump-probe and cw experiments are described in addition. Dressing of the polariton dispersion and bistability of the polariton system due to inter-condensate interactions are discussed. The semiclassical and the quantum theories of these effects are presented and their results analysed. The potential for realisation of devices is also discussed.

Microcavities, Second Edition. Alexey V. Kavokin, Jeremy J. Baumberg, Guillaume Malpuech, Fabrice P. Laussy, Oxford University Press (2017). © Alexey V. Kavokin, Jeremy J. Baumberg, Guillaume Malpuech, Fabrice P. Laussy. DOI 10.1093/oso/9780198782995.001.0001

7.1 Optical properties: background

7.1.1 *Quantum well microcavities*

In 1992, the strong-coupling regime was first identified by Weisbuch et al. (1992) in semiconductor microcavities as an anti-crossing between the two main reflection peaks of their sample (Fig. 7.1). In fact, their goal had been to optimise the superradiant emission of quantum wells inside microcavities and the splitting in reflection that they observed was not really expected, because it was previously thought that the light–matter coupling was too small for strong coupling. The correct identification led to a number of investigations of the emission characteristics of these devices.

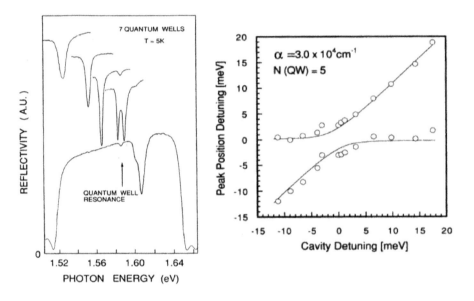

Fig. 7.1: (a) Strong-coupling reflection spectra in a planar semiconductor microcavity. (b) Normal incident polariton energies vs. cavity detuning (scanning across sample), as observed by Weisbuch et al. (1992).

Besides the complete formulation of such a multilayer structure presented in Section 2.7, there are several simple models for the strong coupling that are appropriate for the intuition they provide. A sharp exciton (or atomic) transition produces a characteristic resonant absorption and dispersion lineshape (Fig. 7.2(a)). If this is inserted into a microcavity, then the total round-trip phase as a function of wavelength acquires an extra contribution (Fig. 7.2(b)), which means that there are now three resonant conditions.

The upper and lower resonant conditions occur where the absorption is small and so polariton modes have narrow linewidths, while in this picture the central constructive condition remaining at the resonance energy occurs with strong absorption and is not observed. This simple picture of a net refractive index corresponds to the non-local dielectric susceptibility model presented in Section 4.4.2 so that excitons feel the optical field from the cavity together with the polarisation from all other excitons around them.

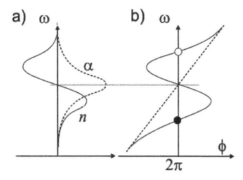

Fig. 7.2: (a) Dispersion and absorption of resonance, producing (b) net microcavity round-trip phase, ϕ, without (dashed) and with (solid) resonant medium. The lower ($\bullet$) and upper ($\circ$) polaritons are at energies where $\phi = 2\pi$.

Claude **Weisbuch** (b. 1945), while visiting the group of Arakawa in Tokyo in 1991, decided to look at QWs in microcavities in the short time (four months) he was there. Thanks to the growth speed varying over the wafer, the sample he got failed to show a resonance in his attempt at observing a luminescence increase when matching the exciton with the cavity, exhibiting instead a permanent doublet. Two Referees of Phys. Rev. Lett. agreed on his interpretation of strong coupling, but diverged on whether this was merely incremental as already observed with atoms, or surprising because observed in the solid state. A third Referee was consulted who liked the "surprise". Weisbuch and Benisty (2005) gives more details on these recollections, such as "cavity polaritons" being christened at the Erice school in a "heated discussion" involving Eli Burstein and Eli Yablonovich. During his PhD with Georges Lampel in the 1970s, Weisbuch developed strong links with the laboratory of Boris Zakharchenia at the Ioffe institute in Leningrad. Despite early sympathies with Soviet and Cuban scientists, he later on took part in several research projects for the French defense ministry. He is now a distinguished professor in the Department of Materials at UC Santa Barbara.

The key characteristic of semiconductor microcavities in the strong-coupling regime is the dispersion relation. This maps how the resonant polariton modes shift with the in-plane wavevector (or angle of incidence). The derivation for the multilayer (eqns 1.9, 2.142 and 4.133) can be simply realised from the resonant condition on the wavevector perpendicular to the planar cavity mirrors (see Sections 1.3, 2.7 and 4.4.4.2). While the angular dispersion near $k = 0$ can be expanded quadratically giving a very light polariton mass, the full dispersion is often crucial for the effects reported and is the solution of eqns (4.129), producing the characteristic shape in Fig. 7.3. We have termed the centre of this dispersion a k-space polariton trap, as the energy of lower-polaritons here is below that of all other electronic excitations and scattering out of the trap is difficult for polaritons if $k_B T < \Omega/2$.

7.1.2 *Variations on a theme*

Quantum well microcavities that incorporate InGaAs microcavities exhibit the clearest polaritonic features because the strain within the InGaAs energetically splits the

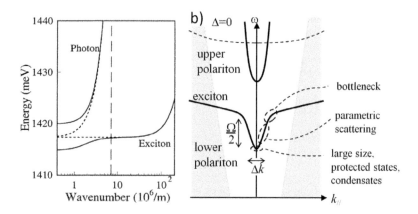

Fig. 7.3: Angular dispersion of a typical semiconductor microcavity at zero detuning, (a) on a log scale showing the contrast between light polaritons and heavier excitons, and (b) showing the critical regions around the polariton trap.

heavy- and light–hole excitons so that only the simple $j = 3/2$ heavy–hole polaritons are resolved. Even in this case the spin-degeneracy and residual lattice strain along [110] produces complicated polariton interactions. Microcavities in which GaAs quantum wells are incorporated have narrower linewidths (typically below 0.1 meV) because of the eliminated alloy disorder, although the light–hole excitons with a third of the oscillator strength are only a few meV to higher energy and also strongly couple to the cavity mode, producing a more complicated polariton dispersion. Microcavities that incorporate GaN are even more complicated, since there are A, B and C excitons that all couple to the cavity mode with different polarisation dependences.

Besides microcavities that use quantum wells for the excitonic coupling to the cavity mode, it is also possible to use wavelength-thickness layers of bulk semiconductors. Because binary semiconductors do not have alloy disorder, their excitonic linewidths can be narrow, although larger thicknesses need to be used to overcome the weaker oscillator strength, see for instance the discussion by Tredicucci et al. (1995). Both reflectivity and luminescence show similar strong coupling to QW microcavities, with clear Rabi splittings of typically 4 meV, as well as extra polaritonic modes from the quantisation of the exciton centre of mass within the finite thickness layer (Fig. 7.4).

It is not even strictly necessary to use a microcavity to produce such polaritonic dispersions, since unwrapped the microcavity looks like a periodic array of quantum wells. The first theoretical study of such "Bragg arranged quantum wells" was undertaken by Ivchenko et al. (1994), who showed that the Bragg arrangement leads to an amplification of the exciton–light coupling strength proportional to the number of quantum wells. Nowadays, this effect is also exploited in 2D and 3D resonant photonic crystals.

Experimentally, "Bragg-arranged quantum wells" have been explored by several groups, for instance by Hübner et al. (1996) and Prineas et al. (2002). They show many analogous features to semiconductor microcavities. However, in practice, it is harder to produce many quantum wells (up to 100 are needed in GaAs) all of exactly the same

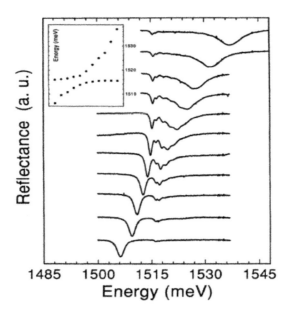

Fig. 7.4: Angular dispersion of a $\lambda/2$ GaAs bulk semiconductor microcavity showing the strong coupling and additional centre-of-mass polariton modes, from Tredicucci et al. (1995).

thickness, spacing and composition, and this is even harder in other material systems. However, even in a small number of closely-spaced quantum wells, polaritonic effects can be observed, as reported by Baumberg et al. (1998).

Typically, it is useful to study different detuning conditions of the microcavity, where the detuning is the energy difference between the normal incidence uncoupled cavity mode and the exciton energy $\Delta = \omega_C - \omega_X$ at $\theta = 0°$. One way in which this is achieved is by increasing the growth variation between different areas of the wafer (typically by eliminating the wafer rotation in the growth reactor), which produces an increasing thickness of the cavity length across the wafer. Hence, it is possible to find areas in which zero detuning is present, and either side of this detunings greater or less than zero. The weaker dependence of the QW energy on the well width means that this method is quite effective. Another possibility is to use temperature to control the detuning, since the expansion of the lattice shifts both exciton energy and cavity mode to lower energy, though the exciton shifts about three times faster. This is seen on Fig. 7.5 from the work of Fisher et al. (1995). At temperatures above 100 K, the thermal ionisation of excitons becomes sufficient to broaden the excitons in conventional III-V semiconductors and wash out strong coupling, limiting the effective range of tuning. However, this technique has been frequently used, often to tune localised excitons in quantum dot microcavities.

However, there are situations in which one would like to remain in a specific position on the sample and tune the cavity mode or exciton energy. Tuning of the exciton

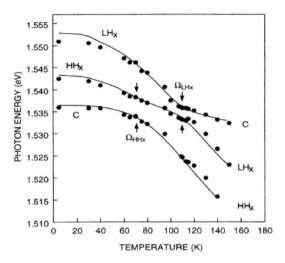

Fig. 7.5: Temperature tuning of an InGaAs microcavity by Fisher et al. (1995).

energy is possible using either electric or magnetic applied fields. By growing the microcavity in a *p-i-n* device, for instance with the DBR mirror stacks doped as in a VCSEL, a vertical electric field can be applied that produces a quantum-confined Stark shift of the exciton energy. Up to $35\,\mathrm{kV\,m^{-1}}$, the oscillator strength of the exciton transition decreases by only 30%, while the excitons redshift by $20\,\mathrm{meV}$, allowing them to be scanned across the cavity mode to demonstrate the strong-coupling regime, as in Fig. 7.6 from Fisher et al. (1995). On the other hand, magnetic fields split the heavy–hole exciton into spin-up and spin-down components, and the magneto-exciton can have a larger oscillator strength due to its more compact binding. This allows a weak to strong coupling transition to be observed with applied magnetic field, as reported by Tignon et al. (1995). In addition, although the magneto-splitting is only $1\,\mathrm{meV}$ for $B = 10\,\mathrm{T}$ and, hence, less than the Rabi splitting, the individual spin-down and spin-up polaritons can be resolved using circularly polarised light, as reported by Fisher et al. (1996).

7.1.3 *Motional narrowing*

Another effect of strong coupling is to change the effect of disorder in the exciton and photon modes. This arises because the length scale over which polaritons average over disorder can be different from the length-scales of disorder in the components. Typically, excitons, even in high-quality quantum wells, are localised on the 10–100 nm length-scales, in the order of 10 Bohr radii. The unavoidable variation in the width of the quantum wells (so-called monolayer fluctuations, which are different on the two sides of the quantum well) means that there is a population of excitons with different energies in different spatial locations. When these excitons are all coupled to the same cavity mode, the resulting polariton averages over all their energies producing an inhomogeneously broadened polariton much narrower than the exciton distribution. This effect has been termed "motional narrowing", as seen in Section 4.4.5. Another way to see this effect

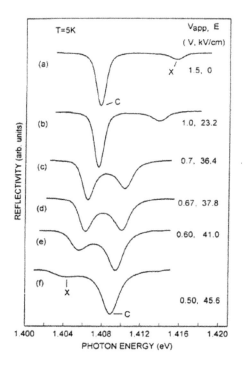

Fig. 7.6: Electric field tuning of an InGaAs microcavity by Fisher et al. (1995).

is that the lower-energy polariton has a linewidth given by the imaginary part of the dielectric constant that is much reduced further away from the centre of the exciton distribution. Measurements of the polariton linewidth in the strong-coupling regime indeed show this effect, with a reduced inhomogeneous distribution for polaritons compared to excitons (see Fig. 4.26).

7.1.4 *Ultra-strong coupling in THz cavities*

Ciuti et al. (2005) were the first to notice that if an intersubband transition in a quantum well is brought into resonance with an optical mode of a relatively large terahertz (THz) waveguide, the resulting Rabi splitting may be comparable with, or higher than, the frequency of the optical (THz) mode. This is a way to realise the so-called ultra-strong coupling regime, introduced in Chapter 5. This regime has been experimentally evidenced by Günter et al. (2009) and is shown in Fig. 7.7. Unlike exciton–polaritons in conventional microcavities, the interband polaritons are polarised perpendicularly to the waveguide plane direction and propagate in the plane of the structure. In the ultra-strong coupling regime, the period of Rabi oscillations may become shorter than the period of oscillations of the electromagnetic field of a THz light mode (the light cycle). In this regime, a class of extremely non-adiabatic phenomena becomes observable. In particular, Günter et al. (2009) have directly monitored how a coherent photon population converts to cavity polaritons during abrupt switching. Ultra-strongly coupled

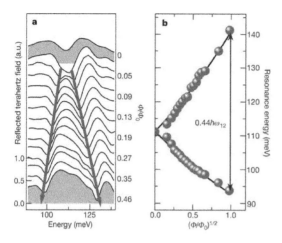

Fig. 7.7: The regime of *ultra-strong coupling*, from Günter et al. (2009). (a) Terahertz reflectance spectra measured at room temperature (293 K) for various excitation conditions of a planar THz waveguide with 50 embedded GaAs/AlGaAs quantum wells. (b) The splitting between eigenmodes of the structure which may be varied from 0 (weak-coupling regime) till 44% of the frequency of the waveguide mode (ultra-strong coupling regime).

light–matter systems form a promising laboratory in which to study novel sub-cycle quantum electrodynamics effects and, potentially, may be used as room-temperature switching devices operating at unprecedented speed.

7.1.5 *Polariton emission*

One of the first observations concerning the polariton radiative emission from InGaAs quantum wells in GaAs semiconductor microcavities was that, although the photoluminescence mapped onto the predicted dispersion relation, the intensity of this luminescence was rather different from the typical thermalised emission seen from bare quantum wells. Similar results were seen in CdTe-based microcavities, for instance by Muller et al. (1999) (see Fig. 7.8). The reason for this is that the excitons generated immediately after non-resonant excitation with a pump laser relax quickly to the high-k part of the lower-polariton dispersion (often termed the "exciton reservoir"). Their cooling to lower energies and lower k, and particularly into the polariton trap, is then restricted by the need to lose large amounts of energy with very little simultaneous reduction in k. Very few quasiparticles exist within the semiconductor that can remove this combination of energy and momentum, and hence the exciton–polaritons collect at the "bottleneck" region in the vicinity of the trap (Fig. 7.3(b)). Hence, instead of the greatest luminescence intensity emerging at the lowest energies, much more luminescence emerges from this bottleneck spectral region. Note also that unlike quantum wells, the luminescence spectrum is also angularly dependent due to the polariton dispersion.

From angular measurements of the luminescence as a function of detuning (which changes the depth of the trap) and temperature, one can estimate that more than five acoustic phonon scattering events are needed to cool a carrier into these 3 meV polariton traps (for GaAs-based microcavities), which is significantly slower than the radiative lifetime in the bottleneck region. Hence, in the linear regime, emission from strong-coupled semiconductor microcavities is reduced, rather than enhanced, besides being strongly angle dependent.

At the end of the 1990s, experiments such as those of Le Si Dang et al. (1998), and Senellart and Bloch (1999) began to show that the bottlenecked luminescence from

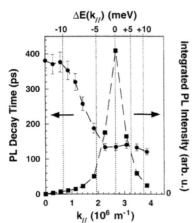

Fig. 7.8: Emission intensity and emission rate as a function of the emission angle from a non-resonantly pumped CdTe-based microcavity, from Muller et al. (1999).

the trap states was highly nonlinear with the injected laser power. Full understanding of this behaviour required an overview of the scattering processes available to exciton–polaritons, which we treat in the next chapter.

7.2   Near-resonant-pumped optical nonlinearities

7.2.1   *Pulsed stimulated scattering*

In quantum wells, many experiments have shown how the injection of excitons or free electron–hole pairs leads to changes in the exciton absorption spectrum. These result from scattering processes between excitons (generally repulsive), and between excitons and free carriers (which can ionise the exciton). However, the experimental difficulty in studying excitons within a quantum well is that the optically accessible states are not distinguishable by changing the angle of incidence (due to their almost flat dispersion), so that the inhomogeneous broadening dominates. Equally problematic is that, despite the best efforts to grow smooth-walled atomically flat interfaces, the disorder from roughness and alloying of quantum wells produces excitonic states that are at most delocalised over a few hundred nanometres, many times the exciton Bohr radius (≈ 15 nm in GaAs), but much less than the optical wavelength. Hence, these exciton states emit and absorb in all directions. Thus, it is not possible to directly observe exciton collision processes using conventional spectroscopy in quantum well (or bulk) samples and one has to resort to indirect methods.

On the other hand, the clear dispersion of exciton–polaritons in semiconductor microcavities allows polaritons at different angles and energies to be distinguished. Hence, it became possible to perform resonant nonlinear experiments on microcavities, by pumping excitons at one angle and measuring how fast they scatter into other states.

The first pump-probe experiments on semiconductor microcavities were performed with two beams under normal or quasinormal incidence with as an underlying objective to modify the system using an intense pump pulse and to record the resulting polarisation by measuring the reflection, transmission, absorption or scattering of a weak probe pulse. The main goal of such preliminary investigations, such as those by Jahnke

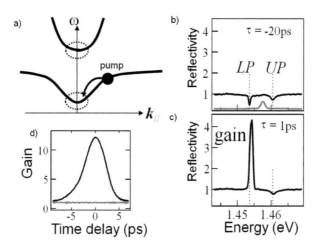

Fig. 7.9: (a) Microcavity dispersion, showing pump pulse injecting polaritons at (k,ω) and scattering to other states on the dispersion. (b,c) Reflection spectrum before and as pump pulse arrives showing strong gain on the lower-polariton at $k=0$. (d) Time response of the gain, from Savvidis et al. (2000).

et al. (1996) (an extensive list of references is given by Khitrova et al. (1999)) was to elucidate the mechanisms responsible for the loss of the strong-coupling regime. After this early stage, the understanding of nonlinear optical properties of microcavities has progressed considerably. This progress has been mainly due to use of advanced spectroscopy techniques, allowing one to tune the angle, energy and time delay between pulses independently. The breakthrough came from an experiment performed by Savvidis et al. (2000) and discussed at length below. This experiment has evidenced the bosonic behaviour of cavity polaritons. It has also shed much light on the main mechanisms governing optical nonlinearity in microcavities. An avalanche of experimental and theoretical studies followed that of Savvidis revealing rich and deep physical phenomena. Most of these results are now being discussed.

One first indication of the peculiarities of the polariton interactions was how the measured optical nonlinearities depended not just on the energies of the polaritons, but on their full dispersion, see the discussion by Baumberg et al. (1998). This confirmed that angular tuning and position tuning (in which the cavity mode energy varies across the sample due to a low-angle wedged thickness variation) were not equivalent. In 2000, the Southampton group's experiments, by Savvidis et al. (2000), definitively showed that the scattering of polaritons was influenced by pre-existing populations of polaritons—in other words, that scattering could be a stimulated process. While it is well known that photons can stimulate photon emission, the process of stimulated scattering is much less studied.

By injecting a pump pulse at a particular angle of incidence (k) and energy (ω), the time-resolved evolution of the scattering of polaritons can be tracked using a weak broadband probe pulse to measure the reflection spectrum at different times (Fig. 7.9), as done by Savvidis et al. (2000). For particular conditions, reflectivities much larger

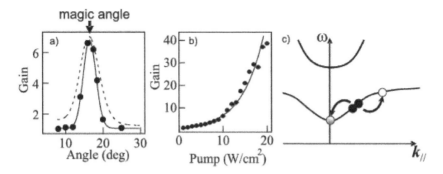

Fig. 7.10: (a) Resonant gain (at k=0) as the pump angle is varied, (b) exponential observed pump-power dependence of the gain and (c) schematic polariton pair scattering, from Savvidis et al. (2000).

than 100%—corresponding to extremely large amplifications—were observed, reaching 10 000%. These gains persisted only while the pump-injected polaritons remained inside the microcavity. Moreover, the gain of the seeded probe pulse is extremely sensitive to the incident pump angle—termed the *magic angle* (see Fig. 7.10)—and pump power. These features are the signature of the polariton pair-scattering process shown in Fig. 7.10(c) in which two polaritons injected by the pump have exactly the right (k, ω) to mutually scatter sending one down to the bottom of the trap (at k=0, often called the "*signal*") and the other to $2k$ (the "*idler*").

Three clear new features are shown in this experiment:

- polaritons can scatter strongly from each other providing that both energy conservation and momentum conservation can be simultaneously satisfied in the two-quasiparticle collision;
- polariton scattering can be enhanced by occupation of the final state. In other words, that polariton scattering can be stimulated, as expected for bosons;
- polaritons are stable at the bottom of the polariton trap, over a time longer than their lifetime (governed by their escape from the cavity).

Stimulated scattering is an example of bosonic behaviour, which has been observed also for spatially indirect excitons in coupled quantum wells, notably by Butov et al. (2001). We note that to observe this effect for excitons, one should cool the system down to about 50 mK; even in this case the observed enhancement of the scattering rate is not as strong as in the case of exciton–polaritons. This is because the exciton dispersion is normally so flat that many other processes can scatter excitons (e.g., disorder, phonons) and, hence, k is not a good quantum number for excitons; a macroscopic population of a single quantum state is unlikely for excitons. On the other hand, for cavity polaritons, because of the tendency for bosons to occupy the same state, the gains measured can exceed 10^6 cm^{-1}, larger than in any other material system.

In the language of nonlinear optics, such processes are said to be *parametric scattering* processes and are commonly observed for parametric down-conversion (where a photon at 2ω transforms into two photons at $\omega + \epsilon$ and $\omega - \epsilon$). Polariton scattering in this experiment is equivalent to a four-wave mixing process (or near-degenerate

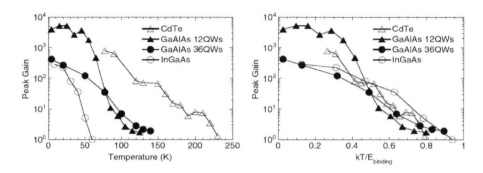

Fig. 7.11: Stimulated polariton scattering peak gain as a function of temperature for different microcavity samples, from Saba et al. (2003).

parametric conversion) where two pump polaritons create a signal and idler polariton, which emerge from the sample at different angles. Because this description via four-wave mixing only deals with the incident and emitted photons, it describes nothing of the solid-state coherence within the semiconductor microcavity, and is thus a limited tool for understanding polariton scattering and coherence.

In a similar way, using the exciton and cavity photon basis for understanding polariton scattering is also limiting. For zero cavity detuning, both the lower and upper-polaritons are composed of half a photon and half an exciton, however the scattering properties of these polaritons are completely different, due to their different energies and the density of states into which they can scatter. In an exciton/photon basis the only difference is the sign with which their wavefunctions are combined.

Further evidence for the coherent nature of the lower-polariton signal state at $k{=}0$ is provided by coherent control experiments that show how the signal polaritons amplified by a first seed pulse may be destroyed by a subsequently oppositely phased reset pulse. Such experiments have been done by Kundermann et al. (2003).

Stimulated scattering at the magic angle has been observed in many different semiconductor microcavities, with single or multiple quantum wells, of different materials, and in patterned mesa microcavities (which have different dispersions with the extra transverse mode). The main effect of using different materials is to change the temperature at which the stimulated scattering process switches off (see Fig. 7.11 and its discussion by Saba et al. (2003)). The current model of temperature-dependence is that because of the parametric process, both signal and idler polaritons together (in a joint coherent state) generate the stimulation (see Section 7.2.2). Scattering of the idler polaritons, which occurs at elevated temperatures thus destroys the polariton stimulation. If the idler is too close energetically to the electron–hole continuum, then the fast scattering (very similar to that of excitons) resumes, which destabilises the idler polaritons. This motivates the current experimental push to building strong-coupling microcavities that are based on ZnSe, GaN and ZnO semiconductors, since these are predicted to provide strong stimulated scattering at room temperature. This would open the way

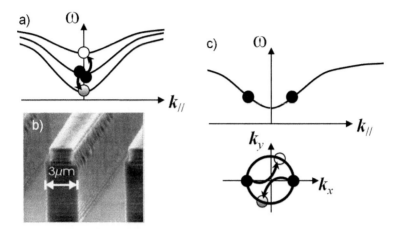

Fig. 7.12: Stimulated scattering processes in (a,b) photonic wires, from Saba et al. (2003), and (c) from two pump beams launched at equal angles either side of $k=0$.

to building more complex optoelectronic devices (such as coherent interferometers and switches) from semiconductor microcavities.

Stimulated scattering also occurs in a variety of geometries. For instance, when a planar microcavity is patterned into photonic wires, quantisation perpendicular to the wire produces a nested series of lower-polariton dispersion curves for k along the wire (see Fig. 7.12 and discussion by Dasbach et al. (2003)). These produce a new range of possibilities for stimulated scattering, involving more than one branch of the dispersion, with final idler states existing at lower energies, thus reducing the scattering that constrains the temperature of operation. Stimulated scattering of polaritons is not limited to the magic-angle condition in which the two initial polaritons are at the same (k,ω). It can also be observed for polaritons that are in initially different states, for instance on either side near the bottom of the polariton trap (Fig. 7.12b), see the discussion by Romanelli et al. (2005). All these schemes have suggested novel ways in which to efficiently produce entangled photons as part of signal and idler beams.

One clear signature of the stimulated scattering process (to be discussed in detail in the theory of Section 7.3) is the rigid blueshift of the whole lower-polariton dispersion as it becomes macroscopically occupied. This results from a self-scattering term to the energy, but it has the effect of modifying dynamically the tuning of incident lasers and dispersion as scattering occurs. This rigid energy shift of the polariton dispersion is only the first-order effect and is proportional to the total population of polaritons. A second-order term means that occupation of the dispersion changes the shape of the dispersion—a highly nonlinear process. One effect of this is that when pump and signal polaritons become macroscopically occupied, new scattering processes appear for which one of the final states is off-branch (see Fig. 7.13 and the discussion by Savvidis et al. (2001)). The polariton dispersion is distorted and produces a flat region around

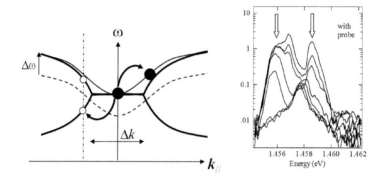

Fig. 7.13: (a) Blueshifted lower-polariton dispersion (dashed to solid thin lines) subsequently produce stimulated scattering. The macroscopic signal and pump polariton occupation (•) generates new off-branch polaritons (◦), observed at the indicated output angle (dash-dot vertical) as (b) new peaks in emission (arrows).

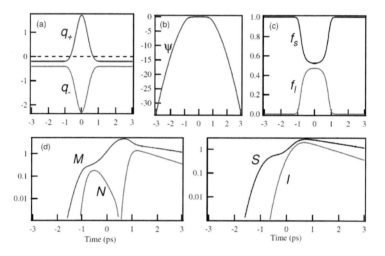

Fig. 7.14: Quasimode calculations as a function of real time (ps) for (a) eigenvalues of M, N, (b) mixing parameter ψ, (c) fractional amount of signal and idler components in M, (d) dynamics of eigenmodes M, N and (e) of signal and idler, when the probe pulse is at $t = -1$ ps, pump at $t = 0$ ps.

k=0, whose onset also signals the destabilisation into spatial solitons (mentioned in Section 6.3.1). Thus, there remain many confusing and novel features about the stimulated scattering process in both space and time that need to be further explored.

7.2.2   Quasimode theory of parametric amplification

In this section, we address the theories of CW parametric scattering in the dynamic regime. This identifies, at each moment in time, the transient eigenstates of the pair polaritons that independently experience the gain or loss. We assume a slowly varying polariton amplitude (a reasonable approximation for narrow spectral linewidth cavities) and also work in the limit of negligible pump depletion (i.e. at low probe powers). In this case the equations governing the slowly varying envelope of signal (S) and idler (I) can be written

$$\frac{\partial S}{\partial t} = -\gamma_S S - \Lambda I^* \,, \tag{7.1a}$$

$$\frac{\partial I^*}{\partial t} = -\gamma_I I^* - \Lambda^* S \,, \tag{7.1b}$$

where $\Lambda(t) = iV P^2(t) e^{i\nu t}$ accounts for the coupling. Here, V is the polariton–polariton interaction constant, P is the dynamic pump polariton occupation and $\nu = 2\omega_P - \omega_S - \omega_I$ is the frequency mismatch from the magic-angle condition. We look for solutions corresponding to gain: $S, I^* \propto e^{qt}$. Solving the determinant of eqn(7.1b) produces the two solutions for the damping:

$$\gamma_\pm = -\frac{\gamma_S + \gamma_I}{2} \pm \sqrt{\alpha^2 + |\Lambda|^2} \,, \tag{7.2}$$

with $\alpha = (\gamma_S - \gamma_I)/2$. These solutions are time dependent, with $q_\pm < 0$ away from the pump pulse corresponding to the individual damping of signal and idler. They repel strongly when the pump arrives, to produce transient gain ($q_+ > 0$, Fig. 7.14(a)). The eigenvectors of these solutions correspond to the two mixed modes (M, N) that experience these gains:

$$M = C\left\{-e^{i\phi} S + e^\psi I^*\right\} \,, \tag{7.3a}$$
$$N = C\left\{e^\psi S + e^{i\phi} I^*\right\} \,, \tag{7.3b}$$

where we have defined $\sinh\psi = \alpha/|\Lambda|$, $\phi = \arg(\Lambda)$ and the normalisation $C = 1/\sqrt{1 + e^{2\psi}}$. This mixed complex transformation of the signal and idler is controlled by the phase mismatch, $\phi = \nu t$, and a mixing parameter, $\psi(t)$. The mode M is amplified when the pump pulse arrives, while the mode N is deamplified. The gain of these modes is given by $q_\pm = \bar{\lambda} \pm \Lambda \cosh\psi$, with the average damping, $\bar{\lambda} = (\gamma_S + \gamma_I)/2$. The incident probe couples into both modes, giving new instantaneously decoupled dynamical equations:

$$\frac{\partial M}{\partial t} = q_+ M - C e^{i\phi} S_{\text{probe}}(t) \,, \tag{7.4a}$$

$$\frac{\partial N}{\partial t} = q_- N - C e^\psi S_{\text{probe}}(t) \,. \tag{7.4b}$$

In the vicinity of the pump pulse, the modes M, N contain roughly equal admixtures of the signal and idler (Fig. 7.14(c)): in other words, when the pump is present, the true modes of the system are not S, I, but M, N. The dynamics of the quasi-uncoupled modes and the signal and idler are shown in Figs. 7.14(d) and (e) for a probe pulse that is 1 ps before the pump, and with damping of signal and idler, $\gamma_{S,I} = 0.2, 0.4\,\text{meV}$ corresponding to the experiments.

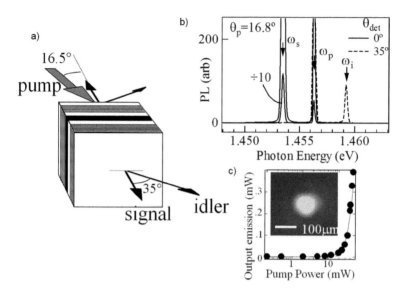

Fig. 7.15: (a) Geometry for microparametric oscillator with (b) spectrum at $k=0$ and (c) power dependence, from Baumberg et al. (2000).

From these equations it can be seen that the amplification of the population of polaritons in the M-mode is roughly given by

$$\left|\frac{M_{\text{out}}}{M_{\text{in}}}\right|^2 = \exp\{2q_+T\} = \exp\{2|\Lambda|T\} = \exp\{2VI_{\text{pump}}T\}, \qquad (7.5)$$

where T is the pulse length and I_{pump} is the pump power. This recovers the experimental result. It is also not what might be intuitively expected from a pair scattering process that in an uncoupled system would have a gain proportional to the *square* of the pump intensity. The completely mixed nature of signal and idler polaritons is what makes the parametric amplification so sensitive to dephasing of the idler component.

7.2.3 *Microcavity parametric oscillators*

While the multiple effects of stimulated scattering are clearest for pulsed excitation, they are also observed in continuous wave (CW) excitation. A pump beam incident at the magic angle first generates spontaneous parametric pairs to signal and idler states, which then act as the seed for further stimulated scattering, see Fig. 7.15 and its discussion by Baumberg et al. (2000). After this threshold (where the signal polariton population exceeds unity), scattering then proceeds exponentially with pump power until saturation occurs. A set of spectra obtained in this configuration by Stevenson et al. (2000) at various angles of detection and at different powers is displayed in Fig. 7.16, where the features at $k=0$ completely dominate above threshold.

The system behaves as a *microparametric oscillator* (μOPO), an integrated equivalent to the cm- to m-scale bulk parametric oscillators (normally based on parametric down-conversion, which is a three-photon and not a four-photon process). Typical

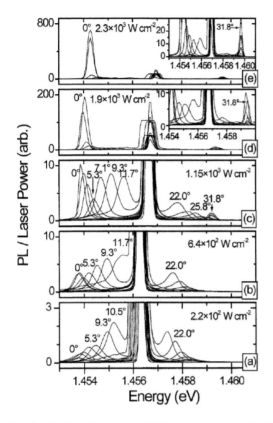

Fig. 7.16: Spectra observed by Stevenson et al. (2000) in the resonant CW pumping of a microcavity, as a function of power. The system is below threshold in (a), approaching threshold in (b), and increasingly above threshold in (c), (d) and (e). Each figure displays spectra collected at different angles. The strong feature around 1.456 eV is induced by the laser. Below threshold, excitations relax to states close to the pump. Above threshold, strong emission is observed at the signal and idler states corresponding to 0° and 32°, respectively. The $k = 0$ emission quickly becomes by far the dominant one (the importance of the idler is seen in the insets).

thresholds for this micro-OPO device are in the 10 mW range, although the physical gain length is some 10 000 times smaller than conventional OPOs.

The output of the μOPO is coherent, narrow spectral linewidth ($\approx$ 1 GHz) and emitted into a narrow angular beam (width $\sim$ 5°). While the signal output phase is independent of the pump laser phase, the sum of signal and idler phases is locked to that of the pump (as in a normal parametric oscillator). However, there are some peculiar novel features of the μOPO system. One of these is that because of the energy shifts possible in the lower-polariton dispersion as it becomes occupied, the device can organise itself to optimise the stimulated scattering. Thus, while pulsed stimulated scattering only occurs close to a magic angle, in the CW case the device adjusts to produce μOPO behaviour over a wide range of pump angles (Fig. 7.17).

One result of the twin-photon production of signal and idler is that they are quantum-mechanically correlated, or entangled. This can be most simply understood from their simultaneous origin with correlated phases from colliding pump polaritons, even though they emerge with different energies in different directions. Such correlations can be extracted from experiments in which the two beams are mixed with a local oscillator on two balanced photodiodes, as has been done by Messin et al. (2001). Theoretically, such experiments can only slightly (by a few %) circumvent the quantum noise limit, due to the degradation of the perfect polariton correlation when they convert into photons on exiting the sample.

More recently, there have been proposals by Ciuti (2004) and Savasta et al. (2005) for generating more useful entangled photon pairs from semiconductor microcavity polariton pair-scattering, using geometries in which both signal and idler are lower in energy than the exciton reservoir. However, it remains a challenge to generate bright high-efficiency correlated photon beams from these devices.

The effect of disorder is getting increasing attention in recent research both in the resonant experiments studied in this Chapter, but also in the off-resonant case exposed in Chapter 8. In the case of the μOPO, a pseudoperiodic potential due to strain—which is present in every sample from CdTe to GaAs—results in local differences in the refractive index of the microcavity. Due to this disorder—or so-called "photonic potential"—the formation of the signal of the μOPO is strongly influenced by the minima of the potential wells. Sanvitto et al. (2006) reported that when the power is increased, the signal occupies different regions in real space.[105] This might prove to be a key point in the formation of OPO and related physics.[106] For instance, Sanvitto et al. (2005) reported that locally, the Q-factor could increase up to 30 000 from a nominal value of 10 000 just by restricting to a region of $5\,\mu m^2$.

In a final review of microcavity parametric scattering, we note that recently a μ-OPO regime has been observed in the weak-coupling regime by Diederichs and Tignon (2005). In these devices, three microcavities are stacked such that their cavity photons can mix between them, while quantum wells still provide the nonlinear scattering process. Instead of relying on the distorted dispersion of the lower-branch polaritons, the three-photon cavity mode branches provide a phase-matched photon-stimulated emission for all pump, signal and idler at k=0. Hence, in a similar way to the strong-coupled microcavities, these cavities lower the emission energies for the participating modes below that of the dissipative excitons, however, at the expense here of returning to photon and not polariton quasiparticle scattering processes.

7.3 Resonant excitation case and parametric amplification

This section presents the theoretical description of microcavity emission for the case of resonant excitation. We focus on the experimental configuration of this chapter where stimulated scattering is observed due to formation of a dynamical condensate of polaritons. Pump-probe and cw experiments are both described. Dressing of the polariton dispersion because of inter-condensate interaction is discussed as well as its main consequence, which is the bistable behaviour of this system. The semiclassical and the quantum theories of these effects are presented and their results analysed.

---

[105] Disorder is an open door to richer physics in relatively well understood systems. In the case of the OPO, for instance, "bistability" (discussed in further detail in Section 7.4.3) manifests strikingly by "switching" the OPO on and off and results in depopulating some regions as others get populated under the influence of potential traps in the disorder.

[106] Effect and potential importance of disorder in the case of spontaneous condensation will be discussed in Chapter 8. The intimate links between these two limits as regards disorder are not yet fully understood.

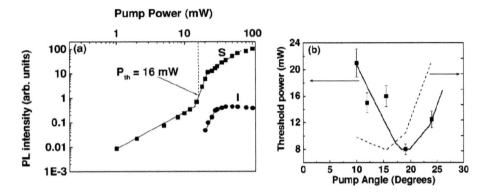

Fig. 7.17: (a) Power dependence of signal and idler near the magic angle. (b) Threshold for signal at different pump angles, from Butté et al. (2003).

7.3.1 *Semiclassical description*

We describe parametric amplification experiments using rate equations.[107] The advantages of such a description with respect to the parametric amplifier model (classical or quantum), which will be presented next, is that it allows us to account for stimulated scattering and to easily include all types of interactions affecting exciton–polariton relaxation. Its disadvantage is that dispersion dressing of polaritonic energies—an important feature of parametric amplification—cannot be easily accounted for in this model. In the resonant configuration, one can single out states where energy-momentum transfer is very efficient and dominates the dynamics. We assume the simplest case of a three-level model:[108] the ground or *signal state*, the *pump state* and the *idler state* (cf. Section 7.3.4). The names arise from similar physics in nonlinear optics. The main loss processes for these states are radiative losses and elastic scattering processes driven by disorder, which can both be included in the same loss constant even if their nature is very different. The radiative loss means the disappearance of the particles, while the disorder scattering implies transfer of a particle towards other states that are neglected in this model. The interaction with phonons in this framework is similar to the disorder interaction and it can also be included as a loss with its appropriate constant. The phonon contribution is often negligible at low temperatures in typically used cavities. It

---

[107] Rate equations of populations are closely linked to so-called "Boltzmann equations", which will be studied in Chapter 8 where relaxation of polaritons will be the central theme of study.

[108] In the complete picture including all states to which we shall return in Chapter 8, the generic Boltzmann equation (see footnote 107) for a state of wavevector $\mathbf{k}$ is given by eqn (8.30). The equation to be solved describing the polariton dynamics is formed by the ensemble of Boltzmann equations written for all allowed values of the in-plane wavevector. It can be solved numerically, choosing suitable initial conditions. These conditions are, for a pump-probe experiment, $n_0(0) = n_{\text{probe}}$, $n_{\mathbf{k}_p}(0) = n_{\text{pump}}$ and $P_{\mathbf{k}} = 0$. For cw experiments, these initials conditions are $n_{\mathbf{k}}(0) = 0$ and $P_{\mathbf{k}_p}(t) = P_0$. In Chapter 8, cylindrical symmetry of the distribution function will be assumed. In the parametric amplification experiments of interest in this chapter, the resonant excitation conditions break this symmetry and a two-dimensional polariton distribution function should be assumed. However, just below and above the amplification threshold a good description of the ground-state population can be performed assuming only the three states of the text.

may cause a significant broadening of the polariton states in experiments performed at high temperatures. The broadening, or loss parameter, can be written as

$$\frac{1}{\Gamma_k} = \frac{|\mathcal{X}_k|^2}{\Delta + \Gamma_{\text{phonons}}} + \frac{|\mathcal{C}_k|^2}{\gamma_c}, \tag{7.6}$$

where $\mathcal{X}_k$ and $\mathcal{C}_k$ are the exciton and photon Hopfield coefficients, respectively, Δ is the exciton inhomogeneous broadening, Γ_{phonons} is the phonon-induced broadening and γ_c is the cavity-photon broadening. At low temperature, $\Gamma_{\text{phonons}} \ll \Delta$. Moreover, in most of the cavity samples studied experimentally, $\Delta \approx \gamma_c$, which yields $\Gamma_0 \approx \Gamma_p \approx \Gamma_i = \Gamma$. In this framework, the system can be described by a set of three coupled equations:

$$\dot{n}_0 = P_0 - \Gamma n_0 - \alpha n_0 n_i (n_p + 1)^2 + \alpha (n_0 + 1)(n_i + 1) n_p^2, \tag{7.7a}$$

$$\dot{n}_p = P_p - \Gamma n_p + 2\alpha n_0 n_i (n_p + 1)^2 - 2\alpha (n_0 + 1)(n_i + 1) n_p^2, \tag{7.7b}$$

$$\dot{n}_i = P_i - \Gamma n_i - \alpha n_0 n_i (n_p + 1)^2 + \alpha (n_0 + 1)(n_i + 1) n_p^2, \tag{7.7c}$$

where

$$\alpha = \frac{2\pi}{\hbar^2} \frac{|M|^2}{\pi \Gamma/2}, \tag{7.8}$$

where M is the polariton–polariton matrix element of interaction, which is here approximately equal to one fourth of the exciton–exciton matrix element of interaction. This system of equations can be easily solved numerically. Moreover, if one considers the CW excitation case, $P_0 = P_{2p} = 0$, this gives $n_0 = n_{2p}$. The system (7.7) thus becomes

$$\dot{n}_0 = -\Gamma n_0 - \alpha n_0 (n_p + 1)^2 + \alpha (n_0 + 1) n_p^2 + \alpha n_p^2, \tag{7.9a}$$

$$\dot{n}_p = P - \Gamma n_p + 2\alpha n_0^2 (n_p + 1)^2 - 2\alpha (n_0 + 1)^2 n_p^2. \tag{7.9b}$$

7.3.2 Stationary solution and threshold

In the stationary regime, $\dot{n}_0 = \dot{n}_p = 0$. Before proceeding further with the formalism, we have to discuss how to define correctly the threshold condition for amplification in the stationary case. Very often, it is believed that a good empirical criterion is that the population of a given state reaches one. Indeed, the evolution equation for the ground-state population formally reads

$$\dot{n}_0 = W_{\text{in}}(n_0 + 1) - W_{\text{out}} n_0. \tag{7.10}$$

W_{in} is supposed to include all channels used by incoming polaritons and W_{out} all channels for their departure from the ground state. The $+1$ in brackets corresponds to the spontaneous scattering process, n_0 in brackets describes the stimulated scattering and $-W_{\text{out}} n_0$ the loss term. Therefore, the condition $n_0 = 1$ means that the stimulation term is as large as the spontaneous scattering term and that the amplification threshold is reached. This point of view can, however, be quite misleading. The equation for the ground-state population can indeed be rewritten as

$$\dot{n}_0 = n_0(W_{\text{in}} - W_{\text{out}}) + W_{\text{in}} \,. \tag{7.11}$$

Thus, the threshold is given by the condition $W_{\text{in}} - W_{\text{out}} = 0$. Equation (7.11) yields in this case

$$-\Gamma - \alpha n_0(n_p + 1)^2 + \alpha(n_0 + 1)n_p^2 = 0 \,, \tag{7.12}$$

which implies

$$n_0 = \frac{\alpha n_p^2 - \Gamma}{\alpha(2n_p + 1)} \,. \tag{7.13}$$

where n_0 is a population, so it should be positive or zero. In the latter case

$$n_p = \sqrt{\frac{\Gamma}{\alpha}} \,. \tag{7.14}$$

Below threshold, $n_p \approx P/\Gamma$, which gives

$$P_{\text{thres}} = \Gamma\sqrt{\frac{\Gamma}{\alpha}} = \Gamma\frac{\hbar\Gamma}{2|M|} \,. \tag{7.15}$$

Using the conventional threshold condition leads to a similar formula for the threshold power (see Exercise 7.1). It is noteworthy that using two apparently independent threshold conditions, one recovers exactly the same value of the amplification threshold. Assuming an exciting laser spot size of 50 μm, $\hbar\Gamma = 1$ meV and for the typical GaAs parameters, $P_{\text{thres}} \approx 10^6\Gamma \approx 50$ μW. This is in good agreement with experimental data.

Exercise 7.1 *Assuming $n_p \gg n_0$, find the solution of the system (7.9). Find the threshold assuming as a threshold condition $n_0 = 1$.*

7.3.3 Theoretical approach: quantum model

Our starting point is the Hamiltonian (5.191). We neglect interactions with phonons or free carriers. The framework used historically to describe this configuration, e.g., by Louisell et al. (1961) for the general problem and by Ciuti et al. (2000), or Ciuti et al. (2001), for microcavities, is the one of the Heisenberg formalism rather than the density matrix approach.

To obtain the equation of motion for polariton operators $a_{\mathbf{k}}$ and $a_{\mathbf{k}}^\dagger$, we write the Heisenberg equation

$$i\hbar\frac{da_{\mathbf{k}}}{dt} = [a_{\mathbf{k}}, H] = E_{\text{LP}}(k)a_{\mathbf{k}} + \sum_{\mathbf{k},\mathbf{k}''} E_{\mathbf{k},\mathbf{k}',\mathbf{k}''}^{\text{int}} a_{\mathbf{k}'+\mathbf{k}''-\mathbf{k}}^\dagger a_{\mathbf{k}'} a_{\mathbf{k}''} + P(\mathbf{k}) \,, \tag{7.16a}$$

$$i\hbar\frac{da_{\mathbf{k}}^\dagger}{dt} = [a_{\mathbf{k}}^\dagger, H^*] = E_{\text{LP}}^*(k)a_{\mathbf{k}}^\dagger - \sum_{\mathbf{k},\mathbf{k}''} E_{\mathbf{k},\mathbf{k}',\mathbf{k}''}^{\text{int}} a_{\mathbf{k}''} a_{\mathbf{k}'} a_{\mathbf{k}'+\mathbf{k}''-\mathbf{k}}^\dagger + P(\mathbf{k}) \,, \tag{7.16b}$$

where E_{LP} is the lower-polariton branch dispersion relation

$$E^{int}_{k,k',k''} = \frac{1}{2}\left(V_{k',k'',k-k'} + V_{k',k'',k''-k}\right),\tag{7.17}$$

and $P(\mathbf{k})$ the polarisation amplitude induced by an external pumping field.

7.3.4 Three-level model

A three-level model has been proposed by Ciuti et al. (2000) to explain the Savvidis–Baumberg experiment described in Section 7.2.1. Its starting point is eqn (7.16), considering only the three most important states, namely the pumped state k_p, the ground or signal state k_0 and the idler state $2k_p$. The authors assumed these three states to be coherently and macroscopically populated. In other words, they assumed these states to behave as classical coherent states and they replaced the operators a_0, a_{k_p}, a_{2k_p} and their adjoint by c-numbers. This ansatz was proposed in the 1950s by Bogoliubov (1947) to describe superfluids (see also Bogoliubov (1970)'s textbook). He diagonalised a Hamiltonian equivalent to eqn (5.191), considering the existence of a macroscopically-occupied ground state (the superfluid). He assumed that only interactions involving the ground state were important and also proposed to neglect fluctuations of the ground-state because of its macroscopic occupation. His argument is that for the ground state $[a, a^\dagger] \gg N$, where N is the ground state population. Therefore, the nonzero value of the commutator can be neglected and the ground-state operators can be replaced by complex numbers. Ciuti et al. proposed a similar approximation but for three condensates instead of one.

In this section, we consider that only the pumped-state operators reduce to complex numbers, keeping the operator nature of signal and idler. In this framework, the system (7.16) can be reduced to just three equations:

$$-i\hbar\dot{a}_0 = \tilde{E}_{LP}(0)a_0 + E_{int}a^\dagger_{2k_p}P^2_{k_p} + P_{probe}(t),\tag{7.18a}$$

$$-i\hbar\dot{P}_{k_p} = \tilde{E}_{LP}(k_p)P_{k_p} + E_{int}P^*_{k_p}a_0a_{2k_p} + P_{pump}(t),\tag{7.18b}$$

$$i\hbar\dot{a}^\dagger_{2k_p} = \tilde{E}_{LP}(2k_p)a^\dagger_{2k_p} + E^*_{int}a_0P^{*2}_{k_p},\tag{7.18c}$$

where

$$\tilde{E}_{LP}(0) = E_{LP}(0) + 2V_{0,k_p,0}|P_{k_p}|^2,\tag{7.19a}$$

$$\tilde{E}_{LP}(k_p) = E_{LP}(k_p) + 2V_{k_p,k_p,k_p}|P_{k_p}|^2,\tag{7.19b}$$

$$\tilde{E}_{LP}(2k_p) = E_{LP}(2k_p) + 2V_{2k_p,k_p,0}|P_{k_p}|^2,\tag{7.19c}$$

and

$$\tilde{E}_{int} = \frac{1}{2}\left(V_{k_p,k_p,k_p} + V_{k_p,k_p,-k_p}\right).\tag{7.20}$$

The advantage of this formalism with respect to rate equations of populations in Section 7.3.1 is that it allows one to account for the energy renormalisation processes driven by interparticle interactions. Here, a blueshift of the three states considered is induced by the pump intensity. Replacing all operators by complex numbers, this equation system can be solved numerically for any pump and probe configuration.

We now consider the steady-state excitation case where a stationary pump of frequency ω_p excites the system, without a probe. This pump drives the pump polarisation given by

$$P_{k_p}(t) = \bar{P}_{k_p} e^{i\omega_p t}, \tag{7.21}$$

with $\bar{P}_{k_p} \in \mathbb{C}$. The system of eqns (7.19) reduces to two coupled equations

$$-i\hbar \dot{a}_0 = \tilde{E}_{LP}(0) a_0 + E_{int} a^\dagger_{2k_p} \bar{P}^2_{k_p} e^{2i\omega_p t}, \tag{7.22a}$$

$$-i\hbar \dot{a}^\dagger_{2k_p} = -\tilde{E}^*_{LP}(2k_p) a^\dagger_{2k_p} - E_{int} a_0 \bar{P}^{*2}_{k_p} e^{-2i\omega_p t}. \tag{7.22b}$$

We define

$$\omega_0 = \frac{1}{\hbar} \mathrm{Re}(\tilde{E}_{LP}(0)), \qquad \omega_i = \frac{1}{\hbar} \mathrm{Re}(\tilde{E}_{LP}(2k_p)), \tag{7.23a}$$

$$\Gamma_0 = \frac{2}{\hbar} \mathrm{Im}(\tilde{E}_{LP}(0)), \qquad \Gamma_i = \frac{2}{\hbar} \mathrm{Im}(\tilde{E}_{LP}(2k_p)), \tag{7.23b}$$

and introduce the two rescaled quantities

$$\tilde{a}_0 = a_0 e^{-i\omega_0 t}, \qquad \tilde{a}^\dagger_{2k_p} = a^\dagger_{2k_p} e^{i\omega_i t},$$

and

$$\beta = |\beta| e^{2i\varphi_p} = E_{int} \bar{P}^2_{k_p}.$$

The two previous equations become

$$-i\dot{\tilde{a}}_0 = -\frac{\Gamma_0}{2} \tilde{a}_0 + \beta \tilde{a}^\dagger_{2k_p} e^{i(2\omega_p - \omega_0)t}, \tag{7.24a}$$

$$-i\dot{\tilde{a}}^\dagger_{2k_p} = -\frac{\Gamma_i}{2} \tilde{a}^\dagger_{2k_p} - \beta^* \tilde{a}_0 e^{i(2\omega_i - \omega_p)t}. \tag{7.24b}$$

This equation is simply a quantum-mechanical equation for parametric processes first written and solved by Louisell et al. (1961). Replacing all quantum operators in this equation system by complex numbers is equivalent to treating the classical parametric oscillator studied in the last century by Faraday and Lord Rayleigh, as has been pointed out by Whittaker (2001). This equation system has been widely studied in recent decades. It can be solved in the Heisenberg representation in the time domain, as detailed in the expositions of Mandel and Wolf (1961) and Louisell et al. (1961) or in the frequency domain, as discussed by Loudon (2000) and in the case of microcavities by Ciuti et al. (2000).

For simplicity, we assume that the resonance conditions are satisfied and that the loss coefficients are the same for signal and idler:

$$\omega_0 + \omega_i - 2\omega_p = 0, \qquad \Gamma_0 = \Gamma_p = \Gamma_i = \Gamma. \tag{7.25}$$

Equations (7.24) become

$$\dot{\tilde{a}}_0 = -\frac{\Gamma}{2}\tilde{a}_0 + i\beta\tilde{a}_{2k_{\mathrm{p}}}^\dagger \,, \tag{7.26a}$$

$$\dot{\tilde{a}}_{2k_{\mathrm{p}}}^\dagger = -\frac{\Gamma}{2}\tilde{a}_{2k_{\mathrm{p}}}^\dagger - i\beta^*\tilde{a}_0 \,. \tag{7.26b}$$

Equations (7.26) imply

$$\dot{\tilde{a}}_{2k_{\mathrm{p}}}^\dagger = \frac{1}{i\beta}\left(\ddot{\tilde{a}}_0 + \frac{\Gamma}{2}\dot{\tilde{a}}_0\right), \tag{7.27a}$$

$$\ddot{\tilde{a}}_0 + \Gamma\dot{\tilde{a}}_0 + \left(\frac{\Gamma^2}{4} - |\beta|^2\right)\tilde{a}_0 = 0 \,. \tag{7.27b}$$

The solutions of the characteristic equation associated with eqn (7.27b) are

$$r_\pm = -\frac{\Gamma}{2} \pm |\beta| \,. \tag{7.28}$$

The solutions of the system (7.27) are therefore

$$\tilde{a}_0(t) = e^{-\frac{\Gamma}{2}t}\left(a_0\cosh(|\beta|t) - ia_{2k_{\mathrm{p}}}^\dagger\sinh(|\beta|t)e^{i2\varphi_{\mathrm{P}}}\right), \tag{7.29a}$$

$$\tilde{a}_{2k_{\mathrm{p}}}(t) = e^{-\frac{\Gamma}{2}t}\left(a_{2k_{\mathrm{p}}}^\dagger\cosh(|\beta|t) + ia_0\sinh(|\beta|t)e^{-i2\varphi_{\mathrm{P}}}\right). \tag{7.29b}$$

Note that the right-hand side of eqn (7.29a) is back in terms of a, rather than $\tilde{a}$ (since at $t=0$, operators coincide).

For $t > 1/|\beta|$, cosh and sinh can be approximated by exponentials with positive argument. Therefore

$$\tilde{a}_0(t \gg |\beta|^{-1}) = \frac{1}{2}e^{(|\beta|-\Gamma/2)t}\left(a_0 - ia_{2k_{\mathrm{p}}}^\dagger e^{i2\varphi_{\mathrm{P}}}\right), \tag{7.30a}$$

$$\tilde{a}_{2k_{\mathrm{p}}}(t \gg |\beta|^{-1}) = \frac{1}{2}e^{(|\beta|-\Gamma/2)t}\left(a_{2k_{\mathrm{p}}}^\dagger + ia_0 e^{-i2\varphi_{\mathrm{P}}}\right). \tag{7.30b}$$

The "particle number" operator for the signal is

$$\tilde{a}_0^\dagger(t)\tilde{a}_0(t) = \frac{1}{4}e^{(2|\beta|-\Gamma)t}\left(a_0^\dagger a_0 + a_{2k_{\mathrm{p}}}^\dagger a_{2k_{\mathrm{p}}} - i(a_0^\dagger a_{2k_{\mathrm{p}}}^\dagger e^{i2\varphi_{\mathrm{P}}} - a_{2k_{\mathrm{p}}}a_0 e^{-2i\phi_{\mathrm{P}}})\right). \tag{7.31}$$

If the signal and idler states are initially in the vacuum state, the average number of particles is, therefore,

$$\langle\tilde{a}_0^\dagger(t)\tilde{a}_0(t)\rangle = \langle 0,0|\,\tilde{a}_0^\dagger(t)\tilde{a}_0(t)\,|0,0\rangle = \frac{1}{4}e^{(2|\beta|-\Gamma)t} \,. \tag{7.32}$$

However,

$$\langle\tilde{a}_0(t)\rangle = \langle 0,0|\,\tilde{a}_0(t)\,|0,0\rangle = 0 \,. \tag{7.33}$$

Equations (7.32) and (7.33) show that a ground-state, initially symmetric in the phase space, will have its population growing exponentially, while its amplitude remains zero. This shows that the symmetry of the ground-state is not broken by the pumping laser. To illustrate our purpose we consider that the system is initially in a state other than the vacuum.

7.3.5   *Threshold*

The threshold condition to stimulated scattering is given by $\Gamma = 2|\beta|$, that is

$$|P_{k_p}|^2 = \frac{\hbar\Gamma}{2E_{\text{int}}}. \tag{7.34}$$

With such a pump polarisation, the energy shift of the signal at threshold is equal to the polariton linewidth. This theoretical result is in good agreement with the available experimental data. It is instructive to compare the criterion (7.34) with the threshold condition obtained in Section 7.3.2 from the population rate equations (Boltzmann equations).

The relation between the pumping power and the coherent polarisation is $\Gamma|P_{k_p}|^2 \approx P$ and the threshold condition for the pump power is thus

$$P = \frac{\hbar\Gamma^2}{2E_{\text{int}}}. \tag{7.35}$$

Assuming, as in Section 7.3.2, that the broadening Γ is independent of the wavevector, and that $E_{\text{int}} \approx |M|$, the polariton–polariton matrix element of interaction, eqn (7.35), becomes

$$P_{\text{thres}} = \Gamma\frac{\hbar\Gamma}{2|M|}. \tag{7.36}$$

This value is exactly the same as the one obtained in Section 7.3.2, illustrating the equivalence of the semiclassical and quantum models in this aspect.

7.4   Two-beam experiment

7.4.1   *One-beam experiment and spontaneous symmetry breaking*

We assume that a cw pump laser excites the sample, together with an ultrashort probe pulse that seeds the probe state. Therefore, at $t = 0$, the probe state is a coherent state $|\alpha_0\rangle$ with $\alpha_0 = |\alpha_0|e^{i\varphi_0}$. The idler state is initially unpopulated (vacuum state). The initial state of the signal⊗idler system is denoted $|\alpha_0, 0\rangle$. With such initial states:

$$\langle\tilde{a}_0^\dagger(t)\tilde{a}_0(t)\rangle = \frac{1}{4}(1 + |\alpha_0|^2)e^{(2|\beta|-\Gamma)t}, \tag{7.37a}$$

$$\langle\tilde{a}_0(t)\rangle = \frac{1}{2}\alpha_0 e^{(|\beta|-\Gamma/2)t}. \tag{7.37b}$$

The phase φ_0 of the order parameter does not depend on the value of the pump phase φ_p and is determined by the probe phase. We define the first order coherence of the system as

$$\eta = \frac{|\langle\tilde{a}_0(t)\rangle|^2}{\langle\tilde{a}_0^\dagger(t)\tilde{a}_0(t)\rangle}. \tag{7.38}$$

We then find

$$\eta = \frac{|\alpha_0|^2}{1 + |\alpha_0|^2}. \tag{7.39}$$

This coherence is constant for any phase relationship between pump and probe. It is close to one if the probe introduces a coherent seed population much larger than one. In this case, the symmetry of the system is broken by the probe.

We have seen in the previous paragraph that the wavefunction of the initially symmetrical system will remain symmetrical during its temporal evolution. Now we are going to artificially break this symmetry, assuming that the initial state is a coherent state characterised by a small, but finite amplitude. Since the signal and idler are now completely identical, we consider that they are both initially in a coherent state with the same amplitude $\alpha_0 = |\alpha|e^{i\varphi_0}$, $\alpha_{2k_p} = |\alpha|e^{i\varphi_{2k_p}}$, but different phases. The average signal polarisation and population are

$$
\begin{aligned}
\langle \tilde{a}_0(t) \rangle &= \frac{1}{2} e^{(|\beta|-\Gamma/2)t} \left(\alpha_0 - i\alpha_{2k_p}^* \right), \\
&= \frac{1}{2} |\alpha| e^{(|\beta|-\Gamma/2)t} e^{i\varphi_0} \left(1 - e^{i(2\varphi_p - \varphi_0 - \varphi_{2k_p})} \right).
\end{aligned}
\tag{7.40}
$$

The signal polarisation strongly depends on the phase relation between pump, probe and idler. Namely, it vanishes if

$$
2\varphi_p - \varphi_0 - \varphi_{2k_p} = 0 \quad \mathrm{mod}\ 2\pi,
\tag{7.41}
$$

and achieves its maximum if

$$
2\varphi_p - \varphi_0 - \varphi_{2k_p} = \pi \quad \mathrm{mod}\ 2\pi.
\tag{7.42}
$$

This last equation is the phase-matching condition for parametric oscillation to take place. A similar equation can be written for the signal population:

$$
\begin{aligned}
\langle \tilde{a}_0^\dagger(t)\tilde{a}_0(t) \rangle &= \frac{1}{4} e^{(2|\beta|-\Gamma)t} \left(1 + |\alpha_0|^2 + |\alpha_{2k_p}|^2 - i(\alpha_0^*\alpha_{2k_p}^* e^{i2\varphi_p} - \alpha_{2k_p}\alpha_0 e^{-i2\varphi_p}) \right), \\
&= \frac{1}{4} e^{(2|\beta|-\Gamma)t} \left(1 + |\alpha|^2 (1 - \cos(2\varphi_p - \varphi_0 - \varphi_{2k_p})) \right).
\end{aligned}
\tag{7.43}
$$

One can see that the population has a minimum (but does not vanish) if condition (7.41) is fulfilled, while it has a maximum if condition (7.42) is fulfilled. The coherence then reads

$$
\eta = \frac{2|\alpha|^2(1 - \cos(2\varphi_p - \varphi_0 - \varphi_{2k_p}))}{1 + 2|\alpha|^2(1 - \cos(2\varphi_p - \varphi_0 - \varphi_{2k_p}))}.
\tag{7.44}
$$

An initial coherent state is expected to appear because of the system fluctuations. It is difficult to describe such fluctuations theoretically and to quantify $|\alpha|$. It is, however, clear that the system will choose to grow on the most "favourable" fluctuation, respecting the constructive phase-matching condition (7.42). It is essential to note that the phase of the signal and idler is not fixed by the phase of the pumping laser together with the phase matching condition, as was proposed by Snoke (2002). One can see that only the quantity $\varphi_0 + \varphi_{2k_p}$ is actually fixed. Therefore, there is a well-defined phase relation between signal and idler but all the values of the signal phase are equivalent for the system. This signal phase is not a priori determined by the pump phase, but it is randomly chosen by the system from experiment to experiment. Choosing its phase, the system "breaks its symmetry". This symmetry-breaking effect is common to the laser phase transition, superconducting phase transition and Bose–Einstein condensation (BEC). To summarise, it is a common feature of phase transitions induced by the bosonic character of the particles involved. This is not a BEC, however, because, as is the case with lasers, it is an out-of-equilibrium phase transition, so that, for example, a chemical potential cannot be defined.

7.4.2 Dressing of the dispersion induced by polariton condensates

As already mentioned, stimulated scattering experiments have shown new emission peaks surprisingly far from the polariton dispersion. This "off-branch emission" is induced by strong interactions taking place between macroscopically populated states, which are the pump, signal and idler states. Interaction between these states is not only a perturbation in the sense that it leads to a dressing of the polariton dispersion. Ciuti et al. (2000) have provided the theoretical interpretation. In this section we briefly summarise this theory. We shall consider all scattering processes that involve two macroscopically populated states as initial states and, as final states, one state on the polariton branch and one off-branch state. We shall require conservation of energy and wavevector. As an example of such a transition, one can consider scattering events having two pump polaritons as initial states. The wavevector- and energy-conservation laws give in this case

$$\{k_p, k_p\} \rightarrow \{k, 2k_p - k\}, \tag{7.45a}$$

$$2E_{\text{LP}}(k_p) = E_{\text{LP}}(k) + E_{\text{pp}}^{\text{off}}(2k_p - k). \tag{7.45b}$$

Equation (7.45b) defines a new dispersion The appearance of a polariton on this branch is possible because the corresponding scattering event is fast enough (see Section 7.2.1). Four other branches corresponding to signal–pump, pump–idler, signal–signal and idler–idler scattering can be defined. The observation of off-branch emission, which can be associated with the existence of macroscopically populated polariton states, is a characteristic feature of phase transitions of weakly interacting bosons. Its experimental observation confirms once again that microcavity polaritons are quasiparticles suitable for the observation of collective bosonic effects.

7.4.3 Bistable behaviour

As discussed in the previous section, an important feature of the resonant excitation scheme is the renormalisation of the polariton energies. This renormalisation is observed in the polariton emission, but it also plays a key role in the absorption of the pump light. Two situations can be distinguished. If the laser is below the bare polariton energy, the absorption is simply reduced by the pump-induced blueshift. If the laser is above the bare polariton energy, the pump gets closer to the absorption energy because of the blueshift that in turn increases the shift that enhances the absorption and so on. There are two different regimes. At low pumping, the pump energy remains above the renormalised polariton energy. At higher pumping, the polariton energy jumps above the pump energy that results in a dramatic increase of the population of the pumped state. The threshold between the two regimes is called a bistable threshold since it comes from

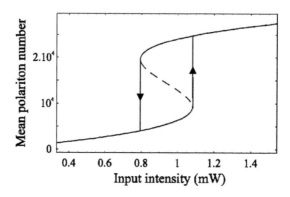

Fig. 7.18: Bistability of the polariton amplifier, from Baas et al. (2004).

the existence of two possible polariton populations and energies for the same pump energy and intensity. Bistability in strongly coupled microcavities was predicted in 1996 by Tredicucci et al. (1996) and observed 8 years later by Baas et al. (2004). This threshold yields a very abrupt jump of the population of the pumped state and can initiate the parametric scattering process. The coexistence of two different nonlinear physical effects (bistability and stimulated parametric scattering) makes this configuration extremely rich to analyse, as shown by Gippius et al. (2004) and Whittaker (2005). In the following we present the formalism describing the pumping of a single state that leads to bistability only.

We can use eqns (7.18b) and (7.21) describing the dynamics of the pump state, but without the coupling term with idler and signal, in other words the nonlinear term of the Hamiltonian reduced to $a_{k_p}^\dagger a_{k_p}^\dagger a_{k_p} a_{k_p}$:

$$\dot{\bar{P}}_{k_p} = i(\omega_{k_p} - \omega_p + i\Gamma_{k_p})\bar{P}_{k_p} + i\frac{2}{\hbar}V_{k_p,k_p,k_p}|\bar{P}_{k_p}|^2\bar{P}_{k_p} + A_p. \quad (7.46)$$

This last quantity should be zero in the stationary regime. Then, multiplying eqn (7.46) by its complex conjugate and replacing $\bar{P}_{k_p}$ by the population of the pump state as well as $|\bar{P}_{k_p}|^2$ by the pump intensity, one gets

$$\left[\left((\omega_{k_p} - \omega_p) + \frac{2}{\hbar}V_{k_p,k_p,k_p}N_p\right)^2 + \Gamma_{k_p}^2\right]N_p = I_p. \quad (7.47)$$

The plot N_p versus I_p is shown in Fig. 7.18 from Baas et al. (2004). The dashed part of the curve is unstable. The plot clearly exhibits the hysteresis cycle taking place when the pump intensity is successively increased and decreased. The position of the two turning points can be found from the condition $dI_p/dN_p = 0$, which yields

$$3(\frac{2}{\hbar}V_{k_p,k_p,k_p})^2N_p^2 + 4(\omega_{k_p} - \omega_p)^2 + \Gamma_{k_p}^2 = 0. \quad (7.48)$$

Bistability takes place if there are two positive different solutions for the quadratic equation (7.48), which gives the condition

$$\omega_p > \omega_{k_p} + \sqrt{3}\Gamma_{k_p}. \tag{7.49}$$

Equation (7.49) says that the only condition to get bistability is to pump about one linewidth above the bare polariton energy. Of course, in a real situation, the pumping cannot be too high in energy since it would require an enormous pump intensity to reach the bistable threshold. If eqn (7.49) is fulfilled the solutions for the turning points read

$$N_p = \frac{2(\omega_p - \omega_{k_p}) \pm \sqrt{(\omega_p - \omega_{k_p})^2 - 3\Gamma_{k_p}^2}}{\frac{6}{\hbar}V_{k_p,k_p,k_p}}. \tag{7.50}$$

The solution with the minus sign corresponds to the turning point with the higher pumping, namely the one that can be found on increasing the pumping power. The solution with the plus sign corresponds, on the other hand, to the turning point that can be found on decreasing the pumping power. Therefore, the minus-sign solution of eqn (7.50) can be injected into eqn (7.47) in order to find the pumping threshold intensity. We note that the large degree of spatial phenomena that can arise from these nonlinear equation, including pattern formation and spatial solitons.

7.5 Propagation of polaritons

7.5.1 *Polariton wavepackets*

Schrödinger (1926) introduced the coherent state (presented in Chapter 3) as a shape-preserving wavepacket in his scientific dispute against Heisenberg (1925) on the nature and meaning of the newly-discovered quantum mechanics. Heisenberg challenged the notion that the wavefunction was the substance for a microscopic mechanical object and advocated his matrix calculus instead to compute observables with no further attachment to some underlying reality. At the heart of the argument was the propagation of the free Gaussian wavepacket through Schrödinger equation. Its solution—which analytical form was probably known to Schrödinger and Heisenberg, but was first published by Darwin (1928) (Márk (1997) gives it in a clearer form)—diffuses and thus eventually results in complete indeterminacy of the position of the particle which is spread over the entire space. This situation is indeed troublesome for associating a particle to the wavefunction and the coherent state was the counter-argument from Schrödinger to prove that his equation could provide localised solutions. Born (1926) provided the (contemporary) interpretation of the wavefunction as a probability amplitude and it soon emerged that Schrödinger and Heisenberg theories were two versions of the same theory. The link between the wavefunction and the particle it describes has remained a blessing question, the problem being taking over by de Broglie (1927) who found particle-like behaviour in the spherical singular region of his wave-theory (from which also emerged the de Broglie wavelength). Bohm (1952a,b) put forward a more elaborate version known as the pilot-wave theory, including many-particle states and highlighting the nonlocal character of the theory. This approach regained considerable interest with Bell (1966)'s work on hidden variables and, recently, was brought to the macroscopic and classical laboratory with the physics of Couder et al. (2005)'s "walkers", i.e., droplets bouncing on a bath of their own liquid sustaining Faraday waves, (see Couder et al. (2010)'s overview for a quantum perspective of these classical experiments). In parallel, the second-quantisation formalism and quantum field theories brought a more robust connection between fields and particles and the interest for the physics of wavepackets took other directions, such as quantum chaos and the formation of quantum scars. The latter does not explain how particles emerge from fields, but how chaos—a feature of nonlinear systems—emerges from the linear Schrödinger equation. As explained by Berry (1987), this can be related to the statistical distribution of the differences between nearest-eigenvalues of the Hamiltonian, leading to linear (Poisson distribution) or chaotic (Wigner distribution) dynamics. This will be taken over in Section 8.9 of Chapter 8. Below, we consider polaritons for their interest as a dream-laboratory to explore wavepackets and fluid dynamics. At low enough driving fields, nonlinearities can be ignored or absorbed in renormalised and blueshifted states while still benefiting from a coherent wavefunction, that can be injected in the sample by resonant (low-power) excitation with some spatial structure, possibly vorticity and at an angle to impart momentum, and thereby set a condensate in motion. This allows to study Schrödinger wavepackets. At higher densities, interactions can be expected to play their full-role in the context of strongly-correlated many-particle fields. This allows to study superfluidity. In all these cases, one can probe continuously the coherent propagation in space and time through the steady emission of photons.

7.5.2   *Self-accelerating and self-interfering wavepackets*

The coherent state is a shape-preserving solution of Schrödinger equation only in presence of an external potential. Yan (1994) showed that the displacement $D(\alpha)$ of any excited state of the harmonic oscillator is also shape-preserving in an harmonic potential, the coherent state being the particular case of displacing the vacuum. Although solutions are many, the necessity of a potential is a considerable limitation as the particle is confined and has an oscillating trajectory. Another drawback of such an externally-enforced cohesion is that, being static in nature, it works for a stiff class of solutions only. For instance, if the Gaussian (coherent state or displaced vacuum) in the harmonic potential has not the correct width, it pulsates (expanding and contracting) as it propagates harmonically in the potential, giving rise to the so-called *squeezed states*.

In contrast, interactions can self-sustain a wavepacket, that can dynamically adapts itself to its environment—including to other wavepackets found on its way—to ensure its integrity. This gives rise to the concept of a "soliton", first reported by Russell (1844) with water in a canal, exhibiting such robustness as to be able to escape far into the sea (although the conditions for the soliton solution were not met anymore). The phenomenon was explained by Korteweg and de Vries (1895) as a balance between dispersion and nonlinearity. The non-diffusion of solitons (and other properties such as their stability under collisions) makes them identifiable as particles in a way Schrödinger would have liked for his wave equation of the nascent quantum mechanics. The connection between fields and particle took another unexpected turn with the discovery by Berry and Balazs (1979) (Nándor Balázs was an assistant of Schrödinger himself) of a peculiar solution to the Schrödinger equation, that is also shape-preserving and exhibits the self-healing effect (reconstructing its shape after passing through a potential). Their solution reads

$$\psi(x,t) = \mathrm{Ai}\left(\frac{B}{\hbar^{2/3}}\left(x - \frac{B^3 t^2}{4m^2}\right)\right) e^{\frac{iB^3 t}{2m\hbar}\left(x - \frac{B^3 t^2}{6m^2}\right)}, \qquad (7.51)$$

with Ai the Airy function—solution of $\partial_x^2 y - xy = 0$—that defines the initial condition $\psi(x,0) = \mathrm{Ai}(Bx/\hbar^{2/3})$, m is the mass of the particle and B a free parameter. The analytical expression shows that such a "Airy packet" (or "Airy beam") propagates without spreading as it is simply translated to the right. A striking feature of the solution, however, is that the packet accelerates, although it is the solution of the free Schrödinger equation with no potential nor any other applied force. This would seem to violate Ehrenfest theorem that connects observables from a quantum wavepacket to classical dynamics and the latter is notoriously adamant since Galileo that in absence of an applied force, a system does not change its velocity. The absence of diffusion is less surprising although this is the sort of quality that Schrödinger had been looking for (his 1926 paper appeared the year Balázs was born). Berry and Balázs' solution achieves this "particle"-like feat without the need of a potential.

All these counter-intuitive features of the Airy beam are understood and all paradoxes are resolved once it is clarified that the solution is unphysical, in the sense that eqn (7.51) is non-square integrable. Ehrenfest theorem, in particular, does not apply, as observables diverge. Berry and Balazs (1979) argued that the lack of normalisation means that the packet represents not one particle in isolation, but a whole family, the Galilean-propagation of which causes an emerging curvature (the caustic) in the space-time trajectory. The Airy beam remained for a long time a mathematical curiosity and did not receive much attention, until the result was born again when Siviloglou and Christodoulides (2007) made the seemingly obvious but genious proposition that had remained pending for almost three decades, of turning to a truncated version of the Airy beam. Multiplying the Airy function by an exponential term, $\mathrm{Ai}(x)\exp ax$, with typically $a \ll 1$, makes the packet normalisable and still retains its striking properties, although now for a finite time (the longer the smaller is a). The peaks of the wavepacket ("sub-packets") do indeed accelerate while also retaining their shape during propagation, but the truncated tail eventually destroys the structure, that thus exists with a finite lifetime. This was readily observed experimentally by Siviloglou and Christodoulides (2007), not directly with a Schrödinger packet but with the optical field, that is ruled by the same equation in the paraxial approximation of diffraction. Demonstrations with other fields followed, for instance Voloch-Bloch et al. (2013) shaped free electrons' wavefunction into Airy wavepackets. Ehrenfest theorem becomes relevant again but is not violated: the total average does not accelerate, only sub-packets do appear to do so, inside a mother packet that moves accordingly to Newtonian dynamics. These fascinating objects, challenging with the purely non-interacting Schrödinger equation one's intuition of the basic laws of physics, call to reconsider one's understanding of fields and particles. They already found striking applications, such as Baumgartl et al. (2008)'s optical "snowblowing", whereby Airy photon-packets drag particles powered by the sub-packets acceleration. It was later shown by Bandres (2009) how Airy packets are but a particular case of a larger family of wavepackets, the so-called "self-accelerating

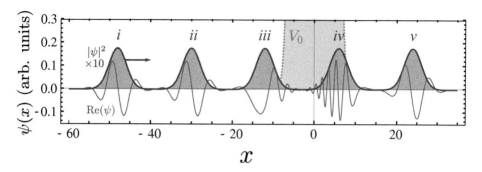

Fig. 7.19: A wavepacket propagating with an exact linear dispersion, as suggested by Laussy (2012) as a toy-model for super-propagation that tunnels through potentials of arbitrary size and strength (dotted wall V_0) without suffering any alteration to its density $|\psi|^2$. The real and imaginary parts, on the other hand, react to the potential by getting compressed as they cross through. Since there are no final states where to scatter on the linear dispersion, these phase distortions have no effect on the wavepacket profile and motion. The labels $i, \ldots, v$ show various instants of time for the packet that propagates from the left to the right.

beams". In these cases, instead of interactions, the acceleration of the packet holds it together against diffusion. This is provided, in the linear regime with no potential nor applied force, by the phase-shaping of the initial condition.

The best fix for wavepacket diffusion is to doctor up the dispersion directly, since it is the source of diffusion in the first place. The kinetic energy of free propagation leads by the correspondence $k \to -i\hbar\nabla$ to a Laplacian form, that spreads and levels-out the bumps in a wavepacket, which are precisely the features that one wishes to attach to a particle. On the other hand, assuming, for instance, an exact linear dispersion $E_k = sk$, with s a real number—the "speed of sound"—and with no mirror symmetry (i.e., it is not $s|k|$), leads to "super-wavepackets" that transport themselves with no distortion of their shape, whatever this may be, but more strikingly, that can also penetrate through any potential of any width and strength without suffering the slightest recoil or even slowing down. This is readily shown by the analytical solution for the wavepacket equation for such an hypothetical dispersion (so-called "transport equation"), $i\partial_t\psi(x) = (-is\partial_x + V(x) - i\gamma/2)\psi(x)$, with, say, a Gaussian wavepacket as the initial condition:

$$\psi(x,t) = \frac{1}{\sigma_x\sqrt{2\pi}}\exp\left(-\frac{(x-x_0-st)^2}{2\sigma_x^2}\right)\exp(ik_0(x-st))$$
$$\exp\left(-\frac{iV_0}{2s}\left\{\mathrm{Erf}\left(\frac{x-x_V}{\sqrt{2}\sigma_V}\right)-\mathrm{Erf}\left(\frac{x-x_V-st}{\sqrt{2}\sigma_V}\right)\right\}\right), \quad (7.52)$$

where we included a Gaussian obstacle on the way (at x_V with strength V_0 and width σ_V). This solution is plotted at various instants of time in Fig. 7.19. As can be seen trivially from the analytical solution, the packet is indeed literally transported as far as its modulus squared is concerned, since the complex exponential disappear. This does not mean that nothing happens, however, and that the potential is transparent for the packet. The complex exponential in the second line shows that, although the modulus is unaffected, the real and imaginary parts are strongly affected, and the phase gets compressed as it crosses the potential. The wider/stronger is the potential, the more energy gets smeared out towards lower k, as the potential attempts to stop the packet, that finds however no other point on the dispersion where to scatter without violating conservation laws. If the dispersion would be mirror symmetric, $E_k = s|k|$, the packet would bounce back, although still retaining its shape-preserving properties. This result is an artificial, but enlightening mathematical toy model that captures the physics of superfluid motion, that suppresses scattering. The role of interactions in superfluidity is basically to distort a quadratic dispersion into a more linear one with similar properties as those just presented. Interestingly, polaritons are known even in the non-interacting regime to feature non-quadratic dispersions. We have already commented on the impact of the complex-valued character of the polariton dispersion. Recently, Pinsker et al. (2015) observed that even when focusing on small ranges of k around the ground state, the polariton dispersion could be more accurately described by non-parabolic models, such as the "fractional Schrödinger equation", introduced by Laskin (2000), that replaces the Laplacian ∇^2 for the kinetic energy term by ∇^{2s} for fractional s (he studies the case $s = 5/6$). Even in absence of interactions, the apparently small deviations from the parabolic case result in notable differences such as shrinking of the condensate, that Pinsker et al. (2015) suggest should be incorporated in models of polariton dynamics.

The effect of a non-parabolic dispersion for wavepacket dynamics was also investigated by Laussy (2012) and in more details by Colas and Laussy (2016) by solving for the dynamics of the coupled exciton-photon fields, which accounts for the non-parabolic dispersion. The polariton as a two-field object comes with important conceptual differences from a single-field particle, of which we overview the most important ones.

The dynamics of the wavefunction $|\psi\rangle$ is ruled by the polariton propagator Π such that $|\psi(t)\rangle = \Pi(t - t_0)|\psi(t_0)\rangle$. In free space, it is diagonal in k space:

$$\langle k'|\,\Pi(t)\,|k\rangle = \exp\left[-i\begin{pmatrix} \frac{\hbar k^2}{2m_C} + \Delta & \Omega_R \\ \Omega_R & \frac{\hbar k^2}{2m_X} \end{pmatrix} t\right]\delta(k - k') \tag{7.53}$$

with $m_{C,X}$ is the photon/exciton mass, Δ their detuning and Ω_R their Rabi coupling. The eigenstates of the propagator

$$\Pi(t)\,\|k\rangle\!\rangle_\pm = \exp(-iE_\pm t)\,\|k\rangle\!\rangle_\pm\,, \tag{7.54}$$

define both the polariton dispersion

$$E_\pm = \hbar k^2 m_+ + 2\Delta \mp k_\Omega^2\,, \tag{7.55}$$

where the notation $\|\rangle\!\rangle_\pm$ stands for upper ("+") and lower ("−") polaritons. They also define the canonical polariton basis

$$\|k\rangle\!\rangle_\pm \propto \begin{pmatrix} E_+(k) \\ 1 \end{pmatrix}|k\rangle\,, \tag{7.56}$$

where $m_\pm = (m_C \pm m_X)/(m_C m_X)$ are the reduced relative masses and $|k\rangle$ the plane wave of well defined momentum k. Another important quantity is the dressed polariton momentum

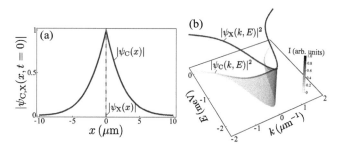

Fig. 7.20: (a) Localising a polariton in space is possible for one of its component only (here in dashed red for the exciton field); the other field smears out to keep the particle on its branch. (b) Counterpart of (a) in energy-momentum space. Computations from Colas and Laussy (2016).

$$k_\Omega^2 = \sqrt{\hbar^2 k^4 m_-^2 - 4\hbar k^2 \Delta m_- + 4(\Delta^2 + 4\Omega_R^2)}\,. \tag{7.57}$$

A general polariton state can thus be expressed as a linear combination

$$\|\psi\rangle\rangle_\pm = \int_{-\infty}^{\infty} \phi_\pm(k) \, \|k\rangle\rangle_\pm \, dk \,, \tag{7.58}$$

with $\phi_\pm(k)$ the scalar-field (upper/lower) polariton wavefunction.

The results that follow come from the impossibility—due to the two-component character—to evolve a general polariton state in time with the complex rotation of free propagation. Except for a well-defined polariton state in k-space, i.e., a completely delocalised polariton in real space, the photon and exciton components of a polariton cannot be jointly defined according to a given wavepacket $\phi(k)$, e.g. a Gaussian packet. Indeed, one component gets modulated by the $E_\pm$ factor needed to maintain the particle on its branch. Gaussian packets for both the photon and the exciton result in populating both polariton branches. The general case obviously admixes the two types of polaritons:

$$|\psi\rangle = \sum_{\sigma=\pm} \int \phi_\sigma(k) \, \|k\rangle\rangle_\pm \, dk \,. \tag{7.59}$$

These results that impose strong constrains on a polariton wavepacket must be contrasted with the conventional picture one has of the polariton as a particle, which is that of states $\|k\rangle\rangle_\pm$ and is, in good approximation, recovered for packets large enough in real space. The composite structure of a polariton therefore forbids its localisation in real space, in the sense that both its photon and exciton components cannot be simultaneously localised. If we choose $\phi(k)$ such that either $\psi_C(x, t = 0)$ or $\psi_X(x, t = 0)$ is $\delta(x)$, the other component is smeared out in a pointed wave function surrounding the singularity of the localised field, as shown in Fig. 7.20(a–b). Such constrains result in a rich phenomenology when involving a large enough set of momenta.

Another consequence of the non-parabolic dispersion is found in the concept of effective mass, a fundamental notion in solid state physics initially introduced in band

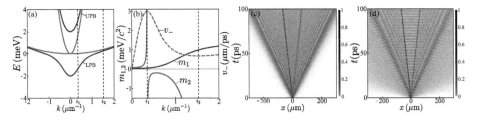

Fig. 7.21: Dynamic of polariton wavepackets, from Colas and Laussy (2016). (a) Dispersion of the exciton-photon and lower–upper polaritons (UPB, LPB). (b) Effective masses for the lower polariton as a function of k: inertial mass m_1, diffusive mass m_2 (negative when $i_1 < k < i_2$) and group velocity v_- (dashed green). (c) Self-interfering packet (SIP) formed out of an initial Gaussian lower-polariton state, exhibiting characteristic fringes as the packet bounce back on a mass wall to fold onto itself. The diffusion is shielded in between two cones, indicated by the dashed lines. (d) Same with a photon as an initial condition, resulting in further interferences from the Rabi oscillations, leading to a spacetime crystal-like structure.

theory to describe the electrons motion in a crystal, providing a useful semiclassical picture. The crystal's band can be locally approximated using a parabolic dispersion:

$$E(k) = E_0 + \frac{\hbar^2 k^2}{2m^*}, \qquad (7.60)$$

where E_0 is a constant giving the band energy at the edge of the Brillouin zone and m^* is the effective mass, which is a constant here. For a crystal electron in presence of an electric field F, with the local parabolic approximation of eqn (7.60), we can obtain the Newton-like equation of motion:

$$\frac{dv(t)}{dt} = \frac{d}{dt}\frac{1}{\hbar}\partial_k E(k) = \frac{1}{\hbar}\frac{d^2 E(k)}{dk^2}\frac{dk}{dt} = \frac{1}{\hbar^2}\frac{d^2 E(k)}{dk^2}(-e)F, \qquad (7.61)$$

where we have used the time derivative of the quasi-momentum of the electron in the crystal $d(\hbar k)/dt = -eF$, with e the elementary charge of the electron. This leads us to the well know expression for the effective mass:

$$m^* = \hbar^2 (\partial_k^2 E)^{-1}. \qquad (7.62)$$

The particle's velocity can be connected to the energy gradient as follows:

$$v(k) = \frac{1}{\hbar}\partial_k E(k). \qquad (7.63)$$

One can note that, in the case of flat bands, this velocity goes to zero, as if we were dealing with extremely heavy classical objects that would stall any motion of the electron.

A notable property of the effective mass is that it can take negative values, as is the case for holes in semiconductors. It can also be the case in a polaritonic system as seen in Fig. 7.21(b). This negative value for the effective mass m^* implies negative velocities when a positive force is applied on the system. In the case of the polariton dispersion, the coexistence of positive and negative masses for the same wavepacket, when it straddles over the inflection point (see Fig. 7.21(b)), leads to an interesting phenomenology of "self-interferences". When this happens, the wavefunction hits a "mass wall" over which the wavepacket bounces back. This results, instead of standard diffusion, in a series of fringes that altogether look alike an Airy beam. In this way, the polariton dispersion affords another mechanism to provide a strong notion of particle (localisable and resilient object) out of a field, but unlike solitons, without interactions and, unlike Airy beams, being normalisable. The resulting structure, the "self-interfering packet" (SIP), is shown in the case of a $k = 0$ initial Gaussian wavepacket in Fig. 7.21(c) with a spatial extent in real-space small enough so that the spread in reciprocal space covers the full polariton dispersion. While for a parabolic dispersion, squeezing the packet in space merely causes a faster diffusion, in the polariton case however, there is a critical diffusion beyond which the packet stops expanding and folds back onto itself.

In the case of relatively small excitons mass, the second inflection point i_2 of the branch becomes accessible and thus, spreading over this point induces another reflection of the wavefunction, shielding from the self-interferences the core of the mother packet. The two diffusion cones of the wavefunction due to the inflection points of the LPB are materialised by blue dashed lines on the space time chart Fig. 7.21(c). They correspond to a propagation with the wavevector related to the inflection points and their expression is thus given by

$$w_{1,2}(t) = \frac{t}{k(i_{1,2})} , \tag{7.64}$$

where the values of the inflection points $k(i_{1,2})$ are obtained by solving $\partial_k E_- = 0$, which is a complex polynomial equation with no closed-form solutions, but numerical solutions can be easily obtained.

These are results for a lower polariton SIP. If considering instead a photon (or exciton) state as an initial condition, then self-interferences coexist with Rabi oscillations and the structure of Fig. 7.21(d) is obtained: there is now an ever-diffusing part beyond the cone due to the upper polariton contribution (whose dispersion is parabolic-like) and the shielded area exhibits simple Rabi oscillations. In between the two cones, one observes an hexagonal lattice, as typical of three interfering beams (two from the lower polariton SIP and one from the upper polariton). This structure in spacetime is reminiscent of a time-crystal.

7.5.3 Superfluid propagation

The possibility to resonantly excite polaritons with high-enough densities for interactions to play a role, imparting them with a momentum and tracking their evolution in spacetime, also makes them ideal candidates to study fundamental effects of interacting condensates, such as superfluidity. Amo et al. (2009b) studied the propagation of polaritons in a planar microcavity in an OPO configuration discussed in Chapter 7, whose signal is perturbed by a short pulse of light, triggering the propagation of a polariton bullet that is observed to be unaltered by small disorder on the one hand and coherently scattered by strong defects on the other hand. The propagation, with no diffusion and keeping its shape and momentum, is shown in Fig. 7.22. It should be noted here that although the lifetime of a single polariton is very short, of the order of picoseconds,

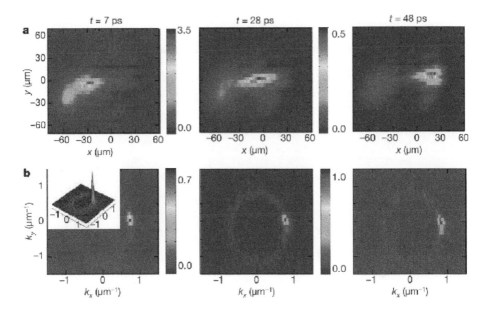

Fig. 7.22: Propagation of a polariton bullet in the experiment of Amo et al. (2009b). The upper row shows the propagation in real space (from left to right) of polaritons moving collectively at a speed of 1.2×10^6 m/s. The images are taken at three different times after the arrival of a pulse that created the wavepacket and imparted it with momentum. Throughout its propagation within the pump spot, the bullet keeps its shape unaltered and is not affected by disorder. The lower row shows the same dynamics in momentum space. A faint Rayleigh circle is observable. The inset displays a three-dimensional view which shows the narrow k distribution of the moving fluid.

the polariton fluid propagates collectively for as long as the spot extends. Also, this configuration does not pin the phase of the signal to that of the laser.

Later, Amo et al. (2009a) investigated the case of a steady flow on a defect and the suppression of disorder scattering as function of pumping power, demonstrating a threshold separating the normal-fluid and the superfluid. The observation was made in transmission. The vanishing scattering is interpreted from the earlier work of Carusotto and Ciuti (2004) who co-author this paper. A defect is found in the sample and the excitation spot is centred on it, resulting—in the normal-fluid regime—in scattering of the individual polaritons. The superposition of the cylindrical wave from the scattered polaritons with the incident polariton plane-wave makes parabolic-like wavefronts around the defect, that are shown in panel I of Fig. 7.23. In momentum space, this scenario produces the Rayleigh elastic circle, displayed in panel IV below. Increasing pumping, polariton–polariton interactions blueshift the polariton dispersion on the one hand and linearise it on the other hand. As a consequence, the fluid finds no scattering channel and collectively prohibits the defect to scatter its constituent particles, making the flow effectively frictionless.

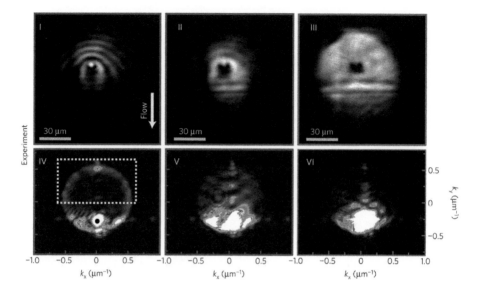

Fig. 7.23: Loss of scattering in the experiment of Amo et al. (2009b). As in Fig. 7.22, the upper row corresponds to real space images and the lower row to their counterparts in reciprocal space, with increasing pumping intensity from left to right. The excitation is now continuous, so that the flow is stationary. In its midst, one can clearly see a defect in the real-space images. This defect scatters polaritons and induces the Cherenkov-like interference pattern in the case of low pumping. In this regime, the Rayleigh elastic circle in the reciprocal space is clearly seen. Increase of the pumping intensity results in shrinkage of this elastic circle, due to the blue shift and distortion of the dispersion of the low polariton branch. In the strong pumping regime, elastic scattering of the exciton polaritons is suppressed (image III), which characterises a superfluid.

These two experiments, from the same first author but in two different groups, have been both realised under resonant optical pumping, where the coherence has been injected into the polariton system by a laser pulse. They demonstrated in two different configurations that scattering of the polaritons due to the static disorder can be essentially suppressed in some given conditions, which is characteristic of a superfluid. The most popular tool used to described such a resonantly pumped polariton system is the driven-dissipative Gross–Pitaevskii equation which reads

$$i\hbar\frac{\partial\psi}{\partial t} = -\frac{\hbar^2}{2m}\Delta\psi + \alpha|\psi|^2\psi - \frac{i\hbar}{2\tau}\psi + P(\mathbf{r},t) - i\hbar\psi(\mathbf{r},t)/2\tau\,. \qquad (7.65)$$

Here, $\psi(\mathbf{r},t)$ is the mean value of polaritons described by a single-particle wavefunction. $P(\mathbf{r},t)$ is the resonant pumping which can have any spatial or temporal dependence. τ is the particle lifetime, which is taken independent of the wavevector. α is the interaction constant (which was called $V_{k_{\mathrm{p}},k_{\mathrm{p}},k_{\mathrm{p}}}$ in the previous section). This equation taken in the limit of a single-mode system is the same as the one describing bistability. Taken in a three-mode approximation, it is similar to the quantum description of an OPO. Also the present formulation assumes massive particles, which corresponds to a constant exciton-photon fraction. This type of specific dependence can be taken into account by using two coupled equations for the photon and exciton parts independently, as done for instance by Carusotto and Ciuti (2004). Also this equation describes scalar particles, neglecting the polarisation/spin degree of freedom, which can be taken into account as well as all sorts of polarisation splittings (see Shelykh et al. (2010)), as detailed in Chapters 9 and 10. The description using this type of equation is at the basis of the idea of describing resonantly pumped polaritons in terms of a quantum fluid of light as introduced by Carusotto and Ciuti (2013), a concept described in Chapter 10. Here, we are going to consider the specific case of homogeneous CW pumping under normal incidence, to compute the renormalised excitations of the system in the presence of this pump and show how it leads indeed to the most basic, but also most striking behaviour of a quantum fluid that is superfluidity.

7.5.4 Elementary excitation of resonantly pumped polaritons

Let us consider a pump $P(\mathbf{r}, t) = A_P \exp[i(\mathbf{k}_P \cdot \mathbf{r} - \omega_P t)]$. This resonant pumping leads to the formation of a macro-occupied mode: $\psi_0(\mathbf{r}, t) = e^{i(\mathbf{k}_p \cdot \mathbf{r} - \omega_p t)} \psi_0$. The next stage consists in finding weak excitations that can develop on top of this macro-occupied mode and to write the total wavefunction as $\psi(r, t) = \psi_0(r, t) + \delta\psi(r, t)$. However, the nonlinear term of the Gross–Pitaevkii Equation GPE couples $\delta\psi(r, t)$ and $\delta\psi^*(r, t)$. This means that a single-particle excitation described by a plane wave $e^{+i(\mathbf{k} \cdot \mathbf{r} - \omega t)}$ gets coupled, through the nonlinear term to the wave with opposite wave vector and energy. The excitation of the pump system therefore take place through some kind of pair-scattering processes. The spectrum of these excitations can be found by inserting the following wave function in the driven dissipative GPE

$$\psi(\mathbf{r}, t) = e^{i(\mathbf{k}_p \cdot \mathbf{r} - \omega_p t)} \left(\psi_0 + A e^{+i(\mathbf{k} \cdot \mathbf{r} - \omega t)} + B^* e^{-i(\mathbf{k} \cdot \mathbf{r} - \omega^* t)} \right). \qquad (7.66)$$

One should then use a linearisation procedure, keeping only first order terms in A and B. It gives the following set of coupled equations:

$$\left[E(k_p) - \hbar\omega_p + \alpha|\psi_0|^2 - \frac{i\hbar}{\tau} \right] \psi_0 + A_P = 0, \qquad (7.67a)$$

$$\left[E(\mathbf{k}_p + \mathbf{k}) - \hbar\omega_p - \hbar\omega + 2\alpha|\psi_0|^2 - \frac{i\hbar}{\tau} \right] A + \alpha\psi_0^2 B = 0, \qquad (7.67b)$$

$$\alpha(\psi_0^2)^* A + \left[E(\mathbf{k}_p - \mathbf{k}) - \hbar\omega_p + \hbar\omega + 2\alpha|\psi_0|^2 - \frac{i\hbar}{\tau} \right] B = 0. \qquad (7.67c)$$

The first equation is similar to the one obtained in the previous section for bistability. It allows to find the macro-occupied mode density $n = |\psi_0|^2$ and the detuning δ_P between the pump energy and the frequency of the excited nonlinear oscillator: $\mu_0 = \omega_p + \alpha|\psi_0|^2$. The dispersion branches can then be found as

$$\hbar\omega_\pm(\mathbf{k}) = \hbar\omega_P(\mathbf{k}_P) \pm \sqrt{[E_L(\mathbf{k}) - \hbar\omega_P + 2\alpha n]^2 - [\alpha n]^2}, \qquad (7.68)$$

Here, $E_L(\mathbf{k})$ is the bare dispersion relation of the lower polariton. The main difference between the macroscopic occupation of a mode driven by an external pump and

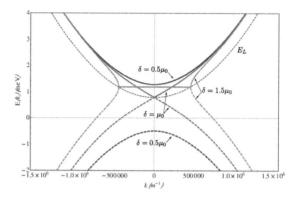

Fig. 7.24: Dispersion of elementary excitations of a driven mode at $k = 0$ for 3 values of the detuning: $\delta = 0.5\mu_0$ (outer lines), $\delta = \mu_0$ (lines intersecting at one point) and $\delta = 1.5\mu_0$ (lines meeting on an horizontal segment). The dashed parts are the negative energy branches. The thin dashed line stands for the bare dispersion of the particles E_L assumed to be parabolic here.

the usual equilibrium situation, described in the next chapter, is that the dispersion of elementary excitations can exhibit diffusive flat parts for $\delta_P > \mu_0$ or a gap for $\delta_P < \mu_0$, and the linear Bogoliubov spectrum of an equilibrium system is recovered only when the detuning is exactly compensated by the interaction energy, which means that $\hbar\delta_P = \alpha n_0 = \mu_0$ as shown in Fig. 7.24. Besides, the imaginary parts of the dispersion branches, which can be positive for large values of δ_P, leads to possible instabilities. Similarly, if one accounts for the non-parabolicity of the polariton dispersion, one finds that pumping at wavevectors beyond the points of inflection, leads to a positive imaginary part of energies. This corresponds to switching-on the OPO process described previously and to a breakdown of the hypothesis that the excitations of the macro-occupied mode are weak. This parametric instability, also called dynamical, or modulational instability is a generic process that takes place when a condensate is accelerated above a critical velocity or in a lattice, as we will see in the next chapters.

7.5.5 Conventional and unconventional polariton superfluidity

Superfluidity occurs because of the linear shape of the elementary excitations of the condensate, the so-called bogolon. The density of states available at the propagating condensate energy becomes vanishing and elastic scattering by random disorder, which is the microscopic origin of viscosity, vanishes as well. Under resonant excitation, a dispersion similar to that of equilibrium systems can occur at a very specific pumping density, for a given pump energy and wavevector. Using a strong homogeneous pumping, which as shown above opens a true gap in the excitation spectrum, can create propagating polaritons that are protected from disorder scattering. This regime, discussed above and shown in Fig. 7.23, cannot be realised in conservative systems and can be viewed as a type of "unconventional superfluidity". This is also the case for the superfluid behaviour of a small signal propagating under the large area excitation spot of a pump in a polariton OPO, as shown in Fig. 7.22, that represents another type of unconventional superfluidity. In Chapter 8, the problem of polariton superfluidity will be revisited from the more conventional viewpoint with no driving, that is to say, under non-resonant excitation. An other alternative, that will be described in Chapter 10, is to use a local pump which injects a polariton flow that is free to evolve and is only affected by the decay (which in good samples can be slow). In such a case, the polariton is described in a quite similar way to an atomic BEC and "conventional superfluidity" occurs for polaritons as well, as shown by Amo et al. (2011).

7.5.6 *High-density effects: the polariton backjet*

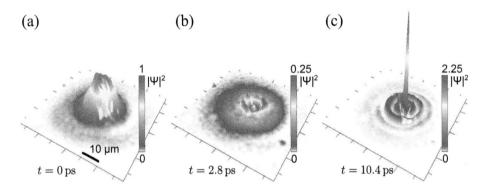

Fig. 7.25: Spatial dynamics of a polariton condensate in the first picoseconds of its creation by an intense laser pulse, from Dominici et al. (2015). The spot with initial spatial extension that of the pumping spot (at $t = 0$ ps) acquires a ring-like structure and develop at longer times a sharp and intense peak at the center, similar to a backjet of a droplet of water impacting a still pond. This striking dynamics has not yet received a compelling explanation for the mechanism that collapses repulsive particles into a sharp peak.

We have considered previously both very low densities, so that one can study the Schrödinger (non-interacting) dynamics, as well as higher densities where polariton interactions play a direct role, allowing a of study quantum fluid dynamics and effects such as superfluidity. We conclude this chapter with the case of very high densities, where new and unexpected phenomenology appears, as reported experimentally by Dominici et al. (2015) in a pulsed resonant experiment, where a bright ultrashort laser is sent at $k = 0$ on a microcavity in the vacuum. The fluid that is suddenly created does not splash, but instead coheres into a very bright and narrow spot, as shown in Fig. 7.25, reminding the backjet of a droplet after its impact at the surface of the liquid, although of course the analogy is limited (there is no gravity for polaritons nor, in this particular experiment, even a fluid before the arrival of the pulse). The radius of this polariton "backjet" is resolution limited and is thus at most of $2\,\mu$m, while the initial spot is 10 times larger. It also gathers a large number of particles, with a local enhancement up to 10 times the original density of polaritons. This striking dynamics takes place in a few picoseconds. Another interesting feature is the generation of a shock wave at early times and concentric rings at later times. It is also reported that the object is self-sustained and does propagate when imparted with a momentum.

The real-space collapse into a sharp peak is at odd with the repulsive interactions of polaritons and their positive mass, suggesting that an unconventional mechanism is at play. There is so far no documented phenomenon in the polariton literature to explain the most striking feature of this experiment: the real-space collapse in the centre of the spot. Real-space localisation could in principle appear under attractive polariton–polariton interactions. While these are possible through various superexchange processes via dark excitons or biexciton states, or due to Van der Waals forces, or involving large exchanged momenta, Dominici et al. (2015) show that this is not consistent with their observations and also exclude other scenarios such as a transition to the weak coupling regime (screening and reduction of coupling), or through effects involving the exciton reservoir.

Instead, they see as the most likely cause for this observation a self-trapping of the polariton condensate by a type of collective polaron effect, involving a correlated coupling to a phonon field. It has been shown by Klembt et al. (2015) that the resonant pumping of polaritons may result in both cooling or heating the crystal lattice depending on the initial lattice temperature and the optical pump power. At the cryogenic temperatures of the backjet experiment and for sufficiently high pump power, one can expect such a local heating of the crystal lattice due to the polariton Auger process, followed by the emission of a cascade of acoustic phonons. The probability of this process is quadratic in the polariton density. The heating results in the local bandgap renormalisation which is responsible for the redshift of the exciton energy. Heating by 20–30 K results in an exciton redshift of 2 meV, which is sufficient for trapping the polariton fluid. In this way, a trap in real space is formed under the pump spot. It becomes deeper as more polaritons are getting trapped, thus providing a positive feedback that stabilises the self-trapping process and explains the robustness of the effect.

This striking effect remains to be further investigated and characterised. It points out at possible future-generations of experiments in very high densities, where the conventional pictures, in particular those rooted in atomic paradigms, are shattered and a new physics altogether takes place. The resulting effects could also be relevant for applications, since the control of ultra-sharp localied light peaks is clearly of technological interest, for instance for high resolution displays or memory units.

8

STRONG COUPLING: POLARITON BOSE CONDENSATION

In this chapter we address the physics of Bose–Einstein condensation and its implications to a driven-dissipative system such as the polariton laser. We discuss the dynamics of exciton-polaritons non-resonantly pumped within a microcavity in the strong coupling regime. It is shown how the stimulated scattering of exciton-polaritons leads to formation of bosonic condensates that may be stable at elevated temperatures, including room temperature.

Microcavities, Second Edition. Alexey V. Kavokin, Jeremy J. Baumberg, Guillaume Malpuech, Fabrice P. Laussy, Oxford University Press (2017). © Alexey V. Kavokin, Jeremy J. Baumberg, Guillaume Malpuech, Fabrice P. Laussy. DOI 10.1093/oso/9780198782995.001.0001

8.1 Introduction

As discussed in Chapter 5, cavity polaritons, although they are a mixture of excitons and photons, behave as bosons in the low-density limit. One would therefore expect them to exhibit bosonic phase transitions such as Bose–Einstein condensation (BEC) or superfluidity. However, polaritons have a finite lifetime and the distribution function achieved by a polariton gas is a steady state distribution resulting from the balance between pumping, which constantly replenishes a high energy excitonic reservoir, scattering processes, which transfer these excitons toward lower energy states (tending to make the distribution function closer to an equilibrium one), and the finite lifetime, which makes particles disappear. The polariton system is therefore intrinsically an open system and polariton condensation is a dynamical out-of-equilibrium process. In fact, as we are going to see in this chapter, how far a polariton gas is from equilibrium depends on system parameters, which can be tuned experimentally and the system can actually be brought relatively close to equilibrium. In such a case the description of the system in terms of equilibrium BEC has some relevance. On the other hand in a vast parameter range polaritons do not accumulate in the state minimising the system energy, but in the state showing the optimal balance between the efficiency of scattering processes and lifetimes, the so-called kinetic regime. There are moreover different parameters, which can characterise a Bose condensate. The shape of the distribution function is one aspect, but spatial or temporal coherences and statistics of the condensed states are other important quantities that experimentalists have tried to measure and theoreticians to calculate, as we are going to see later. An important consequence of polariton condensate formation is the emission of an intense beam of light showing some degree of coherence, typically close to that of a laser, as first pointed out by Ĭmamoğlu and Ram (1996) and this coherent light emitter is very often called a "polariton laser". However, we would like to point out that the use of the label is appropriate only considering the cavity as a "black box", which under some proper excitation condition emits coherent light. The term "laser", however, implies that stimulated emission of photons competes with absorption and that spontaneous emission overcomes the radiative losses of the cavity. In a polariton system, absorption is entirely absent and does not need to be compensated. The population inversion, or gain condition, disappears. Only what remains is the fact that the net scattering toward the condensate should compensate the radiative escape through the mirrors. This means that polariton lasers having potentially a very low threshold, which is attractive from a device application point of view as discussed in detail in Chapter 12.

In this chapter we first give some basic ideas about BEC and bosonic phase transitions in equilibrium systems. We apply these basic concepts to the specific polariton case, including the strongly non-parabolic dispersion, to establish an equilibrium phase diagram. We then introduce several models that allow us to describe the non-equilibrium character of the polariton. We first introduce in detail the semi-classical Boltzmann equations, which allows calculation of the non-equilibrium polariton distribution function. This approach enables the definition of a parameter range where the thermodynamic description of the phase transition works and a parameter range that is completely dominated by the efficiency of relaxation kinetics in the system. We then introduce models inspired from quantum optics methods, in which the quantum fluctuations in a single mode condensate are taken into account, that enables access to the first and second order temporal coherence. We finally introduce a class of models that aim to describe spatial coherence, a key property of BECs and more generally spatial effects. This last class of models is based on a phenomenological coupling between a condensate described by a Gross Pitaevskii equation and a reservoir. Comparison with major experimental results will be given throughout the chapter.

8.2   Basic ideas about Bose–Einstein condensation

8.2.1   *Einstein proposal*

Einstein (1879–1955) and **Bose** (1894–1974) shortly after the publication of "*Plancks Gesetz und Lichtquantenhypothese*" in Zeitschrift für Physik, **26**, 178 (1924), written by Bose (1924) and translated by Einstein, who would later complement it with the concept of what is now known as "Bose–Einstein" condensation, although Bose did not take part in this aspect of the theory.

A fascinating property of bosons is their tendency to accumulate in unlimited quantity in a degenerate state. Einstein (1925) made an insightful proposition based on this property in the case of an ideal Bose gas that led him to the prediction of a new kind of phase transition. Let us consider N non-interacting bosons at a temperature T in a volume R^d, where R is the system size and d its dimensionality. The bosons are distributed in energy following the Bose–Einstein distribution function

$$f_B(\mathbf{k}, T, \mu) = \cfrac{1}{\exp\left(\cfrac{E(\mathbf{k}) - \mu}{k_B T}\right) - 1}, \tag{8.1}$$

where $\mathbf{k}$ is the particle d-dimensional wavevector, $E(\mathbf{k})$ is the dispersion relation for the considered bosons, k_B is Boltzmann constant. The chemical potential, μ, is a negative number if the lowest value of E is zero, so that $-\mu$ is the energy needed to add a particle to the system. Its value is given by the normalisation condition for the fixed total number of particles N:

$$N(T,\mu) = \sum_{\mathbf{k}} f_{\mathrm{B}}(\mathbf{k}, T, \mu) \,. \tag{8.2}$$

Before going to the thermodynamic limit,[109] it is convenient to separate the ground state from the others:

$$N(T,\mu) = \cfrac{1}{\exp\left(-\cfrac{\mu}{k_{\mathrm{B}}T}\right) - 1} + \sum_{\mathbf{k}, \mathbf{k}\neq 0} f_{\mathrm{B}}(\mathbf{k}, T, \mu) \,. \tag{8.3}$$

In the thermodynamic limit, the total polariton density is given by bringing the sum to converge into an integral over the reciprocal space:

$$n(T,\mu) = \lim_{R\to\infty} \frac{N(T,\mu)}{R^d} = n_0 + \frac{1}{(2\pi)^d} \int_0^\infty f_{\mathrm{B}}(\mathbf{k}, T, \mu) d\mathbf{k} \,, \tag{8.4}$$

where

$$n_0(T,\mu) = \lim_{R\to\infty} \frac{1}{R^d} \cfrac{1}{\exp\left(-\cfrac{\mu}{k_{\mathrm{B}}T}\right) - 1} \,. \tag{8.5}$$

If μ is nonzero, the ground-state density vanishes. On the other hand, the integral on the right-hand side of eqn (8.4) is an increasing function of μ. So, if one increases the particle density n in the system, the chemical potential also increases. The maximum particle density that can be accommodated following the Bose distribution function is therefore

$$n_{\mathrm{c}}(T) = \lim_{\mu\to 0} \frac{1}{(2\pi)^d} \int_0^\infty f_{\mathrm{B}}(\mathbf{k}, T) \, d\mathbf{k} \,. \tag{8.6}$$

This function can be calculated analytically in the case of a parabolic dispersion relation. It converges for $d > 2$, but diverges for $d \leq 2$, i.e., in two or less dimension(s), an infinite number of non-interacting bosons can always be accommodated in the system following the Bose distribution, the chemical potential is never zero and there is no phase transition. In higher dimensions, however, n_{c} is a critical density above which it would seem no more particles can be added. Einstein proposed that at such higher densities the extra particles in fact collapse into the ground state, whose density is therefore given by

$$n_0(T) = n(T) - n_{\mathrm{c}}(T) \,. \tag{8.7}$$

This is a phase transition characterised by the accumulation of a macroscopic number of particles—or equivalently by a finite density—in a single quantum state. The order parameter is the chemical potential, which becomes zero at the transition.

8.2.2   Experimental realisation

This proposal was not immediately accepted and understood by the scientific community, principally because of Uhlenbeck's thesis, wherein it was argued that BEC was not

---

[109]The thermodynamic limit is the limiting process by which the system size and the number of particles increase indefinitely, but conjointly so that the density remains constant.

Pyotr Leonidovich **Kapitza** (1894–1984) and Fritz **London** (1900–1954)
The pioneers of quantum hydrodynamics.

Kapitza worked in the Cavendish Laboratory in Cambridge with Ernest Rutherford, whom he called "the crocodile". He had one carved on the outer wall of the Mond Laboratory, now the emblem of the prestigious research centre. The following year he visited the Soviet Union where he was held forcibly to continue his research. He died the only member of the presidium of the Soviet Academy of Sciences who was not a member of the communist party. He was awarded the Nobel prize in 1978 for his work in low-temperature physics.

London (1938)'s proposition to root superfluidity in Bose condensation was highly controversial in his time, if only because he based his argument on an ideal gas, whereas a fluid is a strongly interacting system. He, more than any other, took the Bose condensation seriously, even at the time when Einstein himself seemed to accept its rebuttal by the scientific community.

realistic because it would not occur in a finite system. The interest in BEC saw a revival in 1938 with the first unambiguous report of Helium-4 superfluidity. A few months after this observation, London first proposed the interpretation of this phenomenon as a manifestation of BEC. This link between BEC and ^{4}He superfluidity marked the beginning of an impressive amount of scientific activity throughout the twentieth century, which is still being pursued. Another early field where Einstein's intuition found potential and practical applications is superconductivity. However, the link was only properly understood in the 1950s with the advent of the Bardeen, Cooper and Schrieffer (BCS) theory. Consider the case when the total particle density is fixed. It is thus possible to define a critical temperature T_c given by the solution of

$$n_c(T_c) = n . \tag{8.8}$$

A remarkable indication of the validity of Einstein's hypothesis (as was immediately pointed out by London) is that its direct application yields a BEC critical temperature of 3.14 K for He, very close to the experimental value of 2.17 K. However, a major difficulty is that these systems are strongly interacting, in fact already in their liquid phase, and therefore particle interactions are expected to play a fundamental role, thus making them poor realisations of Einstein's ideal gas. Consequently, the objective of most theoretical efforts of the 1940s–1960s was to describe condensation of strongly interacting bosons. Stimulated emission of light and laser action is also induced by the bosonic nature of the particles involved—the photons—which do not interact. This, however, means that they cannot self-thermalise and photons in a laser represent fundamentally a non-equilibrium system. Consequently, lasing is a non-equilibrium phase transition that cannot be directly interpreted as a BEC. In fact, the first clear manifestation of condensation in a weakly interacting Bose gas was performed recently by Anderson et al. (1995) with trapped alkali atoms. This discovery, crowned by the 2001 Nobel prize for physics, has given a strong revival to this field.

8.2.3 *Modern definition of Bose–Einstein condensation*

Research on BEC was extremely intense in the period 1938–1965, especially on the theoretical side, with several deep advances in understanding. In particular, these efforts led to a new definition of the BEC criterion. BEC is now associated with the appearance of a macroscopic condensate wavefunction $\psi(\mathbf{r})$, which has a nonzero mean value

$$\langle \psi(\mathbf{r}) \rangle = \sqrt{n_{\mathrm{cond}}(\mathbf{r})} e^{i\theta(\mathbf{r})}, \qquad (8.9)$$

where $\langle \psi(\mathbf{r}) \rangle$ is the order parameter for this phase transition. It is a complex number with an amplitude—the square root of the condensate density—and a phase. The system Hamiltonian is invariant under an arbitrary phase change of $\psi(\mathbf{r})$ (a property referred to as *global gauge invariance*). However, at the phase transition, this symmetric solution becomes unstable and the system breaks this symmetry by choosing a specific phase that is assumed throughout the whole condensate, which is therefore completely phase-coherent. Penrose and Onsager (1954) proposed the following criterion for BEC:

$$\langle \psi^\dagger(\mathbf{r})\psi(\mathbf{r}') \rangle \xrightarrow[|\mathbf{r}-\mathbf{r}'|\to\infty]{} \langle \psi(\mathbf{r}') \rangle^* \langle \psi(\mathbf{r}) \rangle, \qquad (8.10)$$

which is now generally accepted. Its significance has emerged gradually through the efforts of many theorists. Goldstone (1961) and Goldstone et al. (1962) advanced the idea of spontaneous symmetry breaking, Yang (1962) termed the phenomenon "*off-diagonal long-range order*" (ODLRO) and Anderson (1966) emphasised the notion of phase coherence. The superfluid velocity can be defined from eqn (8.9) as

$$m v_{\mathrm{s}}(\mathbf{r}, T) = \hbar \nabla \theta(\mathbf{r}, T). \qquad (8.11)$$

A system is therefore "superfluid" if two arbitrary spatial points are connected by a phase-coherent path, allowing for frictionless transport, i.e., no scattering.

8.3 Specificities of excitons and polaritons

Depending on their density and on temperature, excitons behave as either a weakly interacting Bose gas, a metallic liquid, or an electron–hole plasma. It has been understood by Moskalenko (1962) and Blatt et al. (1962) that excitons remain in the gas phase at low densities and low temperatures, and are therefore good candidates for observation of BEC in the way envisioned by Einstein. At that time there were no experimental examples of BEC of a weakly interacting gas and a great deal of research effort was dedicated to the problem of exciton BEC. A number of theoretical works on excitonic condensation and superfluidity have appeared, with major publications such as those by Keldysh and Kozlov (1968), Lozovik and Yudson (1975, 1976a, 1976b), Haug and Hanamura (1975) and Comte and Nozières (1982). In most of these, the fermionic nature of composite exciton is also addressed. The starting point of these models is a system of degenerate electrons and holes of arbitrary densities that is treated in the spirit of the BCS theory. A key point of all the formalisms that have been developed is that they assume an infinite lifetime for the semiconductor excitations. In other words, these theories are looking for steady-state solutions of the Schrödinger equation for interacting excitons. It is indeed clear that to have enough time to Bose-condense, excitons must have a radiative lifetime much longer than their relaxation time. Thus, the use of "dark" (uncoupled to light) excitons seems preferable. This is the case for bulk Cu_2O paraexcitons, whose ground-state spin is 2, or of excitons in coupled quantum wells, where the electron and hole are spatially separated. These two systems have been subject to intense experimental studies that have sometimes claimed achievement of exciton BEC or superfluidity, see for instance the publications by Butov et al. (2001), Snoke et al. (2002) and Butov et al. (2002). However, careful analysis by Tikhodeev (2000) and Lozovik and Ovchinnikov (2002) showed that clear evidence of excitonic BEC has not yet been achieved in such systems. The difficulties of Bose-condensing excitons are two-fold. The first reason is linked to the intrinsic imperfections of semiconductors. Because of unavoidable structural disorder, dark excitons non-resonantly excited are often trapped in local minima of the disorder potential and can hardly be considered as free bosons able to condense. The second source of difficulty is connected with the problem of detection of the condensed phase. The clearest signature of exciton Bose condensation should be the emission of coherent light by spontaneous recombination of condensed excitons. Such emission is a priori forbidden for a system of dark excitons with spin 2. This is why the system of coupled quantum wells is better adapted: it allows for the direct photoluminescence spectroscopy of the condensate. The critical temperature for BEC in coupled quantum wells, estimated by Butov et al. (2002), is below 1 K, which is less than the liquid He temperature, but still much more than T_c for Li or Rb atoms.

On the other hand, "bright" excitons, that are directly coupled to light, might also be good candidates for condensation, despite their short lifetimes. In bulk semiconductors, this coherent coupling gives rise to a polarisation wave that can be considered from a quantum-mechanical point of view as a coherent superposition of pure exciton and photon states (polaritons). Bulk polaritons are stationary states that transform into photons only at surfaces. Polaritons also being bosons, they can, in principle, form condensates that would emit spontaneously coherent light. Typical dispersion curves of bulk polaritons are shown in Fig. 4.14. In the vicinity of the exciton–photon intersection point, the density of states of polaritons is strongly reduced and the excitonic contribution to the polariton is decreased. One should note that strictly speaking, a $k = 0$ photon does not exist and that, consequently, the $k = 0$ polaritonic state of the Lower Polariton Branch (LPB) does not exist either. The polariton dispersion has no minimum so that a true condensation process is strictly forbidden. Polaritons accumulate in a large number of states in the so-called bottleneck region. The situation is drastically different in microcavities. The cavity prevents the escape of photons and allows the formation of long-lifetime cavity polaritons. Conversely to the bulk case, the in-plane cavity polariton dispersion exhibits a well-defined minimum located at $k = 0$, but since they are two-dimensional quasiparticles they cannot exhibit a strict BEC phase transition, but rather a local condensation or so-called *Kosterlitz–Thouless phase transition* towards superfluidity. They have, moreover, an extremely small effective mass around $k = 0$, allowing for polariton lasing at temperatures that could be higher than $300\,\mathrm{K}$. Experimental discovery of stimulated scattering of polaritons in microcavities (see Chapter 7) has proved that a microcavity is a suitable system to observe effects linked to the bosonic nature of polaritons, including BEC. The paper by Kasprzak et al. (2006) is widely accepted as the proof for polariton BEC in CdTe-based microcavities at temperatures up to about $40\,\mathrm{K}$. Moreover, strong indications of BEC of exciton-polaritons at $300\,\mathrm{K}$ have been reported by Baumberg et al. (2008) of which we will give a more detailed overview in a dedicated section. A fundamental peculiarity and difficulty of a microcavity is the finite polariton lifetime that may result in a strongly non-equilibrium polariton distribution function. The relaxation kinetics of polaritons plays a major role in this case.

8.3.1 *Thermodynamic properties of cavity polaritons*

In this section we discuss the thermodynamic properties of microcavity polaritons considered as equilibrium particles, i.e., particles having an infinite lifetime. Even though this approximation is far from the reality, mainly governed by relaxation kinetics, it is instructive to examine the BEC conditions in this limiting case. We shall, moreover, assume that polaritons behave as either an ideal or a weakly interacting boson gas, so that the following analysis is valid only in the low-density limit.

As mentioned in Section 8.2.1, the critical condensation density is finite for nonzero temperature if $d > 2$. However, this density diverges in the two-dimensional case. Thus, a non-interacting Bose gas cannot condense in an infinite two-dimensional system and the same statement turns out to be true when interactions are taken into account. A rigorous proof of the absence of BEC in two dimensions has been given by Hohenberg (1967). An equivalent statement known as the Mermin–Wagner theorem after the work of Mermin and Wagner (1966), asserts that long-range order cannot exist in a system of dimensionality lower than two. Finally, it has been shown that spontaneous symmetry breaking does not occur in two dimensions, see for instance the discussion by Coleman (1973). However, a phase transition between a normal state and a superfluid state can still take place in two dimensions as predicted by Kosterlitz and Thouless (1973) in the framework of the XY spin model. Such a second-order phase transition is forbidden for ideal bosons, but according to Fisher and Hohenberg (1988) can take place in systems of weakly interacting bosons such as low-density excitons or polaritons, as shown by Fisher and Hohenberg (1988). The case of excitons has been especially investigated by Lozovik et al. (1998) and Koinov (2000). In the next section, we introduce the Kosterlitz–Thouless phase transition and its application to the cavity polariton system. We describe the effect of interactions on bosons through a presentation of the Bogoliubov formalism as a guide. The problem of local condensation is addressed in Section 8.3.3.

8.3.2 *Interacting bosons and Bogoliubov model*

Lev Davidovich **Landau** (1908–1968) and Nikolai Nikolaevich **Bogoliubov** (1909–1992) provided the most lasting phenomenological and microscopical interpretations of superfluidity.

Landau made numerous contributions to theoretical physics, especially in Russia where the "Landau School" is still referred to. Alexey Abrikosov, Lev Pitaevskii, Isaak Khalatnikov are among his most famous students. He put special emphasis on broad qualifications for a physicist as opposed to narrow expertise. A child prodigy, he said he couldn't remember a time when he was not familiar with calculus. He kept a list of physicists whom he graded on a logarithmic scale, with his contemporaries Bohr, Heisenberg and Schrödinger falling into the category 1 along with Newton. He made an exception for Einstein who he ranked in a superior category of his own (0.5). He put himself in the category 2.5, but later upgraded himself to 2. He was the victim of a severe car accident in January 1962 that would ultimately claim his life. He abandoned his research activities during this last period. He received the 1962 Nobel Prize in Physics for his work on superfluidity.

Bogoliubov published his first paper at 15. He was one of the first to have studied nonlinearities in physical systems at a time where absence of computers made them forbidding. In the late 1940s and 1950s he studied superfluidity, successfully taking into account the nonlinear terms that others had deemed too complicated. He introduced the Bogoliubov transformation in quantum field theory. In the 1960s he turned his interest to quarks in nuclear physics.

In order to explain the properties of superfluid He, a phenomenological model was developed by Landau (1941) (extended in 1947) who introduced an original energy spectrum, displayed in Fig. 8.1. This spectrum is composed of two kinds of quasiparticles: *phonons* and *rotons*. These quasiparticles are collective modes in the "gas of quasiparticles" and are associated with the first and second sound. Such a spectrum introduced "by hand" allowed Landau to describe most of the peculiar properties of superfluid He. Bogoliubov's work of 1947 was a real breakthrough. He presented a microscopic description of the condensed weakly interacting Bose gas. As we shall see later, he showed how BEC is not much altered in a weakly interacting Bose gas, something that was not obvious at the time. He also showed how interactions completely alter the long-wavelength response of a Bose gas. He recovered qualitatively the spectrum assumed by Landau for the quasiparticle dispersion relation. Most of the further theoretical developments in the field are based on the Bogoliubov approach.

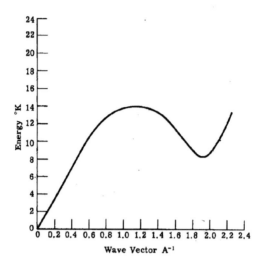

Fig. 8.1: The energy spectrum of liquid Helium II, showing phonons as $k \to 0$ and rotons as the high-k dip.

Bogoliubov considered the Hamiltonian describing an interacting Bose gas:

$$H = \sum_{\mathbf{k}} E(\mathbf{k}) a_{\mathbf{k}}^{\dagger} a_{\mathbf{k}} + \frac{1}{2} \sum_{\mathbf{k},\mathbf{k}',\mathbf{q}} V_{\mathbf{q}} a_{\mathbf{k}+\mathbf{q}}^{\dagger} a_{\mathbf{k}'-\mathbf{q}}^{\dagger} a_{\mathbf{k}} a_{\mathbf{k}'}, \qquad (8.12)$$

with $a_{\mathbf{k}}^{\dagger}$, $a_{\mathbf{k}}$ the creation and annihilation bosonic operators and $V_{\mathbf{q}}$ the Fourier transform of the interaction potential for the boson–boson scattering. His objective was to diagonalise this Hamiltonian making reasonable approximations.

At zero Kelvin, an ideal Bose gas should be completely condensed. Bogoliubov assumed that interactions are only responsible for weak condensate depletion. In other words, most system particles are assumed to be still inside the condensate. This implies $\langle a_0 \rangle, \langle a_0^{\dagger} \rangle \approx \sqrt{N_0}$ and, therefore,

$$[a_0, a_0^{\dagger}] \ll \langle a_0^{\dagger} a_0 \rangle, \qquad (8.13)$$

where N_0 is the condensate population. Bogoliubov proposed to neglect the condensate fluctuations and to replace the operators a_0, $a_0^{\dagger}$ by complex numbers A_0, A_0^*. The condensate is thus treated classically as a particle reservoir. The second Bogoliubov approximation was to keep only the largest contributions in the interacting part of the Hamiltonian. The largest contributions are those that involve the condensate. Therefore, one keeps only the terms that involve a_0 two times or more. Equation (8.12) becomes

$$H = N_0 V_0 + \sum_{\mathbf{k}, \mathbf{k} \neq 0} \left(E(\mathbf{k}) + N_0 (V_{\mathbf{k}} + V_0) \right) a_{\mathbf{k}}^{\dagger} a_{\mathbf{k}}$$

$$+ N_0 \sum_{\mathbf{k}} V_{\mathbf{k}} \left(a_{\mathbf{k}}^{\dagger} a_{-\mathbf{k}}^{\dagger} a_{\mathbf{k}} a_{-\mathbf{k}} \right). \quad (8.14)$$

The Bogoliubov approximations conserve off-diagonal coupling terms that induce the appearance of new eigenmodes. Then, a change of basis is made through the transformations (5.34) that, according to this procedure, diagonalises the Hamiltonian (8.12) into

$$H = N_0 V_0 + \sum_{\mathbf{k} \neq 0} E_{\mathrm{Bog}}(\mathbf{k}) \alpha_\mathbf{k}^\dagger \alpha_\mathbf{k} , \tag{8.15}$$

as a function of new operators $\alpha_\mathbf{k}$, $\alpha_\mathbf{k}^\dagger$, chosen to remain bosonic operators and with, as an all-important consequence:

$$E_{\mathrm{Bog}}(\mathbf{k}) = \sqrt{E(\mathbf{k})[E(\mathbf{k}) + 2N_0 V_\mathbf{k}]} . \tag{8.16}$$

It is this equation that justifies the Landau spectrum. When $V_\mathbf{k}$ is given by the Fourier transform of the Coulomb potential, the spectrum assumes the shape displayed in Fig. 8.1, which is also called the *Bogoliubov spectrum*. The unperturbed dispersion is recovered if N_0 vanishes. On the other hand, if one considers the existence of a condensate, there follows from a quadratic unperturbed dispersion

$$E(\mathbf{k}) = \frac{\hbar^2 k^2}{2m} , \tag{8.17}$$

the renormalised spectrum near $k = 0$ equal to

$$E_{\mathrm{Bog}}(\mathbf{k}) \approx \hbar k \sqrt{\frac{N_0 V_0}{m}} . \tag{8.18}$$

This is the linear part of the Landau spectrum, which displays features of phonon quasiparticles with sound velocity

$$v_s = \sqrt{\frac{N_0 V_0}{m}} . \tag{8.19}$$

Considering a large polariton population of $N_0 = 10^6$ and a system size of $100\,\mu\mathrm{m}$, the typical bogolon velocity in a GaAs microcavity would be $5 \times 10^5\,\mathrm{m\,s^{-1}}$. In the same way one can estimate the reciprocal-space region where the bogolon spectrum is linear as the set of k values such that

$$k < \frac{2}{\hbar} \sqrt{m N_0 V_0} . \tag{8.20}$$

This corresponds to an angular width of about 3 degrees in the above-mentioned conditions.

8.3.3 *Polariton superfluidity*

Superfluidity is a property deeply associated with BEC and at first glance it seems that one cannot exist without the other. This is, however, not true. BEC is linked with the appearance of a Dirac function at $k = 0$ in the distribution function of bosons. The Fourier transform of this Dirac function gives the extension in the direct space of the condensate wavefunction, which is infinite and constant. BEC therefore means the appearance of a homogeneous phase in real space. This homogeneity implies superfluidity. Particles can move throughout space along a phase-coherent, dissipationless path. Superfluidity means that statistically, two points in space are connected by a phase-coherent path, even if the whole space is not covered by a phase-coherent wavefunction. As a conclusion, a superfluid state can occur without the existence of strict BEC. This is the kind of state that arises in two dimensions where a strict BEC is forbidden. We now describe qualitatively how this phase transition takes place and estimate the Kosterlitz–Thouless (KT) transition temperature T_{KT} for the polariton case.

At temperatures higher than the critical temperature T_{KT}, the superfluid number density n_s is zero, but local condensation can take place. Condensate droplets can have quite large sizes as we shall see later, but they are characterised by an exponentially decreasing correlation function and are not connected together. Free vortices prevent long-range ordering, i.e., percolation of the quasicondensate droplets. However, once the critical temperature T_{KT} is reached, single vortices are no longer stable. They bind, forming pairs or clusters with the total *winding number* (or *vorticity*) equal to zero, allowing for a sudden percolation of the quasicondensate droplets that therefore form a superfluid. For temperatures slightly below T_{KT}, the superfluid number density is proportional to T_{KT} with a universal coefficient (see the discussion by Nelson and Kosterlitz (1977)):

$$n_s = \frac{2mk_B T_{KT}}{\pi \hbar^2}, \tag{8.21}$$

where m is the bare polariton mass at $q = 0$. Pairs of vortices remain well below T_{KT} and the correlation function is not constant, but decreases as a power of the distance. The superfluid wavefunction has thus a finite extension in reciprocal space, and consequently is not a BEC wavefunction. Complete homogeneity and true long-range order can be achieved only at $T = 0$, where vortices disappear. Below T_{KT} normal and superfluid phases coexist. The normal fluid can be characterised by a density n_n and a velocity v_n, while the superfluid has a density n_s and velocity v_s. The total fluid density is

$$n = n_n + n_s, \tag{8.22}$$

where n_n can be calculated following, for instance, Koinov (2000). Despite the absence of BEC in two dimensions, the energies of quasiparticles (bogolons) in the superfluid phase are still given by the Bogoliubov expression

$$E_{Bog(\mathbf{k})} = \sqrt{E(\mathbf{k})[E(\mathbf{k}) + 2\mu]}. \tag{8.23}$$

We need to know the chemical potential of interacting polaritons to calculate the quasiparticle dispersion. With the meaning previously given for the chemical potential, when one considers added particles going into the ground state, the associated interaction energy yields

$$\mu = NV_0. \tag{8.24}$$

Once the quasiparticle dispersion is known, one can use the famous Landau formula to calculate the normal mass density. This formula reads, for a two-dimensional system,[110,111]

$$n_{\mathrm{n}} = \frac{1}{(2\pi)^2} \int E(\mathbf{k}, T, \mu) \left(- \frac{\partial f_{\mathrm{B}}(E_{\mathrm{bog}}(\mathbf{k}, T, 0))}{\partial E_{\mathrm{bog}}(\mathbf{k})} \right) d\mathbf{k}. \tag{8.25}$$

Note that the Bose distribution function given by eqn (8.1) that enters this expression is taken at zero bogolon chemical potential, while the nonzero polariton chemical potential is still present in the bogolon dispersion relation (8.23). Equations (8.23)–(8.25) yield $n_{\mathrm{s}}(T, n)$. The substitution of $n_{\mathrm{s}}(T_{\mathrm{KT}}, n)$ into eqn (8.22) allows one to obtain $T_{\mathrm{KT}}(n)$ and therefore to plot a polariton phase diagram. Such a phase diagram is shown in Fig. 8.2.

Solid lines (a–d) show the critical concentration for a KT phase transition according to the above-mentioned procedure, and calculated for typical microcavity structures based on GaAs (a), CdTe (b), GaN (c) and ZnO (d). In all cases we assume zero detuning of the exciton resonance and the cavity photon mode. For GaAs- and CdTe-based microcavities we have used the parameters of the samples studied by Senellart and Bloch (1999), Senellart et al. (2000) and Le Si Dang et al. (1998). The parameters of model GaN and ZnO microcavity structures have been given by Malpuech et al. (2002a) and Zamfirescu et al. (2002). Vertical and horizontal dashed lines show the approximate limit of the strong-coupling regime in a microcavity that come from either exciton screening by a photoinduced electron–hole plasma or from temperature-induced broadening of the exciton resonance. Below the critical density, if still in the strong-coupling regime, a microcavity operates as a polariton diode emitting incoherent light, while in the weak-coupling regime the device behaves like a conventional light-emitting diode. Above the critical density, in the weak-coupling regime, the microcavity acts as a conventional laser. Thin dotted lines in Fig. 8.2 indicate the limit between the latter two phases that cannot be found in the framework of our formalism limited to the strong-coupling regime. One can note that critical temperatures achieved are much higher than those that can be achieved in exciton systems. In existing GaAs- and CdTe-based cavities, these temperatures are high enough for experimental observation of the KT phase transition under laboratory conditions, but do not allow one to produce devices working at room temperature. Record critical temperatures of $T_{\mathrm{KT}} = 450\,\mathrm{K}$ and $560\,\mathrm{K}$ for GaN- and ZnO-based model cavities are given by extremely high exciton dissociation energies in these semiconductors. It is interesting to note that $n - n_{\mathrm{n}}$ does not vanish, which reflects the existence of isolated quasicondensate droplets. As we show below, these

---

[110]The derivation of the Landau formula requires a current-conservation law and therefore is exact for particles with a parabolic spectrum only. In the case of cavity polaritons with a non-parabolic dispersion, eqn (8.25) remains a good approximation at low temperatures, where the exciton-like part of the low polariton dispersion branch is weakly populated. Keeling (2006) has shown that at higher temperatures, the critical density for the KT transition is lower than is predicted by eqn (8.25). See Appendix B for more details.

[111]The integral in eqn (8.25) can be computed approximately in the case of a parabolic dispersion if $\mu/(k_{\mathrm{B}}T) \gg 1$, as shown by Fisher and Hohenberg (1988). This approximation is, for example, well satisfied for excitons having critical temperature in the Kelvin range. It is no longer the case for polaritons because of their small masses. Therefore, numerical integration has to be carried out.

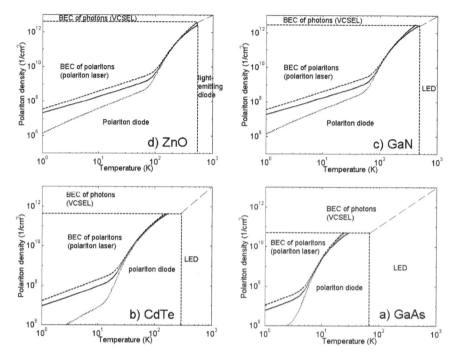

Fig. 8.2: Phase diagrams for GaAs (a), CdTe (b), GaN (c) and ZnO (d) based microcavities at zero detuning, by Kavokin et al. (2003b). Vertical and horizontal dashed lines show the limits of the strong-coupling regime imposed by the exciton thermal broadening and screening, respectively. Solid lines show the critical concentration N_c versus temperature of the polariton KT phase transition. Dotted and dashed lines show the critical concentration N_c for quasicondensation in 100 μm and in one meter lateral size systems, respectively. The thin dashed line (upper right) symbolises the limit between vertical-cavity surface-emitting laser (VCSEL) and light-emitting diode regimes.

droplets can reach substantial size, even above the Kosterlitz–Thouless temperature or density. Their properties may dominate the behaviour of real systems in some cases.

At high excitonic densities the behaviour of the polariton liquid can be affected by fermionic effects linked to the composite nature of the exciton. This may result in the crossover from the quasicondensed or superfluid phase to quantum-correlated plasma (studied by Eastham and Littlewood (2001), Marchetti et al. (2004), Keeling et al. (2005) and subsequent works since then), that is, however, beyond the scope of this monograph.

8.3.4   *Quasicondensation and local effects*

In this section we define a rigorous criterion for boson quasicondensation in finite-size systems. For the sake of simplicity we neglect here all kinds of interactions between particles. Let us consider a system of size R and dimensionality d. The particle density is given by:

$$n(T, R, \mu) = \frac{N_0}{R^d} + \frac{1}{R^d} \sum_{\substack{\mathbf{k} \\ k \geq 2\pi/R}} f_\mathrm{B}(\mathbf{k}, T, \mu), \qquad (8.26)$$

where N_0 is the ground-state population. We define the critical density as the maximum number of bosons that can be accommodated in all the states, but the ground state:

$$n_\mathrm{c}(R, T) = \frac{1}{R^d} \sum_{\substack{\mathbf{k} \\ k \geq 2\pi/R}} f_\mathrm{B}(\mathbf{k}, T, 0). \qquad (8.27)$$

The quasicondensate density is thus given by $n_0 = n - n_\mathrm{c}$. In this case, formally, the chemical potential μ is always strictly negative, but it approaches zero, allowing one to put as many bosons as desired in the ground state, while keeping the concentration of bosons in all other states finite and limited by n_c. The concentration (8.27) can be considered as the critical concentration for local quasi-Bose condensation in two-dimensional systems. Further, we shall refer to T_c defined in this way as the critical temperature of Bose condensation in a finite two-dimensional system. On the other hand, it appears possible, knowing the temperature and density, to deduce the typical coherent droplet size, which is given by the correlation length of the quasicondensate.

From a practical point of view, experiments are performed on samples having a lateral size of about 1 cm. Electron–hole pairs are generated by laser light with a spot area of about 100 μm. These electron–hole pairs rapidly (typically on a timescale less than 10 ps) form excitons, which relax down to the optically active region, where they strongly interact with the light field to form polaritons. Excitons that form polaritons have a finite spatial extension in the plane of the structure, but they are all coupled to each other via light, as illustrated for instance in the applications developed by Malpuech and Kavokin (2001) or Kavokin et al. (2001). The polariton system thus covers the whole surface where excitons are generated. If the KT critical conditions are not fulfilled, but if typical droplet sizes are larger than the light spot size, the whole polariton system can be transiently phase-coherent and thus exhibits local BEC. As we shall show below, this situation is the most likely to happen in current optical experiments performed at low temperature.

Let us underline at this point an important advantage of polaritons with respect to excitons weakly coupled to light for the purposes of BEC or superfluidity. Individual excitons in real structures are subject to strong localisation in inevitable potential fluctuations that prevent them from interacting and forming condensed droplets. Polaritons are basically delocalised, even though the excitons forming them could be localised, a point emphasised in the work of Keeling et al. (2004) and Marchetti et al. (2006). This is why their interactions are expected to be more efficient and bosonic behaviour more pronounced. The dotted and dashed lines in Fig. 8.2(a–d) show the critical concentration for local quasicondensation in microcavity systems of 100 μm and 1 m lateral size, respectively. In the high-temperature (high-concentration) limit, the critical concentrations are very similar for both lateral sizes and they slightly exceed the critical concentrations of the KT phase transition. This means that in this limit the KT transition takes place before the droplet size reaches 100 μm. Conversely, in the low-temperature (low-concentration) limit the KT curve is between the transition curves of the 100 µm and 1 m size systems. This shows that droplets at the KT transition are larger than 100 µm, but smaller than 1 m. Since the typical laser spot size is of about 100 µm, this means that local Bose condensation takes place before the KT transition at low pumping. A detailed analysis could allow one to obtain the percolating droplet size versus temperature, which is beyond the scope of our present discussion.

Note finally that inhomogeneous broadening of the exciton resonance leads to broadening of the polariton ground state in the reciprocal space. Any broadening in the reciprocal space is formally equivalent to localisation in the real space. Such a localisation, present even in an infinite microcavity, allows for quasicondensation, in principle. Agranovich et al. (2003) have recently studied disordered organic semiconductors prone to exhibit such effects.

Experimental results such as those of Richard et al. (2005b) evidenced localisation of the condensate in real space due to the disorder, whereas no superfluid behaviour—like linearisation of the excitation spectrum—was observed. The impact of localisation is therefore extremely strong and the following qualitative picture can be drawn. In the presence of disorder, the lowest-energy states are localised states. In a non-interacting boson picture, only the lower energy state should be filled giving rise to a fully localised condensate. This picture changes drastically in the presence of interactions. Indeed, once a localised state starts to be filled, it blueshifts because of polariton–polariton interactions. The chemical potential increases and reaches the energy of another localised state that in turn starts to be populated and blueshifts. The system therefore assumes an assembly of strongly populated localised states, all with the same chemical potential. This situation, which does not exhibit any superfluidity, is known as a *Bose glass* (see, for instance, the discussion by Fisher et al. (1989)). Once the chemical potential reaches the value of the localisation energy, a delocalisation of the condensate occurs and a standard KT phase transition can take place as shown by Berman et al. (2004). In realistic systems one should therefore expect two successive bosonic phase transitions when increasing the density. A detailed theoretical analysis of the polariton phase diagram in the presence of disorder is given by Malpuech et al. (2007).

8.4 Kinetics of formation of polariton condensates: semiclassical picture

As pointed out in Section 8.3, the condensate of cavity polaritons is an equilibrium state of the polaritonic system for a wide range of external parameters. However, polaritons have a finite lifetime in microcavities and are therefore non-equilibrium particles. Relaxation of polaritons in the steepest zone of the polariton dispersion and therefore towards the ground state, is a slow process compared to the polariton lifetime. The slow energy relaxation of polaritons combined with their fast radiative decay results in the bottleneck effect. The photoluminescence mainly comes from the "bottleneck region" and the population of the ground state remains much lower than what one could expect from the equilibrium distribution function (see Section 7.1.5). A suitable formalism to describe polariton population dynamics is the semiclassical Boltzmann equation. This formalism has been widely used, particularly by Tassone and Yamamoto (1999), Malpuech et al. (2002b), Malpuech et al. (2002a) and Porras et al. (2002). This type of model has then been used to describe experimental data in a large number of works such as Butté et al. (2002), Kasprzak et al. (2008b), Wertz et al. (2009), Levrat et al. (2010), Li et al. (2013). We now discuss qualitatively the main features of polariton photoluminescence, and show how the semiclassical Boltzmann equations can be used to describe polariton relaxation. A major weakness of this approach is, however, that it only allows calculation of the populations of polaritonic quantum states. All other quantities of interest, such as the order parameter and various correlation functions, are beyond its scope and a derivation involving quantum features of the system must be undertaken and will be presented later.

8.4.1 *Qualitative features*

Typical dispersion curves of bulk polaritons are shown in Fig. 4.14. Excitons created by an initial laser excitation relax along the lower-polariton dispersion, which is essentially the bare exciton dispersion, except near the exciton–photon resonance. In this region, the polariton density of states is strongly reduced and the excitonic contribution to the polariton is decreased. One should note that strictly speaking, a photon with $k = 0$ does not exist and that, consequently, the $k = 0$ polaritonic state of the LPB does not exist either. The polariton dispersion has no minimum and polaritons accumulate in a large number of states in the bottleneck region, already mentioned, from where the light is mainly emitted. In this respect, photoluminescence (PL) experiments performed on the bulk can be viewed as being influenced by the polaritonic effect. More simply, the bulk bottleneck effect is induced by the sharpness of the energy/wavevector region where an exciton can emit a photon considering energy and wavevector conservation conditions. In microcavities, the polariton dispersion is completely different and the LPB has a minimum at $k = 0$. A bottleneck effect still arises because of the sharpness of this minimum, but light is clearly emitted from the whole polariton dispersion including the ground state. The polaritonic effect is from this point of view much clearer than in the bulk, as the PL signal comes from polariton modes, which are easily distinguishable from bare exciton and photon modes. States that emit light in a strongly coupled microcavity are polariton states, despite the localisation effect. The consequences are twofold. First, PL gives direct access to the polariton dispersion as pointed out by Houdré et al. (1994). Secondly, a theoretical description of PL experiments should account for the polariton effect and the particle relaxation should be described within the polariton basis.

The initial process is a non-resonant optical excitation or an electrical excitation of the semiconductor. At our level, the differences between the two kinds of excitation are only qualitative. This excitation generates non-equilibrium electron–hole pairs that self-thermalise on a picosecond timescale. A typical temperature of this electron–hole gas is of the order of hundreds or even thousands of Kelvin. This electron–hole gas strongly interacts with optical phonons and is cooled to a temperature smaller than ω_{LO}/k_B on a picosecond timescale. During these few picoseconds excitons may form and populate the exciton dispersion. The exact ratio between excitons and electron–hole pairs and their relative distribution in reciprocal space is still the subject of intense research activity, for example from the work of Selbmann et al. (1996) or from Gurioli et al. (1998). For simplicity, we choose to completely neglect these early-stage processes. Rather, we

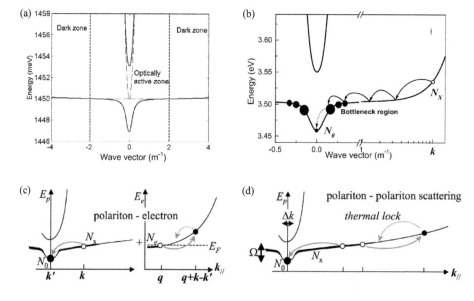

Fig. 8.3: Dynamics of relaxation along the polariton dispersion relation, from Malpuech et al. (2002b) and Kavokin et al. (2003b). (a) Bare exciton and bare cavity photon dispersions (dashed line) and polariton dispersion (solid line) as a function of the wavevector (10×10^7 m^{-1}). In the "dark zone", the exciton wavevector is larger than the light wavevector in the media. (b) Sketch of the polariton relaxation within the lower polariton branch showing the "bottleneck" where polaritons accumulate. (c) Sketch of polariton–electron scattering process (the curve on the right-hand side displays the free-electron dispersion). (d) Sketch of polaritons relaxation due to direct polariton–polariton interaction.

choose to consider as an initial condition the direct injection of excitons in a particular region of reciprocal space. We assume that the typical timescale needed to achieve such a situation is much shorter than the typical relaxation time of polaritons within their dispersion relation. Therefore, our objective is to describe the relaxation of particles (polaritons) moving in a dispersion relation composed of two branches (the upper-polariton mode and the lower-polariton mode), as shown in Fig. 8.3(a). Moreover, we assume for simplicity that the upper branch plays only a minor role since it is degenerate with the high-k LPB and that polaritons only relax within the LPB. The peculiar shape of the dispersion relation plays a fundamental role in the polariton relaxation kinetics.

The LPB is composed of two distinct areas. In the central zone, excitons are coupled to light. In the rest of the reciprocal space, excitons have a wavevector larger than the light wavevector in the vacuum and are therefore dark, as shown in Fig. 8.3(a). In the active zone, the polariton lifetime is mainly associated with radiative decay and is of the order of a few picoseconds. In the dark zone, polaritons only decay non-radiatively with a decay time of the order of hundreds of picoseconds (depending on the temperature). The dark zone has a parabolic dispersion associated with the heavy-hole exciton mass, which is of the order of the free-electron mass. The optically active zone is strongly distorted by strong exciton–light coupling. The central part of this active zone can be associated with a very small effective mass (about $10^{-4}m_0$). This mass rapidly increases with increasing wavevector to reach the exciton mass at the boundary between the optically active and dark zones.

We now list the physical processes involved in polariton relaxation towards lower energy states.

8.4.1.1 *Polariton–acoustic phonon interaction:*   Interactions between excitons and acoustic phonons are much less efficient than optical phonon–exciton interactions. Each relaxation step needs about 10 ps and no more than 1 meV can be exchanged. About 100-200 ps are therefore needed for a polariton to dissipate 10–20 meV of excess kinetic energy and to reach the frontier zone between dark and active areas. This relaxation time is shorter than the particle lifetime within the dark zone and some thermalisation can take place in this region of the reciprocal space. Once polaritons have reached the edge of the active zone, they still need to dissipate about 5–10 meV to reach the bottom of the polariton trap. This process assisted by the acoustic phonon needs about 50 ps, which is at least ten times longer than the polariton lifetime in this region. Therefore, polaritons cannot strongly populate the states of the trap. The distribution function takes larger values in the dark zone and at the edge of the active zone than in the trap. It cannot achieve thermal equilibrium values because of the slow relaxation kinetics. This effect has been called the *bottleneck effect* by Tassone et al. (1997), since it is induced by the existence of a relaxation "neck" in the dispersion relation. Such a phenomenon does not take place in a single QW with a parabolic dispersion. In that case the energy difference between the dark-active zone frontier is only about 0.05 meV and a dark exciton can reach the ground state by a single scattering event.

8.4.1.2 *Polariton–polariton interaction:*   This elastic scattering is a dipole–dipole interaction with a typical timescale of a few ps. It is very likely to happen because of the non-parabolic shape of the dispersion relation. Each scattering event can provide an energy exchange of a few meV. It is the main available process allowing population of the polariton trap and overcoming the "bottleneck effect". All experimental results in this field confirm that polariton–polariton interactions strongly affect polariton relaxation. However, the polariton–polariton interaction does not dissipate energy and does not reduce the net temperature of the entire polariton gas. If one considers the process sketched in Fig. 8.3(b), one polariton drops into the active zone where it will rapidly decay, whereas the other polariton gains energy and stays in a long-living zone. Altogether, this process heats the polariton gas substantially and may generate a non-equilibrium distribution function, as we shall see later.

8.4.1.3 *Polariton–free-carrier interaction:*   This scattering is sketched in Fig. 8.3(c). As stated above, optical pumping generates hot free carriers, which may interact with polaritons. Actually, the formation time of excitons or of strongly correlated electron–hole pairs is much shorter than the polariton lifetime. It is reasonable to assume that they do not play a fundamental role in polariton relaxation. However, a free-carrier excess may exist in modulation-doped structures, or may even be photoinduced if adapted structures are used, as has been done by Harel et al. (1996) and Rapaport et al. (2000).

Ludwig **Boltzmann** (1844–1906) developed the statistical theory of mechanics and thermodynamics. As such he fathered Boltzmann's constant k_B (see footnote 59 on page 126), the Maxwell–Boltzmann distribution and the Boltzmann equation to describe the dynamics of an ideal gas, among others.

Suffering from depression he committed suicide, some say because of the unsatisfying response to his work. His dense and intricate production was later lightened and disseminated by Ehrenfest (who also committed suicide).

A large free-carrier excess destroys excitonic correlations. However, at moderate density it can keep polaritons alive and provide a substantial relaxation mechanism. The polariton–electron interaction is a dipole–charge interaction and the associated scattering process has a sub-picosecond timescale. Electrons are, moreover, quite light particles in semiconductors (typically 4–5 times lighter than heavy-hole excitons). An electron is therefore able to exchange more energy by exchanging a given wavevector than an exciton. This aspect is extremely helpful in providing polariton relaxation in the steepest zone of the polariton dispersion. An electron–polariton scattering event is, moreover, a dissipative process for the polariton gas. It may be argued that the electron system can be heated by such an interaction. This is only partially true and we assume that the two-dimensional electron gas covering the entire sample represents a thermal reservoir. In this framework, an electron gas plays a role similar to acoustic phonons in polariton relaxation, but with a considerably enhanced efficiency. This efficiency allows, in some cases, the achievement of a thermal polariton distribution function, as proposed theoretically by Malpuech et al. (2002b) and later investigated experimentally by Qarry et al. (2003), Tartakovskii et al. (2003) and then by Perrin et al. (2005) and Bajoni et al. (2006).

8.4.2 The semiclassical Boltzmann equation

The classical Boltzmann equation describes the relaxation kinetics of classical particles. In reciprocal space this equation reads

$$\frac{dn_{\mathbf{k}}}{dt} = P_{\mathbf{k}} - \Gamma_{\mathbf{k}} n_{\mathbf{k}} - n_{\mathbf{k}} \sum_{\mathbf{k}'} W_{\mathbf{k} \to \mathbf{k}'} + \sum_{\mathbf{k}'} W_{\mathbf{k}' \to \mathbf{k}} n_{\mathbf{k}'}, \qquad (8.28)$$

where $P_{\mathbf{k}}$ is the generation term, due to optical pumping or to any other physical process, $\Gamma_{\mathbf{k}}$ is the particle decay rate and W is the total scattering rate between the states and due to any kind of physical process. Uhlenbeck and Gropper (1932) first proposed to include the quantum character of the particles by taking into account their fermionic or bosonic nature. Equation (8.28) written for fermions reads

$$\frac{dn_{\mathbf{k}}}{dt} = P_{\mathbf{k}} - \Gamma_{\mathbf{k}} n_{\mathbf{k}} - n_{\mathbf{k}} \sum_{\mathbf{k}'} W_{\mathbf{k} \to \mathbf{k}'}(1 - n_{\mathbf{k}'}) + (1 - n_{\mathbf{k}}) \sum_{\mathbf{k}'} W_{\mathbf{k}' \to \mathbf{k}} n_{\mathbf{k}}', \quad (8.29)$$

whereas for bosons it is

$$\frac{dn_{\mathbf{k}}}{dt} = P_{\mathbf{k}} - \Gamma_{\mathbf{k}} n_{\mathbf{k}} - n_{\mathbf{k}} \sum_{\mathbf{k}'} W_{\mathbf{k} \to \mathbf{k}'} (1 + n_{\mathbf{k}'}) + (1 + n_{\mathbf{k}}) \sum_{\mathbf{k}'} W_{\mathbf{k}' \to \mathbf{k}} n'_{\mathbf{k}}. \quad (8.30)$$

Equations (8.29) and (8.30) are called the *semiclassical Boltzmann equations*. The main task to describe the relaxation kinetics of particles in this framework is to calculate scattering rates. One should first identify the main physical processes that provoke scattering of particles. Then, scattering rates can be calculated using the Fermi golden rule. This procedure is usable only if the scattering processes involved are weak and can be treated in a perturbative way. Interactions should provoke scattering of particles within their dispersion relation and not provoke energy renormalisation. For example, the coupling of particles with the light should be a weak coupling, only responsible for a radiative decay. In a strongly coupled microcavity one cannot describe relaxation of excitons using a Boltzmann equation. One should first treat non-perturbatively the exciton–photon coupling giving rise to the polariton basis. Then, polaritons weakly interact with their environment. This weak interaction provokes scattering of polaritons within their dispersion relation and eqn (8.30) can be used. The scattering rates can indeed be calculated in a perturbative way (Fermi golden rule) because they are induced by weak interactions.

In a semiconductor microcavity the main scattering mechanisms identified are:

- Polariton decay (mainly radiative).
- Polariton–phonon interactions.
- Polariton–free-carrier interactions.
- Polariton–polariton interactions.
- Polariton–structural-disorder interactions.

The calculations of the rates of these scattering mechanisms are presented in Appendix A.

8.4.3 *Numerical solution of Boltzmann equations: practical aspects*

Despite its very simple mathematical form, eqn (8.30) is very complicated to solve[112] and one must use numerical methods. We detail below one possible such approach.

The phase space, in our case the reciprocal polariton space, must first be discretised. As mentioned previously, it is reasonable to assume cylindrical symmetry for the distribution function. The elementary cells of the chosen grid should reflect this cylindrical symmetry and therefore these cells should be annular. The cell number i, $C(i)$ should contain all states with wavevectors $\mathbf{k}$ satisfying $k \in [k_i, k_{k+1}[$. Various choices of scale for the k_i (linear, quadratic or other) have been used in the literature already quoted. The most important requirement being that the distribution function does not vary too abruptly from cell to cell, one should use small cells in the steep zone of the polariton dispersion, whereas very large cells can be used in the flat excitonic area. The nature of the polariton dispersion makes, therefore, the choice of a nonlinear grid a much better candidate. In all cases, one state requires particular attention over all the others: the ground state, especially if one wishes to describe "condensation-like phenomena",

---

[112]Hilbert worked on the problem of integrating the Boltzmann equation, but without success.

namely a discontinuity of the polariton distribution function. If such a discontinuity takes place the actual size of the cells plays a role. We cannot choose infinitely small cells numerically. This means that one cannot solve numerically the Boltzmann equations in the thermodynamic limit in the case of Bose condensation. What can be done is to account for a finite system size R. The spacing between states becomes finite (of the order of $2\pi/R$). The grid size plays a role, but it is no longer arbitrary, but related to a real physical quantity. In such a case the cell size should follow the real state spacing in the region where the polariton distribution function varies abruptly.

8.4.4   Effective scattering rates

The total scattering rate from a discrete state to another discrete state is the sum of all the scattering rates:

$$W_{\mathbf{k}\to\mathbf{k'}} = \frac{w_{\mathbf{k}\to\mathbf{k'}}}{S} = W^{\text{phon}}_{\mathbf{k}\to\mathbf{k'}} + W^{\text{pol}}_{\mathbf{k}\to\mathbf{k'}} + W^{\text{el}}_{\mathbf{k}\to\mathbf{k'}} . \tag{8.31}$$

where $S = R^2$. We now need to calculate two kinds of transition rate. The first is the transition rate between a discrete initial state k and all the states belonging to a cell of the grid, indexed by the integer i:

$$W^{\text{out}}_{\substack{\mathbf{k}\to\mathbf{k'} \\ k'\in C(i)}} = \sum_{k'\in C} \frac{w_{\mathbf{k}\to\mathbf{k'}}}{S} . \tag{8.32}$$

We pass to the thermodynamic limit, changing the sum to an integral

$$W^{\text{out}}_{\substack{\mathbf{k}\to\mathbf{k'} \\ k'\in C(i)}} = \frac{S}{(2\pi)^2} \int_{k'\in C(i)} \frac{w_{\mathbf{k}\to\mathbf{k'}}}{S} \, d\mathbf{k'} = \frac{1}{(2\pi)^2} \int_{k'\in C(i)} w_{\mathbf{k}\to\mathbf{k'}} \, d\mathbf{k'} . \tag{8.33}$$

Here, the integration takes place over final states. The cells have cylindrical symmetry, which means that the scattering rate does not depend on the direction of $\mathbf{k}$. The total scattering rate towards cell i is therefore the same for any state belonging to cell j. So

$$\text{if } k \in C(j), \quad W^{\text{out}}_{i\to j} = W^{\text{out}}_{\substack{\mathbf{k}\to\mathbf{k'} \\ k'\in C(i)}} . \tag{8.34}$$

One also needs to calculate the number of particles reaching a state from the cell $C(i)$:

$$W^{\text{in}}_{\substack{\mathbf{k}\to\mathbf{k'} \\ k'\in C(i)}} = W^{\text{in}}_{i\to j} = \frac{1}{(2\pi)^2} \int_{k'\in C(i)} w_{\mathbf{k}\to\mathbf{k'}} \, d\mathbf{k} . \tag{8.35}$$

Here, as opposed to eqn (8.33), the integration takes place over initial states $\mathbf{k}$.

If one wishes to describe condensation in a finite-size system, the ground-state cell is constituted by a single state and no integration takes place when the final state is the ground state:

$$W^{\text{out}}_{\mathbf{k}\to 0} = W^{\text{out}}_{i\to 0} = \frac{w^{\text{out}}_{\mathbf{k}\to 0}}{S} . \tag{8.36}$$

This scattering rate is inversely proportional to the system size.

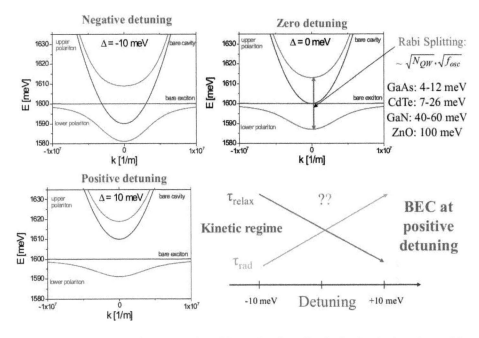

Fig. 8.4: Exciton-polariton dispersion at three different detunings. Sketch, showing the dependence of the polariton relaxation time and radiative lifetime versus detuning.

In this framework, eqn (8.30) becomes

$$\frac{dn_i}{dt} = P_i - \Gamma_i n_i - n_i \sum_j W_{i \to j}(1 + n_j) + (1 + n_i) \sum_j W_{j \to i} n_j. \qquad (8.37)$$

It is really an ensemble of coupled first-order differential equations that can be easily solved numerically. One should point out that despite the cylindrical symmetry hypothesis and, despite the one-dimensional nature of the final equation, all two-dimensional scattering processes are correctly accounted for.

8.4.5   *From thermodynamic to kinetic regime*

Qualitatively, the two parameters that can be modified to tune both the thermodynamic BEC limit and the relaxation kinetics of polaritons are the exciton-photon detuning and the temperature. Figure 8.4 shows polariton dispersions of a typical CdTe microcavity at three different detunings. At positive detuning, the polariton trap is shallow, which typically gives fast relaxation kinetics. The polariton ground state is dominantly excitonic, which typically corresponds to a long lifetime. Fast relaxation and long lifetime are the ideal situation to reach a quasi-thermal equilibrium. On the other hand once the detuning is positive enough for this thermal equilibrium to be reached, a further increase of detuning makes the polariton effective-mass larger, which increases the critical density for BEC as calculated with an infinite lifetime. Above some critical detuning, the

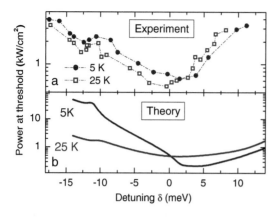

Fig. 8.5: Threshold power for polariton condensation versus exciton-photon detuning at two different temperatures, from Kasprzak et al. (2008b).

BEC threshold should then increase with detuning, as expected from a system at thermal equilibrium. If we now consider the negative detuning case as shown in Fig. 8.4, the polariton trap is deep and sharp and the relaxation kinetics to reach the ground state are slow whereas at the same time the photonic polariton ground state has a short lifetime. In this regime, the system is expected to be strongly out-of-equilibrium and the threshold governed by relaxation kinetics. The threshold is then expected to decrease with increasing detuning. The general behaviour of the relaxation time of polaritons and of their lifetime as a function of detuning is shown on the right lower panel of Fig. 8.4. The rates should cross more or less at the optimal detuning that separates the kinetic regime where the threshold decreases vs detuning and the thermodynamic regime where the threshold increases vs detuning.

Let us now think about the role of temperature. Temperature, in general, makes the kinetics faster and the thermodynamics worse. The optimal detuning should therefore be pushed to lower detuning where the threshold dependencies should cross.

This general behaviour has been described theoretically and observed in all types of microcavities able to support bosonic condensation. It is was demonstrated in CdTe structures by Kasprzak et al. (2008b), in GaAs based structures by Wertz et al. (2009), in GaN based structures by Levrat et al. (2010), allowing access to a larger set of temperatures and in ZnO based structures by Li et al. (2013). Figures 8.5, 8.6 and 8.7 are from Kasprzak et al. (2008b). Figure 8.5 shows the typical U shape with an optimal detuning at 5 K larger than at 25 K. In the thermodynamic region, the measured and calculated distribution functions are indeed thermal as shown in Fig. 8.6. On the opposite, in the kinetic region the distribution function clearly shows a bottleneck behaviour (cf. Fig. 8.7).

8.5 Kinetics of formation of polariton condensates: quantum picture in the Born–Markov approximation

In this section, we upgrade the Boltzmann approach previously discussed, to include some aspects of quantum coherence. We provide a self-consistent microscopic derivation starting with the polariton Hamiltonian. Polaritons are modelled as weakly interacting bosons moving within the lower-polariton branch. They can in principle interact

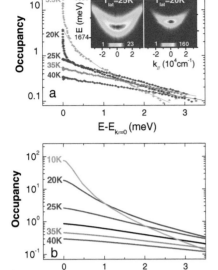

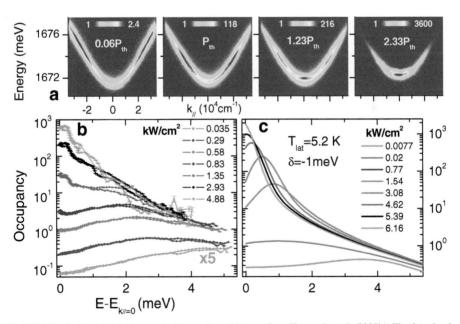

Fig. 8.6: Distribution of polaritons at thermal equilibrium, from Kasprzak et al. (2008b). (a) Experimental distributions of polaritons at different lattice temperature and at a constant pumping power close to threshold. The detuning is positive equal to 6.8 meV. Inset: PL versus in-plane wave vector. (b) Calculated distribution of polaritons under the same conditions using the semiclassical Boltzmann equations.

Fig. 8.7: Distribution of polaritons out of thermal equilibrium, from Kasprzak et al. (2008b). The detuning is negative (-1 meV) at a lattice temperature of 5.2 K. The figure shows (a) the PL and (b,c) the experimental and calculated distribution functions versus pumping power.

with phonons and free carriers (we will limit ourselves to the interaction with electrons). Furthermore, we include in the Hamiltonian the coupling to an external light field that allows for a self-consistent description of pumping and radiative lifetimes. The procedure used can be summarised as follows. We write the equation of motion for the density matrix of the system (von Neumann equation) and simplify it, using the approximations listed below. In all cases we will allow for the Markov approximation, cancelling memory effects. In a second step we use the so-called Born approximation that allows us to decouple the density matrices of different systems and eventually different polariton states. This Born approximation will be applied in any case to decouple the polariton density matrix from the phonon density matrix, the electron density matrix and the external free-photon density matrix. All systems except the polariton one are considered as reservoirs and we trace (average) over them. At the end of this procedure we will find a master equation for the density matrix of the whole polariton system (eqn 8.40) that will depend only on polariton operators and on semiclassical scattering rates. This equation will be our starting point for further approximations on two different levels. First, we describe the case where we fully apply the Born approximation to the polariton system. This allows us to obtain the semiclassical Boltzmann equations, which are therefore rigorously justified. A partial trace applied only on excited states yields a master equation for the ground-state density matrix, depending only on semiclassical quantities that in turn can be calculated, solving the Boltzmann equation. However, in this framework, spontaneous coherence buildup cannot take place, which shows the need to relax some of the approximations. This is done in the following section where we partially relax the Born approximation, keeping the correlations between the ground state and the excited states. As a result we get a quantum Boltzmann master equation that describes the dynamics of the ground-state statistics and allows for the spontaneous coherence buildup in the polariton system.

8.5.1   *Density matrix dynamics of the ground state*

The procedure we are going to outline is standard in quantum optics to describe the dynamics of open systems. The account by Carmichael (2002) is especially readable and insightful. Shen (1967) and Zel'dovich et al. (1968) provide a rather more historical approach. In what follows, we calculate the evolution of the density matrix of the system $\rho(t)$ under the influence of the polariton Hamiltonian (5.191). The tools required to follow the derivation have been introduced in Chapter 3. First, we iterate the Liouville–von Neumann eqn (3.53) to obtain

$$i\hbar\dot{\rho} = [H, \rho(-\infty)] + \int_{-\infty}^{t} [H(t), [H(\tau), \rho(\tau)]] \, d\tau \,. \tag{8.38}$$

Then, we apply the Born approximation to decouple the polariton field and the phonon, electron and photon fields. At this point, the density matrix of the system reads

$$\rho(t) = \rho_{\text{pol}}(t)\rho_{\text{phon}}\rho_{\text{el}}\rho_{\gamma} \,. \tag{8.39}$$

The polariton density matrix evolves in time, while the phonon and electron subsystems are considered as thermal baths at the lattice temperature. ρ_{γ} describes the photon vacuum of the electromagnetic field outside of the cavity (to which is coupled the electromagnetic field from within the cavity). ρ_{phon}, ρ_{el} and ρ_γ are kept equal to their equilibrium value. We then trace over phonon, electron and photon states. After some lengthy, but straightforward algebra, this partial trace gives results in a Lindblad form for the dissipative processes, i.e., with terms of the kind $LL^\dagger\rho + \rho LL^\dagger - 2L^\dagger\rho L$, eqn (5.98), (we now write simply ρ instead of ρ_{pol}), multiplied by time-dependent coefficients and with operators L depending only on polariton operators. It is at this stage that the Markov approximation is invoked: the populations are assumed to vary slowly in time and are taken out of the integral in eqn (8.38) at $t = \tau$. The remaining product of exponentials is then integrated to give the energy-conserving delta functions. The equation of motion for the polariton density matrix then becomes

$$\dot{\rho} = \mathcal{L}_{pump}\rho + \mathcal{L}_{lifetime}\rho + \mathcal{L}_{phon}\rho + \mathcal{L}_{el}\rho + \mathcal{L}_{pol}\rho, \tag{8.40}$$

where $\mathcal{L}$ are Liouvillian superoperators (cf. Section 3.2.1). Due to the linearity of the Liouville equation, one can associate to each part of the Hamiltonian a corresponding Liouvillian that affects the evolution of the density matrix. We define them now. The pump Liouvillian reads

$$\mathcal{L}_{pump}\rho = \frac{1}{2}\sum_{\mathbf{k}} P_{\mathbf{k}}\left(2a_{\mathbf{k}}^\dagger\rho a_{\mathbf{k}} - a_{\mathbf{k}}a_{\mathbf{k}}^\dagger\rho - \rho a_{\mathbf{k}}a_{\mathbf{k}}^\dagger\right), \tag{8.41}$$

where

$$P_{\mathbf{k}} = \frac{2\pi}{\hbar}|g(\mathbf{k})|^2|K_{pump}|. \tag{8.42}$$

The lifetime Liouvillian reads

$$\mathcal{L}_{lifetime}\rho = \frac{1}{2}\sum_{\mathbf{k}} \Gamma_{\mathbf{k}}\left(2a_{\mathbf{k}}\rho a_{\mathbf{k}}^\dagger - a_{\mathbf{k}}^\dagger a_{\mathbf{k}}\rho - \rho a_{\mathbf{k}}^\dagger a_{\mathbf{k}}\right), \tag{8.43}$$

where

$$\Gamma_{\mathbf{k}} = \frac{2\pi}{\hbar}|\gamma(\mathbf{k})|^2 \tag{8.44}$$

is the radiative coupling constant of the state to the external photon field. This result is obtained assuming that the photon modes are empty and therefore unable to replenish the corresponding polariton mode. Only their quantum fluctuations are playing a role, namely bringing a perturbation at the origin of the transition.

There is a clear symmetry between eqns. (8.41) and (8.43), reversing the ordering of a and $a^\dagger$. Note that each expression is equal to its hermitian conjugate. The Liouvillian of interactions with phonons reads

$$\mathcal{L}_{phon} = -\frac{1}{2}\sum_{\mathbf{k}'}\sum_{\mathbf{k}'\neq\mathbf{k}} W_{\mathbf{k}'\to\mathbf{k}}^{phon}\left(2a_{\mathbf{k}}^\dagger a_{\mathbf{k}'}\rho a_{\mathbf{k}'}^\dagger a_{\mathbf{k}} + a_{\mathbf{k}'}^\dagger a_{\mathbf{k}}a_{\mathbf{k}}^\dagger a_{\mathbf{k}'}\rho + \rho a_{\mathbf{k}'}^\dagger a_{\mathbf{k}}a_{\mathbf{k}}^\dagger a_{\mathbf{k}'}\right)$$

$$-\frac{1}{2}\sum_{\mathbf{k}'}\sum_{\mathbf{k}\neq\mathbf{k}'} W_{\mathbf{k}\to\mathbf{k}'}^{phon}\left(2a_{\mathbf{k}'}^\dagger a_{\mathbf{k}}\rho a_{\mathbf{k}}^\dagger a_{\mathbf{k}'} + a_{\mathbf{k}}^\dagger a_{\mathbf{k}'}a_{\mathbf{k}'}^\dagger a_{\mathbf{k}}\rho + \rho a_{\mathbf{k}}^\dagger a_{\mathbf{k}'}a_{\mathbf{k}'}^\dagger a_{\mathbf{k}}\right),$$

$$\tag{8.45}$$

where

$$W^{\text{phon}}_{\mathbf{k}' \to \mathbf{k}} = \frac{2\pi}{\hbar} \sum_{\mathbf{q}} |U(\mathbf{q})|^2 (1 + n_{\mathbf{q}}) \delta(E(\mathbf{k}') - E(\mathbf{k}) \mp \omega_{\mathbf{q}}), \qquad (8.46a)$$

$$W^{\text{phon}}_{\mathbf{k} \to \mathbf{k}'} = \frac{2\pi}{\hbar} \sum_{\mathbf{q}} |U(\mathbf{q})|^2 (\xi_{\pm} + n_{\mathbf{q}}) \delta(E(\mathbf{k}') - E(\mathbf{k}) \mp \omega_{\mathbf{q}}), \qquad (8.46b)$$

where $n_{\mathbf{q}} = \langle b^\dagger_{\mathbf{q}} b_{\mathbf{q}} \rangle$ is the phonon distribution function given by the Bose distribution, and $\xi_{+} = 1$ (with matching $+$ sign in the delta function) corresponds to emission of a phonon, while the case $\xi_{-} = 0$ corresponds to absorption (if $E(\mathbf{k}) < E(\mathbf{k}')$). The electronic Liouvillian $\mathcal{L}_{\text{el}}$ is the same as $\mathcal{L}_{\text{phon}}$, but for the transition rates that get replaced with

$$W^{\text{el}}_{\mathbf{k} \to \mathbf{k}'} = \frac{2\pi}{\hbar} \sum_{\mathbf{q}} |U^{\text{el}}(\mathbf{q}, \mathbf{k}, \mathbf{k}')|^2 n^{\text{e}}_{\mathbf{q}} (1 - n^{\text{e}}_{\mathbf{q}+\mathbf{k}'-\mathbf{k}})$$

$$\times \, \delta\left(E(\mathbf{k}') - E(\mathbf{k}) + \frac{\hbar^2}{2m_e}(q^2 - |\mathbf{q} + \mathbf{k} - \mathbf{k}'|^2)\right), \quad (8.47)$$

with $n^{\text{e}}_{\mathbf{q}}$ the Fermi distribution function.

Exercise 8.1 (*) *Derive the Liouvillian $\mathcal{L}_{\text{pol}}$ for polariton–polariton scattering in the approximation where it can be dealt with perturbatively. Observe the similarity with expressions eqns (8.41), (8.43) and (8.45). Relate it to the Lindblad form, i.e., what is the operator L of eqn (5.98) in this case?*

At this level, making the trace on polariton states in eqn (8.40) produces fourth-order correlators for the phonon and electron terms and even eighth-order correlators for the polariton–polariton terms. These correlators have now to be decoupled in order to get a closed set of kinetic equations. For simplicity we consider only polariton scattering with acoustic phonons. Other cases can be related easily to the following derivation. We now apply the Born approximation to the polariton system itself. This means that we factorise the density matrix of the polariton system as $\rho(t) = \rho_0(t)\rho_{\mathbf{k}_1}(t) \cdots \rho_{\mathbf{k}}(t) \cdots$. This implies that correlators can be decoupled in the following way:

$$\langle a^\dagger_{\mathbf{k}} a_{\mathbf{k}} a^\dagger_{\mathbf{k}'} a_{\mathbf{k}'} \rangle = \langle a^\dagger_{\mathbf{k}} a_{\mathbf{k}} \rangle \langle a^\dagger_{\mathbf{k}'} a_{\mathbf{k}'} \rangle, \qquad (8.48)$$

if $\mathbf{k} \neq \mathbf{k}'$. This quantity is of course the product of polariton populations. We shall note these with uppercase letters, to separate them from phonon populations that we note with lower case:

$$\langle a^\dagger_{\mathbf{k}} a_{\mathbf{k}} \rangle = N_{\mathbf{k}}, \qquad \langle b^\dagger_{\mathbf{k}} b_{\mathbf{k}} \rangle = n_{\mathbf{k}}. \qquad (8.49)$$

In this framework, we make the trace over all polariton states except the ground state. Using this procedure, we get as foretold the semiclassical Boltzmann equations that describe the dynamics of the polariton distribution function:

$$\frac{dN_{\mathbf{k}}}{dt} = P_{\mathbf{k}} - \Gamma_{\mathbf{k}} N_{\mathbf{k}} - N_{\mathbf{k}} \sum_{\mathbf{k}'} W_{\mathbf{k} \to \mathbf{k}'} (1 + N_{\mathbf{k}'}) + (1 + N_{\mathbf{k}}) \sum_{\mathbf{k}'} W_{\mathbf{k}' \to \mathbf{k}} N'_{\mathbf{k}}. \quad (8.50)$$

The dynamics of the ground-state population is also governed by this equation, but we also end up with a master equation for the ground-state density matrix that reads

$$\dot{\rho}_0 = \frac{1}{2}W_{\text{in}}(t)(2a_0^\dagger \rho_0 a_0 - a_0 a_0^\dagger \rho_0 - \rho_0 a_0 a_0^\dagger)$$
$$+ \frac{1}{2}(\Gamma_0 + W_{\text{out}}(t))(2a_0 \rho_0 a_0^\dagger - a_0^\dagger a_0 \rho_0 - \rho_0 a_0^\dagger a_0), \qquad (8.51)$$

where

$$W_{\text{in}}(t) = \sum_{\mathbf{k}} W_{\mathbf{k} \to 0} N_{\mathbf{k}} \quad \text{and} \quad W_{\text{out}}(t) = \sum_{\mathbf{k}} W_{\mathbf{0} \to \mathbf{k}}(1 + N_{\mathbf{k}}), \qquad (8.52)$$

the former is the total scattering rate toward the ground state and the latter the total scattering rate from the ground state toward excited states. This equation is similar to the one used to describe a single-mode linear amplifier—an exhaustive description of which is given by Mandel and Wolf (1995)—except for the time dependence of the transition rates in eqns. (8.52). Equation (8.51) can nevertheless be solved using the Glauber–Sudarshan representation of the density matrix. The derivation and the discussion of this solution have been given by Rubo et al. (2003) and Laussy et al. (2004c). The result is given in Exercise 8.2.

Exercise 8.2 $^{(**)}$ *Show that the solution of eqn (8.51) in Glauber representation with the initial condition* $P(\alpha, \alpha^*, 0) = \delta(\alpha - \alpha_0)$ *is*

$$P(\alpha, \alpha^*, t) = \frac{1}{\pi m(t)} \exp\left(-\frac{|\alpha - G(t)\alpha_0|^2}{m(t)}\right) \qquad (8.53)$$

in terms of time-dependent parameters

$$G(t) = \exp\left[\frac{1}{2}\int_0^t (W_{\text{in}}(\tau) - W_{\text{out}}(\tau)) \, d\tau\right], \quad m(t) = G(t)^2 \int_0^t W_{\text{in}}(\tau)G(\tau)^2 \, d\tau. \qquad (8.54)$$

The exact solution, eqns. (8.53) and (8.54), shows that in the Born–Markov approximation, the state of the condensate is that of a thermalised coherent state, parameters G and m relating to the relative importance of the coherent and thermal fractions, respectively. The quantities of interest that we can derive directly from the solution are the occupation number N_0 (ground-state population), the order parameter $\langle a_0 \rangle$, the ground-state statistics $p_0(n) = \langle n| \rho |n \rangle$ and the second-order coherence $g^{(2)}$. These can be obtained from the complete solution by integrating their equations of motion (derived from eqn (8.51)). First, equations for the scalar quantities (we use for convenience $\eta = 2 - g^{(2)}$, rather than $g^{(2)}$ directly) are given by

$$\dot{N}_0 = (W_{\text{in}} - W_{\text{out}} - \Gamma_0)N_0 + W_{\text{in}}, \qquad (8.55a)$$
$$\langle \dot{a}_0 \rangle = \frac{1}{2}(W_{\text{in}} - (W_{\text{out}} + \Gamma_0))\langle a_0 \rangle, \qquad (8.55b)$$
$$\partial_t(\eta N_0^2) = 2(W_{\text{in}} - (W_{\text{out}} + \Gamma_0))\eta N_0^2. \qquad (8.55c)$$

Secondly, the set of coupled differential equations for the statistics reads

$$\dot{p}_0(n) = (n+1)(W_{\text{out}} + \Gamma_0)p_0(n+1)$$
$$- [n(W_{\text{out}} + \Gamma_0) + (n+1)W_{\text{in}}]p_0(n_0) + nW_{\text{in}}p_0(n-1). \quad (8.56)$$

The quantities $\langle a_0 \rangle^4$ and ηN_0^2 are both extensive and proportional to N_0^2. They are surprisingly described by the same equation. Whereas $\langle a_0 \rangle$ depends on off-diagonal elements of the density matrix, ηN_0^2 depends only on diagonal elements. The equation of motion for the normalised (intensive) quantities η that describe the diagonal coherence and $\langle a_0 \rangle^2/N_0$ that describes the off-diagonal coherence are given by

$$\dot{\eta} = -2\frac{W_{\text{in}}\eta}{N_0} \quad \text{and} \quad \frac{\partial}{\partial t}\left(\frac{\langle a_0 \rangle^2}{N_0}\right) = -\frac{W_{\text{in}}}{N_0}\frac{\langle a_0 \rangle^2}{N_0}. \quad (8.57)$$

8.5.2 Discussion

The set of equations (8.55)–(8.57) is particularly simple and the meaning of the various terms is transparent. Equation (8.55) is inhomogeneous and is composed of a spontaneous scattering term and of a stimulated scattering term. The equations for the order parameter and for the quantity η are both homogeneous and governed by the stimulated terms. This means that if the ground state is initially empty or in a thermal state it will stay thermal forever, even if a large number of particles comes in. However, a coherent seed in the ground state can be amplified if stimulated scattering takes place. As we will see in the next section, this set of equations allows us to describe the transfer of a large number of incoherent reservoir particles within a ground state having a high degree of coherence. This means that the total coherence of the system can increase. However, the coherence degree of the ground state itself can only decay in time as one can see from eqns (8.57), except if the ground-state population becomes infinite, which can be the case only in infinite systems.

In the steady-state regime the equilibrium value for N_0 is

$$N_0(\infty) = \frac{W_{\text{in}}}{W_{\text{out}}(\infty) + \Gamma_0 - W_{\text{in}}(\infty)}, \quad (8.58)$$

and the system is in a thermal state with zero diagonal and off-diagonal coherence. The characteristic decay constant of the order parameter is called the phase diffusion coefficient that is equal to the emission linewidth of the ground state

$$D = \frac{1}{2}[W_{\text{in}}(\infty) - W_{\text{out}}(\infty) - \Gamma_0]a = \frac{W_{\text{in}}}{2N_0(\infty)}. \quad (8.59)$$

The energy broadening of the condensate is no longer given by the radiative lifetime, but by the balance between incoming and outgoing scattering rates. In the low-temperature limit W_{out} is small and $W_{\text{in}} \approx \Gamma_0$ that allows recovery of the well-known diffusion coefficient from laser theory

$$D \approx \frac{\Gamma_0}{2N_0}. \quad (8.60)$$

However, one should notice that the polariton–polariton interaction that is not included in the former equations plays an important role in the phase diffusion coefficient, as discussed by Laussy et al. (2006b) and in more detail later.

8.5.3 Coherence dynamics

We now give a qualitative analysis of the polariton condensate formation. The kinetics are characterised by a transient regime, during which the polaritons arrive to the condensate, after being excited in some $\mathbf{k} \neq \mathbf{0}$ state at $t = 0$. Their relaxation rate depends nonlinearly on the pumping intensity. For strong enough pumping the stimulated scattering of polaritons into the condensate flares up at a time $t > 0$, so that the in-scattering rate increases drastically and becomes greater than the out-scattering rate. In the time domain, where $W_{\text{in}}(t) > W_{\text{out}}(t) + \Gamma_0$, the solution becomes unstable. This instability allows the condensation to develop. The formation of the condensate with $t \neq 0$ implies breaking the symmetry of the system, which cannot happen spontaneously in the framework of the formalism used. Therefore, to study the possibility of coherence buildup, we introduce an initial seed (a coherent state with a small number of polaritons). This initial coherence can survive and be amplified for the high relaxation rates, as long as a time window exists in which $W_{\text{in}}(t) > W_{\text{out}}(t) + \Gamma_0$. After the steady-state regime is reached, the point $\langle a_0 \rangle$ becomes stable again, since the rates reach the time-independent values $W_{\text{in}}(\infty)$ and $W_{\text{out}}(\infty)$, with $W_{\text{out}}(\infty) + \Gamma_0 > W_{\text{in}}(\infty)$. However, the difference between the stationary rates is very small, inversely proportional to the system area, which corresponds to a large stationary number of condensed polaritons. Once a coherence is formed its decrease due to spontaneous scattering is slow in large cavities. Rubo et al. (2003) studied this mechanism and found two regimes: one below threshold, where the seed coherence is rapidly washed out and that of the entire system remains zero. The other regime, in contrast, leads to a coherent amplification for the seed, leading to a macroscopic number of particles populating the coherent fraction of the initially small seed. In this case, the whole system coherence strongly increases from 0 to more than 90%. At intermediate pumping densities the steady-state coherence degree depends noticeably on the seed's characteristics.

8.6 Kinetics of formation of polariton condensates: quantum picture beyond the Born–Markov approximation

8.6.1 Two-oscillator toy theory

In this section, we retain particle-number correlations between states, which allows to dispense from the seed of the previous section to describe coherence in the system. To gain insight into the mechanisms at work, we first revert to a toy model that reduces all the relevant physics to its bare minimum. Later, we present numerical results describing a realistic microcavity.

Since dimensionality is not an issue because it is not the accommodation of a population in phase space, but dynamical effects that are responsible for populating the ground state, we describe the system by a zero-dimensional two-oscillator model: one oscillator representing the ground state, the other an excited state (or assembly of excited states combined as a whole). We also neglect interparticle interactions, which will clearly show that efficient relaxation is required (conserving particle number), but that intrinsic interparticle interactions are not necessary. The number of polaritons in the entire system fluctuates, but we shall see that the correlations implied by conservation of polaritons in their relaxation are at the heart of our mechanism. We label the states as 1 and 2. There is only one parameter to distinguish them that is the ratio ξ of the rate of transitions $w_{1\to2}$ and $w_{2\to1}$ between these states:

$$\xi \equiv \frac{w_{2\to1}}{w_{1\to2}}. \tag{8.61}$$

These w are constants. In particular, they have no time dependence coming from the populations included in these scattering terms. We assume $\xi > 1$, which identifies state 1 as the ground state (i.e., state of lower energy), since from elementary statistics:

$$\frac{w_{2\to1}}{w_{1\to2}} = e^{(E_2 - E_1)/k_B T}, \tag{8.62}$$

with E_i the energy of state i (by definition of the ground state, $E_2 > E_1$) and T is the temperature of the system once it has reached equilibrium. The Hamiltonian for the two-oscillator system coupled through an intermediary oscillator (depicting a phonon) in the rotating wave approximation reads in the interaction picture

$$H = V e^{i(\omega_1 - \omega_2 + \omega)t} a_0 a_1^\dagger b + \text{h.c.}, \tag{8.63}$$

where a_1, a_2 and b are (time-independent) annihilation operators with bosonic algebra for first oscillator (ground state), second oscillator (excited state) and the "phonon" respectively, with free propagation energy $\hbar\omega_1$, $\hbar\omega_2$, $\hbar\omega$ and coupling strength V. Carrying through the same procedure as previously for ϱ, i.e., evaluating the double commutator gives

$$\partial_t \rho = -\tfrac{1}{2}[w_{1\to2}(a_1^\dagger a_1 a_2 a_2^\dagger \rho + \rho a_1^\dagger a_1 a_2 a_2^\dagger - 2a_1 a_2^\dagger \rho a_1^\dagger a_2)$$
$$+ w_{2\to1}(a_1 a_1^\dagger a_2^\dagger a_2 \rho + \rho a_1 a_1^\dagger a_2^\dagger a_2 - 2a_1^\dagger a_2 \rho a_1 a_2^\dagger)], \tag{8.64}$$

after the Markov approximation ($\varrho(\tau) \approx \varrho(t)$) and factorisation of the entire system density matrix ϱ into $\rho\rho_{\text{ph}}$, with ρ, ρ_{ph} the density matrices describing the two oscillators and the phonons, respectively. Of course at this stage the whole construct is very close to our previous considerations, but note that no Born approximation is made on ρ so that correlations between the two oscillators are fully taken into account. The transition rates are given by

$$w_{1\to2} = 2\pi|V|^2 \langle b^\dagger b\rangle / (\hbar\omega_2 - \hbar\omega_1), \tag{8.65a}$$
$$w_{2\to1} = 2\pi|V|^2 (1 + \langle b^\dagger b\rangle) / (\hbar\omega_2 - \hbar\omega_1). \tag{8.65b}$$

We now obtain from eqn (8.64) the equation for diagonal elements

$$p(n, m) \equiv \langle n, m|\rho|n, m\rangle, \tag{8.66}$$

where $p(n, m)$ is the joint probability distribution to have n particles in state 1 and m in 2. This equation reads

$$\partial_t p(n, m) = (n + 1)m[w_{1\to2}p(n + 1, m - 1) - w_{2\to1}p(n, m)]$$
$$+ n(m + 1)[w_{2\to1}p(n - 1, m + 1) - w_{1\to2}p(n, m)]. \qquad (8.67)$$

This equation for a probability distribution parallelling the Boltzmann equation is the quantum Boltzmann master equation (QBME) for the two-oscillator model.

The QBME can be derived rigorously from a microscopic Hamiltonian, as shown by Gardiner and Zoller (1997) in a series of papers on quantum kinetics. However, in the two-oscillator toy model, the physical picture is so straightforward that one hardly needs this approach (investigated below). In a similar spirit, Scully (1999) applied a fixed number of particles to the case of Bose condensation of atoms. Coming back to our two modes and focusing, for instance, on the first term on the right-hand side, one sees that it expresses how the probability can be increased to have (n, m) particles in states $(1, 2)$ through the process where starting from an $(n + 1, m - 1)$ configuration, one reaches (n, m) by transfer of one particle from state 1 to the other state. This is proportional to $n + 1$, the number of particles in state 1 and is stimulated by $m - 1$ the number of particles in state 2, to which we add one for spontaneous emission, whence the factor $(n + 1)m$. We repeat that $w_{1\to2}$ and $w_{2\to1}$ are constants and should not be confused with the *bosonic transition rate* defined as $w_{1\to2}(1 + m)$ and $w_{2\to1}(1 + n)$ to account in a transparent way for stimulation.

For all quantities Ω that pertain to a single state only, say the ground state (so we can write $\Omega(n)$), it suffices to know the reduced probability distribution for this state, i.e., for the ground state:

$$p_1(n) \equiv \sum_{m=0}^{\infty} p(n, m), \qquad (8.68)$$

and vice versa, i.e., for the excited state $p_2(m) \equiv \sum_{n=0}^{\infty} p(n, m)$. So that indeed

$$\langle \Omega(n) \rangle = \sum_{n=0}^{\infty} \sum_{m=0}^{\infty} \Omega(n)p(n, m) = \sum_{n=0}^{\infty} \Omega(n)p_1(n). \qquad (8.69)$$

We will soon undertake to solve exactly eqn (8.67), but to explain how coherence arises in the system, we first show that if we make the approximation to neglect correlations between the two states, i.e., if we assume the factorisation

$$p(n, m) = p_1(n)p_2(m), \qquad (8.70)$$

then the system at equilibrium never displays any coherence, i.e., in accord with our previous discussion, both states remain in a thermal state no matter the initial conditions, the transition rates or any other parameters describing the system. Indeed putting eqn (8.70) into eqn (8.67) and summing over m, we obtain

$$\partial_t p_1(n) = p_1(n + 1)w_{1\to2}(n + 1)(\langle m \rangle + 1)$$
$$- p_1(n)\big(w_{2\to1}(n + 1)\langle m \rangle + w_{1\to2}n(\langle m \rangle + 1)\big) \qquad (8.71)$$
$$+ p_1(n - 1)w_{2\to1}n\langle m \rangle,$$

with $\langle m \rangle \equiv \sum_m m p_2(m)$ the average number of bosons in state 2. At equilibrium the detailed balance of these two states gives the solution

$$p_1(n+1) = \frac{\langle m \rangle}{\langle m \rangle + 1} \frac{w_{2 \to 1}}{w_{1 \to 2}} p_1(n) . \tag{8.72}$$

The same procedure for state 2 yields likewise:

$$p_2(m+1) = \frac{\langle n \rangle}{\langle n \rangle + 1} \frac{w_{1 \to 2}}{w_{2 \to 1}} p_2(m) , \tag{8.73}$$

with the notational shortcuts

$$\theta \equiv \frac{\langle n \rangle}{\langle n \rangle + 1} , \qquad \nu \equiv \frac{\langle m \rangle}{\langle m \rangle + 1} , \tag{8.74}$$

eqns (8.72) and (8.73) read after normalisation

$$p_1(n) = (1 - \nu\xi)(\nu\xi)^n , \tag{8.75a}$$
$$p_2(m) = (1 - \theta/\xi)(\theta/\xi)^m , \tag{8.75b}$$

so that $\langle n \rangle \equiv \sum n p_1(n) = \nu\xi/(1 - \nu\xi)$, which inserted back into eqn (8.74) yields

$$\xi = \frac{\theta}{\nu} , \tag{8.76}$$

or, written back in terms of occupancy numbers and transition rates,

$$\frac{w_{2 \to 1}}{w_{1 \to 2}} = \frac{\langle n \rangle}{\langle n \rangle + 1} \frac{\langle m \rangle + 1}{\langle m \rangle} , \tag{8.77}$$

which give in eqns. (8.72) and (8.73)

$$p_1(n+1) = \frac{\langle n \rangle}{\langle n \rangle + 1} p_1(n) \quad \text{and} \quad p_2(m+1) = \frac{\langle m \rangle}{\langle m \rangle + 1} p_2(m) , \tag{8.78}$$

achieving the proof that both states are (exact) thermal states under the hypothesis (8.70) that we will now relax. This can still give rise to a regime where both states are thermal states, but also to another regime where the excited state (state 2) is still in a thermal state, but the ground state (state 1) is non-thermal (and in some limit, has the statistics of a coherent state). This is possible if one takes into account correlations between states. In our case these correlations come from the conservation of particle number, so that the knowledge of particle number in one state determines the number in the other state. In fact, observe how the QBME connects elements of $p(n, m)$ that lie on antidiagonals of the plane (n, m). One such antidiagonal obeys the equation

$$n + m = N, \tag{8.79}$$

where N is a constant, namely, the distance of the antidiagonal to the origin from the geometrical point of view and the number of particles from the physical point of view.

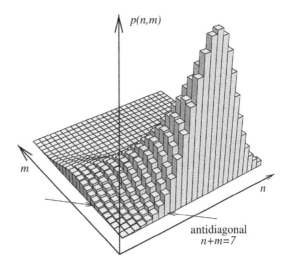

Fig. 8.8: $p(n,m)$ steady state solution from Laussy et al. (2004a) in the case where the distribution function for the number of particles in the entire system $P(N)$ is a Gaussian of mean (and variance) 15 and $\xi = 1.2$. One "antidiagonal", $n + m = 7$, is shown for illustration. The projection on the n-axis displays a coherent state, whereas the projection on the m-axis displays a thermal state.

One such antidiagonal is sketched in Fig. 8.8. The equation can be readily solved if only one antidiagonal is concerned, i.e., if there are exactly N particles in the system. In the other cases, one can still decouple the equation into its antidiagonal projections, solve for them individually and add up afterwards weighting each antidiagonal with the probability to have the corresponding particle number. We hence focus on one antidiagonal N whose *conditional* probability distribution is given by $d(n|N) \equiv p(n, N - n)$, with an equation of motion given by eqn (8.67) as

$$\partial_t d(n|N) = (n+1)(N-n)[w_{1\to2}d(n+1|N) - w_{2\to1}d(n|N)]$$
$$+n(N-n+1)[w_{2\to1}d(n-1|N) - w_{1\to2}d(n|N)]. \tag{8.80}$$

The equation is well behaved within its domain of definition $0 \le n \le N$ since it secures that $d(n|N) = 0$ for $n > N$. This also ensures uniqueness of solution despite the recurrence solution being of order 2, for $d(1|N)$ is determined uniquely by $d(0|N)$, itself determined by normalisation. The stationary solution is obtained in this way (or from detailed balance)

$$d(n+1|N) = \frac{w_{2\to1}}{w_{1\to2}}d(n|N), \tag{8.81}$$

with solution

$$d(n|N) = d(0|N) \left(\frac{w_{2\to1}}{w_{1\to2}} \right)^n, \tag{8.82}$$

where $d(0|N)$ is defined for normalisation as

$$d(0|N) = \frac{\xi - 1}{\xi^{N+1} - 1}, \qquad \xi \equiv \frac{w_{2\to1}}{w_{1\to2}}. \tag{8.83}$$

Technically solving for d resembles the procedure already encountered to solve the equation under assumption (8.70). However, we are now paying full attention to correlations between the two states, which turns detailed balancing of eqns (8.72) and (8.73) into one of an altogether different type, eqn (8.82). This gives by weighting eqn (8.82) the solution to the QBME

$$p(n, m) = \frac{\xi - 1}{\xi^{n+m+1} - 1} \xi^n P(n + m),$$ (8.84)

where

$$P(N) \equiv \sum_{n+m=N} p(n, m),$$ (8.85)

is the distribution of total particle number, i.e., the probability to have N particles in the *entire* system. $P(N)$ is time independent since the microscopic mechanism involved conserves particle number for any transition (one can also check that $\partial_t P(N) = 0$). This allows us to derive the statistics of separate states

$$p_1(n) = \xi^n \sum_{N=n}^{\infty} \frac{\xi - 1}{\xi^{N+1} - 1} P(N) \quad \text{and} \quad p_2(n) = \xi^{-n} \sum_{N=n}^{\infty} \frac{\xi - 1}{\xi^{N+1} - 1} \xi^N P(N).$$ (8.86)

Observe how the n dependence of the sum index prevents a trivial relationship between p_1 and p_2 of the kind $p_1(n) = p_2(N - n)$. Also, the asymmetry between ground and excited state is obvious from eqn (8.84). It is this feature that allows two states with drastically different characteristics, typically a thermal and a coherent state. Indeed, p_1 (resp. p_2) is the product of a sum with an exponentially diverging (resp. converging to zero) function of ξ. In both cases, the sum of positive terms is a decreasing function of n, so that clearly no coherence can ever survive in the excited state whose fate is always thermal equilibrium, or at least, in accord with our definitions,

$$p_2(n) > p_2(n + n_0) \quad \text{for all } n, n_0 \text{ in } \mathbf{N}.$$ (8.87)

For p_1, however, ξ^n diverges with n, what leaves open the possibility of a peak not centred about zero in this distribution, while it can still be a decreasing function if the sum converges faster still. It is to $P(N)$ to settle this issue, which as a constant of motion is completely determined by the initial condition. The solution for the case where $P(N)$ is a Gaussian of mean (and average) 15 is displayed in Fig. 8.8. $p(n, m)$ is in this case manifestly not of the type $p_1(n)p_2(m)$ and there is always coherence in the system. In the next section we investigate the more interesting situation where coherence does not exist a priori in the system.

8.6.1.1 *Growth of the condensate at equilibrium* By growth at equilibrium we mean that, still in the approximation of infinite lifetime, coherence can arise when one lowers the temperature, i.e., increases ξ, in a system where initially all states are thermal states. In this case the initial condition for the system is the thermal equilibrium

$$p(n, m) = (1 - \theta)(1 - \nu)\theta^n \nu^m,$$ (8.88)

where θ, ν are the thermal parameters for ground and excited states, respectively. They link to $\langle n \rangle$, the mean number of particles in the ground state, through

$$\langle n \rangle = \frac{\theta}{1 - \theta},\tag{8.89}$$

or, the other way around

$$\theta = \frac{\langle n \rangle}{1 + \langle n \rangle}.\tag{8.90}$$

Similar relations hold for ν and m. This is one possible steady solution of eqn (8.67) and we discuss how it arises from eqn (8.84) below. The thermal state is wildly fluctuating. Once in a while, thermal kicks transfer into the state one or many particles, which, however, do not stay for long before the state is emptied again or replaced by other, unrelated particles. This accounts for the chaotic, or incoherent, properties of such a state. This essentially empty, but greatly fluctuating statistics brings no conceptual problem for small populations, but one might enquire whether it is conceivable to have a thermal distribution with high mean number. This is possible for a single state, but not for the system as a whole. A macroscopic population can distribute itself in a vast collection of states so that each has thermal statistics, constantly exchanging particles with other states and displaying great fluctuations, but as expected from physical grounds, the whole system does not fluctuate greatly in its number of particles. Therefore, we expect $P(N)$, the distribution of particles in the *entire system*, to be peaked about a nonzero value, typically to be a Gaussian of mean and variance equal to N. In the pure Boltzmann case, this results from the central limit theorem since the total number of particles is the sum of a large number of uncorrelated random variables and thus is itself a Gaussian random variable. Remembering our previous definitions, this, however, does not qualify the system as a coherent emitter, since the statistics must refer to a *single state*, not to a vast assembly of differing emitters. Thus, not surprisingly, coherence arises when a *single* quantum mode models or copies features of a macroscopic system, typically its population distribution. The two-oscillator system, which is a rather coarse approximation to a macroscopic system, will, however, display very clearly this mechanism. In the limit where $\xi \gg 1$ it is already clear from eqn (8.86) that $p_1(n) \approx P(n)$, so that the statistics of the entire system indeed serves as a blueprint for the ground state (and it alone, excited states being always decreasing as already shown). At equilibrium, with two thermal states, the distribution for the whole system reads

$$P(N) = \sum_{n+m=N} p(n, m) = (1 - \theta)(1 - \nu)\frac{\theta^{N+1} - \nu^{N+1}}{\theta - \nu}.\tag{8.91}$$

This exhibits a peak at a nonzero value provided that

$$\nu + \theta > 1.\tag{8.92}$$

This criterion refers to a first necessary condition: there must be enough particles in the system. The fewest particles available so that eqn (8.92) is fulfilled, is two. This minimum required to grow coherence fits nicely with the Bose–Einstein condensation picture (one needs at least two bosons to condense). It is not a necessary condition, though; also the dynamical aspect is important as shown by the key role of ξ. Indeed, if the system is steady in configuration (8.88), ξ is not a free parameter, but is related to θ and ν by

$$\xi = \frac{\theta}{\nu}, \tag{8.93}$$

and in this case the distribution of the ground state

$$p_1(n) = (1 - \theta)(1 - \nu)\xi^n \sum_{N=n}^{\infty} \frac{\theta^{N+1} - \nu^{N+1}}{\xi^{N+1} - 1} \frac{\xi - 1}{\theta - \nu}, \tag{8.94}$$

reduces by straightforward algebra to $p_1(n) = (1 - \theta)\theta^n$, i.e., as should be for consistency, the ground state is in a thermal state, independently of the value of θ (i.e., no matter what is the number of particles in the ground state). This can come as a surprise, but it must be borne in mind that this two-oscillator model is an extreme simplification that cannot dispense from some pathological features, namely, the ability to sustain a thermal macroscopic population, an ability that we understand easily since the ground state accounts for half of the system! With increasing number of states, dimensionality forbids such an artifact. Also, the shape of $P(N)$ hardly resembles a Gaussian, but already in this limiting case it is able to display a peak at a nonzero value provided there are enough particles. With increasing number of states, the central limit theorem will turn this distribution into an actual Gaussian. Once again, $P(N)$ is time independent because the relaxation mechanism conserves particle number, which results in correlations between the two states. By increasing ξ to ξ', one might search new values of θ, ν, say θ', ν', so that $\theta/(1-\theta)+\nu/(1-\nu) = \theta'/(1-\theta')+\nu'/(1-\nu')$ (conservation of particle number) and $\xi' = \theta'/\nu'$. This is possible if one allows $P(N)$ to change, in which case the two new states are also thermal states. If $P(N)$ is constrained by correlations induced by strict conservation of particle number—so that the uncertainty is not shifted as the system evolves—then eqn (8.93) breaks down and this allows eqn (8.84) to grow a coherent state in the ground state. In the two-oscillator model, coherence grown out of thermal states is maximum in the limiting case $\xi \to 0$ for which the ground-state distribution reduces to $P(N)$ from eqn (8.91), i.e., $p_1(n) = (1-\theta)(1-\nu)(\theta^{n+1}-\nu^{n+1})/(\theta-\nu)$. This is obvious on physical grounds (one particular realisation is the one for which $w_{1\to2} = 0$ and thus with all particles eventually reaching the ground state) and reinforces our understanding of Bose condensation as the ground-state distribution function coming close to the macroscopic distribution, with complete condensation corresponding to identification of $p_1(n)$ with $P(N)$.

Laussy (2012) has extended the two-oscillator toy model to $M \in \mathbb{N}$ oscillators as excited state plus one for the ground state, with the approximation of a single relaxation rate from any excited state to the ground state, which would correspond, for instance, to relaxation from a Rayleigh circle of degenerate states. The analytical expression he obtains (not given here) confirms the mechanism detailed above, as shown in Fig. 8.9 where the number n_0 of bosons in the ground state is shown as a function of the inverse effective temperature ξ. As temperature is decreased, the particles distributed according to a thermal distribution gather in the ground state with Poissonian fluctuations. As a result, the $g^{(2)}$ goes from 2 to 1.

8.6.1.2 *Growth of the condensate out of equilibrium*   The previous case holds in an equilibrium picture and for that matter refers to coherence buildup in systems like cold atom BEC. To address the polariton laser case, it is necessary to extend the model with the additional complication of finite lifetime τ of particles in state 1, with a balance in the total population provided by a pump that injects particles in state 2 at a rate Γ. We still focus on the case of the two-oscillator toy model for simplicity (we will not crucially need a finite lifetime in the excited state and thus neglect it, which is a good approximation in microcavities, where the radiative lifetime in the exciton reservoir is two to three orders of magnitude longer than the polariton lifetime, state 2 representing the reservoir in our model.) Although the QBME can be readily extended phenomenologically to take these into account:

$$
\begin{aligned}
\dot{p}(n, m) = {}& (n+1)m[w_{1\to2}p(n+1, m-1) - w_{2\to1}p(n, m)] \\
& + n(m+1)[w_{2\to1}p(n-1, m+1) - w_{1\to2}p(n, m)] \\
& + \frac{1}{\tau}(n+1)p(n+1, m) - \frac{1}{\tau}np(n, m) \\
& + \Gamma p(n, m-1) - \Gamma p(n, m),
\end{aligned}
\tag{8.95}
$$

the couplings between different particle numbers forbid solving this new equation along the same analytical lines as previously, even for the simple case of two oscillators only. However, numerical solutions can be obtained straightforwardly. Introducing $\langle m \rangle_n$ the mean number of polaritons in the excited state given that there are n in the ground state, i.e.,

$$
\langle m \rangle_n\, p_0(n) = \sum_{m=0}^{\infty} mp(n, m),
\tag{8.96}
$$

with $p_0(n) \equiv \sum_m p(n, m)$ is the reduced ground-state statistics, we obtain, by averaging eqn (8.67) over excited states, an equation for the ground-state statistics only:

$$
\begin{aligned}
\partial_t p_1(n) = {}& (n+1)\big(w_{1\to2}(\langle m \rangle_{n+1} + 1) + 1/\tau\big)p_1(n+1) \\
& - \Big\{n\big((\langle m \rangle_n + 1)w_{1\to2} + 1/\tau\big) \\
& \quad + (n+1)\langle m \rangle_n w_{2\to1}\Big\}p_1(n) \\
& + n\langle m \rangle_{n-1}w_{2\to1}p_1(n-1).
\end{aligned}
\tag{8.97}
$$

However, in this out-of-equilibrium regime, the excited state is not as important as in the equilibrium case where it must be thermal and whose configuration is of utmost consequence on the ground state. Thus, we can dispense with the actual distribution of the excited state and simply use the mean $\langle m \rangle_n$ obtained from $\sum_m mp(n, m) =$

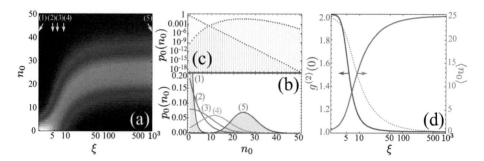

Fig. 8.9: Dynamics of condensation of an ideal Bose gas of $N = 25$ particles as described through QBME by Laussy (2012). In (a), the density plot $p_0(n_0)$ of particles in the ground state shows the evolution from a thermal distribution on the left (with $\langle n_0 \rangle = 0.3$ particles initially in the ground state with the monotically decreasing distribution shown in (c)) to a coherent distribution on the right (with $\langle n_0 \rangle$ close to 25 particles with peaked fluctuations also shown in (c)) as temperature is decreased. Slices at the points indicated by the arrows are shown in (b). In (d) $\langle n_0 \rangle$ (and $N - \langle n_0 \rangle$ dotted) are shown together with $g^{(2)}$ as temperature is decreased. Particles accumulate in the ground state as coherence grows.

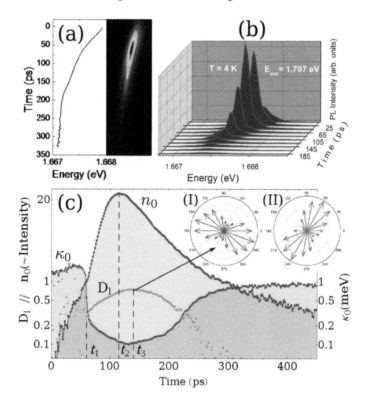

Fig. 8.10: From del Valle et al. (2009b). Time-resolved formation and decay of a polariton BEC after the arrival of a non-resonant pulse, as reported by del Valle et al. (2009a). (a) Emission energy from the ground state ($k = 0$) as a function of time (left) and the corresponding photoluminescence intensity (right) recorded by a streak camera. The colour code goes from black, no emission, to red, intense emission, saturating in the black central region. (b) Spectral line shapes extracted from (a). (c) In blue dots, average condensate population n_0 (normalised photoluminescence intensity); in purple squares, linewidth κ_0 (meV) of the emission peak, spectral-resolution limited resolution; in brown rhombus, degree of linear polarisation D_l. Three times are highlited: t_1 when the coherence build-up ignites, t_2 when the population is maximum and t_3 when the coherence is maximum. Inset: Polarisation polar plot at $t = t_3$ for two points in the sample. The polarisation pinning differs from defect to defect, proving indirectly the symmetry breaking.

$\langle m \rangle_n p_2(n)$. In this case eqn (8.95) can be decoupled to give an equation for $p_1(n)$ alone and in the "dynamical" steady-state, the detailed balance reads

$$p_1(n+1) = \frac{w_{2\to1}\langle m \rangle_n}{w_{1\to2}(\langle m \rangle_{n+1}+1)+1/\tau}p_1(n).\qquad(8.98)$$

Up to this point it is still exact, and also in the out-of-equilibrium regime we grant the conservation of particle number as the origin of correlations between the two states, but because of lifetime and pumping, it can now be secured only in the mean, leading us to the approximation

$$\langle m \rangle_n = N - n.\qquad(8.99)$$

The pump, which has quantitatively disappeared from the formula, is implicitly taken into account through this assumption, since even though particles have a finite lifetime, their number is constant on average. In the coherent case, $p_1(n)$ is a Poisson distribution with maximum at $N - N_c$, so that this dependency of N on the pump is in this case $N = \tau P + N_c$, with N_c the critical population defined by

$$N_c = \frac{1}{\tau(w_{2\to1} - w_{1\to2})},\qquad(8.100)$$

obtained from eqns (8.98) and (8.99) with the requirement that $p_1(1) > p_1(0)$. When the population has stabilised in the ground state by equilibrium of radiative lifetime and pumping, it is found in a coherent state if $N > N_c$. If this population is exceeded, coherence builds up in the system along with the population, which stabilises at an average given by the maximum of the Gaussian-like distribution:

$$\langle n \rangle = n_{\max} = N - N_c,\qquad(8.101)$$

obtained from $p_1(n) = p_1(n+1)$. Effectively if $N < N_c$ there is no such Gaussian and coherence remains low with a thermal-like state whose maximum is for zero occupancy. If $N > N_c$ the state is a Gaussian whose mean increases with increasing departure of population from the critical population. Thus, the higher the number of particles, the less the particle number fluctuations of the state and the better its coherence.

8.6.2 *Coherence of polariton laser emission*

The above toy model embeds the main ingredients than those to describe a more realistic microcavity, i.e., counting with a large number of states quantising a two-dimensional structure with a full dispersion and momentum-dependent relaxation rates, mediated, say, still by phonon-mediated scattering only (neglecting polariton–polariton scattering) for simplicity.

The dynamics is still of the master equation type in the Lindblad form for the density matrix ρ of polariton states in the reciprocal space: $\partial_t \rho = (\mathcal{L}_{\text{pol}-\text{ph}} + \mathcal{L}_\tau + \mathcal{L}_{\text{pump}})\rho$, accounting for scattering (through phonons), lifetime and pumping. The polariton–polariton scattering would add a unitary contribution $\mathcal{L}_{\text{pol}-\text{pol}} = -\frac{i}{\hbar}[H_{\text{pol}-\text{pol}}, \rho]$. In the following, we undertake the derivation of $\mathcal{L}_{\text{pol}-\text{ph}}$ from the microscopic Hamiltonian $H_{\text{pol}-\text{ph}}$ for polariton–phonon scattering. Exactly the same procedures can be

carried out for $\mathcal{L}_\tau$ and $\mathcal{L}_{\text{pump}}$ to yield $\mathcal{L}_\tau \rho = -\sum_{\mathbf{k}} \frac{1}{2\tau_{\mathbf{k}}}(a_{\mathbf{k}}^\dagger a_{\mathbf{k}}\rho + \rho a_{\mathbf{k}}a_{\mathbf{k}}^\dagger - 2a_{\mathbf{k}}\rho a_{\mathbf{k}}^\dagger)$
and $\mathcal{L}_{\text{pump}}\rho = -\sum_{\mathbf{k}} \frac{P_{\mathbf{k}}}{2}(a_{\mathbf{k}}a_{\mathbf{k}}^\dagger \rho + \rho a_{\mathbf{k}}^\dagger a_{\mathbf{k}} - 2a_{\mathbf{k}}^\dagger \rho a_{\mathbf{k}})$ with $\tau_{\mathbf{k}}$ the lifetime, $P_{\mathbf{k}}$ the pump intensity in the state with momentum $\mathbf{k}$ and $a_{\mathbf{k}}$ the Bose annihilation operator for a polariton in this state. Pumping, if incoherent, injects excitons well above the bottom of the bare exciton band, what is modeled by nonzero values of $P_{\mathbf{k}}$ for a collection of $\mathbf{k}$-states normally distributed about a high momentum mean value. This describes an incoherent pumping provided by a reservoir that pours particles into the system, but does not allow their return, thereby populating the system with incoherent polaritons, which relax towards the ground state where they can join in a coherent phase before escaping the cavity by spontaneous emission.

We give a more detailed derivation for $\mathcal{L}_{\text{pol-ph}}$ that contains the key aspects of the technique. We repeat here the interaction picture polariton–phonon scattering term from the polariton Hamiltonian

$$H_{\text{pol-ph}} = \sum_{\mathbf{k},\mathbf{q}\neq 0} \mathcal{V}_{\mathbf{q}} e^{\frac{i}{\hbar}(E_{\text{pol}}(\mathbf{k}+\mathbf{q})-E_{\text{pol}}(\mathbf{k})-\hbar\omega_{\mathbf{q}})t} a_{\mathbf{k}+\mathbf{q}} a_{\mathbf{k}}^\dagger b_{\mathbf{q}}^\dagger + \text{h.c.}, \qquad (8.102)$$

with $\mathcal{V}_{\mathbf{q}}$ the interaction strength, E_{pol} the lower-polariton-branch dispersion, $\hbar\omega_{\mathbf{q}}$ the phonon dispersion and $a_{\mathbf{q}}$ (resp. $b_{\mathbf{q}}$), the Bose annihilation operator for a polariton (resp. a phonon), in state $\mathbf{q}$. As previously, we obtain $\mathcal{L}_{\text{pol-ph}}$ starting with the Liouville–von Neumann equation for polariton–phonon scattering, $\partial_t \varrho = -\frac{i}{\hbar}[H_{\text{pol-ph}}, \varrho]$ where ϱ is the density matrix for polaritons and phonons. The polariton density matrix, is obtained by tracing over phonons, $\rho \equiv \text{Tr}_{\text{ph}}\varrho$, assumed as a reservoir in equilibrium with no phase coherence nor correlations with ρ. To dispense from this reservoir we write the equation for ϱ to order two in the commutator and trace it out, to get

$$\partial_t \rho(t) = -\frac{1}{\hbar^2} \int_{-\infty}^t \text{Tr}_{\text{ph}}[H_{\text{pol-ph}}(t), [H_{\text{pol-ph}}(\tau), \varrho]] \, d\tau. \qquad (8.103)$$

We define

$$\mathcal{E}_{\mathbf{k},\mathbf{q}}(t) \equiv \mathcal{V}_{\mathbf{q}} e^{\frac{i}{\hbar}(E_{\text{pol}}(\mathbf{k}+\mathbf{q})-E_{\text{pol}}(\mathbf{k})-\hbar\omega_{\mathbf{q}})t}, \qquad (8.104)$$

and for convenience we write

$$\mathcal{H}(t) \equiv \sum_{\mathbf{k},\mathbf{q}} \mathcal{E}_{\mathbf{k},\mathbf{q}}(t) a_{\mathbf{k}+\mathbf{q}} a_{\mathbf{k}}^\dagger b_{\mathbf{q}}^\dagger, \qquad (8.105)$$

so that $H_{\text{pol-ph}} = \mathcal{H} + \mathcal{H}^\dagger$. Operators are time independent. Because the phonon density matrix is diagonal, $[H_{\text{pol-ph}}(t), [H_{\text{pol-ph}}(\tau), \varrho]]$ reduces to $[\mathcal{H}(t), [\mathcal{H}^\dagger(\tau), \varrho(\tau)]]$+h.c., which halves the algebra. Also, the conjugate hermitian follows straightforwardly, so we are left only with explicit computation of two terms, of which the first reads

$$[\mathcal{H}(t), \mathcal{H}^\dagger(\tau)\varrho] = \sum_{\mathbf{k},\mathbf{q}\neq 0} \sum_{\mathbf{l},\mathbf{r}\neq 0} [\mathcal{E}_{\mathbf{k},\mathbf{q}}(t) a_{\mathbf{k}+\mathbf{q}} a_{\mathbf{k}}^\dagger b_{\mathbf{q}}^\dagger, \mathcal{E}_{\mathbf{l},\mathbf{r}}^*(\tau) a_{\mathbf{l}} a_{\mathbf{l}+\mathbf{r}}^\dagger b_{\mathbf{r}} \varrho(\tau)], \qquad (8.106)$$

which, taking the trace over phonons and calling $\theta_{\mathbf{q}}\rho \equiv \text{Tr}_{\text{ph}}(\varrho b_{\mathbf{q}}^\dagger b_{\mathbf{q}})$, becomes

$$\mathrm{Tr_{ph}}[\mathcal{H}(t), \mathcal{H}^\dagger(\tau)\varrho(\tau)]$$

$$= \sum_{k,l,q\neq 0} \mathcal{E}_{k,q}(t)\mathcal{E}^*_{l,q}(\tau)\theta_q(a_{k+q}a^\dagger_k a_l a^\dagger_{l+q}\rho(\tau) - a_l a^\dagger_{l+q}\rho(\tau)a_{k+q}a^\dagger_k). \quad (8.107)$$

Solving numerically this equation is a considerable task, which, however, has already been carried out for a similar equation by Jaksch et al. (1997), using quantum Monte-Carlo simulations. We prefer to make further approximations to reduce its simulation to a level of complexity of the same order as for the Boltzmann equations: we take into account correlations between the ground state and excited states only, neglecting all correlations between excited states. This is justified by the fast particle redistribution between excited states and their rapid loss of phase correlations. Physically this means that if a particle reaches the ground state, its absence is felt to some extent in the collection of excited states in a way that ensures particle number conservation. On the contrary, redistribution of particles between excited states obey the usual Boltzmann equations that pertain to averages only. Formally, we thus neglect terms like $\langle a_{k_1} a^\dagger_{k_2} a_{k_3} a^\dagger_{k_4}\rangle$ if k_i involve non-diagonal elements in the excited state. For non-vanishing terms, we further assume $\langle a_{k_1} a^\dagger_{k_1} a_{k_2} a^\dagger_{k_2}\rangle = \langle a_{k_1} a^\dagger_{k_1}\rangle\langle a_{k_2} a^\dagger_{k_2}\rangle$ if neither k_1 nor k_2 equal 0, while otherwise retaining the unfactored expression. Terms from eqn (8.107) featuring the ground-state are

$$\sum_{k\neq 0} \mathcal{E}_{k,-k}(t)\mathcal{E}_{k,-k}(\tau)^*\theta_k(a_0 a^\dagger_0 a^\dagger_k a_k\rho(\tau) - a_k a^\dagger_0\rho(\tau)a_0 a^\dagger_k) \quad (8.108a)$$

$$+ \sum_{k\neq 0} \mathcal{E}_{0,k}(t)\mathcal{E}_{0,k}(\tau)^*\theta_k(a^\dagger_0 a_0 a_k a^\dagger_k\rho(\tau) - a_0 a^\dagger_k\rho(\tau)a^\dagger_0 a_k). \quad (8.108b)$$

This expression (8.108) is one part of the term inside the time integral that gives $\partial_t\rho(t)$ evolution. Since

$$\mathcal{E}_{k,q}(t)\mathcal{E}_{k,q}(\tau)^* = |V_q|^2 \exp\left(-\frac{i}{\hbar}(E_{\mathrm{pol}}(k+q) - E_{\mathrm{pol}}(k) - \hbar\omega_q)(t-\tau)\right),$$

the time integration would yield a delta function of energy if ρ in eqn (8.108) was τ-independent. This delta would itself provide selection rules for allowed scattering processes through the sum over k. That $\rho(\tau)$ time evolution is slow enough compared with this exponential to allow this (Markov) approximation can be checked through evaluation of the phonon reservoir correlation time, which, when the reservoir has a broadband spectrum as in our case, is short enough to allow the approximation of $\rho(\tau)$ by $\rho(t)$. In this case, eqn (8.108b) vanishes as a non-conserving energy term. Gathering other terms similar to eqn (8.107) eventually gives (from now on we no longer write ρ time dependence, which is t everywhere)

$$\partial_t\rho = -\frac{1}{2}\sum_{k\neq 0} W_{0\to k}(a^\dagger_0 a_0 a_k a^\dagger_k\rho + \rho a^\dagger_0 a_0 a_k a^\dagger_k - 2a_0 a^\dagger_k\rho a^\dagger_0 a_k) \quad (8.109a)$$

$$-\frac{1}{2}\sum_{k\neq 0} W_{k\to 0}(a_0 a^\dagger_0 a^\dagger_k a_k\rho + \rho a_0 a^\dagger_0 a^\dagger_k a_k - 2a^\dagger_0 a_k\rho a_0 a^\dagger_k), \quad (8.109b)$$

where

$$W_{0 \to \mathbf{k}} \equiv \frac{2\pi}{\hbar} |\mathcal{V}_{\mathbf{k}}|^2 \theta_{\mathbf{k}} \delta(E_{\text{pol}}(\mathbf{k}) - E_{\text{pol}}(\mathbf{0}) - \hbar\omega_{\mathbf{k}}) \tag{8.110a}$$

$$W_{\mathbf{k} \to 0} \equiv \frac{2\pi}{\hbar} |\mathcal{V}_{\mathbf{k}}|^2 (1 + \theta_{\mathbf{k}}) \delta(E_{\text{pol}}(\mathbf{k}) - E_{\text{pol}}(\mathbf{0}) - \hbar\omega_{\mathbf{k}}) . \tag{8.110b}$$

We call $p(\{n_{\mathbf{k}}\})$ the diagonal of the polariton density matrix, i.e., the dotting of ρ with $|\{n_{\mathbf{k}}\}\rangle = |n_0, n_{\mathbf{k}_1}, \cdots, n_{\mathbf{k}_i}, \cdots\rangle$ the Fock state with $n_{\mathbf{k}_i}$ polaritons in state $\mathbf{k}_i$:

$$p(\{n_{\mathbf{k}}\}) \equiv \langle \cdots, n_{\mathbf{k}_i}, \cdots, n_{\mathbf{k}_1}, n_0 | \rho | n_0, n_{\mathbf{k}_1}, \cdots, n_{\mathbf{k}_i}, \cdots \rangle . \tag{8.111}$$

This is the probability that the system be found in configuration $\{n_{\mathbf{k}}\}$ whose equation of motion is the master equation obtained from eqn (8.109) as

$$\dot{p}(\{n_{\mathbf{k}}\}) = - \sum_{\mathbf{k}} (W_{0 \to \mathbf{k}} n_0 (n_{\mathbf{k}} + 1) + W_{\mathbf{k} \to 0} (n_0 + 1) n_{\mathbf{k}}) p(\{n_{\mathbf{k}}\})$$

$$+ \sum_{\mathbf{k}} W_{0 \to \mathbf{k}} (n_0 + 1) n_{\mathbf{k}} p(\{n_0 + 1, \ldots, n_{\mathbf{k}} - 1, \ldots\}) \tag{8.112}$$

$$+ \sum_{\mathbf{k}} W_{\mathbf{k} \to 0} n_0 (n_{\mathbf{k}} + 1) p(\{n_0 - 1, \ldots, n_{\mathbf{k}} + 1, \ldots\}) .$$

This is the counterpart to eqn (8.67), which also parallels closely the Boltzmann equation with which it shares the same transition rates (8.110) given by Fermi's golden rule and so it represents the QBME for polariton lasers. In its spirit, we do not solve it for the entire joint probability $p(\{n_{\mathbf{k}}\})$, but average over all excited states to retain the statistical character for the ground state only. Excited states will be described with a Boltzmann equation, thus with thermal statistics. Calling

$$p_0(n_0) \equiv \sum_{i=1}^{\infty} \sum_{n_{\mathbf{k}_i}=0}^{\infty} p(n_0, n_{\mathbf{k}_1}, n_{\mathbf{k}_2}, \cdots, n_{\mathbf{k}_j}, \cdots) , \tag{8.113}$$

the ground state reduced probability (the sum is over all states but the ground state, cf. eqn (8.68)), and

$$\langle n_{\mathbf{k}} \rangle_{n_0} p_0(n_0) \equiv \sum_{n_{\mathbf{k}_1}, n_{\mathbf{k}_2}, \ldots} n_{\mathbf{k}} p(\{n_{\mathbf{k}}\}) , \tag{8.114}$$

cf. eqn (8.96), we get the ground-state QBME equation

$$\dot{p}_0(n_0) = (n_0 + 1)(W_{\text{out}}^{n_0+1} + 1/\tau_0) p_0(n_0 + 1)$$

$$- \left(n_0(W_{\text{out}}^{n_0} + 1/\tau_0) + (n_0 + 1) W_{\text{in}}^{n_0} \right) p_0(n_0) \tag{8.115}$$

$$+ n_0 W_{\text{in}}^{n_0-1} p_0(n_0 - 1) ,$$

with rate transitions now a function of the ground-state population number n_0,

$$W_{\text{in}}^{n_0}(t) \equiv \sum_{\mathbf{k}} W_{\mathbf{k}\to\mathbf{0}}\langle n_{\mathbf{k}}(t)\rangle_{n_0}, \tag{8.116a}$$

$$W_{\text{out}}^{n_0}(t) \equiv \sum_{\mathbf{k}} W_{\mathbf{0}\to\mathbf{k}}(1 + \langle n_{\mathbf{k}}(t)\rangle_{n_0}), \tag{8.116b}$$

while for excited states, in the Born–Markov approximation, we indeed recover the Boltzmann equations

$$\langle \dot{n}_{\mathbf{k}}\rangle = \langle n_{\mathbf{k}}\rangle \sum_{\mathbf{q}\neq\mathbf{0}} W_{\mathbf{k}\to\mathbf{q}}(\langle n_{\mathbf{q}}\rangle+1) - (\langle n_{\mathbf{k}}\rangle+1)\sum_{\mathbf{q}\neq\mathbf{0}} W_{\mathbf{q}\to\mathbf{k}}\langle n_{\mathbf{q}}\rangle, \qquad \mathbf{k}\neq\mathbf{0}. \tag{8.117}$$

Inclusion of $\mathcal{L}_\tau$ and $\mathcal{L}_{\text{pump}}$ for the above adds $-\langle n_{\mathbf{k}}\rangle/\tau_{\mathbf{k}} + P_{\mathbf{k}}$ to this expression. Observe that in this case, the transition rates are constants.

Cast in this form, eqn (8.115) has the same transparent physical meaning in terms of a rate equation for the probability of a given configuration, much like the usual rate equations for occupation numbers in the framework of the Boltzmann equations. The difference is that transitions from one configuration to a neighbouring one occur at rates that depend on the configuration itself, through the population of the state. $\langle n_{\mathbf{k}}\rangle_{n_0}$ is a function of n_0 that we estimate through a first-order expansion about $\langle n_0\rangle$. This implies that fluctuations of excited states are proportional (with opposite sign) to fluctuations of the ground state:

$$\langle n_{\mathbf{k}}\rangle_{n_0} \approx \langle n_{\mathbf{k}}\rangle_{\langle n_0\rangle} + \left.\frac{\partial\langle n_{\mathbf{k}}\rangle_{n_0}}{\partial n_0}\right|_{\langle n_0\rangle}(n_0 - \langle n_0\rangle), \tag{8.118}$$

where $\langle n_{\mathbf{k}}\rangle_{\langle n_0\rangle}$ is given by a Boltzmann equation. Since the derivative does not depend on n_0 (it is evaluated at $\langle n_0\rangle$), we compute it by evaluation of both sides at a known value, for instance $n_0 = N$ with N the total particle number in the entire system, ground and excited states together. This gives

$$\left.\frac{\partial\langle n_{\mathbf{k}}\rangle_{n_0}}{\partial n_0}\right|_{\langle n_0\rangle} = \frac{\langle n_{\mathbf{k}}\rangle}{\langle n_0\rangle - N}, \tag{8.119}$$

since $\langle n_{\mathbf{k}}\rangle_N = 0$ (no particles are left in excited states when they are all in the ground-state). With the knowledge of eqn (8.118) and (8.119), which are known from semiclassical Boltzmann equations, this is now only a matter of numerical simulations.

8.6.3 Numerical simulations

Quantum Boltzmann master equations for polaritons have been numerically solved by Laussy et al. (2004b), with parameters of a CdTe microcavity excited by a light spot of $10\,\mu\text{m}$ lateral size with one quantum well and a Rabi splitting of $7\,\text{meV}$ at zero detuning. The spot size (the system size for the model) is an important parameter as correlations increase with decreasing size of the system. Scattering is mediated by a bath of phonons at a temperature of $6\,\text{K}$ and with a thermal gas of electrons of density $10^{11}\,\text{cm}^{-2}$ accounted for to speed up relaxation. This is below the exciton bleaching density. The

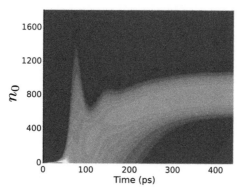

Fig. 8.11: Density plot of $p_0(n_0)$ (ground-state polaritons distribution) as a function of time for a realistic microcavity (with parameters given in the text), from Laussy et al. (2004b). Darker colours correspond to smaller values. Initial state is the vacuum. A coherent state is obtained in the timescale of hundreds of ps.

cavity is initially empty and pumped non-resonantly from $t = 0$ onwards. Figure 8.11 shows the density plot for $p_0(n_0)$ that results, starting from the vacuum. While the calculation is much more involved and the dynamics exhibits some specificities, qualitatively, the behaviour is very similar to the simple toy model of Fig. 8.9. In the realistic polariton case, the function $p_0(n_0)$, which at first varies wildly, quickly flattens with a large number of particles in a thermal state, then a nonzero maximum appears and the statistics evolves as Gaussian-like towards a Poisson distribution of mean $\langle n_0 \rangle$ in the steady-state. Consequently, coherence steadily rises in the system as more polaritons enter the ground state. Interestingly, even though the dynamics can give rise to a temporary decrease in the number of ground-state polaritons, the coherence does not drop in concert but continues its ascent. The polariton density of 10^{10} cm^{-2} achieved in the simulation of is more than one order of magnitude smaller than the strong/weak coupling transition density in CdTe.

8.6.4   *Order parameter and phase diffusion coefficient*

As a final remark on the mechanism of spontaneous coherence buildup, Laussy et al. (2004b) observe that if the system is spin degenerate, the polariton condensate forming in both spin components would result in a spontaneous buildup of the linear polarisation as well, due to the nonzero order parameter for each condensate ($\langle a_{0\uparrow}^\dagger \rangle$ for spin-up and $\langle a_{0\downarrow}^\dagger \rangle$ for spin-down) resulting in a well-defined value of their product, which is, for coherent states, equal to $\langle a_{0\uparrow}^\dagger a_{0\downarrow} \rangle$, the latter being linked to the degree of linear polarisation $D_l \equiv 2|\mathrm{Re}(\langle a_{0\uparrow}^\dagger a_{0\downarrow} \rangle)|/(\langle a_{0\uparrow}^\dagger a_{0\uparrow} \rangle + \langle a_{0\downarrow}^\dagger a_{0\downarrow} \rangle)$. This mechanism has been theoretically studied in more details by Laussy et al. (2006b), Shelykh et al. (2006) and Read et al. (2009) and was also used in experiments, by del Valle et al. (2009a), to track coherence under pulsed excitation with the aim of following the dynamics of formation and decay of a polariton BEC. This is shown in Fig. 8.10. This experiment was made possible by a pinning of the linear polarisation to a crystallographic axis, without which the symmetry breaking, as demonstrated by Baumberg et al. (2008), would result in averaging out the effect. The degree of linear polarisation D_l is directly related to the degree of second order coherence $D_l \approx \sqrt{2 - g^{(2)}(0)}$, so that it provides a direct measurement of $g^{(2)}$ without direct quantum-optical measurements. By detecting the time-resolved photoluminescence together with the ground state degree of linear polarisation, del Valle et al. (2009a) could thus monitor with a streak camera all the aspects of the the formation and decay of a polariton condensate following a circularly polarised pulsed excitation at high energy. The intensity and energy of the polariton ground state are reproduced in Fig. 8.10(a,b), along with the degree of linear polarisation D_l and the linewidth κ_0, shown in Fig. 8.10(c) (also with the total ground state population n_0, linked to the intensity). This is a neat observation of condensation through the several interconnected key variables that are population, coherence (here through polarisation, adding the phase locking between the spin-up and spin-down condensates) and linewidth. The time-resolved experiment also clearly shows how the ground-state energy is already blueshifted even when n_0 is small (at early times), confirming that the blueshift is mainly due to exciton screening and carrier-carrier interaction in the exciton reservoir. Finally, the experiment reveals that coherence (as measured by the polarisation) and the condensate populations are not matched in time, with the former lagging behind the latter. All this phenomenology is well accounted for by a QBME model similar to the one presented above, upgraded to take into account polarisation.

The factorisation $\langle a_{0\uparrow}^\dagger a_{0\downarrow} \rangle = \langle a_{0\uparrow}^\dagger \rangle \langle a_{0\downarrow} \rangle$ that supports the linking of polarisation with condensation is not strictly required as the former can be nonzero even when each $\langle a_{0\uparrow}^\dagger \rangle$ is zero and the former has in fact a more robust meaning and existence than the case of isolated order parameters. Still, the consideration of a well defined $\langle a_0 \rangle$ is an important and useful one as it connects directly with the phase of classical physics. The dynamics of the phase can in turn be related to several observables. It is ruled by the phase diffusion coefficient D, that is a decay constant for $\langle a_0(t) \rangle = \langle a_0(0) \rangle e^{i\omega_0 t} e^{-Dt}$ where ω_0 is the free propagation energy. This quantity has already been introduced in Chapter 6 and discussed in Section 8.5.2. It is related to the laser linewidth and inversely proportional to its coherence length. For normal lasers, this phase diffusion coefficient is due to the finite ratio that exists between spontaneous and stimulated emission of light. D is therefore inversely proportional to the number of photons in the lasing mode. This number is finite in any finite size system and realistic system.

For polariton lasers, as we have seen in Sections 8.3.1 and 8.3.2, this decay constant has, for non-interacting particles, the same physical origin as for normal lasers. It is due to the spontaneous scattering toward the ground state that provokes a dephasing. This dephasing is responsible for a broadening of the emission linewidth. The calculated emission spectra is a classical Lorentzian shape. This aspect has been widely discussed in laser theory. However, polaritons are interacting particles and this may lead to a substantial shift of the emission energy and to an increase of the linewidth. Formally, this shift and broadening originate from the terms $\sum_k V_k a_0^\dagger a_0 a_k^\dagger a_k$ of the Hamiltonian. These do not invoke any real scattering and they do not modify either the populations or the statistics. They are called *dephasing terms*. Their impact has been analysed by Porras and Tejedor (2003) in the steady-state and, as discussed previously, by Laussy et al. (2006b) on the polarisation. Porras and Tejedor considered the broadening of a thermal state. Figure 8.12 shows the full-width at half-maximum of the ground-state line versus

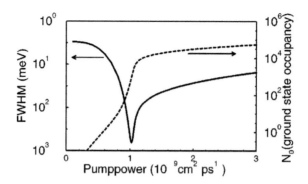

Fig. 8.12: Ground-state occupancy of a polariton laser versus pumping power (dashed line) and full-width at half-maximum of the ground-state emission (solid line), as predicted by Porras and Tejedor (2003).

pumping power. The width first decays above the threshold because of the decrease of dephasing induced by spontaneous scattering. However, the dephasing induced by interactions increases proportionally to the ground-state population and becomes dominant for large pumping powers. In what follows we give an intuitive derivation that gives the same results as the quantum formalism developed by Laussy et al. (2006b).

We consider again a polariton system with N particles in total and N_0 in the ground state. We also define $N_1 = N - N_0$ the number of particles in the excited states. The interparticle interaction constant is fixed and given by V. Therefore, the average interaction energy of the ground state is VNN_0. The energy per particle is equal to the energy shift of the line:

$$E_{\text{shift}} = \frac{VNN_0}{N_0} = VN.$$ (8.120)

This shift is a rigid shift of the complete dispersion, if one neglects the dependence of V on the wavevector. N, however, is a fluctuating quantity governed by a statistical distribution. The uncertainty of the number of particles leads to an uncertainty in the energy. We are going to separate the uncertainty from N_1 and the uncertainty from N_0. If one considers a large ensemble of independent states, all having thermal statistics and a small average number of particles, the statistics of the total number of particles in this system is Poissonian. This is a consequence of the central limit theorem. Therefore, the number of particles in excited states follows a Poissonian distribution $P_1(n_1)$. The root mean square of this distribution is $\sigma_1 = \sqrt{N_1}$. The emission line resulting from the interaction between the ground state and the excited states is therefore Poissonian with a width at half-maximum of $V2\sqrt{2\ln 2}\sqrt{N_1}$. On the other hand, the uncertainty of the number of particles in the condensate is sensitive to the statistics of exciton-polaritons in this state. For a state having the thermal statistics, $\sigma_0 = \sqrt{N_0^2 + N_0} \approx N_0$. The associated line has the shape of a thermal distribution, namely exponential on the high-energy tail and with an abrupt cut-off corresponding to the case $n_0 = 0$. Such statistics yields an asymmetric broadening as shown in Fig. 8.12. The total line is in this case a superposition of two very different functions. If, however, the ground state is coherent the uncertainty in the number of particles for the coherent statistics is only $\sigma_0 = \sqrt{N_0}$. In general, the condensate is in a mixed state having both coherent and thermal fractions. Its statistics is described by the second-order coherence function $g^{(2)}(0)$, which varies from 2, for the thermal state, to 1 for the coherent state. The root mean square of such statistics is given by $\sigma_0 = \sqrt{N_0 + N_0^2(g^{(2)}(0) - 1)}$. If $g^{(2)}(0)$ is close enough to 1, the associated statistics is well described by a Gaussian. Making the assumption that the two random variables N_0 and N_1 are not correlated, the statistics of their sum is given by a Gaussian of mean $N_0 + N_1$ and of root mean square $\sigma = \sqrt{\sigma_0^2 + \sigma_1^2}$. Now, if we neglect the spontaneous broadening that is very small if N_0 is large, the linewidth of the emission line is

$$\Delta E \approx V 2\sqrt{2 \ln 2}\sqrt{N_1 + N_0 + N_0^2(g^{(2)}(0) - 1)}. \qquad (8.121)$$

Now, we suppose that N_0 can be measured experimentally. We use eqn (8.120) to get N_1 as a function of N_0 and E_{shift}. The knowledge of these two quantities—V combined with a measurement of ΔE—finally allows one to deduce the second-order coherence as

$$g^{(2)}(0) = 1 + \frac{1}{8 \ln 2}\left(\frac{\Delta E}{V N_0}\right)^2 - \frac{E_{\text{shift}}}{V N_0^2}. \qquad (8.122)$$

This formula has a weak precision if σ_1 is larger than σ_0. In the opposite limit it represents a clear and direct way to measure $g^{(2)}(0)$. This technique can only be applied to the case of interacting particles. It is specific to the polariton system as opposed to a photonic system.

8.7 Spatial dynamics of polariton condensates

8.7.1 *Gross–Pitaevskii equation*

In the previous sections, we considered the Boltzmann and quantum dynamics of a homogeneous system, with full spatial coherence. In this Section, we address the spatial dynamics of polariton condensates. Experimental results obtained since 2005 have revealed that spatial dynamics is a crucial aspect of polariton condensation, in particular as it manifests through the first-order spatial coherence. A powerful theoretical tool typically used in equilibrium systems is the Gross–Pitaeskii equation (GPE) for the condensate wavefunction $\psi(r, t)$, that we introduced in previous chapters:

$$i\hbar\frac{\partial\psi(r, t)}{\partial t} = -\frac{\hbar^2}{2m}\nabla^2\psi(r, t) + \alpha|\psi(r, t)|^2\psi(r, t). \qquad (8.123)$$

Here, m is the particle mass and α is the interaction constant. Indeed, in the presence of a homogeneous propagating condensate $\psi_0 e^{i\mathbf{k}_0 \cdot \mathbf{r}} e^{-i\mu t}$, single plane-wave excitations are not eigenstates of the GPE (contrary to the linear Schrödinger equation). The plane-wave terms $e^{i\mathbf{k}\cdot\mathbf{r}}$ and $c^{i(2\mathbf{k}_0 - \mathbf{k})\cdot\mathbf{r}}$ interfere due to the nonlinear term $\alpha|\psi(r, t)|^2\psi(r, t)$. As a result, the dispersion of the true elementary excitations (bogolons), containing both complex exponents, is linear at low k, verifying the Landau criterion of superfluidity as previously discussed. The state $\psi_0 e^{i\mathbf{k}_0 \cdot \mathbf{r}} e^{-i\mu t}$ is thus protected against excitations despite not being the ground state of the system.

Nevertheless, there is still not a fully satisfactory model to describe all the aspects of condensation in an open-dissipative system. The need for such a model arose long ago, since all superfluids, including liquid Helium or atomic BECs, are only approximately conservative. A fundamental aspect of BECs and superfluids is the presence of a normal (non-condensed) fraction. This aspect has been first considered by Landau (1941) who developed a two-fluid hydrodynamic theory for liquid ^{4}He. The two-fluid model is only valid when the collisions between the excitations occur so often that they lead to thermalisation within the gas of the excitations. The evaporation of helium is simply neglected in this model and the conservation of the total number of particles is thus required. Pitaevskii (1959) introduced a phenomenological damping term in eqn (8.123) for the superfluid fraction, which drives the system towards a stationary state. This damping accounts for the decay of the excitations of the superfluid fraction due to various thermalisation mechanisms. The modified GPE is obtained by transformation of the Hamiltonian as $H' = (1 - i\Lambda) H$ where Λ is a small dimensionless parameter controlling the strength of dissipation. Pitaevskii used this Hamiltonian in the hydrodynamic limit, where the modification of the local density associated with bogolons is averaged out, since their density is very high. The dissipation term in this simplest approximation is therefore linearly proportional to the deviation of the energy from the equilibrium value (which is the chemical potential). This dissipative model has later been used to describe relaxation processes in atomic condensates by Choi et al. (1998), beyond the limits of the hydrodynamic approximation. One of the most representative finite temperature particle-conserving theories developed later for atomic BECs is the kinetic theory of Griffin (1996) and Hutchinson et al. (1997), based on the equation for the condensate within the time-dependent Hartree–Fock–Popov approximation, where the dynamics of the condensate wavefunction is coupled to the thermal gas. The latter is described in terms of a semi-classical Boltzmann equation for the distribution function. Two-fluid hydrodynamic theories are successful in describing some finite-temperature features of atomic BECs, as shown by Zaremba et al. (1998), for instance.

8.7.2 *Modified Gross–Pitaevskii equations*

A proper description of BEC in an open dissipative system, where the particle density results from a balance between pumping and decay, such as polariton BECs, requires to include the two above-mentioned ingredients: the relaxation and dissipation processes for both the condensed and normal parts. Most of the efforts have been put into models where an equation describing a pumped reservoir is coupled to a GPE-like equation first by Wouters and Carusotto (2007) and Keeling and Berloff (2008). These phenomenological models, which we will refer to as "diffusive" models, became popular and have been developed and used under several versions, for instance by Wouters and Carusotto (2010), Wouters et al. (2010), Larré et al. (2012), Chiocchetta and Carusotto (2013), Altman et al. (2015) and others. Within these models, the condensate dynamics can be mostly described in two different forms, known as i) the diffusive Ginzburg–Landau model:

$$i\hbar\frac{\partial\psi}{\partial t} = -\frac{\hbar^2}{2m}\nabla^2\psi + \alpha|\psi|^2\psi - i\Lambda\alpha\left(|\psi|^2 - n_c\right)\psi, \qquad (8.124)$$

and ii) the diffusive Pitaevskii model:

$$i\hbar\frac{\partial\psi}{\partial t} = (1 - i\Lambda)\left(-\frac{\hbar^2}{2m}\nabla^2\psi + \alpha|\psi|^2\psi - \mu\right)\psi. \qquad (8.125)$$

As we shall see, these models provide a diffusive dispersion for the elementary excitations of a condensate at rest, which does not fulfill the Landau criterion of superfluidity and can show some instabilities. As a result, disorder and defects provoke a finite dissipation of a flow (friction), even for velocities below the condensate speed of sound. The use of these models therefore leads to the conclusion that a condensate in an open dissipative system must not be superfluid according to Landau's criterion, even if their wavefunction exhibit long-range order. A third form of modified GPE, where the coupling between the condensate is written in reciprocal space using semi-classical Boltzmann rates, has been derived by Solnyshkov et al. (2014) and is called the hybrid Boltzmann Gross–Pitaevskii equation. It yields, in the limit of fast relaxation rates, to the following equation for the condensate:

$$i\hbar\frac{\partial\psi}{\partial t} = -(1 - i\beta)\frac{\hbar^2}{2m}\nabla^2\psi + \alpha\left(|\psi|^2 + n_R\right)\psi, \qquad (8.126)$$

where β is a parameter that mostly depends on the temperature and exciton density, with an exact expression that is given by Solnyshkov et al. (2014). The key difference with respect to the diffusive models, is that the damping term acts only on the kinetic energy and not on the interaction energy. Next, we briefly summarise what type of excitation spectrum is found for a condensate described by each of the above equations and what is the resulting first order coherence. As noted earlier in this section, the nonlinear term of the GPE couples, at first order, a plane wave with its complex conjugate, which forces us to look for elementary excitations of the condensate of the form

$$\psi(\mathbf{r}, t) = \psi_0 e^{-i\mu t}\left\{1 + A_{\mathbf{k}}\exp\left(i\left(\mathbf{k}\cdot\mathbf{r} - \omega_{\mathbf{k}}t\right)\right) + B_{\mathbf{k}}^*\exp\left(-i\left(\mathbf{k}\cdot\mathbf{r} - \omega_{\mathbf{k}}^*t\right)\right)\right\},$$
$$(8.127)$$

such that $n(\mathbf{k} = 0, t) = n_0 = |\psi_0|^2$ is the condensate density. This wavefunction can then be introduced in the different versions of the GPE, keeping only the first order terms in A_k and B_k. This approach allows one to obtain the dispersion of these excitations ω_k and the A_k and B_k coefficients using the normalisation condition $|A_k|^2 - |B_{-k}|^2 = 1$. This condition physically means that $|A_k|^2 + |B_{-k}|^2$ particles are scattered out from the condensate to become $|A_k|^2$ particles of energy ω_k and $|B_k|^2$ of energy $-\omega_k$. In total, with the above mentioned normalisation condition, the process involves an energy ω_k: the energy of a bogolon.

8.7.3 *Bogolon dispersion*

Applied to the conservative GPE, the resulting bogolon dispersion is similar to that found in the previous sections by direct diagonalisation of the Hamiltonian, and which fulfills the Landau criterion of superfluidity. This procedure can also be applied to condensates put in motion described by a wavefunction $\psi_0 e^{i\mathbf{k}_0 \cdot \mathbf{r}}$. This wavefunction is not

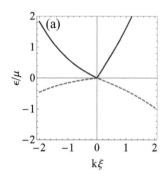 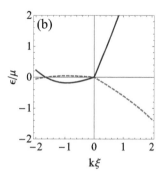

Fig. 8.13: Bogoliubov spectra calculated using the conservative GP equation in (a) the subsonic case and (b) the supersonic case, from Solnyshkov et al. (2014).

that of the ground state, which obviously corresponds to having all particles at rest. The Bogolon dispersion in such a case becomes

$$\hbar\omega_k = \hbar\mathbf{k}_r \cdot \mathbf{v}_0 + \sqrt{\left(\frac{\hbar^2 k_r^2}{2m}\right)^2 + 2\alpha n_c \frac{\hbar^2 k_r^2}{2m}}, \qquad (8.128)$$

where $\mathbf{k}_r = \mathbf{k} - \mathbf{k}_0$. The corresponding dispersions are shown as solid lines in Fig. 8.13. Panel (a) shows the superfluid case, when the condensate velocity is smaller than c_s, the speed of bogolons, also called the speed of sound. The depletion of the flow by elastic scattering and real phonon emission is entirely suppressed, if one considers only single particle depletion. The flow is metastable and therefore superfluid. Some depletion of the condensate takes place because of quantum fluctuations and absorption of phonons. This depletion leads to a finite occupation of the bogolon states and to a spatial decay of the 1st-order coherence of the condensate as we will see in next sections. Panel (d) shows the dispersion for a supersonic flow. Bogolon states with zero energy do exist and allow depletion of the condensate by resonant elastic scattering. Also some bogolons have a negative energy that provokes an energetic instability of the propagating condensate.

The Bogoliubov dispersion obtained using the hybrid Boltzmann Gross–Pitaevskii equation reads

$$\hbar\omega_k = -i\Lambda\left(\alpha n_c + \frac{\hbar^2 k^2}{2m}\right) \pm \sqrt{-\alpha^2 n_c^2 \Lambda^2 + \alpha n_c \frac{\hbar^2 k^2}{m} + \frac{\hbar^4 k^4}{4m^2}}. \qquad (8.129)$$

This is just a damped Bogoliubov mode as shown in Fig. 8.14a. This should be compared to the bogolon disperion using the diffusive Pitaevskii model, which reads

$$\hbar\omega_k = -i\beta\frac{\hbar^2 k^2}{2m} + \sqrt{\left(\frac{\hbar^2 k^2}{2m}\right)^2 + 2\alpha n_c \left(\frac{\hbar^2 k^2}{2m}\right)}. \qquad (8.130)$$

The corresponding dispersion is shown in figure 8.14b. The dispersion at low k is diffusive and the corresponding system is not a superfluid.

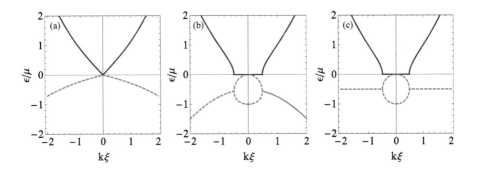

Fig. 8.14: Bogoliubov spectra calculated by Solnyshkov et al. (2014) using: (a) the Hybrid Boltzmann–Gross Pitaevskii equation, (b) the diffusive Pitaevskii model and (c) the diffusive Ginzburg Landau model. The solid lines show the real part of the energy and the dashed lines the imaginary part.

8.7.4   *Spatial coherence. The thermal fluctuation effect*

To calculate the spatial coherence, one considers the wavefunction as being a superposition of the condensate wavefunction and a thermal mixture of bogolons. The first order coherence is then calculated as $g_1(r) = \langle \psi(r)^* \psi(0) \rangle$, the mean value being taken on the statistical mixture. In order to make explicitly the calculation, the condensate wavefunction is written as $\psi(\mathbf{r}, t) = \sqrt{n(\mathbf{r}, t)} e^{iS(\mathbf{r}, t)}$ and the coherence decay is considered as a consequence of phase fluctuations. The quantity of importance is the mean value of the condensate wavefunctions at low k, which governs the coherence at long distances and which reads $\langle \psi_k \psi_k^* \rangle = N_b(k) \left(|A_k|^2 + |B_k|^2 \right)$ where $N_b(k)$ is the bogolon occupation. In a conservative system, the bogolon occupation is just given by a Bose distribution function. In an open system, it has to be found as the solution of a kinetic equation. It turns out that the wavevector dependence of $\langle \psi_k \psi_k^* \rangle$ is $1/k^2$ for all models, as shown by Solnyshkov et al. (2014) (though with different prefactors). Then the first order coherence depends only on dimensionality, exhibiting an exponential decay in 1D and a power law decay in 2D. The different shapes of the bogolon dispersion, however, affect the stability of the condensate against different types of perturbation. The dynamical stability relates to the imaginary part of the bogolon energy. If it becomes positive, it means that the corresponding mode amplifies with time, which will lead to a condensate instability. The energetic instability occurs if bogolon modes with negative energies are present, which again will lead to a decay of the condensate. The superfluid behaviour relies on the decay of the flow induced by disorder scattering.

As one can see in Table 8.1, the stability properties that are extracted are different for all the models. A specific of the diffusive models is that they do not describe a superfluid because the dissipation term acts on the nonlinear term. So in principle these models are unable to describe evaporating superfluid Helium for instance.

	Conservative	Open hybrid BGP	Open diffusive Ginzburg–Landau	Open diffusive Pitaevskii
quasicondensation temperature range	$T < T_c^d$	$T_{c_1} < T < T_{c_2}$	Not applicable	Not applicable
$N_b(k)$	$1/k$	$1/k$	$1/k^2$	$1/k^2$
$\langle \psi_k \psi_k^* \rangle$	$1/k^2$	$1/k^2$	$1/k^2$	$1/k^2$
$g_1^{1D}(r)$ for $v_c = 0$	$\exp(-r/r_0)$ $r_0 = \dfrac{2n_c\hbar^2}{mk_BT}$	$\exp(-r/r_0')$ $r_0' = r_0\left(1 + \dfrac{\hbar W_b}{\beta k_BT}\right)$	$\exp(-r/r_0'')$ $r_0'' = \dfrac{n_c\hbar^2}{mW_b\hbar\Lambda}$	$\exp(-r/r_0'')$ $r_0'' = \dfrac{n_c\hbar^2}{mW_b\hbar\Lambda}$
$g_1^{2D}(r)$ for $v_c = 0$	$\left(r_0^{2D}/r\right)^{\nu}$ $r_0^{2D} = \dfrac{\hbar c_s}{k_BT}$ $\nu = \dfrac{mk_BT}{2\pi\hbar^2 n_c^{2D}}$	$\left(r_0^{2D}/r\right)^{\nu'}$ $r_0^{2D} = \dfrac{\hbar c_s}{k_BT}$ $\nu' = \nu\dfrac{W_b}{W_b + \beta k_BT/\hbar}$	$\left(r_0'^{2D}/r\right)^{\nu''}$ $r_0'^{2D} = \dfrac{\hbar c_s}{2mW_b\hbar\Lambda}$ $\nu'' = \dfrac{mW_b\hbar\Lambda}{\pi\hbar^2 n_c^{2D}}$	$\left(r_0'^{2D}/r\right)^{\nu''}$ $r_0'^{2D} = \dfrac{\hbar c_s}{2mW_b\hbar\Lambda}$ $\nu'' = \dfrac{mW_b\hbar\Lambda}{\pi\hbar^2 n_c^{2D}}$
dynamic stability for $v_c < c_s$	Yes	Yes	Yes	Yes
for $v_c > c_s$	Yes	No	Yes	No
energic stability for $v_c < c_s$	Yes	Yes	No	Yes
for $v_c > c_s$	No	No	No	No
superfluid for $v_c < c_s$	Yes	Yes	No	No
for $v_c > c_s$	No	No	No	No

Table 8.1 Summary of the main condensate features described by the conservative, hybrid Boltzmann Gross–Pitaevskii and diffusive models, from Solnyshkov et al. (2014).

8.8 Experiments on Bose–Einstein condensation, superfluidity and lasing of polaritons

8.8.1 *Experimental observation*

We complete this chapter by discussing some experimental findings, reporting a positive outcome to the quest for solid state BEC. Indeed, a number of groups have found that thermalisation of polaritons on the lower-polariton branch is possible in the conditions of a positive cavity detuning. In II-VI microcavities, Kasprzak et al. (2006) could obtain a direct fitting of the experimentally observed polariton occupations using the Bose–Einstein distribution above and below threshold, as seen in Fig. 8.15 and show that this is different from Maxwell–Boltzmann statistics, i.e., evidencing strongly the bosonic nature of polaritons. In addition, it is clear that despite the strong spatial fluctuations and disorder-induced localisation of the emission, the different emitting "hot" spots are coherently locked in phase, i.e., they are part of the same state. The transition to this polariton BEC state is found around 20 K, where the polariton temperature is close to the lattice temperature. Similarly, in III-V microcavities, evidence has emerged that polariton BECs are indeed thermalised at low temperatures, whether for positive cavity detuning when excited at resonant energies of the excitons, as done by Deng et al. (2006), or using localised stress to form a real-space trap for polaritons, as done by Balili et al. (2006) and Balili et al. (2007). All these experiments show the appearance of bosonic-induced coherence in the semiconductor microcavity system in the strong coupling regime and pave the way to coherent matter–wave monolithic devices.

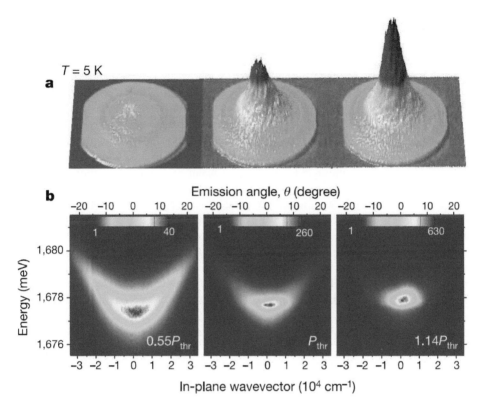

Fig. 8.15: Far-field angular emission below and above threshold showing the formation of a polariton Bose–Einstein condensate, as claimed by Kasprzak et al. (2006).

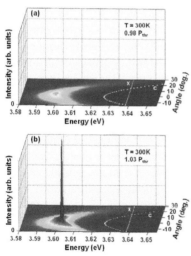

Fig. 8.16: Room temperature polariton lasing in an In-GaN/GaN microcavity, as reported by Christmann et al. (2008). The polariton distribution as a function of wavevector and energy is shown (a) below and (b) above the threshold for lasing.

8.8.2   *Polariton lasing vs Bose–Einstein condensation*

In the late 90s, polariton lasing was observed by Le Si Dang et al. (1998) in a CdTe-based microcavity and by Senellart and Bloch (1999) in GaAs-based microcavities. Polariton lasing at room temperature has been reported by Christopoulos et al. (2007) and Christmann et al. (2008) in wide-band gap microcavities based on GaN, as shown in Fig. 8.16. These works suggest that BEC of exciton-polaritons may take place in the studied samples and polariton lasing is one of its indications. Polariton lasing implies appearance of a macroscopically occupied coherent polariton state that can be called a "condensate". On the other hand, polariton lasing does not imply thermal equilibrium in the system and is not necessarily a result of a phase transition like BEC. Therefore, the observation of polariton lasing is not a sufficient condition to claim BEC. In works that followed, a number of additional criteria have been proposed including non-resonant excitation to ensure that the condensate is formed spontaneously, the thermal distribution of exciton-polaritons in the excited states, the spatial coherence of the condensate and spontaneous vector polarisation buildup. Strictly speaking, each of these criteria for polariton BEC can be criticised. The thermal distribution of exciton-polaritons can only be demonstrated at very low wavevectors, inside the "polariton trap". On the other hand, a huge sub-population of the polaritons remains in the exciton-like part of the dispersion, probably, out of thermal equilibrium. The spatial coherence can be checked only within the spot occupied by the condensate, whose size is usually not more than 30–50 μm. Measurements within such a small spot can hardly allow one to verify if the Onsager–Penrose criterion of BEC (spatial correlator having a finite value for the infinite distance) is satisfied. The polarisation criterion[113] may not work because of the pinning of linear polarisation, as discussed in detail in the next chapter. The question whether the observed effects manifest a BEC phase transition or not can be discussed indefinitely and, probably, does not make much sense. What really matters is the experimentally documented spontaneous formation of a coherent ensemble of exciton-polaritons at high temperatures. Be it polariton lasing or Bose–Einstein condensation, it is important, though, that the system remain in strong-coupling and that polaritons are involved. Among many positive reports, let us finish with the word of caution of Bajoni et al. (2007), who observe similarities with Bose–Einstein condensation in a system that is in weak-coupling. The authors suggest the observation of two thresholds to prove that the first one is associated with polaritons.

8.8.3   *Polariton diodes*

In 2007–2008, three groups have demonstrated electrically pumped polariton light-emitting diodes, with reports from Bajoni et al. (2008a), Khalifa et al. (2008) and Tsintzos et al. (2008). In these experiments a microcavity has been embedded in a p-i-n junction, and injection of carriers has been realised through p- and n-doped Bragg mirrors. The strong-coupling regime has been maintained up to 220 K by Tsintzos et al. (2008), while no threshold to polariton lasing has been observed.

---

[113] Spontaneous symmetry breaking in a GaN microcavity at room temperature has been evidenced by single-shot polarisation measurements under non-resonant pulsed pumping by Baumberg et al. (2008). This observation may be considered a signature of BEC of exciton-polaritons at room temperature.

8.8.4   *Experiments on superfluidity*

The experimental quest for polariton superfluidity naturally started as soon as their Bose–Einstein condensation was reported. Several groups started to look for features of superfluidity in these systems that, hosting a two-dimensional gas of weakly interacting bosons, are susceptible to undergo a Kosterlitz–Thouless phase transition and form a superfluid. Polariton lasing would still be observed in this case as a superfluid is a macroscopically coherent quantum state and in the case of polaritons, is therefore expected to emit coherent light. The question thus remains how one can distinguish experimentally between a two-dimensional superfluid and a localised Bose–Einstein condensate (or quasicondensate)? Among specific features of superfluidity, one should mention zero viscosity and, as a consequence, suppression of scattering by defects, formation of quantised vortices and appearance of a linear long-wavelength part in the dispersion of quasiparticles. All these features have been recently observed in exciton-polariton fluids in microcavities.

In particular, one landmark of superfluidity—quantised vortices—has been reported, first in non-resonantly pumped condensates in CdTe microcavities, by Lagoudakis et al. (2008), and later also half-quantum vortices, by Lagoudakis et al. (2009).[114] Vortices have also been found under resonant excitation, by Sanvitto et al. (2010) with a pulse (allowing the study of their dynamics, in particular those multiply charged) and with continuous excitation, by Krizhanovskii et al. (2010). More involved, a Bogoliubov-like dispersion mode has been reported by the group of Y. Yamamoto by Utsunomiya et al. (2008), in a GaAs microcavity pumped at a large oblique angle.

Does this conjunction of characteristic features represent a definite evidence for superfluidity of exciton-polaritons? The first report of quantised vorticity, by Lagoudakis et al. (2008), carefully noted that this was not sufficient to establish superfluidity, although it suggests parallels. Regarding the dissipationless and unperturbed propagation, one can argue that optical solitons propagating in a nonlinear medium would demonstrate a similar behaviour to the resonantly excited fluids studied by Amo et al. (2009b) or that nonlinear effects such as screening cast doubts on the phenomenology observed by Amo et al. (2009a). Indeed, a wavepacket of polaritons resonantly created by a laser pulse resembles a soliton. On the other hand, in a system with a parabolic dispersion, solitons exist when interactions are attractive, to balance the diffusion.[115] This is, however, not the case for polaritons, where it is established, for instance by Krizhanovskii et al. (2006), that polaritons with parallel spins repel each other. Also, the dimensionality goes against the soliton picture, which holds strictly in 1D, whereas polaritons are 2D. Egorov et al. (2009) studied the solitonic solutions with a confining potential along the direction of motion, finding many similarities with Amo et al. (2009b)'s polariton bullet. On the other hand, theoretical works by Wouters and Savona (2010) and Wouters and Carusotto (2010) bring new questions on what defines superfluidity in a dissipative system. Along with the ongoing theoretical investigations, stronger evidence

---

[114]More details about half-quantum vortices specific to polariton superfluids are given in Chapter 9.

[115]Or the soliton has to be "dark", meaning that a depletion of the fluid is what propagates.

in favour of or against superfluid phenomenology would be given when one can observe propagation of polaritons under non-resonant optical pumping.

8.9 Polariton billiard

Gao et al. (2015b) used polaritons under non-resonant pumping to implement a chaotic polariton billiard. This allowed them to observe features of so-called "exceptional point", introduced by Kato (1980) (Heiss (2012) give a good account). This occurs when a parameter makes both the real and imaginary parts of two eigenvalues from a non-Hermitian Hamiltonian become equal, therefore lifting simultaneously their anticrossing in energy (the usual case for Hermitian systems), but also in broadening. These points correspond to a singularity in the parameter space that tune the eigenvalues, with a square-root type of branch discontinuity between two Riemann sheets. They have peculiar topological features, such that encircling such a point leads to a swapping for its eigenmodes $|\psi_i\rangle$ as

$$|\psi_i\rangle \to -|\psi_{\bar{i}}\rangle \to -|\psi_i\rangle \to |\psi_{\bar{i}}\rangle \to |\psi_i\rangle \,, \qquad (8.131)$$

where $i = 1, 2$ and $\bar{i} = 3 - i$, with the acquisition of Berry phase resulting in the $-$ sign, that requires to go round the singularity four times to recover the initial state. This was first demonstrated by Dembowski et al. (2001) in a microwave cavity and implemented with polaritons by Gao et al. (2015b) who enforced a complex potential by combining pumping and decay, $V(\mathbf{r}) = P(\mathbf{r}) + i(P(\mathbf{r}) - \gamma)$, allowing them to control the dynamics of (non-interacting) states ruled by the non-Hermitian Hamiltonian

$$H = \begin{pmatrix} E_1 - i\Gamma & q \\ q^* & E_2 - i\Gamma \end{pmatrix} \,, \qquad (8.132)$$

with eigenvalues

$$\lambda_{1,2} = \frac{E_1 + E_2 - i(\Gamma_1 + \Gamma_2)}{2} \pm \sqrt{\left(\frac{E_1 - E_2 - i(\Gamma_1 - \Gamma_2)}{2}\right)^2 + |q|^2} \,, \qquad (8.133)$$

that "coalesce" (as the terminology of exceptional points calls the crossing of real and energy parts, i.e., $\lambda_1 = \lambda_2$), when $E_1 = E_2$ and $\Gamma_1 - \Gamma_2 = \pm|q|$. The differences in energies and linewidth can be tuned by varying two parameters of the billiard, namely the radius R of a quarter-disk sitting on one of its side, and the thickness d of the billiard wall. In this way, it is possible to go round the exceptional point and track the evolution of the mode, as shown in Fig. 8.17

8.10 Superconductivity mediated by exciton-polaritons

BEC at room temperature is now widely accepted by the microcavity community. One can logically inquire whether another striking effect based on the Bose statistics, *superconductivity*, may be realised at equally high temperatures in optically coupled systems. Superconductivity is nowadays understood as a Bose–Einstein condensation of so-called *Cooper pairs*, which are bound states of two electrons, with a total spin of 0

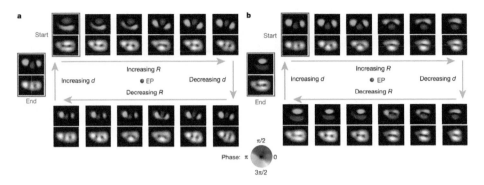

Fig. 8.17: *Going round a polariton exceptional point*, with Gao et al. (2015b). The exceptional point, labelled "EP", is a discontinuity in the parameter space controlled essentially by the radius R of a quarter-disk in the Billiard and its walls width d. Circling round it leads to swapping the eigenmodes, so that the "Start" real-space image (black & white snapshots) differs from the "End" one. Circling another time, as done in (b), where "Start" is now the "End" of the previous round, swaps again the mode, but the phase acquired a minus sign as shown in the colour-panel (numerical simulations). Another two-rounds would be necessary to recover the initial condition.

or 1, therefore following Bose–Einstein statistics. As shown by Cooper (1956), on top of the Fermi sea, two electrons will always form a bound state, no matter how weak the attractive interaction between them. In the famous Bardeen et al. (1957) (BCS) theory, phonons are the mediator of the attractive interaction between the electrons. They can overcome Coulomb repulsion (which is, moreover, screened), thanks to the so-called retardation effect: their slow dynamics (vibration of the lattice) extends over long times and one electron can influence another with a retarded interaction. The BCS theory leads to a simple estimate for the critical temperature of the superconducting phase transition:

$$T_C = 1.14 T_D \exp\left(\frac{1}{gV_{ee}}\right), \tag{8.134}$$

where T_D is the Debye temperature (of the order of hundred Kelvin in typical metals), g is the electron density of states at the surface of the Fermi sphere and V_{ee} is the effective electron–electron interaction constant. In conventional superconductors, $gV_{ee} \ll 1$, which is referred to as the *weak-coupling* regime (not to be confused with the weak-coupling regime for exciton-polaritons). While the Debye temperature is of the order of a few hundred Kelvin in many metals, the exponential factor in eqn (8.134) limits the critical temperature of the BCS transition to no more than $40\,\mathrm{K}$ in conventional superconductors. Research towards a better combination of parameters allowing for higher T_c in new systems or materials has been initiated in the 1950s and is still going on.

Exciton-polaritons being electrically neutral, cannot carry an electric current. On the other hand, one can perhaps use polaritons as the "glue" to bind together electrons in a metallic layer embedded in a microcavity. Laussy et al. (2010) have studied a scenario where a Bose–Einstein condensate of polaritons assumes this role.

An exciton-mechanism for superconductivity has been proposed by several authors in the 1960–1970s, starting with Little (1964), revisited by Allender et al. (1973) (see

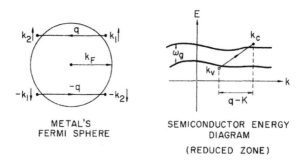

METAL'S
FERMI SPHERE

SEMICONDUCTOR ENERGY
DIAGRAM
(REDUCED ZONE)

Fig. 8.18: Electron–electron pairing in a metal via excitation of virtual excitons in the neighbouring semiconductor layer, as sketched by Allender et al. (1973).

Fig. 8.18) and championed by Ginzburg (1972). The idea is simply to replace phonons by excitons as mediators of the Cooper pairing of electrons. In this scenario, free electrons in a metal are expected to interact with the valence band electrons in a semiconductor. Creation and subsequent annihilation of virtual excitons in a semiconductor due to this interaction constitutes a second-order scattering process, which—in a perfect analogy with the acoustic phonon scattering—leads to the decrease of the ground state energy for free electrons at the Fermi level. This results in the formation of the Cooper pairs. The adepts of exciton-mediated superconductivity argue that the Debye temperature in the BCS formula should be replaced by the exciton energy in the case of the exciton mechanism, which corresponds to tens of thousands Kelvin in most popular direct bandgap semiconductors. On the other hand, the electron–exciton interaction constant could be very small, because of the spatial separation of the electrons in a metal and a semiconductor. In order to increase it, Ginzburg (1972) proposed to use metal–insulator sandwiches composed of two-dimensional metal and insulator (semiconductor) layers. Such sandwich structures were fabricated and studied experimentally by Gozar et al. (2008), showing superconductivity at temperatures up to 50 K. There is, however, no evidence yet that the observed superconductivity in sandwiches is caused by the exciton mechanism. One of the arguments against this mechanism is the high speed of the excitons that can participate in the exchange interaction leading to Cooper pairing. They have wavevectors comparable to the Fermi wavevector in metals. While acoustic phonons are slower than electrons at the Fermi level by three orders of magnitude, the velocity of excitons exchanging their wavevectors with electrons is typically only 3 to 5 times slower than the Fermi velocity. This means that the retardation effect is strongly reduced in the case of the exciton mechanism. This reduced retardation can only be compensated if the exciton-mediated attractive interaction is orders of magnitude stronger than the phonon mediated interaction. Usually, this is not the case, as even in the best quality structures the Cooper coupling constant V_{ee} remains quite small. An immediate advantage of the superconductivity based on the free electron scattering with a Bose–Einstein condensate (BEC) of exciton-polaritons is in the possibility to tune the coupling constant V_{ee} by tuning the optical pumping power. The Cooper coupling constant scales as the occupation number of the BEC, which can be directly controlled by optical pumping, as discussed by Laussy et al. (2010). This opens a perspective for the realisation of light-mediated superconductivity with very high T_c. The superconducting gap width and the Cooper pair size in this case would be also controlled by the intensity of pumping light, which would pave the way towards optical band engineering and other interesting applications. In the limit of very strong pumping (at the onset of the Mott transition for excitons) the system may be brought to the strong electron–electron coupling regime, where the Cooper pair size is comparable with the average distance between electrons. The superconducting phase transition may be described in this case as the Bose condensation of quatron-polaritons, as discussed by Kavokin et al. (2007).

8.11 Further reading

Grifin et al. (1996) edited a proceedings volume for the first Levico conference on BEC in 1993, held before its experimental realisation 2 years later and announced "hot off the press" at the following conference. Its lucky timing and contributions from eminences of the time made it an important publication in the field. The review papers give insightful and personal accounts by pioneers of the field, while brief reports provide an interesting historical snapshot of the time. It also includes some unusual systems for the condensed-matter physicist, such as BEC of mesons or Cooper pairing in nuclei. Griffin remembers the conference, now an historical one, and the story of this book is in J. Phys. B: At. Mol. Opt. Phys. **37** (2004).

More pedagogical and unified texts have flourished since. Especially notable is the excellent text by Pitaevskii and Stringari (2003). Pethick and Smith (2001) provide a good introductory description of the atomic case that became the "hero" of BEC as its first experimental realisation and a source of inspiration for the condensed-matter community. Another useful proceedings is the Enrico Fermi's Varenna summer school volume edited by Inguscio et al. (1999).

Moskalenko and Snoke (2000) provide a good overview of theoretical work (with many discussions of experimental results) related to BEC in semiconductors. The first chapter opens with *"Many people seem to have trouble with the concept of an exciton..."* and gives a gentle introduction to an otherwise involved exposition. The content indeed borrows a lot from research papers of the Russian literature. It is therefore a useful window to the non-Russian-speaking reader into the extensive amount of theoretical work made by the Soviet school on this topic.

An enduring classic dealing with superfluidity is the textbook by Khalatnikov (1965), which covers all the main characteristics of Bose gas and liquids, especially the excitation spectrum, hydrodynamics and kinetics of the problem. The discussion of the Landau spectrum and its two-liquid model is one of the best expositions. The Bogoliubov model is reviewed in full details in the review by Zagrebnov and Bru (2001).

We also recommend readers interested in the controversial subject of exciton interactions to the textbook by Combescot and Shiau (2015), where polaritons are described beyond the bosonic approximation. This contains a large list of references to the previous papers by Combescot and co-authors on interacting systems of composite bosons, where the so-called "Shiva diagrams" were proposed and a number of interesting conclusions of general relevance for the many-body physics have been drawn.

Being perfectly aware that the bosonic approximation for composite particles—unlike exciton-polaritons—has its limitations, we nevertheless use it largely throughout this book for two reasons: because it allows a qualitative understanding of most of the new nonlinear optical effects observed in microcavities and because the full theory of cavity polaritons treated as composite particles is still under construction. In particular, the important issue of spin and polarisation-related effects has only been treated within the bosonic approximation, as discussed in the next chapter.

9

SPIN AND POLARISATION

In this chapter we consider a complex set of optical phenomena linked to the spin dynamics of exciton-polaritons in semiconductor microcavities. We review a few important experiments that reveal the main mechanisms of the exciton-polariton spin dynamics and present the theoretical model of polariton spin relaxation based on the density matrix formalism. We also discuss the polarisation properties of the condensate and the superfluid phase transitions for polarised exciton-polaritons. We briefly address the polarisation multistability and switching in polariton lasers. Finally, the optical spin-Hall and spin-Meissner effects are described.[116]

[116] We acknowledge enlightening discussions with I. Shelykh, Yu. Rubo and K. Kavokin who contributed substantially to the content of this chapter.

Microcavities, Second Edition. Alexey V. Kavokin, Jeremy J. Baumberg, Guillaume Malpuech, Fabrice P. Laussy, Oxford University Press (2017). © Alexey V. Kavokin, Jeremy J. Baumberg, Guillaume Malpuech, Fabrice P. Laussy. DOI 10.1093/oso/9780198782995.001.0001

9.1 Introduction

When optically created, polaritons inherit their spin and dipole moment from the exciting light. Their polarisation properties can be fully characterised by a Stokes vector or—using the language of quantum physics—a pseudospin accounting for both spin and dipole moment orientation of a polariton. From the very beginning of their creation in a microcavity, polaritons start changing their pseudospin state under the effect of effective magnetic fields of various natures and due to scattering with acoustic phonons, defects, and other polaritons. This makes pseudospin dynamics of exciton-polaritons rich and complex. It manifests itself in non-trivial changes in the polarisation of light emitted by the cavity versus time, pumping energy, pumping intensity and polarisation. Experimental studies of the polarisation properties of microcavities have given evidence of a set of unusual effects (giant Faraday rotation, polarisation beats in photoluminescence, polarisation inversion in the parametric oscillation regime, etc.). Linear optical effects (like Faraday rotation) have been interpreted in terms of the theory of non-local dielectric response (developed in Chapter 3). The analysis of the data of non-linear optical experiments require more substantial theoretical effort. The experiments unambiguously indicate that energy and momentum-relaxation of exciton-polaritons are spin dependent, in general. This is typically the case in the regime of stimulated scattering when the spin polarisations of initial and final polariton states have a huge effect on the scattering rate between these states. It appears that critical conditions for polariton Bose-condensation are also polarisation dependent. In particular, the stimulation threshold (i.e., the pumping power needed to have a population exceeding 1 at the ground-state of the lower-polariton branch) has been experimentally shown to be lower under linear than under circular pumping, but this swaps when trapped condensates are used. These experimental observations have stimulated theoretical research toward understanding the mutually dependent polarisation- and energy-relaxation mechanisms in microcavities.

9.2 Spin relaxation of electrons, holes and excitons in semiconductors

The concept of spin was introduced by Dutch-born US physicists Samuel Abraham Goudsmit and George Eugene Uhlenbeck in 1925. In the same year, the Austrian theorist Wolfgang Pauli proposed his exclusion principle that states that two electrons cannot occupy the same quantum state. Later, it was understood that this principle applies to all particles and quasiparticles with a semi-integer spin (fermions) including electrons and holes in semiconductors. On the other hand, it is not valid for particles (quasiparticles) having an integer spin, in particular for excitons.

Spins of electrons and holes govern the polarisation of light generated due to their recombination. The conservation of spin in photoabsorbtion allows for spin-orientation of excitons by polarised light beams (optical orientation), an effect that manifests itself also in the polarisation of photoluminescence. σ^+ and σ^- circularly polarised light excites $J = +1$ and -1 excitons, respectively. Linearly polarised light excites a linear combination of $+1$ and -1 exciton states, so that the total exciton spin projection on the structure axis is zero in this case. Optical orientation of carrier spins in bulk semiconductors was discovered by a French physicist Georges Lampel in 1968. In quantum

George **Uhlenbeck** (1900–1988), Hans A. **Kramers** and Samuel A. **Goudsmit** (1902–1978) on the left. Wolfgang **Pauli** (1900–1958) on the right.

Uhlenbeck is mainly known for his introduction (with Goudsmit) of the concept of spin of the electron, although he favoured and was most active in statistical physics, where lies his work on quantum kinetics (see on page 338). Known to prefer rigour to originality, one of his student comments about him that "*He felt that something really original one did only once—like the electron-spin—the rest of one's time one spent on clarifying the basics.*"

Pauli's first talk in Sommerfeld's "Wednesday Colloquium" impressed the venerable professor so much that he entrusted the young student to prepare an article on relativity for the "Encyklopädie der mathematischen Wissenschaften." He produced a 237-pages article that impressed Einstein himself who reviewed the work as "grandly conceived." He remained widely praised for the mastery and clarity of his expositions, with his two review articles on quantum mechanics in the Handbuch der Physik known as the "Old and New Testament." Oppenheimer called the later "the only adult introduction to quantum mechanics." However, he published little as compared to his actual scientific production, especially as he aimed for a thorough understanding. He onced commented "*The fact that the author thinks slowly is not serious, but the fact that he publishes faster than he thinks is inexcusable.*" Many of his own results were confined to private correspondence, like the equivalence of Heisenberg and Schrödinger pictures (in a letter to Jordan), the time–energy uncertainty relation (to Heisenberg), or the neutrino to rescue energy conservation in radioactivity. He almost never cared about recognition. He developed the "Ausschliessungsprinzip" or exclusion principle in 1924, already known as the Pauli principle when it earned him the Nobel prize in 1945. Despite a strong opposition to the initial idea of the spin of the electron (which Kronig never published after his idea was ridiculed by Pauli), he formalised the concept of spin following the young theory of Heisenberg, culminating with Pauli matrices. He later laid the foundations of quantum field theory with the spin-statistics theorem, linking bosons and fermions to integer and half-integer spins, respectively. After his first marriage to a dancer broke down in less than a year and when his mother committed suicide, he suffered a deep personal crisis and started to drink. He consulted the psychologist Carl Jung (who first delegated a female assistant thinking that Pauli's problem were linked to women) for whom he detailed over a thousand dreams and established a deep relationship as a strong believer in psychology. He wrote that "*in the science of the future reality will neither be 'psychic' nor 'physical' but somehow both and somehow neither.*" Numerous anecdotes are in circulation about him, like the "Pauli effect" dooming an experiment if he was in its vicinity. He was also known for his severity and scathing comments, but also for his wit, illustrated by his famous remark "*This isn't right. This isn't even wrong.*" One of the most brilliant theorists of all times, Albert Einstein described him as his intellectual successor during Pauli's Nobel celebration in Princeton.

wells, this has been extensively studied since the 1980s by many groups. For good reviews we address the reader to the famous volume "Optical orientation" edited by Meier and Zakharchenia (1984).

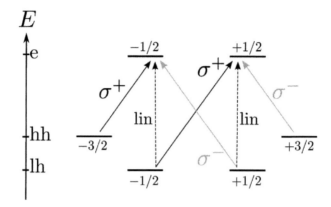

Fig. 9.1: Polarisation of optical transitions in zincblende semiconductor quantum wells. Black, gray and dashed lines show σ_+, σ_- and linearly polarised transitions, respectively.

An exciton is formed by an electron and hole, i.e., by two fermions having projections of the angular momenta on a given axis equal to $J_z^e = S_z^e = \pm\frac{1}{2}$ for an electron in the conduction band with S-symmetry and $J_z^h = S_z^h + M_z^h = \pm\frac{1}{2}, \pm\frac{3}{2}$ for a hole in the valence band with P-symmetry (in zincblende semiconductor crystals). The states with $J_z^h = \pm\frac{1}{2}$ are formed if the spin projection of the hole S_z^h is antiparallel to the projection of its orbital momentum M_z^h. These states are called *light holes*. If the spin and orbital momentum are parallel, the heavy holes with $J_z^h = \pm\frac{3}{2}$ are formed (see Section 4.2). In bulk samples, at $\mathbf{k} = 0$ the light- and heavy-hole states are degenerate. However, in quantum wells the quantum confinement in the direction of the structure growth axis lifts this degeneracy so that energy levels of the heavy holes lie typically closer to the bottom of the well than the light-hole levels (see Fig. 9.1).

The ground-state exciton is thus formed by an electron and a heavy hole. The total exciton angular momentum[117] J has projections ±1 and ±2 on the structure axis. Bearing in mind that the photon spin is 0 or ±1 and that the spin is conserved in the process of photoabsorption, excitons with spin projections ±2 cannot be optically excited. These are spin-forbidden states that, since they are not coupled to light, are also called *dark states*. This vivid terminology extends to states that couple to light (with spin projection ±1) calling them *bright states*. Since, in quantum microcavities, $J = 2$ states are not coupled to the photonic modes, we shall neglect them in the following consideration, but it should be borne in mind, however, that in some cases these dark states still come into play, as Combescot et al. (2007b) emphasise. They can, for instance, be

---

[117]In this text, as is common practice in the field, the total exciton angular momentum is referred to for convenience as the exciton *spin*.

mixed with the bright states by an in-plane magnetic field or be populated due to the polariton–polariton scattering.

Georges **Lampel** Professor at the Ecole Polytechnique, Paris and Boris P. **Zakharchenia** (1928–2005).

Lampel is the father of "optical orientation", an important branch of crystal optics studying the optically induced spin effects in solids. One of Lampel's PhD students was Claude Weisbuch, who was the first to observe the strong light–matter coupling in semiconductor microcavities. At present, Lampel and Weisbuch work in the same laboratory of the famous École Polytechnique in Paris.

Zakharchenia was a PhD Student of Evgenii Gross and participated in the early studies on the Wannier–Mott excitons in CuO_2. Later, he headed a laboratory of the Ioffe Institute in Saint Petersburg that published the pioneering works on what is now called "spintronics". Zakharchenia is also known for his theory on the Pushkin's duel place.

The polarisation of exciting light cannot be retained infinitely long by excitons. Sooner or later they lose polarisation due to inevitable spin and dipole moment relaxations. Excitons, being composed of electrons and holes, are subject to mechanisms of spin-relaxation for free carriers in semiconductors. The main important ones being:

1. Elliot (1954)–Yaffet mechanism involving the mixing of the different spin wavefunctions with $k \neq 0$ as a result of the kp interaction with other bands. In quantum wells this effect plays a major role in the spin relaxation of holes and can induce transitions between the optically active and dark exciton states $|+1\rangle \rightarrow |-2\rangle$ and $|-1\rangle \rightarrow |+2\rangle$.

2. Dyakonov and Perel (1971) mechanism caused by the spin-orbit interaction induced spin splitting of the conduction band in non-centrosymmetric crystals (like zincblende crystals) and asymmetric quantum wells (Dresselhaus and Rashba terms, respectively) at $k \neq 0$. This mechanism is predominant for the electrons and also leads to transitions between the optically active and dark exciton states, $|+1\rangle \rightarrow |+2\rangle$ and $|-1\rangle \rightarrow |-2\rangle$.

3. The Bir–Aronov–Pikus (BAP) mechanism first published by Pikus and Bir (1971), involving the spin-flip exchange interaction of electrons and holes. For excitons the efficiency of this mechanism is enhanced, as the electron and the hole form a bound state. The exchange interaction consists of so-called "short-range" and

"long-range" parts of the Coulomb interaction[118]. The short-range part leads to the coupling between heavy-hole (hhe) and light-hole excitons (lhe) and thus is suppressed in the quantum wells where the degeneracy of lhe and hhe is lifted. The long-range part leads to transitions within the optically active exciton doublet[119].

4. Spin-flip scattering between carriers and magnetic ions in diluted magnetic semiconductors. As an example, paramagnetic semiconductors containing Mn^{2+} ions (spin $5/2$) allow for efficient spin relaxation of electrons, heavy and light holes flipping their spins with the magnetic ion spins.[120]

Vladimir Idelevich **Perel** (1928–2007) and Mikhail Igorevich **Dyakonov** (b. 1940) in 1976.

Dyakonov and Perel personalise the excellence of the Soviet theoretical physics. Their names are associated with numerous physical effects and theories, including the non-radiative recombination and spin-relaxation models, spin-Hall effect, among others. Their elegant and seemingly simple analytical models helped the understanding of extremely complex phenomena of modern solid-state physics. The famous "Tea seminar" created by Dyakonov and Perel still runs at the Ioffe institue of Saint Petersburg every Thursday at 5pm.

---

[118]For details see E.L. Ivchenko, Optical Spectroscopy of Semiconductor Nanostructures, Alpha, Harrow (2005), pages 252–253.

[119]The ability of the long-range part of the exchange interaction to couple the exciton doublet can lead to the inversion of the circular polarisation in time-resolved polarisation measurements, as shown by Kavokin et al. (2003a).

[120]The spin-flip scattering is discussed at length by P. A. Wolff, in Semiconductors and Semimetals, edited by J. K. Furdyna and J. Kossut, Diluted Magnetic Semiconductors, Vol. 25. (Academic Press, Boston, 1988).

In a key paper, Maialle et al. (1993) have shown that the third (BAP) mechanism is predominant for the quantum-confined excitons in non-magnetic semiconductors. The long-range electron–hole interaction leads to the longitudinal-transverse splitting of exciton states, i.e., energy splitting between excitons having a dipole moment parallel and perpendicular to the wavevector. This splitting is responsible for rapid spin-relaxation of excitons in quantum wells. For description of exciton-polaritons in microcavities it has a very important consequence: the dark states can be neglected, which allows us to consider an exciton as a two-level system and use the well-developed pseudospin formalism for its description. From the formal point of view, the exciton can be thus described by a 2×2 spin density matrix that is completely analogous to the spin density matrix for electrons.

Exercise 9.1 [*] *In 1924, British physicists Wood and Ellett (1924) reported an amazing polarisation effect: they measured the circular polarisation degree of fluorescence of the mercury vapour resonantly excited by a circularly polarised light. In their initial experiments a high degree of circular polarisation was observed, while it significantly diminished in later experiments. They wrote: "It was then observed that the apparatus was oriented in a different direction from that which obtained in the earlier work, and on turning the table on which everything was mounted through ninety degrees, bringing the observation direction East and West, we at once obtained a much higher value of the polarisation." Explain this effect.*

9.3 Microcavities in the presence of a magnetic field

A magnetic field strongly affects excitons in quantum wells and thus it also affects exciton-polaritons in microcavities. One can distinguish between three kinds of linear magneto-exciton effects in cavities:

- The energies of electron and hole quantum-confined levels change as a function of the magnetic field, following the so-called *Landau fan diagram*. As a result, the energies of electron–hole transitions increase by

$$\delta E = (l + 1/2)\hbar\omega_c , \qquad (9.1)$$

where the cyclotron frequency $\omega_c = eB/(\mu c)$ depends on the reduced effective mass of electron–hole motion $\mu = m_e m_h^{\|}/(m_e + m_h^{\|})$ and magnetic field B. $m_h^{\|}$ is the in-plane hole mass, m_e is the electron effective mass and $l = 0, 1, 2, \cdots$ is the Landau quantum index. Landau quantisation of exciton energies takes place due to the magnetic field effect on the orbital motion of electrons and holes. This effect is polarisation independent. In microcavities, Landau quantisation results in the appearance of a fine structure of polariton eigenstates and gives the possibility of tuning of different Landau levels into resonance with the cavity mode, as observed by Tignon et al. (1995).

- Zeeman splitting of the exciton resonance. Excitons with spins parallel or antiparallel to the magnetic field have different energies. The splitting is given by

$$\Delta E = \mu_B g B , \qquad (9.2)$$

where $\mu_B \approx 0.062\,\text{meV}\,\text{T}^{-1}$ is the Bohr magneton and g is the exciton g-factor that depends on the materials composing the quantum well and the magnetic-field orientation with respect to the QW. In the following we shall consider the so-called *Faraday geometry*, i.e. the magnetic field parallel to the wavevector of incident light and, consequently, normal to the QW plane (normal-incidence case). Note that g can be positive or negative; in most semiconductor materials it varies between -2 and 2, and in GaAs/AlGaAs QWs it changes sign as the QW width changes. Exciton Zeeman splitting leads to an effect known as *resonant Faraday rotation*, i.e., rotation of the polarisation plane of linearly polarised light passing through a QW in the vicinity of the exciton resonance. In microcavities, this effect is strongly amplified due to the fact that light makes a series of round-trips, each time crossing the QW before escaping from the cavity. This is to be discussed in detail below.

- An increase of the exciton binding energy and oscillator strength due to the shrinkage of the exciton wavefunction in a magnetic field. This effect is important for strong enough magnetic fields, for which the magnetic length $L = \sqrt{\hbar/(eB)}$ is comparable to the exciton Bohr diameter (typically about 200 Å). An increase of the exciton oscillator strength in a magnetic field enhances the vacuum-field Rabi splitting.

Figure 9.2 shows the relative exciton radiative broadening, Rabi splitting and period of Rabi oscillations (i.e. oscillations in time-resolved reflection due to beats between two exciton-polariton modes in the cavity) measured experimentally and calculated with a variational approach of Berger et al. (1996). One can see that the oscillator strength increases by about 80% with the magnetic field changing from 0 to 12 T, which results in the Rabi splitting increasing by more than 30% and a corresponding decrease in the period of Rabi oscillations. Note that due to the exciton diamagnetic shift $\delta E - E_B$ (with δE given by eqn (9.1) and E_B being the exciton binding energy, the energy of the polariton ground-state shifts up. This shift is parabolically dependent on the magnetic field at low fields. On the other hand, due to the increase of the exciton oscillator strength and resulting increase of the Rabi splitting, the polariton ground-state energy is pushed down. In realistic cavities, the latter effect dominates at low fields and the ground-state energy shifts down, while in the limit of strong fields, when the magnetic lengths become comparable to the exciton Bohr radius, the ground state is expected to start moving up. Moreover, due to the increase of the Rabi splitting, the exciton-polariton dispersion gets slightly steeper near the ground-state, so that the polariton effective mass decreases with the increase of the magnetic field. The variational calculation of the change of the Rabi splitting in presence of a magnetic field applied to a microcavity is given in Appendix C.

9.4   Resonant Faraday rotation

Faraday rotation is rotation of the electric field vector of a linearly polarised light wave propagating in a medium in the presence of a magnetic field oriented along the

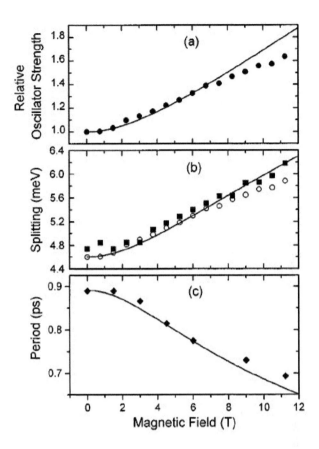

Fig. 9.2: Relative exciton radiative broadening, Rabi splitting and period of Rabi oscillations, i.e., oscillations in time-resolved reflection due to beats between two exciton-polariton modes in the cavity, measured experimentally from a GaAs-based microcavity with InGaAs/GaAs QWs (points) and calculated (lines), from Berger et al. (1996).

light propagation direction (see Chapter 2). Consider the propagation of linearly (X) polarised light, whose initial polarisation is described by a Jones vector

$$\begin{pmatrix} 1 \\ 0 \end{pmatrix} = \frac{1}{2} \begin{pmatrix} 1 \\ i \end{pmatrix} + \frac{1}{2} \begin{pmatrix} 1 \\ -i \end{pmatrix} . \tag{9.3}$$

Here, the upper and lower components of the vectors correspond to the electric-field projections in the x- and y-directions, respectively, and the two terms on the right-hand side describe σ^+ (right-circular) and σ^- (left-circular) polarised waves, respectively. If the structure is placed in a magnetic field, the transmission coefficients for σ^+ and σ^- polarised waves become different, so that the transmitted light can be represented as

$$\hat{t} = \frac{1}{2} A \exp(i\varphi_+) \begin{pmatrix} 1 \\ i \end{pmatrix} + \frac{1}{2} C \exp(i\varphi_-) \begin{pmatrix} 1 \\ -i \end{pmatrix} , \tag{9.4}$$

where the amplitudes A and C and phases φ_+ and φ_- coincide in the absence of the field, but may be different in the presence of the field. $\hat{t}$ is the polarisation vector of elliptically polarised light, which can be conveniently represented as a sum of waves linearly polarised along the main axes of the ellipse

$$t = \frac{i}{2}(B - C)\exp(i\psi)\begin{pmatrix} \sin\phi \\ -\cos\phi \end{pmatrix} + \frac{1}{2}(A + C)\exp(i\psi)\begin{pmatrix} \cos\phi \\ \sin\phi \end{pmatrix}, \qquad (9.5)$$

where $\psi = (\varphi_+ + \varphi_-)/2$ and $\phi = (\varphi_- - \varphi_+)/2$ is the *Faraday rotation angle*.

The transmission coefficient of the structure for light detected in X-polarisation is

$$T_x = \frac{1}{4}|A\exp(i\varphi_+) + C\exp(i\varphi_-)|^2, \qquad (9.6)$$

and in Y-polarisation, it is

$$T_y = \frac{1}{4}|A\exp(i\varphi_+) - C\exp(i\varphi_-)|^2. \qquad (9.7)$$

In σ^+ and σ^- polarisation, the amplitude of light transmitted across the quantum well at the exciton resonance frequency is given by

$$t_{\sigma^+,\sigma^-} = 1 + \frac{i\Gamma_0}{\omega_0^{\sigma^+,\sigma^-} - \omega - i(\gamma + \Gamma_0)}, \qquad (9.8)$$

where $\omega_0^{\sigma^+,\sigma^-}$ is the exciton resonance frequency in the two circular polarisations, whose splitting in a magnetic field is referred to as exciton Zeeman splitting, Γ_0 is the exciton radiative decay rate and γ is the exciton non-radiative decay rate.

Hereafter, we shall neglect the exciton inhomogeneous broadening. The polarisation plane of linearly polarised light passing through the QW rotates by the angle

$$\phi = \frac{1}{2}\left[\arctan\frac{(\omega_0^- - \omega)\Gamma_0}{(\omega_0^- - \omega)^2 + (\gamma + \Gamma_0)^2} - \arctan\frac{(\omega_0^+ - \omega)\Gamma_0}{(\omega_0^+ - \omega)^2 + (\gamma + \Gamma_0)^2}\right],$$

$$\approx \frac{(\omega_0^{\sigma^-} - \omega_0^{\sigma^+})\Gamma_0}{(\gamma + \Gamma_0)^2}. \qquad (9.9)$$

In the case of a microcavity, the Faraday rotation can be greatly amplified. Let us first analyse the expected effects in the framework of ray optics. Consider a cavity-polariton mode as a ray of light travelling backwards and forwards inside the cavity within its lifetime. At the anticrossing point of the exciton and photon modes, the lifetime of cavity polaritons τ is

$$\tau = \frac{1}{2(\kappa + \gamma)}, \qquad (9.10)$$

where κ is the cavity decay rate, dependent on the reflectivity of the Bragg mirrors. The average number of round-trips of light inside the cavity is

$$N = \frac{\tau c}{2n_c L_c} = \frac{Q}{2}, \qquad (9.11)$$

where n_c is the cavity refractive index and L_c is its length, Q is the quality factor of the cavity (see Chapter 2). In standard-quality GaAs-based microcavities this factor reaches a few hundred. While circulating between the mirrors the light accumulates a rotation, before escaping the cavity. The amplitude of the x- and y-polarised components of the emitted light can be found as $E_x = \Re(E)$ and $E_y = \Im(E)$ with

$$E = t_B + t_B r_B e^{i\phi} + t_B r_B^2 e^{2i\phi} + \cdots = \frac{t_B}{1 - r_B e^{i\phi}}, \tag{9.12}$$

where r_B and t_B are the amplitude reflection and transmission coefficients of the Bragg mirror, respectively. The angle of the resulting rotation of the linear polarisation is

$$\theta = \arg(E) = \arctan \frac{r_B \sin \phi}{1 - r_B \cos \phi}. \tag{9.13}$$

Note that this consideration neglects reflection of light by a QW exciton, since the amplitude of the QW reflection coefficient is more than an order of magnitude smaller than the transmission amplitude.

To observe a giant Faraday rotation, a peculiar experimental configuration is needed. In the reflection geometry only a very small rotation of the polarisation plane can be observed (Kerr effect). Actually, the reflection signal is dominated by a surface reflection from the upper Bragg mirror that does not experience any polarisation rotation. To observe the giant effect predicted by eqn (9.13), either the measurement should be carried out in the transmission geometry, which would imply etching any absorbing substrate, or a pump-probe technique should be used to introduce the light into the cavity at an oblique angle and then to probe emission at the normal angle. The Faraday rotation described above is a linear optical effect induced by an external magnetic field applied to the cavity. There exists also an optically induced Faraday rotation in microcavities that is a nonlinear effect having a strong influence on the polarisation of emission of the cavities. It will be considered in detail below. Experimentally, the magnetic field induced Faraday rotation in microcavities has been studied by Kavokin (1997) while the rotation induced by a circularly polarised light has been reported by Brunetti et al. (2006).

Exercise 9.2 [***] *Describe the resonant Faraday rotation of TE-polarised light incident on a QW at oblique angle. A magnetic field is oriented normally to the QW plane.*

9.5   Spin relaxation of exciton-polaritons in microcavities: experiment

The spin dynamics of cavity polaritons has been experimentally studied by measurement of time-resolved polarisation from quantum microcavities in the strong-coupling regime. At the beginning of the twenty-first century, a series of experimental works appeared in this field, which reported unexpected results. Let us briefly summarise the most important of them.

In the experimental work by Martin et al. (2002), the dynamics of the circular polarisation degree ϱ_c of the photoemission from the ground-state of a CdTe-based microcavity was measured. ϱ_c was determined as

$$\varrho_c = \frac{I^+ - I^-}{I^+ + I^-} = \frac{N_{k=0}^+(t) - N_{k=0}^-(t)}{N_{k=0}(t)}, \tag{9.14}$$

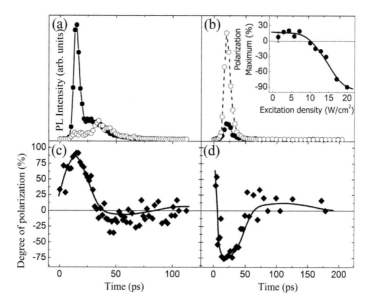

Fig. 9.3: Experimentally measured temporal evolution of the photoluminescence of a CdTe-based microcavity excited by circularly polarised light at the positive detuning, upper polariton branch (a, $\delta = 10\,\text{meV}$ and negative detuning, lower polariton branch (b, $\delta = -10\,\text{meV}$). The filled circles/solid line (open circles/dashed line) denote the σ^+ (σ^-) emission. The deduced time evolution of the circular polarisation degree for positive and negative detunings is shown in (c) and (d), respectively. The inset shows the maximum value of the polarisation degree at 20 ps in the negative detuning case. From Martin et al. (2002).

where $I^\pm$ denotes the measured circularly polarised intensities, $N_{k=0}^\pm$ represent the population of the ground-state ($k = 0$) with polaritons having spin projections on the structure axis equal to ± 1, respectively, $N_{k=0}(t) = N_{k=0}^+(t) + N_{k=0}^-(t)$. The pump pulse was circularly polarised and centred on the energy above the stop-band of the Bragg mirrors composing the microcavity (non-resonant excitation). In the linear regime, when the stimulated scattering of the exciton-polaritons did not play any role, an exponential decay of the circular polarisation degree of photoemission was observed. However, above the stimulation threshold, ρ_c exhibited a non-monotonic temporal dependence. At positive detuning, the initial polarisation of $\approx 30\%$ first increased up to $\approx 90\%$ and then showed damped oscillations (see Fig. 9.3). For negative detuning, the polarisation degree started from a positive value $\approx 50\%$, rapidly decreased down to strongly negative values and then increased showing attenuated oscillations with a period of about 50 ps. As we show in the next section this effect is caused by the TE–TM splitting of polariton states with $\mathbf{k} \neq \mathbf{0}$.

In the experiments carried out by the group of Y. Yamamoto and published by Shelykh et al. (2004), the microcavity was pumped resonantly at an oblique angle ($\approx 50°$). The dependence of the intensity and of the polarisation of light emitted by the ground-state on pumping polarisation and power was studied, with all experiments carried out in the continuous-pumping regime. The pump intensity corresponding to the stimulation threshold appeared to be almost twice as high in the case of circular pumping as in the

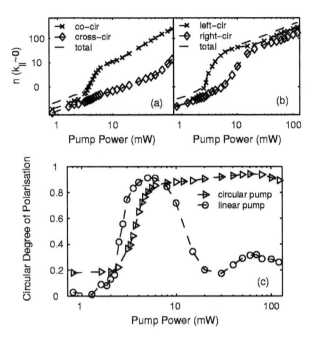

Fig. 9.4: a) Emission from $k_\parallel \approx 0$ polaritons vs. pump power under the circular pump. The two circular-polarisation components of the emission and their total intensity are plotted. b) Same as a) but for the linear pump. c) Circular polarisation degree ϱ_c vs. pump power with circular (triangles) and linear (circles) pumps. From Shelykh et al. (2004).

case of linear pumping as shown in Fig. 9.4(a) and (b). Figure 9.4(c) shows the dependence of the circular polarisation degree of the emitted light on the pumping intensity for different pumping polarisations. For both polarisations, $\varrho_c \approx 0$ is observed far below the threshold and $\varrho_c \approx 0.9$ near threshold. This indicates that spin relaxation is complete at low excitation density and that stimulated scattering into a definite spin component (say, spin-up) of the ground-state takes place at high densities. Above threshold, ϱ_c remains large in the case of a circular pump. In the case of a linear pump, however, ϱ_c decreases sharply at high pumping intensities. This effect was interpreted in terms of the interplay between ultrafast scattering and spin-relaxation of exciton-polaritons due to TE–TM splitting (see also below).

The polarisation dynamics of polariton parametric amplifiers was the focus of experimental research carried out by several groups. The parametric amplifier is realised if polaritons are created by resonant optical pumping close to the inflection point of the lower dispersion branch at the "magic angle" (see Chapter 7). In this configuration, the resonant scattering of two polaritons excited by the pump pulse toward the signal ($k = 0$) and the idler state is the dominant relaxation process. The scattering can be stimulated by an additional probe pulse used to create the seed of polaritons in the signal state, or it can be strong enough to be self-stimulated.

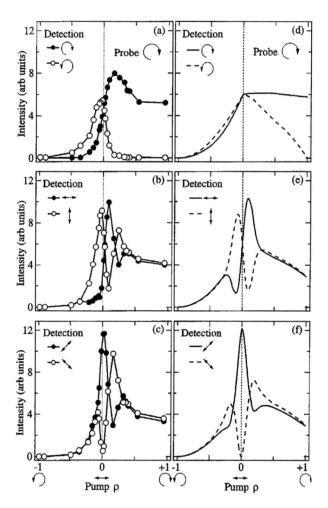

Fig. 9.5: Experimental (a)–(c) and theoretical intensities of circularly (a)–(d) and linearly (b), (c), (e) and (f), polarised components of the light emitted by a microcavity ground-state as a function of the circular polarisation degree of the pumping light. From Kavokin et al. (2003a).

In the experiments carried out by Lagoudakis et al. (2002), the polarisation of the probe pulse was kept right-circular, whereas the pump polarisation was changed from right- to left-circular passing through elliptical and linear polarisation. Consequently, the spin-up and spin-down populations of the pump-injected polaritons were varied while the pump intensity was kept constant. The two circular components of polarisation of light emitted by the ground-state and four in-plane linear components (vertical, horizontal and the two diagonal ones) were detected as functions of the circular polarisation degree of the excitation. To briefly summarise the results of these measurements— shown in Fig. 9.5(a), (b) and (c): in the case of a linearly polarised pump pulse and circularly polarised probe pulse the observed signal was linearly polarised, but with a plane of polarisation rotated by 45 degrees with respect to the pump polarisation. In the case of elliptically polarised pump pulses, the signal also became elliptical, while the direction of the main axis of the ellipse rotated as a function of the circular polarisation degree of the pump. In the case of a purely circular pump, the polarisation of the signal was also circular, but its intensity was half that found for a linear pump. The polarisation of the idler emission emerging at roughly twice the magic angle showed a similar behaviour, although in the case of a linearly polarised pump the idler polarisation was rotated by 90 degrees with respect to the pump polarisation. As we show below, the rotation of the polarisation plane in this experiment is a manifestation of the optically induced Faraday effect: the imbalance of populations of spin-up and spin-down polariton states produces a spin-splitting of the polariton eigenstates. This imbalance has been introduced by the elliptically polarised pump. The whole effect, referred to as "self-induced Larmor precession" will be discussed in the next section.

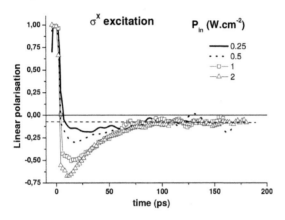

Fig. 9.6: Experimentally measured degree of linear polarisation of emission from the ground exciton-polariton state in a microcavity measured under linearly polarised pumping at the magic angle. Different curves correspond to different pumping intensities. One can see that the linear polarisation degree quickly becomes negative, which corresponds to 90° of rotation of the polarisation plane with respect to the pumping light polarisation. From Renucci et al. (2005).

The polarisation dynamics in a parametric oscillator without a probe has been considered in the experimental work of Renucci et al. (2005). Quite surprisingly, it was observed that for a linear pump, the polarisation of the signal is also linear, but rotated by 90° (see Fig. 9.6). This effect is connected with the anisotropy of the spin-dependent polariton–polariton scattering.

An interesting polarisation effect in a microcavity was observed in 2002 by Langbein et al. (2007) (Fig. 9.7) who illuminated a microcavity with a spot of linearly polarised light (say, along the x-axis) and detected spatially and temporally resolved emission from the sample in co- and cross-linear polarisations. In the cross-polarisation he observed a characteristic cross (*Langbein cross*), showing that the conversion of co- to cross-inear polarisation is efficient within the quarters of the reciprocal space separated by the x- and y-axes, but not in the vicinity of the axes. These striking patterns

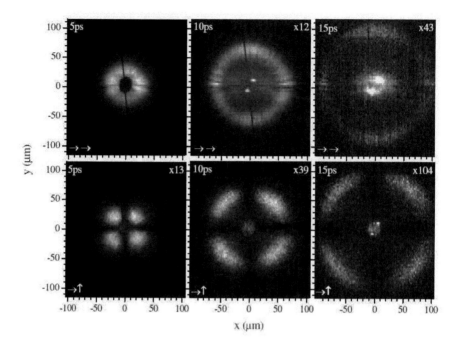

Fig. 9.7: Spatial imaging of polariton propagation (linear grayscale, all images are from W. Langbein in the Proceedings of the 26th International Conference on Physics of Semiconductors, Institute of Physics, Bristol, 2003, p. 112). Top: colinearly polarised polaritons. Bottom: crosslinearly polarised polaritons.

appear due to precession of the pseudospin of propagating exciton-polaritons in the effective magnetic field induced by TE–TM splitting (see next section).

9.6 Spin relaxation of exciton-polaritons in microcavities: theory

As we mentioned in Section 9.2, the main mechanisms for spin relaxation of exciton-polaritons in the linear regime are the transitions within the optically active doublet due to the longitudinal transverse TE–TM splitting of exciton-polaritons. Consequently, the dark states can be neglected in most cases and an exciton-polariton state with a given in-plane wavevector $\mathbf{k}$ can be treated as a two-level system. It can be described by the 2×2 density matrix ρ_k that is completely analogous to the spin density matrix of an electron.

It is convenient to decompose the polariton pseudospin density matrix as

$$\rho_k = \frac{N_k}{2}\mathbf{1} + \mathbf{S_k} \cdot \boldsymbol{\sigma}, \tag{9.15}$$

where $\mathbf{1}$ is the identity matrix, $\boldsymbol{\sigma}$ is the Pauli-matrix vector,[121] $\mathbf{S_k}$ is the mean *pseudospin* of the polariton state characterised by the wavevector $\mathbf{k}$. It describes both the exciton spin state and its dipole moment orientation (see Fig. 9.8).

[121]The *Pauli matrix vector* is a vector $\boldsymbol{\sigma} = (\sigma_x, \sigma_y, \sigma_z)$ whose components σ_i are matrices, namely the Pauli matrices (3.11) on page 85.

The pseudospin is the quantum analogue of the Stokes vector (see Section 2.2): the mean value of the pseudospin operator coincides with the Stokes vector of partially polarised light in our notation. Note that here and further we use the basis of circularly polarised states, i.e., associate the states having definite S_z with the polariton radiative states with their spin projection on the structure axis equal to ± 1. Their linear combinations correspond to eigenstates of S_x and S_y yielding linearly polarised emission. The pseudospin parallel to the x-axis corresponds to X-polarised light, the pseudospin antiparallel to the x-axis corresponds to Y-polarised light and the pseudospin oriented along the y-axis describes diagonal linear polarisations.

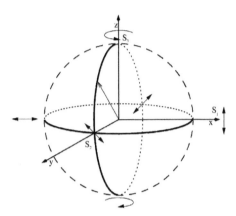

Fig. 9.8: Poincaré sphere with pseudospin (identical to the Stokes vector in this case, cf. Fig. 2.2). The equator of the sphere corresponds to different linear polarisations, while the poles correspond to two circular polarisations.

For non-interacting polaritons the temporal evolution of each density matrix (9.15) is governed by its Liouville–von Neumann equation (3.50)

$$i\hbar \partial_t \rho_k = [H_k, \rho_k], \tag{9.16}$$

where the Hamiltonian H_k reads in terms of the pseudospin

$$H_k = E_n(\mathbf{k}) - g\mu_B \mathbf{B}_{\text{eff},k} \cdot \mathbf{S}_k. \tag{9.17}$$

Here, $E_n(\mathbf{k})$ is the energy of the nth polariton branch, g is the effective polariton g-factor, μ_B is the Bohr magneton and $\mathbf{B}_{\text{eff},k}$ is an effective magnetic field. Unlike real magnetic fields the effective field only applies to the radiatively active doublet and does not mix optically active and dark states. We do not consider here the effects of a real magnetic field on the polariton spin dynamics.

The x- and y-components of $\mathbf{B}_{\text{eff},k}$ are nonzero if the exciton states whose dipole moments are oriented in, say, x- and y-directions have different energies. This always happens for excitons having nonzero in-plane wavevectors. The splitting of exciton states with dipole moments parallel and perpendicular to the exciton in-plane wavevector is called *longitudinal-transverse splitting* (LT-splitting, see Chapter 4 for the bulk polariton LT-splitting). The longitudinal-transverse splitting of excitons in quantum wells is a result of the long-range exchange interaction. It can be described by the reduced spin Hamiltonian of Pikus and Bir (1971) as

$$H_{\text{ex}} = \frac{3}{16} \frac{|\Phi_{\text{ex}}(0)|}{|\Phi_{\text{ex}}^{\text{bulk}}(0)|^2} \hbar \omega_{\text{LT}} \frac{f(k)}{k} \left[\begin{matrix} k^2 & (k_x - ik_y)^2 \\ (k_x + ik_y)^2 & k^2 \end{matrix} \right], \tag{9.18}$$

where $\hbar \omega_{\text{LT}}$ is the longitudinal-transverse splitting of a bulk exciton (see Chapter 4) and $\Phi_{\text{ex}}(0)$, $\Phi_{\text{ex}}^{\text{bulk}}(0)$ are bulk and QW exciton envelope functions, respectively, taken with equal electron and hole coordinates. The form-factor $f(k)$ is given by

$$f(k) = \iint U_e(z) U_{hh}(z) e^{-k|z-z'|} U_e(z') U_{hh}(z') \, dz dz', \tag{9.19}$$

with $U_e(z)$, $U_{hh}(z)$ being electron and heavy-hole envelope functions normal to the QW plane direction. Off-diagonal terms of the Hamiltonian (9.17) lead to polariton spin flips and thus create an effective in-plane magnetic field $\mathbf{B}_{\text{eff},k}$ referred to below as the *Maialle field*. The Maialle field is in general not parallel to $\mathbf{k}$, but makes with the x-axis twice the angle as $\mathbf{k}$. The effective magnetic field is zero for $\mathbf{k} = \mathbf{0}$ and increases as a function of k following a square root law at large k. For more details on the longitudinal-transverse splitting the reader can refer to the work of Tassone et al. (1992). In microcavities, splitting of longitudinal and transverse polariton states is amplified due to the exciton coupling with the cavity mode. Note that the cavity mode frequency is also split in TE- and TM- light polarisations (see Chapter 2). The resulting polariton splitting strongly depends on the detuning between the cavity mode and the exciton resonance and, in general, depends non-monotically on k. Figure 9.9 shows the TE–TM polariton splitting calculated for a CdTe-based microcavity sample for different detunings between the bare cavity mode and the exciton resonance. For these calculations, polariton eigenfrequencies in two linear polarisations have been found numerically by the transfer matrix method. One can see that the splitting is very sensitive to the detuning and may achieve 1 meV, which exceeds by an order of magnitude the bare-exciton LT splitting.

Using the pseudospin representation, one can rewrite the Liouville–von Neumann equation for the density matrix as

$$\partial_t \mathbf{S_k} = -\frac{g \mu_{\text{B}}}{\hbar} \mathbf{S_k} \times \mathbf{B}_{\text{eff},k} . \tag{9.20}$$

The Maialle field thus induces precession of the pseudospin of the ensemble of circularly polarised polaritons and can cause oscillations of the circular polarisation degree of the emitted light in time as was observed experimentally by Martin et al. (2002) (Fig. 9.3) and conversion from linear to circular polarisation observed by the Yamamoto group (Fig. 9.4). In the latter case, the pseudospin of exciton-polaritons excited at the oblique angle rotates about the Maialle field, while in the ground state ($\mathbf{k} = \mathbf{0}$) the rotation vanishes. Thus, linear-to-circular polarisation conversion depends on the ratio of the rotation period at the oblique angle and energy relaxation time of exciton-polaritons,

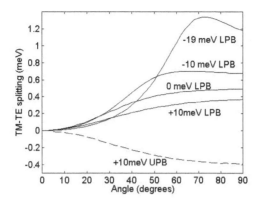

Fig. 9.9: Longitudinal-transverse polariton splitting calculated by Kavokin et al. (2004) for detunings: $+10$ meV, 0 meV, -0 meV and -19 meV. Solid lines: lower polariton branch, dashed lines: upper-polariton branch.

i.e., the time needed for a polariton to relax to the ground-state. At low pumping intensity, the relaxation time is longer than the rotation period, at some intermediate pumping the two times coincide and at high pumping the relaxation time becomes shorter than the period of pseudospin rotation. This explains the non-monotonic dependence of the circular-polarisation degree of emission on the pumping intensity in the linear pumping case (cf. Fig. 9.4).

The precession of the polariton pseudospin about the Maialle field resembles the precession of electron spins about the Rashba effective magnetic field. The Rashba effect, predicted by Bychkov and Rashba (1984), has stimulated development of spintronics, a new area of semiconductor physics studying propagation of spin-polarised electrons. The physical origin of these two fields is, however, different. The Rashba field is created by the spin-orbit interaction in asymmetric quantum wells, whereas the Maialle field appears because of the long-range electron–hole interaction and TE–TM splitting of the cavity modes. The orientation of these two fields is also different. The Rashba field is perpendicular to the electron wavevector, whereas the Maialle field is neither perpendicular nor parallel to the exciton (polariton) wavevector, in general.

9.7 Optical spin Hall effect

In high-quality microcavities, exciton-polaritons can propagate *ballistically*, i.e., without scattering, over a few picoseconds. This presents an opportunity of observing the pseudospin precession of individual exciton-polaritons under the effect of the Maialle field. The pseudospin precession in the ballistic regime leads to the *optical spin Hall effect* predicted by Kavokin et al. (2005c) and experimentally observed by Leyder et al. (2007).

The spin Hall effect is the appearance of a spin flux due to the direct current flow in a semiconductor. This effect predicted by Dyakonov and Perel (1971) found its experimental proof only recently in the work of Kato et al. (2004). It has a remarkable analogy in semiconductor optics, namely, in Rayleigh scattering of light in microcavities. The spin polarisation in a scattered state can be positive or negative, dependent on the orientation of the linear polarisation of the initial state and on the angle of rotation of the polariton wavevector during the act of scattering. Very surprisingly, spin polarisations of the polaritons scattered clockwise and anticlockwise have different signs.

Consider a semiconductor microcavity in the strong-coupling regime. We suppose that one of the $\mathbf{k} \neq \mathbf{0}$ states of the lower-polariton dispersion branch is resonantly excited by linearly polarised light (here, $\mathbf{k} = (k_x, k_y)$ is the in-plane polariton wavevector). The Rayleigh-scattered signal comes from the quantum states whose wavevector is rotated with respect to the initial $\mathbf{k}$ by some nonzero angle θ. We study polarisation of the scattered light as a function of θ and $\mathbf{k}$ taking into account the pseudospin rotation under the effect of the Maialle field (9.20).

Let us assume that light is incident in the (x, z)-plane and is polarised along the x-axis (TM-polarisation). It excites the polariton state with a pseudospin parallel to the x-axis. As the pseudospin is parallel to the effective field, it does not experience any precession at the initial point (see Fig. 9.10(b)). Consider now the scattering act that brings our polariton into the state (k'_x, k'_y) with $k'_x = k \cos \theta$ and $k'_y = k \sin \theta$. Following the classical theory of Rayleigh scattering, we assume that the polarisation does not change during the scattering act, so that at the beginning the pseudospin of the scattered state remains oriented in the x-direction (Fig. 9.10(b)). As the effective field is no longer parallel to the pseudospin, it starts precessing. Due to precession, the pseudospin acquires a z-component and circular polarisation emerges. The polarisation of emission depends on the ratio between the period of precession (given by the TE–TM splitting of the polariton state) and the polariton lifetime. If both quantities are of the same order, a peculiar angle dependence of the circular polarisation of scattered light appears.

Figure 9.11 shows schematically the dependence of the circular polarisation degree of light scattered by the cavity in different directions. Dark areas correspond to the right (left) circularly polarised light. One can notice an inequivalence of clockwise and anticlockwise scattering: if the spin-up majority of polaritons (right-circular polarisation) dominate scattering at the angle θ, the signal at angle $-\theta$ is mostly emitted by spin-down polaritons (left-circular polarisation) and vice versa. In order to obtain the polarisation distribution in scattering of TE-polarised light (incident electric field in the yz-plane) one should simply interchange areas in Fig. 9.11.

Rotation of the polariton pseudospin around the Maialle field is responsible for the appearance of the Langbein cross (Fig. 9.7). In his experiment, in x-, $-x$-, y- and $-y$-directions, X-polarised light corresponds to one of the eigenstates of the system (TE- or TM-polarised). The effective field in these points is parallel (antiparallel) to the pseudospin of the exciting light, thus no precession takes place and the polarisation is conserved. On the other hand, within the quarters, the effective field is inclined at some angle to the pseudospin, so that the precession takes place and the emission signal in cross-linear polarisation appears.

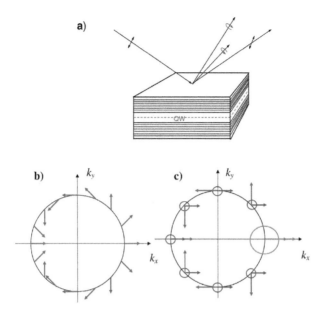

Fig. 9.10: (a) experimental configuration proposed by Kavokin et al. (2005a) allowing for observation of the optical spin Hall effect: linearly polarised light is incident at an oblique angle, circular polarisation of the scattered light is analysed, (b) orientations of the effective magnetic field induced by the LT splitting of the polaritons for different orientations of the in-plane wavevector, (c) pseudospin of polaritons created by the TM-polarised pump pulse at the point shown by the large circle, arrows show the pseudospins of polaritons just after the Rayleigh-scattering act and the effective field orientation.

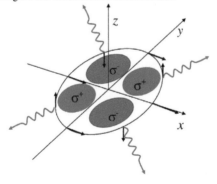

Fig. 9.11: Schematic diagram of the angular dependence of circular polarisation of emission of the microcavity. Dark areas correspond to the right-(left-) circularly polarised light. From Kavokin et al. (2005a).

9.8   Full Poincaré beams and polarisation shaping in microcavities

Exciton-polariton Rabi oscillations induced by cross-circularly polarised laser pulses arriving successively give rise to a peculiar effect that embeds a particular case of "polarisation shaping". In the regime of Rabi oscillations, photon and exciton populations in the system oscillate in time with the same frequency, but opposite phases. These

402   SPIN AND POLARISATION

oscillations can be measured independently, e.g., by the pump-probe Kerr rotation technique, as demonstrated by Brunetti et al. (2006). From the point of view of classical optics, they can be viewed as the effect of interference of two coherent electromagnetic waves emitted at different frequencies corresponding to the lower and upper polariton branches (beating). The intensity of the emitted light oscillates with a terahertz frequency corresponding to the splitting between upper and lower polariton frequencies. In optical experiments, Rabi oscillations are usually excited by short pulses of light. The excitation initially places the system in a purely photonic state that may be expanded in a polariton basis as

$$|C\rangle = \frac{1}{\sqrt{2}} \left(|LP\rangle + |UP\rangle \right), \tag{9.21}$$

where $|LP\rangle$ and $|UP\rangle$ correspond to the lower and upper polariton branches, respectively. The system evolves with time as

$$\frac{1}{\sqrt{2}} \left(|LP\rangle \exp\left(i\omega_{LP}t\right) + |UP\rangle \exp\left(i\omega_{UP}t\right) \right), \tag{9.22}$$

where ω_{LP} and ω_{UP} are the eigenfrequencies of lower and upper branch polariton states, respectively. After a time $t_x = \frac{\pi/2}{\omega_{UP}-\omega_{LP}}$, the system is brought to the purely excitonic state

$$|X\rangle = \exp\left(i\frac{\omega_{UP}+\omega_{LP}}{\omega_{UP}-\omega_{LP}}\frac{\pi}{4}\right)\frac{i}{\sqrt{2}}\left(|UP\rangle - |LP\rangle\right). \tag{9.23}$$

At the time $2t_x$, the system comes back to the purely photonic state

$$-|C\rangle = -\frac{1}{\sqrt{2}}\left(|LP\rangle + |UP\rangle\right). \tag{9.24}$$

And the cycle is repeated, the system bouncing between exciton and photon states until the dephasing and dissipation eventually suppress the oscillations.

Polarisation shaping is the extension to polarisation of the concept of "pulse shaping", that consists in sculpting desired profiles of a laser pulse (for instance for quantum control). The output of more complex devices can shape in this way the quantum states of light, for instance, Nisbet-Jones et al. (2011) engineered the shape of one photon to follow the skyline of the Tower Bridge, the Bodleian Library or the Giza pyramids. The shaping of the light field's amplitude is therefore an advanced discipline. Doing so at the vectorial level, to control also the state of polarisation, is a more recent concept, cast in a general framework by Brixner and Gerber (2001), who can shape arbitrary trajectories of polarisations thanks to a computer-controlled pulse shaper that modulates the degree of ellipticity and orientation of the elliptical axis through a liquid-crystal display. At the same time, another interest for peculiar states of polarised light arose in the framework of cylindrical-vector beams with so-called "Full Poincaré beams", introduced by Beckley et al. (2010), that are beams of light that takes all the states of polarisation in the cross-section of the beam. Polariton Rabi oscillations allow to merge these two ideas to realised Full Poincaré beams in time, that is, a pulse of light that takes all the states of polarisation during its duration, and shown in Fig. 9.12. This is a linear optical effect based on the coherent superposition of fields of different polarisations, each in the regime of Rabi oscillations. Suppose indeed that the polariton Rabi oscillator is excited by two short pulses of light of the same amplitude, but with opposite circular polarisations, the second one (left-circularly polarised) arriving with a delay τ after the first one (right-circularly polarised). The Jones vector of light emitted by the oscillator, in the limit of no dissipation, depends on time as

$$\begin{bmatrix} E_x \\ E_y \end{bmatrix} \propto \begin{bmatrix} 1 \\ i \end{bmatrix} \cos(\Omega t) + \begin{bmatrix} 1 \\ -i \end{bmatrix} \cos(\Omega(t-\tau)), \qquad (9.25)$$

where $\Omega = \omega_{UP} - \omega_{LP}$. In the particular case of $\Omega \tau = \pi/2$, we obtain

$$\begin{bmatrix} E_x \\ E_y \end{bmatrix} \propto \begin{bmatrix} \cos(\Omega t) + \sin(\Omega t) \\ i(\cos(\Omega t) - \sin(\Omega t)) \end{bmatrix} = \sqrt{2} \begin{bmatrix} \sin(\Omega t + \pi/4) \\ i\cos(\Omega t + \pi/4) \end{bmatrix}. \qquad (9.26)$$

One can see that the polarisation of light periodically changes with time: at $t = 0$ light is right-circularly polarised, at $t = \pi/4\Omega$ it is x-polarised, at $t = \pi/2\Omega$, it is left-circularly polarised, at $t = 3\pi/4\Omega$ it is y-polarised, etc. The resulting polarisation dynamics can be described as a circular motion of the Stokes vector of light at the surface of the Poincaré sphere. Note that the overall intensity of the emitted light remains time-independent in this regime. Now, accounting for the decay of Rabi-oscillations, we obtain the more complex polarisation dynamics

$$\begin{bmatrix} E_x \\ E_y \end{bmatrix} \propto \begin{bmatrix} 1 \\ i \end{bmatrix} \cos(\Omega t)\exp(-\gamma t) + \begin{bmatrix} 1 \\ -i \end{bmatrix} \cos(\Omega(t-\tau))\exp(-\gamma(t-\tau)), \quad (9.27)$$

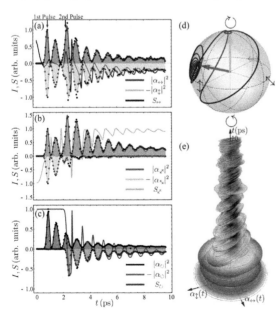

Fig. 9.12: Experimentally measured dynamics of the three components of the Stokes vector of light emitted by a semiconductor microcavity excited by two optical pulses with opposite circular polarisations, from Colas et al. (2015). The arrival times of two pulses are indicated by arrows (top left panel). Three left panels show the degrees of linear polarisation measured in xy, diagonal axes and to the circular polarisation degree, respectively. The theoretical fit is shown by lines. The dynamics of the Stokes vector on the surface of the Poincaré sphere corresponding to this experiment is shown by a line in the upper right panel. The vertical line shows the initial state of the polarisation, the horizontal line shows its final state. The three-dymensional plot in the right bottom panel shows the time-dependence of $|E_x^2|$ and $|E_y^2|$, taking all states of polarisation in time.

where γ is the decay rate. This dynamics corresponds to the spiral ending up in the pole of the Poincaré sphere corresponding to the left-circular polarisation. Varying the polarisations of the two pump pulses and delay between them, one can realise various trajectories of the Stokes vector, that can span the whole surface of the Poincaré sphere within several picoseconds. Polarisation shaping with polariton Rabi-oscillations has been studied in details by Colas et al. (2015) who also reports its experimental realisation. The authors provide closed-form expressions for the trajectories on the Poincaré sphere, that consist of spirals between an initial and final points whose parametrisation is specified in terms of sequence of two exciting pulses (see Exercise 9.3). This opens the way to the ultrafast coherent control of light-polarisation, that may be used in laser microprocessing.

Exercise 9.3 $^{(*)}$ *To parametrise the initial and final points on the Poincaré sphere, one needs to be able to excite the joint coherent state $|\alpha\beta\rangle$ for the photon (α) and exciton (β) fields. Show that a Gaussian pulse of temporal width σ achieves that for fields with a Rabi coupling Ω_R at resonance, provided that it is sent at a frequency ω_L given by*

$$\omega_L = \frac{\sigma^2}{\Omega_R} \ln \frac{|\alpha+\beta|}{|\alpha-\beta|}. \tag{9.28}$$

9.9 Optically-induced Faraday rotation

The z-component of $\mathbf{B}_{\mathrm{eff},k}$ splits $J=+1$ and $J=-1$ exciton states. It would be zero in the absence of polariton–polariton interactions. However, in the nonlinear regime it can arise due to the difference of concentrations of spin-up and spin-down polaritons, which leads to the optically induced Faraday rotation in microcavities. To show this, let us first consider the connection of the pseudospin formalism with the second quantisation representation.

If the polariton concentration is lower than the saturation density, the polaritons behave as good bosons and thus a pair of bosonic annihilation operators $a_{k,\uparrow}$, $a_{k,\downarrow}$ can be introduced to describe the polariton quantum states having a wavevector $\mathbf{k}$. The occupation numbers of spin-up, spin-down polariton states and the z-component of the corresponding pseudospin can be found from

$$N_{\mathbf{k}\uparrow} = \mathrm{Tr}\left[\rho_{\mathbf{k}} a_{\mathbf{k},\uparrow}^\dagger a_{\mathbf{k},\uparrow}\right] = \langle a_{\mathbf{k},\uparrow}^\dagger a_{\mathbf{k},\uparrow}\rangle, \tag{9.29a}$$

$$N_{\mathbf{k}\downarrow} = \mathrm{Tr}\left[\rho_{\mathbf{k}} a_{\mathbf{k},\downarrow}^\dagger a_{\mathbf{k},\downarrow}\right] = \langle a_{\mathbf{k},\downarrow}^\dagger a_{\mathbf{k},\downarrow}\rangle, \tag{9.29b}$$

$$S_{\mathbf{k},z} = \frac{1}{2}\left[\langle a_{\mathbf{k},\uparrow}^\dagger a_{\mathbf{k},\uparrow}\rangle - \langle a_{\mathbf{k},\downarrow}^\dagger a_{\mathbf{k},\downarrow}\rangle\right]. \tag{9.29c}$$

To find the dynamics of in-plane components of the pseudospin, one should introduce the bosonic operators for linear-polarised polaritons as:

$$a_{k,x} = \tfrac{1}{\sqrt{2}}\left(a_{k,\uparrow} + a_{k,\downarrow}\right), \qquad a_{k,-x} = \frac{1}{\sqrt{2}}\left(a_{k,\uparrow} - a_{k,\downarrow}\right), \tag{9.30a}$$

$$a_{k,y} = \tfrac{1}{\sqrt{2}}\left(a_{k,\uparrow} + ia_{k,\downarrow}\right), \qquad a_{k,-y} = \frac{1}{\sqrt{2}}\left(a_{k,\uparrow} - ia_{k,\downarrow}\right). \tag{9.30b}$$

Knowing the dynamics of these operators, $S_{\mathbf{k},x}$ and $S_{\mathbf{k},y}$ are expressed as

$$S_{\mathbf{k},x} = \frac{1}{2}\left[\langle a^\dagger_{\mathbf{k},x}a_{\mathbf{k},x}\rangle - \langle a^\dagger_{\mathbf{k},-x}a_{\mathbf{k},-x}\rangle\right] = \Re\langle a^\dagger_{\mathbf{k},\downarrow}a_{\mathbf{k},\uparrow}\rangle, \qquad (9.31a)$$

$$S_{\mathbf{k},y} = \frac{1}{2}\left[\langle a^\dagger_{\mathbf{k},y}a_{\mathbf{k},y}\rangle - \langle a^\dagger_{\mathbf{k},-y}a_{\mathbf{k},-y}\rangle\right] = \Im\langle a^\dagger_{\mathbf{k},\downarrow}a_{\mathbf{k},\uparrow}\rangle. \qquad (9.31b)$$

We shall consider the polariton pseudospin dynamics in the nonlinear regime. Let us start from the simplest case where all the polaritons are in the same quantum state, i.e., form a "condensate", so that the scattering to the other states is completely suppressed. Of course, this is true only at zero temperature. The general form of the interaction Hamiltonian reads[122]

$$H = \varepsilon(a^\dagger_\uparrow a_\uparrow + a^\dagger_\downarrow a_\downarrow) + V_1(a^\dagger_\uparrow a^\dagger_\uparrow a_\uparrow a_\uparrow + a^\dagger_\downarrow a^\dagger_\downarrow a_\downarrow a_\downarrow) + 2V_2(a^\dagger_\uparrow a_\uparrow a^\dagger_\downarrow a_\downarrow), \qquad (9.32)$$

where the index corresponding to the polariton state $\mathbf{k}$ in the reciprocal space is omitted.[123] In eqn (9.32), ε is the free polariton energy, while the matrix elements V_1 and V_2 correspond to the forward scattering of the polaritons in the triplet configuration (parallel spins) and in the singlet configuration (antiparallel spins). If the polariton–polariton interactions were spin-isotropic, i.e., the Hamiltonian (9.32) were covariant with respect to the linear transformation of the operators (9.30a), the matrix elements would be interdependent, so that

$$V_1 = V_2. \qquad (9.33)$$

However, this situation is not realised in semiconductor microcavities, where the major contribution to polariton–polariton interaction is given by the exchange term, so that the polariton–polariton interaction is in fact anisotropic (dependent on the mutual orientation of spins of interacting polaritons) and $|V_1| \gg |V_2|$. This anisotropy manifests itself experimentally in the optically induced splitting of the spin-up and spin-down polariton states, which has been experimentally measured as a function of the circular polarisation degree of the excitation (directly linked to the imbalance of populations of spin-up and spin-down states).

The Hamiltonian (9.32) conserves $N_\uparrow$ and $N_\downarrow$, but *not* the in-plane components of the pseudospin. It is straightforward to directly evaluate the commutator of this Hamiltonian with the operators governing linear polarisation. This commutator is zero only if

---

[122]Note that a different notation is used in a number of publications on polariton-polariton interactions. The constant describing the interaction of polaritons with parallel spins is denoted α_1, in which case $\alpha_1 = V_1$ according to our current definition. The constant describing the coupling of polaritons with antiparallel spins is introduced as $\alpha_2 = V_1 - 2V_2$. We will use the α notation where it is convenient and in agreement with the notations adopted in the literature (mostly in Chapter 10).

[123]The Hamiltonian without notation shortcut reads

$$H = \varepsilon(a^\dagger_{\mathbf{k},\uparrow}a_{\mathbf{k},\uparrow}+a^\dagger_{\mathbf{k},\downarrow}a_{\mathbf{k},\downarrow})+V_1(a^\dagger_{\mathbf{k},\uparrow}a^\dagger_{\mathbf{k},\uparrow}a_{\mathbf{k},\uparrow}a_{\mathbf{k},\uparrow}+a^\dagger_{\mathbf{k},\downarrow}a^\dagger_{\mathbf{k},\downarrow}a_{\mathbf{k},\downarrow}a_{\mathbf{k},\downarrow})+2V_2(a^\dagger_{\mathbf{k},\uparrow}a_{\mathbf{k},\uparrow}a^\dagger_{\mathbf{k},\downarrow}a_{\mathbf{k},\downarrow}).$$

the condition (9.33) is satisfied, which is not the case experimentally. The equation of motion for $\langle a_\downarrow^\dagger a_\uparrow \rangle$ can be obtained from eqns (9.16) and (9.32) to read

$$\partial_t \langle a_\downarrow^\dagger a_\uparrow \rangle = \frac{i}{\hbar} 2(V_1 - V_2) \left[\langle a_\downarrow^\dagger a_\downarrow a_\downarrow^\dagger a_\uparrow \rangle - \langle a_\uparrow^\dagger a_\uparrow a_\uparrow a_\downarrow \rangle \right] . \tag{9.34}$$

In the mean-field approximation, the fourth-order correlators in the right side of eqn (9.34) can be decoupled and eqn (9.34) can be transformed into an equation of precession of the pseudospin in an effective magnetic field $\mathbf{B}_{int}$ oriented along the structure growth axis. The absolute value of the field is determined by the difference between the populations of spin-up and spin-down polaritons as

$$g\mu_B |\mathbf{B}_{int}| = 2(V_1 - V_2)(N_\downarrow - N_\uparrow) . \tag{9.35}$$

Experimentally, the effect manifests itself as a rotation of the polarisation plane of emission if the σ_+ and σ_- populations are imbalanced, i.e., in the case of optically induced Faraday rotation. This effect has been observed by Brunetti et al. (2006). The imbalance of spin-up and spin-down polaritons is characteristic for elliptically polarised polariton condensates. The rotation of the main axis of the polarisation ellipse due to the effective field $\mathbf{B}_{int}$ induced by the condensate itself is referred to as self-induced Larmor precession. An example is given by Krizhanovskii et al. (2006).

To summarise, in general the effective magnetic field acting on the polariton pseudospin has two components. The in-plane component is governed by the TE–TM splitting of the cavity modes and the long-range electron–hole interaction in the exciton. It is concentration independent and leads to the beats between the circularly polarised components of the photoemission. The component parallel to the structure growth axis arises because of the anisotropy of the polariton–polariton interaction and depends on the imbalance between spin-up and spin-down polaritons. It leads to the Faraday rotation of the polarisation plane of light propagating through the cavity.

9.10   Interplay between spin and energy relaxation of exciton-polaritons

The linear model developed above neglects the inelastic scattering of polaritons leading to their relaxation in reciprocal space and their energy relaxation. Interpretation of nonlinear polarisation effects in microcavities requires its generalisation. Formally, kinetic equations for the occupation numbers and pseudospins can be decoupled only in the linear regime. Thus, a theoretical description of nonlinear processes requires the self-consistent accounting of both energy and spin relaxation processes.

Our starting point is the Hamiltonian of the system in the interaction representation:

$$H = H_{shift} + H_{scat} . \tag{9.36}$$

Polariton interaction with acoustic phonons and polariton–polariton scattering are taken into account here. Only the lower-polariton branch is considered, coupling with the upper branch and dark exciton states are neglected. The "shift" term describes interaction of exciton-polaritons having the same momentum, but possibly different spins, with Hamiltonian

$$H_{\text{shift}} = \sum_{\substack{\mathbf{k},\sigma=\uparrow,\downarrow}} \left(g_{\text{B}}\mu_{\text{B}}B_{\text{eff}}a_{\sigma,\mathbf{k}}^{\dagger}a_{-\sigma,\mathbf{k}} + V_{\mathbf{k},\mathbf{k},0}^{(1)}(a_{\sigma,\mathbf{k}}^{\dagger}a_{\sigma,\mathbf{k}})^{2} + V_{\mathbf{k},\mathbf{k},0}^{(2)}(a_{\sigma,\mathbf{k}}^{\dagger}a_{-\sigma,\mathbf{k}})^{2} \right)$$

$$+ \sum_{\substack{\mathbf{k},\mathbf{k}'\neq\mathbf{k} \\ \sigma=\uparrow,\downarrow}} \left(V_{\mathbf{k},\mathbf{k}',0}^{(1)}a_{\sigma,\mathbf{k}}^{\dagger}a_{\sigma,\mathbf{k}'}^{\dagger}a_{\sigma,\mathbf{k}}a_{\sigma,\mathbf{k}'} + V_{\mathbf{k},\mathbf{k}',0}^{(2)}a_{\sigma,\mathbf{k}}^{\dagger}a_{-\sigma,\mathbf{k}'}^{\dagger}a_{\sigma,\mathbf{k}}a_{-\sigma,\mathbf{k}'} \right). \quad (9.37)$$

Here, $a_{\uparrow,\mathbf{k}}$, $a_{\downarrow,\mathbf{k}}$ are annihilation operators of the spin-up and spin-down polaritons, $B_{\text{eff},\mathbf{k}} = B_{\text{eff},\mathbf{k},x} + iB_{\text{eff},\mathbf{k},y}$ ($B_{\text{eff},\mathbf{k},x}$ and $B_{\text{eff},\mathbf{k},y}$ are the x- and y-projections of the effective magnetic field). The "scattering term" H_{scat} describes scattering between states with different momenta

$$H_{\text{scat}} = \frac{1}{4} \sum_{\substack{\mathbf{k},\mathbf{k}'\neq\mathbf{k} \\ \sigma=\uparrow,\downarrow}} \exp\left(\frac{i}{\hbar}(E(\mathbf{k}) + E(\mathbf{k}') - E(\mathbf{k}+\mathbf{q}) - E(\mathbf{k}'-\mathbf{q}))t \right)$$

$$\times \left(V_{\mathbf{k},\mathbf{k}',\mathbf{q}}^{(1)}a_{\sigma,\mathbf{k}+\mathbf{q}}^{\dagger}a_{\sigma,\mathbf{k}'-\mathbf{q}}^{\dagger}a_{\sigma,\mathbf{k}}a_{\sigma,\mathbf{k}'} + 2V_{\mathbf{k},\mathbf{k}',\mathbf{q}}^{(2)}a_{\sigma,\mathbf{k}+\mathbf{q}}^{\dagger}a_{-\sigma,\mathbf{k}'-\mathbf{q}}^{\dagger}a_{\sigma,\mathbf{k}}a_{-\sigma,\mathbf{k}'} + \text{h.c.} \right)$$

$$+ \frac{1}{2} \sum_{\substack{\mathbf{k},\mathbf{q}\neq 0 \\ \sigma=\uparrow,\downarrow}} \exp\left(\frac{i}{\hbar}(E(\mathbf{k}) + \hbar\omega_{\mathbf{q}} - E(\mathbf{k}+\mathbf{q})) \right) U_{\mathbf{k},\mathbf{q}}a_{\sigma,\mathbf{k}+\mathbf{q}}^{\dagger}a_{\sigma,\mathbf{k}}b_{\mathbf{q}} + \text{h.c.} \quad (9.38)$$

In eqn (9.38), $b_{\mathbf{q}}$ is an acoustic phonon operator, $U_{\mathbf{k},\mathbf{q}}$ is the polariton–phonon coupling constant, $E(\mathbf{k})$ is the dispersion of the low polariton branch. The matrix elements $V_{\mathbf{k},\mathbf{k}',\mathbf{q}}^{(1)}$ and $V_{\mathbf{k},\mathbf{k}',\mathbf{q}}^{(2)}$ describe scattering of two polaritons in the triplet and the singlet configurations, respectively. As has been discussed above, in real microcavities the polariton–polariton interaction is strongly anisotropic: the triplet scattering is usually much stronger than the singlet one. Moreover, the matrix elements $V_{\mathbf{k},\mathbf{k}',\mathbf{q}}^{(1)}$ and $V_{\mathbf{k},\mathbf{k}',\mathbf{q}}^{(2)}$ can have opposite signs, as shown by Kavokin et al. (2005c). Indeed, the interaction between two polaritons with parallel spins is always repulsive, while polaritons with opposite spins may experience an attractive interaction and can even form a bound state (bipolariton), as discussed by Ivanov et al. (2004).

We use the Hamiltonian (9.36) to write the Liouville equation

$$i\hbar\frac{d\rho}{dt} = [H(t),\rho(t)] = [H_{\text{shift}} + H_{\text{scat}}(t),\rho(t)], \quad (9.39)$$

for the total density matrix of the system. We solve eqn (9.39) within the Born–Markov approximation already used in previous chapters. The Markov approximation means that the system has no phase memory. This is, in general, not true for the coherent processes described by the H_{shift} part of the Hamiltonian but, is a reasonable approximation for the scattering processes involving the momentum transfer. We apply the Markov approximation to the scattering part of eqn (9.39), which reduces to

$$\frac{d\rho}{dt} = -\frac{i}{\hbar}[H_{\text{shift}}, \rho] - \frac{1}{\hbar^2}\int_{-\infty}^{t}[H_{\text{scat}}(t), [H_{\text{scat}}(\tau), \rho(\tau)]]\, d\tau. \tag{9.40}$$

The next step is to perform the Born approximation $\rho = \rho_{\text{phon}} \otimes \prod_{\mathbf{k}} \rho_{\mathbf{k}}$. The phonons are then traced out with their occupation numbers being treated as fixed parameters determined by the temperature. The density matrices are given by eqn (9.15). They contain information about both occupation numbers and pseudospin components of all states in the reciprocal space. Equations. (9.36–9.40) together with formulas (9.29a) for occupation numbers and pseudospins are sufficient to derive a closed set of dynamics equations for $N_{\uparrow,\mathbf{k}}$, $N_{\downarrow,\mathbf{k}}$ and $S_{\mathbf{k},\perp} = \mathbf{e}_x S_{\mathbf{k},x} + \mathbf{e}_y S_{\mathbf{k},y}$, as has been done by Shelykh et al. (2005). Maialle spin-relaxation because of the TE–TM splitting and self-induced Larmor precession are reduced in this model to the precession of the polariton pseudospin about an effective magnetic field $\mathbf{B}_{\text{eff},\mathbf{k}}$ that arises from the Hamiltonian term H_{shift}. Once polariton populations and pseudospins are known, the intensities of the circular and linear components of photoemission are given by

$$I_{\mathbf{k}}^{+} = N_{\uparrow,\mathbf{k}}, \tag{9.41a}$$

$$I_{\mathbf{k}}^{-} = N_{\downarrow,\mathbf{k}}, \tag{9.41b}$$

$$I_{\mathbf{k}}^{x} = \frac{N_{\uparrow,\mathbf{k}} + N_{\downarrow,\mathbf{k}}}{2} + S_{x,\mathbf{k}}, \tag{9.41c}$$

$$I_{\mathbf{k}}^{y} = \frac{N_{\uparrow,\mathbf{k}} + N_{\downarrow,\mathbf{k}}}{2} - S_{x,\mathbf{k}}. \tag{9.41d}$$

Note that light polarisations parallel to the x- and y-axes correspond to the pseudospin parallel and antiparallel to the x-axis, respectively.

The general form of the kinetic equations (9.36)–(9.40) is complicated and their solution requires heavy numerical calculations. However, some particular cases can be considered analytically. If the pump intensity is weak, the polariton–polariton scattering is dominated by the scattering with acoustic phonons. The polariton–polariton interaction terms can be thus neglected and the system of kinetic equations becomes much simpler. For the occupation numbers and pseudospins, we have in this case

$$\frac{dN_{\mathbf{k}}}{dt} = -\frac{1}{\tau_{\mathbf{k}}}N_{\mathbf{k}} + \sum_{\mathbf{k}'}\left[(W_{\mathbf{k}\to\mathbf{k}'} - W_{\mathbf{k}'\to\mathbf{k}})(\frac{1}{2}N_{\mathbf{k}}N_{\mathbf{k}'} + 2\mathbf{S}_{\mathbf{k}}\cdot\mathbf{S}_{\mathbf{k}'}) \right.$$
$$\left. + (W_{\mathbf{k}\to\mathbf{k}'}N_{\mathbf{k}'} - W_{\mathbf{k}'\to\mathbf{k}}N_{\mathbf{k}})\right], \tag{9.42a}$$

$$\frac{d\mathbf{S}_{\mathbf{k}}}{dt} = -\frac{1}{\tau_{\text{sk}}}\mathbf{S}_{\mathbf{k}} + \sum_{\mathbf{k}'}\left[(W_{\mathbf{k}\to\mathbf{k}'} - W_{\mathbf{k}'\to\mathbf{k}})(N_{\mathbf{k}}\mathbf{S}_{\mathbf{k}'} + N_{\mathbf{k}'}\mathbf{S}_{\mathbf{k}}) \right.$$
$$\left. + (W_{\mathbf{k}\to\mathbf{k}'}\mathbf{S}_{\mathbf{k}'} - W_{\mathbf{k}'\to\mathbf{k}}\mathbf{S}_{\mathbf{k}})\right] + \frac{g_s\mu_B}{\hbar}[\mathbf{S}_{\mathbf{k}} \times \mathbf{B}_{\text{eff},\mathbf{k}}], \tag{9.42b}$$

where the transition rates are

$$
W_{\mathbf{k}\rightarrow\mathbf{k}'} = \begin{cases} \dfrac{2\pi}{\hbar}|U_{\mathbf{k},\mathbf{k}'-\mathbf{k}}|^2 n_{\mathrm{phon},\mathbf{k}'-\mathbf{k}}\delta(E(\mathbf{k}') - E(\mathbf{k}) - \hbar\omega_{\mathbf{k}'-\mathbf{k}}) \\[2ex] \dfrac{2\pi}{\hbar}|U_{\mathbf{k},\mathbf{k}'-\mathbf{k}}|^2 (n_{\mathrm{phon},\mathbf{k}'-\mathbf{k}} + 1)\delta(E(\mathbf{k}') - E(\mathbf{k}) - \hbar\omega_{\mathbf{k}'-\mathbf{k}}). \end{cases} \tag{9.43}
$$

n_{phon} are acoustic phonon occupation numbers. The Dirac delta functions account for energy conservation during the scattering processes. Mathematically, they appear from the integration of the time-dependent exponents in the second term of eqn (9.40). Writing the energy-conserving delta functions in eqn (9.43) we assume that the polariton longitudinal transverse splitting does not modify strongly the polariton dispersion curve. In numerical calculations, the delta functions may be replaced by resonant functions, e.g., Lorentzians, having a finite amplitude that can be estimated as an inverse energy broadening of the polariton state. The polariton lifetime has been introduced in eqn (9.42) to take into account the radiative decay of polaritons. The pseudospin lifetime is in general less than $\tau_{\mathbf{k}}$, it can be estimated as $\tau_{\mathrm{sk}}^{-1} = \tau_{\mathbf{k}}^{-1} + \tau_{\mathrm{sl}}^{-1}$, where τ_{sl} is the characteristic time of the spin-lattice relaxation (which accounts for all the processes of relaxation within the polariton spin doublet apart from one due to the LT splitting). The last term in eqn (9.42b) is the same as in eqn (9.20). It describes the pseudospin precession about an effective in-plane magnetic field given by the polariton LT splitting. This term is responsible for oscillations of the circular polarisation degree of the emitted light.

Qualitatively, the spin relaxation of exciton-polaritons can be understood from the following arguments. Below the threshold, the spin system is in the collision-dominated regime, i.e., relaxation of polaritons down to the ground-state goes through a huge number of random paths, each corresponding to the scattering process with an acoustic phonon. The polarisation degree displays a monotonic decay in this case. This is easy to understand, as in each scattering event the direction of the effective magnetic field changes randomly, so that on average no oscillation can be observed. The situation is completely analogous to one for electrons undergoing spin relaxation while moving in an effective Rashba field. Spin relaxation of spin-up and spin-down polaritons proceeds with the same rate, in general.

The situation changes dramatically above the stimulation threshold. In this case, the relaxation rates for spin-up and spin-down polaritons become different, in general. Once the ground-state is populated preferentially by polaritons having a given spin orientation, the relaxation rate of polaritons having this spin orientation is enhanced. This stimulated scattering process makes the circular polarisation degree of the emission increase in time, as was experimentally observed. Also, the polarisation degree is found to oscillate with a period sensitive to the pumping power and the detuning. The detuning (difference between bare exciton and cavity mode energies) has an important effect on the polariton spin relaxation. First, the LT splitting at a given value of the wavevector strongly depends on the detuning, as one can see from Fig. 9.9. Also, the energy relaxation is extremely sensitive to the detuning. At positive detuning, polaritons relax to the ground-state with more random steps than at negative detuning.

Exercise 9.4 $^{(**)}$ *Polaritons linearly polarised in the y-direction propagate ballisti-cally under the effect of an effective magnetic field orientated (a) in the x-direction, (b) in the y-direction, (c) along the z-axis (normal to the cavity plane). Find the time-dependent intensity of light emitted by the cavity in X,Y and circular polarisations if the polariton lifetime is τ.*

9.11   Polarisation of Bose condensates and polariton superfluids

In this section we address an important issue of polarisation of the condensates of exciton-polaritons in microcavities. Of course, the condensate polarisation is crucially dependent on the dynamics of its formation, i.e., pumping polarisation, polariton re-laxation mechanisms, etc. A realistic polariton system is typically out of equilibrium, since polaritons have a finite lifetime and the system should be pumped from outside to compensate the losses. The thermodynamic limit where the polariton distribution is given by an equilibrium Bose–Einstein distribution function is hardly achievable in re-ality.[124] Nevertheless, consideration of this limit is very important for understanding the fundamental effects of Bose condensation of polaritons. The characteristics of polariton condensates out of equilibrium deviate from the parameters predicted assuming thermal equilibrium, but the main qualitative tendencies remain valid.

Here, we assume that the Bose condensation of exciton-polaritons has already taken place and the polariton system is fully thermalised. The condensate is in a purely co-herent state that can be characterised by a wavefunction. The lifetime of polaritons is assumed infinite and there is no pumping. Moreover, in this section we shall assume zero temperature. However, we take into account the spin structure of the condensate. An exciton-polariton can have a +1 or −1 spin projection on the structure axis, thus we have a two-component (or spinor) Bose condensate. This is an important peculiarity of the polariton system with respect to various known 0-spin bosonic systems (atoms, Cooper pairs, He^4) and 1-spin systems (He^3). We consider the heavy-hole exciton-polaritons polarised in the plane of the cavity. In this case the condensate can be described by a spinor

$$\boldsymbol{\phi} = \begin{pmatrix} \phi_x \\ \phi_y \end{pmatrix}, \tag{9.44}$$

where ϕ_x, ϕ_y are complex functions of time and coordinates describing projections of the polarisation of the condensate on two corresponding in-plane axes. In this basis, the free energy of the system can be represented as

$$F = \boldsymbol{\phi}^* \hat{T}(-i\nabla)\boldsymbol{\phi} - \mu(\boldsymbol{\phi} \cdot \boldsymbol{\phi}^*)^2 + \frac{1}{4}(V_1 + V_2)(\boldsymbol{\phi} \cdot \boldsymbol{\phi}^*)^2 - \frac{1}{4}(V_1 - V_2)|\boldsymbol{\phi} \times \boldsymbol{\phi}^*|^2, \tag{9.45}$$

where $\hat{T}(-i\nabla)$ is the kinetic energy operator, μ is the chemical potential, which cor-responds to the experimentally measurable blueshift of the photoluminescence line of the microcavity due to formation of the condensate, V_1 and V_2 are interaction constants

---

[124]Experiments by the groups of Le Si Dang (Grenoble), Deveaud (Lausanne), Snoke (Pittsburg) and Bloch (Paris) show, however, that under sufficiently strong pumping the polariton gas quickly thermalises, so that its distribution function becomes very close to the Bose–Einstein function.

of polaritons with parallel and antiparallel spins, respectively. The two last terms of eqn (9.45) describe contributions of polariton–polariton interactions to the free energy. It can be shown that no other terms of this order can exist in the free energy if the cavity is isotropic in the xy-plane.

At zero temperature, only the ground state is occupied, so that substitutions can be made in eqn (9.45),

$$n = \boldsymbol{\phi} \cdot \boldsymbol{\phi}^*, \tag{9.46a}$$

$$S_z = \frac{i}{2} |\boldsymbol{\phi} \times \boldsymbol{\phi}^*|, \tag{9.46b}$$

where n is the occupation number of the condensate and S_z is the normal-to-plane component of the pseudospin of the condensate, which characterises the imbalance of populations of spin-up, $n_\uparrow$, and spin-down, $n_\downarrow$ polaritons

$$n = n_\uparrow + n_\downarrow, \tag{9.47a}$$

$$S_z = \frac{1}{2}(n_\uparrow - n_\downarrow), \tag{9.47b}$$

and is related to the circular polarisation degree of the emitted light ρ by

$$\rho = \frac{2S_z}{n}. \tag{9.48}$$

The free energy therefore reads

$$F = -\mu n + \frac{1}{4}(V_1 + V_2)n^2 + (V_1 - V_2)S_z^2. \tag{9.49}$$

The parameters V_1 and V_2 have a crucial impact on the critical conditions for the condensation of exciton-polaritons in microcavities. Figure 9.13 from Vladimirova et al. (2010) shows the phase diagram of a uniform polariton gas in the coordinates of interaction constants V_1 and V_2. In order to interpret it, let us recall that the system must have a minimum free energy at equilibrium. Four regions can be delimited depending on the signs of the quantities V_1, $V_1 + V_2$ and $V_1 - V_2$.

If $V_1 - V_2 > 0$, the minimum of free energy is achieved at $S_z = 0$, so that the polariton gas is linearly polarised (dense hatched regions in Fig. 9.13). The energy shift of the polariton gas is found by minimisation of the free energy over concentration:

$$E_{\text{lin}} = \mu = n(V_1 + V_2)/2. \tag{9.50}$$

Depending on the sign of $V_1 + V_2$, the minimum of the free energy F is achieved either at the minimum or at the maximum of the polariton concentration n.

1. If $V_1 + V_2 > 0$ (region I), the minimum of free energy is achieved at the minimum polariton concentration. The uniform polariton gas is stable and linearly polarised. The uniform distribution of the condensate in the real space is equivalent to its condensation in the reciprocal space. The Bose–Einstein condensate of polaritons is formed in the lowest energy quantum state of the system corresponding to zero in-plane wavevector.

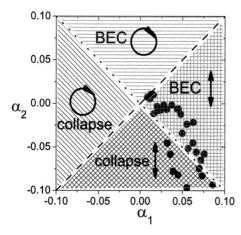

Fig. 9.13: Phase diagram of a gas of interacting exciton-polaritons, from Vladimirova et al. (2010). Regions of the parameter space corresponding to the real (reciprocal) space condensation are shown by tilted (horizontal or vertical) lines. The polarisation of the polariton gas is shown by hatching. Dense hatch stands for linearly and rare hatch for circularly polarised states. The values of interaction constants measured by Vladimirova et al. (2010) in a GaAs based microcavity with a single quantum well for different values of the exciton–photon detuning are shown by circles.

2. If $V_1 + V_2 > 0$ (region II), the minimum of free energy is achieved at the maximum polariton concentration. There are two possible scenarios of maximisation of the polariton concentration. First of all, in a system with a finite number of polaritons, the condensation in real space may take place, in which case the local concentration of polaritons would increase in certain spots of the sample while in the other spots the concentration would decrease. This would lead to the fragmentation of the condensate, similar to what has been observed by Kasprzak et al. (2006). The location of real space condensates would be pinned if a lateral potential disorder is present in the cavity. We note that the fragmentation of the condensate is only possible if the increase of its kinetic energy due to localisation is weaker than the decrease of polariton-polariton interaction energy. According to eqn (9.49), once started, the condensation in real space may go on until the collapse of the condensate. In reality, the increase of polariton concentration would be saturated due to the higher order positive terms in the free energy (omitted in eqn (9.49), which would eventually compensate the decrease of the free energy. If the increase of kinetic energy prevents fragmentation of the condensate and if the condensate is fed by a reservoir of exciton-polaritons situated in the exciton-like part of the lower polariton branch, another scenario would be realised. The condensate in this case would accumulate the polaritons while remaining spatially homogeneous. We note that the increase of polariton concentration is actually limited in this regime as well, due to the higher order terms in the free energy.

If $V_1 - V_2 < 0$, the minimum of free energy is achieved at $S_z = n/2$, which means that the polariton gas is circularly polarised (rare hatched regions in Fig. 9.13). In this case the shift of the polariton energy is

$$E_{\text{circ}} = \mu = nV_1 \,. \tag{9.51}$$

Depending on the sign of V_1, the minimum of the free energy F is achieved either at the maximum ($V_1 < 0$, region III) or at the minimum ($V_1 > 0$, region IV) of the polariton concentration n. In the region IV the polariton gas condenses in the reciprocal space

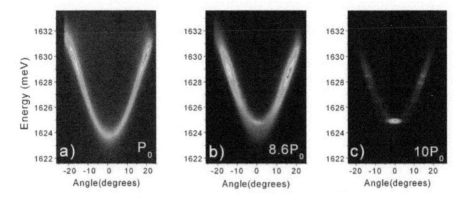

Fig. 9.14: Angle-dependent photoluminescence (PL) from a microcavity in the strong-coupling regime at different pumping intensities (in units of the threshold pumping P_0) (a, b, c) by Richard et al. (2005a). One can clearly see formation of the condensate at $\mathbf{k} = 0$ point (c) as well as the blueshift of the ground-state energy by about 1 meV between (a) and (c).

and Bose–Einstein condensation takes place. In the region III classical condensation of exciton-polaritons and fragmentation of the condensate in real space are expected for the systems with a finite number of exciton-polaritons. If the number of polaritons is unlimited, the virtually unlimited accumulation of polaritons in the condensate would take place, as we discussed above for the region II.

There have been various attempts to estimate V_1 and V_2 from the experimental data and to calculate them theoretically. Most authors believe that these constants should not be strongly dependent on the in-plane wavevector of interacting polaritons if it remains smaller than the inverse exciton Bohr radius. Despite the deviations between different works, they agree in two essential conclusions:

$$V_1 > 0 \quad \text{and} \quad V_1 > |V_2|. \tag{9.52}$$

These conditions are satisfied in the region I of the phase diagram in Fig. 9.13, where a linearly polarised Bose–Einstein condensate is the ground state of the system. In perfectly isotropic microcavities, the orientation of linear polarisation would be chosen spontaneously by the system and would randomly change from one experiment to another. In real structures it is generally *pinned* to one of the crystallographic axes. The pinning of linear polarisation can be caused by various factors, including the exchange splitting of the exciton state and the photon mode splitting at $\mathbf{k} = 0$. The second scenario is the most likely one, as the slightest birefringence in the cavity (say, in-plane variation of the refractive index by 0.01%) would yield the splitting of the cavity mode by about 0.1 meV, which is quite sufficient for pinning of the condensate polarisation. Here, and later, we shall assume the condensate is x-polarised in the absence of any magnetic field. The chemical potential of the condensate is given by eqn (9.50) in this case. As we mentioned above, this value corresponds to the blueshift of the photoluminescence line due to formation of the condensate. Such a blueshift, linearly dependent on the occupation number of the condensate, has been experimentally observed by the group of Le Si Dang (Fig. 9.14).

The dependence of the blueshift of the condensate on the polarisation of polaritons in the system allows for polarisation multistability and optical spin switching, as demonstrated by Amo et al. (2010a). The effect takes place if the frequency of the excitation laser is slightly detuned from the polariton mode toward higher energies. The blueshift induced by excitation may bring the polaritons of one or another spin orientation into resonance with the pump laser frequency, which "switches on" emission of light of a circular polarisation corresponding to the selected spin state. From the theoretical point of view, the nonlinear behaviour of optically pumped microcavities is excellently described by the spin-dependent Gross-Pitaevskii equations (9.45) for the condensates of exciton-polaritons.

9.12   Magnetic-field effect and superfluidity

We consider an exciton-polariton condensate in a microcavity subject to a magnetic field normal to the plane ($\mathbf{B} \parallel z$). We neglect the field effect on electron–hole relative motion and the resulting diamagnetic shift of the condensate and only consider the spin splitting of exciton-polaritons resulting from the Zeeman effect. The diamagnetic shift would result in a parabolic dependence of the energy of the condensate on the amplitude of the applied field, but would not influence the superfluidity and spin structure of the polaritons.

The free energy of the system can be represented as

$$F_\Omega = F - i\Omega|\boldsymbol{\phi} \times \boldsymbol{\phi}^*|, \tag{9.53}$$

where F is the free energy in the absence of the magnetic field given by eqn (9.45), $\Omega = \mu_B g_B/2$ with g_B being the exciton-polariton g-factor, μ_B being the Bohr magneton, B being the amplitude of the applied magnetic field.

At zero temperature, one can use eqn (9.49) for F so that eqn (9.53) yields

$$F_\Omega = -\mu n - 2\Omega S_z + \frac{1}{4}(V_1 + V_2)n^2 + (V_1 - V_2)S_z^2. \tag{9.54}$$

If a weak magnetic field is applied to the cavity (we shall define later what "weak" means in this case), the pseudospin projection and the number of polaritons n can still be considered as independent variables. In this regime, further referred to as the *weak-field regime*, the free-energy minimisation over S_z yields

$$S_z = \frac{\Omega}{V_1 - V_2}. \tag{9.55}$$

The condensate in this case emits elliptically polarised light with

$$\rho = \frac{\mu_B g B}{(V_1 - V_2)n}. \tag{9.56}$$

Interestingly, in this regime the chemical potential is still given by eqn (9.50) as immediately follows from minimisation of the free energy. Thus, the condensate emits light at the same energy. The redshift of the polariton energy due to the Zeeman effect is exactly compensated by an increase of the blueshift due to polarisation of the condensate. The minimum free energy of the system decreases quadratically with the field

$$F_{\Omega,\mathrm{min}} = -\frac{\Omega^2}{V_1 - V_2} - \frac{V_1 + V_2}{4}n^2 . \tag{9.57}$$

At the critical magnetic field

$$B_{\mathrm{c}} = \frac{(V_1 - V_2)n}{\mu_B g} , \tag{9.58}$$

the condensate becomes fully circularly polarised ($\rho = 1$). Beyond this point, S_z and n are no longer independent parameters, and $S_z = n/2$. Therefore, minimisation of the free energy eqn (9.53) over n yields a different result:

$$\mu = -\Omega + V_1 n , \tag{9.59a}$$

$$F_{\Omega,\mathrm{min}} = -\frac{V_1}{2}n^2 . \tag{9.59b}$$

In this regime, further referred to as the *strong-field regime*, the emission energy of the condensate decreases linearly with the field, so that the normal Zeeman effect can be observed in photoluminescence. The magnetic susceptibility of the condensate is discontinuous at the critical field.

In order to reveal the superfluid properties of exciton-polaritons we consider the Gross–Pitaevskii equation

$$i\hbar\frac{\partial \phi_j}{\partial t} = \frac{\partial F_\Omega}{\partial \phi_j^*} . \tag{9.60}$$

Here, the free energy F is given by eqn (9.45), $j = x, y$. The dynamics of the in-plane components of the polarisation of the system is given by

$$i\hbar\frac{\partial \phi_x}{\partial t} = T_{xy}(-i\nabla)\phi_x - \mu\phi_x + i\Omega\phi_y + \frac{1}{2}(V_1 + V_2)(\boldsymbol{\phi} \cdot \boldsymbol{\phi}^*)\phi_x$$
$$+ \frac{1}{2}(V_1 - V_2)|\boldsymbol{\phi} \times \boldsymbol{\phi}^*|\phi_y , \tag{9.61a}$$

$$i\hbar\frac{\partial \phi_y}{\partial t} = T_{yx}(-i\nabla)\phi_y - \mu\phi_y - i\Omega\phi_x + \frac{1}{2}(V_1 + V_2)(\boldsymbol{\phi} \cdot \boldsymbol{\phi}^*)\phi_y$$
$$- \frac{1}{2}(V_1 - V_2)|\boldsymbol{\phi} \times \boldsymbol{\phi}^*|\phi_x . \tag{9.61b}$$

The last terms in eqns (9.61) describe the self-induced Larmor precession of the polariton pseudospin that exactly compensates precession induced by the external magnetic field weaker than the critical field (9.58). This follows from eqns (9.46b) and (9.55). The exact compensation of the Larmor precession is one of the manifestations of the *full paramagnetic screening* or *spin Meissner effect* in polariton condensates (discussed in detail below).

Landau and his group in Moscow (1956).
Back row: S.S. Gershtein, L.P. Pitaevskii, L.A. Vainshtein, R.G. Arkhipov, I.E. Dzyaloshinskii.
Front row: L.A. Prozorova, A.A. Abrikosov, I.M. Khalatnikov, L.D. Landau, E.M. Lifshitz.

Following Lifshitz and Pitaevskii (1980) one can obtain the dispersion of excited states of the system from eqns (9.61) by the substitution

$$\phi(\mathbf{r}, t) = \sqrt{n}\mathbf{e} + \mathbf{A}e^{i(\mathbf{k}\cdot\mathbf{r}-\omega t)} + \mathbf{C}^* e^{-i(\mathbf{k}\cdot\mathbf{r}-\omega t)}, \qquad (9.62)$$

where

$$\mathbf{e} = \begin{cases} \mathbf{x}\cos\theta + i\mathbf{y}\sin\theta, & \text{if } B < B_{\mathrm{c}}, \\ \dfrac{1}{\sqrt{2}}(\mathbf{x} + i\mathbf{y}), & \text{if } B > B_{\mathrm{c}}, \end{cases} \qquad (9.63)$$

and $\theta = \frac{1}{2}\arcsin\left(\Omega/[n(V_1 - V_2)]\right)$.

Retaining only the terms linear in $\mathbf{A}$ and $\mathbf{C}$, and separating the terms with different complex exponential functions, one can obtain a system of four linear algebraic equations for A_x, A_y, C_x and C_y. The condition for existence of a non-trivial solution of this system allows one to obtain the dispersion of excited polariton states. At magnetic fields below B_{c} it reads

$$\omega^2 = \omega_0^2 + nV_1\omega_0 \pm \omega_0\sqrt{(nV_2)^2 + (V_1^2 - V_2^2)n^2\frac{\Omega^2}{\Omega_{\mathrm{c}}^2}}, \qquad (9.64)$$

where $\omega_0(k)$ is the energy of the lower-branch polariton state as a function of the in-plane wavevector $\mathbf{k}$ in the absence of the condensate, $\omega_0(k = 0) = 0$, $\Omega_{\mathrm{c}} = \mu_{\mathrm{B}}gB_{\mathrm{c}}/2$.

One can see that both branches start at the same point at $k = 0$, which corresponds to the bottom of the lowest polariton band blueshifted by μ. This shows that the ground state of the system remains two-fold degenerate in the presence of the external magnetic field below B_c, i.e., the Zeeman effect is fully suppressed (see inset in Fig. 9.15). This suppression results from the full paramagnetic screening of the magnetic field. The polarisation plane of light going through the microcavity does not experience any Faraday rotation in this case. The polariton spins orient along the field, so that the energy of the system decreases as in any paramagnetic material, but this decrease is exactly compensated by the increase of polariton–polariton interaction energy. This effect can be considered as a spin analogue of the *Meissner effect*, familiar in superconductors.

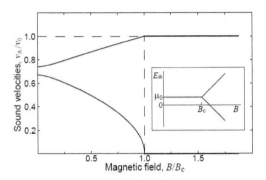

Fig. 9.15: Sound velocities of two branches of the excitations of a polariton condensate in a microcavity as a function of applied magnetic field, as predicted by Rubo et al. (2006). The inset shows Zeeman splitting of the polariton ground state, $\mu_0 = n(V_1 + V_2)/2$.

Both dispersion branches described by eqn (9.62) have a linear part characteristic of a superfluid. The peculiarity of our two-component superfluid consists in the difference of the sound velocities for the two branches

$$v_{\pm} = \left.\frac{\partial \omega}{\partial k}\right|_{k=0} = \sqrt{\frac{V_1 n}{2m^*} \pm \frac{1}{2m^*}\sqrt{(nV_2)^2 + (V_1^2 - V_2^2)n^2\frac{\Omega^2}{\Omega_c^2}}}, \qquad (9.65)$$

where m^* is the effective mass of the lowest polariton band. Note that at zero magnetic field the difference of two sound velocities persists, it is given by $\sqrt{n/(2m^*)}(\sqrt{V_1 + V_2} - \sqrt{V_1 - V_2})$. The sound velocity of the branch polarised with the condensate increases with magnetic field and achieves $\sqrt{V_1 n/m^*}$ at the critical field (Fig. 9.15). On the other hand, the sound velocity of the other branch decreases with increasing field and vanishes at $B = B_c$.

At strong fields, the ground-state of the lowest polariton band is split into the right-circularly and left-circularly polarised doublet. The lowest (right-circular, if $g > 0$) branch remains superfluid. Its dispersion is given by

$$\omega^2 = \omega_0^2 + 2nV_1\omega_0 . \qquad (9.66)$$

The sound velocity is independent of the field and always equals $\sqrt{V_1 n / m^*}$ in this regime. On the contrary, the higher (left-circular) branch has the same shape as the bare-polariton band

$$\omega = \omega_0 + 2(\Omega - \Omega_c). \tag{9.67}$$

The second term in the right part of eqn (9.66) describes the Zeeman splitting of the ground-state.

The superfluidity of exciton-polaritons at zero temperature exists below and above the critical field: at $B < B_c$ both branches of excitations have a linear dispersion and they touch each other at $k = 0$ (zero Zeeman splitting). At $B = B_c$ the Zeeman splitting is still zero, but one of two branches of excitations no longer has linear component in the dispersion. The condensate of polaritons exists only at $T = 0$ in two-dimensional systems with a parabolic spectrum, thus at any finite temperature, the polariton superfluidity disappears at the critical field. Interestingly, at $B > B_c$ the superfluidity reappears again due to the Zeeman gap that opens between the two branches of excitations: the polariton dispersion in the vicinity of the lowest-energy state again becomes linear (see the phase diagram in Fig. 9.16). With further increase of the field the critical temperature of the superfluid transition is expected to increase and approach the Kosterlitz–Thouless transition temperature, i.e., the critical temperature of the superfluid transition in a one-component interacting Bose gas. One can see that the two-component (spinor) nature of the polariton condensate leads to its peculiar behaviour in a magnetic field, in particular, to the full paramagnetic screening (or spin Meissner effect) and the appearance of a bicritical point in the phase diagram at $B = B_c$.

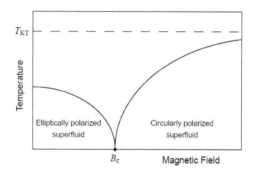

Fig. 9.16: Phase diagram from Rubo et al. (2006) of the polariton superfluidity: at zero field the superfluid is linearly polarised, in the weak-field regime it is elliptically polarised and Zeeman splitting is equal to zero, at the critical field no superfluidity exists, the condensate disappears at any finite temperature, at strong fields the superfluidity exists again, the condensate is circularly polarised.

9.13 Finite temperature case

At nonzero temperature, the excited polariton states are not empty. The temperature increase leads to depletion of the condensate, i.e., departure of the polaritons from the

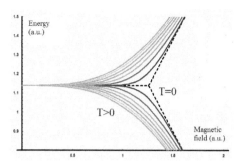

Fig. 9.17: Zeeman splitting of a polariton condensate at zero temperature (dashed lines) and finite temperatures increasing from right to left. One can see that the spin Meissner effect is relaxed as the temperature increases.

condensate to the states having a nonzero in-plane wavevector) and its depolarisation (as some polaritons can flip their spins). The depletion results in the decrease of n with increasing temperature. Let us analyse the depolarisation effect assuming a given occupation number of the condensate n and finite temperature T.

We shall consider only the $k = 0$ state assuming that all other states are weakly occupied and do not contribute to the polarisation of emission. We take into account all the excited states of the condensate characterised by different projections of its pseudospin on the magnetic field. The chemical potential of the system can be found from:

$$\mu = \frac{\partial}{\partial n}\langle E\rangle, \tag{9.68}$$

where the average energy of the condensate is

$$\langle E\rangle = -2\Omega\langle S_z\rangle + \frac{1}{4}(V_1 + V_2)n^2 + (V_1 - V_2)\langle S_z^2\rangle, \tag{9.69}$$

with

$$\langle S_z\rangle = \frac{\sum_{j=-n/2}^{n/2} j\exp\left(-E_j/(k_BT)\right)}{\sum_{j=-n/2}^{n/2}\exp\left(-E_j/(k_BT)\right)}, \tag{9.70a}$$

$$\langle S_z^2\rangle = \frac{\sum_{j=-n/2}^{n/2} j^2\exp\left(-E_j/(k_BT)\right)}{\sum_{j=-n/2}^{n/2}\exp\left(-E_j/(k_BT)\right)}, \tag{9.70b}$$

where $E_j = -2\Omega j + (V_1 - V_2)j^2$. One can see that with $T \to 0$, $S_z \to \Omega/(V_1 - V_2)$ if $\Omega < (V_1 - V_2)n/2$ and $S_z \to n/2$ if $\Omega > (V_1 - V_2)n/2$, which yields the results obtained above (eqns (9.55) and (9.58)). In order to obtain a compact expression for the chemical potential, we substitute the sums (9.70) by integrals. This is perfectly valid for large enough occupation numbers n. Now, the derivatives are easily calculated and the chemical potential reads

$$\mu = \frac{1}{2}(V_1+V_2)n - 2\Omega\frac{\exp(-(V_1-V_2)n^2/4k_BT)}{\Sigma_n}\left(\frac{n}{2}\sinh\frac{\Omega n}{k_BT} - \langle S_z\rangle\cosh\frac{\Omega n}{k_BT}\right)$$
$$+ (V_1 - V_2)\frac{\exp(-(V_1-V_2)n^2/4k_BT)}{\Sigma_n}\cosh\frac{\Omega n}{k_BT}\left(\frac{n^2}{4} - \langle S_z^2\rangle\right), \quad (9.71)$$

where

$$\Sigma_n = \int_{-\frac{n}{2}}^{\frac{n}{2}}\exp\left(-\frac{E(x)}{k_BT}\right)dx, \qquad E(x) = -2\Omega x + (V_1 - V_2)x^2. \quad (9.72)$$

Figure 9.17 shows schematically the behaviour of the chemical potential of a condensate as a function of the magnetic field at different temperatures (the lower branches in
the figure show the chemical potential behaviour as a function of the magnetic field). At
nonzero temperature, the chemical potential decreases with the field increase even if the
field is lower than the critical one. The condensate is never fully circularly polarised,
however, its magnetic susceptibility $\chi = \partial\mu/\partial B$ remains strongly field dependent. The
upper set of curves shows the excitation energy of the polariton having its spin opposite
to the magnetic field, i.e., the upper component of the Zeeman doublet observable in
reflection or transmission experiments. One can see that at finite temperatures, the Zeeman splitting is never exactly zero. It is, however, strongly suppressed below the critical
field due to the paramagnetic screening (spin Meissner effect).

Exercise 9.5 (**) *Obtain the Bogoliubov spectrum of excitations (dispersion relation)
from the Gross–Pitaevskii equation for a scalar wavefunction.*

9.14 Stationary states of spinor condensates

By definition, stationary states are those that do not evolve with time. Hence, for the
pseudospin vector **S** of a polariton condensate in a stationary state,

$$\frac{d\mathbf{S}}{dt} = 0. \quad (9.73)$$

Having in mind that, in general, the pseudospin dynamics is described by (see eqn 9.20)

$$\hbar\frac{d\mathbf{S}}{dt} = \mu_B g\left(\mathbf{B}_{\text{eff}}\times\mathbf{S}\right), \quad (9.74)$$

where $\mathbf{B}_{\text{eff}}$ is an effective field acting on the pseudospin, μ_B in the Bohr magneton, g
is the polariton g-factor, we arrive to a simple condition:

$$\mathbf{B}_{\text{eff}}\times\mathbf{S} = 0, \quad (9.75)$$

that describes the whole variety of stationary states of spinor condensates.

Now, the effective magnetic field $\mathbf{B}_{\text{eff}}$ is dependent on the linear polarisation splittings (e.g., TE-TM splitting), circular polarisation splitting induced by an external magnetic field $\mathbf{B}^{ext}$ and spin-dependent polariton-polariton interactions. Due to the interactions, eqn (9.75) becomes nonlinear. Its z-component becomes pseudospin-dependent:

$$B_z = B_z^{ext} + \frac{\alpha_1 - \alpha_2}{\mu_B g} S_z \, , \tag{9.76}$$

where α_1, α_2 are interaction constants of polaritons with parallel and antiparallel spins, respectively. Normalising the pseudospin to the number of particles in the condensate, N, as:

$$S_x^2 + S_y^2 + S_z^2 = N^2/4 \, , \tag{9.77}$$

one obtains four solutions of eqn (9.75) for each particular N, in general. Note that these solutions may correspond to different energies of the condensate. The energy of a stationary state may be expressed as:

$$E = -\mu_B g \left(\mathbf{B}_{\text{eff}} \cdot \mathbf{S} \right) \, . \tag{9.78}$$

Switching between stationary states of a polariton condensate has been experimentally demonstrated by Ohadi et al. (2015).

9.15 Conclusion

Spin-dependent phenomena in polariton gases and liquids are rich and unique. Due to the specific band structure of zinc-blend semiconductors, in microcavities, exciton-polaritons have just two allowed spin projections to the structure axis, which is highly untypical for integer spin particles. Resulting spinor condensates demonstrate a peculiar spin-dependent dispersion (modified Bogoliubov dispersion), give rise to specific magnetic effects, e.g. the spin Meissner effect. Microcavities offer several tools for the control of the polariton spin dynamics. All of them can be described in terms of effective magnetic fields acting upon polariton pseudospins. Such fields may be originated by TE-TM splitting of optical modes in microcavities, that leads to the optical spin Hall effect, in particular. Also, spin-dependent polariton-polariton interactions contribute to the effective magnetic fields giving rise to the self-induced Larmor precession and polarisation multistability. The spinor structure of polariton fluids is responsible for unusual topological phenomena including generation of half-vortices and half-solitons as discussed in detail in the next chapter.

9.16 Further reading

Hanle (1924) provides excellent additional reading on spin relaxation. We already mentioned the volume "Optical orientation" edited by Meyer and Zakharchenia (1984), which contains good experimental and theoretical chapters on the polarised optics of semiconductors. For those interested in spintronics, we propose to compare the optimistic review of Wolf et al. (2001) with the skeptical one of Dyakonov (2004).

10

QUANTUM FLUIDS OF LIGHT

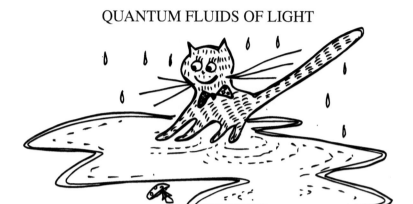

In this chapter, we deal with polaritons as a "quantum fluid of light", described by variants of the Gross–Pitaevskii equation. We discuss how interactions between flowing polaritons and a defect allow to study their superfluid regime and generate topological defects. Including spin gives rise to an effective magnetic field (polariton spin-orbit coupling) that acts on the topological defects—half-solitons and half-vortices—behaving as effective magnetic monopoles. We describe various techniques to create periodic potentials, that can lead to the formation of polaritonic bands and gaps with a unique flexibility. Special focus is given to topologically nontrivial bands, leading to a polariton topological insulator, based on a polariton graphene analog.[125]

---

[125] We thank A. Nalitov and H. Flayac for their several contributions to this chapter.

Microcavities, Second Edition. Alexey V. Kavokin, Jeremy J. Baumberg, Guillaume Malpuech, Fabrice P. Laussy, Oxford University Press (2017). © Alexey V. Kavokin, Jeremy J. Baumberg, Guillaume Malpuech, Fabrice P. Laussy. DOI 10.1093/oso/9780198782995.001.0001

10.1   Introduction

When the temperature of a system becomes lower than the quantisation energy, but particles remain free to exchange places, both their quantum nature and statistics merge to create a *quantum fluid*. Famous examples are superfluid Helium and the Bose–Einstein condensate (Leggett (2008) gives a succinct review). Such a fluid has qualitatively different properties than a classical fluid, in many cases due to the restrictions on the allowed excitations. Vortices in such fluids are quantised and topologically protected, while single-particle excitations are fully replaced by collective sound-like modes. Considering vortices as "particles" and sound modes as "electromagnetic waves", one can emulate many interesting and otherwise inaccessible systems, such as the early Universe, as shown by Zurek (1996), the Standard Model, by Volovik (2003), or black holes, by Unruh (1981). In microcavities, a coherent polariton gas of large density can be obtained through either Bose–Einstein condensation or resonant laser excitation as seen in previous chapters. The coherent evolution of this polariton gas is well described by various versions of the Gross–Pitaevskii equation and represents an ideal realisation of a quantum fluid. This fluid, made out of photonic quasi-particles, has been called a "Quantum Fluid of Light" by Carusotto and Ciuti (2013), a helpful epithet that we will also use in this chapter. Resonant laser excitation allows one to probe specific states of the polariton dispersion. The success encountered by this research field is predominantly due to the easy experimental access to several quantum fluid effects using the powerful and versatile tools of optics. A wide range and types of flow can be created by resonant laser excitation, whereas density, phase, temporal and spatial coherence, and motion, can be directly accessed by well-developed optical spectroscopy techniques. Several approaches allow one to model many types of in-plane potential landscape, among which are lateral patterning of microcavities, (see for example Jacqmin et al. (2014)), or using optical potentials induced by exciton-exciton interactions, as shown by Wertz et al. (2010) and Tosi et al. (2012). Moreover, polaritons are spinor particles with spin-anisotropic interactions, a feature used by Amo et al. (2010a) to design spin switches. Their specific spin-orbit coupling is at the origin of the optical spin Hall effect proposed by Kavokin et al. (2005a) and observed by Leyder et al. (2007).

All these ingredients make the polariton system a unique platform to study spinor quantum fluid effects. Their semiconductor implementation allows one to evidence on a chip properties that are otherwise very difficult to access. This has been clearly demonstrated by the avalanche of results obtained since the first observation of polariton BEC by Kasprzak et al. (2006). Further evidence of phase coherence has been given through the observation of quantised vortices by Lagoudakis et al. (2008). Dissipation-less flow of a signal in a polariton OPO has been reported by Amo et al. (2009a) whereas superfluidity and the Cerenkov regime of polaritons were claimed the same year by Amo et al. (2009b). This was followed by an original experimental scheme based on the interaction between a dense mono-kinetic flow of polaritons and a defect, allowing Amo et al. (2011) to confirm the superfluid nature of the polariton flow and to observe, for the first time, oblique solitons. In a similar scheme, Nardin et al. (2011) observed the dynamical creation of vortices and vortex–anti-vortex pairs. When the spin degree of freedom is taken into account, different types of spin polarised topological defect can

be formed. Spin-isotropic interactions, typical of atomic systems lead to the formation of skyrmions, whereas the spin-anisotropic interactions of the polaritonic system
lead to the formation of so called half-vortices (studiedby Rubo (2007) and Lagoudakis
et al. (2009)) and half-solitons (by Flayac et al. (2011) and Hivet et al. (2012)). Moreover, the energy splitting between linearly polarised eigenstates in photonic systems can
be interpreted as an effective magnetic field, which couples the polariton spin components. This effective magnetic field interacts with topological defects with half-integer
quantum numbers as if these defects would be magnetic charges interacting with a real
magnetic field as shown by Hivet et al. (2012) and Solnyshkov et al. (2012). Quantum
fluid phenomena present a significant applied interest as well, thanks to the flow properties for future optical devices: on the one hand, the backscattering is suppressed by
superfluidity. On the other hand, the quantum turbulence dramatically modifies the results of the scattering on possible defects with respect to the classical case, as discussed
by Barenghi et al. (2014).

The rest of this chapter is organised as follow. In section 10.2, we consider topological excitations of a homogeneous quantum fluid described by the Gross–Pitaevskii
equation with repulsive interactions, namely grey solitons in 1D and quantum vortices
in 2D. We address the specific case of the interaction between a static defect and the
polariton flow. In section 10.3, we consider the spinor case, introducing the concepts of
half-vortices, half-solitons and their effective magnetic monopole behaviour. The interaction between a spinor flow and a defect will be described in details in section 10.2.2.
In Section 10.5, we discuss the role played by the change of the linear particle dispersion through the application of an in-plane potential. We first consider the basic cases
of quasi-1D and 0D systems, the case of coupled 0D systems, which allowed the implementation of bosonic Josephson junctions and of the so-called self-trapping effect.
We then move to periodic systems, first 1D, which allows bandgap opening and allows,
in the nonlinear regime to obtain nonlinear gap states, the so-called "gap solitons" (see
their discussion by Chen and Mills (1987) and Eggleton et al. (1996)). Two-dimensional
lattices can also be designed and we will put the focus on the case of a honeycomb periodic potential. We will consider the polarisation degree of freedom and show how
the interplay between TE and TM splitted photonic modes and a finite Zeeman splitting allows the opening of topologically nontrivial bandgaps and of a polaritonic analog
of topological insulators (for recent reviews on topological insulators see for example
Hasan and Kane (2010) and Qi and Zhang (2011)).

10.2   Topological excitations in quantum fluids of light

10.2.1   *Topological defects in scalar condensates*

Apart from the small-amplitude excitations (Bogolons) described previously, a significant perturbation of the BEC allows the appearance of nontrivial modifications of its
wavefunction, including topological defects as detailed in the book of Kevrekidis et al.
(2008). Depending on the dimensionality of the system and on the nature of the interactions between particles, these defects can be of various types. In 1D, they manifest
themselves as solitons that are density dips/humps (so-called dark/bright solitons) for

the case of repulsive/attractive interactions, respectively. This is accompanied by a maximum phase shift of π through the soliton. Such objects are stable thanks to the interplay between nonlinear interactions and the dispersion, that compensate each other. However, since a 1D soliton can be continuously unfolded into a homogeneous solution, it should rather be called a pseudo-topological defect. A vortex—the soliton's counterpart in two-dimensional (2D) systems—carries a quantum of angular momentum. Such objects cannot be continuously transformed (at least in scalar condensates) into a vortex-free solution and are said to be topologically stable. Soliton excitations can also occur in 2D in the form of oblique solitons, as first shown by El et al. (2006) and as we will discuss in more detail later.

A BEC being a system of bosonic particles that occupy, at low temperature, the same lowest energy (ground) state, these particles can be described collectively by the same single-particle wavefunction. It is referred to, for this reason, as a macroscopic wavefunction or order parameter of the condensate. It is convenient to write it as

$$\Psi\left(\mathbf{r}, t\right) = \sqrt{n\left(\mathbf{r}, t\right)} e^{i\theta(\mathbf{r}, t)} . \tag{10.1}$$

This wavefunction is complex-valued and thus possesses a phase θ, that can possibly contain a propagation term $\mathbf{k} \cdot \mathbf{r}$. The wavefunction's amplitude is governed by the density $n = |\Psi|^2$ of the particles in the BEC. The evolution of the wavefunction in the mean-field approximation is well described by the Gross–Pitaevskii equation (GPE)

$$i\hbar\frac{\partial\Psi}{\partial t} = -\frac{\hbar^2}{2m}\nabla^2\Psi + \alpha_1|\Psi|^2\Psi , \tag{10.2}$$

where m is the mass of the particles. In what follows, we will concentrate on the case $\alpha_1 > 0$, which corresponds to repulsive interactions, as is the case with polaritons. The GPE requires the normalisation condition $\int |\Psi|^2 d\mathbf{r} = N$ where N is the total number of particles in the system. Stationary solutions are found upon writing $\Psi\left(\mathbf{r}, t\right) = \psi\left(\mathbf{r}\right) e^{-i\mu t}$ where μ is the chemical potential. This yields the stationary Gross–Pitaevskii equation

$$\mu\psi = -\frac{\hbar^2}{2m}\nabla^2\psi + \alpha_1|\psi|^2\psi . \tag{10.3}$$

10.2.1.1 *Dark solitons in 1D Bose–Einstein condensates* In a 1D system ($\mathbf{r} \to x$), the GPE remains valid according to the so-called multiple-scale expansion, as shown by Konotop and Salerno (1997). This allows special solutions to occur: the grey solitons, namely, dips in the density that remain stable (do not spread with time) even if they propagate, provided that the interactions are repulsive. This dip is associated with a local shift of the condensate's phase. The grey soliton solution reads

$$\Psi_S\left(x, t\right) = \sqrt{n_\infty}\left[\sqrt{1 - \frac{v_s^2}{c^2}}\tanh\left(\frac{x - v_s t}{\xi\sqrt{2}}\sqrt{1 - \frac{v_s^2}{c^2}}\right) + i\frac{v_s}{c}\right] . \tag{10.4}$$

Here n_∞ is the density far away from the soliton's core, v_s is the speed of the soliton related to its depth via $v_s = c\sqrt{n(0)/n_\infty}$ and $\xi = \hbar/\sqrt{2m\mu}$ is the healing length of the BEC, which defines the size of the soliton's core, the latter being modulated by the relativistic Lorentz factor $1/\gamma = \sqrt{1 - v^2/c^2}$. The faster a soliton moves, the shallower and the larger it becomes, behaving as a relativistic particle with respect to the speed of sound. The energy of a grey soliton is obtained by injecting eqn (10.4) into the condensate's energy, to obtain

$$E_s = \frac{4}{3}\hbar c n_\infty \left(1 - \frac{v^2}{c^2}\right)^{\frac{3}{2}} \tag{10.5}$$

which also coincides with a relativistic equation. One interesting feature is that assuming $v \ll c$, the effective mass of the soliton is given by

$$m_s = -\frac{4\hbar n_\infty}{c} \tag{10.6}$$

and is, therefore, *negative*. This is actually not so surprising since we are considering a particle-like density dip. This is similar to effects in semiconductors for example where holes also have a negative effective mass. A direct consequence is that a grey soliton minimises its energy by moving to higher density regions, and consequently two grey solitons (forming local density minima) see each other as potential barriers and *repel* at short range. The phase shift through the soliton is given by $\Delta\theta = \arccos\left(v_s/c\right)$, varying between 0 and π. For $v_s = 0$, the wavefunction is real-valued, the density at the soliton's core is exactly zero and the phase is the discontinuous Heaviside function of amplitude π, undefined at $x = 0$. This solution is called a "dark soliton". Grey solitons remain stable in the system, because interactions are present to compensate the dispersion that would disperse the wavepacket under the free propagation of the linear Schrödinger equation. Figure 10.1 on page 429 shows the normalised density profile $n_s\left(x\right) = \left|\Psi_s\left(x\right)\right|^2$ [panel (a)] of several grey solitons, together with their phase [panel (b)]. Notably, the phase is discontinuous at the core of a dark soliton (as a Heaviside function) and is undefined (singular) at this point. This is an important feature of topological defects, that governs their stability.

10.2.1.2 *Vortices in 2D scalar Bose–Einstein condensates*   We now turn to the 2D case. In the previous chapters, we have seen that the superfluid is irrotational due to the link between its velocity field and the spatial variations of its phase: $\mathbf{v} = \hbar/m\nabla\theta$, and therefore $\nabla \times \mathbf{v} = \mathbf{0}$. One would therefore, at first sight, conclude that the fluid would not respond to an imposed rotation. This is only partially true. Indeed, the phase θ of the superfluid flow is defined up to 2π, which means that any spatial variation of θ around a closed path is $2\pi l$ for integer l:

$$\oint \nabla\theta\left(\mathbf{r}, t\right) - 2\pi l. \tag{10.7}$$

The case $l = 0$ corresponds to a homogeneous condensate or a slightly perturbed one with a continuous flow. Cases where $l \neq 0$ are more intriguing. The only way to have a phase that changes, e.g., by 2π on a closed loop is to make it wind around a central point, where it is singular. This phase winding defines the so-called *quantised vortex* introduced by Onsager (1949) in the context of liquid helium, the "quantum" corresponding to the integer l. One direct consequence is that the circulation of the velocity on a closed loop is quantised in units of h/m (h being Planck's constant):

$$\oint \mathbf{v} \cdot d\mathbf{l} = \frac{\hbar}{m} \oint \boldsymbol{\nabla}\theta\,(\mathbf{r}, t) \cdot d\mathbf{l} = l\frac{h}{m} \,. \qquad (10.8)$$

The central point where the phase is singular corresponds to a vanishing density of the fluid $n(0) = 0$, just like in the dark soliton case. The velocity field around the vortex core, given that $\theta(\phi) = l\phi$ in polar coordinates, reads

$$\mathbf{v} = \frac{\hbar}{m} \boldsymbol{\nabla}\theta\,(\mathbf{r}, t) = l\frac{\hbar}{mr}\mathbf{u}_\theta \,. \qquad (10.9)$$

As one can see, the particles rotate faster and faster while approaching the density minimum (the vortex core) like in a classical whirlpool [see the right panel of Fig. 10.2 (arrows)]. But their nucleation within the fluid is radically different. The quantisation of the velocity field means that vortices appear in a superfluid put in rotation above some critical velocity below which the rotation is ignored by the flow. This is a strongly counter-intuitive property of quantum fluids.

In the thermodynamic limit, the wavefunction of a condensate carrying a vortex is, according to Pitaevskii and Stringari (2003):

$$\psi_V\,(r, \phi) = f_l\,(r)\,e^{il\phi} \,. \qquad (10.10)$$

The radial part $f_l(r)$ can be evaluated variationally, which yields, e.g., for $l = 1$,

$$f_1\,(r) = \sqrt{n_\infty}\frac{r/\xi}{\sqrt{(r/\xi)^2 + 2}} \,, \qquad (10.11)$$

where n_∞ is the density at infinity and $\xi = \hbar/\sqrt{2m\mu}$ is the healing length of the condensate, determining the density-dependent vortex (core) size. The corresponding density profile $n_V = |\psi_V|^2$ is plotted in Fig. 10.2 [right panel] together with the velocity field, and the associated phase [left panel]. In the Born approximation, the free energy of the condensate for the stationary solution $\psi(\mathbf{r})$ reads

$$E = \int \left\{ \frac{\hbar^2}{2m}|\boldsymbol{\nabla}\psi\,(\mathbf{r})|^2 + \frac{\alpha}{2}|\psi\,(\mathbf{r})|^4 \right\} d\mathbf{r} \,. \qquad (10.12)$$

Injecting eqn (10.10) into eqn (10.12) gives the energy of a condensate that contains a single vortex. To find the vortex energy, one should subtract the energy of the uniform state $\int \frac{\alpha}{2}|\psi\,(\mathbf{r})|^2 d\mathbf{r}$ with the same number N of particles. Following this procedure, the energy of a vortex is separated into a core energy part that should be evaluated

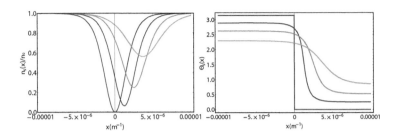

Fig. 10.1: Grey solitons solutions plotted with the polariton parameters at $t = 5\,\text{ps}$ for the values $v_s = \{0, 0.25, 0.5, 0.75\}c$: (a) Density profiles and (b) phases profiles. When the minimum of the density becomes zero, the grey soliton is called "dark".

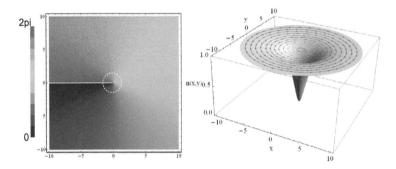

Fig. 10.2: Left panel: Phase of the wavefunction (10.10), the singularity is marked with a white/dashed circle and the phase jump with solid white line. Right panel: Vortex density profile together with its velocity field (arrows).

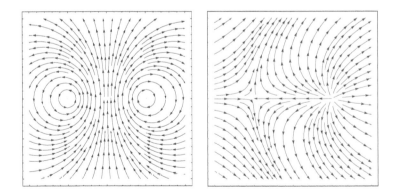

Fig. 10.3: Vortex-antivortex pair ($l = \pm 1$). Left: velocity field. Right: phase field $(\cos(\theta), \sin(\theta))$.

numerically and a kinetic energy part E_k, which dominates for large system sizes. The latter is given by

$$E_k = \frac{l^2 \pi n_\infty \hbar^2}{m} \ln \left(\frac{R}{\xi} \right) , \qquad (10.13)$$

where R is the radial size of the system, which has to be finite, otherwise the vortex energy, growing logarithmically, would become infinite. The single quantised vortex ($l = 1$) is of course the lowest energy (vortex) state. Since the energy of a vortex is proportional to the condensate density, it moves along density gradients to minimise its energy.

In the case of multiple vortices, the corresponding kinetic energy is found from

$$E_k = \frac{\hbar^2}{2m} \int \left[\sum_i \nabla \theta_i (\mathbf{r}_i) \right]^2 d\mathbf{r} , \qquad (10.14)$$

where $\theta_i(\mathbf{r}_i)$ is the phase of the vortex i at the position $\mathbf{r}_i$. In the simplest case of a vortex pair, the energy is found to be

$$E_k = E_{k1} + E_{k2} + E_{int} , \qquad (10.15a)$$

$$E_{int} = \frac{2l_1 l_2 \pi n_\infty \hbar^2}{m} \ln (R/d) , \qquad (10.15b)$$

l_1 and l_2 being the winding number associated to each vortex and d the distance between them. The interaction energy E_{int} is negative (positive) for vortices having a winding number with an opposite (identical) sign, respectively, which means that they attract (repel) each other. Furthermore, a vortex with a large winding number is unstable (Exercise 10.1) and thus decays into a corresponding number of singly-charged vortices. Another different configuration is the vortex-antivortex pair.

Exercise 10.1 (*) *Show that forming a vortex with a winding number $l = \pm 2$ is more energetically costly than forming a pair of vortices with $(l_1, l_2) = (\pm 1, \pm 1)$. Generalise to higher winding numbers.*

10.2.1.3   *The Berezinskii–Kosterlitz–Thouless transition*   In 2D, vortices behave differently from other excitations (non-topological ones). Indeed, their energy grows logarithmically with the size of the system. They can consequently be activated thermally, but only above a critical temperature T_{BKT}, named after Berezinskii (1971) and Kosterlitz and Thouless (1973). In order for the vortex solution to be profitable, it has to lower the free energy F_V of the system. Taking into account the vortex entropy S_V induced by thermal fluctuations at temperature T, the free energy reads

$$F_V = E_V - T S_V . \qquad (10.16)$$

Here E_V is the energy of a vortex at zero temperature as established previously. The vortex entropy depends on the logarithm of accessible positions for the vortex as

$$S_V = k_B \ln \left(\frac{R^2}{\xi^2} \right) , \qquad (10.17)$$

where k_B is the Boltzmann constant, and assuming a size ξ^2 for the vortex. Finally, we obtain

$$F_V = \left(\frac{l^2 \pi n_\infty \hbar^2}{m^2} - 2k_B T \right) \ln \left(\frac{R}{\xi} \right) . \tag{10.18}$$

We can immediately derive a critical temperature above which the energy of vortices become smaller than the thermal energy $k_B T$:

$$T_{\text{BKT}} = \frac{l^2 \pi n_\infty \hbar^2}{2 k_B m^2} . \tag{10.19}$$

In this case, single vortices can be thermally activated. Below T_{BKT}, they are bound into vortex-antivortex pairs that only perturb the fluid locally, as shown in Fig. 10.3 on page 429. The appearance of vortices results in a breakdown of superfluidity, as they lead to a friction between the normal and superfluid fractions of the fluid. The temperature T_{BKT}, therefore, sets the frontier to be crossed to enter a superfluid regime in 2D. This defines the so-called Berezinskii–Kosterlitz–Thouless transition. This was evidenced by Hadzibabic et al. (2006) with a 2D gas of Rubidium atoms. It is clear from eqn (10.19) that singly-quantised vortices ($|l| = 1$), which are favored energetically anyway, give the lowest critical temperature with respect to other winding numbers. In spinor condensates, we will see that half-quantum vortices, with half-integer winding numbers, result in a T_{BKT} that is two times smaller.

Vortices are the elementary *topological* excitations of superfluids and are well-known in the context of superconductors, since their prediction by Abrikosov (1957). Vortices and vortex lattices were first nucleated putting an atomic BEC into rotation with a stirring laser, as done by Anderson et al. (1995) and Matthews et al. (1999). In polariton condensates, the spontaneous formation of vortices has been first observed by Lagoudakis et al. (2008) at deterministic positions pinned to disorder, as shown in Fig. 10.4. Next, an artificial phase-imprinting method was proposed by Marchetti et al. (2010), Tosi et al. (2011) and Roumpos et al. (2011), based on a Gauss-Laguerre probe in the optical parametric oscillator regime. Later, vortices have been detected by Amo et al. (2011), Nardin et al. (2011) and Sanvitto et al. (2011) in the turbulence of a polariton fluid propagating past an obstacle. We will focus on this kind of configuration for the discussion on oblique (half) solitons later on. Such a large number of experimental observations reflects again the convenience offered by the polaritonic system for the investigation of hydrodynamic-related effects. One of the advantages is that vortices are much larger in spatial size than in atomic condensates ($\xi \sim 1\,\mu\text{m}$ for polaritons) and, therefore, are more easily observable. The other advantage is that using the techniques of classical optics, it is possible to reconstruct the entire wavefunction of the condensate, namely its density, from the intensity of light that escapes from the microcavity, and its phase (mod 2π), from interferograms. Furthermore, the control of the condensate parameters such as its density or its wavevector can be performed by simply changing the pumping intensity.

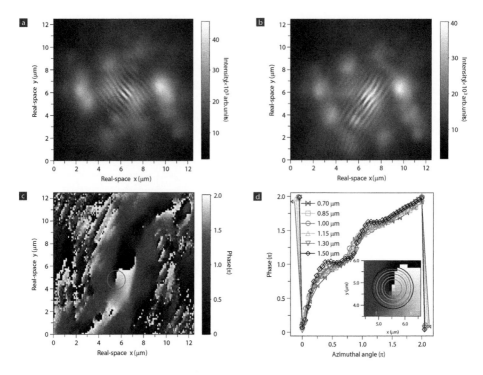

Fig. 10.4: Observation of vortices in a polariton condensate from Lagoudakis et al. (2008). (a) Interferogram evidencing a vortex: the typical fork-like dislocation (phase singularity) is visible inside the circle. (b) Same information, but this time the vortex is overlapped with a different region of the condensate. (c) Real-space phase profile extracted from the interferogram of panel (a). The circle highlights the vortex (same real-space area as in (a) and (b)). (d) Phase as a function of the azimuthal angle for different radii as shown in the inset of (d) (magnification of (c)).

10.2.2 *Interaction with a static defect; superfluidity and topology*

One fruitful configuration, first described by El et al. (2006), consists of a defect (potential barrier) larger than the healing length and crossing the flow of a BEC. This experiment is the quantum analogue of, e.g., a jet aircraft flying at supersonic velocity in a classical fluid (air). It revealed considerable differences between the two cases. Depending on the velocity of the fluid with respect to the speed of Bogolons in the media, the flow can be unperturbed (superfluid regime) or give rise in the wake of the obstacle to turbulence, vortex streets and oblique solitons, that would extend without deformation. One can expect that this kind of solitonic solution in 2D should become unstable against perturbations. While the stability analysis performed later by Kamchatnov and Pitaevskii (2008) showed that the soliton is indeed unstable, the instability is only convective, which means that it is damped while being dragged away downstream from the defect. The oblique solitons are nucleated from the shock waves because of the dispersion of the excitations (which is not linear), and because of the interactions between the particles, which favor the formation of solitons as stable structures. Another way to describe the generation of solitons is the following: the fluid tends to accelerate locally

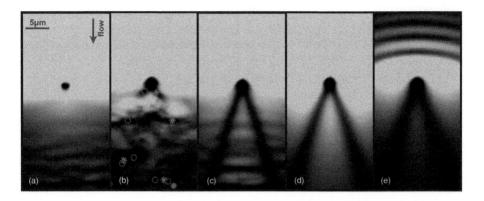

Fig. 10.5: Interaction between a polariton flow and a defect, for decreasing polariton densities from left to right, as simulated by Pigeon et al. (2011). (a) Superfluid regime. (b) Turbulent regime, the circles and crosses show at a given time the vortex/antivortex positions, respectively. (c,d,e) Soliton regime.

close to the defect (see Frisch et al. (1992)), and since the velocity field and the phase of the fluid are related via $\mathbf{v} = \hbar\nabla\theta/m$, an important local phase shift is acquired, giving birth to negative interference that produce density dips, which are the solitons.

This experimental scheme is, however, difficult to implement using atomic BECs, whereas it revealed to be quite easy to implement by resonant excitation of strongly coupled cavities. Pigeon et al. (2011) first considered theoretically this problem by direct numerical solution of the GPE including pumping and lifetime. Their result is well summarised by Fig. 10.5, which shows different regimes depending on the velocity of the fluid: superfluidity, turbulence and oblique solitons. This phenomenology has been demonstrated by Amo et al. (2011) and without being exhaustive, this work has been followed by several others, e.g., by Nardin et al. (2011) and Sanvitto et al. (2011), where vortices and turbulence have also been evidenced. More quantitatively, a stationary BEC flow containing a soliton is described by the solution

$$v_x = \frac{M\left(1 + a^2 n\right)}{\left(1 + a^2\right) n}, \qquad v_y = -\frac{aM\left(1 - n\right)}{\left(1 + a^2\right) n}, \tag{10.20a}$$

$$n\left(\chi\right) = 1 - \left(1 - p\right)\operatorname{sech}\left[\frac{\sqrt{1 - p}}{\sqrt{1 + a^2}}\chi\right]^2, \tag{10.20b}$$

where $\chi = x - ay$ is a tilted coordinate perpendicular to the oblique soliton (with a the slope of the soliton with respect to the y-axis), $M = v/c$ is the so-called Mach number and $p = M^2/(1 + a^2)$. In 1D systems, we have seen that the speed of a soliton is related to its depth and this is still true in 2D. Indeed, at fixed value of M, increasing a increases the depth of the soliton and thus reduces its speed. In other words, the more the soliton is tilted with respect to the x-axis, the faster it moves with respect to the condensate, and the shallower and larger it becomes. The phase of the wavefunction can be recovered by writing

$$\theta\left(\mathbf{r}\right) = \frac{m}{\hbar}\int \mathbf{v}\left(\mathbf{r}\right) d\mathbf{r}, \tag{10.21}$$

and as in the one dimensional case, the presence of the soliton (and the related density dip) is accompanied by a local phase shift lying between 0 and π. Indeed, the oblique soliton can be seen as a 1D soliton for which the second spatial coordinate plays the role of time. However, the strict analogue of the 1D dark soliton does not exist in 2D, since an oblique soliton cannot have zero velocity ($p = 0$) with respect to the flow. Interestingly, it has been shown by Cilibrizzi et al. (2014) and discussed by Amo et al. (2015) and Cilibrizzi et al. (2015) that interference patterns seen in the wake of a defect, in the linear regime, can lead to features showing similarities with oblique solitons, such as density dips accompanied with phase shifts. It is indeed a general fact that complex quantum flows can lead even in the linear regime to the appearance of different types of phase-singularities as demonstrated theoretically by Flayac et al. (2013) or Kamchatnov and Pavloff (2015), for instance.

10.3   Half-integer topological defects in spinor quantum fluids

10.3.1   *Introduction*

Multicomponent (spinor) condensates allow more complex topological excitations than scalar condensates, as they indeed allow the mixing of both the phase and the spin topologies. One can cite, as illustrative examples, solitons in spinor 1D condensates from the works of Öhberg and Santos (2001) and Kevrekidis et al. (2004) (vector solitons) and oblique solitons in spinor 2D systems, already considered theoretically, for instance by Gladush et al. (2009). Kawaguchia and Ueda (2012) give a thorough review of spinor condensates. In 1D, many possible configurations have been described, depending on the strength and type of the particle interactions (repulsive or attractive). In particular, a solution where the kink lies in only one component was reported by Salomaa and Volovik (1989) known as the dark-antidark soliton or half-soliton (HS). The counterpart of such a defect in 2D systems is the so-called half-vortex or skyrmion, introduced by Khawaja and Stoof (2001) and Kasamatsu et al. (2005), depending on the intercomponent interactions type. Half-integer topological defects have been originally predicted by Volovik and Mineev (1976) in the context of superfluid Helium 3, and their experimental observation was reported at the intersection of three grain boundaries of cuprate superconductors by Kirtley et al. (1996) in the form of half-vortices. With the reports of polariton condensation, it appeared quite natural to investigate the possibility of having quantised vortices in the systems. Rubo (2007) predicted that due to the spontaneous formation of the polariton condensate with a well defined linear polarisation, as we have seen previously, the elementary topological excitations are not regular, but half-integer vortices. This statement, being at first quite mysterious for the polariton community, has attracted much attention. The first experimental observation of vortices in a polariton condensate by Lagoudakis et al. (2008) followed quickly under non-resonant pulsed excitation. In that work, the sample had strong enough structural imperfections to induce a significant disorder landscape. The separate condensates forming islands (due to the disorder) at early times have different phases, and their reconnection at higher densities induces phase dislocations at their interface in the form of vortices. Therefore, vortices were observed pinned to defects at deterministic positions. Since this very first experiment was not polarisation resolved, it could not resolve

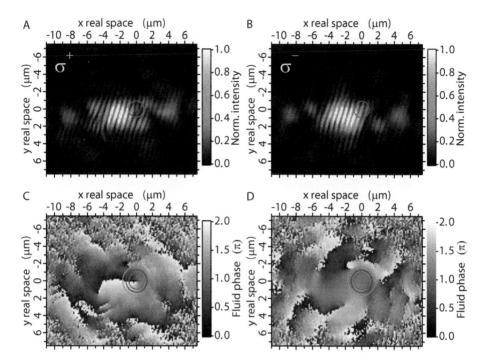

Fig. 10.6: Observation of half-quantum vortices in a polariton condensate from Lagoudakis et al. (2009). (a,b) Reconstructed interferograms for σ_+ and σ_- polarisations at the energy of the condensate. We see the typical forklike dislocation appearing in only one component, highlighted by the circles, evidencing the half-vortex. (c,d) Associated real space phase map extracted from the interferograms.

half-vortices. The next year, the same group, Lagoudakis et al. (2009), revisited their experiment [see Fig. 10.6] this time polarisation-resolved and confirmed that regular vortices actually co-existed with the expected half-quantum vortices. These experiments have paved the way for a large number of proposals and experimental reports on vortices in semiconductor microcavities.

We now study in more details these half-integer topological defects, namely half-vortices and half-solitons. A two-component polariton condensate at $0\,\mathrm{K}$ can be described by a vectorial macroscopic wavefunction $\boldsymbol{\Psi} = (\Psi_+, \Psi_-)^T$, whose evolution follows a set of (coupled) GPEs. Assuming first a parabolic dispersion with an effective mass m^* and an infinite lifetime of the particles, their equation of motion reads

$$i\hbar\frac{\partial\Psi_+}{\partial t} = -\frac{\hbar^2}{2m^*}\nabla^2\Psi_+ + \left(\alpha_1|\Psi_+|^2 + \alpha_2|\Psi_-|^2\right)\Psi_+,\qquad(10.22\mathrm{a})$$

$$i\hbar\frac{\partial\Psi_-}{\partial t} = -\frac{\hbar^2}{2m^*}\nabla^2\Psi_- + \left(\alpha_1|\Psi_-|^2 + \alpha_2|\Psi_+|^2\right)\Psi_-.\qquad(10.22\mathrm{b})$$

Time-independent solutions are obtained by expressing the BEC's wavefunctions as $\Psi_\pm(\mathbf{r},t) = \psi_\pm(\mathbf{r})\exp(-i\mu t)$, where $\mu = (\alpha_1 + \alpha_2)n_0/2$ is the chemical potential related to the density of the homogeneous condensate $n_0 = n_{0+}/2 = n_{0-}/2$, consistent with the linearly polarised ground state for a spin anisotropy $\alpha_2 \approx -0.1\alpha_1$. For a single-kink solution such as a vortex or a soliton, the asymptotic behaviour can be found analytically. For this, let us assume that far away from the defect the condensate density is unperturbed by the presence of this defect and is, therefore, constant in each component.

10.3.2 Half-vortices

The original description of half-vortices proposed by Rubo (2007) was performed in the linear polarisation basis, that neglects the TE-TM splitting. The stationary vectorial order parameter $\psi(\mathbf{r})$ of the two component spinor condensate reads

$$\psi = \begin{pmatrix} \psi_x \\ \psi_y \end{pmatrix} = \begin{pmatrix} \sqrt{n_x}e^{i\theta_x} \\ \sqrt{n_y}e^{i\theta_y} \end{pmatrix}. \tag{10.23}$$

Indeed, for an arbitrary polarisation of the condensate, the phases $\theta_{x,y}(\mathbf{r})$ and the densities $n_{x,y}(\mathbf{r})$ are defined independently for each component. However, since the condensate forms with a well defined polarisation, the previous expression can be rewritten as:

$$\psi_{lin} = \begin{pmatrix} \psi_x \\ \psi_y \end{pmatrix} = \sqrt{n_0} \begin{pmatrix} \cos{(\eta)}\,e^{i\theta} \\ \sin{(\eta)}\,e^{i\theta} \end{pmatrix}. \tag{10.24}$$

This means that $\theta_x = \theta_y = \theta$, $\sqrt{n_x} = \sqrt{n_0}\cos\eta$ and $\sqrt{n_y} = \sqrt{n_0}\sin\eta$, where n_0 is the total density, $\eta(\mathbf{r})$ is the linear polarisation angle and $\theta(\mathbf{r})$ is a global phase of the spinor condensate, namely the phase that would be measured in an experiment where the polarisation is not resolved. This representation, where the phase and polarisation angle are separated, means that we consider the particles being part of a global fluid, each particle having a linear polarisation defined by η. It is possible here to define a global velocity for the spinor condensate $\mathbf{v} = \hbar/m\nabla\theta$. This representation is valid only for linear polarisation states, indeed no global phase can be correctly defined for elliptic condensates (i.e. forming under applied magnetic field as in the case of Rubo et al. (2006)).

Therefore, for the two-component spinor condensate, one is considering two phases: η and θ. From the same argumentation than in Section 10.2.1.2 devoted to vortices, it follows that $\theta(\mathbf{r})$ being defined modulus 2π is allowed to wind around a central point to form a global phase vortex, that actually coincides with a usual integer vortex, provided that the orientation of η remains homogeneous in space. Furthermore, η can wind, in its turn keeping θ constant to form the so-called polarisation vortex. To describe a vortex state in spinor condensates, we, therefore, need two winding numbers, denoted as k and m for the polarisation and phase respectively. The latter cases correspond to an integer value of one winding number while the other one is zero. Since a linear polarisation direction (corresponding to the oscillation direction of the electric field) is defined only up to π, it should therefore be possible for η to wind, e.g., by only π around the vortex core preserving the continuity of the polarisation texture. However, the whole order parameter defined by eqn (10.24) is not invariant under the transformation $\eta \to \eta + p\pi$

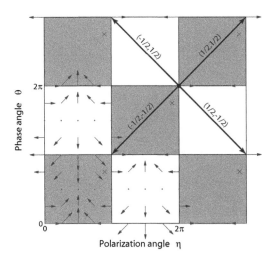

Fig. 10.7: Representation of the spinor polariton condensate's order parameter, from Rubo (2007). A vortex corresponds to a diagonal line connecting two points of the chessboard. This line is associated with the transformation $\eta \to \eta + 2k\pi$ and $\theta \to \theta + 2m\pi$. The blue arrows show the vector field associated with η, namely $\text{Re}(\boldsymbol{\psi}_{\text{lin}})$.

Phase angle θ

Polarization angle η

(with $p \in \mathbb{Z}$), so something more is needed and since the orientation of η is oscillating with time anyway, the continuity is broken. Indeed, we have

$$\begin{pmatrix} \psi_x \\ \psi_y \end{pmatrix} = \begin{pmatrix} \cos\left(\eta + p\pi\right) e^{i(\theta + q\pi)} \\ \sin\left(\eta + p\pi\right) e^{i(\theta + q\pi)} \end{pmatrix} = \begin{pmatrix} (-1)^p (-1)^q \cos\left(\eta\right) e^{i\theta} \\ (-1)^p (-1)^q \sin\left(\eta\right) e^{i\theta} \end{pmatrix}, \qquad (10.25)$$

which means that the order parameter is invariant under the *combined* transformations $\{\eta, \theta\} \to \{\eta + p\pi, \theta + q\pi\}$ where $\{p, q\} \in \mathbb{Z}$ provided that p and q have the same parity. In the winding number representation, given that $\{k, m\} = \{p/2, q/2\}$, we immediately obtain that k and m are allowed to take both integer and *half-integer* values, defining the so-called *half-quantum vortices* in the latter case. The π winding of a quantity on a closed loop appears counter-intuitive since the particles need to wind two times to recover the initial phase for a Möebius band. However, note that, for example, the $\{k, m\} = \{1/2, 0\}$ half-vortex is forbidden, since the global continuity of the order parameter is violated in that case. The half-winding of one of the two phases imposes the other one to behave in a fractional fashion as well, to preserve this global continuity. A global winding number could be defined as $\kappa = k + m$ and the continuity would require κ to be an integer in order to recover physical phase windings of $2\kappa\pi$ around the vortex core.

The order parameter of the spinor polariton condensate can be seen as an infinite chessboard. A vortex is defined by a diagonal line connecting two equivalent points on this chessboard characterised by the winding numbers k and m, as shown in Fig. 10.7.

10.3.2.1 *Energy* We know from Section (10.2.1.2) that the vortex energy is dominated by its kinetic part (or elastic part) growing logarithmically with the system size. This statement should hold for half-vortices as well. Since for the linearly polarised condensate far from the vortex core, we are allowed to separate θ and η, the kinetic energy of the condensate reads

$$E_k = \frac{\hbar^2 n_0}{2m^*} \int \left[\nabla\eta\left(\mathbf{r}\right) + \nabla\theta\left(\mathbf{r}\right)\right]^2 d\mathbf{r}. \qquad (10.26)$$

Having a single vortex in the system means that:

$$\theta = k\phi \quad \text{and} \quad \eta = m\phi, \tag{10.27}$$

where ϕ is the polar angle. E_k is, therefore, reduced to

$$E_k = \frac{\hbar^2 n_0}{2m^*}(k^2 + m^2)\ln\left(\frac{R}{\xi}\right), \tag{10.28}$$

where $\xi = \hbar/\sqrt{2m^*\mu}$ is the healing length of the condensate and the chemical potential is defined by $\mu = (U_0 - U_1)n_0$. As one would have expected, the lowest energy half-vortices possess winding numbers equal to one-half. The four building blocks are thus defined by

$$\{k, m\} = \begin{cases} \{+1/2, +1/2\} \\ \{-1/2, -1/2\} \\ \{+1/2, -1/2\} \\ \{-1/2, +1/2\} \end{cases} . \tag{10.29}$$

Noticing moreover that if, e.g., $l = 1$ and $k, m = \pm 1/2$, one has $l^2 = 2(k^2 + m^2)$, which means that the energy of an integer (phase or polarisation) vortex is twice larger than that of a half-vortex. The latter, therefore, embodies the elementary topological excitation of a polariton condensate. In addition, the Berezinskii–Kosterlitz–Thouless (discussed in Sec. 10.2.1.3) is reduced by a factor 2 if half-vortices are involved.

10.3.2.2 *Polarisation texture* So far we have not yet discussed the core structure of the half-vortex, which is especially interesting. This aspect is, however, far more transparent in the circular polarisation basis.

Let us rewrite the order parameter (10.24) in the circular polarisation basis. Using the transformation $\psi_\pm = (\psi_x \mp i\psi_y)/\sqrt{2}$, we obtain:

$$\psi_{lin} = \sqrt{\frac{n_0}{2}}\begin{pmatrix} e^{i(\theta+\eta)} \\ e^{i(\theta-\eta)} \end{pmatrix}. \tag{10.30}$$

Each component possesses its own phase $\theta_+ = \theta + \eta$ and $\theta_- = \theta - \eta$. As well, the linear polarisation angle is defined by the phase difference between the two circular components $\eta = (\theta_+ - \theta_-)/2$ and the global phase is the sum of the two $\theta = (\theta_+ + \theta_-)/2$. Considering a vortex state characterised by the winding numbers k and m, we can define a new set of winding numbers $(l_+, l_-) = (k + m, k - m)$, which are bound to be integers. The corresponding vortex order parameter reads

$$\psi_{HV} = \sqrt{\frac{n_0}{2}}\begin{pmatrix} e^{il_+\phi} \\ e^{il_-\phi} \end{pmatrix}. \tag{10.31}$$

This makes the half-vortex representation much clearer and intuitive. Indeed, we see that from the phase point of view, each component takes separately the form a vortex wave-function with the winding numbers (l_+, l_-). In this representation, the four elementary half-vortices are characterised by:

$$\{l_+, l_-\} = \begin{cases} \{+1, 0\} \\ \{-1, 0\} \\ \{0, +1\} \\ \{0, -1\} \end{cases} . \tag{10.32}$$

This means that a half-vortex solution corresponds to having a vortex in one component while the other one remains unperturbed (at least if $\alpha_2 = 0$), which makes their interpretation more natural.

Let us now discuss the half-vortex as a whole, including its core structure. To do so, we need to rewrite the order parameter out of the linear polarisation approximation, which was found by an asymptotic analysis far away from the vortex core, including the radial profiles $f_+(r) = \sqrt{n_+(r)}$ and $f_-(r) = \sqrt{n_-(r)}$:

$$\psi_{HV}(r, \phi) = \begin{pmatrix} f_+(r)\, e^{il_+ \phi} \\ f_-(r)\, e^{il_- \phi} \end{pmatrix} . \tag{10.33}$$

The stationary ($\partial_t \to -i\mu$) spinor GPEs (10.22b) including further the TE-TM splitting can be rewritten in the following dimensionless form:

$$\begin{pmatrix} -\frac{\nabla^2}{2} - 1 + A_1|\psi_+|^2 + A_2|\psi_-|^2 & \chi(\partial_y + i\partial_x)^2 \\ \chi(\partial_y - i\partial_x)^2 & -\frac{\nabla^2}{2} - 1 + A_1|\psi_-|^2 + A_2|\psi_-|^2 \end{pmatrix} \begin{pmatrix} \psi_+ \\ \psi_- \end{pmatrix}$$
$$= \begin{pmatrix} 0 \\ 0 \end{pmatrix} . \tag{10.34}$$

We have used the scaling relations $\psi_\pm \to \sqrt{\mu/(\alpha_1 + \alpha_2)}\psi_\pm$, $\mathbf{r} \to \sqrt{\hbar^2/(m^*\mu)}\mathbf{r}$ and $t \to (\hbar/\mu)\, t$. Here, $\mu = (\alpha_1 + \alpha_2)\, n_\infty/2$ with $n_\infty = |\psi_+(\infty)|^2 + |\psi_-(\infty)|^2$ is the condensate density far away from the vortex core, $A_{1,2} = \alpha_{1,2}/(\alpha_1 + \alpha_2)$ and $\chi = \beta m^*/\hbar^2$. Injecting the ansatz (10.33) in eqn (10.34), and neglecting first the TE-TM splitting ($\chi = 0$), yields the equations for the radial functions:

$$f_+'' + f_+' + \left(2 - 2A_1 f_+^2 - 2A_2 f_-^2 - \frac{l_+^2}{r^2}\right) f_+ = 0 \tag{10.35a}$$

$$f_-'' + f_-' + \left(2 - 2A_1 f_-^2 - 2A_2 f_+^2 - \frac{l_-^2}{r^2}\right) f_- = 0 \tag{10.35b}$$

In the simplest case, where the circular polarised components do not interact ($A_2 = 0$), the half-vortex with e.g. $l_+ = +1$, $l_- = 0$ corresponds to a homogeneous distribution of σ_+ component and a simple vortex in σ_-. Clearly, in the centre of such a half-vortex, the density is nonzero (due to the σ_+ component) and the polarisation is circular, since the density of the σ_- component vanishes in the core of the vortex. This kind of vortex is referred to as a *coreless vortex* in atomic condensates, but in that case, the spin isotropy of the interactions rather gives birth to skyrmions, as discussed by Khawaja and Stoof (2001) and Kasamatsu et al. (2005), that are asymptotically circularly polarised. In our case, moving from the centre of the vortex, the polarisation of the half-vortex changes from circular to linear in a continuous manner [see Fig. 10.8].

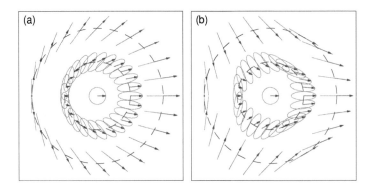

Fig. 10.8: Two half-vortex polarisation textures, from Rubo (2007). (a) $(+1/2, +1/2)$ or $(+1/2, -1/2)$ and (b) $(-1/2, +1/2)$ or $(-1/2, -1/2)$ half-vortices. The red arrows show the orientation of the linear polarisation at a specific time and the red circles show the trajectories of the arrows displaying the degree of circular polarisation.

It is interesting to note that at the core position neither the σ_+ nor the σ_- particles contribute to the condensate's motion since one component is completely static and the other one is absent. As well, far away from the vortex core, particles of the component carrying the vortex are almost immobile since $v(r) \sim 1/r$. It is, therefore, tempting to define a global hydrodynamic velocity

$$\mathbf{v}_g = \frac{n_+}{n_0}\mathbf{v}_+ + \frac{n_-}{n_0}\mathbf{v}_- , \qquad (10.36)$$

for the spinor condensate, which describes well this behavior. However, apart from a linear polarisation case (which is definitely not the case of the vortex core) for which $\mathbf{v}_g = \hbar/m^*\theta$ (θ the global phase defined above), this velocity cannot be linked with any phase in the system and its circulation is, in general, not quantised.

The set of equations (10.35) can only be solved numerically. Some results are shown in Fig. 10.9 in the case $(l_+, l_-) = (+1, 0)$ for $A_2 = 0$ [panel(a)] and $A_2 = -0.1A_1$ [panel (b)]. In the first case, the normalised radial function in the σ_+ component is obviously that of a regular vortex and can be well approximated by the function $f_+(r) = r/\sqrt{r^2 + 2}$, while $f_-(r) = 1/2$. In the second case, the intercomponent interactions being attractive ($A_2 < 0$) the presence of the dip in the σ_+ density is seen as a weak potential barrier by the σ_- component, which, therefore, exhibits a shallow density minimum at the vortex position as well [see Fig. 10.9]. Additionally, the vortex size is slightly reduced: the healing length ξ is renormalised in the σ_+ component.

10.3.3 Half-solitons

So far, we have introduced the half-vortex solution in a two-dimensional system. We now analyse its one-dimensional counterpart that has been called a half-soliton (HS) by Flayac et al. (2011), Solnyshkov et al. (2012), Flayac et al. (2012) and Hivet et al. (2012) as a continuation of the works of Salomaa and Volovik (1989). We shall see how the behaviour of half-quantum topological defects can be described in terms of relativistic "material points" and "point charges" (an easy way) or in terms of underlying local spin

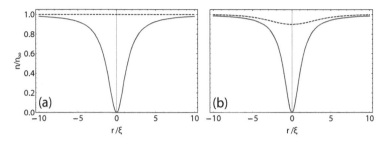

Fig. 10.9: Half-vortex normalised density slice $n_\pm(r) = f_\pm^2(r)$. (a) $A_2 = 0$ and (b) $A_2 = -0.1A_1$. The solid red (dashed/blue) curve shows the σ_+ (σ_-) component. The vortex, therefore, lies here in the σ_+ component here: $(\pm 1, 0)$ half-vortex.

dynamics (a harder way). The importance of such analogies as "magnetic charges" or magnetic monopoles introduced by Dirac (1931) will thus become especially clear.

Analogously to the half-vortex discussion, the half-soliton corresponds to its integer counterpart in one component while the other component remains homogeneous (at least for $\alpha_2 = 0$). Therefore, the polarisation of the condensate is circular (elliptic if the soliton is moving) at the half-soliton core and linear at $\pm\infty$. Since the scalar dark soliton solution [see 10.2.1.1] is simply given by $\psi_S(x) = \sqrt{n_0}\tanh(x)$ and its phase is a Heaviside function of amplitude π, the dark half-soliton order parameter reads (in the circular polarisation basis and when $\alpha_2 = 0$)

$$\begin{pmatrix} \psi_+(x) \\ \psi_-(x) \end{pmatrix} = \sqrt{\frac{n_0}{2}} \begin{pmatrix} 1 \\ \tanh(x) \end{pmatrix}, \tag{10.37}$$

where the dark soliton lies here in the σ_- component. Obviously the σ_- phase displays a π phase shift while it remains constant in the σ_+ component. Rewriting eqn(10.37) in the linear polarisation basis yields

$$\begin{pmatrix} \psi_x(x) \\ \psi_y(x) \end{pmatrix} = \frac{\sqrt{n_0}}{2} \begin{pmatrix} 1 + \tanh(x) \\ i - i\tanh(x) \end{pmatrix}. \tag{10.38}$$

Looking at asymptotic forms, one can easily obtain

$$\psi_x^{HS}(+\infty) = \sqrt{n_0}e^{2ih\pi}\cos(2s\pi), \tag{10.39a}$$
$$\psi_y^{HS}(+\infty) = \sqrt{n_0}e^{2ih\pi}\sin(2s\pi), \tag{10.39b}$$
$$\psi_x^{HS}(-\infty) = \sqrt{n_0}e^{ih\pi}\cos(s\pi), \tag{10.39c}$$
$$\psi_y^{HS}(-\infty) = \sqrt{n_0}e^{ih\pi}\sin(s\pi), \tag{10.39d}$$

where h and s are half-integer numbers that can be seen as topological charges. Elementary dark half-solitons appear for $\{h, s\} = \{\pm 1/2, \pm 1/2\}$ and their phase and polarisation angle are shifted from 0 to $\pi/2$ going through their circularly polarised core. This topological defect can also be seen as a domain wall with respect to x- and

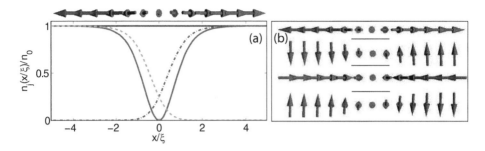

Fig. 10.10: (a) Half-soliton density profiles scaled to n_0 and ξ. The solid (blue) and (red) curves represent the σ_+ (σ_-) density profiles, while the dashed-dotted purple and dashed cyan curves show the x and y components respectively. The red arrows show the pseudospin vector field. (b) ϕ_0 dependent half-soliton pseudospin textures. From the top to the bottom: $\phi_0 = 0, \pi/2, \pi, 3\pi/2$.

y-polarised particles. A plot of the HS density profiles ($n_j = \left|\psi_j^{HS}\right|^2, j = \pm, x, y$) is given in Fig. 10.10(a). The corresponding pseudospin projections are found as

$$\begin{pmatrix} S_x \\ S_y \\ S_z \end{pmatrix} = \frac{n_0}{4} \begin{pmatrix} 2\tanh(x) \\ 0 \\ 1 - \tanh(x)^2 \end{pmatrix}, \tag{10.40}$$

and it is not surprising that the pseudospin points in opposite directions at the opposite sides of the soliton, as shown in Fig. 10.10 (red arrows), since it is a domain wall between x and y polarisations. It is possible to tune the global orientation of the linear polarisation introducing a constant relative phase between the σ_+ and σ_- components appearing as an $\exp(i\phi_0)$ factor in one of the two wavefunctions $\psi_\pm$. Four configurations are shown in Fig. 10.10(b) changing ϕ_0 from 0 to 2π with a $\pi/2$ step. Remarkably, the pseudospin field of the half-soliton is similar to the field created by a point charge being either divergent or convergent for $\phi_0 = 0$ or $\phi_0 = \pi$ respectively.

10.3.3.1 *Half-solitons acceleration*   Let us first consider the interaction between a half-soliton and a constant in-plane effective magnetic field $\mathbf{H}_{LT} = H_x \mathbf{u}_x$ pointing along x. This splitting relates to the energy splitting between linearly-polarised eigenmodes parallel and perpendicular to the orientation of the crystallographic axes of a planar cavity or the orientation of a microwire. Its interpretation in terms of a constant effective in-plane magnetic field is even more direct than in the case of the interaction-induced field in the z direction: if the x and y polarisations have different energies when the Gross-Pitaevskii equation is written in the xy basis, this splitting transforms into a term $-H_x\psi_\mp/2$ in the circular polarisation basis with the usual coordinate transformation rules $\psi_\pm = (\psi_x \mp i\psi_y)/\sqrt{2}$.

Since the existence of a soliton requires a significant interaction energy and, therefore, a significant condensate density, the intrinsic Zeeman splitting behaving as an effective magnetic field $\mathbf{H}_Z = H_z \mathbf{u}_z$ is expected to play an important role in circularly polarised regions such as a half-soliton (or a half-vortex) core. Indeed, due to the spin anisotropy of polariton interactions, the effective field is amplified and tends to lock circularly polarised states (strong density imbalance) in the system. It provides the natural stability of half-integer topological defects against an effective magnetic field preventing the precession of the polarisation at their core up to the critical value

$$H_c = \frac{(\alpha_1 - \alpha_2)\, n}{4}, \tag{10.41}$$

derived by Solnyshkov et al. (2012) and that would lead to their destruction.

Assuming here $\phi_0 = 0$, let us consider the pseudospin dynamics at each point, since the other terms in the Gross–Pitaevskii equation (the kinetic energy and the interaction energy) compensate each other at $t = 0$. For the linearly polarised regions far from the soliton's core, $|x| \gg \xi$, the pseudospin is aligned or anti-aligned with the magnetic field (which contains only a x component) and there is no evolution: $\partial \mathbf{S}/\partial t = \mathbf{0}$. For the core region, there is a nonzero pseudospin projection on the z axis, and, therefore, the pseudospin will rotate (precess) around the magnetic field (which now contains both x and z components). Let us consider the initial moments of this rotation for the pseudospin in the centre of the soliton: $\mathbf{S}(x = 0, t = 0) = \mathbf{u}_z$, while the total magnetic field is:

$$\mathbf{H}\,(x = 0, t = 0) = H_x \mathbf{u}_x - \frac{\alpha_1 - \alpha_2}{2} \left(-\frac{n}{2} \right) \mathbf{u}_z. \tag{10.42}$$

Rotating around the (positive) H_x component, the pseudospin (initially negative along z) gains a positive S_y projection and starts to precess around the positive H_z field, turning towards the negative direction of the x-axis. This is the main result of our qualitative vectorial consideration of the polarisation dynamics: the pseudospin in the centre of the soliton gains a *negative* x-projection. Therefore, the domain of negative x pseudospin projection becomes larger, the domain of positive x-projection smaller, and the wall between these domains moves to the right. However, once the soliton core starts to propagate, the kinetic and interaction energy terms are no longer compensated everywhere, and one cannot discuss the evolution of the system using the qualitative arguments based on polarisation dynamics. At first sight, one may even think that the nonlinear system in question can only be solved numerically.

However, an important insight into the behaviour of the system can be gained by "changing the zoom". Forgetting about the internal structure, a vectorial grey soliton in a Bose–Einstein condensate can be considered as a particle (see for instance Volovik (2003)) with a negative effective mass (at least at low velocities). Moreover, the pseudospin pattern of this particle is the same as the field of a point magnetic charge in 1D. The magnetic energy of the system can be found from the Hamiltonian as the usual scalar product of the field and the spin, and this magnetic energy depends on the position of the soliton because of the finite system size. Thus, one can evaluate the force acting on the magnetic charge from the magnetic field as a gradient of the magnetic energy with respect to the position of the soliton. This force will consequently accelerate the soliton.

Considering the soliton as an elementary particle without internal structure means passing to the limit $L \gg \xi$, where L is the system size (for example, the length of a wire-shaped cavity, which is usually of the order of 100 μm). The healing length of a polariton condensate for a reasonable blueshift of 1 meV expected for GaAs or CdTe cavities and a polariton mass of 5×10^{-5} of a free electron mass is $\xi \sim 1\,\mu m$. In this limit, the wavefunction of the soliton at $t = 0$ becomes simply $\psi_+ = \sqrt{n/2}\,\mathrm{sign}(x - x_0)$, $\psi_- = \sqrt{n/2}$, where x_0 is the position of the soliton. In general, the tanh function is replaced by the sign function. The magnetic energy of a condensate containing a half-soliton in an external in-plane magnetic field is:

$$E_{mag} = -\int \mathbf{H} \cdot \mathbf{S}\, dx\,, \qquad (10.43)$$

Here $\mathbf{S}(x - x_0) = n\,\mathrm{sign}(x - x_0)\mathbf{u}_x$ in the limit we consider (where x_0 is the soliton position), giving:

$$E_{mag}(x_0) = H_x n x_0\,, \qquad (10.44a)$$

$$F_{mag} = -\frac{dE_{mag}}{dx_0} = -nH_x\,. \qquad (10.44b)$$

The force in eqn (10.44b) is, therefore, acting opposite to the direction of the magnetic field, but the acceleration occurs in the direction opposite to the force, because the effective mass of the soliton is negative, at least at low velocities.

For a grey soliton propagating at speed v, the phase shift induced by the soliton in the σ_+ component is $\Delta\theta = 2\arccos(v/c) < \pi$ and the pseudospin projection S_x is reduced, which can be expressed as a renormalisation of the magnetic charge. The correction to the charge is found as

$$q = q_0\left(1 - \frac{v^2}{c^2}\right)\,, \qquad (10.45)$$

where $q_0 = \alpha n/2$ is the charge at rest for a dark half-soliton. The total correction for the mass of the soliton and its charge gives the equation of motion

$$a = q_0 \frac{nH_x}{m_0}\left(1 - \frac{v^2}{c^2}\right)^{3/2}\,, \qquad (10.46)$$

which is the same as in relativistic physics, whose integration yields

$$v(t) = c\tanh\left(\frac{q_0 H_x n}{c}t\right)\,, \qquad (10.47)$$

assuming zero initial velocity. This trajectory is confirmed by numerical simulations as shown in Fig. 10.11. In panel (b), the soliton is introduced as an initial condition and no lifetime is accounted for. In the configuration of panel (c), we have included the polariton lifetime and the pumping term, while the half-solitons are created with a pulsed potential acting on only one component of the condensate (σ_+ here). We see that, using the abstraction of a point magnetic charge appears particularly useful, since it allows to solve the nonlinear spinor Gross–Pitaevskii equation analytically and gives

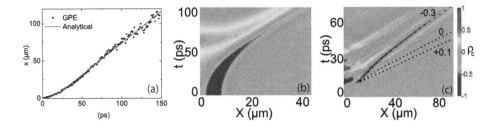

Fig. 10.11: Half-soliton acceleration. (a) Trajectory of a half soliton accelerated in a constant in-plane effective magnetic field calculated analytically (red solid line) and numerically (black dots). (b,c) Circular polarisation degree ρ_c as a function of coordinate and time calculated using the spinor Gross-Pitaevskii equation, including the constant in-plane effective magnetic field. The half-soliton trajectory is visible as the deep blue minimum (the core is filled with σ_- particles). (b) Half-soliton as an initial condition, no lifetime. (c) Pumping and lifetime the half-soliton is created here by a pulsed potential. The black/dotted lines are guides for the eyes showing trajectories for the splitting values indicated on the figure (in meV).

a good qualitative understanding of the observed phenomena. Several theoretical works have followed the description of half-solitons in terms of effective magnetic monopoles. One can mention the study of the stability of a half-soliton gas, and its evolution toward an ordered phase similar to a Wigner crystal, by Terças et al. (2013). One can also mention the study of the transport properties of a half-soliton gas under an accelerating force provided by an effective magnetic field, theoretically analysed by Terças et al. (2014), which allowed to put forward the concept of magnetricity in photonic systems.

10.4   Hydrodynamic generation of oblique half-solitons and half-vortices

In the previous sections, we have introduced the oblique dark soliton solution of the scalar Gross–Pitaevskii equation. Its experimental evidence was obtained in an exciton-polariton condensate by Amo et al. (2011) thanks to the high degree of control offered by the system [see Fig. 10.5]. The experiment involved a propagating polariton fluid injected resonantly against an immobile structural defect in the microcavity. In this initial observation, Amo *et al.* did not resolve the polarisation of the condensate and the solitons were "scalar", since the excitation was performed with a circularly polarised pump.

In the following, we discuss the hydrodynamic generation of oblique half-solitons (OHS) as described theoretically by Flayac et al. (2011) and half-vortices, that were later observed by Hivet et al. (2012).

Let us now focus on the possibility of creating 2D *oblique* half-solitons. First of all, it is clear that in the case where the two components of a spinor BEC do not interact ($\alpha_2 = 0$), if they are initially equally populated, a significant perturbation in only *one* of the components will lead to the formation of half-integer topological excitations. Next, what happens if the interaction between the two components is no longer negligible? To answer this question, we return to the spinor Gross–Pitaevskii equation, following Gladush et al. (2009), we rescale the two fields as in eqns (10.34), and look for stationary solutions, where the phase of each component is expressed by means of their stationary and irrotational velocity fields via $\mathbf{v}_{\pm}(\mathbf{r}) = \hbar/m^* \boldsymbol{\nabla} \theta_{\pm}(\mathbf{r})$, with $\mathbf{r} = (x, y)$. We look

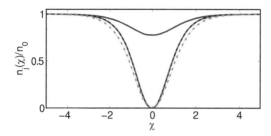

Fig. 10.12: An oblique half-soliton density slice normal to its axis. The solid curves show numerical profile for the two spin components. The dashed black curve is the perturbative solution described in the text, which cannot be distinguished from the numerical solution. The dashed dotted curve shows the scalar soliton solution.

for oblique solutions that depend only on the tilted coordinate $\chi = (x - ay)/\sqrt{1 + a^2}$, which leads to the set of equations

$$\left(n_+'^2/4 - n_+ n_+''/2\right) + 2n_+^2 \left(\Lambda_1 n_+ + 2\Lambda_2 n_-\right) = (q + 2\mu) n_+^2 - q n_0^2, \qquad (10.48a)$$

$$\left(n_-'^2/4 - n_- n_-''/2\right) + 2n_-^2 \left(\Lambda_1 n_- + 2\Lambda_2 n_+\right) = (q + 2\mu) n_-^2 - q n_0^2, \qquad (10.48b)$$

where $\Lambda_{1,2} = \alpha_{1,2}/(\alpha_1 + \alpha_2)$ and $q = U^2/(1+a^2)$ (U is the velocity of the flow). This system has to be solved numerically, but we can first consider some simple arguments. The density profile of an integer oblique soliton in a spinor fluid is given by $n_{ODS} = 1 - (1 - q/\mu) \operatorname{sech}[\chi\sqrt{\mu - q}]^2$ with $\mu = (\Lambda_1 + \Lambda_2)n_0/2 = \Lambda n_0/2$. Now, for the case of the OHS, the density notch in the σ_- component that contains the defect, is seen as an external potential by the initially unperturbed σ_+ component, because of the interactions between the particles of different spins. We suppose that the σ_+ component fits the shape of this potential, which is simply $\Lambda_2 n_-$. Then, this perturbation creates in turn a potential for the σ_- component given by $-\Lambda_2 n_+ = -\Lambda_2^2 n_-$. Therefore, the density profile is modified as $\widetilde{n}_- \leftarrow \left(\Lambda - \Lambda_2^2\right) n_-/\Lambda$. Iterating this procedure leads to a geometric series and to a renormalisation of the interaction constant seen by the component containing the soliton $\widetilde{\Lambda} \leftarrow \Lambda - \Lambda_2^2/\Lambda_1$. Consequently, the OHS solution is approximated by

$$n_{OHS} = 1 - (1 - q/\widetilde{\mu}) \operatorname{sech}\left[\chi\sqrt{\widetilde{\mu} - q}\right]^2, \qquad (10.49)$$

with $\widetilde{\mu} = \widetilde{\Lambda} n_0$. In this description, the sound velocity is changed like $c_s \to \widetilde{c}_s = \sqrt{\widetilde{\mu}/m^*}$ and the healing length like $\xi \to \widetilde{\xi} = \hbar/\sqrt{2m^*\widetilde{\mu}}$. In the case where $\Lambda_2 < 0$ (resp. > 0), which corresponds to an attractive (resp. repulsive) interaction, c_s is slightly increased (resp. decreased) and inversely for ξ. The component without a soliton obviously presents a minimum (resp. maximum). This argumentation is compared to direct numerical solutions of eqns (10.48a–10.48b) in Fig. 10.12 showing excellent accuracy provided that Λ_2 remains small.

Now let us see how half-vortices and oblique half-solitons can be generated in a propagating exciton-polariton fluid. The setup that has been proposed is basically the same as the one required to generate the integer oblique solitons described previously, namely a continuous and resonant pumping scheme locally upstream from a defect, imposing a supersonic flow. However, we will now focus on the spin degree of freedom of the condensate, namely the polarisation of the pump and the polarisation of the emission along the propagation. To describe more accurately the spinor polariton BEC, we take into account the real non-parabolic dispersion of the particles, their decay and injection, via the set of four coupled spin-dependent equations:

$$
i\hbar\frac{\partial\phi_\pm}{\partial t} = -\frac{\hbar^2}{2m_\phi}\nabla^2\phi_\pm + \Omega_R\chi_\pm + D_\pm\phi_\pm + \beta\left(\frac{\partial}{\partial x}\mp i\frac{\partial}{\partial y}\right)^2\phi_\mp
$$

$$
+P_\pm e^{i(\mathbf{k}_P\cdot\mathbf{r}-\omega_P t)} - \frac{i\hbar}{2\tau_\phi}\phi_\pm,
\tag{10.50a}
$$

$$
i\hbar\frac{\partial\chi_\pm}{\partial t} = -\frac{\hbar^2}{2m_\chi}\nabla^2\chi_\pm + \Omega_R\phi_\pm + \left(\alpha_1|\chi_\pm|^2 + \alpha_2|\chi_\mp|^2\right)\chi_\pm - \frac{i\hbar}{2\tau_\chi}.
\tag{10.50b}
$$

The pump terms of amplitude $P_\pm$ allow us to select the polarisation of injected photons and thus of polaritons. $D_\pm$ is the impenetrable potential barrier, that can affect each component independently. We have chosen a bar-shaped pump spot upstream from the defect. The effective mass m_ϕ is, therefore, approximated by $m_\phi = m_\phi^{TM} m_\phi^{TE}/(m_\phi^{TM} + m_\phi^{TE})$. To generate half-integer topological defects past the obstacle we need to be able to break the symmetry of the flow not only with respect to the density (integer topological defects), but also with respect to its polarisation, which can be realised using the polarisation separation brought by the TE-TM. One needs to select the polarisation of the pump laser to avoid pseudospin rotation before the fluid reaches the defect. We choose the latter to be linear in a TM state, which corresponds to a polarisation along the direction of propagation (x-axis) and to $\mathbf{S}$ pointing along $\mathbf{H}_{LT}$.

Arriving at the obstacle, the supersonic fluid is split into two parts, propagating in opposite oblique directions around the obstacle. Before the defect, the pseudospin was aligned with the effective field, but when the propagation direction changes, the angle between the pseudospin and the field starts to increase. It induces an antisymmetric rotation of the pseudospin: the particles going up (down) will gain a σ_+ (σ_-) component, providing the seed for the OHS/HV generation (Fig. 10.13).

Downstream from the defect, the flows are complex, but globally the fluid is moving along the x-axis. Since the half-solitons have been slightly separated, they start to behave as magnetic monopoles having opposite charges, just like in the 1D case, because their pseudospin texture is divergent. They start to feel the TE-TM induced effective magnetic field $\mathbf{H}_{LT}$ that is pointing along the flow. Since the solitons are oblique, the scalar product $\mathbf{H}_{LT} \cdot \mathbf{S}$ is not zero, providing their separation and acceleration, increasing with the distance from the defect. One of them is bent slightly towards the axis of symmetry of the flow, becoming deeper, while the other one is moved in the opposite direction, becoming shallower and larger and possibly hardly visible for larger values of H_{LT} (larger $\mathbf{k}_P$). The situation is totally antisymmetric for the lower oblique soliton. Moreover, this separation effect is emphasised for large density regions (close to the defect) if α_2 is negative as it induces the half-soliton repulsion (see Fig. 10.13).

Once again, the intrinsic Zeeman splitting arising from the polariton spin-anisotropy protects the half-soliton up to the critical magnetic field H_c [see eq.(10.41)]. One should keep in mind, however, that due to the finite polariton lifetime, the total density is decaying away from the defect and the Zeeman splitting is decaying as well. At some

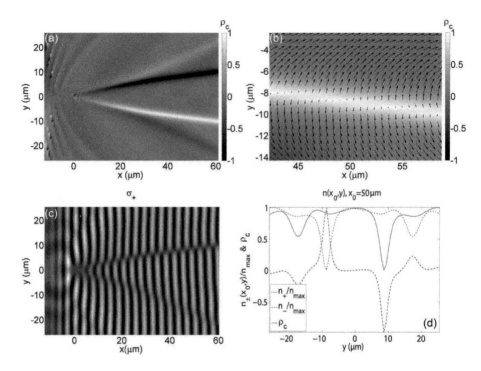

Fig. 10.13: Stationary 2D oblique half-soliton, the flow is going from left to the right. (a) Degree of circular polarisation ρ_c: one clearly sees the antisymmetric pattern imposed by the effective field's geometry and the OHS separation. (b) Zoom on the lower σ_- soliton (white one in (a)) with the in-plane pseudospin component $\mathbf{S}_\parallel = (S_x, S_y)^T$ (black arrows) exhibiting a rotation of almost π through the OHS ($\pi/2$ rotation of η). (c) Interference pattern in the σ_+ component showing the phase shifts at the soliton position. The situation is obviously antisymmetric in the other component (not shown). The repelled soliton is shallower, which corresponds to smaller phase shifts. (d) Density slices $50\,\mu m$ downstream from the defect together with a ρ_c slice.

point, it is unavoidable that the effective magnetic field felt by the half-solitons exceeds the critical value. At this point, they start to be converted into the other component, and thus their extension is finite.

Let us remember that 1D half-solitons are the domain walls between linear polarisations, which means that the polarisation angle η rotates by $\pi/2$ and the in-plane projection of $\mathbf{S}_\parallel = (S_x, S_y)^T$ rotates by π, going through the HS. This rotation of η is also expected for a the 2D system, nevertheless, the *oblique* half-solitons possess a nonzero velocity with respect to the flow, and, therefore, have nonzero density in the component supporting the soliton. Thus, the rotation of η as well as the shift of the global phase θ are bound to be smaller than $\pi/2$.

We know now that in a scalar condensate, increasing the fluid density (or reducing its speed) leads to the dissociation of oblique solitons into vortex streets and eventually to the onset of superfluidity. We obviously expect the same behavior to occur for the spinor system. The half-vortex generation at higher densities can be understood similarly to the half-soliton nucleation: integer vortices are split into HVs by the effective magnetic field

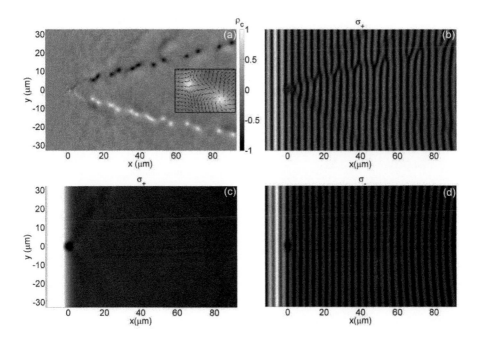

Fig. 10.14: (a) Degree of circular polarisation: HV trains generation increasing the pump intensity with respect to Fig. 10.13. The inset displays **S**$\parallel$ for a σ_- HV pair (black arrows) and shows the opposite winding
of the polarisation around the HVs core. (b) The interference pattern in the σ_+ component shows the typical
forklike dislocations at the HV position. The positions corresponding to σ_- vortices show no phase modification. (c) Superfluid regime: The flow ignores the presence of the obstacle and shows no phase perturbation
as one can see in the panel (d) showing the complementary antisymmetric σ_- component.

around the defect and they become accelerated. The difference is that vortices cannot be
more or less shallow, like solitons, they can only appear or not, being real topological
defects, which explains why only one species of half-vortices appears in each half-
plane in our simulation, but the situation can be different depending on the strength
of the field. We show the corresponding numerical stationary solutions in Figs.10.13
and 10.14, demonstrating the three density dependent hydrodynamic regimes: Oblique
Half-Solitons [Fig. 10.13], streets of half-vortex dipoles [Fig. 10.14(a, b)], and finally a
superfluid regime [Fig. 10.14(c, d)].

10.5   Spin bifurcation theory (broken parity)

Spontaneous buildup of circular polarisation was first demonstrated by Ohadi et al.
(2015) under nonresonant pumping in the presence of a linear polarisation splitting.
In this experiment, optically trapped polariton condensates were formed in the centre of a four-spot confining potential by spatially patterning a nonresonant excitation
beam (Fig. 10.15). Separation of the condensate and excitonic reservoir, which proved
to be crucial in this experiment, allowed the stochastic formation of long-lived left-

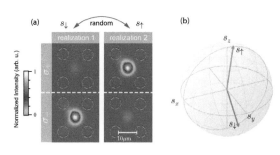

Fig. 10.15: (a) Polarisation-resolved spatial image of the only two realisations observed by Ohadi et al. (2015) in a single trapped condensate in a 4-spot trapping geometry. (b) Simultaneously measured components of the pseudospin for 1000 realisations. The total degree of polarisation is 0.93 ± 0.03 and the two average pseudospin states are $s_\downarrow = [-0.22, 0.19, -0.94]$ (blue vector) and $s_\uparrow = [-0.22, -0.14, 0.96]$ (orange vector).

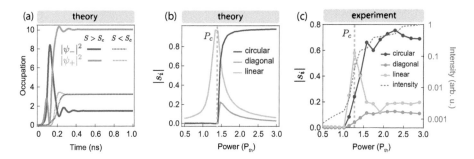

Fig. 10.16: (a) Numerical calculations using eqn 10.53 for the case when the condensate occupation is below the critical occupation S_c (dotted lines) and when above S_c (solid lines). Blue lines show occupation of spin-down component and orange lines show spin-up component. (b) Numerical calculations of the Stokes parameters at different powers using eqn 10.53. (c) Measured power dependence of the condensate Stokes parameters.

or right-circularly polarised condensates under linear excitation, demonstrating spontaneous breaking of the parity symmetry. This phenomenon was described by the "spin bifurcation theory", which is a development of Aleiner et al. (2012)'s theory of polariton weak lasing in two coupled condensation centres, but for the case of the spin degree of freedom. In the spin bifurcation theory there are the right and left circular polarisations instead of two separated condensates, and the gain-saturation nonlinearity is included in the system.

Similar to Section 10.3, the order parameter for an exciton-polariton condensate is a two-component complex vector $\Psi = [\psi_{+1}, \psi_{-1}]^T$, where ψ_{+1} and ψ_{-1} are the spin-up and spin-down wavefunctions. The components of the order parameter define the measurable condensate pseudospin $\mathbf{S} = (1/2)(\Psi^\dagger \cdot \sigma \cdot \Psi)$, or the normalised vector $\hat{\mathbf{s}} = \mathbf{S}/S$, where $\sigma_{x,y,z}$ are the Pauli matrices. The components of this vector contain information about the intensities and relative phases of the emitted light. The order parameter evolves according to the driven dissipative equation

$$i\frac{d\Psi}{dt} = -\frac{i}{2}g(S)\Psi - \frac{i}{2}(\gamma - i\varepsilon)\sigma_x\Psi + \frac{1}{2}\left[(\alpha_1 + \alpha_2)S + (\alpha_1 - \alpha_2)S_z\sigma_z\right]\Psi, \quad (10.51)$$

or in components:

$$\dot{\psi}_{+1} = -\frac{1}{2}g(S)\psi_{+1} - \frac{1}{2}(\gamma - i\varepsilon)\psi_{-1} - \frac{i}{2}(\alpha_1|\psi_{+1}|^2 + \alpha_2|\psi_{-1}|^2)\psi_{+1}, \quad (10.52a)$$

$$\dot{\psi}_{-1} = -\frac{1}{2}g(S)\psi_{-1} - \frac{1}{2}(\gamma - i\varepsilon)\psi_{+1} - \frac{i}{2}(\alpha_1|\psi_{-1}|^2 + \alpha_2|\psi_{+1}|^2)\psi_{-1}. \quad (10.52b)$$

Following Keeling and Berloff (2008), $g(S) = \Gamma - W + \eta S$ is the pumping-dissipation balance, Γ is the (average) dissipation rate, W is the incoherent in-scattering (or "harvest" rate), and η captures the gain-saturation term with $S = (|\psi_{+1}|^2 + |\psi_{-1}|^2)/2$. Here, this gain saturation depends on the total occupation of the condensate (this is more generally treated by Ohadi et al. (2015)). It is assumed now that X (horizontal) and Y (vertical) linearly-polarised single-polariton states have different energies and dissipation rates. The energy of the X-polarised state is shifted by $-\varepsilon/2$, and the energy of the Y-polarised state by $+\varepsilon/2$. The dissipation rate from the X-polarised state is $\Gamma + \gamma$, while the dissipation rate from the Y-polarised state is $\Gamma - \gamma$. Finally, α_1 is the repulsive interaction constant for polaritons with the same spin, and α_2 is the interaction constant for polaritons with opposite spins. From eqn (10.51) we obtain for the components of the pseudospin vector ($\alpha = \alpha_1 - \alpha_2$) as

$$\dot{S}_x = -g(S)S_x - \gamma S - \alpha S_z S_y, \quad (10.53a)$$

$$\dot{S}_y = -g(S)S_y + \varepsilon S_z + \alpha S_z S_x, \quad (10.53b)$$

$$\dot{S}_z = -g(S)S_z - \varepsilon S_y, \quad (10.53c)$$

and the related equation for the total spin $\dot{S} = -g(S)S - \gamma S_x$. There are two sets of solutions, which we call here the paramagnetic and ferromagnetic solutions.

10.5.1 *Paramagnetic solutions.*

These give simple condensation into either X or Y linearly-polarised states. The Y state possesses the longest lifetime, and the condensation threshold is reached for this state first at $W_1 = \Gamma - \gamma$. There is no parity breaking for this condensate: $S_y = S_z = 0$, $S_x = -S$ with $S = (W - W_1)/\eta$, so that the occupations of +1 and −1 components are equal. However, this condensate solution becomes unstable for $W > W_2$. The values of the critical occupation S_c and the critical pumping rate are:

$$S_c = \frac{\gamma^2 + \varepsilon^2}{\alpha\varepsilon}, \qquad W_2 = W_1 + \eta S_c. \quad (10.54)$$

This instability is present also for equal dissipation rates, i.e. when $\gamma = 0$. In this case, the system (eqn 10.53a-c) describes the self-induced Larmor precession of the pseudospin vector. Incorporating energy relaxation (e.g., using small negative γ) then leads to the formation of the X-polarised condensate—an intuitively expected result.

10.5.2 *Ferromagnetic solutions.*

The key ingredient of the theory is the presence of the $\gamma > 0$ parameter describing the variation of dissipation rates. This parameter allows the formation of the "weak lasing" regime of Aleiner et al. (2012), which is characterised by two important features: (i) the X-polarised condensate is also unstable, and (ii) when the Y-polarised condensate loses stability at the critical occupation S_c, it continuously transforms into one of the two ferromagnetic states. While eqns (10.52) are parity symmetric, i.e., they are not affected by the interchange of left and right circular polarisation, the new solutions are characterised by broken parity symmetry and by spontaneous formation of either left or right elliptical polarisation. These solutions are

$$S_x = -\frac{g(S)}{\gamma} S \, , \qquad S_y = -\frac{g(S)}{\varepsilon} S_z \, , \tag{10.55a}$$

$$S_z = \pm \frac{\varepsilon}{\gamma} \sqrt{\frac{\gamma^2 - g(S)^2}{\varepsilon^2 + g(S)^2}} \, S \, , \qquad S = \frac{\gamma[\varepsilon^2 + g(S)^2]}{\alpha \varepsilon g(S)} \, , \tag{10.55b}$$

where the positive root of the second equation in eqn (10.55b) should be taken. We note that while the sign of S_x is always negative, S_y and S_z have opposite signs for the two solutions. This means that the left-circular component is accompanied by a diagonal component, and the right-circular by anti-diagonal. Moreover, if these components change for some reason, they mirror each other as long as the total condensate occupation stays fixed. We label these two solutions as the $s_\downarrow$ and $s_\uparrow$ spin states.

Numerical calculations for the occupation of the two circular components of the wavefunction when $S < S_c$ (dotted lines) and when $S > S_c$ are shown in Fig. 10.16(a). Here, the condensate is initialised with a small asymmetry in spin-up and spin down occupations ($< 1\%$). Below the critical occupation S_c, the condensate is linearly polarised, but when the occupation is increased above the threshold S_c, the condensate adopts one of two elliptically polarised configurations depending on the initial conditions. In the experiment, the stochastic behaviour is due to random spin fluctuations at the onset of the condensation. In theory, this can be reproduced by randomly setting the initial conditions. Numerical calculations of the condensate polarisation versus excitation power are shown in Fig. 10.16(b). Directly at the condensation threshold, the condensate is linearly polarised, but once it reaches the critical occupation ($P_c = 1.3 P_{th}$, marked by a dashed grey line), the linear component quenches and the circular polarisation builds up. This behavior reproduces the experimental data, as shown in Fig. 10.16(c). An initial buildup and subsequent quenching of linear polarisation with the continuing increase of circular polarisation at $P_c = 1.25 P_{th}$ (marked by dashed grey line, total intensity marked by red dotted line) is observed experimentally. Once circular polarisation is achieved, the orientation of the condensate circular polarisation becomes stochastic under linearly polarised pumping.

10.6   Engineering of the polariton band structure

10.6.1   *Introduction*

Starting from 2D massive quasi-particles, it is possible to engineer the polariton band structure by creating an in-plane potential. This strategy has been explored already at the end of the 90s by Gutbrod et al. (1999) and Tartakovskii et al. (1998) through the creation of photonic dots and photonic wires by etching of planar cavities. Then, more complicated real periodic potentials have been implemented by Kim et al. (2011) with metallic deposition or by Cerda-Méndez et al. (2013) through the use of Surface Acoustic Waves. Optical potential can also be created and combined with the static ones imposed by the structure design as shown by Wertz et al. (2010) and Tosi et al. (2012). This possibility was then enhanced with the use of Spatial Light Modulators as shown by Boulier et al. (2015). The most recent etched structures combine exceptional optical properties and a complete flexibility of design. A remarkable aspect is that eigenmodes of the structures can be found with an excellent precision using a particle-like description, namely, Schrödinger equation. An important specificity relies on the polarisation degree of freedom. It is naturally taken into account while solving Maxwell equations. It has to be re-introduced with care while going to a Schrödinger-like description. We should already underline a key difference of the systems we are describing with photonic crystal slabs, in which a 2D periodic potential acts on the propagating modes of a waveguide, which are characterised with a large in-plane wave vector. In cavities, the confinement along the z-direction is provided by the mirrors, and the in-plane motion is small. As a result, in a lattice of coupled pillars for instance, the tight-binding approximation is an excellent approximation.

In this Section, we focus first on the band engineering in the linear regime. In that case the results are essentially the same considering pure photonic modes and polaritonic states except for the sensitivity to external fields, to which excitons are more sensitive than photons. We review the basic cases of 1D (wire) and 0D (pillar) confinement, of a photonic molecule composed by two coupled pillars. We then describe periodic potentials which allow in principle the opening of energy gaps in the photonic spectrum. We focus on the specific case of a photonic lattice with a honeycomb periodicity as fabricated by Jacqmin et al. (2014). We show that this design combined with the intrinsic chirality of TE and TM photonic modes and the Zeeman splitting of cavity polaritons allows one to build topologically nontrivial polariton bands and chiral edge states as shown by Nalitov et al. (2015b) .

10.6.2 Wire cavities

The 1D etching of a planar cavity adds a 1D lateral confinement that transforms the 2D dispersion of polaritons into a set of quantised bands. Within a particle-like description (parabolic dispersion, scalar case, use of the Schrödinger equation instead of Maxwell's equations), the new quantised dispersion reads

$$E_n = \frac{\hbar^2}{2m_{pol}} \left(\left(n\frac{2\pi}{L} \right)^2 + k_x^2 \right), \tag{10.56}$$

where n is the quantisation number and m_{pol} the polariton mass. Individual lines are well separated if their splitting is larger than the linewidth. If we take $m_{pol} = 5 \times 10^{-5}m_0$, a splitting of 1.5 meV corresponds to a wire width of 3 μm. However, polarisation effects cannot be completely neglected. Each confined mode is typically made out of two polarisation eigenstates having the electric field parallel (Longitudinal mode) or perpendicular (Transverse mode) to the wire axis X. As discussed in Chapter 9, the TE and TM modes are characterised by two different effective masses (m_{TE}, m_{TM}).

The X-polarised mode is TE when the wavevector points along one axis, and TM when it points along the other, a decomposition reversed for the Y- polarised mode. As a result the polarisation mode with the lighter mass along the wire axis is quantised with the heavier mass along the confinement direction. The polarisation eigenmodes read:

$$E_n^X = \frac{\hbar^2}{2} \left(\frac{1}{m_{pol}^{TE}} \left(n\frac{2\pi}{L} \right)^2 + \frac{1}{m_{pol}^{TM}} k_x^2 \right), \tag{10.57a}$$

$$E_n^Y = \frac{\hbar^2}{2} \left(\frac{1}{m_{pol}^{TM}} \left(n\frac{2\pi}{L} \right)^2 + \frac{1}{m_{pol}^{TE}} k_x^2 \right). \tag{10.57b}$$

As a result the two polarisation modes are split at $k_x = 0$ by values that could be of the order of 50 to 100 μeV. The lower mode in energy having the lighter mass, the modes are expected to cross at $k_x = \frac{n\pi}{L}$. Figure 10.17 from Sturm et al. (2015) shows the PL emission of a 3 μm wire, where one can observe the splitting between the two first confined subbands, the polarisation splitting, and the crossing of the X and Y (labelled TE and TM here) polarised lines around 2.6 μm^{-1}, whereas the value given by the formula above is 2.1 μm^{-1}. One should notice that the polarisation splitting mentioned above can be strongly modified by strain and many other effects.

10.6.3 Single pillars and molecules

Pillars have been fabricated for many years by etching cavities and can be square, circular, elliptical and other general shapes. A 3 μm-size pillar typically shows an energy splitting of 3 meV between the two first quantised states. It therefore represents a very good implementation of atomic-like photonic or polaritonic states. In a symmetric circular or square pillar, the polarisation splitting associated with the TE-TM splitting cancels out, the only remaining source of splitting being structural anisotropies. As we are going to see below, these pillars can be coupled to form molecules, or lattices, which are ideally described by the tight-binding method.

The simplest case that can be organised is a double pillar molecule whose study has been reported in details by Galbiati et al. (2012). Neglecting the polarisation degree of freedom, the coupling between the atomic like orbital of each of the pillars can be described by a 2×2 matrix

$$H = \begin{pmatrix} E_i & J_i \\ J_i & E_i \end{pmatrix}, \tag{10.58}$$

which, for each orbital i, gives rise to a bound and anti-bound state. Figure 10.18(a) shows an image of double-pillars structures and the PL emission of one molecule. The localised polariton states overlap and form bonding and anti-bonding states, evidenced by photoluminescence spectra shown in (b). The same panel demonstrates that one can tune the tunnelling coefficient by changing the distance between pillars. In (c) is shown the measured and calculated bonding and anti-bonding states emitted by the two first orbitals. The consequence of the action of the TE-TM splitting on the simplest polariton molecule, consisting of A and B pillars, is the dependence of the tunnelling probability on polarisation. The potential barrier due to the size quantisation

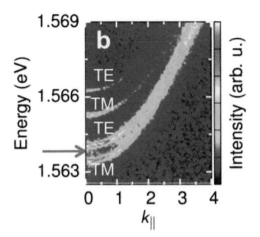

Fig. 10.17: Photoluminescence of a 3 μm wide polariton wire at slight negative detuning as reported by Sturm et al. (2015). A neat quantisation of the energy levels is observed.

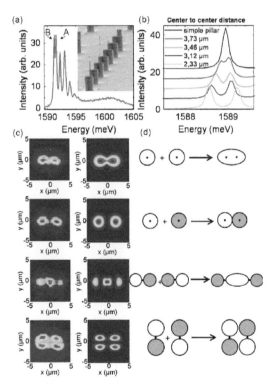

Fig. 10.18: (a) Emission spectrum measured by Galbiati et al. (2012) on a single molecule at low excitation power ($d = 4\,\mu m$ and $d_{CC} = 3.73\,\mu m$); A and B indicate the binding and anti-binding states. Inset : scanning electron micrography of an array of pillars and molecules. b) Emission spectra measured on a 4 μm round micropillar (black line) and on photonic molecules with $d = 4\,\mu m$ and various values of d_{CC}. c) left (resp. right) column : measured (resp. calculated) emission pattern of the four lowest energy modes of a photonic molecule with $d = 4\,\mu m$ and $d_{CC} = 3.73\,\mu m$; d) Schematic of the hybridisation of the individual pillar modes within a photonic molecule..

at the narrow junction between the pillars for a given polarisation may be estimated as $V_{L(T)} = \pi^2\hbar^2/(2m_{L(T)}w^2)$, where w is the junction width. Note that the barrier height depends on the mass of the same polarisation, but in the direction orthogonal to the axis between the pillars. For the state $|A, L(T)\rangle$, localised in pillar A and having

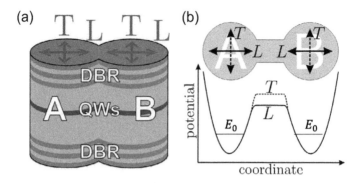

Fig. 10.19: (a) A scheme of the simplest polariton molecule: a pair of overlapping pillar microcavities. The basis of linearly polarised polariton modes, localised in each cavity, is chosen according to the axis between the pillar centres. (b) A sketch of effective potential for polaritons. Potential barrier dependence on polarisation is due to TE-TM coupling.

$L(T)$ polarisation (see Fig. 10.19), the wavefunction tail in pillar B is determined by the tunnel coefficient $\kappa_{L(T)} = \hbar^{-1}\sqrt{2m_{L(T)}(V_{T(L)} - E)}$. Therefore, we find that the nondiagonal blocks of the Hamiltonian matrix have different matrix elements $-J_{L,T}$, $J_L > J_T$ at their diagonals, whose values depend on the overlaps of corresponding wavefunctions.

The off-diagonal terms are zero, because all sources of spin conversion other than LT splitting are neglected. In the basis of states $|A, L\rangle$, $|A, T\rangle$, $|B, L\rangle$ and $|B, T\rangle$ the Hamiltonian therefore reads

$$H_{AB} = - \begin{pmatrix} 0 & 0 & J_L & 0 \\ 0 & 0 & 0 & J_T \\ J_L & 0 & 0 & 0 \\ 0 & J_T & 0 & 0 \end{pmatrix}. \tag{10.59}$$

A transition to the basis of circularly polarised polaritons $|A, +\rangle$, $|A, -\rangle$, $|B, +\rangle$ and $|B, -\rangle$, where the polarisation state $|\pm\rangle = |x\rangle \pm i|y\rangle$ is defined with reference to x, y axes, only affects nondiagonal blocks of the Hamiltonian (10.59)

$$H_{AB} = - \begin{pmatrix} 0 & 0 & J & \delta J e^{-2i\varphi} \\ 0 & 0 & \delta J e^{2i\varphi} & J \\ J & \delta J e^{-2i\varphi} & 0 & 0 \\ \delta J e^{2i\varphi} & J & 0 & 0 \end{pmatrix}, \tag{10.60}$$

where $J = (J_L + J_T)/2$, $\delta J = (J_L - J_T)/2$ and φ is the angle between the AB and x axes. The diagonalisation of Hamiltonian (10.60) gives the eigenstates of the polariton molecule. In the absence of the TE-TM splitting ($\delta J = 0$) the eigenstates are double degenerate pairs of symmetric and anti-symmetric polariton states. Effective spin-orbit coupling (SOC) therefore lifts the degeneracy for both bonding and anti-bonding states and splits them by $2\delta J$.

10.6.4   *Lattices: a few basics about 1D lattices*

The simplest possible case is constituted by a 1D periodic lattice, which straightfor-wardly shows gap-opening at the edge of the first Brillouin zone. The unique flexibility of the photonic platform has been further evidenced by emulating quasi-periodic lat-tices using Fibonacci series, which allowed Tanese et al. (2014) to demonstrate fractal band structures. In this subsection, we describe several basic results based on the text-book theory of linear waves in a periodic potential. We shall consider here that particles experience a cosine potential

$$U\left(x\right) = A\cos\left(k_0 x\right)^2, \tag{10.61}$$

where $k_0 = \pi/d$ with d the period of the potential. Assuming that in the absence of U the dispersion is parabolic, the stationary Schrödinger equation reads

$$E\psi\left(x\right) = -\frac{\hbar^2}{2m^*}\frac{\partial^2\psi\left(x\right)}{\partial x^2} + A\cos\left(k_0 x\right)^2\psi\left(x\right). \tag{10.62}$$

Applying Bloch's theorem, we look for eigenfunctions in the form

$$\psi_{p,q}\left(x\right) = u_{p,q}\left(x\right)e^{iqx} \tag{10.63}$$

where q is the so-called quasimomentum and p is the band index. The function $u_{p,q}$ is d-periodic. We are, therefore, allowed to rewrite $\psi(x)$ and $U(x)$ in terms of Fourier series:

$$\psi_{p,q}\left(x\right) = e^{iqx}\sum_m c_m^n e^{2imk_0 x}, \tag{10.64a}$$

$$U\left(x\right) = \sum_m U_m e^{2imk_0 x}. \tag{10.64b}$$

Injecting this anzatz into eqn(10.62) considering $2N+1$ modes, which means truncating the sums to $|m| = N$, leads to the set of $2(2N + 1)$ linear equations:

$$\left[\frac{\hbar^2}{2m^*}\left(q - 2mk_0\right)^2 + V_0\right]c_{q-mG} + U_{+2k_0}c_{q-2(m+1)k_0} + U_{-G}c_{q-2(m-1)k_0}$$
$$= Ec_{q-2mk_0}. \tag{10.65}$$

The choice of our periodic potential imposes $U_{\pm 2k_0} = A/4$ and $U_{m=0} = A/2$. For a given quasimomentum q, we obtain $2N + 1$ eigenenergies forming the so-called en-ergy bands separated by forbidden energy gaps. To each eigenvalue E_p corresponds an eigenfunction defined by the Fourier component c_{q-2mk_0}. The total dispersion relation obviously strongly depends on both the amplitude A of the potential and the quasimo-mentum q. The recoil energy $E_R = \hbar^2 k_0^2/2m^*$ is the characteristic scale in the system and we can consider two opposite limits depending on the ratio A/E_R.

In the case $A \simeq E_R$ the potential is said to be shallow and the eigenenergies crucially depend on q. Since the gap widths E_g^p scale like A^{p+1} the magnitude is only significant between the first and second bands. Therefore, particles having an energy larger than E_g^1 are well described as free particles. The band structure can be estimated as

$$\frac{E\left(\widetilde{q}\right)}{E_R} = \widetilde{q}^2 \mp \sqrt{4\widetilde{q}^2 + \frac{A^2}{16E_R^2}}, \tag{10.66}$$

where $\widetilde{q} = q/k - 1$ and $s = A/E_R$. This approximation gives the first (minus sign) and second (plus sign) bands. In Fig. 10.20, the energies are shown in real space. In the same graph are also shown the real space probability distributions of the eigenfunctions. The eigenfunctions at the lowest and highest energy are almost constant, which implies that the wavefunction is mainly given by a plane wave corresponding to an almost free particle. It is important to note that for energies near the upper band edge of the lowest band, the probability distribution is periodic and its maxima coincide with the potential minima. For this energy we additionally depict the wavefunction, which reveals that the relative phase in the adjacent potential minima is π. This is the well-known sinusoidal *Bloch state* at the band edge defining the Brillouin zone. From this graph one can also see that the Bloch state in the first excited band is also sinusoidal, but it is in-phase with the periodic potential. Thus the energy of this state is higher due to the bigger overlap with the periodic potential.

In the opposite limit, where $A \gg E_R$, we enter the well known tight-binding treatment. In that case, the eigenenergies of the first bands are only weakly dependent on q and the energy gaps are large. The first band can be found analytically to be

$$\frac{E\left(q\right)}{E_R} = \sqrt{\frac{A}{E_R}} - 2J\cos\left(qd\right), \tag{10.67}$$

where J is the coupling energy between adjacent wells (Josephson coupling constant), which can be found as

$$J = \frac{4}{\sqrt{\pi}}\left(\frac{A}{E_R}\right)^{3/4}\exp(-2\sqrt{A/E_R}). \tag{10.68}$$

The expression (10.67) is plotted in Fig. 10.20(c) as a dotted line and reveals the good agreement with the numerically calculated eigenenergies. The corresponding eigenfunctions are depicted on the right-hand side. Although the absolute value of the eigenfunctions for the lowest band shows no significant dependence on the quasimomentum, the wavefunctions at $q = 0$ and at $q = \pi/d$ differ by the relative phase between adjacent potential minima [solid lines in Fig. 10.20(d)]. As in the weak periodic potential limit, the wavefunction at the upper band edge of the lowest band is staggered, i.e., there is a π phase jump between different sites. Phenomena studied in this regime only involve the lowest band, which is well described by localised wavefunctions at each site. Therefore, in this limit the dynamics can be described using the localised Wannier functions, which are superpositions of the Bloch functions

$$\psi_p\left(R, x\right) = \frac{1}{d}\int \psi_{p,q}\left(x\right)e^{-iRq}dq, \tag{10.69}$$

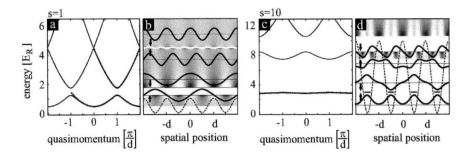

Fig. 10.20: Band structure for two different potential depths: (a) weak potential $A = E_R$, (b) deep potential with $A = 10E_R$ ($s = A/E_R$). In both cases, the analytical expressions (10.66) and (10.67) are depicted with dotted lines. In (b) and (d) we show the spatial dependence of the corresponding Bloch states. The periodic potentials are represented by the dashed lines. For each energy, the absolute square value of the corresponding Bloch states is depicted in the grey scale plot, where high probabilities are darker. Additionally, the wavefunctions are shown for the energies at the gaps indicated with the arrows. One clearly sees that the wavefunctions at the first gap change their sign from well to well, i.e., there is a phase slip of π. These modes are also known as staggered modes.

where R is the centre of the function. The dynamics is simply described via inter-well tunnelling. The characteristic energy scale of tunnelling coupling between two sites is given by the width of the band, which is $4J$.

We have seen that the linear properties of the periodic potential are defined by the potential amplitude A. The transition between weak and strong potential is continuous and thus no well-defined boundary can be given. A characteristic potential modulation for this transition may be found by equating the bandwidth and the gap energy, which have the same magnitude at a potential modulation depth of $A = 1.4E_R$.

10.6.5 *Bright- and gap-solitons in 1D polariton systems*

We have seen that the 1D GPE (10.2) describing particles with a repulsive interaction $\alpha > 0$ and positive mass $m > 0$ allows for dark and grey soliton solutions, which correspond to density dips occurring in a constant condensate. The qualitative explanation for the existence of this stationary density dip is the interplay between kinetic energy, which tends to spread the dip, and interaction energy, which tends to make it sharper. In fact another type of solitonic solution of the GPE exists if the product αm is negative. In such a case, a bright soliton solution, namely a stationary density peak over an empty system, does exist and the reader will be guided to find this stationary wavefunction by solving Exercise 10.2. The condition $\alpha m < 0$ means that, for a positive mass, there should be an attractive interaction between particles. The density peak should spread because of the kinetic energy dispersion of massive particles, but it is compensated by their attraction. Another option, a priori less expected, is to stick with the normal repulsive interactions of polaritons, but turn to a negative mass for the particles. This can be in fact naturally realised in periodic lattices where the effective mass of the valence band at the edge of the Brillouin zone is indeed negative, while particles still have repulsive interactions. This should allow for the formation of nonlinear "gap solitons". Gap solitons have been first described by Chen and Mills (1987) and then studied and observed in photonic systems by Eggleton et al. (1996) and in Bose–Einstein Condensates by Eiermann et al. (2004). In the framework of this book, it is worth mentioning the self-localised structures that exist in a dissipative system with an external supply of energy and that are often called dissipative solitons (see the book by Akhmediev and Ankiewicz-Kik (2005)). The balance between dispersion and nonlinearity for the dissipative solitons is also accompanied by a loss-gain balance. Dissipative solitons in optical resonators have been observed, e.g., in the vertical cavity surface emitting semiconductor lasers (VCSELs) operating in the bistable regime as shown by Barland et al. (2002). VCSEL solitons can be switched on and off using an external writing pulse, which makes them potentially suitable for applications in optical information processing. The first proposal of creating bright polariton solitons was formulated by Egorov et al. (2009). The authors proposed to use, not a periodic lattice, but the negative curvature of the lower polariton branch above the inflection point of the dispersion, an effect evidenced experimentally soon after by Sich et al. (2012), while polarisation effects have been investigated by Sich et al. (2014) and in the guided geometry by Walker et al. (2015). The first studies of polariton gap solitons have been jointly theoretical and experimental and were performed by Tanese et al. (2013) in a 1D lattice and by Cerda-Méndez et al. (2013) in a 2D lattice. An important conclusion that came out from the work of Tanese et al. (2013) and that was then further confirmed by Jacqmin et al. (2014) is that under non-resonant focused excitation, polariton condensation occurs preferentially in solitonic states bound to the potential created by the excitonic reservoir present at the position of the excitation. This type of state indeed shows a very good spatial overlap with the reservoir and is strongly favoured by relaxation kinetics with respect to a homogeneous ground state.

Exercise 10.2 *Stability of a bright soliton in 1D. The Gross-Pitaevskii equation in 1D has a bright soliton solution, if the product αm is negative, then*

$$i\hbar \frac{\partial \psi}{\partial t} = -\frac{\hbar^2}{2m} \nabla^2 \psi + \alpha |\psi|^2 \psi - \mu \psi. \tag{10.70}$$

In this case, particles attract each other and are able to form a droplet, described by the exact wavefunction

$$\psi(x) = \frac{\sqrt{n_0}}{\cosh(x/\xi)}. \tag{10.71}$$

1. *By inserting the above wavefunction into the Gross-Pitaevskii equation, find the expression for the chemical potential and the healing length.*

2. *Check the stability of the bright soliton by using the trial Gaussian wavefunction*

$$\frac{1}{\sqrt{w\sqrt{\pi}}} \exp\left(-\frac{x^2}{2w^2}\right), \tag{10.72}$$

(instead of the exact wavefunction) and checking if the total energy of the system

$$E(w) = -\frac{\hbar^2}{2m} \int_{-\infty}^{+\infty} \psi^*(x) \nabla^2 \psi(x) \, dx + \frac{\alpha}{2} \int_{-\infty}^{+\infty} |\psi(x)|^4 dx, \tag{10.73}$$

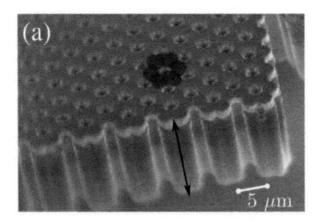

Fig. 10.21: Jacqmin et al. (2014)'s SEM image of polariton graphene.

as a function of the variational parameter w has a minimum.

If necessary, use the integrals

$$I_n = \int_0^\infty x^n e^{-x^2}\, dx = \frac{(n-1)\, I_{n-2}}{2}, \quad I_0 = \frac{\sqrt{\pi}}{2}, \quad I_1 = \frac{1}{2}. \tag{10.74}$$

Draw a sketch of this dependence (taking into account the negative sign of αm) and conclude on the stability of a bright soliton in 1D.

10.6.6 Honeycomb lattice (scalar approximation)

Two-dimensional lattices have also been implemented through the various methods mentioned previously. The emblematic example on which we will focus is that of a honeycomb lattice for polaritons, the so-called "polariton graphene", introduced by Jacqmin et al. (2014) and shown in the SEM of Fig. 10.21. In the rest of this Section, we explain in detail how the different ingredients at hand allow to engineer a photonic system showing topologically nontrivial bulk band structure and the corresponding chiral edge states.

Let us start by considering the spin-less case, which is the standard problem of the tight-binding description of a honeycomb lattice. The 2D graphene lattice contains two different types of inequivalent atoms A and B (cf. Fig. 10.22a). An atom A is surrounded only by atoms B and an atom B is surrounded only by atoms A. Therefore, in the basis of A and B orbitals, in a tight-binding approach and neglecting second-neighbour interactions, the Hamiltonian takes the form

$$H_{\text{graphene}} = \begin{pmatrix} 0 & J f_k \\ J f_k^+ & 0 \end{pmatrix} = J \left(\text{Re}(f_k)\sigma_x - \text{Im}(f_k)\sigma_y \right), \tag{10.75}$$

where

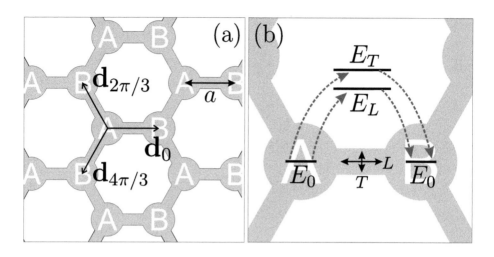

Fig. 10.22: A sketch of the tight-binding model. (a) Photon tunnelling between microcavity pillars is described as photon propagation through "waveguide"-like links. (b) Polarisation dependence of tunnelling probability due to TE-TM energy splitting: the L state, which is polarised longitudinally to the "waveguide" link, is closer in energy to the degenerate pillar-pinned states than the transversely-polarised state T, resulting in higher L-photon tunnelling probability through the link.

$$f_k = \sum_{j=0}^{2} \exp\left(ia\vec{k} \cdot \left(\cos\left(\frac{2j\pi}{3}\right) \vec{u}_x + \sin\left(\frac{2j\pi}{3}\right) \vec{u}_y \right) \right). \qquad (10.76)$$

This system of equations describes the coupling between a pseudospin associated with the pseudo-spinor $(\psi_A, \psi_B)^T$ and an effective magnetic field

$$\vec{\Omega}_{\text{graphene}} = J(\text{Re}(f_k), -\text{Im}(f_k), 0)^T. \qquad (10.77)$$

It is not related with the real spin of the particles: it defines the correlations between A and B atoms. The eigenstates of the system are found through the alignment and anti-alignment of the isospin with a $\vec{k}$-dependent effective magnetic field $\vec{\Omega}_{\text{graphene}}$. The states with negative energy have their pseudospin aligned with this field, and their energy lies below the Dirac points (valence band). The ones with their pseudospin anti-aligned have a positive energy and are above the Dirac points. The effective field cancels at the Dirac points and depends linearly on k around it, which provides the famous linear Dirac dispersion for electrons $\Omega_{\text{graphene}} \equiv (\tau_z k_x, k_y, 0)^T$, with $\tau_z = \pm 1$ being the valley index. One can observe that the field is monopolar for $\tau_z = +1$ and Dresselhaus-like for $\tau_z = -1$, and, therefore, has an opposite winding at K and K'.

10.6.7 Honeycomb lattice (polarised)

We follow the approach presented in the previous Section on double pillars to construct the polariton graphene Hamiltonian.

The zeroth approximation of the model is the basis of σ-polarised polariton states, localised on one of the pillars $\varphi^\sigma(\mathbf{r} - \mathbf{a}_i(\mathbf{b}_i))$, where $\sigma = \pm 1$ and $\mathbf{a}_i(\mathbf{b}_i)$ is the radius vector of the ith type-A(B) pillar. The general form of the polariton wavefunction may be represented as a linear combination of the basis states:

$$|\Psi(\mathbf{r})\rangle \equiv \sum_{i,\sigma} \left[c_{i,\sigma}^A |\varphi^\sigma(\mathbf{r} - \mathbf{a}_i)\rangle + c_{i,\sigma}^B |\varphi^\sigma(\mathbf{r} - \mathbf{b}_i)\rangle \right]. \tag{10.78}$$

We assume a slight difference in the tunnelling coefficients between two neighbouring pillars for a polariton linearly polarised along and transverse to the axis between them (due to the TE-TM splitting, as discussed above):

$$\langle \varphi^{L(T)}(\mathbf{r} - \mathbf{a}_i)|\hat{H}|\varphi^{L(T)}(\mathbf{r} - \mathbf{a}_i - \boldsymbol{\delta}_j)\rangle = -J_{L(T)}, \tag{10.79}$$

where the vector $\boldsymbol{\delta}_j$ links a type-A pillar to its jth neighbor, as shown in Fig. 10.22. We further put the energy of the degenerate basis states to zero and define the Hamiltonian non-diagonal matrix elements (see eqn (10.60))

$$H_j^{\sigma,\sigma} \equiv \langle \varphi^\sigma(\mathbf{r} - \mathbf{a}_i)|\hat{H}|\varphi^\sigma(\mathbf{r} - \mathbf{a}_i - \boldsymbol{\delta}_j)\rangle = -J, \tag{10.80a}$$

$$H_j^{+,-} \equiv \langle \varphi^+(\mathbf{r} - \mathbf{a}_i)|\hat{H}|\varphi^-(\mathbf{r} - \mathbf{a}_i - \boldsymbol{\delta}_j)\rangle = -\delta J e^{ij\frac{2\pi}{3}}, \tag{10.80b}$$

where $J = (J_L + J_T)/2$ and $\delta J = (J_L - J_T)/2$. Note that the presence of δJ results in a nonzero probability for a σ-polarised polariton to tunnel to a neighboring pillar, simultaneously flipping its polarisation. Moreover, the phase gained in such tunnelling is different for each of the three tunnelling directions. Consequently, dotting Schrödinger's equation $\hat{H}\Psi = E\Psi$ on $\langle \varphi^\pm(\mathbf{r} - \mathbf{a}_n)|$ and $\langle \varphi^\pm(\mathbf{r} - \mathbf{b}_n)|$ and substituting the general form (10.79) with $c_{i,\sigma}^{A(B)} = c_\sigma^{A(B)}(\mathbf{k})e^{-i\mathbf{k}\mathbf{a}_i(\mathbf{b}_i)}$, we obtain a system of four coupled equations on the coefficients $c_\sigma^{A(B)}(\mathbf{k})$:

$$\sum_{j,\sigma} c_\sigma^{B(A)} e^{-i\mathbf{k}\boldsymbol{\delta}_j} H_j^{+(-),\sigma} = E c_{+(-)}^{A(B)}, \tag{10.81}$$

which is more clearly presented in a matrix form as

$$\begin{pmatrix} E_k & 0 & f_k J & f_k' \delta J \\ 0 & E_k & f_k'' \delta J & f_k J \\ f_k^* J & f_k''^* \delta J & E_k & 0 \\ f_k'^* \delta J & f_k^* J & 0 & E_k \end{pmatrix} \begin{pmatrix} c_+^A \\ c_-^A \\ c_+^B \\ c_-^B \end{pmatrix} = 0, \tag{10.82}$$

where $f_k = \sum_{j=0}^2 e^{-i\mathbf{k}\boldsymbol{\delta}_j}$, $f_k' = \sum_{j=0}^2 e^{-i\mathbf{k}\boldsymbol{\delta}_j + ij\frac{2\pi}{3}}$ and $f_k'' = \sum_{j=0}^2 e^{-i\mathbf{k}\boldsymbol{\delta}_j - ij\frac{2\pi}{3}}$. The polariton dispersion is obtained from the block determinant of the matrix M_k in the left-hand part of eqn (10.82):

$$\det(M_k) = \det(\hat{E}_k^2 - \hat{F}_k^\dagger \hat{F}) = 0, \tag{10.83}$$

where $\hat{E}_k$ and $\hat{F}_k$ are the upper-left and upper-right 2×2 blocks of M. Equation (10.83) presents a biquadratic equation on E, with a set of solutions:

$$E_k^2 = |f_k|^2 J^2 + \frac{|f_k'|^2 + |f_k''|^2}{2} \delta J^2 \pm \frac{\sqrt{(|f_k'|^2 - |f_k''|^2)^2 \delta J^4 + 4|f_k f_k'^* + f_k^* f_k''|^2 J^2 \delta J^2}}{2}.$$

(10.84)

Note that in absence of SOC, $\delta J = 0$ and eqn (10.84) is reduced to the conventional equation for graphene dispersion. The effective Hamiltonian of polariton graphene may be formulated in terms of pseudospin operators σ and s, having the same matrix form of Pauli matrices vector and corresponding to sublattice (A/B) and polarisation degrees of freedom. It may be separated into a polarisation-independent part $H_k^{(0)}$, coupling σ with momentum and giving a standard graphene dispersion with two Dirac valleys K and K', and a spin-orbit term H_k^{SO}, coupling s with σ and momentum:

$$H_k^{(0)} = -J\sigma_+ f_k \otimes 1 + \text{H.c.},$$

(10.85a)

$$H_k^{SO} = -\delta J\sigma_+ \otimes \left(f_k^+ s_+ + f_k^- s_-\right) + \text{H.c.},$$

(10.85b)

where $\sigma_\pm = (\sigma_x \pm i\sigma_y)/2$, $s_\pm = (s_x \pm is_y)/2$, $\otimes$ is the Kronecker product and 1 is the identity operator in the spin subspace. Expanding eqns (10.85) and keeping the first order in $q = k - K$, we isolate the momentum-independent and dependent parts H_K^{SO} and H_K^{SO} to finally rewrite all terms in the low-energy approximation:

$$H_q^{(0)} = \hbar v_F \left(\tau_z q_x \sigma_x + q_y \sigma_y\right),$$

(10.86a)

$$H_K^{SO} = \Delta \left(\tau_z \sigma_y s_y - \sigma_x s_x\right),$$

(10.86b)

$$H_q^{SO} = \frac{\Delta a}{2} \left[s_x \left(\tau_z q_y \sigma_y - q_x \sigma_x\right) - s_y \left(\tau_z q_x \sigma_y + q_y \sigma_x\right)\right],$$

(10.86c)

where $v_F = 3Ja/(2\hbar)$, $\Delta = 3\delta J/2$ and τ_z equals $+1$ and -1 for K and K' valleys respectively. Here, the same basis as the one of Kane and Mele (2005a) is used in order to allow a direct comparison with their Hamiltonian. This basis is different from the original basis used by Wallace (1947), which is used in eqn (10.90). The passage from Wallace's basis to that of Kane and Mele (2005a,b) is obtained by substituting $q_x \to q_y$ and $q_y \to -q_x$. The term in eqn (10.86b), similar to the Rashba term introduced by Kane and Mele (2005a,b), is dominant in the region of reciprocal space where $qa \ll \delta J/J$ and is responsible for the band-splitting at K and K'. The interplay between eqns (10.86a) and (10.86b) produces an effective photon mass $m^* = (2c\hbar^2\delta J)/(3a^2 J^2)$ in this region.

Provided that $\delta J \ll J$, the term in eqn (10.86b) dominates over that of eqn (10.86c) in the region $\delta J/J \ll qa \ll 1$, but plays the role of a perturbation over the polarisation-independent term (10.86a), splitting its linearly-polarised eigenstates in energy. It may, therefore, be interpreted as a pseudospin interaction with an emergent in-plane effective magnetic field in this region. Considering either positive ($c = +1$) or negative ($c = -1$) energy states, the spin-orbit term (10.86b) transforms to a symmetry-allowed Dresselhaus-like emergent field

$$H_c^{SO} = -\Delta c\tau_z \left(q_x s_x - q_y s_y\right)/q.$$

(10.87)

The SOC effective-field Hamiltonian in the highest order in momentum may thus be expressed in the minimal coupling form

$$H_{\mathbf{q}}^{(0)} + H_{\mathbf{K}}^{SO} = \hbar v_F \left(\tau_z \sigma_x \left[q_x - \frac{\Delta}{\hbar v_F} \tau_z^{-1} s_x \right] + \sigma_y \left[q_x + \frac{\Delta}{\hbar v_F} \tau_z s_y \right] \right). \quad (10.88)$$

The effective-field x and y components are not commutative, inheriting this property from the Pauli matrices s_x and s_y. Therefore, it can be represented as a non-Abelian gauge field. The effective field follows the direction of the lowest branch pseudospin direction map, plotted in Fig. 10.23.

Figure 10.24(b–d) show the degree of circular polarisation as a function of coordinates, taken at $t = 30$ ps. Panel (b) shows the polarisation degree for the excitation at the Γ point, where the field has the typical TE-TM texture, featuring the typical four polarisation domains (see Kavokin et al. (2005a); Leyder et al. (2007); Flayac et al. (2013)). Panels (c–d) demonstrate the optical spin Hall effect at the K and K' points, respectively, where the field has the texture of the Dresselhaus SOC. This is evidenced by two polarisation domains in real space as shown by Vishnevsky et al. (2013) and Terças et al. (2014), being inverted between the K and K' points, as the fields around K and K' have opposite signs. The OSHE texture is a demonstration of the different nature of the effective SOC field at the two Dirac points K and K'. From this numerical experiment, one may clearly see the advantage of photonic systems in comparison to their solid state counterparts, which allow us to excite and analyse any point of the dispersion with much more facility. Another very interesting consequence of this work relies on the possibilities offered i) by the manipulation of the lattice geometry and ii) by the hybrid exciton-photon nature of the polaritons. Combined with SOC, it paves the stage for a broad range of applications. For example, the mixed nature of exciton-polaritons provides a magnetic response of the system at optical frequencies, which is crucial for realisation a photonic topological insulator as first shown by Haldane and Raghu (2008) and Wang et al. (2009), and later by Karzig et al. (2015) and Nalitov et al. (2015b). This is discussed in more details in the following.

10.6.8 *Polariton topological insulators*

10.6.8.1 *Introduction*   The history of topological insulators (TI) dates back to the discovery of the Quantum Hall Effect (QHE) by von Klitzing et al. (1980). In his experiment, a strong magnetic field pinned the conduction band electrons to the Landau levels, opening a bandgap in the bulk and thus turning an electron conductor into an insulator. The edge electron states, on the contrary, carried one-way currents, protected from backscattering and responsible for the integer Hall conductance as shown by Thouless et al. (1982). This is an example of a $\mathbb{Z}$ topological insulator. The classification that allows one to distinguish this phase from a conventional band insulator, is based on the Chern topological invariant, introduced by Simon (1983). This is an integer number that characterise the band structure in terms of the Berry phase and a $\mathbb{Z}$ topological insulator is often called a Chern Insulator.

A whole new family of TI materials with different sets of topological invariants and symmetries were later proposed and discovered (see the review from Hasan and Kane (2010)). Graphene has a special place in this family, allowing Novoselov et al. (2007) to observe the QHE at room temperature. It played the role of a model system for the description by Haldane (1988) of the Quantum Anomalous Hall effect (QHE without net

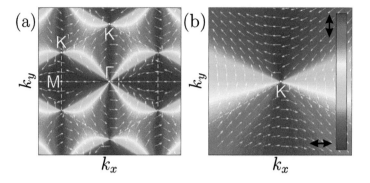

Fig. 10.23: The eigenstate pseudospin map of the lowest energy dispersion branch, as calculated by Nalitov et al. (2015b). The pseudospin direction is plotted with white arrows, and its projection on the x axis is additionally plotted with colour. Blue corresponds to negative projection and horizontal linear polarisation, red to positive sign of projection and vertical linear polarisation. (b) Zoom of the whole Brillouin zone pseudospin polarisation map, plotted in (a). It demonstrates Dresselhaus SOC field configuration due to TE–TM field in the vicinity of the Dirac points, as the eigenstate pseudospin is directed along the effective field.

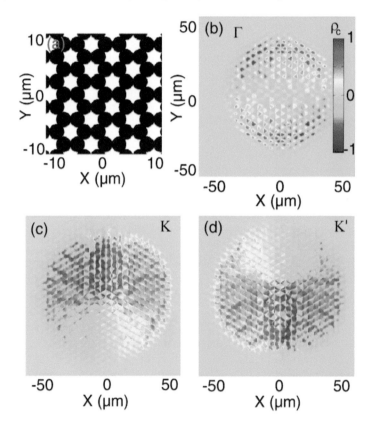

Fig. 10.24: Optical spin Hall effect (OSHE) modelling in polariton graphene by Nalitov et al. (2015b). (a) Potential profile: black regions correspond to negative potential and simulate photon confinement. (b) OSHE modelling with resonant linearly polarised excitation of the Γ point of the lowest energy dispersion branch (ground state). The circular polarisation ρ_c, color-coded, has four domains in space, like the conventional OSHE in planar microcavities. (c) and (d) panels demonstrate OSHE simulation with resonant excitation of K and K' points (excitation frequency corresponds to the centre of the lowest band). They have two spatial domains of circular polarisation sign, revealing the gauge nature of the effective spin-orbit interaction field.

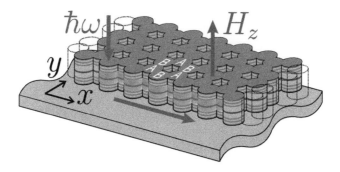

Fig. 10.25: Topologically protected light propagation along the edge of a polariton topological insulator. The polariton graphene considered is based on an etched planar microcavity. The cavity is constituted by two Distributed Bragg Reflectors (DBRs) sandwiching a cavity with embedded Quantum Wells (QWs). The energy splitting existing between TE and TM polarised modes provides the photonic SOC. The application of a real magnetic field perpendicular to the x-y plane of the structure opens the nontrivial gap. Edge modes are one way propagative modes that cannot elastically scatter to the bulk states. In a stripe geometry, normally incident light is guided either clockwise or anti-clockwise, depending on the external magnetic field sign.

magnetic field), and for the description by Kane and Mele (2005a) of the Quantum Spin Hall effect. The latter is associated with topologically-protected boundary spin currents and is characterised by a nonzero $\mathbb{Z}_2$ invariant stemming from the SOC for electrons (see Kane and Mele (2005b)) (without external magnetic field). These spin currents in the $\mathbb{Z}_2$ TI are formed from two spin components propagating in opposite directions, contrary to the $\mathbb{Z}$ TI, where both spin components propagate in the same direction, the other being forbidden. Although the extremely small SOC has not allowed the observation of QSHE in graphene, it was later demonstrated in various 2D structures by König et al. (2007) and in 3D by Hsieh et al. (2008). A Floquet TI having a topologically nontrivial gap was realised in a 2D heterostructure under microwave electromagnetic irradiation by Lindner et al. (2011).

Many promising implementations of topological phases in bosonic systems have been proposed, including photonic systems. One can mention honeycomb photonic gyromagnetic waveguides by Haldane and Raghu (2008) and Wang et al. (2008), coupled microcavities by Umucalılar and Carusotto (2012), coupled cavity rings by Hafezi et al. (2011), coupled helical waveguides by Rechtsman et al. (2013), and photonic waveguides based on metamaterials with bi-anisotropic behaviour by Khanikaev et al. (2013) and Chen et al. (2014). Finally, optical QHE due to artificial gauge fields was predicted by Ozawa and Carusotto (2014) in microcavity lattices. An excellent review of the emergent field now known as topological photonics has been published by Lu et al. (2014), before several proposals to use polaritons for topological photonics (by Karzig et al. (2015), Nalitov et al. (2015b) and Bardyn et al. (2015)) appeared. The microcavity platform provides an intrinsic susceptibility of polaritons to magnetic field due to the excitonic component and the effective SOC due to the photonic part. This Section is devoted to the demonstration of the $\mathbb{Z}$ topological insulator behaviour of polariton graphene in an external magnetic field, as sketched in Fig. 10.25

10.6.8.2 *Energy polariton bandgap in external magnetic field* We construct the polaritonic graphene effective Hamiltonian in the nearest neighbour approximation. A state of the polariton graphene can be described by a bispinor $\Phi = \left(\Psi_A^+, \Psi_A^-, \Psi_B^+, \Psi_B^-\right)^{\mathrm{T}}$, with $\Psi_{A(B)}^{\pm}$ the wavefunctions of the two sublattices and two spin components. We account for the magnetic field via the Zeeman splitting Δ of the states localised at each pillar. It affects the diagonal blocks of the Hamiltonian matrix form (10.82). The effective Hamiltonian in the presence of a real magnetic field applied along the z-direction reads

$$H_{\mathbf{k}} = \begin{pmatrix} \Delta\sigma_z & F_{\mathbf{k}} \\ F_{\mathbf{k}}^\dagger & \Delta\sigma_z \end{pmatrix} \quad \text{and} \quad \Delta = |x|^2 g_X \mu_B H_z / 2\,, \tag{10.89}$$

where σ_z is the Pauli matrix, x is the excitonic Hopfield coefficient, g_X is the effective g-factor for the 2D exciton, μ_B is the Bohr magneton, and H_z is the applied magnetic field, giving rise to polariton Zeeman splitting Δ. The matrix that appears on the off-diagonal reads

$$F_{\mathbf{k}} = -\begin{pmatrix} f_{\mathbf{k}} J & f_{\mathbf{k}}^+ \delta J \\ f_{\mathbf{k}}^- \delta J & f_{\mathbf{k}} J \end{pmatrix}, \tag{10.90}$$

where the complex coefficients $f_{\mathbf{k}}, f_{\mathbf{k}}^{\pm}$ are defined by

$$f_{\mathbf{k}} = \sum_{j=1}^{3} \exp(-i\mathbf{k}\mathbf{d}_{\varphi_j}), \quad f_{\mathbf{k}}^{\pm} = \sum_{j=1}^{3} \exp(-i\left[\mathbf{k}\mathbf{d}_{\varphi_j} \mp 2\varphi_j\right])\,, \tag{10.91}$$

and $\varphi_j = 2\pi(j-1)/3$ is the angle between the horizontal axis and the direction to the jth nearest neighbor of a type-A pillar. J is the polarisation independent tunnelling coefficient and δJ is the SOC-induced polarisation dependent term. Without the magnetic field, the Hamitonian (10.89) can be exactly diagonalised, as shown by Nalitov et al. (2015a). The energy dispersions obtained are relatively close to those of bilayer graphene (see McCann and Koshino (2013)) and of a monolayer graphene in presence of Rashba SOC (see Rakyta et al. (2010)). The polarisation texture of the eigenstates is, however, different.

The dispersion close to the **K** point is shown as a dashed line on Fig. 10.26(a). Under the effect of SOC, the Dirac point transforms into four inverted parabolas. Two parabolas are split off, while the two central ones cross each other. It is instructive to consider the eigenstates exactly at the Dirac points. At the K point, the eigenstates of the two central parabolas are fully projected on Ψ_A^- and Ψ_B^+ respectively, whereas at the K' point they project on Ψ_A^+ and Ψ_B^- respectively. Let us now qualitatively consider the consequence of a finite Zeeman splitting. As sketched on Fig. 10.26(b), the degeneracy between the states in the crossing points of two branches (dashed lines) is lifted by 2Δ, and the states split off by the SOC are further shifted. At the K (K') point, the "valence" band is formed from the B(A)-pillars and the "conduction" band from the A(B)-pillars. The reversed order of the band in the basis of the sublattices signifies a topological nontriviality of the gap. In the spin basis, however, the valence and conduction bands are equivalent at K and K', unlike the $\mathbb{Z}_2$-topological insulator (see the discussion by Hasan and Kane (2010)).

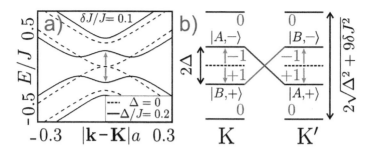

Fig. 10.26: Nontrivial bulk band structure of the polariton graphene in an external magnetic field. (a) bulk energy dispersion without (dashed) and with (solid) magnetic field. (b) illustration of degeneracy lifting in K and K' points. Due to coupling between sublattice and polarisation, the states, localised on one sublattice, go up in energy at one point and down in the other. Chern numbers at each point are shown in green. From Nalitov et al. (2015b), with parameters: $\delta J/J = 0.1$, $\Delta/J = 0.01$.

The result of the complete diagonalisation of the Hamiltonian (10.89) is plotted with solid lines in Fig. 10.26(a). As expected, it shows an energy gap, saturating to $E_g \sim 3\delta J$ at $|\Delta| \sim \delta J$. Without SOC ($\delta J = 0$), the application of a magnetic field does not not open any gap, as it keeps the symmetry between K and K' valleys, while breaking the symmetry between the spin projections on the z axis. The band structure in this latter case is constituted by two graphene dispersions shifted in energy by the polariton Zeeman splitting.

10.6.8.3 *Topological invariant analysis*   The topological invariants are integer numbers characterising the band structure, which cannot be changed with continuous transformations of the bands. Nonzero topological invariants distinguish topologically trivial band insulators from nontrivial ones. In particular, nonzero Chern numbers correspond to a topological insulator of the $\mathbb{Z}$ type. Band Chern numbers may be calculated from the Berry connection over the Brillouin zone (see Hasan and Kane (2010)) as

$$C_m = \frac{1}{2\pi} \iint\limits_{\mathrm{BZ}} \mathbf{B}_{\mathbf{k},m} \mathrm{d}^2 \mathbf{k} , \qquad (10.92)$$

where m is the branch index, and the Berry curvature $\mathbf{B}_{\mathbf{k},m}$ is expressed in the effective Hamiltonian (10.89) and its eigenstates $|\Phi_{\mathbf{k},m}\rangle$ with corresponding energies $E_{\mathbf{k},m}$ as

$$\mathbf{B}_{\mathbf{k},m} = i \sum_{l \neq m} \frac{\langle \Phi_{\mathbf{k},m}|\nabla_{\mathbf{k}} H_{\mathbf{k}}|\Phi_{\mathbf{k},l}\rangle \times \langle \Phi_{\mathbf{k},l}|\nabla_{\mathbf{k}} H_{\mathbf{k}}|\Phi_{\mathbf{k},m}\rangle}{(E_{\mathbf{k},m} - E_{\mathbf{k},l})^2} . \qquad (10.93)$$

A direct numerical calculation of the Berry curvature for the polariton graphene bands shows that it is nonzero in the vicinity of the Dirac points. The two inner branches, split by the interplay of the external magnetic field and effective SOC, have nonzero Berry connections around the K and K' points, each giving ± 1 contribution to the total band Chern number ± 2 (marked in green in Fig. 10.26(b)). Outer branches, on the contrary, have zero Berry curvature over all reciprocal space.

Nonzero Chern numbers demonstrate the topological nontriviality of the polariton graphene band structure in the presence of an external magnetic field. The bulk-boundary correspondence yields the existence of one-way edge states on an interface between topologically trivial and nontrivial insulators. The energy of the states must lie in the energy gap, and their dispersion connects the lower and upper bands. The vacuum itself is a trivial band insulator, therefore, any edge of the polariton graphene, either zigzag or armchair, conducts polaritons in a direction specified by the magnetic field sign.

10.6.8.4 *Topologically protected edge states*

To demonstrate the consequence of the bulk-boundary correspondence with one-way edge states, we used the same tight-binding approach to model a quasi-1D stripe of microcavity pillars, consisting of $N = 50$ zig-zag chains. We derive and diagonalise the $4N \times 4N$ Hamiltonian to obtain the band structure. For this, we set a basis of Bloch waves $\Psi^{\pm}_{A/B,n}(k_x)$, where the index n enumerates the stripes, and k_x is the quasi-wavevector in the infinite zig-zag direction. The 4×4 diagonal blocks D_{k_x} describe coupling within one stripe and are derived in the same fashion as Hamiltonian (10.89), while the coupling between the stripes is accounted for in subdiagonal and superdiagonal 4×4 blocks $S^{\dagger}_{k_x}$ and S_{k_x}. The whole matrix reads

$$
H_{k_x} = \underbrace{\begin{pmatrix}
D_{k_x} & S_{k_x} & 0 & \cdots & 0 & 0 & 0 \\
S^{\dagger}_{k_x} & D_{k_x} & S_{k_x} & \cdots & 0 & 0 & 0 \\
0 & S^{\dagger}_{k_x} & D_{k_x} & \cdots & 0 & 0 & 0 \\
\vdots & \vdots & \vdots & \ddots & \vdots & \vdots & \vdots \\
0 & 0 & 0 & \cdots & D_{k_x} & S_{k_x} & 0 \\
0 & 0 & 0 & \cdots & S^{\dagger}_{k_x} & D_{k_x} & S_{k_x} \\
0 & 0 & 0 & \cdots & 0 & S^{\dagger}_{k_x} & D_{k_x}
\end{pmatrix}}_{\text{50 blocks } 4 \times 4}, \tag{10.94}
$$

where

$$
D_{k_x} = - \begin{pmatrix}
-\Delta & 0 & Jf_{k_x} & \delta Jf^{+}_{k_x} \\
0 & \Delta & \delta Jf^{-}_{k_x} & Jf_{k_x} \\
Jf^{*}_{k_x} & (\delta Jf^{-}_{k_x})^{*} & -\Delta & 0 \\
(\delta Jf^{+}_{k_x})^{*} & Jf^{*}_{k_x} & 0 & \Delta
\end{pmatrix} \quad \text{and} \quad S_{k_x} = - \begin{pmatrix}
0 & 0 & 0 & 0 \\
0 & 0 & 0 & 0 \\
J & \delta J & 0 & 0 \\
\delta J & J & 0 & 0
\end{pmatrix}, \tag{10.95}
$$

with complex coefficients f_{k_x} and $f^{\pm}_{k_x}$ defined as

$$
f_{\mathbf{k}} = \sum_{j=1}^{2} \exp(-i\mathbf{k}\mathbf{d}_{\varphi_j}), \quad f^{\pm}_{\mathbf{k}} = \sum_{j=1}^{2} \exp(-i\left[\mathbf{k}\mathbf{d}_{\varphi_j} \mp 2\varphi_j\right]). \tag{10.96}
$$

Note that here the summation is only taken over two terms, corresponding to the two nearest neighbors within a zigzag chain. Figure 10.27(c) shows the result of the band structure calculation. The degree of localisation on edges is calculated from the wave-function densities on the very left and the very right zigzag chains $|\Psi_L|^2$ and $|\Psi_R|^2$

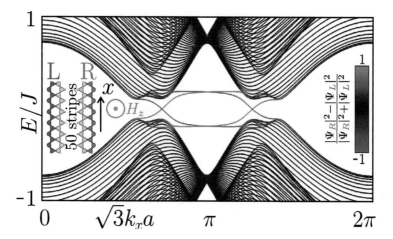

Fig. 10.27: Numerical calculation of eigenstates of a zig-zag chain by Nalitov et al. (2015b): edge states are marked with colour. The direction of their propagation is given by the sign of the product $H_z g_X$ and is protected from both backscattering and scattering into the bulk.

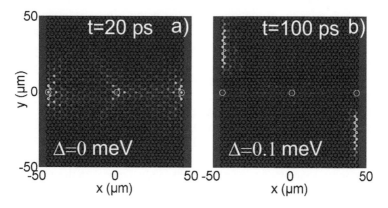

Fig. 10.28: Propagation of light in the conducting ($\Delta = 0$) and topological insulator phase ($\Delta \neq 0$), as calculated by Nalitov et al. (2015b) in the spatial distribution of emission intensity. a) rapid expansion of the bulk propagative states after 20 ps at $\Delta=0$; b) surface states after 100 ps at $\Delta=0.1$ meV. White circles show the pumping spots located at $\mathbf{r}_j$, and black line traces the contours of the potential. The parameters are: $\beta = \hbar^2 \left(m_l^{-1} - m_t^{-1} \right) / 4m$ where $m_{l,t}$ are the effective masses of TM and TE polarised particles respectively and $m = 2\left(m_t - m_l \right) / m_t m_l$; $m_t = 5 \times 10^{-5} m_0$, $m_l = 0.95 m_t$, where m_0 is the free electron mass.$\tau_0 = 35$ ps, $\sigma = 1\,\mu\text{m}$, $\omega = 1.6$ meV and lies within the bulk energy gap, $\tau = 25$ ps. Pumping P is circularly polarised.

(see inset), and is shown with colour, so that the edge states are blue and red. The propagation direction of these edge states is related to the direction of the external magnetic field: the photon edge current is either clockwise or anti-clockwise depending on the signs of H_z and g_X. One should insist on the fact that we deal with a real polariton current and not with a spin current.

The presented results were obtained for the most realistic case of rather small SOC ($\delta J/J = 0.1$). In this case, the dispersion topology in the absence of magnetic field is characterised by the trigonal warping effect, that consists in the emergence of three additional Dirac cones in the vicinity of each Brillouin zone corner. This is typical for bilayer graphene (see McCann and Koshino (2013)) and monolayer graphene with Rashba SOC (see Rakyta et al. (2010)). However, at a critical strength of the spin-orbit interaction $\delta J = J/2$, a transition occurs in the topology of the dispersion: additional Dirac cones with opposite Chern numbers meet in pairs at the centres of Brillouin zone edges and recombine. This leads to a change of the band Chern number set from $C_m = \pm 2$ to $C_m = \pm 1$. Figure 10.28, calculated by direct solution of the Schrödinger equation including the honeycomb lattice potential and the TE-TM splitting, demonstrates qualitatively different behaviour with and without the magnetic field (parameters of the simulations are given in the caption). Without the field ($\Delta = 0$), the excitation energy corresponds to propagating states, and the resulting expansion of polaritons is visible on panel (a). No gap is opened, and the particles created on the surface by the two corresponding pumps are rapidly expanding into the bulk: after only 20 ps, the intensity is distributed over a significant part of the sample. However, under an applied field giving $\Delta = 0.1$ meV, the excitation energy lies within a gap, which makes the injection in the centre ineffective: no particles created by the pump are visible in the centre after 100 ps. But the spots on the edges become now resonant with the surface states that have appeared there, and their one-way propagation is visible in panel (b): after 100 ps, particles are about 25 μm away from the pump spots. Although particles are created with both positive and negative wavevectors along a given edge, they propagate in only one direction, which proves the one-way nature of the surface states. The transverse profile of these states shows an exponential decay with a characteristic length of $(2\kappa)^{-1} = 3.1(1)$ μm, corresponding to the analytical estimate of the extinction coefficient $\kappa \approx \sqrt{2mE_g/\hbar^2}$ determined by the size of the gap E_g. Thus, the full numerical simulation confirms the predictions of the tight-binding model on the appearance of the gap in the presence of magnetic field and the formation of one-way surface states.

Exercise 10.3 *Consider a tight-binding description of a Honeycomb lattice made out of two different atoms, neglecting the real spin degree of freedom. The Hamiltonian of the system will be a 2×2 matrix similar to (10.75), but with two different elements on the diagonal. Find the eigenenergies and eigenstates of the Hamiltonian. Calculate the Valley and total Chern numbers for the conduction and valence band.*

10.7   Further reading

This chapter comes with an extensive bibliography, including several books and reviews. The basics of Quantum fluid physics is described in the Bose–Einstein Condensation book by Pitaevskii and Stringari (2003) already recommended in previous chapters. A more advanced description is provided by Leggett (2006) in his book "Quantum liquids". A highly inspiring monograph, which details how the unique properties of liquid helium can be used to emulate an enormous quantity of physical situations is the "Universe in a Helium Droplet" by Volovik (2003). As mentioned in the introduction, the epithet "Quantum fluids of Light" has been introduced in the comprehensive review of Carusotto and Ciuti (2013). A very complete description of topological defects in spinor polariton quantum fluids can be found in the PhD thesis of Flayac (2012). The physics of nonlinear states in periodic systems is addressed in detail for both atomic and photonic systems by Kevrekidis et al. (2008) and Kivshar and Agrawal (2001), respectively. Finally the physics of topological insulators and topologically nontrivial systems has been addressed in many reviews and we suggest the popular ones of Hasan and Kane (2010) and of Qi and Zhang (2011). The field of topological photonics is even more recent, but already counts with an excellent review, from Lu et al. (2014). Amo and Bloch (2016) make a review of polaritons in lattices and their prospects as simulators.

11

QUANTUM POLARITONIC

Microcavity polaritons have demonstrated their unique propensity to host macroscopic quantum phenomena. While they appear to be highly promising for applications in a classical realm, they are still far from competing even with decade old electronics. Another playground where polaritons could emerge as strong contenders is the microscopic quantum regime with single-particle effects and nonlinearities at the one-polariton level. Several theoretical proposals exist to explore polariton blockade mechanisms, realise sophisticated quantum phase transitions, implement quantum simulations and/or quantum information processing, thereby opening a new page of the polariton physics when such ideas will be implemented in the laboratory.[126]

[126] We acknowledge enlightening discussions with E. del Valle, C. Sánchez Muñoz and J. C. López Carreño, who contributed substantially to the content of this chapter.

Microcavities, Second Edition. Alexey V. Kavokin, Jeremy J. Baumberg, Guillaume Malpuech, Fabrice P. Laussy, Oxford University Press (2017). © Alexey V. Kavokin, Jeremy J. Baumberg, Guillaume Malpuech, Fabrice P. Laussy. DOI 10.1093/oso/9780198782995.001.0001

11.1   Microcavity QED

11.1.1   *Quantum vs classical polaritons*

Haroche (2013)'s Nobel Prize honored the field of cavity QED, that is, the quantum interaction of light and matter in the confines of a cavity. These laboratories of the extreme bring together two realms of physics:

1. *Classical physics* through the light field,
2. *Quantum physics* through the quantum emitter.

The light-field can inherit the quantum features of the quantum emitter and stray beyond a purely classical description, but even in a fully-quantised description, it bears all the features of a classical field (at least in quantum optics where intrinsic nonlinearities such as a photon creating an electron-positron pair are not possible). This was the view of Jaynes and Cummings (1963) who brought this dichotomy between light and matter to some extreme conclusions by proposing to replace quantum mechanics by their own substitute (the so-called neo-classical theory) where the light-field is not quantised, solving, according to Jaynes (1977), the famous problem of divergences in QED.

Semiconductor technology brought this physics of light-matter interactions down to the micro- and even nano-world, and many of the observed effects have some analogue in cavity QED, in this respect, "microcavity QED". The analogy can be fairly exact by substituting, on the one hand, an atomic transition by a 0D solid-state two-level system, such as a quantum dot transition, a nitrogen vacancy, a single dye molecule, or a superconducting qubit, and on the other hand a dilute atomic gas by a 2D excitonic gas. In the case of 2D exciton-polaritons, extra-care must be taken by the fact that excitons are bosons, in which case the need of quantising anything at all might not always be compelling. Nevertheless, the polariton is popularly described as a quantum superposition of light and matter, which comes from the one-particle Hamiltonian eigensolution, i.e., in Dirac's notation:

$$\|1, 0\rangle = \cos\theta\,|0, 1\rangle - \sin\theta\,|1, 0\rangle \,, \tag{11.1a}$$

$$\|0, 1\rangle = \sin\theta\,|0, 1\rangle + \cos\theta\,|1, 0\rangle \,, \tag{11.1b}$$

where, with the conventions of Chapter 5, $|0, 1\rangle$ is the one-exciton state with no photon and $|1, 0\rangle$ the one-photon state with no exciton. The corresponding dressed states are those of a lower-polariton $\|1, 0\rangle$ with no upper-polariton, and vice-versa, $\|0, 1\rangle$. This status of a quantum superposition was proclaimed from the beginning by the discoverers of QW polaritons, Weisbuch et al. (1992), who saw them as the "*solid state analog of the vacuum-field Rabi splitting*". Even in atomic cavity QED, discussions soon started on the genuine quantum character of strong-coupling. Zhu et al. (1990) summarise the appeal of this identification by noting that "*the vacuum Rabi splitting has attracted the attention of the quantum-optics community because it is considered to be an important manifestation of the quantum nature of the electromagnetic field*", but then proceed to show that a fully classical model of linear absorption and dispersion provides the same results and that, therefore, "*contrary to popular belief, the vacuum Rabi splitting is not an inherently quantum phenomenon.*"

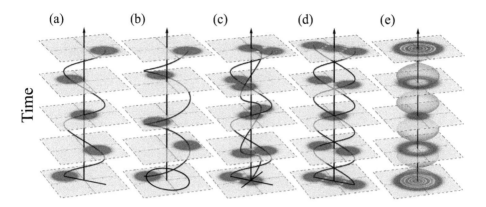

Fig. 11.1: Quantum states of two coupled harmonic oscillator pictured in the Wigner representation of one of the oscillators (this would be the cavity field for polaritons). In each case, the x and y axis represent the real and imaginary parts (a) Rabi oscillations, with motion along a diagonal of the space. (b) Polariton state, with motion along a circle, in the direction of the rotating frame for the upper polariton (case shown here) and counter-clockwise for the lower polariton. (c) Superposition of coherent states $|\alpha 0\rangle + |0\beta\rangle$, combining the two previous types of motions simultaneously, still with only positive values of the Wigner function. (d) Cat state $|\alpha\rangle + |-\alpha\rangle$, with Rabi-like oscillations of the two lobes, but exhibiting deep quantum interferences when far apart. (e) Fock states of polaritons, breathing around the centre from the vacuum to the characteristic Fock Wigner functions. "Negative probabilities" occur only in the two last cases.

The resolution of this apparent paradox—eqns (11.1) describing unmistakably an entangled state—is found in the concept of the "cebit" introduced by Spreeuw (1998) that packages in suitable notations and a fitting conceptual approach the description of classical fields to feature their common and distinctive attributes from quantum fields. Instead of Dirac's $\langle\text{bra}|\text{ket}\rangle$, the scalar products of the classical fields are expressed as a $(\text{parent}|\text{thesis})$ and two orthogonal classical states provide the bit states $|0\rangle$ and $|1\rangle$. In this conceptual framework, one can show that entanglement is not a genuine quantum feature, since classical fields, with an ad hoc interpretation, can be shown to be likewise intermingled and one can even design quantum-like gates and protocols for them, such as teleportations of cebits. This does not mean, however, that there is nothing peculiar about quantum states. Only that they can be "simulated" or "spoofed" by classical fields. To do so, however, requires exponentially growing resources, and the interest for quantum information processing (QIP) with genuine qubits remains total. Conceptually, the cebits also allow to identify, not the entanglement per se, but nonlocality as the defining attribute of "quantum weirdness". Entangling classically two objects is possible if it is done through internal degrees of freedom that remain attached to the same physical reality, while quantum entanglement can likewise bind two separate objects (e.g., particles) very far apart and, therefore, not susceptible to Einsteinian mediation.

The states in eqns (11.1) can be argued to be genuine quantum states because they are single-particle objects. They would, therefore, exhibit antibunching or any other criterion to characterise quantum light (here, we will not enter the discussion, started by Jaynes and Cummings (1963), whether even single-particle quantum states of the light field can be cast in a classical theory). These states are popular for polaritons mainly as

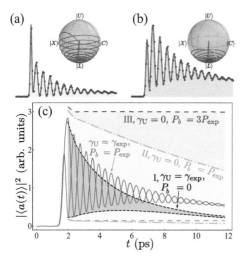

(a)   (b)

(c)

$|\langle a(t)\rangle|^2$ (arb. units)

t (ps)

Fig. 11.2: **Rabi oscillations in the laboratory.**
(a) Oscillations of the cavity photon intensity
observed as a result of preparing the initial condition close to a photon state with a resonant
pulse, and its reconstruction on the Bloch sphere.
(b) Same for excitation at a different energy, resulting in a different initial condition closer to
the lower polariton. The contrast in the oscillations is not due to dephasing, but to the polariton
lifetimes imbalance. Points are experimental data
and the lines theory fit. (c) Theory full reconstruction of the dynamics of the two fields for the
case (b) exciton (blue) and photon (red). The effect of reservoir pumping on the lifetime of the
oscillations is shown with the several envelopes.
Adapted from Dominici et al. (2014).

a mathematical convenience, rather than to describe such actual states of the polariton
field. They are indeed particular cases of the basis states $|n,m\rangle = p^{n\dagger}q^{m\dagger}|0,0\rangle$ that
make up the most general quantum state, described by a density matrix:

$$\rho = \sum_{m,n,\mu,\nu\in\mathbb{N}} \rho^{mn}_{\mu\nu} |\mu,\nu\rangle\langle m,n|. \qquad (11.2)$$

A typical representation for quantum states is the Wigner representation, defined as

$$P(x,p) \equiv \frac{1}{\pi\hbar} \int_{-\infty}^{\infty} \psi^*(x+y)\psi(x-y)e^{2ipy/\hbar}\,dy \qquad (11.3)$$

and several cases, both classical and genuinely quantum, are shown in Fig. 11.1 for
the photon fraction. While they all display a type of Rabi oscillation, their dynamics
varies greatly from one case to the other. Note that the case of Fock states is one of the
most remote from one's intuition. Also the fact that Fock states of n lower polaritons
(eqn 5.48), m upper polaritons or the combined states of n lower and m upper polaritons
together (eqn 5.49) are seldom encountered in the literature is a clue that one-particle
polariton states are in fact not realized in the laboratory. Instead, their correspondence
at the level of one-photon observables with classical states makes them so popular. In
next Section, we discuss how this similitude extends to their time dynamics.

11.1.2   *Control of polariton Rabi oscillations*

In the same way as the polariton eigenstates can be interpreted in a quantum (eqns 11.1)
or a classical (eqn 5.56) way, the time-dynamics has two possible descriptions, the
quantum one at the single-particle level

$$|\psi(t)\rangle = \alpha(t)|1_a,0_b\rangle + \beta(t)|0_a,1_b\rangle\,, \qquad (11.4)$$

and the classical one, that can still be written in a quantum mechanical framework:

$$|\psi(t)\rangle = |\alpha(t)\rangle \, |\beta(t)\rangle \, . \tag{11.5}$$

The paradigm shift is clear for variables α and β that otherwise are of the same nature (complex numbers) and also have the same dynamics: in the quantum version they correspond to probability amplitudes for the respective states, while in the classical version they are the amplitudes of coherent fields. The latter type of dynamics, eqn (11.5), has been observed in the laboratory, as shown in Fig. 11.2.

With the coherent states $|\alpha_0\rangle \, |\beta_0\rangle$ as the initial condition, the evolution is given by

$$|\psi(t)\rangle = |\alpha_0 \cos(gt) - i\beta_0 \sin(gt)\rangle \, |-i\alpha_0 \sin(gt) + \beta_0 \cos(gt)\rangle \, . \tag{11.6}$$

Two extreme cases are of interest: on the one hand, $\alpha_0 = \mp\beta_0$, in which case the dynamics becomes

$$|\psi(t)\rangle = \left|\alpha_0 e^{\pm igt}\right\rangle \left|\mp\alpha_0 e^{\pm igt}\right\rangle \, , \tag{11.7}$$

corresponding to the freely propagating lower $\left\| -\sqrt{2}\alpha e^{igt}, 0\right\rangle$ and upper $\left\| 0, \sqrt{2}\alpha e^{-igt}\right\rangle$ polariton coherent states, respectively. On the other hand, for α nonzero and $\beta = 0$ (or vice versa),

$$|\psi(t)\rangle = |\alpha_0 \cos(gt)\rangle \, |-i\alpha_0 \sin(gt)\rangle \, , \tag{11.8}$$

corresponding to Rabi oscillations. These dynamics are shown for the photon field in Fig. 11.1 (first two columns). This is the Hamiltonian dynamics. Including decay makes the solution slightly more heavy:

$$|\psi(t)\rangle = \left\| \left[a_0 \cosh(\tfrac{1}{4}Rt) - \left(\frac{b_0 G + a_0\Gamma}{R} \right) \sinh(\tfrac{1}{4}Rt) \right] \exp(-\tfrac{1}{4}\gamma t) \right\rangle \otimes$$

$$\left\| \left[b_0 \cosh(\tfrac{1}{4}Rt) + \left(\frac{-a_0 G + b_0\Gamma}{R} \right) \sinh(\tfrac{1}{4}Rt) \right] \exp(-\tfrac{1}{4}\gamma t) \right\rangle \tag{11.9}$$

where we have introduced:

$$G = i4g \, , \qquad\qquad R = \sqrt{G^2 + \Gamma^2} \, , \tag{11.10}$$
$$\gamma = \gamma_a + \gamma_b \, , \qquad\qquad \Gamma = \gamma_a - \gamma_b \, .$$

With decay only, the state remains pure throughout and thus one can write it through its wavefunction as above. Experimentally, there are typically sources of pure dephasing and incoherent excitations of the modes that turn the pure state into a mixed state. In such cases, one has to write a density matrix. Interestingly, however, the coherent fractions of both the photon and exciton fields, $\langle a(t)\rangle$ and $\langle b(t)\rangle$, are still given by the expressions featured in the respective kets of eqn (11.9), but with the following expressions for the coefficients instead:

$$G = i4g + \gamma_U \, , \qquad\qquad R = \sqrt{G^2 + \Gamma^2} \, , \tag{11.11}$$
$$\gamma = \gamma_a + \gamma_b + \gamma_U - P_b \, , \qquad\qquad \Gamma = P_b + \gamma_a - \gamma_b \, ,$$

where, to match with experimental observations, an explicit lifetime for the upper polariton γ_U has been included as well as an incoherent rate of excitation P_b for the excitons. Since only the sum of radiative decay γ_U^R and pure dephasing γ_U^ϕ of the upper polariton plays a role in the coherent dynamics, the result does not depend on which factor is actually responsible for the polariton dephasing: $\gamma_U = \gamma_U^R + \gamma_U^\phi$. Such analytical formulae can fit perfectly the observations by Dominici et al. (2014) in Fig. 11.2, of controlling the polariton dynamics by sequences of pulses, allowing to switch on and off the Rabi oscillations and even annihilate the fields. The examples shown compare the data (blue points) and the fit by eqn (11.9) (actually, also including the pulsed excitation, which, however, yields too cumbersome formulae to write here). The evolution is also shown on a Bloch sphere, that is suitable to picture both the quantum (eqns 11.1) and classical (eqn 5.56) dynamics of two coupled oscillators. A feature of interest of the above solutions is that incoherent pumping counteracts decay for the coherent fraction, as seen in Fig. 11.2(c) when comparing the observed oscillations (red and blue lines) and the envelope of the same system with $P_b = 0$. Larger values of incoherent pumping would allow to extend the lifetimes of these oscillations of the coherent fraction, as seen through the yellow and blue envelope, the latter case corresponding essentially to infinitely lived Rabi oscillations. There is, however, an incoherent fraction that is not shown here, as indeed coherent imaging of polaritons would not detect it. Such facts have been used by Chestnov et al. (2016) in a related mechanism to achieve permanent Rabi oscillations and by Demirchyan et al. (2014) to propose a polariton qubit (more accurately, a cebit as the field is classical and incoherent pumping would not be able to sustain the coherence of a one-particle state). Applications are, therefore, so far restricted to the classical realm. Colas et al. (2015), for instance, have proposed a pulsed source of full Poincaré beams in time by combining the above ideas with Rabi oscillations in two polarisations, as presented in Section 9.8.

11.1.3   *Polariton squeezing*

There has been many attempts since early stages of the field to bring polaritons into genuine quantum states. At the root of all such attempts are the polariton-polariton interactions, which are strong as compared to other optical systems, where they take the form of a Kerr nonlinearity. The "Kerr effect" is the change of the refractive index of an optical material with an applied electric field, more precisely, varying proportionally to the square of an applied electric field (while a linear dependence goes by the name the "Pockels effect"). Such Kerr nonlinearities are known to produce a popular family of non-classical states in the form of "squeezed states", first observed with light by Slusher et al. (1985) and in semiconductors by Fox et al. (1995). Such states have fluctuations in a specific quadrature of the field reduced (as compared to the fluctuations of the quantum vacuum) at the expense of the other. In this way, fluctuations can be brought below the quantum noise limit while still satisfying the Heisenberg uncertainty relation. Breitenbach et al. (1997) achieved a comprehensive collection (squeezed vacuum, amplitude squeezing, phase squeezing and in arbitrary quadratures). Since a squeezed state is basically making a quantum state resemble an ideal classical field, with no uncertainty, in one of its quadratures, this is a less striking manifestation than states with negative Wigner distribution, that have no classical counterpart. Still, squeezing polaritons is a worthwhile enterprise that has been first contemplated by

Tassone and Yamamoto (2000) who proposed a theoretical scheme to generate number-squeezed state using correlations between frequency and number noise. Karr et al. (2004b) from the quantum-optical background brought the quantum problem into microcavities early-on and reported a 6% polariton squeezing (4% in the output light) based on polariton-polariton interactions under coherent excitation. Romanelli et al. (2010) made an in-depth theoretical analysis predicting that photon-like polaritons are more suitable since they are filtering out the excitonic noise. Improvements upon these values limited to a few percent have been impaired mainly by the continuous nature of the polariton field and in particular its coupling with lattice phonons. Additional limitations such as disorder have been identified and analysed by Bamba et al. (2010) who also proposed a way to recover ideal quadrature squeezing in the strong localisation limit. This was implemented by Boulier et al. (2014) by turning to a system of reduced-dimensionality, namely, a micropillar, along with other favourable factors, such as the negative detuning advocated by Romanelli et al. (2010) and an improved mode matching between the pump and the polariton field. In this way, significant improvement ($\approx 20\%$) of squeezing in the amplitude quadrature was observed (in a reproducible way across several samples) at the bistability turning points, where theory predicts it to be maximum. These findings show that although some basic quantum features such as squeezing may be difficult to observe with polaritons, with dedicated effort and attention to all details, the promised behaviors are confirmed beyond ambiguity. The same should be true for deeper incursions into the quantum, such as genuine quantum states (beyond convex mixtures of Gaussian states) or entangled polaritons.

11.1.4 *Polariton statistics*

The particle statistics $g^{(2)}$ is one of the key quantities that goes beyond the classical description of a field, although there is a rich physics also for classical fields brought by stochasticity. In the context of polariton physics, interest in Glauber's correlators has been largely with the aim of describing coherence buildup as a transition from thermal statistics ($g^{(2)} = 2$) to that of a coherent state (or condensate, $g^{(2)} = 1$), and there has been a sustained theoretical effort to describe and then detail this scenario, from Laussy et al. (2004a), Schwendimann and Quattropani (2006), Schwendimann and Quattropani (2008), Doan et al. (2008), Sarchi et al. (2008b) and others. The first claim for polariton condensation by Deng et al. (2002) was in fact based on the decrease above threshold from $g^{(2)} \approx 1.8$ to slightly below 1.5, as shown in Fig. 11.3(a). Interestingly, no such criterion of polariton second-order coherence was retained by Kasprzak et al. (2006) whose work is regarded as the first definite confirmation of polariton condensation. Such an analysis followed, by Kasprzak et al. (2008a) on the same sample as the 2006 condensing one, with both cw and pulsed excitation results, and reporting the opposite phenomenology as Deng et al. (2002), shown in Fig. 11.3(b), although with some arguments to make it match. In more detail, well below threshold, a small $g_m^{(2)} \approx 1.03$ is measured (the m subscript denotes "measurement"), that is accounted for by the small time resolution of the setup $\tau_{AC} = 120$ ps as compared to the polariton lifetime $\tau_C = 2$ ps, leading to a corrected $g^{(2)} = 1 + (g_m^{(2)} - 1)\tau_{AC}/\tau_C \approx 2.8$, well-above a coherent state. Above threshold, however, the bunching is lost within noise

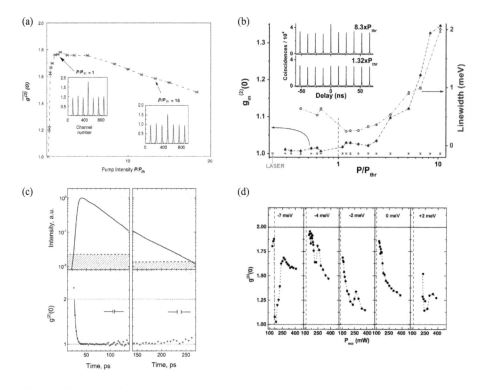

Fig. 11.3: Photon statistics from photons emitted by a polariton condensate, as seen by (a) Deng et al. (2002), (b) Kasprzak et al. (2008a), (c) Adiyatullin et al. (2015) and (d) Aßmann et al. (2011). All these works support the idea that the second-order coherence goes from 2 to 1 with condensation, although with much room for complications in the details.

until its reappearance at larger pumping, attributed this time to polariton-polariton interactions. The same sample was used again by Adiyatullin et al. (2015) this time with a streak-camera setup to track the time-resolved $g^{(2)}$ in the process of condensation. The transition from slightly above 2 to 1 as the population builds up was clearly observed and is shown in Fig. 11.3(c). A remarkable finding of this work is that the almost ideal value $g^{(2)} = 1$ for the condensate, as seen on the figure, is obtained by spatial filtering of the core of the BEC, without which a typical slight bunching departure is found that had been attributed in previous works to the effect of interactions. Adiyatullin et al. (2015), instead, suggest an incomplete phase synchronisation of the condensate, ruled by the Kibble (1976)–Zurek (1985) mechanism. The data at long times also gives some insight into the unexplored regime of loss of condensation, when most of the polaritons have evaporated and not enough are left to maintain coherence. In this regime, expectedly, the reverse trend should occur from 1 to 2 and although the little signal then available makes this observation difficult, this is indeed the trend as seen in Fig. 11.3(c) right panel. Other experimental reports of polariton statistics have been made available, e.g., from Love et al. (2008) and Horikiri et al. (2010), sometimes with contradictory

observations with the background literature. A comprehensive study was made by Aßmann et al. (2011) as function of pumping power and detuning, and it was reported that
a wealth of different behaviours can indeed be obtained depending on the configurations. A subset of their systematic analysis is shown in Fig. 11.3(d) with various level
of variations of the overall trend of decrease from 2 to 1. In particular, a pronounced
dip in $g^{(2)}$ is resolved at negative detunings not unlike previous observations around
threshold (cf. Fig. 11.3) that had been ruled out as artifacts. Such an effect could be
attributed to non-resonant scattering processes and the presence of a bottleneck of relaxation such as descibed by Sarchi et al. (2008b)'s two-reservoir model, that predicts
such a succession of a sharp drop of $g^{(2)}$ from 2 to 1 and a sharp increase in the opposite
direction followed by a weak decrease towards 1 with pumping. Recently, on the basis of
quantum Boltzmann equations, Kavokin et al. (2015) have analysed the cross-polariton
correlations in the presence of an incoherent (stochastic) exciton-photon coupling, in
addition to the coherent (Rabi) coupling, and find some trends similar to those observed
by the experiments, showing that a definite picture of the polariton statistics is far from
established.

　　While the bulk of the work has been devoted to the condensation process turning a
thermal cloud into a condensate, polaritons statistics are increasingly queried for their
antibunching, that is, for their indication of a nonclassical behaviour. This will be discussed in the next sections, but mainly at a theoretical level since experiments have not
yet reached this stage. Also of interest is the other limit, where $g^{(2)} \gg 2$, a so-called
"superbunching" that also indicates departures from a conventional scenario, albeit with
no guarantee as to their quantum character. This will also be discussed in the following
sections as well as in Chapter 12 with the statistical properties of the Bosonic cascades.

11.1.5   Polariton entanglement

A monumental finding in the physics of polaritons, maybe charting the second milestone in the field after their discovery by Weisbuch et al. (1992), is the observation
by Savvidis et al. (2000) of parametric amplification and oscillation at the magic angle, demonstrating compellingly their coherent and nonlinear aspect as well as their
bosonic character, that triggered the hunt for polariton lasing. The models from Ciuti
et al. (2000) and Ciuti et al. (2001) have been foundational in the quantum-optical
treatments of the polariton field and the description of correlations between different modes. In their work on polariton statistics, Schwendimann et al. (2003) tackled
explicitly the case of parametric amplification. At about the same time as they were
squeezing polaritons, Karr et al. (2004a) were also chasing twin polaritons. Parametric
oscillators are known in nonlinear optics to provide the best performances as sources
of entangled-photon pairs, e.g., already from the seminal work of Kwiat et al. (1995),
with violation of Bell's inequality by over 100 standard deviations from data acquired
in less than 5 minutes. Consequently, polaritons have long been regarded as potential
candidates to implement similar schemes. This led Ciuti (2004) to propose entangling
of the form $|j_1, k_s\rangle |j_2, k_i\rangle + |j_1, k_i\rangle |j_2, k_s\rangle$ with $|j, k\rangle$ a polariton in the branch $j \in$
$\{\mathrm{Lower}, \mathrm{Upper}\}$ and with momentum k at the signal or idler points of an interbranch

parametric process, leading to the emission of frequency-entangled photons. A similar idea was explored (cf. Fig. 11.4) and brought to the laboratory by Savasta et al. (2005) who went further into claiming a quantum effect by invoking the which-path indeterminacy, in their case, of the possible scattering events on two intersecting 8-shaped final-state curves on the lower branch dispersion. It was observed that polaritons from distinct idler modes interfere if and only if they share the same signal mode. This was interpreted as a quantum erasing of the which-way information, similarly to the experiment of Herzog et al. (1995). Special triple-cavity designs have been engineered to provide a better energy-conserving phase-matching by Diederichs et al. (2006), which, from several calculations, e.g., by Portolan et al. (2014), should result in the expected entanglement, although it has not been directly and unambiguously experimentally observed so far, despite unambiguous demonstration of parametric correlations at the classical level by Romanelli et al. (2007). To date, however, strong correlations indicative of entangled polaritons have failed to show up. As representative of the state-of-the-art, Ardizzone et al. (2012) studied the cross-correlations between the signal and idler modes of a polariton OPO in a 1D wire, and despite excellent parametric condition at the classical level (classical correlations and mode-matching), the $g^{(2)}$ only showed a weak dependence and did not lead to violation of classical inequalities, showing that further work is required to reach the regime of entangled polaritons.

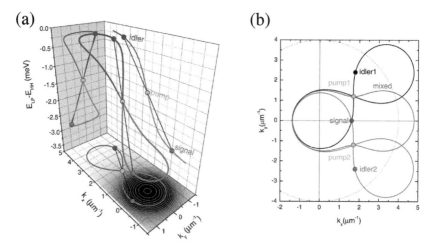

Fig. 11.4: (a) 8-shape trajectory on the polariton dispersion that conserves energy and momentum in a parametric scattering process, as observed by Langbein (2004) and Langbein (2005). (b) When using two pumps, the processes that share the same signal lead to idler pairs that cannot be distinguished, leading to a which-path indeterminacy observed through characteristic interferences. The dotted line is the border of the k range detected in the experiment.

11.2   Polariton blockade

The "one-particle blockade" effect originates with Fulton and Dolan (1987)'s observation of the strong impact of a single electron on the current-voltage characteristic

curves of small tunnel junctions. Geerligs et al. (1990) and others (see Grabert and Devoret (1992) for a review) established the picture of a "Coulomb blockade", with one
electron transiting from a source to a drain in a microscopic structure, passing by an island between them. This one electron can interrupt the current flow by its mere presence
on the island, since its electric charge is enough to shift the voltage for the other electrons. Once this blocking electron has tunneled through completely, the next electron
repeats the process, resulting in time-correlations in the transport and allowing control
at the single-electron level. Isolating successive electrons has important applications
in quantum transport, but mainly amplifies or at best harnesses a natural tendency of
fermionic particles. More dramatic is the counterpart of this effect with bosons, which
have the opposite tendency of clustering together. This started the field of "photon
blockade", as termed by Ĭmamoğlu et al. (1997). Ĭmamoğlu and Yamamoto (1994) had
previously brought the two concepts together by powering a single-photon source from
an emitting Coulomb-blocked heterojunction, but the mechanism became integral to
the nanophotonic device with Ĭmamoğlu et al. (1997)'s proposal of relying directly on
large photon-photon interactions (in their case through the mechanism of Schmidt and
Ĭmamoğlu (1996) of a giant Kerr nonlinearity obtained by electromagnetically induced
transparency). Various mechanisms for the interactions were studied, and we will focus
on the two major ones, namely, when the interactions originate from a two-level system
(Sec. 11.2.1) and when it comes from particle self-interactions (Sec. 11.2.2).

11.2.1 *Jaynes–Cummings blockade*

Photon blockade from the Jaynes–Cummings scenario relies on the anharmonic energy
spacing due to strong-coupling, as shown in Fig 5.5 and reproduced in Fig. 11.5(a).
At zero-detuning, the $\sqrt{2}$ splitting of the two-photon dressing of the two-level system
results in a shift that bring two photons from the laser out of resonance with the polari-
tonic $|2, -\rangle$ state when one photon from the laser is resonant with $|1, -\rangle$ (idem with +).
As a result, the Poissonian fluctuations of the laser that visit the states with two or
more photons get blocked by the ladder, letting the one-photon state alone pass and resulting in an overall antibunching (single-photon states). The observation of this effect
was first made by Birnbaum et al. (2005) in the light transmitted by an optical cavity
in strong-coupling with one cesium atom's transition, resulting in an antibunching (of
$g^{(2)} \approx 0.13 \pm 0.11$), shown in Fig. 11.5(b). This principle has been implemented in
a large family of systems, including superconducting qubits by Fink et al. (2008) and
quantum dots in photonic crystals by Faraon et al. (2008) and Reinhard et al. (2012).

All of these schemes have been studied at or close to exciton-photon resonance,
where the mechanism follows the principle of Fig. 11.5. It is better, however, to take advantage of detuning, which brings several benefits. First is that the resonances are better
resolved with detuning, even in presence of large dissipation rates, by the very nature of
detuning, which is to bring apart the bare energies. The dressed energies also inherit this
separation, which makes it easier to target a given manifold. In fact, although detuning
weakens the effective coupling, it strengthens its splitting, through a contribution that
affects both coupling and dissipation. This is seen in the expression for the complex
energies $E_{\pm}^k$ of the Jaynes–Cummings ladder given in eqn (5.78) that one can rewrite as

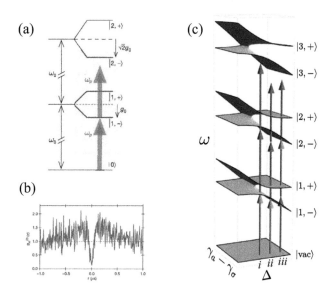

Fig. 11.5: **Jaynes–Cummings blockade** from Birnbaum et al. (2005), showing how (a) while one photon from the impinging laser is resonant with the lower polariton $|1, -\rangle$, two of them get out-of-resonance due to the anharmonic splitting. (b) The $g^{(2)}(\tau)$ observed from this experiment exhibits a clear antibunching dip at zero-delay ($\tau = 0$). (c) Detuned dissipative Jaynes–Cummings ladder, with three configurations of resonant excitation: (i) photon-blockade at resonance, similar to Fig. 11.5 (showing also the blockade by the third manifold), (ii) detuned photon-blockade and (iii) the case of resonant excitation of the second manifold with blockade by the first and third.

$$E^k_\pm = k\omega_a - \frac{\Delta}{2} - i\frac{(2k-1)\gamma_a + \gamma_\sigma}{4}$$
$$\pm \sqrt{\left[(\sqrt{k}g)^2 + \Delta^2/4 - \left(\frac{\gamma_a - \gamma_\sigma}{4}\right)^2\right] - i\Delta\frac{\gamma_a - \gamma_\sigma}{4}}, \quad (11.12)$$

with the square bracket explicitly showing how, while dissipation harms the splitting, detuning reinforces it. Also, which component of the system is driven is important. While most schemes drive the cavity, it is more efficient to drive the emitter, resulting in a much stronger antibunching and a better response of the system, as shown through comparisons between the two schemes by Laussy et al. (2012). This is expected once it is reminded that the emitter bears the quantum character while the cavity bears the classical one. Detuning helps to separate the two modes and to excite one in isolation. The lifetime of the emitter is also typically much longer than that of the cavity, which makes the broadening of the targeted transition sharper, that is, better resolved. This also makes its coupling to the exciting light weaker, hence the advantage of having the emitter remote from the cavity. Together, these variations on the paradigm of photon blockade (sketched in Fig. 11.5(a)) lead to the general configuration shown in Fig. 11.5(c) on the dissipative Jaynes–Cummings ladder. The soundness of detuning the system can be demonstrated by exciting the higher manifolds, as shown in the case iii of Fig. 11.5(c) where two photons from the laser are resonant with $|2, -\rangle$ while, for the same reasons as photon blockade at the one-photon level, other states are not excited. The response of the system, in particular through its photon statistics at such resonances, proves that there is an efficient quantum coupling between the modes despite their detuning. The resonant excitation of the first manifold leads to much stronger and robust antibunching. This configuration has been investigated and exploited by Müller et al. (2015) who, thanks to the "detuned photon blockade", could achieve much higher figures of merit from a quantum dot in a photonic crystal cavity, in terms of repetition rates, purity and, of course, antibunching. Blockade by the higher manifolds is studied in Section 11.4.

11.2.2   *Kerr blockade*

Photon blockade was initially proposed with a Kerr type of nonlinearity, which is some anharmonic term for an otherwise bosonic field. This was the approach of Carusotto (2001) who modelled an atom laser as a classical field Ω_L driving an anharmonic oscillator serving as a cavity accommodating the atoms, in this way describing an "atomic blockade". The resulting formalism is thus that of Drummond and Walls (1980), who provide exact solutions. A more complex model came from Verger et al. (2006) who include two coupled fields, exciton and photon, to describe the "polariton blockade". The Hamiltonian then reads

$$H = \hbar\omega_a a^\dagger a + \hbar\omega_b b^\dagger b + U b^\dagger b^\dagger bb + g(a^\dagger b + ab^\dagger),\qquad(11.13)$$

and is supplemented with coherent driving $\Omega e^{-i\omega_L t} a^\dagger + $ h.c. and Lindblad terms to describe decay. The underlying idea remains the same: polariton-polariton interactions result in a shift of the energy level that, if strong-enough with a single polariton, subsequently blocks the others and thus achieves a one-polariton quantum nonlinearity, resulting in antibunching just like the other types of blockade. This proposal became one of the chief objectives for quantum polaritonics, but has remained elusive so far due to too small values of polariton-polariton interaction as compared to their line-broadening. The lower-polariton case as first obtained by Verger et al. (2006) is shown on Fig. 11.6(a). Here, a quantum box was invoked for confining the excitons instead of the planar geometry, so as to boost the interaction constant to levels necessary for the effect to take place. The authors scanned around the lower polariton resonance only, since the upper polariton provides identical results in theory and much worse ones in practice due to the much higher dephasing rates and smaller lifetimes of the upper branch. A full-scan, however, reveals that the system has a strong response to the impinging light not only at the polariton resonances, but also at the bare modes, where light is super-bunched, a phenomenon also observed in the other types of blockades, where it comes with the name of "photon tunnelling". This is shown in Fig. 11.6(b). Interestingly, the strong bunching arises in this case because the bare states are very efficiently suppressed at resonance while other types of excitation remain as at other detunings. This results in comparatively large values of the correlators, including $\langle a^{\dagger 2} a^2 \rangle$, as compared to the population $n_a = \langle a^\dagger a \rangle$, which ratio yields the large values of $g^{(2)}$, showing that "blockade" does not naively imply antibunching. Even more remarkable is a much stronger antibunching resonance in between the bare states and the upper polariton, that differs

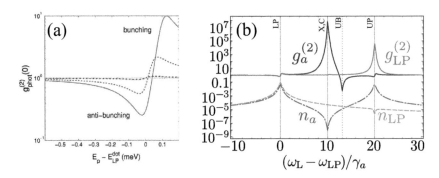

Fig. 11.6: **Polariton blockade** as seen through the antibunching of coherent light passing through the cavity. (a) The effect as observed for three types of exciton-photon detunings [positive (solid), resonance (dashed) and negative (dotted)] when exciting nearby the lower polariton resonance, from Verger et al. (2006). (b) Zoomed out version of (a) showing the response also of the upper polariton. Superbunching is observed at resonance. A much stronger antibunching—so-called "unconventional blockade"—occurs between the bare mode and the upper polariton.

so much from the previous types of blockade and became so popular in the community that it acquired its own name—of "unconventional photon blockade"—and spurred a new line of research. This is discussed separately in the next Section. Meanwhile, the quest for Kerr blockade is still ongoing and new designs of fiber-optics microcavities by Besga et al. (2015) may allow one to reach this regime by providing smaller confinements required for interactions to overcome broadening.

11.2.3 *Unconventional blockade*

The strong antibunching peak observed in Fig. 11.6(b) between the polaritons resonances for the same value of the Kerr nonlinearity came as a surprise for the polariton community and had a strong impact since it promises strong quantum effects with polaritons despite their weak interactions at the single-particle level. The effect, sitting nearby the conventional blockade, had remained hidden until it was discovered numerically by Liew and Savona (2010) in a different and more complex configuration, namely, of two-coupled interacting quantum boxes (or other variations, e.g., two photonic-crystal cavities). The mechanism was later identified by Bamba et al. (2011) as a destructive quantum interference for probability amplitude of two-photon states in the driven cavity and ultimately understood as a particular case of optimised Gaussian amplitude squeezing by Lemonde et al. (2014). Such interferences and their effects were known in the cavity QED literature, and in fact "unconventional photon blockade" had already been observed, e.g., by Rempe et al. (1991), following the work from Carmichael et al. (1991) who showed how weak driving of a cavity containing a collection of two-level atoms results in characteristic strongly oscillating photon correlations, with values at $\tau = 0$ below one, thus producing antibunching different from a single-photon source that decays smoothly (exponentially) from zero to unity with time. The unconventional blockade $g^{(2)}$ for N atoms is shown in Fig. 11.7(a) from an experimental observation by Foster et al. (2000). The physical mechanism itself follows from the cancellation of the probability amplitude for the transmission of two photons, that can

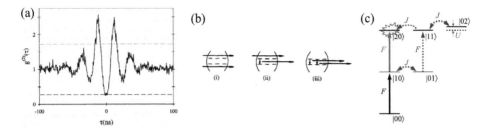

Fig. 11.7: **Unconventional blockade.** (a) Antibunching observed by Foster et al. (2000) with $N \approx 13$ atoms in a cavity under conditions of weak driving. (b) The general principle causing the antibunching as explained by Carmichael et al. (1991) consists in a destructive interference of probability amplitudes for two-photon states. (c) The particular case of two-couple modes revived by Liew and Savona (2010) as explained by Bamba et al. (2011): the two paths to $|20\rangle$ interfere destructively.

be decomposed as shown in Fig. 11.7(b) into (i) a transmission without absorption of two photons through the cavity, (ii) a transmission of one unabsorbed photon and an emission of a photon by an excited atom and (iii) a transmission of two photons from the excited atoms. Adapting this scheme to the case of two-coupled modes, Bamba et al. (2011) approximate the steady state of a system described by eqn (11.13) with also interaction for the mode a as

$$|\psi\rangle = C_{00}|00\rangle + e^{-i\omega_p t} \sum_{i\in\{01,10\}} C_i|i\rangle + e^{-2i\omega_p t} \sum_{i\in\{02,11,20\}} C_i|i\rangle + \cdots \quad (11.14)$$

and solving Schrödinger equation for the condition $C_{20} = 0$, they find an interference between the transition $|10\rangle \rightarrow |20\rangle$ driven by the laser (only the mode a is driven) and the transition $(|10\rangle \leftrightarrow |01\rangle) \rightarrow (|11\rangle \leftrightarrow |02\rangle) \rightarrow |20\rangle$ that results from both the laser Ω and the coupling g. For the simpler case where levels are resonant, $\omega = \omega_a = \omega_b$ and with identical lifetimes $\gamma_a = \gamma_b = \gamma$, the optimum values for the detuning with the driving laser $\Delta\omega = \omega - \hbar\omega_L$ and the nonlinearity U are given by

$$\Delta\omega_{\text{opt}} = \pm\frac{1}{2}\sqrt{\sqrt{9g^4 + 8\gamma^2 g^2} - \gamma^2 - 3g^2}, \quad (11.15a)$$

$$U_{\text{opt}} = \frac{\Delta\omega_{\text{opt}}(5\gamma^2 + 4\Delta\omega_{\text{opt}}^2)}{2(2g^2 - \gamma^2)}, \quad (11.15b)$$

which proves the great advantage of the unconventional mechanism, namely, that antibunching survives the case $U < \gamma$—that is fatal to the conventional mechanism—provided that $g > \gamma/\sqrt{2}$, which is typically much more straightforward to achieve. Only one mode needs to interact for the interference to take place, making unconventional polariton blockade operate for polaritons as modeled by eqn (11.13), and as shown in Fig. 11.6(b). The unconventional mechanism can also be linked to other types of blockade, as also discussed by Bamba et al. (2011) for the case of Jaynes–Cummings blockade, where it is found that two cavities in the presence of a two-level system in strong-coupling can also produce much stronger antibunching thanks to these

interferences. The question has enjoyed a considerable followup to explore its ramifications and variations, e.g., in other platforms (e.g., photonic crystals by Majumdar
et al. (2012) or Ferretti et al. (2013), optomechanical resonators by Xu and Li (2013),
etc.), with other types of nonlinearities (e.g., χ^2 by Zhou et al. (2015) or $ab^{\dagger 2}$ by Gerace and Savona (2014)), under pulsed excitation regime (by Flayac et al. (2015)) or
with more complex engineering such as involving parametric scattering (by Kyriienko
et al. (2014)). Despite its limitations for applications,[127] unconventional blockade has
generated a thriving branch of research and its laboratory realisation is one of the most
awaited for milestone in the field.

11.3   Frequency-resolved photon correlations

One approach to boost quantum correlations is to select photons that are the most
correlated. This can be achieved by frequency filtering, as we discuss in this Section.

11.3.1   *Photo-detection theory*

Eberly and Wódkiewicz (1977) introduced the concept of the physical spectrum of light
following the observation that the Wiener–Khinchin theorem assumes abstract and unphysical properties of the light field, such as stationarity. A bridge between the quantum
system and the observer can be made with the so-called input-output formalism: the
photons *inside* the system, say with operator a (we consider a single mode for simplicity), are weakly coupled to an *outside* continuum of modes, with operators A_ω
(corresponding to their frequency ω). In the Heisenberg picture, the output field allows one to compute the time-dependent power spectrum of emission as the density of
output photons with frequency ω_1 at time T_1, i.e., $S^{(1)}(\omega_1, T_1) = \langle A_{\omega_1}^\dagger(T_1) A_{\omega_1}(T_1)\rangle$.
This quantity is physical only if the uncertainties of detection in both time and fre-
quency are jointly taken into account. Mathematically, this amounts to adding two
exponential decays in the Fourier transform of the time-autocorrelation $S_{\Gamma_1}^{(1)}(\omega_1, T_1) =$
$\frac{\Gamma_1}{2\pi} \iint_{-\infty}^{T_1} dt'_1 dt'_4 e^{-\frac{\Gamma_1}{2}(T_1-t'_1)} e^{-\frac{\Gamma_1}{2}(T_1-t'_4)} e^{i\omega_1(t'_4-t'_1)} \langle a^\dagger(t'_1)a(t'_4)\rangle$, where Γ_1 has the interpretation of the linewidth of the detector. This shows how the physics of the detector
needs to be included if a more realistic description of the light field is required. To
compute a single-particle power spectrum (e.g., photoluminescence), the impact is conceptually modest as it basically requires a broadening from the detector, which is physically obvious and conveniently added by hand whenever required. Besides, the limit of
an ideal detector $\Gamma_1 \to 0$ can be taken and still arrive to a nontrivial result, which is,
in the steady-state, the Wiener–Khinchin expression. The situation becomes more complex when turning to observables involving more than one-photon, in particular those
dealing with correlations, where a full theory of photo-detection is required (Vogel and
Welsch (2006) give an excellent textbook account of the the general principles). For
instance, photon correlations combining both their frequency and time information are
now routinely measured in the laboratory, where they have proven extremely powerful

---

[127]Weaknesses of unconventional blockade include i) a small signal since the interference occurs for weak
driving, and ii) the nature of antibunching that is strong at $\tau = 0$, but rapidly oscillates in time, linked to the
fact that the state belongs to the family of convex mixture of Gaussian states.

to characterise quantum systems, starting with the Mollow (1969) triplet that has been investigated in this way throughout the last decades (by Aspect et al. (1980); Schrama et al. (1991); Ulhaq et al. (2012) among others). Press et al. (2007); Hennessy et al. (2007); Kaniber et al. (2008) have studied in this way the Rabi doublet. Sallen et al. (2010) proposed it as a way to access spectral diffusion of single emitters with unsurpassed accuracy. We will consider with particular attention the Mollow triplet, that is the drosophila of quantum optics. Its peculiar lineshape calls naturally to ask what are the correlations between the peaks. This question has been first addressed experimentally by Aspect et al. (1980) who found the peaks of the triplet to exhibit strong intensity correlations. A more general version of this pioneering experiment is sketched in Fig. 11.10 (on page 495) where a Monte Carlo simulation was used to generate photons emitted in the highlighted frequency windows. The first notable feature is a trivial one: the different intensities. There are twice as many detection events from the central peak than from the satellites and considerably more so than in the tails or between the peaks. What is of interest to the quantum optician, however, is the correlation between these photons. The figure gives the autocorrelation values as computed from these clicks in each window. One finds, as observed by Aspect et al. (1980), that the central peak is bunched while satellites are antibunched. Although this was not contemplated by Aspect *et al.*, one can also consider correlations in off-peak windows, and find, as could be expected, that the emission there is essentially uncorrelated. However, cross-correlations between these photons are extremely strong. They are much stronger, in particular, than the correlations between peaks, shown on the figure as well and, as was also observed in 1980, displaying a positive correlation between satellites and an anticorrelation between a satellite and the main peak.

Such correlations of photons from different frequencies of an emitter were first described theoretically by dedicated methods for the problem at hand, from Cohen-Tannoudji and Reynaud (1979) and Reynaud (1983) (dressed atom picture) and Dalibard and Reynaud (1983) (diagrammatic expansion). A more general extension of photo-detection in the spirit of Eberly and Wódkiewicz (1977) was impulsed by Knöll et al. (1984) and Arnoldus and Nienhuis (1984), considering two detectors with respective linewidths Γ_1 and Γ_2. Due to the complexity of the integrals they obtained, they also focused on the particular case of resonance fluorescence for illustration. These mathematical foundations, shaky in their initial development, were firmly established in the course of the following years by Knöll and Weber (1986), Knöll et al. (1986) and Cresser (1987). The multiplicity of photons requires a careful time ($\mathcal{T}_\pm$) and normal (:) ordering of the operators, and it was eventually realised that it is the time ordering of $\langle :A_{\omega_1}^\dagger(T_1)A_{\omega_1}(T_1)A_{\omega_2}^\dagger(T_2)A_{\omega_2}(T_2): \rangle$, which provides the physical two-photon spectrum

$$
S_{\Gamma_1\Gamma_2}^{(2)}(\omega_1, T_1; \omega_2, T_2) = \frac{\Gamma_1\Gamma_2}{(2\pi)^2} \iint_{-\infty}^{T_1} dt_1' dt_4' e^{-\frac{\Gamma_1}{2}(T_1-t_1')} e^{-\frac{\Gamma_1}{2}(T_1-t_4')} \iint_{-\infty}^{T_2} dt_2' dt_3'
$$
$$
e^{-\frac{\Gamma_2}{2}(T_2-t_2')} e^{-\frac{\Gamma_2}{2}(T_2-t_3')} e^{i\omega_1(t_4'-t_1')} e^{i\omega_2(t_3'-t_2')} \times \langle \mathcal{T}_-[a^\dagger(t_1')a^\dagger(t_2')]\mathcal{T}_+[a(t_3')a(t_4')] \rangle .
$$
$$(11.16)$$

Here, we have defined $\mathcal{T}_+$ (resp. $\mathcal{T}_-$) to order the operators in a product with the latest time to the far left (resp. far right) following, e.g., Vogel and Welsch (2006). Normalising this expression yields the sought time- and frequency-resolved two-photon correlation function:

$$g^{(2)}_{\Gamma_1\Gamma_2}(\omega_1, T_1; \omega_2, T_2) = S^{(2)}_{\Gamma_1\Gamma_2}(\omega_1, T_1; \omega_2, T_2)/\left[S^{(1)}_{\Gamma_1}(\omega_1, T_1)S^{(1)}_{\Gamma_2}(\omega_2, T_2)\right]. \quad (11.17)$$

It is positive and finite, and suitably reflects that the frequency and time of emission cannot be both measured with arbitrary precision, in accordance with Heisenberg's uncertainty principle. The limiting behaviours of $g^{(2)}_{\Gamma_1\Gamma_2}$ defined in this way are those expected on physical grounds: photons are uncorrelated at infinite delays, and color-blind detectors recover the standard two-time correlators:

$$\lim_{|T_2-T_1|\to\infty} g^{(2)}_{\Gamma_1\Gamma_2}(\omega_1, T_1; \omega_2, T_2) = 1, \quad (11.18a)$$

$$\lim_{\Gamma_1,\Gamma_2\to\infty} g^{(2)}_{\Gamma_1\Gamma_2}(\omega_1, T_1; \omega_2, T_2) = g^{(2)}(T_1; T_2). \quad (11.18b)$$

Further generalisation to N-photon correlations follow in this way, adding pairs of operators with their corresponding integrals, as detailed by Knöll and Weber (1986) and Knöll et al. (1990). The actual computation of such $g^{(N)}_{\Gamma_1...\Gamma_N}$ by computation of these integrals has, however, proved intractable for $N > 2$, even for simple single-mode systems, such as resonance fluorescence or the single mode laser, treated by Centeno Neelen et al. (1993). The case $N = 2$ is already demanding and Nienhuis (1993) and Joosten and Nienhuis (2000) who treated this problem, made some approximations to simplify the algebra. More recently, the resonance fluorescence problem was revisited without approximations by Bel and Brown (2009), but still for two photons and at zero time delay only. The main reason for such limitations is that all the possible time orderings of the $2N$-time correlator $\langle\mathcal{T}_-[a^\dagger(t'_1)\ldots a^\dagger(t'_N)]\mathcal{T}_+[a(t'_{N+1})\ldots a(t'_{2N})]\rangle$ result in $(2N-1)!!2^{N-1}$ independent terms. Furthermore, each of these correlators requires the application of the quantum regression theorem $2N-1$ times. This growth of the complexity makes a direct computation hopeless for a quantity that is otherwise straightforward to measure experimentally, merely by detecting photon clicks as function of time and energy, a technology provided for instance by a streak camera, as demonstrated by Wiersig et al. (2009). In the next Section, we introduce a method to compute them easily.

11.3.2   The sensor method

We now present del Valle et al. (2012)'s theory of N-photon correlations that allows one to compute exactly the integrals of eqn (11.16) effortlessly. The method comes with several advantages: i) it allows for arbitrary time delays and frequencies, ii) it is applicable to any open quantum system and iii) it relies on the elementary toolbox of quantum optics. This technique simply consists in the introduction of N two-level systems to the dynamics of the open quantum system (noted Q in Fig. 11.8), each with annihilation operator ς_i and transition frequency ω_i, that is matched to the frequency to be probed in

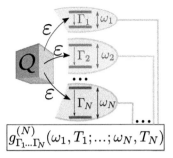

Fig. 11.8: **The sensors technique.** N two-level systems of determined frequencies are weakly coupled to the quantum system to be characterised (here modeled as a black box Q) and serve as correlation sensors at these frequencies, with their decay rate providing the detector linewidth. The calculation is exact and considerably easier than through the corresponding integral expression.

the system. Its lifetime $1/\Gamma_i$ corresponds to the inverse linewidth of detection. The coupling ε_i to each system is small enough so that the dynamics of the system is unaltered by their presence, with $\langle n_i \rangle = \langle \varsigma_i^\dagger \varsigma_i \rangle \ll 1$. For this reason, these systems are called "sensors", as they do not affect the dynamics. More precisely, calling γ_Q any transition rate within Q (either with internal or external degrees of freedom) linked to the field of interest a, the tunnelling rates ε_i must be such that losses into the sensors and their back action are negligible, leading to $\varepsilon_i \ll \sqrt{\Gamma_i \gamma_Q/2}$. Under this condition, one then solves the full quantum dynamics of the system supplemented with the N sensors. The latter play the role of the output fields $A_{\omega_i}(t)$, but instead of formally solving the Heisenberg equations and expressing their correlations in terms of the system operators (as in the standard method exposed above), it is enough to compute directly intensity-intensity correlations between sensors, which is a considerably simpler task. The main result of this technique is thus that

$$g^{(N)}_{\Gamma_1 \dots \Gamma_N}(\omega_1, T_1; \dots; \omega_N, T_N) = \lim_{\varepsilon_1, \dots, \varepsilon_N \to 0} \frac{\langle : n_1(T_1) \dots n_N(T_N) : \rangle}{\langle n_1(T_1) \rangle \dots \langle n_N(T_N) \rangle}, \qquad (11.19)$$

where the left hand side is the time- and frequency-resolved N-photon correlation function as defined previously. The right-hand side is simply intensity correlations, i.e., at identical times, a standard quantum average that can be easily obtained through anybody's favourite formalism in any convenient picture (Heisengerg, Shrödinger, etc.) The proof of this statement is given by del Valle et al. (2012) in the supplementary material of the text.

With this technique, we are now in a position to compute frequency correlations exactly (with no approximations). We compare in Fig. 11.9 the frequency-resolved photon correlations of the Mollow triplet with the sensor method of del Valle et al. (2012) (solid lines) and through an approximation introduced by Nienhuis (1993) to tackle the integrals (11.16), that consists in introducing auxiliary modes corresponding to the transitions of the Mollow triplet. As such, the results are expected to be accurate at the frequencies of the peaks, which is indeed the case on the figure that compares both auto-correlations—corresponding to the $g^{(2)}$ of filtered light—and cross-correlations between frequencies symmetric about the central peak. The approximation is typically excellent at $\omega/\omega_S = 0$ (central peak) and ± 1 (satellites) and can even bear some frequency dependence, except that it may predict exact antibunching while the exact calculation always find it finite. More importantly, however, is that some strong

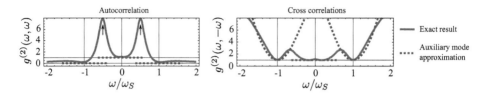

Fig. 11.9: **Frequency-resolved photon correlations of the Mollow triplet**, as calculated exactly with the sensors method of del Valle et al. (2012) [solid lines] and through the auxiliary operator approximation [dashed lines] of Schrama et al. (1992) and Nienhuis (1993) *et al.* The approximation is accurate around the peaks at 0 and ± 1, but fails to access important features, most importantly the leapfrog processes, highlighted by the arrows, where one gets value of 0 and 1 depending on around what point one is making the approximation, instead of $g^{(2)} \gg 2$, characteristic of superbunching.

features occur outside of the peaks, where the approximation fails considerably (and is even multi-valued). Namely, one observes two strong bunching peaks between the peaks (marked with arrows) in the autocorrelation, that will turn out to be features of very high importance, which we explain in the next Section.

11.3.3 *Two-photon spectra*

The easiness of the sensor method as well as its ability to consider any frequency window, with no restriction to special spectral features such as transitions between the states, make it possible to compute a comprehensive map of all the possible frequency-frequency correlations. This is done in Fig. 11.11 for the Mollow triplet, where the type of statistics is encoded as a color code: blue for antibunching, white for no-correlation, red for bunching. There appears a striking landscape of correlations, in particular for the Mollow triplet, two half-circles of antibunching (they close into full-circles when detuning the driving laser from the two-level system) clutching three antidiagonal lines of strong bunching. The two peaks in Fig. 11.9 are the outer lines in Fig. 11.11. These features vanish when they intersect with the frequencies of any transition (central peak or a satellite). On the diagonal (case of auto-correlations), this is the case at the origin where the gap in correlation is obvious. This figure makes particularly clear that the structure of the correlation lies precisely outside of the peaks. The antidiagonal lines are the most significant. They come from two-photon transitions one manifold to another with two-quanta of excitations less, jumping over the intermediate manifold. The jump is allowed outside of the usual one-photon transitions by involving a virtual state. This is why the features vanish when they cross with a transition, since in such a case the system simply emits two independent photons in succession, say, $|+\rangle_N \rightarrow |+\rangle_{N-1}$ and then $|+\rangle_{N-1} \rightarrow |+\rangle_{N-2}$. Any correlation in this case is of a classical character (bunching, in this case, as this corresponds to a cascade, while antibunching if mixing the $+$ and $-$ states when the system has to change its internal state in some way). In stark contrast, a transition between two manifolds separated by two quanta of excitation without transiting by the manifold in-between is still possible through a process $|+\rangle_N \overset{\rightarrow}{\rightarrow} |+\rangle_{N-2}$ that involves a direct two-photon process, or similarly, a virtual intermediate process. For this reason, these states are called "leapfrog processes". Clearly, these two photons are strongly correlated. Sánchez Muñoz et al. (2014a) has shown that

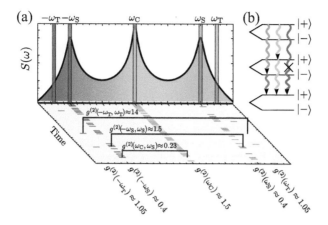

Fig. 11.10: **Photons emitted by the Mollow triplet** (a) in five frequency windows, from Monte Carlo simulations by Sánchez Muñoz et al. (2014a). The strongest cross-correlations are observed in frequency windows far from the peaks. (b) Transitions between levels from the Mollow ladder. The peaks in the spectrum originate from transitions between neighboring manifolds. Direct two-photon transitions jumping over an intermediate manifold (red arrow), the so-called "*leapfrog processes*", are quantum correlated.

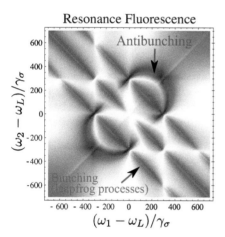

Fig. 11.11: **Two-photon correlation spectrum of the Mollow triplet.** As calculated by González-Tudela et al. (2013), with color code ranging from blue (antibunching) to red (bunching) passing by white (uncorrelated). The antidiagonals correspond to the "*leapfrog processes*", whereby the system undergoes a two-photon transition from one manifold of states to that two rungs below, jumping over the intermediate manifold. The two-photon emitted in the process are strongly quantum-correlated.

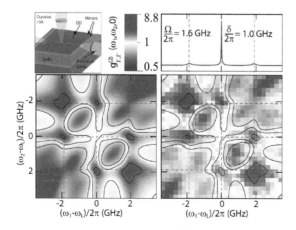

Fig. 11.12: **Two-photon correlation spectrum of the Mollow triplet.** As calculated (left) and measured (right) by Peiris et al. (2015) from the resonance fluorescence of a quantum dot in the Mollow triplet regime (upper row). The leapfrog processes, here observed for the first time experimentally, are neatly resolved. The excellent agreement also shows that the quantum dot behaves as a two-level system under coherent driving, which is an important result for semiconductor QIP.

these processes have a quantum character, namely, they violate Cauchy–Schwarz inequalities and Bell's inequalities, contrarily to photons emanating from the peaks, that satisfy them. Such processes are tightly related to the two-photon processes studied by Göppert-Mayer (1931) except that their observation through photon correlations allows one to bring them in full-view in any system. Two-photon processes are usually smothered by much stronger single-photon signals. For instance, the hydrogen atom has such a two-photon transition $2S_{1/2} \vec{\rightarrow} 1S_{1/2}$, as worked out by Breit and Teller (1940), but the long lifetime of the metastable 2S state, of the order of a tenth of a second, makes it unlikely that no other de-excitation (in particular inelastic collisions) takes away the excitation through another channel. A notable exception is in planetary nebulae, where densities of hydrogen are so small, but present in such large quantities, that two-photon processes are directly observable, in the form of a continuous background instead of the characteristic lines of the hydrogen atom, as first explained by Spitzer Jr. and Greenstein (1951). Barring such exceptions of cosmological proportions, two-photon processes are difficult to observe directly, and require specific attention, such as that provided by Hayat et al. (2008). Photon-correlations in a two-photon spectrum, on the other hand, allow their direct access, along with higher-order processes (N photon leapfrogs).

The antidiagonals of the two-photon spectrum (2PS) based on their interpretation as two-photon emission processes are, therefore, given by the simple equation

$$\omega_1 + \omega_2 = E_{\pm,n} - E_{\pm,n-2}, \tag{11.20}$$

where $E_{\pm,n}$ are the energy of the Mollow manifolds (transitions are quasi-degenerate for large enough n). In this way, Fig. 11.9 literally shows a two-photon Mollow triplet. These processes can be filtered and distilled to be exploited, as discussed by del Valle (2013) and González-Tudela et al. (2015), and also shown later in this chapter. The mechanism is moreover not limited to two photons and by going to $g^{(3)}$, one can then identify the transitions $|+\rangle_N \vec{\rightarrow} |+\rangle_{N-3}$, with equation (in a 3D space, not shown)

$$\omega_1 + \omega_2 + \omega_3 = E_{\pm,n} - E_{\pm,n-3}. \tag{11.21}$$

This remarkable and fundamental structure, predicted by the theory, thus shows that the important quantum features are to be looked for far from the real-state transitions (peaks of the spectrum). This has been observed experimentally by Peiris et al. (2015) in the emission of a strongly-driven quantum dot. Their observation along with their calculation of the expected 2PS for their parameters are shown in Fig. 11.12 (note that they use an opposite direction for the y-axis so the leapfrog lines are diagonal for them). The agreement is remarkable. In particular, the two-photon processes are clearly observed.

The ease of use and power of the sensor technique makes it possible to undertake similar analysis for a vast variety of quantum emitters, and not only on the Mollow triplet, that had been focused on in the literature (in some approximations). Such a program has been initiated by González-Tudela et al. (2013) and we refer to this text for several variations of coupling two modes, with bosonic and fermionic statistics (see Exercise 11.1 for these cases in isolation), and under various types of pumping. del Valle (2013) has also dealt with the biexciton case in great detail, revisiting a

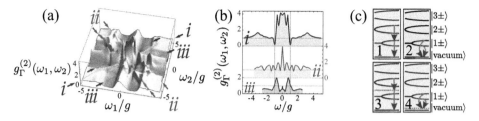

Fig. 11.13: **Frequency-resolved photon correlations in the Jaynes–Cummings model.** (a) Landscape of two-photon correlations, here as a 3D plot. (b) Cuts along the lines indicated in panel (a), showing (i) the filtered autocorrelations, (ii) the leapfrog processes and (iii) transitions involving a virtual state as the final state. (c) Processes in the dissipative Jaynes–Cummings ladder corresponding to the highlighted features, where 3 is the leapfrog process.

configuration that is extremely popular for the radiative cascade in two polarisations leading to entanglement by which-way indeterminacy. Silva et al. (2016) observed a structure in the spontaneous emission from a polariton condensate that they attributed to a generalized Hanbury Brown–Twiss effect. Second to the Mollow triplet is the Jaynes–Cummings model. The frequency-filtered correlations in this case are considerably more complicated owing to the much richer structure of this system. We give a brief overview of the main results in Fig. 11.13 and refer to González-Tudela et al. (2013) and González-Tudela et al. (2015) for more details.

Exercise 11.1 $^{(**)}$ *Derive the analytical two-photon correlation spectra for the harmonic oscillator and the two-level system when pumped incoherently.*
   Solution: Calling $\Gamma_a = \gamma_a - P_a$ *and* $\Gamma_\sigma = \gamma_\sigma + P_\sigma$, *we find*

$$g_\Gamma^{(2)}[O](\omega_1,\omega_2) = \frac{\Gamma_O}{(\Gamma+\Gamma_O)^2}\left[\Gamma_O + \frac{\Gamma^2(2\Gamma+\Gamma_O)}{\Gamma^2+(\omega_1-\omega_2)^2} + \tilde{g}_\Gamma^{(2)}[O](\omega_1,\omega_2)\right] \qquad (11.22)$$

with a common expression regardless of the mode $O = a, \sigma$ *spelt out above, and a term* $\tilde{g}_\Gamma^{(2)}[O]$ *specific to each case, given by*

$$\tilde{g}_\Gamma^{(2)}[a] = 2\Gamma\left(1+\frac{\Gamma}{\Gamma_a}\right), \qquad (11.23)$$

for the harmonic oscillator (note that it is frequency independent), and by

$$\tilde{g}_\Gamma^{(2)}[\sigma](\omega_1,\omega_2) = 4\Gamma - 2\Gamma(2\Gamma+\Gamma_\sigma)\left(\frac{\frac{3\Gamma+\Gamma_\sigma}{2}}{(\frac{3\Gamma+\Gamma_\sigma}{2})^2+\omega_1^2} + \frac{\frac{3\Gamma+\Gamma_\sigma}{2}}{(\frac{3\Gamma+\Gamma_\sigma}{2})^2+\omega_2^2}\right),$$

for the two-level system. These expressions have been first reported by González-Tudela et al. (2013) who further discuss them.

11.4   N–photon emitters

We bring together the concepts of photon-blockade and frequency filtering presented in the previous Section to show how they can drive new approaches of quantum engineering. We focus in particular on a family of *N-photon emitters* introduced by Sánchez Muñoz et al. (2014b) that release their energy exclusively in groups, or *bundles*, of N photons (for integer N) and in effect provide us with light made up from building blocks that are not single photons anymore. This ability to substitute the quantum of light by a bundle has unforeseeable consequences for both applications and fundamental physics. We will concentrate on the design itself in this Section and provide grounds for some applications in the remaining part of this chapter.

11.4.1 *Super-Rabi oscillations*

In the configuration of detuned polariton-blockade, scanning the laser in frequency between the bare modes leads to a series of resonances as shown in Fig. 11.14, that correspond to the manifolds of the Jaynes–Cummings ladder. The figure shows the time-averaged probabilities of the vacuum (in red) and the excited-state dressed by N photons for $N \in \mathbb{N}$ going from 0 on the right-hand side when driving at the frequency of two-level system, to $N \to \infty$ on the left-hand side when driving at the frequencies close to the cavity. The time dynamics is that of perfect (sinusoidal) full-amplitude Rabi oscillations of frequency $\Omega_n \propto 10^{-3N}$ (Sánchez Muñoz et al. (2014b) give the exact expression in their supplemental material for these "super-Rabi oscillations", the latter terminology was proposed by Strekalov (2014)), meaning that the system in this configuration sustains "super-polaritons" of the type:

$$|\psi\rangle = \frac{1}{\sqrt{2}}(|0g\rangle + |Ne\rangle). \qquad (11.24)$$

These superpositions of the vacuum with a N-photon state occur in the Hamiltonian system, with exponentially sharper resonances as the superposition involves higher excitations. In the presence of dissipation, Laussy et al. (2012) pointed out that the resonances are extremely robust and can be observed even in systems that are in weak-coupling at the one-particle level, thanks to the enhancement of the coupling by the field itself for the higher manifolds. This allows one to climb a Jaynes–Cumming ladder whose bottom rungs are missing. Figure 11.15 shows the counterpart of Fig. 11.14 for different values of γ_a/g for a system wih very small γ_σ such that the former ratio essentially quantifies strong-coupling of the bottom rung. The detuned blockade results in unambiguous Jaynes–Cummings features even for systems well into the weak-coupling, up to fourth-order correlation $g^{(4)}$ and even higher for larger detunings. While Müller et al. (2015) have started to climb the ladder, they tiptoed on the first rung to explore the next one with three-photon measurements and the higher rungs remain so far unexplored. Importantly, a smooth ascension is through direct coupling to the emitter, rather than to the cavity.

11.4.2 *Robust Jaynes–Cummings resonances*

The remarkable structure displayed in Fig. 11.15, linked to the resonances of eqn (11.24) involving N photons in the cavity, invites one to think that in a similar way that conventional blockade serves as a single-photon source, its higher-rung counterparts could power N photon sources. This is, however, mis-reading the information provided by $g^{(n)}$ and so as to provide a resolute answer to this question, Sánchez Muñoz et al. (2014b) undertook Monte Carlo simulations of this system to get access to the photons actually emitted, recorded as wavefunction-collapses and thus producing a simulation of an actual experiment, from which photon correlations can be reconstructed rather than

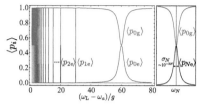

Fig. 11.14: N-photon resonances in the detuned Jaynes–Cumming system, as observed by the time-averaged amplitudes of Rabi oscillations $\langle p_i \rangle$ between manifolds $|0g\rangle$ and $|ie\rangle$. The right inset is a zoom of any of the resonance on the left. The transition from quantised modes towards a continuum as one gets closer to the cavity is nicely observed.

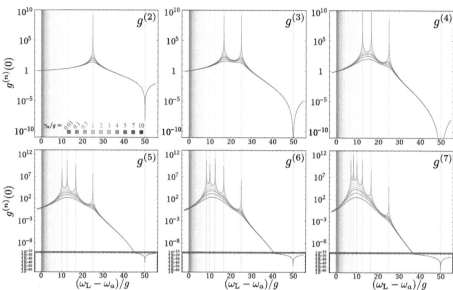

Fig. 11.15: **Robust Jaynes–Cummings resonances** as observed through the nth order correlator $g^{(n)}$ of the cavity photons for $n = 2$ up to $n = 7$ for various values of γ_a/g (color coded) when driving a detuned two-level system (here at $50g$ from the cavity) with a laser of frequency ω_L. Even when the system is in weak coupling, $\gamma_a/g > 4$, clear resonances at the Jaynes–Cummings transitions are observed, that consist in strong antibunching when resonant with the two-level system, and strong bunching when resonant with a Jaynes–Cummings transition. The nth order transitions and all those below are seen in $g^{(n)}$.

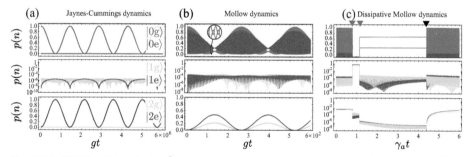

Fig. 11.16: **Wavefunctions of a two-level system in a cavity driven at the two-photon resonance.** (a) At low pumping, only the manifolds $|0g\rangle$ and $|2e\rangle$ have sizable amplitudes, resulting in a "super-polariton". (b) At high pumping, the dynamics enters the Mollow regime with various timescales and, (c) in presence of dissipation, leads to a pure two-photon emitter. From Sánchez Muñoz et al. (2014b).

computed with a density matrix. Such simulations are shown in Fig. 11.16(c). The former case shows the evolution of the wavefunction in presence of decay for a single Monte Carlo run (so-called trajectory), in which case instead of a smooth decay that results from the averaging of several trajectories, one witnesses abrupt changes in the probability amplitudes, that correspond to a wavefunction collapse and to which one can attribute photon emission events, displayed as triangles (red for the cavity, black for the emitter). The other panels in Fig. 11.16 show the Hamiltonian dynamics (no decay) for two intensities of the driving laser at the two-photon resonance. The low-driving case on panel (a) realises the state (11.24) with $N = 2$. It is seen in particular how the manifold $N = 1$ is largely suppressed (notice the log scale), as the state $|2e\rangle$ gets coupled to the vacuum. Observe as well the full-amplitude oscillation with 100% probability of the system to accommodate the two-photon Fock state at regular intervals, averaging in time to 1/2 as shown in Fig. 11.14. Adding decay to this Hamiltonian dynamics, one finds that while the system produces super-bunching in this regime, as in Fig. 11.15, the Monte Carlo simulation (not shown) reveals that the emission still consists predominantly of single-photon events, with more occurences of pairs of photons than from a coherent state, but without any clear pattern. As driving is increased, the system enters the Mollow regime, with faster oscillations and a more complicated dynamics, as shown in panel (b). Adding decay to this case turns the dynamics to that shown in panel (c). This time, the Monte Carlo simulation reveals a striking change of regime: the cavity emission now consists exclusively of strongly-correlated two-photon emissions, that come in rapid succession one after the other and, when involving more than two-photons, in harmonic progression in time.

11.4.3 *Bundles of photons*

The dynamics of Fig. 11.16(c) and its resulting emission powers a new type of light-source. At the level of the observer, the device emits two-photon packets, or "bundles" following Sánchez Muñoz et al. (2014b)'s terminology, that, at the timescale of the cavity, consists in two successive clicks. The photons are emitted well within their life-time and overlap in time, making the bundle a two-photon object different from two photons coming together, as discussed by del Valle et al. (2011) in another, but related two-photon emission process. Interestingly, in this regime, the conventional $g^{(2)}$ exhibits a local minimum and superbunching is not a good measure of N-photon emission. Instead, Sánchez Muñoz et al. (2014b) introduce a generalised N-photon nth order correlator $g_N^{(n)}$ defined as

$$g_N^{(n)}(t_1, \ldots, t_n) = \frac{\langle \mathcal{T}_-\{\prod_{i=1}^n a^{\dagger N}(t_i)\} \mathcal{T}_+\{\prod_{i=1}^n a^N(t_i)\}\rangle}{\prod_{i=1}^n \langle a^{\dagger N} a^N\rangle(t_i)} \tag{11.25}$$

with $\mathcal{T}_\pm$ the time ordering operators. The case $N = 1$ recovers the definition of the standard $g^{(n)}$. This expression describes the statistics of the bundles, or pairs of red triangles at the top of Fig. 11.16. The case shown corresponds to $g_2^2 = 1$ and, therefore, characterises a two-photon laser. The lifetime of the emitter allows one to tune the bundle-statistics and two-photon guns or two-photon thermal sources can also be realised. Single photons have little physical meaning in such a regime of emission, as

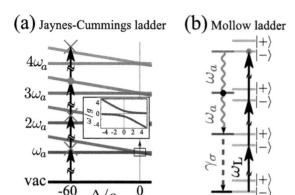

(a) Jaynes-Cummings ladder (b) Mollow ladder

Fig. 11.17: Jaynes–Cummings blockade and Purcell enhancement of Leapfrog processes. From Sánchez Muñoz et al. (2014b)

they lost relevance in describing the dynamics. The transition from panels (a) to (c) in Fig. 11.16 is a far-reaching one that shows how complex (and rewarding) a simple-idea such as polariton blockade can become when bringing it in a regime of high-driving. In this case, a more adequate picture shifts from a driven Jaynes-Cummings system (laser on the one hand, and cavity-emitter on the other) in the wake of polariton blockade, to that of a frequency-filtered Mollow triplet (cavity as the filter on the one hand, emitter strongly-driven by the laser on the other hand). These two viewpoints are contrasted in Fig. 11.17. The left-hand side is the now familiar detuned-polariton blockade with the laser impinging on one manifold in isolation (here the 3rd one); the right-hand side shows the laser in resonance with all the $|-\rangle$ states of the dots and the cavity in resonance with a two-photon transition, being accordingly out-of-resonance with single-photon ones. When compared to Fig. 11.11, one can understand the two-photon bundle emission (100% of two-photon states) as placing a cavity on the antidiagonal where the leapfrog processes are maximum. This is, therefore, simply Purcell enhancing them by opening a channel of escape for an otherwise weak process. Placing the cavity at the *N*-leapfrog process frequency likewise results in "bundling" with the corresponding number of photons. Such an understanding allows one to easily engineer other types of quantum sources, as illustrated for instance in the two-photon emission from a biexciton, theorised on these grounds by Sánchez Muñoz et al. (2015) and implemented in the laboratory by Hargart et al. (2016). Such quantum sources are still to be better developed and characterised as devices. A natural question, though one to which intuition strongly suggests a multitude of answers, is what applications such sources would have. We will offer some suggestions in Section 11.5.

11.4.4 *Yudson representation*

The *N*-photon states emitted by the prospective device from the previous Section have an internal structure, e.g., a five-photon bundle, in terms of its constituting photons, looks like ● ● ●●● with darker photons at the front (right) emitted first. This is the pattern that follows from a Fock state $|5\rangle$ emitted by spontaneous emission from a cavity where it is put as the initial condition, and this is basically what the "bundler" achieves when the state (11.24) collapses into its $|Ne\rangle$ component. However, since a Fock state is a well-defined mathematical object whose closest physical realisation is as the quantum state of some conservative bosonic field, Sánchez Muñoz et al. (2014b) instead called the dynamical output a "bundle" of photons. It will need some further investigations to establish if the terminology is redundant and whether the device is accurately described as an emitter of Fock states. A theoretical background to do so has been proposed in connection to the problem of Dicke's superradiance by Rupasov and Yudson (1984a) and Rupasov and Yudson (1984b), who obtain the wavefunction of the N-particle states that arise in the superradiant dynamics using Bethe (1931)'s Ansatz, that allows one to find some exact solutions to certain one-dimensional quantum many-body models, in particular useful for the quantum inverse scattering method (see Korepin and Bololiubov (1997)'s textbook for an introduction). The method was developed and generalised by Yudson (1985) and was applied to a variety of systems, including the Mollow physics (but kept in a Russian preprint form). This formalism allows one to describe the internal structure of the N-photon state, and along with other dynamical studies such as indistinguishability of two bundles, this would allow one to describe the structure of quantum light that is strongly correlated.

11.5 Exciting with quantum light

11.5.1 *Cascaded formalism*

Much of this book, as indeed much of optics, is concerned with the excitation by classical light, which already leads to many possibilities (cw or pulsed excitation, deterministic or stochastic fields, etc.) Section 11.2, for instance, treats the response of polaritons to impinging coherent light, a problem that still leaves much room for original research to this day. In this section, we address the emerging question of excitation with quantum light. The situation is sketched in Fig. 11.18. The ever increasing availability of quantum sources—so far mainly single-photon sources, but including also new devices like the N-photon emitter of Section 11.4—makes it necessary to enlarge the type of excitations employed. Until now, this has been dominated by lasers and, e.g., for cases of non-resonant excitation, incoherent sources that correspond to excitation with thermal light (for instance from a mercury lamp).

Exciting with classical light   Exciting with quantum light

(a) 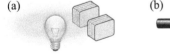 (b)   (c)

Incoherent excitation   Coherent excitation   Coherent SPS

Fig. 11.18: Various types of excitations. (a) and (b) correspond to classical light, which can be of different types requiring different formalisms, but always correspond to a classical field, even if the target itself is quantum. Incoherent light for instance is typically described by Boltzmann types of equations or stochastic fields, while coherent excitation is described by an oscillating deterministic field, in the simplest case, of the form $\exp(i\omega t)$. (c) describes the excitation by quantum light. For instance, a single photon source (SPS) can be used as the source, described for instance by a two-level system driven coherently (by a classical field, but the output of the 2LS is indeed quantum).

Already at this level, some quantum features of the light such as the underlying photon statistics, is largely ignored. Aßmann and Bayer (2011) addressed this point and discussed "photon-statistics excitation spectroscopy" based on the differing statistical features of thermal and coherent light, as discussed in Chap. 4 (see Fig. 11.19 for a recent application to a polariton laser). Even before such semi-classical calculations, it had been foreseen that quantum light could be of value as a source. In particular, Kira and Koch (2006) proposed the concept of "Quantum-optical spectroscopy" to generate and detect quasi-particles, a notion that culminated with their discovery of a new semiconductor quasi-particle, the dropleton with Almand-Hunter et al. (2014), although they still had to recourse to classical excitation in the experiment. Other proposals along similar lines have been advanced, for instance by Mukamel and Dorfman (2015) or Kazimierczuk et al. (2015). In all theses cases, some hypotheses and restrictions are made on the nature of the quantum light. In some respects, such treatments are related to the setting of an initial condition, a configuration that has been thoroughly studied, including for genuinely quantum states. The drawback is that one is limited in such a case to a small set of particular cases, bounded by one's imagination and tolerance for specifying the details of the state under consideration. As discussed in Section 11.3, on the other hand, the two-photon spectrum reveals that even the simplest quantum emitter actually embeds features of very high complexity.

To treat the problem fully and self-consistently, one must, therefore, describe the source completely and independently, and couple its output to the target that is to be excited. It should not be surprising to the quantum physicist who has been sometimes asked to include the observer along with the system in the description of the latter, that one should also include the excitation source in the process. Here, a subtlety is that a simple Hamiltonian model is not enough as it is often desirable for a realistic description that the source remains unaffected by the target, which is in contradiction with the Hamiltonian description that, by hermiticity, allows i) all reversible processes and ii) only reversible processes. Therefore, it also results in the excitation of the source by its target. Note that this problem does not pose itself with classical excitation, which appears in the model as mere parameters (e.g., intensity, frequency, etc.), while a quantum description requires operators with their own dynamics. The necessity to fully include the source as part of the problem was realised by Gardiner (1993) and Carmichael (1993), motivated by the development of squeezed light (see Walls (1983) for a contemporary review). Their (independent) treatment of the problem in these two consecutives papers provided a formalism—christened the "*cascaded formalism*" by Carmichael— that allows one to excite a system (which we will call the "target") by another (the "source") without back-action from the target to the source. This allows one to think separately of the quantum source, in the line of Fig. 11.18, whose properties can be first studied separately. The formalism can be cast in the master equation form, for the Hamiltonians H_1 of the source and H_2 of the target, and reads

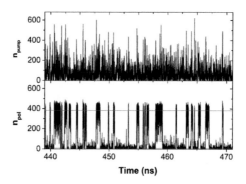

Fig. 11.19: Exciting with classical light with different statistical fluctuations, as computed by Aßmann and Bayer (2015). The upper panel shows the fluctuation of the number n_{pump} of photons from the source, that is constant (horizontal line) for a coherent field or greatly fluctuating for a thermal field, both with otherwise the same average intensity. The bottom panel shows the response of a polariton laser driven slightly off-resonantly to these two types of light, with no fluctuation in the former case and only depending on initial conditions, or exhibiting blinking in the later case.

$$\frac{d\rho}{dt} = i[\rho, H_1 \otimes H_2] + \frac{\gamma_1}{2}\left\{2c_1\rho c_1^\dagger - \rho c_1^\dagger c_1 - c_1^\dagger c_1 \rho\right\}$$
$$+ \frac{\gamma_2}{2}\left\{2c_2\rho c_2^\dagger - \rho c_2^\dagger c_2 - c_2^\dagger c_2\rho\right\} - \sqrt{\gamma_1\gamma_2}\left\{[c_2^\dagger, c_1\rho] + [\rho c_1^\dagger, c_2]\right\}, \quad (11.26)$$

for ρ the density matrix of the combined (source and target) system, with c_i operators corresponding to ouput ($i = 1$) an input ($i = 2$) operators for the source and target, respectively. This master equation (11.26) is such that any process back from the target to the source is cancelled by quantum interferences and while the target is driven by the source through the coupling term $\sqrt{\gamma_1\gamma_2}$, the source itself is unaffected by the target. The derivation of this master equation is given by Gardiner and Zoller (2000) in detail. We leave as an exercise (Exercise 11.2) the correction of a mistake that appears in this text.

Exercise 11.2 (*) *Show that eqn (11.26) is in the Lindblad form (compare with the form given by Gardiner and Zoller (2000) and correct your local copy).*[128]

11.5.2 *Exciting simple targets*

It is insightful to get an intuition of the physics of cascaded systems to consider the case of a two-level system in its excited state exciting without feedback a cavity with $n_a(0)$ photons at $t = 0$. In this case, $H_1 = H_2 = 0$ (in the rotating wave approximation with the free energies of both the source and target at resonant set at zero) and eqn (11.26) with, from our notations throughout, $\gamma_1 \equiv \gamma_\sigma$ and $\gamma_2 \equiv \gamma_a$, can be solved exactly for both the source and its target:

---

[128]Solution: the Lindblad operator is a superposition of the target and the source $c = \sqrt{\gamma_\sigma}\sigma + \sqrt{\gamma_a}a$ in which case $\partial_t\rho = i\left[\rho, H_a + H_\sigma + i\frac{\sqrt{\gamma_a\gamma_\sigma}}{2}(\sigma^\dagger a - a^\dagger\sigma)\right] + \frac{1}{2}\mathcal{L}_c\rho$.

$$n_\sigma(t) = e^{-\gamma_\sigma t}, \tag{11.27a}$$

$$n_a(t) = n_a(0)e^{-\gamma_a t} + \frac{4\gamma_a \gamma_\sigma}{(\gamma_a - \gamma_\sigma)^2} \left[e^{-\gamma_\sigma t/2} - e^{-\gamma_a t/2} \right]^2. \tag{11.27b}$$

As should be—this is the main point of this formalism—the source is unaffected by its coupling to the target so eqn (11.27a) is simply the (trivial) solution of the decaying two-level system. The target, however, is affected by the source, beyond the decay of its initial state (first term). Note that while there is a strong asymmetry in this sense (there is no target parameter in the dynamics of the source), the target displays, on the other hand, a remarkable symmetry between γ_a and γ_σ (that can be interchanged without affecting the result). This shows how the target behaves either as itself, or as the source feeding it, or as a mixture of both depending on the ratio of the pumping and decay rates. The most efficient coupling is achieved when both rates are equal, in which case the maximum population of the target reaches the value $4/e^2 \approx 0.54$ at the time given by twice the inverse decay rate. The factor 4 is reminiscent of the Purcell enhancement in weak-coupling. Note also that the coupling is through dissipative channels in the Lindblad form, which means that an infinitely-lived system cannot be excited. This also puts some strict constraints on the regime that the source can bring its target into, since coupling and decay are not independent, unlike the usual scenario of ligth-matter coupling (see, e.g., eqn (5.81) where g is independent of γ_a and γ_σ). This could appear to be, at first, a detrimental limitation. However, López Carreño and Laussy (2016) and López Carreño et al. (2016) show that this allows, instead, to reach a new regime of light-matter coupling that optimises the excitation thanks to the one-way coherent transfer. Before that, note that there is no Bose-stimulation effect from the source to the target, since, by construction, the source is not affected by the target and cannot, therefore, increase its decay rate by final-state stimulation.

Exercise 11.3 (**) *Solve the general case of a first cavity, A, in the arbitrary quantum state ρ_A, exciting without feedback another cavity, B, in the arbitrary quantum state ρ_B. What happens when the decay rates are equal?*

Spontaneous emission is a basic case that leaves aside the dynamics of correlations that arises when driving the system, as discussed in Section 11.3. We now consider the excitation of a target by a driven SPS, contrasting both the incoherent and coherent excitation of the source. The incoherent excitation is the simplest configuration. Here, the two-level system that acts as the source is excited at a rate P_σ, at which the 2LS is put in its excited state, and is otherwise left to decay. The system is thus described by eqn (11.26) with $c_1 = \sigma$ and $c_2 = a$. The 2LS pumping is described by a Lindblad term $(P_\sigma/2)\mathcal{L}_{\sigma^\dagger}\rho$. Thanks to the absence of feedback, the dynamics is ruled by closed equations for which exact solutions can be found for the observables of interest. Namely, for the cavity (target) population n_a and statistics $g^{(2)}$, one finds

$$n_a = \frac{4P_\sigma \gamma_\sigma}{\Gamma_\sigma (\Gamma_\sigma + \gamma_a)}, \qquad g^{(2)} = \frac{2\Gamma_\sigma}{\Gamma_\sigma + 3\gamma_a}, \tag{11.28}$$

where $\Gamma_\sigma = \gamma_\sigma + P_\sigma$. Eliminating P_σ from eqn (11.28), provides the equation for the trajectory in the $(n_a, g^{(2)})$ as a function of the parameter γ_a/γ_σ:

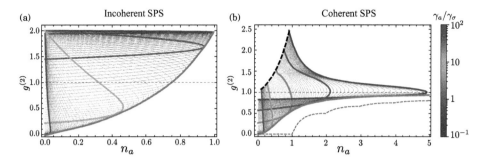

Fig. 11.20: **Exciting an harmonic oscillator with a SPS.** (a) incoherent and (b) coherent excitation of a two-level system, in turn exciting an harmonic oscillator. The curves correspond to various ratios γ_a/γ_σ of the target and source decay and emission rates, respectively. Highlighted are the cases $\gamma_a/\gamma_\sigma = 10^{-2}$ (red), 10^{-1} (yellow), 1 (green), 10 (chartreuse) and 10^2 (blue). In both cases, the lower envelopes show the closest one can get to the ideal limit of Fock-duos, shown in dashed. From López Carreño and Laussy (2016).

$$n_a = \frac{2}{3} \frac{(2 - g^{(2)})(3g^{(2)}(\gamma_a/\gamma_\sigma) - (2 - g^{(2)}))}{g^{(2)}(1 + g^{(2)})(\gamma_a/\gamma_\sigma)} . \tag{11.29}$$

These trajectories in the Hilbert space are plotted in Fig. 11.20. The curves start from the point $\left(n_a = 0, g^{(2)} = \frac{2}{1+3\gamma_a/\gamma_\sigma}\right)$ at vanishing pumping and tend to $(n_a = 0, g^{(2)} = 2)$ with increasing pumpings, where the source gets quenched. These solutions show that low pumping, with a source that has a smaller decay rate than its target, leads to best antibunching/population figures of merits. The curve from eqn (11.29) is fairly constant till the turning point and leads to antibunching as long as $\gamma_a/\gamma_\sigma > 1/3$. From the analytic solution, one can derive the expression for the lower envelope that shows the points accessible in the Hilbert space charted by both population and statistics, by varying the ratio γ_a/γ_σ. One finds

$$g^{(2)} = \frac{2n_a}{3 - 2n_a} \xrightarrow[P_\sigma \to 0]{} \frac{2}{1 + 3\gamma_a/\gamma_\sigma} , \tag{11.30}$$

where we also gave the case of vanishing excitation.

Similar results can be obtained for the coherent driving of the two-level systems, of which we only provide the plotted version, in Fig. 11.20(b), for comparison and discussion, as analytical expressions, when they exist, are cumbersome. The main difference is that while the incoherent case remains bounded in the area delimited by eqn (11.30) from below and $g^{(2)} \leq 2$, $n_a \leq 1$ from above, the coherent excitation leads to a greater selection of states, in particular with $g^{(2)}$ now bounded by 3 (at large driving rates) and with populations of the target extending much beyond unity (at intermediate driving rates). The $g^{(2)}$ at low driving, corresponding to vanishing coherent excitation, can also be found in closed form (but not the general expression for arbitrary n_a):

$$g^{(2)} \xrightarrow[\Omega_\sigma \to 0]{} \frac{1}{(1 + (\gamma_a/\gamma_\sigma))^2} , \tag{11.31}$$

which, when compared to the corresponding limit in eqn (11.30) shows again that coherent excitation provides a much better antibunching than its incoherent counterpart.

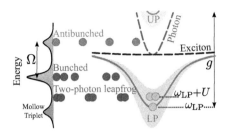

Fig. 11.21: **Exciting polaritons with quantum light.** Mollow spectroscopy, as proposed by López Carreño et al. (2015): the photoluminescence of a strongly-driven two-level sytem provides the Mollow triplet, shown on the left with energy on the vertical axis. Various spectral windows provide different types of photon correlations, sketched here as photon balls with different temporal spacing. Exciting the Lower Polariton (LP) with leapfrog photon pairs allows one to measure accurately very small values of the interaction.

If we also compare these results to reversible dynamics, i.e., in the conventional Liouvillian form, one finds that the cascaded excitation provides better antibunching for a given population, despite the apparent shortcoming already mentioned of the coupling not being independent in this case from the decay rates. As such, cascaded coupling provides a better quantum driving than the Hamiltonian coupling brought in the strong-coupling regime, because in the latter case, the reversible process from the "target" back to the "source" acts like an effective decay. The simple cases treated here for illustration already show that the cascaded excitation, i.e., exciting a cavity with quantum light and without feedback, realises quantum states closer to the physical limits than conventional mechanisms. López Carreño and Laussy (2016) and López Carreño et al. (2016) have considered these cases and a plethora of other configurations in more details. We refer the interested reader to these texts and references therein for further details.

11.5.3 *Mollow spectroscopy*

We conclude this Section with another illustration of the benefits of exciting with quantum light, namely, for spectroscopic purposes, by measuring weak nonlinearities in strongly dissipative systems, a problem that is still unresolved for polaritons. The concept has been introduced by López Carreño et al. (2015) and is sketched in Fig. 11.21. It relies on the correlations of the Mollow triplet, discussed in Section 11.3, applied to the Hamiltonian (11.13).

By scanning the rich variety of quantum light provided by the various frequencies of the Mollow triplet, one can probe the response of a target to all types of input, from single-photon light to super-bunched, strongly-correlated photon pairs. This addition to the emerging field of photon-statistics spectroscopy has been called *"Mollow Spectroscopy"*. In contrast to the excitation by a laser, the use of quantum light allows a strong response in statistics with small populations for all ranges of excitation and values of U/γ_a. The measurement is thus clean from all complications of high densities, such as heating, phase-space filling (as described by Schmitt-Rink et al. (1985)), loss of strong-coupling (as described by Houdré et al. (1995)), population of an exciton reservoir contributing the bulk of the blueshift (see the discussions from Ferrier et al. (2011) and Christmann et al. (2012)), and exciton-photon fraction deviating from the Hopfield coefficients (an effect pointed out by Kartashov et al. (2012)). Since it recourses to the minimum amount of polaritons required to poke the interaction (two), Mollow spectroscopy acts as a "probe" in the ultimate sense of the term, with as little disturbance as possible. The information is retrieved, e.g., through the auto-convolution f of the target's statistics, $f = g^{(2)} * g^{(2)}$, i.e., $f(\omega_0) = \int_{-\infty}^{+\infty} g^{(2)}(\omega_0 - \omega) g^{(2)}(\omega)\, d\omega$. The triangle inequality places the maximum of $f(\omega)$ at the value ω_0^* that minimises the asymmetry of $g^{(2)}(\omega)$ around $\omega_0 - \omega_{\mathrm{LP}}$. Provided that the leapfrog processes are strong enough (that is, the Mollow splitting is sufficiently larger than the target's spectral broadening), one finds that the shift is precisely given by the two-polariton interaction

$$\omega_0^* = \omega_{\mathrm{LP}} + 2U\chi_{02}^2, \qquad (11.32)$$

where χ_{02} is the two-polariton Hopfield coefficient for the state $|02\rangle$ of two excitons and zero photon. The closed-form expression of χ_{02} in terms of detuning is too cumbersome to be given here, but the value is straightforwardly obtained by diagonalisation of the polariton Hamiltonian and is in excellent approximation given by χ_{01}^2. In the case of an anharmonic oscillator with no exciton-photon structure, the shift would be directly given by $2U$. There is thus no dependence of the measurement ω_0^* on the population or other dynamical variables and the measurement is thus absolute, unlike the blueshift from a classical driving that requires knowledge of the effective laser intensity.

These results suggest numerous other applications: including the spin, performing quantum pump-probe and two-tone spectroscopy with delay between quantum correlated beams, probing and exciting continuous fields or their resonant configurations such as optical parametric oscillators, or driving other quantum systems, e.g., two-level systems, coupled cavities, polariton circuits, etc., each of them amenable to several types of quantum light. The emitter of Fock states discussed in Section 11.4, releasing all its energy in bundles of N photons, should also be a key resource for quantum excitation. Beyond the exotic states of light already alluded to, this may open the door to new classes of excitations, such as the already mentionned correlated electron-hole clusters: the dropletons.

11.6   Quantum information processing

11.6.1   *Quantum computation*

The processing of information by quantum means is highly topical, mainly for the prospects of a so-called *"quantum computer"*, a concept developed by Deutsch (1985) that exploits quantum mechanical features such as superposition and entanglement to power new computing techniques. Deutsch (1997) is a proponent of the many-worlds interpretation theory of Everett III (1957) that states that all possible variations of an event are real and take place in a wider "multiverse", of which we experience, as observers, only a subset (in this intepretation, we therefore exist as zillions of parallel copies in different parallel universes). What a quantum computer achieves is (according to this view) to distribute various inputs in these parallel universes that each performs a classical computation, and subsequently collect the results back to a single universe, that of the observer who can benefit from their single device having been duplicated a large number of times. The first example of how this is possible from the laws of quantum mechanics regardless of its interpretation, came from Deutsch (1985) who shows how one can decide if a binary function f of the binary variable x is constant or not with

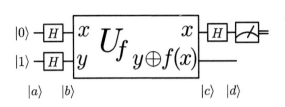

David **Deutsch** (born in 1953, Haïfa) established the conceptual foundations of quantum computation by formulating a quantum Turing machine. He also designed the first quantum algorithm (now bearing his name), shown left, to be run on a prospective quantum hardware. In "The Fabric of Reality", he writes *"The theory of computation has traditionally been studied almost entirely in the abstract, as a topic in pure mathematics. This is to miss the point of it. [...] What computers can or cannot compute is determined by the laws of physics alone, and not by pure mathematics."* A thought-provoking natural philosopher, founding member of the "Taking Children Seriously" pedagogical method, formulator of Deutsch's Law—*"Every problem that is interesting is also soluble"*—he is currently working on a theory of everything known as "constructor theory".

only one evaluation of the function, although it takes two input values. This is achieved with the algorithm shown together with David Deutsch's photograph and detailed in Exercise 11.4. The problem is highly specific and relies on strong properties such as the function satisfying $f(0) \oplus f(1)$ (where $\oplus$ is the sum modulo 2) being zero if it is constant and equal to 1 otherwise. The U_f implementation is, therefore, simply a XOR gate.

Exercise 11.4 $^{(*)}$ *Propagate the state $|a\rangle = |01\rangle$ in the quantum logic sketched on this page and show that the result is 0 if and only if f is constant.*

The algorithm was subsequently extended by Deutsch and Jozsa (1992) to binary function, but with n qubits as an input, and deciding if the function is constant or balanced (rather than not-constant, "balanced" meaning that it returns 0 for half the input domain). Rather than a gain of only a factor two, in its definite present form (designed by Cleve et al. (1998), but still under the name of "Deutsch–Josza algorithm"), the quantum computation is, therefore, exponentially faster than any possible deterministic classical algorithm, since it makes the economy of $2^n - 1$ evaluations of the function. This is an important result in the classification of complexity theory as it establishes a paradigm of computation that separates the P class—of problems that can be solved by a deterministic Turing machine in polynomial time—from the so-called EQP class—of problems that can be solved exactly in polynomial time on a quantum computer. Non-deterministic algorithms for such a problem are, however, efficient, and the real breakthrough came from an algorithm that tackles a problem of actual importance and for which there is no known efficient classical substitute. Shor (1997)'s algorithm provides a non-deterministic way to factor integers (it was first experimentally implemented by Vandersypen et al. (2001) with seven qubits to factorise $15 = 3 \times 5$). Rather than interfering values of the qubits through Boolean algebra, Shor's algorithm interferes complex amplitudes of the wavefunction. The algorithm relies on a *quantum Fourier Transform* (in the words of Shor (1997), "a quantum version of the FFT") that transforms the input state $|a\rangle$ for some $0 \leq a < q$ into

$$|a\rangle \rightarrow \frac{1}{\sqrt{q}} \sum_{c=0}^{q-1} |c\rangle \exp(2i\pi ac/q). \qquad (11.33)$$

If the scheme could be applied to thousands of qubits, one could then factorise in this way large integers, which is a problem of enormous importance as it provides the basis for the public key circulation in the RSA scheme of encryption, that is widely used in all areas of communication, civil and military (one needs 1 154 ideal qubits to break RSA with a key of 768 bits, and 30 002 ideal qubits for a key of 20 000 bits; actual implementations including error corrections, however, bring these numbers into the millions). Such an algorithm turned the quantum computer from an abstract contender to Church-Turing machines into an actually desirable device in a laboratory. It made selected problems related to quantum information processing of prime importance and in its wake, other applications were discovered, such as Grover (1996)'s algorithm for database searching, but also—although they are less known (they are also more recent)—the HHL algorithm (named after Harrow, Hassidim, and Lloyd (2009)), that solves systems of linear equations, Wiebe et al. (2012)'s algorithm that performs data fitting, or Garnerone et al. (2012)'s algorithm that computes PageRanks (the indexing used by Google to classify webpages). Such applications clearly make the issue more pressing as they suggest QIP as a seemingly obvious technology of the future.

While the technological requirements to control interactions between so large numbers of quantum systems are daunting, some breakthroughs were made that offered alternative schemes to implement a quantum computer. An influential one is the so-called KLM, after Knill, Laflamme, and Milburn (2001), that use only linear optical elements. It established a paradigm of quantum computation, known as LOQC (Linear Optical Quantum Computing). Photons are used as "flying qubits" and are processed in quantum superposition states through beam splitters, phase shifters, mirrors and likewise passive optics elements. The major drawback of LOQC is the number of optical elements needed to "wire" a quantum circuit. In LOQC KLM scheme, one requires intermediate measurements on ancilla photons with a feed-forward mechanism. Interactions would solve this problem, but photons are weakly interacting. Polaritons could be candidates for such implementations since they have, through their excitonic fraction, the opportunity for tunable and potentially strong interactions, being in effect strongly interacting photons.

11.6.2 *Limits of quantum computation*

Despite considerable efforts to build a working and scalable quantum computer, the problem (which was known to be extremely challenging from the start) remains stalled in terms of technological added-value. Several critics have tampered the enthusiasm for both what has been accomplished so far as well as to what can be expected in the light of two decades of intense dedication. Viamontes et al. (2005), for instance, detail how an in-depth benchmarking of actual quantum computation (using the case of Grover's quantum search algorithm) fails to compete with optimised classical solutions. Smolin et al. (2013) draws attention to the fact that the experimental implementations so far take advantage of already knowing the solution and do not really explicitly show the theoretical power of Shor's algorithm (interestingly the title of their text was changed from "*pretending to factor large numbers on a quantum computer*" in arXiv to "*Oversimplifying quantum factoring*" in Nature). Commenting on the HHL algorithm, Aaronson (2015) similarly identifies several caveats that could ultimately spoil the claim of an actual gain over classical computation (it is not proven, for instance, that in the condition of well-conditioned sparse matrices required by HHL, a classical algorithm does not exist that performs equally well) and he invites us to "*maintain a sober understanding of what these algorithms would and wouldn't do, an understanding that the original papers typically convey, but that often gets lost in second-hand accounts*". Our second-hand account above is guilty in this sense, incidentally, since we implied that HHL finds the solutions of linear systems while what it actually achieves is, in the constrains already listed, the solution $|x\rangle$ to be averaged over an observable, rather than the solution x itself (whose mere listing already brings it to the level of the classical computation). The quantum computer, although being a universal computing machine able to tackle any problem, could thus be actually suitable for a very limited subset of practical problems. Besides, quantum computation might be useful for classical computation, but it remains hopelessly frustrating for quantum manipulations. For instance, Alvarez-Rodriguez et al. (2015) observe how the no-cloning theorem forbids a quantum operation that, given two inputs $|\psi_1\rangle$ and $|\psi_2\rangle$, returns their superposition $|\psi_1\rangle + |\psi_2\rangle$ (this forbidden "quantum adder" does not forbid classical addition which, as for any Turing machine, must be able to implement all the elementary arithmetic operations, as has been pointed out by Vedral et al. (1996)). The best a quantum computer could achieve is, therefore, a redefinition of the complexity classes for classical computation.

While some of the most brilliant minds and successful entrepreneurs of our times accepted the challenge of building a quantum computer, others, beyond mere restrains, are outright opponents to the basic idea, such as 't Hooft (1999), Dyakonov (2003), Levin (2003), Davies (2007), Kalai (2011) and others (the opinion of these sceptics is often published in secondary channels such as blogs, side notes in papers or conference proceedings; Aaronson (2013)'s Chapter 15 provides a good panorama of arguments from both sides, see also Aaronson (2004) for a more technical discussion). Even before quantum computers got a chance to jeopardise RSA if they could achieve their grail application—factorisation—cryptography was strengthened by using the same tools and developed its own quantum protocols. The most famous one, BB84, after Bennett and Brassard (1984), communicates the full key (rather than encoding it as a product of two large integers in the classical scheme), ensuring privacy by the fact that each bit is encoded by a single photon, making the reading of the information secure by its immediate destruction (assuming photons are not entangled; more secure protocols followed such as the one based on Bell's theorem, developed by Ekert (1991)). Overall, quantum computers based on quantum gates in a circuit-model could result in a dead-end and the idea of a universal quantum computer remain, in Dyakonov (2013)'s words, a *"love-song for military sponsors"*.[129] In the pursuit of the initial ideas, however, other branches of quantum information processing have emerged, that are similar in spirit and could turn out to be more fruitful in the immediate future. Quantum cryptography, for instance, has been much more successful in terms of practical applications, even though it started amid greater skepticism still already at the abstract idea itself, as illustrated by the difficulty of Wiesner (1983)'s seminal idea (of quantum money) to be published, well over a decade after its first submission in 1970 with several rounds of rejection in many journals, for what would later become the pillar of quantum public key distribution.

One of the current actively sought implementation of quantum computation is the so-called *boson sampling* problem, proposed by Aaronson and Arkhipov (2011) who observed that the computational complexity of photon Fock states propagating through a network of beam-splitters—completely passive linear optical elements (without intermediate measurements on ancilla photons as in KLM)—is exponential (#P hard), that is, out of reach of a classical computer. The computation involved is related to the permanent of a large matrix, which, unlike the determinant, is not conserved by changes of basis and thus collides head first with the combinatorics involved by the boson indistinguishability. Setting up the device in the laboratory is comparatively easy (the difficulty is mainly related to preparing the input Fock state of N photons) and there is no drawback of interactions being too weaks as the scheme is fully linear anyway. While this does not allow to compute, e.g., a permanent (this would require exponential sampling), but only to sample the output, this is, however, a sampling that can be performed by a physical system that outperforms its classical counterpart. The practical interest in such a sampling is not yet known and might never surface. However, the demonstration of its feasibility by a physical device that can produce a number according to an algorithmic procedure that no computer similation could compute, would already be a concrete demonstration of the basic principles of quantum information processing. A good review of boson sampling is given by Gard et al. (2015). In the next Section, we turn to other routes that the field is exploring while circuit-based quantum computers struggle to progress, in the search for an escape route towards technology.

11.6.3 *Quantum annealing*

Quantum Annealing is one of the hot topics and trendiest directions in the field of quantum information processing. It brings to a quantum computer the technique of simulated annealing. This technique is, at the classical level, the algorithmic implementation of a method used in metallurgy to reach the ground state of a metal, that can be trapped in local minima of energy if simply cooled, but is found to reach its absolute ground state more quickly when alternating cycles of heating and slow cooling. Annealing,

---

[129]The love song: *"The transistors in our classical computers are becoming smaller and smaller, approaching the atomic scale. The functioning of future devices will be governed by quantum laws. However, quantum behavior cannot be efficiently simulated by digital computers. Hence, the enormous power of quantum computers will help us to design the future quantum technology"*. Dyakonov (2013)'s essay is a highly recommended read.

whose rudiments have been empirically known since antiquity, allows atoms to mi-
grate in the crystal lattice and decreases the number of dislocations. The simulation
of this process in a (classical) algorithmic form comes from Kirkpatrick et al. (1983)
who applied it to combinatorial optimisation problems (such as the travelling sales-
man problem) by introducing an effective temperature for the optimisation, that acts
as the probability of accepting higher-energy configurations, which allows the system
to hop to states that increase the cost function (energy) to be optimised. The insight is
that accepting only changes that lower the cost function will trap the system into local minima, so it is beneficial to let the system temporarily worsen its configuration
to explore global solutions. Kadowaki and Nishimori (1998) allowed such fluctuations
to be quantum, and considered tunnelling through high but shallow potential barriers
that surround a local minimum, which are difficult to escape by classical annealing.
Quantum annealing primarily addresses the discrete optimisation problem, and is thus
not a universal computer aiming to tackle all that is computable. This still remains of
extreme importance since many topical problems—such as machine learning and artificial intelligence—are ultimately problems of combinatorial optimisation. Besides, it has
been shown by Aharonov et al. (2007) that quantum annealing (under the related variation of "adiabatic quantum algorithms" that implements a more coherent version[130]) is
polynomially equivalent to conventional quantum computation. Peng et al. (2008), for
instance, have shown how it could also tackle factorisation (factorising 21 with fewer
qubits than Shor's algorithm). Contrary to circuit-based quantum computation, it has
also been shown (by Somma et al. (2012)) that quantum annealing provides exponential
speedup, albeit on a problem of no known practical importance. To perform quantum annealing, one writes the cost function in terms of the Ising model of statistical mechanics
as

$$H = -\sum_{i,j} J_{ij}\sigma_i^z\sigma_j^z - \Gamma(t)\sum_i \sigma_i^x, \tag{11.34}$$

with the Hamiltonian (energy function) of the Ising model chosen such that its lowest-
energy state (ground state) represents the solution to the combinatorial optimisation
problem. A term representing quantum-mechanical fluctuations is added to the Hamiltonian to induce quantum transitions between the states. The idea is to encode a problem
by specifying the hundreds of interactions within the chip and solve it by finding the
qubits' lowest energy "ground state", which is the sought solution. The problem took
a particular impetus when it involved corporate companies, eventually attracting major
players such as Google and NASA. At the forefront of the commercial line lies "D-
Wave Systems" (the name comes from the d-wave superconducting technology used to
implement the qubits), a company that made the claim for the world's first commercial quantum computer, in the form, first of a 28-qubit quantum annealing processor in
2007, boosted to 128-qubit in 2011, 512 qubits in 2013 and, with the D-Wave TwoX,
a 2 kilo-qubit processor in late 2015 (operated with only half of the qubits activated).
Initial demonstrations by the company involved searching closest matches in a database

---

[130] In the sense of "quantum coherence".

of proteins, optimising guest seating in a wedding reception and solving a Sudoku puzzle. More academic focus was subsequently targeted on simply finding the ground state for a binary Ising Hamiltonian. These results stirred considerable media attention, but also controversy from leading academics, sometimes scoffing at the claims. After years of debates, a first recognition that D-Wave had achieved relevant results to support their claims was made with their report, by Johnson et al. (2011), of quantum tunnelling—at the heart of the quantum version of annealing—occuring in an eight-qubit processor, but yet with no consensus whether this quantum character was playing any role in the actual optimisation. Significant progress was unanimously agreed upon with the report by Boixo et al. (2014) (working on a D-Wave machine, but not affiliated to the company) that the machines did perform genuine "quantum annealing" at the hardware level as compared to "classical annealing", based on comparison with quantum Monte Carlo and classical algorithms (that exhibit bimodal or unimodal distributions of success rates, respectively). Still, there was no agreement on the performance this implied, with suspicion that the machine was lacking the degree of coherence necessary for useful quantum operation and was exhibiting mean-field quantum effects. Google's Researchers, Boixo et al. (2016), recently made another step forward by proving that quantum tunnelling was playing an active role in the optimisation when working on a problem specifically designed to take advantage of quantum annealing strengths. There is currently still no consensus whether the D-Wave machine can, or even could, provide useful quantum computation in the sense of outperforming a classical computer, as classical alternatives have so far always been shown to keep the upper hand, and despite the considerable progresses and achievements made by D-Wave, it is still not recognised whether any of its machines qualifies as a quantum computer. Ironically, the circuit-based model could come back on the scene (with its own design features such as error-correction), since the limitations might be in the algorithmic technique itself rather than on the underlying hardware.

11.6.4 *Polariton simulator*

A universal quantum simulator was formalised by Lloyd (1996) based on Feynman (1982)'s proposal to simulate quantum systems by a programmable quantum system. The field got much impetus from the cold-atoms community that could manipulate them in lattices, thereby implementing Bose–Hubbard physics in a controlled environment, as demonstrated by Greiner et al. (2002) and reviewed by Bloch et al. (2012) (Georgescu et al. (2014) give a more thorough review of the general principles of quantum simulation). Similar schemes based on the BEC/superfluid to Mott insulator transition have been transposed to polaritons, e.g., by Byrnes et al. (2010) and Na and Yamamoto (2010). Polaritons bring several assets to implement quantum simulators: they strongly interact and are easily controlled as well as observed by standard optical methods. Some quantum information processing methods, such as quantum annealing, are particularly resilient to noise. One asset of quantum annealing over the standard quantum computer is that the "computation" is performed in an explicitly open dissipative system, possibly tolerating high degrees of decoherence in some bases that do not hurt the computation. The idea to take advantage of dissipation to drive the computation rather than to be

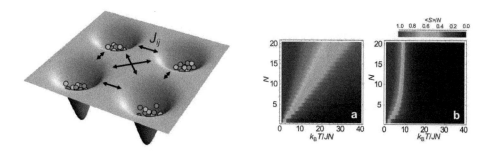

Fig. 11.22: A boson lattice as proposed by Byrnes et al. (2011), that can be implemented with polaritons, boost the principle of quantum optimisation with Bose stimulation. The insets (a) and (b) compare the average spin for a two-site lattice in the case of N distinguishable and bosonic particles, respectively. Such a speedup in reaching the ground state remains in larger lattices and also occurs in the temporal domain, accelerating considerably the computation time.

detrimental to its quantum character emerged in a variety of fields. One original im-plementation of quantum annealing exploiting this scheme and that takes advantage of the bosonic properties of polaritons (it is also relevant for other Bose gases) came from Byrnes et al. (2011), who proposes to boost quantum optimisation with Bose stimula-tion. This is related to boson sampling in that it relies on particles' indistinguishability, rather than quantum superposition or entanglement. The speedup brought by Bose stim-ulation is shown in Fig. 11.22(b) as compared to the conventional implementation with spinor instead of bosonic sites for the particular case of a two-site system, with the Hamiltonian involving N particles:

$$H = -JS_1S_2 - \lambda N(S_1 + S_2). \tag{11.35}$$

The figure shows the average spin $\langle S \rangle \equiv \sum_{S_1,S_2} S_i \exp[-H/(k_{\mathrm{B}}T)]/Z$ when as-suming distinguishable (panel a) or indistinguishable particles following bosonic statis-tics (b). The result is the same for $N = 1$, but as N grows, smaller values (corresponding to the ground state) are obtained independently from the temperature instead of the lin-ear increase in the classical case. Byrnes et al. (2011) have shown how this basic result remains true for any size of the lattice and how the equilibration time likewise benefits from this bosonic speedup. As annealing procedures will develop, it is likely that such a scheme will prove extremely popular and favourable. Byrnes et al. (2013) have also proposed neural networks based on similar ideas, to be contrasted to a similar proposal from Espinosa-Ortega and Liew (2015).

At the heart of polariton simulators lies their capacity to be hosted in lattices, as sketched in Fig. 11.22. There are several approaches to trap polaritons, but they consist basically in confining either the exciton or the photon. A particular versatile tool to con-fine the exciton is through Surface Acoustic Waves (SAW), implemented by M. M. de Lima et al. (2006) although heating is a severe problem, while confinement of the pho-ton can be achieved most directly by etching, as developed by Daïf et al. (2006) and Kaitouni et al. (2006). Lai et al. (2007) also demonstrated the potential of metallic arrays deposited on the surface of the sample. These authors also discussed the unconventional condensation that resulted in these geometries. Directly shaping the potential by light is also possible and convenient, as demonstrated recently by Tosi et al. (2012) and Amo et al. (2010b). As these various techniques will mature, one can expect to see in the polariton community the blossoming of Bose–Hubbard and Tonks–Girardeau physics, and bring to polaritons the rich literature of strongly-correlated coupled cavity arrays, with representative works given by Hartmann et al. (2006), Greentree et al. (2006), Angelakis et al. (2011) and others (see Hartmann et al. (2008) for a review).

11.6.5   *Other paradigms*

Since the quantum-gate based model of quantum computation, several other paradigms have emerged, passing by quantum simulation and quantum annealing. We conclude this Section with a brief overview of some popular trends that are useful to keep in mind in the wider landscape of QIP.

11.6.5.1   *Continuous variables quantum computation.*   This brings to the quantum realm the classical counterpart of analogue computation where real variables of infinite precision are used instead of discrete bits. Even at the classical level, computing with continuous variables is an arcane field whose foundations remain uncertain. One application that is known to be sound is the simulation of one analogue system by another, more conveniently operated. This is the idea that Feynman (1982) brought to its quantum conclusion, laying down the premises for quantum information processing that would later fully resurface as quantum simulation. Continuous variable quantum computation itself was introduced by Lloyd and Braunstein (1999) with variables such as the amplitudes of the electromagnetic field or the position of a particle instead of a countable collection of two-level systems. While this seems to provide a concept of "quantum floating point" computation, in practice, the same difficulty as in the classical case of the intractability of infinite precision variables ultimately limit the performances of a continuous variables computer to those of its discrete counterpart. However, the possibility to work with continuous fields allows the implementation of computation with physical objects such as superfluids, Josephson circuits, Bose–Einstein condensates, etc. This makes this scheme particularly relevant for polaritons. This is the approach taken by Kyriienko and Liew (2016), who observe that—since polariton-polariton interactions are still too small at the few-particles level—the continuous variable scheme would allow one to take advantage of the currently demonstrated polariton condensates to exploit quantum polaritonics with currently available technology. In the continuous variable scheme, two gates of fundamental importance are the NOT and CNOT gates, defined, say, on the state $|x\rangle$ with $x \in \mathbb{R}$ the real variable that could be the position of a (one-polariton) particle. The continuous gates are thus defined as:

$$\text{NOT}\,|x\rangle = |-x\rangle \,, \tag{11.36a}$$

$$\text{CNOT}\,|x\rangle_1\,|y\rangle_2 = |x\rangle_1\,|x+y\rangle_2 \,, \tag{11.36b}$$

Equation (11.36b) is the generalisation to the continuum of the two-level version of the gate, $\text{CNOT}\,|i\rangle\,|j\rangle = |i\rangle\,|i \oplus j\rangle$ where $\oplus$ is the addition modulo 2 and $i, j \in \{0, 1\}$. The modulo drops in the continuous case as the dimension goes to infinity. Wang (2001) introduces such continuous-gates and also hybrid versions where a discrete qubit controls

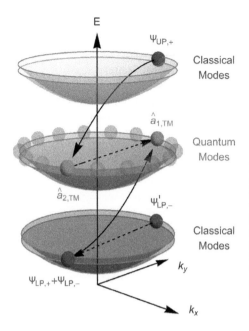

Fig. 11.23: Continuous-variable quantum computation as proposed by Kyriienko and Liew (2016). Two modes are wired in a continuous-variable CNOT gate mediated by the coupling to a macroscopic condensate, allowing entanglement at the few-polariton level despite the weakness of the polariton-polariton interaction. By using a system of dipolaritons split in polarisation with opposite momenta on the middle-branch and classical excitation of the outer branches, the scheme can be scaled to several qubits.

a continuous variable. In the position/momentum space, a microscopic implementation of the CNOT gate eqn (11.36b) on a two-particle input is $\mathrm{CNOT} = \exp(-iq_1p_2)$ with $q_i = (a_i + a_i^\dagger)/\sqrt{2}$ and $p_i = -i(a_i - a_i^\dagger)/\sqrt{2}$ being the quadratures of the mode. Kyriienko and Liew (2016) bring as an implementation the mean-field Hamiltonian of parametric scattering between the polaritonic modes a_k and a_{-k} of opposite momentum and the condensate ψ with

$$H = \alpha(\psi^{*2}a_k a_{-k} + \psi^2 a_k^\dagger a_{-k}^\dagger) - J(a_k^\dagger a_{-k} + a_k^\dagger a_{k-}),\qquad(11.37)$$

that is further tuned such that $\alpha\psi^2 = -J$ (which can be enforced by coherent control of the condensate) and making a $\pi/2$ phase shift of the a_{-k} mode, so that it gets a factor i in the equation above. This reduces H to $2Jq_1p_2$, which implements the propagator $\exp(-i2Jq_1p_2t/\hbar)$ and, therefore, the CNOT continuous-gate simply by driving the microcavity for a duration $\tau_0 = \hbar/2J$ (or multiple of this). The scheme is robust to lifetime as the authors find that the fidelity of such a polaritonic continuous-variable CNOT gate can be made arbitrarily high in the limit where the coupling J dominates the decay, with values as high as 99% with realistic parameters. It also appears to offer good scalability prospects. For instance, using dipolaritons (with coupling to direct and indirect excitons as done by Cristofolini et al. (2012)) to get three-branch polaritons as shown in Fig. 11.23, Kyriienko and Liew (2016) propose to sandwich quantum modes with two branches excited externally and sustaining condensates (classical modes at the mean field level) that can provide a version of eqn (11.37) for several quantum states, identified by different momenta (instead of positions).

11.6.5.2 *Cluster computation.* Also known as a "one-way quantum computer" or "measurement based quantum computer", acts on a prepared so-called "cluster state", that is a highly entangled state within a lattice of qubits with Ising type interactions (so-called "graph states" achieve similar purposes in more general geometries). The quantum algorithm is specified by the choice of bases for single-qubit measurements alone, performed in a given order. The computation is "one-way" because projective measurements destroy the state as the computation progresses. Nielsen (2006) gives a good review of this scheme. A continuous variable version of cluster quantum computation has been proposed by Zhang and Braunstein (2006) and Menicucci et al. (2006) and a proposal for polaritons made by Liew and Savona (2011) for the simplest possible case of a four-mode cluster state. The underlying idea is similar in some aspects to that later formulated by Kyriienko and Liew (2016) in that it combines a mean-field polariton condensate and parametric scattering, with theoretical prediction of four-body entanglement (as evidenced by the violation of so-called van Loock–Furusawa quadripartite inequalities) even in presence of lifetime and dephasing.

11.6.5.3 *Surface codes.* Introduced by Kitaev (2003) as "topological quantum error-correcting" codes (Fowler et al. (2012) give a good introduction), surface codes implement a logical qubit (the one used in a quantum circuit to implement a quantum algorithm) out of a collection of physical qubits (the quantum states implemented on a physical object). The latter are arranged in a two-dimensional array. In this scheme, the logical qubit performs better than the underlying individual physical qubits. While this scheme has not yet been proposed with a polariton implementation, the rapid progress with topological insulator analogues should make such a scheme of QIP also within reach.

11.7   Future reading

The field of quantum polaritonic is one of the fastest evolving in microcavities, and this Chapter is likely to rapidly see its content out of date with some of the most spectacular developments of polariton physics. The field of quantum information processing is itself a thriving and very active one, so keeping updated with the literature is mandatory. Beside the practical problem of building any type of QIP device, a fundamental problem is whether these will be ultimately needed for technology. The problem is known as that of "quantum supremacy", that is to say, whether quantum aspects do really result in a net gain of performances, when all other factors are taken into account. At a fundamental level, the problem is one of complexity theory. Stephen Jordan maintains a "quantum algorithm zoo" at `http://math.nist.gov/quantum/zoo` that aims at being a comprehensive listing of quantum algorithms believed to outperform their best classical counterpart. There is also a complexity zoo (from Scott Aaronson) at `https://complexityzoo.uwaterloo.ca`. Montanaro (2016) gives an overview of quantum algorithms. Kim and Yamamoto (2015) gives a thorough review of the topic of polariton simulators.

12

POLARITON DEVICES

Polariton devices offer multiple advantages compared to conventional semiconductor devices. The bosonic nature of exciton polaritons offers opportunity of realisation of polariton lasers: coherent light sources based on bosonic condensates of polaritons. The final state stimulation of any transition feeding a polariton condensate has been used in many proposals such as for terahertz lasers based on polariton lasers. Furthermore, large coherence lengths of exciton-polaritons in microcavities open the way to realisation of polariton transport devices including transistors and logic gates. Being bosonic spin carriers, exciton-polaritons may be used in spintronic devices and polarisation switches. This chapter offers an overview on the existing proposals for polariton devices.

Microcavities, Second Edition. Alexey V. Kavokin, Jeremy J. Baumberg, Guillaume Malpuech, Fabrice P. Laussy, Oxford University Press (2017). © Alexey V. Kavokin, Jeremy J. Baumberg, Guillaume Malpuech, Fabrice P. Laussy. DOI 10.1093/oso/9780198782995.001.0001

12.1   Polariton lasers

12.1.1   *Concept of polariton lasing*

LASER stands for Light Amplification by Stimulated Emission of Radiation. The mechanism is a direct application of the Einstein (1916) coefficients, with amplification of light when stimulated emission exceeds absorption. This condition cannot be fulfilled at thermal equilibrium: it requires the inversion of electronic population, i.e., a negative temperature. To invert its electronic population the system needs external pumping of energy above some critical value, referred to as the *"lasing threshold"*. Nowadays, the term "laser" is applied to any device that produces coherent, monochromatic and unidirectional light (see, e.g., Coldren and Corzine (1995)'s textbook for an applied perspective). Conventional lasers are fermionic devices: they take advantage of optical transitions in electronic systems, such as in semiconductor crystals. If the inversion of electronic population (negative temperature) is achieved, the optical emission in such systems can be amplified by external optical cavities or waveguides that provide the positive feedback, which results in the laser generation.

It turns out that stimulated emission of radiation is not the only way to generate laser light. In this chapter, we discuss polariton lasers, which are an example of bosonic lasers, based on optical transitions in a gas or liquid of bosonic particles. In bosonic lasers, light is emitted spontaneously by a condensate of particles accumulated in a single quantum state. Bosonic lasers do not require negative temperatures: they may even operate at thermal equilibrium. They still need pumping, but, theoretically, can have vanishing thresholds.

Which particles are good for forming condensates able to emit light? Bose–Einstein condensates of atoms have been realised by Bloch et al. (1999) at extremely low temperatures with a cw coupler so as realise an atom laser. The condensed atoms are usually in their ground state, incapable of emitting light. This makes atomic condensates impractical for the generation of light (the term "atom laser" refers to a coherent flow of atoms, not of photons). On the other hand, condensates of polaritons emit light very well. Historically, the first hint to bosonic lasing based on excitons belongs to Gergel' et al. (1968a), who have calculated the dielectric response function of a bulk semiconductor in the presence of a Bose–Einstein condensate of excitons. It turned out that the imaginary part of the dielectric susceptibility becomes negative in a narrow frequency range below the frequency of the exciton transition. This corresponds to "negative absorption" or gain.

The paper by Gergel' et al. (1968a) was submitted on August 17, 1966 and resubmitted on February 10, 1967, to be finally published in February 1968. Between the two submission dates, the seminal paper of Keldysh and Kozlov (1967) on the Bose–Einstein condensation of excitons was submitted (on the 10th of January 1967). Gergel' *et al.*'s cite the paper by Keldysh and Kozlov as well as the earlier papers by Blatt et al. (1962) and Moskalenko (1962). The modern concept of polariton lasers based on semiconductor microcavities was formulated nearly 30 years later by Ĭmamoḡlu et al. (1996). Bosonic condensates of exciton-polaritons may be realised in semiconductor microcavities at relatively high temperatures, even at room temperature. This is why

Robert Arnol'dovich **Suris** (born in 1936, Moscow) is a charismatic Russian theoretical physicist. Graduated from the Moscow Institute of Steel and Alloys, he received his PhD in physics from the same institution in 1964. The title of his thesis was *"Application of Green Functions Method to Some Problems of the Solid State Physics"*. In 1971, together with Rudolf Kazarinov, he developed the concept of quantum cascade lasers, implemented by Faist et al. (1994). The pioneering works of Gergel, Suris and Kazarinov on the Bose–Einstein condensation and superfluidity of excitons have escaped the attention of the excitonic community for several decades and are now only getting rediscovered. In 1988, Suris moved to the Ioffe Institute of the Russian Academy of Sciences in St-Petersburg where he currently leads the Solid State Theory sector. He is a notorious amateur of pipe smoking.

exciton-polariton lasers (polariton lasers) are most likely to become the first practical bosonic lasers.

The seminal paper by Weisbuch et al. (1992) reporting the observation of a strong exciton-photon coupling in semiconductor microcavities opened to a wide scientific community the world of exciton-polaritons: mixed light-matter quasiparticles with very peculiar properties. More than a decade afterward, the first unambiguous experimental observation of the Bose–Einstein condensation (BEC) of exciton polaritons in microcavities was reported by Kasprzak et al. (2006). Earlier works by Le Si Dang et al. (1998) reported the stimulation of polariton photoluminescence and Deng et al. (2002) the onset of second order coherence, but the measurement of the key feature of spontaneous spatial coherence was reported later, by Kasprzak et al. (2006), using a CdTe-based microcavity. Soon after, two other groups, Balili et al. (2007) and Lai et al. (2007), reported the transient BEC of exciton-polaritons in GaAs-based microcavities. In a polariton laser, the microcavity is excited non-resonantly, either optically or electronically. A gas of electrons and holes created in the cavity forms excitons, which subsequently thermalise amongst themselves, mainly through exciton-exciton interactions.

Their kinetic energy can be lowered by interactions with phonons, as evidenced by Tassone et al. (1997) and Maragkou et al. (2010), or by excitons, as suggested by Porras et al. (2002), and they relax along the lower polariton dispersion branch, as shown in Fig. 12.1. They finally scatter to their lowest energy state, where they accumulate because of stimulated scattering. The coherence of the condensate therefore builds up from an incoherent equilibrium reservoir and a BEC phase transition takes place. The condensates of exciton-polaritons emit light that decays through the Bragg mirrors. This emission is spontaneous, however, the light going out has the properties of laser light: it is coherent, monochromatic, polarised and unidirectional. This spontaneous but coherent emission by a mixed light-matter condensate constitutes the polariton lasing effect.

12.1.1.1   *Coupled semi-classical Boltzmann equations for an exciton-polariton laser*
Kinetics of exciton-polaritons in a polariton laser can be described by the set of rate equations

$$\frac{dn_k}{dt} = P_k - \frac{n_k}{\tau_k} + \sum_{k' \neq k} \left(W_{k'k} n_{k'} \left(n_k + 1 \right) - W_{kk'} n_k \left(n_{k'} + 1 \right) \right). \qquad (12.1)$$

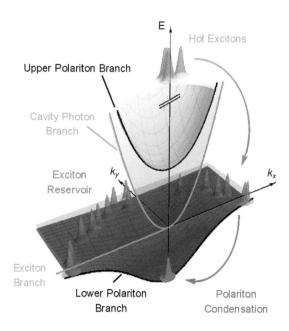

Fig. 12.1: Polariton dispersion and schematic of polariton condensation. Exciton and cavity photon modes couple to generate two new dispersion branches: the upper and lower polariton branches. Hot excitons injected at high energies relax their energies, through states with high in-plane wavevector, before undergoing stimulated scattering to the lowest energy state. The macroscopic buildup of polaritons in a low energy state forms a polariton condensate.

Fig. 12.2: Sketch of the electrically driven polariton laser, from Arash Rahimi-Iman. Electrically injected electrons and holes attract each other, which results in the formation of excitons: hydrogen like quasiparticles, which emit and reabsorb light inside the microcavity. This leads to the formation of exciton-polaritons, quasiparticles that spend a part of their time as photons and another part as excitons. The exciton-polaritons accumulate in a single quantum state called a condensate, which spontaneously emits light going away by decaying through the mirrors.

Here n_k is the concentration of exciton-polaritons with in-plane wavevector k, and P_k describes the pumping rate. τ_k is the polariton radiative lifetime, $W_{kk'}$ is the total scattering rate between quantum states indicated by k and k' with $k = (k_x, k_y)$, $k_{x,y} = \pm\frac{2\pi j}{L}$, $j = 0, 1, 2, \ldots$, $|k| < \frac{\omega}{c}$, where L is the lateral size of the system, ω is the frequency of the exciton resonance, and c is the speed of light. The condensate of polaritons in a state k_0 starts forming if the incoming flux of polaritons to this state exceeds the losses, i.e.,

$$\frac{dn_{k_0}}{dt} > 0.$$ (12.2)

This condition cannot be fulfilled permanently: the saturation of condensate formation takes place because of the depletion of the exciton reservoir that feeds polaritons to the system in the cw non-resonant pumping regime.

In a stationary regime, $dn_{k_0}/dt = 0$ and the lasing threshold is characterised by a superlinear increase of the population n_{k_0} as a function of pumping $P = \sum_k P_k$ and

$$\frac{\partial^2 n_{k_0}}{\partial P^2} > 0.$$

12.1.2 Realisation of polariton lasers in semiconductor microcavities

Polariton lasers with optical pumping have been realised in GaAs, CdTe, GaN and ZnO-based planar cavities by Christopoulos et al. (2007) and Daskalakis et al. (2013), in pillars by Bajoni et al. (2008b) and Zhang et al. (2014), and in photonic crystal by Azzini et al. (2011). They represent the first class of opto-electronic devices based on exciton-polaritons and possess unique characteristics including ultra-low threshold power, controllable polarisation of emission, and peculiar statistics of emitted photons. Room temperature operation has been demonstrated in GaN and ZnO-based polariton lasers by Christopoulos et al. (2007), Xie et al. (2012), Duan et al. (2013) and Li et al. (2013).

While a majority of research has been focused on inorganic semiconductor-based systems, significant advances have also been made in the development of organic-based semiconductor systems, where polariton BEC was reported by Kéna-Cohen and Forrest (2010) and Plumhof et al. (2014), and the spatial coherence of room temperature polariton lasers was demonstrated by Plumhof et al. (2014) and Daskalakis et al. (2014).

12.2 Polariton lasers with electrical injection

For more than a decade until 2013, only polariton lasers with optical pumping had been successfully implemented and demonstrated. However, a laser that needs optical pumping by another laser is inherently limited in its applicability as bulky optical setups are required to operate the device. Electrical injection and control of exciton-polaritons naturally is at the heart of any direct device application, and has motivated the community to work towards the realisation of such a device.

In order to describe the direct electrical pumping geometry, it is convenient to introduce the pumping term in eqn (12.1) as a function of the energy of the lower polariton state E_k, which reads

$$P_k = \begin{cases} 0 & \text{if } E_k - E_0 < \Delta, \\ W n_{eh}/\tilde{N} & \text{if } E_k - E_0 \geq \Delta, \end{cases} \qquad (12.3)$$

where Δ is a cut-off energy below which polariton states cannot be resonantly pumped by the electron-hole plasma, $E_0 + \Delta$ being the QW interband absorption edge. W is the exciton formation rate from the electron-hole plasma, $\tilde{N}$ is the number of states within the light cone, which satisfy the condition $E_k \geq E_0 + \Delta$, n_{eh} is the time-dependent density of the electron-hole plasma. It is given by

$$\frac{dn_{eh}}{dt} = \frac{J}{q} - \frac{n_{eh}}{\tau_{eh}} - W n_{eh}, \qquad (12.4)$$

where J is the electric pumping rate, q is the elementary charge, and τ_{eh} is the decay rate of the electron-hole plasma. Equation (12.4) can be solved analytically, which yields the simple dependence

$$n_{eh}(t) = \frac{J}{q} \frac{\tau_{eh}}{1 + W\tau_{eh}} \left[1 - \exp\left(-Wt - \frac{t}{\tau_{eh}} \right) \right], \qquad (12.5)$$

assuming an electron-hole plasma density equal to zero at $t = 0$. Equation (12.4) coupled to eqn (12.1) describes the action of an electrically pumped polariton laser. The threshold to lasing is achieved if

$$\frac{\partial^2 n_{k_0}}{\partial J^2} > 0.$$

12.2.1   *Experimental manifestations*

The first demonstration of polariton electroluminescence appeared in a resonant cavity LED based on organic polyelectrolyte/J aggregate dye-bilayers in 2005 by Tischler et al. (2005). This is rather surprising, given the exhaustive research activity in the III/V and II/VI inorganic systems preceding this important step. This organic polariton LED could be readily operated at room temperature, which paves the way for real life applications of such a device. This remarkable accomplishment was facilitated by the very large oscillator strength of the Frenkel-excitons in the organic compound, leading to a Rabi-splitting as large as 265 meV in the studied device.

The first demonstration of electroluminescence from polariton states in an inorganic microcavity was almost simultaneously reported by three groups: Khalifa et al. (2008), Tsintzos et al. (2008) and Bajoni et al. (2008a). All the samples contained GaInAs quantum wells as excitonic emitters, integrated in an (Al)GaAs microcavity, which was surrounded by doped (Al)GaAs/Al(Ga)As distributed Bragg reflectors. These devices strongly resemble conventional VCSEL lasers, and share many similarities with the samples previously exploited to observe polariton lasing under optical pumping.

While these experiments were all carried out in the linear regime, a few years later two research groups, Schneider et al. (2013) and Bhattacharya et al. (2013), reported the successful demonstration of polariton lasing with electrical injection in GaAs based vertically emitting microcavities (Fig. 12.2). Initially, the operation of these devices was restricted to cryogenic temperatures, and magnetic fields had to be applied to stabilise the excitons and optimise polariton scattering. This work has been extended one year later by Bhattacharya et al. (2014) to a room temperature prototype, which is based on an edge emitting GaN cavity.

These technological breakthroughs open the way to a new generation of coherent, compact and low energy consuming coherent light emitting devices, but also form a platform for developing a new class of opto-electronic devices based on Bose–Einstein condensates of mixed light-matter quasi-particles. This emerging field of polaritonics will strongly benefit from the capability to switch and tune the energy of (electrically and optically injected) polariton condensates via applied electric fields, such as demonstrated by Amthor et al. (2014), Tsotsis et al. (2014) or Brodbeck et al. (2015), which puts a whole zoo of new polariton devices within reach.

12.2.2   *Weak lasing*

Due to their finite lifetime, exciton-polaritons exhibit condensation and lasing in those quantum states where their influx exceeds the dissipative losses. This defines the threshold for "bosonic lasing", in contrast to the threshold of "fermionic lasing" governed by the condition of inversion of electronic population. In general, different one-particle states decay differently, so that the state with the longest lifetime has better chances to host a polariton lasing mode than short-living states. A supplementary complication comes from interactions between bosons that mix different single-particle quantum states. The weak-lasing regime, recently described by Aleiner et al. (2012), is formed by the spontaneously symmetry breaking and phase-locking self-organisation of bosonic modes, which results in an essentially many-body state with a stable balance between gains and losses. The physics of weak lasing may be understood considering the problem of phase locking in a system of two localised condensates of exciton polaritons, as shown in Fig. 12.3. Let us assume that due to some stationary disorder potential, two localised condensates of exciton-polaritons are formed close to each other. They are separated by a shallow potential barrier that allows for tunnelling. The tunnelling of polaritons leads to the coherent coupling (Josephson coupling) of two condensates. The Josephson coupling forces two condensates to adapt the same phase. In this case the energy of the system is minimised. One can easily understand the physics of Josephson phase-locking considering a single quantum particle in a double quantum well structure. Its lowest energy state is characterised by a real wavefunction having no zeros, that corresponds to the same phase in both quantum wells.

Now, what happens if we take into account the dissipation? Polaritons in both condensates have finite lifetimes. The polariton lasing-threshold is the lowest for the phase relation between the two condensates that maximises their lifetime. If two condensates are close to each other, the lifetime is maximised if they emit light with opposite phases. In this case, the negative interference of electromagnetic fields emitted by the two condensates minimises the radiative loss. This is the dissipative coupling effect known since the famous work of Huygens (1665). Huygens was the first to notice that two pendulum clocks mounted on a common beam antisynchronise after a while. Bennett et al. (2002) have explained this effect by a dissipative coupling mechanism. If the two pendulums

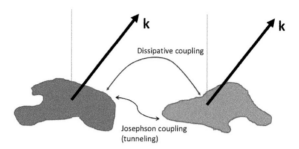

Fig. 12.3: Spontaneous symmetry breaking in the weak lasing regime. The interplay between coherent (Josephson) and dissipative coupling of two polariton condensates results in the specific phase locking that corresponds to a nonzero value the mean wavevector. As a result, two condensates emit light at an oblique angle.

swing in phase, they tend to push the beam in the same direction resulting in frictional forces that eventually dampen the motion of the pendulums. If, however, they swing in anti-phase, the back actions cancel out and the beam does not move. After a while, the two pendulums swing in the state that has the smallest dissipation.

One can see that the coherent (Josephson) and the dissipative coupling between the two condensates have opposite effects on their relative phase: while the coherent coupling imposes the same phase, the dissipative coupling favors an anti-phase configuration. In a specific range of pump intensities, the interplay between coherent and dissipative coupling makes favourable a state where the phases of two condensates are not equal, but not opposite either. Such a state would break a mirror reflection symmetry with respect to the plane separating two condensates. The k-space image of the system becomes strongly asymmetric in this case.

Experimentally, the asymmetric pattern of emission of light by an array of localised condensates has been observed by Krizhanovskii et al. (2009). Another manifestation of weak lasing has been reported by Zhang et al. (2015) who studied room-temperature polariton lasing in a ZnO microwire deposited on a silicon grating that produced a 1D periodical potential for exciton polaritons. Real and reciprocal space photoluminescence images demonstrate that in a certain pump power range, the spatial period of the condensate is twice as large as the period of the underlying periodic potential. This is a consequence of the mixing of polariton modes corresponding to the bottom and to the edge of the 1D Brillouin zone due to the dissipative coupling in the nonlinear regime. It is argued that weak lasing may also lead to spontaneous formation of persistent polariton fluids.

12.3   Polariton terahertz lasers

12.3.1   *Variety of proposals*

The realisation of efficient terahertz (THz) radiation sources and detectors is an important objective of modern applied physics. THz emitters and detectors are promising for applications in biology, medicine, security and non-destructive in-depth imaging. Dragoman and Dragoman (2004) describe several other generic applications. Wireless data transfer using THz radiation is likely to provide a higher transfer rate for indoor short distance or high altitude communication (propagation of THz radiation in the atmosphere is limited by water vapour absorption). Federici and Moeller (2010) provide a review of terahertz and subterahertz wireless communications. For these multiple applications, the creation of cheap, reliable, scalable and portable THz radiation sources and detectors is extremely important. None of the existing THz emitters universally satisfies the application requirements. For example, emitters based on nonlinear-optical frequency down-conversion, gas THz lasers, vacuum tubes and systems based on short-pulse spectroscopy are bulky, expensive and power consuming. Various semiconductor devices based on intersubband optical transitions are compact, but have a limited wavelength tunability range, low quantum efficiency (except for quantum cascade lasers), and require cryogenic cooling. Among the factors that limit the efficiency of semiconductor THz sources is the short lifetime of the electronic states involved (typically, fractions of a nanosecond) compared to the time for spontaneous emission of a THz photon (typically a few milliseconds). The methods of reducing this mismatch include the use of the Purcell effect in THz cavities, implemented by Walther et al. (2010) and/or the cascade effect in quantum cascade lasers (QCL), by Faist et al. (1994). Nevertheless, until now QCLs in the spectral region about 1 THz remain costly and short-lived and still show quantum efficiencies of less than 1%. Moreover, so far, there are no commercially available, reliable, compact and cheap detectors of THz radiation, which are in a great need for Information and Communication Technology (ICT) applications.

Recently, novel approaches to THz generation have been formulated by Kavokin et al. (2010), Savenko et al. (2011), Kavokin et al. (2012), Liberato et al. (2013) and Huppert et al. (2014), solving the problem of lifetime mismatch by exploiting polariton-polariton scattering, which allow to increase the emission rate through stimulation of the THz emission by the final state polaritonic population. The idea behind polariton terahertz lasing is the bosonic amplification of the emission rate of terahertz radiation by the final state population, with the final state being a Bose–Einstein condensate of exciton-polaritons. Some of the concepts on which these proposals are based are being tested experimentally at the time of writing. Proposals published so far differ in one important aspect. For instance, there is the choice of the initial state for the optical transition that brings new quasiparticles to a polariton condensate. In the work of Kavokin et al. (2010) and Huppert et al. (2014), this is the upper polariton branch, in Liberato et al. (2013)'s, the initial state is a plasma state, while Kavokin et al. (2012) propose $2p$-exciton states that may be pumped by a two-photon absorption process.

12.3.2 *Polariton terahertz lasers with two-photon excitation*

As direct optical creation of the $2p$ exciton under single photon absorption is prohibited, it was proposed to use two-photon pumping of a $2p$ exciton state, as has been realised already in GaAs based quantum well structures by Catalano et al. (1989) and Kaindl et al. (2009). After creation, a $2p$ exciton can radiatively decay to the $1s$ exciton state emitting a THz photon. The inverse process (THz absorption by a lower polariton mode with excitation of a $2p$ exciton) has been recently observed experimentally by Tomaino et al. (2012). The THz transition from the $2p$ state pumps the lowest energy exciton

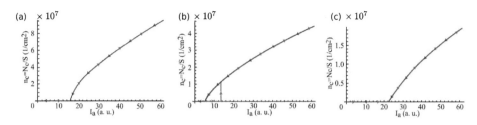

Fig. 12.4: The dependence of the concentration of THz photons on the pumping intensity for the case where the decay rate of the THz resonator $\gamma_c = 100\,\mathrm{eV}$. The branches of hysteresis corresponding to increasing and decreasing pumping intensities are indicated by arrows. Adapted from Pervishko et al. (2014).

state, which can have macroscopic occupation and thus stimulate the THz emission. The situation can be further improved by replacing of the 1s exciton state by an exciton-polariton. In this case, THz emission eventually leads to polariton lasing.

As an example of a THz laser based on the $2p$ to $1s$ optical transition let us consider a quantum well placed inside a microcavity tuned close to the resonant frequency of two-photon absorption $\hbar\omega_{2p}$. The device operates in a polariton lasing regime, so that a condensate of exciton-polaritons is formed in the $1s$ state at the energy $\hbar\omega_s$. As $2p$ and $1s$ exciton states have different parities, the radiative transition from the $2p$ exciton level to the $1s$ exciton level is possible and is accompanied by the emission of a THz photon. The terahertz transition from $2p$ to $1s$ state is stimulated by population of the $1s$ state, which represents the bosonic laser action. The structure is then placed in a THz cavity of larger size, which further increases the efficiency of THz emission by confining THz photons and thus leading to the double stimulation at the terahertz mode frequency $\hbar\omega_{THz}$. The system is excited resonantly by an external continuous wave laser beam of frequency ω_c close to $\omega_{2p}/2$.

The model Hamiltonian of this system in the rotating wave approximation reads

$$\hat{\mathcal{H}} = \hbar\omega_p \hat{p}^+\hat{p} + \hbar\omega_a \hat{a}^+\hat{a} + \hbar\omega_s \hat{s}^+\hat{s} + \hbar\omega_{THz}\hat{c}^+\hat{c} + \\ + g(\hat{p}\hat{a}^+\hat{a}^+ + \hat{p}^+\hat{a}\hat{a}) + G\hat{p}^+\hat{s}\hat{c} + G^*\hat{p}\hat{s}^+\hat{c}^+ , \quad (12.6)$$

where $\hat{p}$, $\hat{a}$, $\hat{s}$ and $\hat{c}$ are annihilation operators of $2p$ exciton, cavity photon, $1s$ exciton and terahertz photon, respectively, which satisfy bosonic commutation relations. The first four terms describe the energies of the free modes. The fifth term corresponds to the interaction between $2p$ excitons and photons. The last two terms describe THz emission associated with the $2p$-$1s$ transition. g and G are the coupling constants describing the two-photon excitation and the THz transition, respectively. The constant of interaction between $2p$ and $1s$ exciton levels with THz lasing can be found from the matrix element of this transition in the dipole approximation (cf. Scully and Zubairy (2002)), using the wavefunctions of the 2D hydrogen atom of Yang et al. (1991):

$$G^* = \frac{1}{\sqrt{2}}\left(\frac{eA_0}{\mu}\right)\frac{im_0}{\hbar}\langle 1s|r|2p\rangle , \\ = \frac{A_0}{\sqrt{2}\mu}\frac{27im_0\pi\epsilon\epsilon_0\hbar}{8\sqrt{6}\mu e} . \quad (12.7)$$

Here A_0 is the vector potential of the THz photon mode, μ is the magnetic susceptibility e is the electron charge and m_0 is the free electron mass.

12.3.2.1 *Classical field approximation* The dynamics of the system can be found by writing the Heisenberg equations of motion for the Hamiltonian in eqn (12.6):

$$i\hbar\frac{d\hat{p}}{dt} = \left[\hat{p}, \hat{\mathcal{H}}\right] = \hbar\omega_p\hat{p} + g\hat{a}^2 + G\hat{s}\hat{c}, \tag{12.8a}$$

$$i\hbar\frac{d\hat{a}}{dt} = \left[\hat{a}, \hat{\mathcal{H}}\right] = \hbar\omega_a\hat{a} + 2g\hat{p}\hat{a}^+, \tag{12.8b}$$

$$i\hbar\frac{d\hat{s}}{dt} = \left[\hat{s}, \hat{\mathcal{H}}\right] = \hbar\omega_s\hat{s} + G^*\hat{p}\hat{c}^+, \tag{12.8c}$$

$$i\hbar\frac{d\hat{c}}{dt} = \left[\hat{c}, \hat{\mathcal{H}}\right] = \hbar\omega_{THz}\hat{c} + G^*\hat{p}\hat{s}^+. \tag{12.8d}$$

In the mean-field approximation one replaces second-quantisation operators by their expectation values. Terms to describe the finite lifetime of the excitons and photons and external coherent pumping can be added to the equations phenomenologically. As a result, the system of nonlinear differential equations of motion takes the form

$$i\hbar\frac{dp}{dt} = \hbar\omega_p p + ga^2 + Gsc - i\gamma_p p, \tag{12.9a}$$

$$i\hbar\frac{da}{dt} = \hbar\omega_a a + 2gpa^+ + Pe^{-i\omega t} - i\gamma_a a, \tag{12.9b}$$

$$i\hbar\frac{ds}{dt} = \hbar\omega_s s + G^*pc^+ - i\gamma_s s, \tag{12.9c}$$

$$i\hbar\frac{dc}{dt} = \hbar\omega_{THz}c + G^*ps^+ - i\gamma_c c. \tag{12.9d}$$

Here $p = \langle\hat{p}\rangle$, $a = \langle\hat{a}\rangle$, $s = \langle\hat{s}\rangle$, and $c = \langle\hat{c}\rangle$ are the mean values of the annihilation operators of $2p$ dark excitons, cavity photons, $1s$ bright excitons and terahertz photons, respectively. γ_p, γ_a, γ_s and γ_c are the decay rates for $2p$ excitons, cavity photons, $1s$ excitons and terahertz photons, respectively. The corresponding lifetimes of the modes are $\tau_x = 1/\gamma_x$ for $x = p, a, s, c$. P is the amplitude of the external coherent pump of the cavity photons with frequency ω.

Figure 12.4 shows the population of the THz mode as a function of the optical pumping intensity calculated for different decay rates γ_c in the THz cavity. One can observe the hysteresis curve that appears at the intermediate decay and characterises the bistable regime of THz lasing. This regime is especially important for realisation of low-power terahertz switches.

12.3.3 *Superradiant emission of terahertz radiation by dipolaritons*

The work of Kyriienko et al. (2013) stands apart from the main stream of theoretical proposals of polariton terahertz lasing. It argues that terahertz radiation may be produced due to the classical Hertz dipole oscillations in the superradiant regime. This regime may be realised in biased microcavities containing double quantum wells, where

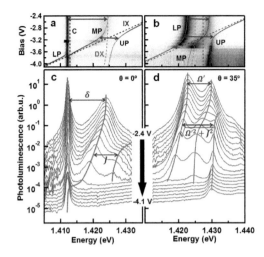

Fig. 12.5: Angle-resolved photoluminescence spectra of a biased semiconductor microcavity containing asymmetric double quantum wells, from Cristofolini et al. (2012). One can see the double anticrossing feature that evidences strong coupling of direct (DX) and indirect (IX) excitons with a cavity mode (C). One of the resulting exciton-polariton states has a large static dipole moment due to the strong contribution of IX, while other states are dominated by DX and C.

Rabi oscillations between a direct exciton-polariton state having zero stationary dipole moment and a spatially indirect exciton state having a large dipole moment, so called "dipolaritons", introduced by Cristofolini et al. (2012), may be induced by a short pulse of light (see Fig. 12.5). The correct treatment of a real dipolariton system involves both coherent and decoherent parts for the Hamiltonian, $\hat{\mathcal{H}} = \hat{\mathcal{H}}_{coh} + \hat{\mathcal{H}}_{dec}$. The Hamiltonian corresponding to the coherent part reads

$$\hat{\mathcal{H}}_{coh} = \hbar\omega_C \hat{a}^\dagger \hat{a} + \hbar\omega_{DX} \hat{b}^\dagger \hat{b} + \hbar\omega_{IX} \hat{c}^\dagger \hat{c} + \frac{\hbar\Omega}{2}(\hat{a}^\dagger \hat{b} + \hat{b}^\dagger \hat{a})$$
$$- \frac{\hbar J}{2}(\hat{b}^\dagger \hat{c} + \hat{c}^\dagger \hat{b}) + P\hat{a}^\dagger + P^*\hat{a},$$

where $\hat{a}^\dagger$, $\hat{a}$, $\hat{b}^\dagger$, $\hat{b}$ and $\hat{c}^\dagger$, $\hat{c}$ are creation and annihilation operators for cavity photons, direct excitons and indirect excitons, respectively. Here $\hbar\omega_C$, $\hbar\omega_{DX}$ and $\hbar\omega_{IX}$ denote cavity mode, direct exciton and indirect exciton energies, and first three terms of Hamiltonian describe energy of bare modes. The fourth and fifth terms describe coupling between modes, where Ω denotes the coupling constant between photon and direct exciton, and the tunnelling rate corresponding to DX–IX coupling is J. The last two terms in Hamiltonian describe the optical pumping of cavity mode with rate $P(t) = P_0(t)e^{i\omega_P t}$, where ω_P is a pumping frequency.

The incoherent part of the dynamics of the system is mainly governed by the radiative lifetimes of the modes and phonon-scattering processes, which lead to population of thermalised exciton reservoirs. They can be described by the exciton-phonon interaction Hamiltonian $\hat{\mathcal{H}}_{int} = \hat{\mathcal{H}}^+ + \hat{\mathcal{H}}^-$, where

$$\hat{\mathcal{H}}^+ = D_{DX} \sum_k \hat{b}^\dagger \hat{b}_{R,k} \hat{d}_k^\dagger + D_{IX} \sum_k \hat{c}^\dagger \hat{c}_{R,k} \hat{d}_k^\dagger, \qquad (12.10a)$$

$$\hat{\mathcal{H}}^- = D_{DX} \sum_k \hat{b}_{R,k}^\dagger \hat{b}\hat{d}_k + D_{IX} \sum_k \hat{c}_{R,k}^\dagger \hat{c}\hat{d}_k, \qquad (12.10b)$$

correspond to processes with emission ($\hat{d}_k^\dagger$) and absorption ($\hat{d}_k$) of phonon with in-plane momentum k, respectively. Here $\hat{b}_{R,k}^\dagger$, $\hat{b}_{R,k}$, $\hat{c}_{R,k}^\dagger$ and $\hat{c}_{R,k}$ are creation and annihilation operators for direct and indirect excitons in reservoirs. $D_{DX,IX}$ denote the exciton-photon interaction constants. These incoherent processes can be conveniently treated using a master equation for the density matrix

$$i\hbar \partial \rho / \partial t = [\hat{\mathcal{H}}_{coh}, \rho] + \hat{\mathcal{L}}\rho^{(dis)} + \hat{\mathcal{L}}\rho^{(th)}, \qquad (12.11)$$

where $\hat{\mathcal{L}}\rho^{(dis)}$ are Lindblad superoperator describing dissipation while $\hat{\mathcal{L}}\rho^{(th)}$ is responsible for phonon-assisted processes.

Final-state bosonic amplification or polariton superradiance effects help to accelerate the rate of emission of radiation in the terahertz frequency range by several orders of magnitude, which is crucial for the realisation of a terahertz semiconductor laser. Kyriienko et al. (2013) have shown that in a microcavity excited with a short pulse of light, the densities of IX and DX oscillate with a THz frequency. This infers that a dipolariton is an oscillating dipole, with a dipole moment in z direction d_z changing periodically from value $d_z = d_0$ of IX to $d_z = 0$ corresponding to DX. One can define the total dipole moment of the system as $D_z = N_{IX}d_z$, where $N_{IX} = n_{IX}A$ is the number of indirect excitons within the area A illuminated by the pumping light. Since after the initial transient regime the density of indirect excitons $n_{IX}(t)$ is a decaying harmonic function of time, the total dipole moment of the system can be found as $D_z(t) = d_0 n_{IX}^0 A \cos^2(\omega t/2)e^{-t/\tau}$, where n_{IX}^0 is the maximum density of indirect excitons.

The mechanism of THz radiation by an oscillating density of indirect excitons is similar to the photo-Dember effect where oscillations of electron-hole plasma on a surface of semiconductor generate THz signals, as in the case of Shan and Heinz (2004) and Johnston et al. (2002) and other lasing driven terahertz emitters. However, the system of dipolaritons has some important advantages over previously studied systems, namely:

1. The better tunability of the system allowing for fast modulation of the THz emission frequency.

2. Improved spectral characteristics of THz signal which can be controlled using applied voltage and pumping conditions.

3. Possibility to achieve a high output power as described below.

The total intensity of the far field radiation emitted by a classical Hertzian dipole can be found (for instance in Landau and Lifshitz (1980)) as $I = \ddot{D}_z^2 / 6\pi\epsilon_0 c^3$, where D_z is the total dipole moment of the dipolariton array, ϵ_0 is vacuum permittivity and c is the speed of light. For the particular case of an array of harmonic dipole oscillations the intensity is $I = N_{IX}^2 d_0^2 \omega^4 / 3\pi\epsilon_0 c^3$, where $d_0 = eL$ is the dipole moment of the indirect exciton. To achieve stable cw radiation one needs using a sequence of pump pulses. Similarly to the conventional case of an elementary dipole emitter, the polar pattern is given by $I_\theta \sim \sin^2 \theta$, where θ is an angle between direction of radiation and the axis of the structure.

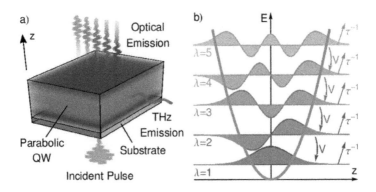

Fig. 12.6: Illustration of a bosonic cascade, from Liew et al. (2016). a) Structure schematic: a parabolic quantum well is excited with an optical pulse and emits light at a range of frequencies. b) Energy level diagram: the parabolic trapping potential engineers equidistant energy levels and transitions occur between neighbouring levels with transition element V. Bosons in each level can decay radiatively, with characteristic decay time τ.

It is important to note that the total emitted intensity is proportional to the square of the density of indirect excitons and is dependent on the pump intensity in a nonlinear way. This is a manifestation of the *superradiance* introduced by Dicke (1954) and put to dramatic effects by Bohnet et al. (2012) and Yukalov and Yukalova (2010): due to the interference of coherent in-phase oscillations of elementary dipoles the output power is enhanced. This effect is sensitive to the quality factor of the cavity: the longer the Rabi oscillations persist, the stronger the amplification effect is. Superradiance is a specific feature of the dipolariton THz emitter, which makes it more efficient than any existing laser-to-THz converter. The efficiency of emission would be further increased by the Purcell factor of the external cavity.

12.4 Bosonic cascade lasers

12.4.1 *The Boltzmann dynamics of bosonic cascades*

Bosonic cascade lasers are promising systems for efficient terahertz generation and also represent a significant fundamental interest. In bosonic cascade lasers (BCLs), excitons or exciton-polaritons are resonantly injected into one of the upper quantum confined states in a parabolic potential trap, whose frequency lies in the THz region (see Fig. 12.6). Examples of potential experimental realisations of bosonic cascades are shown in Fig. 12.7. In the original scheme for bosonic cascades, as proposed by Liew et al. (2013), excitons, being neutral particles, cannot directly emit THz photons by falling from one level to the next. Still the harmonic confining potential, being different for electrons and holes, couples the internal and center-of-mass degrees of freedom of the exciton, allowing for the radiative cascade depicted in Fig. 12.6, in which a THz photon is emitted each time an exciton falls down a step of the harmonic ladder. This relaxation process would be stimulated by the occupation numbers of the quantum states that are subsequently visited by relaxing excitons. The estimations from Liew et al.

(2013) show that one can realistically expect $\approx 7\,\mathrm{THz}$ photons to be emitted per exciton (or exciton-polariton) injected into the system, which results in a quantum efficiency of several hundred per cent. BCLs would be a valuable alternative to fermionic QCLs as, in principle, they do not require a THz cavity, may be assembled in matrices, and are expected to be easily tunable in a wide spectral range by external electric/magnetic field, as shown by Christmann et al. (2010), Zhang et al. (2010) and Amthor et al. (2014) or by tuning the depth of the optical parabolic trap. Terahertz cavities, would, however, further increase efficiency by double bosonic stimulation, as discussed by Kaliteevski et al. (2014). Furthermore, condensates at certain steps of the cascade develop peculiar superbunching of highly populated states, as analysed by Liew et al. (2016) and discussed below, which could have further applications such as for imaging and quantum lithography.

The occupation numbers of exciton quantum confined states and the THz optical mode in a Bosonic Cascade Laser can be found from the following set of kinetic equations (0 is the state with the lowest energy, m is the state with the highest energy, which is resonantly pumped):

$$\frac{dN_m}{dt} = P - \frac{N_m}{\tau} - WN_m(N_{m-1}+1) + W'N_{m-1}(N_m+1)\,, \tag{12.12a}$$

$$\frac{dN_k}{dt} = -\frac{N_k}{\tau} + W\left[N_{k+1}(N_k+1) - N_k(N_{k-1}+1)\right] \tag{12.12b}$$
$$+ W'\left[N_{k-1}(N_k+1) - N_k(N_{k+1}+1)\right]\,, \qquad \text{for } 1 \le k \le m-1,$$

$$\frac{dN_0}{dt} = -\frac{N_0}{\tau} + WN_1(N_0+1) - W'N_0(N_1+1)\,, \tag{12.12c}$$

$$\frac{dn_{\mathrm{THz}}}{dt} = -\frac{n_{\mathrm{THz}}}{\tau_{\mathrm{THz}}} + W\sum_{1}^{m} N_k\,(N_{k-1}+1) - W'\sum_{1}^{m} N_{k-1}\,(N_k+1)\,. \tag{12.12d}$$

Here $W = W_0(n_{\mathrm{THz}}+1)$ is the THz emission and $W' = W_0 n_{\mathrm{THz}}$ is the THz absorption rate, n_{THz} is the THz mode occupation, and τ_{THz} is the THz mode lifetime. The lifetime of the cascade levels, τ, should include both their radiative lifetime and non-radiative lifetime that includes losses due to phonon scattering to states with nonzero in-plane wavevector, $k_\parallel \ne 0$. Above a threshold pump power it is not necessary to calculate explicitly the dynamics of these states. We note that any phonon or exciton-exciton scattering to $k_\parallel \ne 0$ states would not be stimulated and any population in a particular sub-band can be expected to return to the $k_\parallel = 0$ state of the same sub-band by stimulated scattering. Phonon-assisted relaxation between subbands is also expected to have a limited rate, according to Golub et al. (1995). We assume that the matrix element of THz transition is nonzero only for neighbouring stairs of the cascade and that it is the same for all neighbouring pairs. This simplifying assumption allows to solve eqns (12.12) exactly.

We first consider the case where there is no THz cavity, and assume that THz photons leave the system immediately such that $n_{\mathrm{THz}} = 0$. The solid curves in Fig. 12.8 show the dependence of the mode occupations on the pump power in this case, which

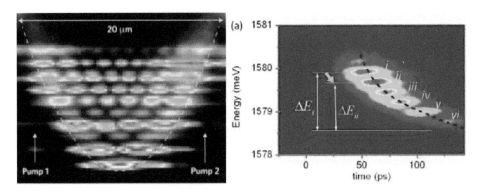

Fig. 12.7: Examples of Bosonic cascades. Left: Equidistant polariton condensates in a parabolic trap observed by near-field photoluminescence in a planar GaAs-based microcavity. The image is adapted from Tosi et al. (2012). Right: Time resolved relaxation dynamics of a polariton condensate in a cascade of quantum confined states of a one-dimensional trap realised in a polariton microwire (a long and narrow stripe etched from a GaAs based planar microcavity). The image is adapted from Wertz et al. (2012).

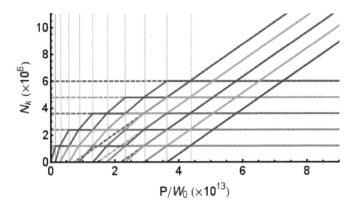

Fig. 12.8: Dependence of the mode occupations in the absence of a THz cavity on pump intensity, calculated numerically by Liew et al. (2013) from the kinetic eqns (12.12a)–(12.12c) (solid curves) and analytically (dashed curves). The vertical gray lines correspond to the step locations given by $P/W_0 = n^2/(W_0\tau)^2$ where n is a half-integer. Parameters: $W_0\tau = 3 \times 10^{-6}$, $m = 15$, $n_{\text{THz}} = 0$.

were calculated by numerical solution of eqns (12.12) for the steady state. For increasing pump power we see that the modes become occupied one-by-one and a series of steps appears, each corresponding to the occupation of an additional mode. In the limit, $W_0\tau \ll 1$, the position of the steps is given by $P/W_0 = n^2/(W_0\tau)^2$ where n is a half-integer. For high pump powers, where all modes are occupied two different behaviours of the modes can be identified: 0th, 2nd, 4th, etc., modes continue to increase their occupation with increasing pump power, while 1st, 3rd, 5th, etc., modes have a limited occupation. This effect persists independently of whether an even or odd number of modes is considered in the system.

Qualitatively, these results can be understood as follows. Every mode in the chain experiences both a gain and a loss. The last mode in the chain is unique since it only experiences loss due to the finite lifetime rather than THz emission. Since it experiences loss only due to the lifetime, we can expect that the last mode is strongly occupied in the limit of high pump power. This means that the second-to-last mode experiences a strong loss due to stimulated scattering to a highly occupied state. Thus the second-to-last mode has a much smaller occupation. The third-to-last mode then experiences only a small loss due to stimulated scattering and so can again have a large occupation. The series repeats such that alternate modes have high and low intensity, with the highly occupied modes introducing a fast loss rate that limits the occupation of low intensity modes.

12.4.2 *Quantum model of a bosonic cascade laser*

In order to analyse the quantum statistics of the ladder of polariton condensates forming the Bosonic Cascade Laser, one can describe the bosonic cascade of N states with a set of master Boltzmann equations for the probabilities $P(n, j)$ to find n excitons $(n = 0, 1, 2, \dots)$ at the level j, where $j = 1, 2, \dots, N$:

$$\frac{dP(n, j)}{dt} = V \sum_{k=0}^{\infty} P(n-1, j) P(k, j+1) nk$$

$$-V \sum_{k=0}^{\infty} P(n, j) P(k, j-1) n(k+1)$$

$$+V \sum_{k=0}^{\infty} P(n+1, j) P(k, j-1)(n+1)(k+1)$$

$$-V \sum_{k=0}^{\infty} P(n, j) P(k, j+1)(n+1)k$$

$$-\frac{n}{\tau} P(n, j) + \frac{n+1}{\tau} P(n+1, j) . \tag{12.13}$$

Here the terms containing $P(k, j)$ with $j < 1$ or $j > N$ should be taken equal to zero. Let us assume that the upper level $(j = N)$ is resonantly excited at $t = 0$. Various pumping conditions correspond to different initial conditions, namely for particular cases of interest, we set $P(n, N) = e^{-\bar{n}} \bar{n}^n / n!$ at the time $t = 0$ for the coherent excitation $P(n, N) = \bar{n}^n / (\bar{n} + 1)^{n+1}$ for incoherent pumping and, for a more academic case, $P(n, N) = \delta_{n\bar{n}}$ for a well-defined polariton number (Fock state). Here $\bar{n}$ is the average number of excitons in the initial state. For all other energy levels $(j < N)$ we set $P(n, j) = 0$ as an initial condition. The quantities of interest are: the time dependent average number of excitons $\langle n_j \rangle = \sum_{n=0}^{\infty} nP(n, j)$ and the average squared number of excitons $\langle n_j^2 \rangle = \sum_{n=0}^{\infty} n^2 P(n, j)$. From these two values we easily find the second order coherence for each energy level: $g_2^j(0) = (\langle n_j^2 \rangle - \langle n_j \rangle)/\langle n_j \rangle^2$. Figure 12.9 shows the time dependence of $g_2^j(0)$ calculated for different polariton condensates forming a bosonic cascade localised in a parabolic trap. It is assumed that an infinitely short

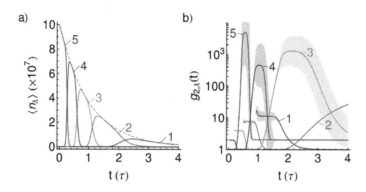

Fig. 12.9: Evolution of five bosonic levels following a coherent pulsed excitation. a) Average level occupations $\langle n \rangle$. The dashed curve shows the exponential decay of the total population. b) Second-order correlation function $g^{(2)}(t)$.

pulse of coherent light excites the Nth condensate at zero time. The mean polariton populations of different energy levels are shown for comparison. One can see that levels are filled and emptied successively, from top to bottom. Emptying of a level j is accompanied by a strong increase of $g_2^j(0)$ that greatly exceeds 2 in many cases. This demonstrates formation of super-bunched polariton states in the bosonic cascades.

12.5 Spatial dynamics of polariton lasing structures

A key characteristic of polariton condensates and lasers is their spatial coherence, of which Love et al. (2008) provided an in-depth study. Belykh et al. (2013) later showed how it develops spontaneously and spreads both in time and spatially across the microcavity plane. The coherence also spreads and synchronises across regions in the presence of disorder, as reported by Baas et al. (2008). The resulting coherent state of polaritons is well described by the Gross–Pitaevskii equation (GPE), amply described in the previous chapters. For polaritons, we remind that this equation is modified to account for gain and loss in the system, following for instance Wouters and Carusotto (2007)[131]. Such equations support a variety of spatially non-trivial structures, including vortices and vortex lattices as revealed by theoretical studies from Keeling and Berloff (2008) and Rodrigues et al. (2014), solitons, by Ostrovskaya et al. (2012), Smirnov et al. (2014), Xue and Matuszewski (2014), Pinsker and Flayac (2014) and Karpov et al. (2015), and various other patterns, by Borgh et al. (2010), Ge et al. (2013) or Liew et al. (2015). For more details on the specific solutions of Gross–Pitaevskii equations for exciton-polariton condensates, we refer to Chapter 10.

12.5.1 Pattern formation

Experimentally, one of the first examples of non-trivial spatial structures of polariton BECs was the observation of quantised vortices by Lagoudakis et al. (2008) and

---

[131] Similar equations were considered earlier in purely excitonic systems with uncoupled quantum wells by Gergel' et al. (1968b).

Lagoudakis et al. (2009). In principle, vortices can form out of phase fluctuations during the formation of the polariton condensate, in a Kibble-Zürek-like mechanism (after and Kibble (1976) and Zurek (1985))[132]. In practice, the observation of vortices at well defined positions, after averaging over multiple laser shots, implies that disorder in the system also plays an important role in pinning vortex paths, as studied by Lagoudakis et al. (2011) and Manni et al. (2012) (see also Fig. 12.10). In addition, polariton condensates being two-dimensional can also exhibit a Berezinskii-Kosterlitz-Thouless (BKT) transition, characterised by the binding of vortex-antivortex pairs. Indeed this effect is challenging to distinguish as disorder tends to make microcavities spatially inhomogeneous. Still experiments using spatially inhomogeneous pumping have successfully demonstrated the formation of vortex-antivortex pairs, studied by Roumpos et al. (2010).

In addition to the potential to form vortex lattices, such as introduced by Keeling and Berloff (2008), polariton condensates and lasers can form a variety of other spatial patterns. Expanding condensates were shown by Christmann et al. (2012) to lead to "sunflower ripples". Ring-shaped optical excitation has also been shown to allow the spontaneous formation of patterns Manni et al. (2011), which can be interpreted as a modulational instability. An example simulation of such a situation from the driven-dissipative Gross-Pitaevskii equation is shown in Fig. 12.11.

12.5.2 *Control of lasing modes in structured potentials*

Spatially-structured excitation of planar microcavities has been considered by several groups, allowing the trapping of polariton condensates in optically controlled potentials, e.g., by Askitopoulos et al. (2013) and Cristofolini et al. (2013). Chiral trapping potentials can also be engineered to generate vortices, as shown by Dall et al. (2014). Experiments with Mexican hat shaped profiles from Manni et al. (2013) and Cristofolini et al. (2013) also showed how it was possible to trap vortex-antivortex pairs. The optical trapping of polaritons can also be achieved based on the balance between optically controlled gain and loss, a mechanism considered by Roumpos et al. (2011) that Ostrovskaya et al. (2012) cast in the framework of dissipative solitons. Aside from optically induced potentials, polaritons can also be confined by growing structures of reduced dimensionality, such as microwires, as was done by Wertz et al. (2010), or micropillars, by Gutbrod et al. (1998), Bajoni et al. (2008c) or Fischer et al. (2014). This can lead to the condensation of polaritons in artificial photonic molecules, as shown by Galbiati et al. (2012) and discussed in greater details in Chapter 10. Several works have also focused on the formation of polariton condensates in periodic potentials including those from Lai et al. (2007), Kim et al. (2011), Kim et al. (2013), Jacqmin et al. (2014) or Winkler et al. (2015). Here interesting effects include: i) the condensation into solitonic gap states, reported by Tanese et al. (2013), both in shallow and deep lattices, see Winkler et al. (2016), ii) weak lasing reported by Zhang et al. (2015) and iii) stochastic

---

[132]Being a non-equilibrium system, exciton-polaritons do not strictly follow the standard Kibble-Zürek mechanism, but nevertheless there are similarities in the spreading of coherence throughout the system and the existence of non-equilibrium universal scaling laws was predicted theoretically Matuszewski and Witkowska (2014).

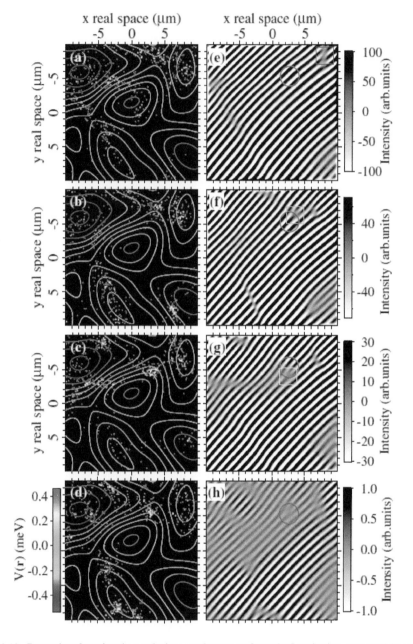

Fig. 12.10: **Dynamics of vortices in a polariton condensate under pulsed excitation,** simulated from the driven-dissipative Gross-Pitaevskii equation. (a-d) Shot-integrated locations of vortices at subsequent time frames (40, 50, 60 and 90 ps) and contours of a disorder potential. The vortices can be seen to preferentially form near local minima of the potential. (e-h) Corresponding shot-integrated interference patterns, which are comparable to experimental measurements. Figure reproduced from Lagoudakis et al. (2011).

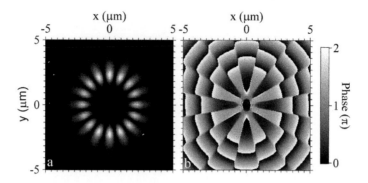

Fig. 12.11: **Polariton stationary state under continuous-wave ring-shaped excitation**, simulated from the driven-dissipative Gross-Pitaevskii equation. Left: The polariton density breaks up into a series of lobes. Right: Corresponding polariton phase distribution in space. The figure has been adapted from Manni et al. (2012).

condensation at different symmetry points in the reciprocal lattice, reported by Kusudo et al. (2013), which could be relevant for an exciton-polariton analogue of valleytronics.

12.5.3   *Bistability and polariton condensate memories*

Resonantly excited microcavities, which rely on the well-defined wavelength of an external laser, are well-known for demonstrating polariton bistability, reported by Baas et al. (2004), and multistability, by Gippius et al. (2007), Paraïso et al. (2010) or Cancellieri et al. (2011) among others. This can lead to very long-lived optical memory elements, given by the duration of application of a control laser as shown by Cerna et al. (2013), as well as switches, as shown by Giorgi et al. (2012) and Amo et al. (2010a) for optical circuits. To interface such devices with electronics, one ideally aims to construct mechanisms of bistability based on non-resonant/incoherent excitation. Here, a few options are available. First, using long-lifetime cavities and ring-shaped excitation, Li et al. (2015a) have shown that polaritons can form stochastically in one of two orbital angular momentum states, representing a stable persistent current. Theoretically, taking orbital angular momentum states as a basis, one can consider the copying and inversion of binary information in a lattice, see Sigurdsson et al. (2014). Alternatively, the difference in polariton linewidths for different linear polarisations (transverse-electric and transverse-magnetic) was shown experimentally and theoretically by Ohadi et al. (2015) to lead to a form of bistability between spin polarised states. Mechanisms of electrically controlled bistability were also demonstrated by Amthor et al. (2015) and several theoretical mechanisms of bistability with incoherent excitation remain open, with preliminary investigations from Kyriienko et al. (2014), Liew et al. (2015) and Li et al. (2015b), etc.

12.5.4   *Polariton quantum random number generators*

Recent discoveries that polariton condensates can adopt spin states in a stochastic fashion open up possibilities for other types of polariton devices, including low-energy spin

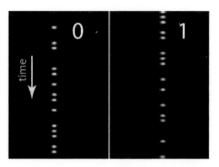

Fig. 12.12: Polariton condensate random number generation (RNG) excited with a non-resonant pump repeatedly turned on for $2\,\mu$s and then turned off again, as reported by Ohadi et al. (2015). The two spin channels are separated onto different detectors, and set to correspond to '0' and '1'.

switches, as well as high bandwidth random number generators. The observation of spin bistability by Ohadi et al. (2015) immediately suggests that small injections of optical or electrical energy can perturb the system from one stable state to another.

When using polariton trapped configurations, with carefully non-polarised incoherent pumping conditions, a condensate will predominantly polarise either spin-up or spin-down once the power is above a critical threshold. In etched microcavity structures, additional residual strain from the sidewalls seems to modify this behaviour. However, in planar microcavities, this phenomena is independent of location, and every time the condensate is switched on, the spin is randomly chosen seeded from the quantum noise in the condensation process (Fig. 12.12). Coding the spin up/down states as 0/1 allows the use of this as a random number generator (RNG), with several advantages to current technologies. Since readout is not a single photon readout, but has been amplified up from quantum noise to give a large macroscopic coherent emission, the detectors of the bit stream can be cheap and high speed (in comparison to the dead time and expense of single-photon detectors). The production of each bit $100\,$ps after the pumping is switched on, allows random number generation on demand. The use of electrically biased microcavities allows the percentage of 0 and 1 bits to be exactly balanced to 50%, which is highly desirable in RNG, and this can be simply monitored in real time and maintained. The high speed for condensate formation (typically $< 100\,$ps) provides single condensate $> 10\,$gb/s data rates, but array generation can easily allow Tb/s RNG rates to be achieved.

12.6 Polariton condensate transistors and optical circuits

The development of practical optical integrated circuits using ultralow powers would represent a revolutionary breakthrough in modern optoelectronics, making possible ultrafast information processing with extremely low losses. However, the realisation of optical integrated circuits is a complex task, which cannot be achieved without exploiting new physical effects, device concepts and technological approaches. Several works have considered the control of propagating polariton condensates in incoherently controlled interferometers, such as Sturm et al. (2014), and condensate transistors by Gao

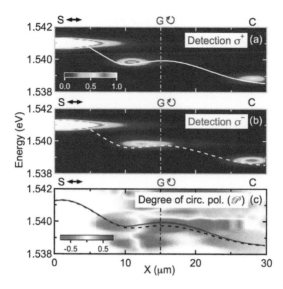

Fig. 12.13: **Polariton spin transistor**, from Gao et al. (2015a). Spatially resolved spectrum of a polariton condensate spin transistor calculated from Wouters (2012)'s driven-dissipative Gross-Pitaevskii equation with energy relaxation. Polaritons are injected incoherently from a source S, with a linear polarisation. A circularly polarised gate, G, creates a barrier in the polariton potential (white curve), which partially inhibits the propagation of polaritons with the same circular polarisation (a), while allowing those of opposite polarisation to propagate (b) to the collector region C. c) Degree of circular polarisation.

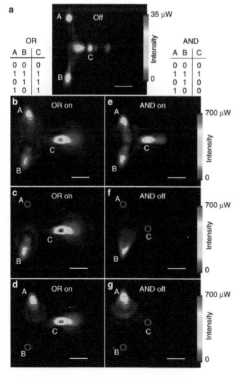

Fig. 12.14: **Polariton logic.** Two polariton-transistors cascaded by Ballarini et al. (2013) to build logic gates: (a–d) an OR gate and (a;e–g) an AND gate, depending on the intensities of the control beams. In each panel, showing a real-space image of the polariton field (with a distance ≈ 30 μm between the spots), A and B show the input of the gate and C the output, indeed satisfying the logic tables of the implemented gate. The top panel, common to both, has rescaled energy, but would appear all-black on the scale of the other images.

et al. (2012) or Antón et al. (2012) (see also Fig. 12.13). A proposal has been made
by Ballarini et al. (2013) also for realisation of a polariton transistor operating under
resonant optical pumping (Fig. 12.14).

12.6.1 *Polariton transistors*

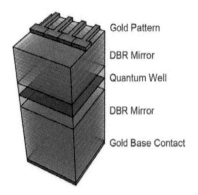

Gold Pattern

DBR Mirror

Quantum Well

DBR Mirror

Gold Base Contact

Fig. 12.15: Scheme of one of possible implementation of a polariton transistor, here using the spin-degree of freedom. From source to drain, polaritons are channeled below the metallic stripes.

The transistor is the core of our electronic-based technology and its implementation
in one form or another in other platforms opens the possibility of bringing fully-fledged
computing resources in possibly cheaper, smaller, faster and/or lower-consumption machines.

One of the early proposals of a polariton transistor, by Gao et al. (2015a), exploits the effects of polariton-mediated optical bistability, plasmon-polariton coupling,
and propagation of polariton domain walls in microcavities in a concept of an optical
integrated circuit. It is based on several ideas, namely:

1. bistable switching of the optically driven polariton state due to nonlinear polariton-
   polariton interactions, which has been experimentally observed recently,

2. channeling of polaritons below metallic stripes due to the hybridisation of the
   Tamm-plasmon states and cavity exciton-polaritons,

3. electrical control of the detuning between the exciton-polariton mode and the
   pumping laser due to the quantum confined Stark effect,

4. propagation of domain walls between different bistable states.

Figure 12.15 shows schematically a proposed polariton spin transistor. The figure illustrates a conventional planar microcavity in which Bragg mirrors confine light so that it
interacts strongly with excitons in a quantum well. However, in addition the structure
has a patterned metal structure on the top surface, comprising a chain of four segments.
The band structure of exciton-polariton modes in a microcavity can be strongly affected
by the presence of a metallic layer on the top of the structure, and this can be used
to provide the lateral confinement necessary for integrated circuits. The thicknesses of
the metallic layer and of the semiconductor layer next to it determine the energy of
the Tamm-plasmon mode localised at the metal-Bragg mirror interface. If the Tamm

plasmon is put in resonance with the cavity polaritons, a new hybrid so-called Tamm-plasmon-exciton-polariton mode (TPEPM) can appear, which has components of the exciton state, cavity photon and Tamm plasmon. The lowest TPEPM is redshifted with respect to the lowest bare cavity polariton mode, with the size of the shift dependent on the exciton oscillator strength and the parameters of the Bragg mirror between the cavity and metal. Selective optical excitation of the TPEPM is achieved by tuning the photon energy of a continuous-wave cw laser to lie between the lowest TPEPM and the lowest cavity polariton mode outside the channel. In this case, there is no absorption of light in the regions of the sample not covered by metal. TPEPMs are confined in the channels and the lowest TPEPM absorbs laser light only if its energy is tuned to the laser energy. This tuning can be achieved and controlled by the application of an electric field across the cavity in the direction normal to the cavity plane. An electric field affects the exciton energy and oscillator strength due to the quantum confined Stark effect, which leads to the shift of the TPEPMs as well. Electric fields to control the local TPEPM energy selectively can be produced by applying a potential difference relative to the structure's back contact to one or more of the four metallic segments.

One particularly demanding requirement for optical technology is cascadability, with the output of one stage being suitable as the input for another stage. This is for instance hardly met in the previous proposal as the output signal is strongly attenuated. Building on the work of Giorgi et al. (2012), Ballarini et al. (2013) have demonstrated an all-optical polariton transistor and its cascading with another polariton transistor with effect of providing an OR or AND gate, suitable for further cascading to assemble a full polariton-circuit. The transistor is based on the renormalisation of the polariton energies with polariton densities: remaining either unperturbed when a driving laser lies far from the polariton branch and abruptly getting into resonance and being efficiently excited passed a given threshold. In contrast to the previous design, here bistability is avoided to prevent any hysteresis that would spoil the switching on and off of the device. Crucially for cascading, the system is robust in its output, with small variations of intensities regardless on how much polaritons from how many branches did trigger the on-state. This allows to feed with one output several inputs (fan-out, or gain) and vice versa (fan-in). Indeed, by bringing together several polariton beams propagating in different directions, Ballarini et al. (2013) are able to build logic gates, as shown in Fig. 12.14. External lasers serve as the input with polariton flowing and triggering or not the logic states depending only, as far as the ON/AND mode of operation is concerned, on the intensities at which the device is operated. The authors estimate the power for operating such a polariton transistor at the level of a few attojoules per μm^2. The speed of the device is given by the inverse polariton lifetime, which is of the order of the ps. The device has less appealing merits regarding its size, which is, however, a problem for all optical devices, that can be addressed in the future would the technology prove to be useful by, e.g., turning to plasmonic counterparts of these ideas, which passed the proof-of-principle demonstration with polaritons.

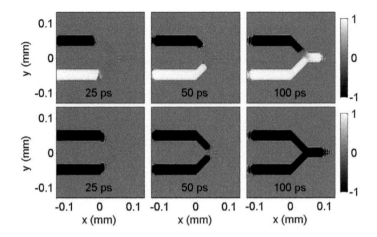

Fig. 12.16: Calculated circular polarisation degree of polariton condensates in the region where two "polariton neurons" merge, at different instants after the pulse arrival time, as calculated by Liew et al. (2008). In the top row the system is excited with oppositely circularly polarised pulses. In the bottom row it is excited with two pulses of identical circular polarisation.

12.6.2   *Polariton neurons*

By exploiting the polarisation multistability of polaritons, polarised signals can be conducted in the plane of a semiconductor microcavity along controlled channels or "neurons." Furthermore, because of the interaction of polaritons with opposite spins, it is possible to realise binary logic gates operating on the polarisation degree of freedom. Multiple gates can be integrated together to form an optical circuit contained in a single semiconductor microcavity. In polariton neurons, polaritons themselves do not move the whole distance from one end of the channel to the other; rather it is the motion of a circularly polarised domain wall or switching of successive parts of the channel caused by very short propagation of polaritons that results in a long signal propagation. In this sense the channel bears a loose analogy to biological neurons, which is why they are called "polariton neurons." Since one does not have to rely on single polaritons traveling the full length of the channels, the short lifetime of polaritons does not limit the length of signal propagation.

12.7   Conclusions

Over the past decade, polaritonics has made a huge step forward, and expanded from a purely fundamental research field into an interdisciplinary research area with promising applications suggested in diverse fields that include solid-state lighting, information processing and medicine. The most significant progress has been achieved in the realisation of polariton lasers: the first realisation of the bosonic lasing concept proposed in 1996 by Ĭmamoḡlu et al. (1996). In terms of practical applications, polariton lasers still need to find their niche. Their undoubted advantage over conventional lasers is in the significantly lower threshold power, as convincingly demonstrated by Schneider et al.

(2013) and Bhattacharya et al. (2014). On the other hand, polariton condensates are fragile: they lose coherence as soon as pumped a bit more strongly. This is why polariton lasers are not good for high-power operation. On the other hand, bosonic condensates of exciton-polaritons may be manipulated by applying external electric and magnetic fields and by external laser beams. The polarisation and intensity of light emitted by polariton lasers can be switched from one value to another within several tens of picoseconds, as consistently reported by several groups, e.g., Amo et al. (2010a); Giorgi et al. (2012); Cerna et al. (2013); Antón et al. (2014). This high controllability of the most essential characteristics of the emitted light make polariton lasers most promising for applications in optical integrated circuits and at the interface between electronic devices and optical communication lines. Another application area that remains to be explored is the stimulation of terahertz frequency generation by polariton condensates. Given a high demand for compact and reliable sources of coherent terahertz radiation, bosonic cascade lasers based on excitons or exciton-polaritons would offer a valuable alternative to quantum cascade lasers based on electronic transitions in semiconductor superlattices. Potentially, they could operate at room temperature, emit terahertz light in the vertical direction (normal to the plane of the structure) and be as small as any vertical cavity surface emitting semiconductor laser.

12.8   Further reading

Further information on perspective polariton devices can be found in Chapters 12–15 of Sanvitto and Timofeev (2012)'s book. In particular, Chapter 14 of this book nicely presents the progress achieved in the realisation of polariton devices based on organic microcavities, which remained out of the scope of the present text.

APPENDIX A

SCATTERING RATES OF POLARITON RELAXATION

This appendix complements Chapter 8 with detailed calculations of the rates of polariton scattering with phonons, electrons and other polaritons in microcavities.

A.1 Polariton–phonon interaction

The theoretical description of carrier–phonon or of exciton–phonon interaction has received considerable attention throughout the history of semiconductor heterostructures. Here we present a simplified picture that is, however, well suited to our problem. Cavity polaritons are two-dimensional particles with only an in-plane dispersion. They are scattered by phonons that are in the QWs we consider, mainly three-dimensional (acoustic phonons) or two-dimensional (optical phonons). Scattering events should conserve the wavevector in the plane. We call $\mathbf{q}$ the phonon wavevector and $q_\parallel$, q_z the in-plane and z-component of $\mathbf{q}$, i.e.,

$$\mathbf{q} = (q_\parallel, q_z). \tag{A.1}$$

$\omega_\mathbf{q}$ will denote the phonon energy.

Using Fermi's golden rule, the scattering rate between two discrete polariton states of wavevector $\mathbf{k}$ and $\mathbf{k}'$ reads

$$W^{\text{phon}}_{\mathbf{k} \to \mathbf{k}'} = \frac{2\pi}{\hbar} \sum_\mathbf{q} |M(\mathbf{q})|^2 (\theta_\pm + N_{\mathbf{q}=\mathbf{k}-\mathbf{k}'+q_z}) \delta(E(\mathbf{k}') - E(\mathbf{k}) \pm \omega_\mathbf{q}), \tag{A.2}$$

where $N_\mathbf{q}$ is the phonon distribution function and $\theta_\pm$ is a quantity whose sign matches the one in the delta function corresponding to phonon emission and absorption, and is defined as $\theta_+ = 1$ and $\theta_- = 0$. In the case of an equilibrium phonon distribution, $N_\mathbf{q}$ follows the Bose distribution. The sum of eqn (A.2) is over phonon states.

M is the matrix element of interaction between phonons and polaritons. If one considers polariton states with a finite energy width $\gamma_\mathbf{k}$, the function can be replaced by a Lorentzian and eqn (A.2) becomes

$$W^{\text{phon}}_{\mathbf{k} \to \mathbf{k}'} = \frac{2\pi}{\hbar} \sum_\mathbf{q} |M(\mathbf{q})|^2 (\theta_\pm + N_{\mathbf{q}=\mathbf{k}-\mathbf{k}'+q_z}) \frac{\gamma_{\mathbf{k}'}/\pi}{(E(\mathbf{k}') - E(\mathbf{k}) \pm \omega_\mathbf{q})^2 + \gamma'^2_\mathbf{k}/\pi^2}. \tag{A.3}$$

Wavevector conservation in the plane actually limits the sum of eqn (A.3) to the z-direction. In the framework of the Born approximation the matrix element of interaction reads

$$|M(\mathbf{q})|^2 = \left| \left\langle \psi^{\text{pol}}_\mathbf{k} \middle| H^{\text{pol−phon}}_\mathbf{q} \middle| \psi^{\text{pol}}_{\mathbf{k}'} \right\rangle \right|^2 = |\mathcal{X}_\mathbf{k}|^2 |\mathcal{X}_{\mathbf{k}'}|^2 \left| \left\langle \psi^{\text{ex}}_\mathbf{k} \middle| H^{\text{ex−phon}}_\mathbf{q} \middle| \psi^{\text{ex}}_{\mathbf{k}'} \right\rangle \right|^2, \tag{A.4}$$

where $\left|\psi_{\mathbf{k}}^{\text{pol}}\right\rangle$ is the polariton wavefunction and $\left|\psi_{\mathbf{k}}^{\text{ex}}\right\rangle$ the exciton wavefunction, and $\mathcal{X}_{\mathbf{k}}$ is the exciton Hopfield coefficient (which squares to the exciton fraction). The exciton wavefunction reads

$$\psi_{\mathbf{k}}^{\text{ex}}(\mathbf{r}_{\text{e}}, \mathbf{r}_{\text{h}}) = U_{\text{e}}(z_{\text{e}})U_{\text{h}}(z_{\text{h}})\frac{1}{\sqrt{S}}e^{i\mathbf{k}(\beta_{\text{e}}\mathbf{r}_{\text{e}} + \beta_{\text{h}}\mathbf{r}_{\text{h}})}\sqrt{\frac{2}{\pi}\frac{1}{a_{\text{B}}^{2\text{D}}}}\exp\left(-\frac{|\mathbf{r}_{\text{e}} - \mathbf{r}_{\text{h}}|}{a_{\text{B}}^{2\text{D}}}\right), \quad (A.5)$$

where z_{e}, z_{h} are electron and hole coordinates along the growth axis and $\mathbf{r}_{\text{e}}$ and $\mathbf{r}_{\text{h}}$ their coordinates in the plane, U_{e} and U_{h} are the electron and hole wavefunctions in the growth direction, $a_{\text{B}}^{2\text{D}}$ is the two-dimensional exciton Bohr radius, $\beta_{\text{e,h}} = m_{\text{e,h}}/(m_{\text{e}} + m_{\text{h}})$.

A.1.1 *Interaction with longitudinal optical phonons*

This interaction is mainly mediated by the Frölich interaction Frohlich (1937). In three dimensions the exciton–LO-phonon matrix element reads

$$M^{\text{LO}}(\mathbf{q}) = -\frac{e}{q}\sqrt{\frac{4\pi\hbar\omega_{\text{LO}}}{SL\varepsilon_0}\left(\frac{1}{\varepsilon_\infty} - \frac{1}{\varepsilon_{\text{s}}}\right)} = \frac{M_0^{\text{LO}}}{q\sqrt{SL}}, \quad (A.6)$$

where ω_{LO} is the energy for creation of a LO-phonon, ε_∞ is the optical dielectric constant, ε_{s} the static dielectric constant, L is the dimension along the growth axis and S the normalisation area. In two dimensions one should consider confined optical phonons with quantised wavevector in the z-direction. L becomes the QW width and $q_z^m = m\pi/L$ with m an integer. Moreover, the overlap integral between exciton and phonon wavefunctions quickly vanishes while m increases. Therefore, we consider only the first confined phonon state and the matrix element (A.6) becomes

$$M^{\text{LO}}(\mathbf{q}) = \frac{M_0^{\text{LO}}}{\sqrt{|\mathbf{q}_\parallel|^2 + (\pi/L)^2}\sqrt{SL}}. \quad (A.7)$$

The wavevectors exchanged in the plane are typically much smaller than π/L and eqn (A.7) can be approximated by

$$M^{\text{LO}}(\mathbf{q}) = \frac{M_0^{\text{LO}}}{\pi}\sqrt{\frac{L}{S}}. \quad (A.8)$$

Considering a dispersionless phonon dispersion for LO-phonons, the LO-phonon contribution to eqn (A.3) reads

$$W_{\mathbf{k}\rightarrow\mathbf{k}'}^{\text{phon-LO}} = \frac{2L}{\pi^2 S}|\mathcal{X}_{\mathbf{k}}|^2|\mathcal{X}_{\mathbf{k}'}|^2|M_0^{\text{LO}}|^2 \times$$

$$\left(\theta_\pm + \frac{1}{\exp(-\omega_{\text{LO}}/k_{\text{B}}T) - 1}\right)\frac{\gamma_{\mathbf{k}'}}{(E(\mathbf{k}') - E(\mathbf{k}) \pm \omega_{\text{LO}})^2 + \gamma_{\mathbf{k}}'^2}. \quad (A.9)$$

Optical phonons interact very strongly with carriers. They allow fast exciton formation. Their energy of formation ω_{LO} is, however, of the order of 20 to 90 meV, depending on the nature of the semiconductor involved. An exciton with a kinetic energy smaller than 20 meV can no longer emit an optical phonon. The probability of absorbing an optical phonon remains extremely small at low temperature. This implies that an exciton gas cannot cool down to temperature lower than 100–200 K by interacting only with optical phonons. Optical phonons are therefore extremely efficient at relaxing a hot-carrier gas (optically or electrically created) towards an exciton gas with a temperature 100–300 K in a few picoseconds. The final cooling of this exciton gas towards the lattice temperature should, however, be assisted by acoustical phonons or other scattering mechanisms. The semiconductor currently used to grow microcavities, and where optical phonons play the largest role, is CdTe. In such a material, ω_{LO} is only 21 meV, namely larger than the exciton binding energy in CdTe-based QWs. Moreover, the Rabi splitting is of the order of 10–20 meV in CdTe-based cavities. This means that the direct scattering of a reservoir exciton towards the polariton ground-state is a possible process that may play an important role.

A.1.2 *Interaction with acoustic phonons*

This interaction is mainly mediated by the deformation potential. The exciton–acoustic phonon matrix element reads

$$M^{\mathrm{ac}}(\mathbf{q}) = \sqrt{\frac{q}{\mu 2\rho c_s SL}} G(\mathbf{q}_{\|}, q_z) \,, \tag{A.10}$$

where μ is the reduced mass of electron–hole relative motion, ρ is the density and c_s is the speed of sound in the medium. Assuming isotropic bands, G reads

$$G(\mathbf{q}_{\|}, q_z) = D_e I_e^{\perp}(q_z) I_e^{\|}(q_{\|}) D_h I_h^{\perp}(q_z) I_h^{\|}(q_{\|}) \approx D_e I_e^{\|}(q_{\|}) D_h I_h^{\|}(q_{\|}) \,. \tag{A.11}$$

D_e, D_h are the deformation coefficients of the conduction and valence band, respectively, and $I_{e(h)}^{\perp(\|)}$ are the overlap integrals between the exciton and phonon mode in the growth direction and in the plane, respectively,

$$I_{e(h)}^{\|}(q_{\|}) = \left(\sqrt{\frac{2}{\pi}} \frac{1}{a_B^{2D}} \right)^2 \int \exp\left(-\frac{2r}{a_B^{2D}} \right) \exp\left(i \frac{m_{h(e)}}{m_e + m_h} \mathbf{q}_{\|} \cdot \mathbf{r} \right) d\mathbf{r}$$

$$= \left(1 + \left(\frac{m_{h(e)} q_{\|} a_B^{2D}}{2M_x} \right) \right)^2 \,, \tag{A.12a}$$

$$I_{e(h)}^{\perp}(q_z) = \int |f_{e(h)}(z)|^2 e^{iq_z z} \, dz \approx 1 \,. \tag{A.12b}$$

Using this matrix element and moving to the thermodynamic limit in the growth direction, the scattering rate (A.3) becomes

$$W_{\mathbf{k} \to \mathbf{k}'}^{\mathrm{phon}} = \frac{2\pi}{\hbar} \frac{L}{2\pi} |\mathcal{X}_{\mathbf{k}}|^2 |\mathcal{X}_{\mathbf{k}'}|^2 \int_{q_z} \frac{q}{2\rho c_s SL} |G(\mathbf{k} - \mathbf{k}')|^2 (\theta_{\pm} + N_{\mathbf{k} - \mathbf{k}' + \mathbf{q}_z}^{\mathrm{phon}})$$

$$\frac{\gamma_{\mathbf{k}'}/\pi}{(E(\mathbf{k}') - E(\mathbf{k}) \pm \omega_{\mathbf{k} - \mathbf{k}' + \mathbf{q}_z})^2 + \gamma_{\mathbf{k}'}^2} \,. \tag{A.13}$$

Moving to the thermodynamic limit means that we let the system size in a given direction (here the z-direction) go to infinity, substituting the summation with an integral, using the formula $\sum_{q_z} \rightarrow (L/2\pi) \int dq_z$.

Equation (A.13) can be easily simplified as

$$
W_{\mathbf{k}\rightarrow\mathbf{k}'}^{\text{phon}} = \frac{|G(\mathbf{k}-\mathbf{k}'|^2}{2\pi Z \rho c_s}|\mathcal{X}_{\mathbf{k}}|^2|\mathcal{X}_{\mathbf{k}'}|^2 \int_{q_z} |\mathbf{k}-\mathbf{k}'+\mathbf{q}_z|(\theta_\pm + N_{\mathbf{k}-\mathbf{k}'+\mathbf{q}_z}^{\text{phon}})
$$

$$
\frac{\gamma_{\mathbf{k}'}/\pi}{(E(\mathbf{k}')-E(\mathbf{k})\pm\omega_{\mathbf{k}-\mathbf{k}'+\mathbf{q}_z})^2+\gamma_{\mathbf{k}'}^2}. \qquad \text{(A.14)}
$$

A.2 Polariton–electron interaction

The polariton–electron scattering rate is calculated using Fermi's golden rule as

$$
W_{\mathbf{k}\rightarrow\mathbf{k}'}^{\text{el}} = \frac{2\pi}{\hbar}\sum_{\mathbf{q}}|M_{\mathbf{q},\mathbf{k},\mathbf{k}'}^{\text{el}}|^2|\mathcal{X}_{\mathbf{k}}|^2|\mathcal{X}_{\mathbf{k}'}|^2 N_{\mathbf{q}}^{\text{e}}(1-N_{\mathbf{q}+\mathbf{k}'-\mathbf{k}}^{\text{e}})
$$

$$
\frac{\gamma_{\mathbf{k}'}}{\left(E(\mathbf{k}')-E(\mathbf{k})+\frac{\hbar^2}{2m_{\text{e}}}(q^2-|\mathbf{q}+\mathbf{k}-\mathbf{k}'|^2)\right)^2+\gamma_{\mathbf{k}'}^2}, \qquad \text{(A.15)}
$$

where $N_{\mathbf{q}}^{\text{e}}$ is the electron distribution function and m_{e} the electron mass. If one considers electrons at thermal equilibrium, it is given by the Fermi–Dirac electron distribution function with a chemical potential

$$
\mu_{\text{e}} = k_B T \ln\left(\exp\left(\frac{\hbar^2 n_{\text{e}}}{\pi k_B T m_{\text{e}}}\right)-1\right), \qquad \text{(A.16)}
$$

where n_{e} is the electron concentration. M^{el} is the matrix element of interaction between an electron and an exciton. A detailed calculation of the electron–exciton matrix element has been given by Ramon et al. (2003). M^{el} is composed of a direct contribution and of an exchange contribution:

$$
M^{\text{el}} = M_{\text{dir}}^{\text{el}} \pm M_{\text{exc}}^{\text{el}}. \qquad \text{(A.17)}
$$

The $+$ sign corresponds to a triplet configuration (parallel electron spins) and the $-$ to a singlet configuration (antiparallel electron spins). If both electrons have the same spin, the total exciton spin is conserved through the exchange process. However, if both electron spins are opposite, an active exciton state of spin $+1$, for example, will be scattered towards a dark state of spin $+2$ through the exchange process. Here, and in what follows, we shall consider only the triplet configuration for simplicity.

In order to calculate M^{el} we adopt the Born approximation and obtain

$$M_{\text{dir}}^{\text{el}} = \iiint \psi_{\mathbf{k}}^*(\mathbf{r}_e, \mathbf{r}_h) f_{\mathbf{q}}^*(\mathbf{r}_e') \Big[V(|\mathbf{r}_e - \mathbf{r}_e'|)$$

$$- V(|\mathbf{r}_h - \mathbf{r}_e'|) \Big] \psi_{\mathbf{k}'}(\mathbf{r}_e, \mathbf{r}_h) f_{\mathbf{q}+\mathbf{k}-\mathbf{k}'}^*(\mathbf{r}_e') d\mathbf{r}_e d\mathbf{r}_h d\mathbf{r}_{e'} ,$$

$$\text{(A.18a)}$$

$$M_{\text{exc}}^{\text{el}} = \iiint \psi_{\mathbf{k}}^*(\mathbf{r}_e, \mathbf{r}_h) f_{\mathbf{q}}^*(\mathbf{r}_e') \Big[V(|\mathbf{r}_e - \mathbf{r}_e'|) - V(|\mathbf{r}_h - \mathbf{r}_e|)$$

$$- V(|\mathbf{r}_h - \mathbf{r}_e'|) \Big] \psi_{\mathbf{k}'}(\mathbf{r}_e, \mathbf{r}_h) f_{\mathbf{q}+\mathbf{k}-\mathbf{k}'}^*(\mathbf{r}_e') d\mathbf{r}_e d\mathbf{r}_h d\mathbf{r}_{e'} ,$$

$$\text{(A.18b)}$$

with Coulomb potential $V(r) = e^2/(4\pi\varepsilon\varepsilon_0 r)$ with ε the dielectric susceptibility of the QW. The free-electron wavefunction f is given by

$$f_{\mathbf{q}}(\mathbf{r}_e') = \frac{1}{\sqrt{S}} e^{i\mathbf{q}\cdot\mathbf{r}_e'} . \tag{A.19}$$

Integrals (A.18) can be calculated analytically. One finds

$$M_{\text{dir}}^{\text{el}} = \frac{e^2}{2S\varepsilon\varepsilon_0 |\mathbf{k} - \mathbf{k}'|} \Big[(1 + \xi_h^2)^{-3/2} - (1 + \xi_e^2)^{-3/2} \Big] , \tag{A.20a}$$

$$M_{\text{dir}}^{\text{dir}} = \frac{2e^2}{S\varepsilon\varepsilon_0} \left[\frac{(1 + \xi_c^2)^{-3/2} - (1 + 4\xi_h^2)^{-3/2}}{\left(a^{-2} + |\mathbf{q} - \beta_e\mathbf{k}'|^2\right)^{1/2}} \right.$$

$$\left. - \frac{(1 + 4\xi_h^2)^{-3/2}}{\left(a^{-2} + |\mathbf{k}' - \mathbf{k} - \mathbf{q} - \beta_e\mathbf{k}'|^2\right)^{1/2}} \right] , \tag{A.20b}$$

where

$$\xi_{e,h} = \frac{1}{2}\beta_{e,h} |\mathbf{k}' - \mathbf{k}| a_B^{2D} \quad \text{and} \quad \xi_c = |\beta_e\mathbf{k} + \mathbf{k}' - \mathbf{k} - \mathbf{q}| a_B^{2D} . \tag{A.21}$$

Passing to the thermodynamic limit, eqn (A.15) becomes

$$W_{\mathbf{k}\to\mathbf{k}'}^{\text{el}} = \frac{S}{2\pi} \int_{\mathbf{q}} |M_{\text{dir}}^{\text{el}} + M_{\text{exc}}^{\text{el}}|^2 |\mathcal{X}_{\mathbf{k}}|^2 |\mathcal{X}_{\mathbf{k}'}|^2 N_{\mathbf{q}}^e (1 - N_{\mathbf{q}+\mathbf{k}'-\mathbf{k}}^e)$$

$$\times \frac{\gamma_{\mathbf{k}'}}{\left(E(\mathbf{k}') - E(\mathbf{k}) + \frac{\hbar^2}{2m_e}(q^2 - |\mathbf{q} + \mathbf{k} - \mathbf{k}'|^2) \right)^2 + \gamma_{\mathbf{k}'}^2} . \tag{A.22}$$

The polariton–electron interaction is a dipole-charge interaction that takes place on a picosecond time scale. An equilibrium electron gas can thermalise a polariton gas quite efficiently. A more complex effect may, however, take place such as trion formation or exciton dephasing.

A.3 Polariton–polariton interaction

The polariton–polariton scattering rate reads

$$W^{\text{pol}}_{\mathbf{k}\to\mathbf{k}'} = \frac{2\pi}{\hbar} \sum_{\mathbf{q}} |M^{\text{ex}}|^2 |\mathcal{X}_{\mathbf{k}}|^2 |\mathcal{X}_{\mathbf{k}'}|^2 |\mathcal{X}_{\mathbf{q}}|^2 |\mathcal{X}_{\mathbf{q}+\mathbf{k}'-\mathbf{k}}|^2 N^{\text{pol}}_{\mathbf{q}} (1 + N^{\text{pol}}_{\mathbf{q}+\mathbf{k}'-\mathbf{k}})$$

$$\times \frac{\gamma_{\mathbf{k}'}}{\left(E(\mathbf{k}') - E(\mathbf{k}) + E(\mathbf{q}+\mathbf{k}'-\mathbf{k}) - E(\mathbf{q})\right)^2 + \gamma_{\mathbf{k}'}^2}. \qquad (A.23)$$

The exciton–exciton matrix element of interaction is also composed of a direct and an exchange term. It has been investigated and calculated by Ciuti et al. (1998) and recently by Combescot et al. (2007a). Here, and in what follows, we use a numerical estimate provided by Tassone and Yamamoto (1999) that we further assume constant over the whole reciprocal space:

$$M_{\text{ex}} \approx 6 \frac{(a_{\text{B}}^{\text{2D}})^2}{S} E_{\text{b}} = \frac{1}{S} M^0_{\text{exc}}, \qquad (A.24)$$

where E_{b} is the exciton binding energy. Passing to the thermodynamic limit in the plane, eqn (A.23) becomes

$$W^{\text{pol}}_{\mathbf{k}\to\mathbf{k}'} = \frac{1}{2\pi S} \int |M^0_{\text{ex}}|^2 |\mathcal{X}_{\mathbf{k}}|^2 |\mathcal{X}_{\mathbf{k}'}|^2 |\mathcal{X}_{\mathbf{q}}|^2 |\mathcal{X}_{\mathbf{q}+\mathbf{k}'-\mathbf{k}}|^2 N^{\text{pol}}_{\mathbf{q}} (1 + N^{\text{pol}}_{\mathbf{q}+\mathbf{k}'-\mathbf{k}})$$

$$\times \frac{\gamma_{\mathbf{k}'}}{\left(E(\mathbf{k}') - E(\mathbf{k}) + E(\mathbf{q}+\mathbf{k}'-\mathbf{k}) - E(\mathbf{q})\right)^2 + \gamma_{\mathbf{k}'}^2}. \qquad (A.25)$$

As one can see, the a priori unknown polariton distribution function is needed to calculate scattering rates. This means that in any simulation these scattering rates should be updated dynamically throughout the simulation time, which can be extremely time consuming.

Polariton–polariton scattering has been shown to be extremely efficient when a microcavity is resonantly excited. It also plays a fundamental role in the case of non-resonant excitation. Depending on the excitation condition and on the nature of the semiconductor used, the exciton–exciton interaction may be strong enough to self-thermalise the exciton reservoir at a given temperature.

A.3.1 *Polariton decay*

There are three different regions in reciprocal space: $[0, k_{\text{sc}}]$, $]k_{\text{sc}}, k_{\text{L}}]$ and $]k_{\text{L}}, \infty[$.

- $[0, k_{\text{sc}}]$ is the region where the exciton–photon anticrossing takes place. Cavity mirrors reflect the light only within a finite angular cone, which corresponds to an in-plane wavevector k_{sc} that depends on the detuning. In this central region the polariton decay is mainly due to the finite cavity photon lifetime $\Gamma_k = |\mathcal{C}_k|^2/\tau_{\text{c}}$, where $\mathcal{C}_k$ is the photon Hopfield coefficient (which squares to the photon fraction of the polariton) and τ_{c} is the cavity photon lifetime. k_{sc} values are typically of the order of 4 to 8×10^6 m^{-1} and τ_{c} is in the range between 1 and 10 ps.

- $]k_{sc}, k_L]$ where k_L is the wavevector of light in the medium. In this region excitons are only weakly coupled to the light and polariton decay is $\Gamma_k = \Gamma_0$, which is the radiative decay rate of QW excitons.
- $]k_L, \infty[$. Beyond k_L excitons are no longer coupled to light. They only decay non-radiatively with a decay rate Γ_{nr}. We do not wish to enter into the details of the mechanism involved in this decay, which we consider as constant in the whole reciprocal space. This quantity is given by the decay time measured in time-resolved luminescence experiments, and is typically in the range between 100 ps and 1 ns.

A.4 Polariton–structural-disorder interaction

This scattering process is mainly associated with the excitonic part of polaritons. Structural disorder induces coherent elastic (Rayleigh) scattering with a typical timescale of about 1ps. It couples very efficiently all polaritons situated on the same "elastic circle" in reciprocal space (see the results of Freixanet et al. (1999) and Langbein and Hvam (2002)). This allows us to simplify the description of polariton relaxation by assuming cylindrical symmetry of the polariton distribution function. Also, disorder induces a broadening of the polariton states, which should be accounted for when scattering rates are calculated.

APPENDIX   B

DERIVATION OF THE LANDAU CRITERION OF
SUPERFLUIDITY AND LANDAU FORMULA

This appendix presents the derivation of the important "Landau formula" used to esti-
mate the critical temperature of the superfluid phase transition.

Let us consider a uniform fluid at zero temperature flowing along a capillary at a
constant velocity $\mathbf{v}$. The only dissipative process assumed is the creation of elementary
excitations due to the interaction between the fluid and the boundaries of the capillary.
If this process is allowed by the conservation laws and Galilean invariance, the flow will
demonstrate viscosity; otherwise it will be superfluid, i.e., dissipativeless. The basic idea
of the derivation is to calculate energy and momentum in the reference frame moving
with the fluid and in the static frame, then making the link between the two frames by a
Galilean transformation. If a single excitation with momentum $p = \hbar k$ appears, the total
energy in the moving frame is $E = E_0 + \epsilon(\mathbf{k})$, where E_0 is the energy of the ground-
state and $\epsilon(\mathbf{k})$ is the dispersion of the fluid excitations. In the static frame, however, the
energy and momentum of the fluid read

$$E = E_0 + \epsilon(\mathbf{k}) + \hbar\mathbf{k} \cdot \mathbf{v} + \frac{1}{2}Mv^2 , \tag{B.1a}$$
$$\mathbf{P} = \mathbf{p} + M\mathbf{v} , \tag{B.1b}$$

where M is the total mass of the fluid.

Equation (B.1a) shows that the energy of the elementary excitations in the static
system is $\epsilon(\mathbf{k}) + \hbar\mathbf{k} \cdot \mathbf{v}$. Dissipation is possible, only if the creation of elementary
excitations is profitable energetically, which means

$$\epsilon(\mathbf{k}) + \hbar\mathbf{k} \cdot \mathbf{v} < 0 . \tag{B.2}$$

Dissipation can, therefore, take place only if $v > \epsilon(\mathbf{k})/(\hbar k)$. On the other hand the flux
is stable if the velocity is smaller than

$$v_c = \min\left(\frac{\epsilon(\mathbf{k})}{\hbar k}\right) . \tag{B.3}$$

Formula (B.3) is the Landau criterion of superfluidity. In the case of a parabolic dispersion, v_c is zero and there is no superfluid motion. In the opposite case of a Bogoliubov
dispersion (eqn (8.16)), v_c is simply the speed of sound, which means that the fluid can
move without dissipation with any velocity less than the speed of sound.

We now consider the finite-temperature case. In such a situation it is natural to
assume that part of the thermally excited particles does not remain superfluid. We therefore consider the coexistence of a superfluid component of velocity $\mathbf{v}_s$ and a normal

555

component of velocity v_n. In the frame moving with the normal fluid, the energy of an elementary excitation reads $\epsilon(\mathbf{k}) + \hbar\mathbf{k}\cdot(\mathbf{v_s}-\mathbf{v_n})$. The occupation number of elementary excitations is

$$f_B\big(\epsilon(\mathbf{k}) + \hbar\mathbf{k}\cdot(\mathbf{v_s}-\mathbf{v_n})\big)\,, \tag{B.4}$$

where f_B is the Bose distribution function, eqn (8.1). The total mass density of the fluid can be written as $\rho = \rho_n + \rho_s$, where ρ_n and ρ_s are the mass densities of the normal fluid and superfluid. At this stage, we will assume that all particles of the fluid have the same mass m, which is not correct, in principle, for polaritons. In the static frame the mass current of the liquid reads

$$m\mathbf{j} = \rho_s\mathbf{v_s} + \rho_n\mathbf{v_n}\,. \tag{B.5}$$

Following eqn (B.1b), the total momentum of the fluid in the static frame can also be written as $\mathbf{P} = M\mathbf{v_s} + \sum_i \hbar\mathbf{k}_i$ where the sum is taken over all the excitations. The mass current therefore reads

$$m\mathbf{j} = \rho\mathbf{v}_s + \frac{1}{S}\sum_i \hbar\mathbf{k}_i = \rho\mathbf{v}_s + \frac{\hbar}{(2\pi)^2}\int f_B\big(\epsilon(\mathbf{k}) + \hbar\mathbf{k}\cdot(\mathbf{v_s}-\mathbf{v_n})\big)\mathbf{k}\,d\mathbf{k}. \tag{B.6}$$

Comparing eqns (B.5) and (B.6), one gets

$$\rho_s(\mathbf{v_s}-\mathbf{v_n}) = \frac{\hbar}{(2\pi)^2}\int f_B\big(\epsilon(\mathbf{k}) + \hbar\mathbf{k}\cdot(\mathbf{v_s}-\mathbf{v_n})\big)\mathbf{k}\,d\mathbf{k}. \tag{B.7}$$

We are now going to assume that $|\mathbf{v_s}-\mathbf{v_n}|$ is small with respect to the speed of sound and develop eqn (B.4) in power series of $\mathbf{v_s}-\mathbf{v_n}$, which gives

$$f_B(\epsilon(\mathbf{k}) + \hbar\mathbf{k}\cdot(\mathbf{v_s}-\mathbf{v_n})) \approx f_B(\epsilon(\mathbf{k})) + \hbar\mathbf{k}\cdot(\mathbf{v_s}-\mathbf{v_n})\frac{df_B(\epsilon(\mathbf{k}))}{d\epsilon}\,. \tag{B.8}$$

We can now insert eqn (B.8) in eqn (B.7) to get

$$\rho_n = -\frac{\hbar^2}{(2\pi)^2}\int \frac{df_B(\epsilon(\mathbf{k}))}{d\epsilon}k^2 d\mathbf{k}. \tag{B.9}$$

This formula is known as the *Landau formula of superfluidity*. It is here written for a 2D system. It is valid only for the case of a parabolic dispersion. In Chapter 8, we use this result for polaritons, replacing $\hbar^2\mathbf{k}^2/(2m)$ by the polariton dispersion. The formula used remains, however, fundamentally inexact since the hypothesis of the constant particle mass is needed in order to derive eqn (B.9).

APPENDIX C

LANDAU QUANTISATION AND RENORMALISATION OF RABI SPLITTING

*This appendix addresses the renormalisation of the exciton binding energy and oscilla-
tor strength in a quantum well subjected to an external magnetic field, resulting in the
variation of the Rabi splitting in a microcavity.*

Consider an exciton confined in a QW and subject to a magnetic field normal to
the QW plane. Separating the exciton centre of mass motion and relative electron–hole
motion in the QW plane, and assuming that the exciton does not move as a whole in the
QW plane, we obtain the exciton Hamiltonian

$$\hat{H} = \hat{H}_{\mathrm{e}} + \hat{H}_{\mathrm{h}} + \hat{H}_{\mathrm{ex}}, \tag{C.1}$$

where

$$\hat{H}_\nu = -\frac{\hbar^2}{2m_\nu}\frac{\partial^2}{\partial z_\nu^2} + V_\nu(z_\nu) - \mu_{\mathrm{B}} g_\nu \mathbf{s}_\nu \mathbf{B} + (l_\nu + 1/2)\hbar\omega_{\mathrm{c}}^\nu, \quad \text{for } \nu = \mathrm{e, h}, \tag{C.2a}$$

$$\hat{H}_{\mathrm{ex}} = -\frac{\hbar^2}{2\mu}\left[\frac{1}{\rho}\frac{\partial}{\partial\rho}\left(\rho\frac{\partial}{\partial\rho}\right) - \frac{\rho^2}{4L^4}\right] - \frac{e^2}{4\pi\varepsilon_0\epsilon\sqrt{\rho^2 + (z_{\mathrm{e}} - z_{\mathrm{h}})^2}}, \tag{C.2b}$$

ρ is the coordinate of electron–hole relative motion in the QW plane, $L = \sqrt{\hbar/(eB)}$
is the so-called magnetic length, B is the magnetic field, $\mathbf{s}_{\mathrm{e(h)}}$ is the electron (hole)
spin, $m_{\mathrm{e(h)}}$ is the electron (hole) effective mass in normal to the plane direction, μ is the
reduced mass of electron–hole motion in the QW plane, $V_{\mathrm{e(h)}}$ is the QW potential for an
electron (hole), $g_{\mathrm{e(h)}}$ and $\omega_{\mathrm{c}}^{\mathrm{e(h)}}$ are the electron (hole) g-factor and cyclotron frequency,
respectively, $l = 0, 1, 2, \cdots$ and ϵ is the dielectric constant. Hereafter, we neglect the
heavy–light hole mixing.

The excitonic Hamiltonian (C.2b) was first derived by Russian theorists Gor'kov
and Dzialoshinskii (1967). It contains a parabolic term dependent on the magnetic field.
If the field increases, the magnetic length L decreases, which leads to the shrinkage of
the wavefunction of the electron–hole relative motion. Thus, the probability of finding
the electron and hole at the same point increases, leading to an increase of the exciton
oscillator strength. In order to estimate this effect, let us solve the Schrödinger equa-
tion (3.1) variationally for the wavefunction $\Psi_{\mathrm{exc}}(z_{\mathrm{e}}, z_h, \rho)$ chosen as an approximate
(or *trial*) expression equal to

$$\Psi_{\mathrm{exc}}(z_{\mathrm{e}}, z_{\mathrm{h}}, \rho) = U_{\mathrm{e}}(z_{\mathrm{e}})U_{\mathrm{h}}(z_{\mathrm{h}})f(\rho); \tag{C.3}$$

where $z_{\mathrm{e(h)}}$ is the electron (hole) coordinate in the direction normal to the plane and ρ is
the coordinate of electron–hole in-plane relative motion. If the conduction-band offsets

are large in comparison to the exciton binding energy, which is the case in conventional GaAs/AlGaAs QWs, we find $U_{e,h}(z_{e,h})$ as a solution of single-particle problems in a rectangular QW. We separate variables in the excitonic Schrödinger equation, and choose

$$f(\rho) = \sqrt{\frac{2}{\pi} \frac{1}{a_\perp}} e^{-\rho/a_\perp}, \tag{C.4}$$

where $a_\perp$ is a variational parameter. Substituting this trial function into the Schrödinger equation for electron–hole relative motion with Hamiltonian (C.1), we obtain the exciton binding energy

$$E_B = -\frac{3}{16} \frac{\hbar^2 a_\perp^2}{\mu L^4} - \frac{\hbar^2}{2\mu a_\perp^2} + \frac{4}{a_\perp^2} \int_0^\infty \rho d\rho e^{-2\rho/a_\perp} V(\rho) - (l_e + 1/2)\hbar\omega_c^e - (l_h + 1/2)\hbar\omega_c^h, \tag{C.5}$$

where

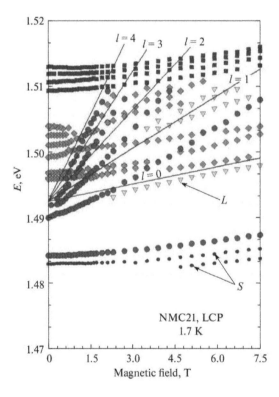

Fig. C.1: Typical "fan-diagram" of an InGaAs/GaAs QW. Circles show the resonances in transmission spectra of the sample associated with the heavy-hole exciton transition. In the limit of strong fields Landau quantisation dominates over the Coulomb interaction of electron and hole, and the energies of excitonic transitions increase linearly with field. Square and diamond correspond to the light-hole exciton transitions. From Seisyan (2016).

$$V(\rho) = \frac{e^2}{4\pi\varepsilon_0\epsilon} \int_{-\infty}^{\infty} \int_{-\infty}^{\infty} dz_e dz_h \frac{U_e^2(z_e)U_h^2(z_h)}{\sqrt{\rho^2 + (z_e - z_h)^2}}.$$ (C.6)

The parameter $a_\perp$ should maximise the binding energy. Differentiating eqn (C.5) with respect to $a_\perp$ we obtain

$$\frac{\hbar^2}{8\mu}\left[1 - \frac{a_\perp{}^4}{2L}\right] = \int_0^\infty \rho d\rho \left(1 - \frac{\rho}{a_\perp}\right)e^{-2\rho/a_\perp}V(\rho).$$ (C.7)

The exciton radiative damping rate Γ_0, defined in Chapter 3, can be expressed in terms of exciton parameters as

$$\Gamma_0 = \frac{\omega_0}{c}\omega_{LT}\sqrt{\epsilon}a_B^2 a_\perp^{-2} J_{eh}^2.$$ (C.8)

Here, $J_{eh} = \int U_e(z)U_h(z)\,dz$, ω_0 is the exciton resonance frequency and ω_{LT} and a_B are the longitudinal–transverse splitting and Bohr radius of the bulk exciton, respectively.[133] The vacuum-field Rabi splitting in a microcavity is

$$\Omega \propto \sqrt{\Gamma_0}\frac{1}{a_\perp}.$$ (C.9)

Shrinkage of the wavefunction of electron–hole relative motion in the magnetic field becomes essential if the magnetic length L is comparable to the exciton Bohr radius a_B, i.e., for magnetic fields of about 3 T and more in the case of GaAs QWs. Taking into account the fact that $L \approx 70$ Å at $B = 10$ T, the exciton Bohr radius can be realistically reduced by a factor of two.

---

[133] Typical parameters for GaAs are $\omega_{LT} = 0.08$ meV and $a_B = 14$ nm.

BIBLIOGRAPHY

S. Aaronson. Multilinear formulas and skepticism of quantum computing. *Proceedings of the Thirty-sixth Annual ACM Symposium on Theory of Computing*, 2004. (Cited on page 511).

S. Aaronson. *Quantum Computing since Democritus*. Cambridge University Press, Cambridge, 2013. (Cited on page 511).

S. Aaronson. Read the fine print. *Nat. Phys.*, 11:291, 2015. (Cited on page 511).

S. Aaronson and A. Arkhipov. The computational complexity of linear optics. *Proceedings of the 43rd Annual ACM Symposium on Theory of Computing*, page 333, 2011. (Cited on page 512).

M. Abbarchi, A. Amo, V. G. Sala, D. D. Solnyshkov, H. Flayac, L. Ferrier, I. Sagnes, E. Galopin, A. Lemaître, G. Malpuech, and J. Bloch. Macroscopic quantum self-trapping and Josephson oscillations of exciton polaritons. *Nat. Phys.*, 9:275, 2013. (Cited on page 194).

A. A. Abrikosov. Ordering and metastability and phase transitions in two-dimensional systems. *Zh. Eksp. Teor. Fiz*, 32:1442, 1957. (Cited on page 431).

A. A. Abrikosov, L. P. Gorkov, and I. E. Dzyaloshinski. *Methods of quantum field theory in statistical physics*. Dover, 1963. (Cited on page 122).

A. F. Adiyatullin, M. D. Anderson, P. V. Busi, H. Abbaspour, R. André, M. T. Portella-Oberli, and B. Deveaud. Temporally resolved second-order photon correlations of exciton-polariton Bose–Einstein condensate formation. *Appl. Phys. Lett.*, 107:221107, 2015. (Cited on page 482).

G. S. Agarwal and S. Dutta Gupta. Steady states in cavity QED due to incoherent pumping. *Phys. Rev. A*, 42:1737, 1990. (Cited on page 229).

V. Agranovich, H. Benisty, and C. Weisbuch. Organic and inorganic quantum wells in a microcavity: Frenkel–Wannier–Mott excitons hybridization and energy transformation. *Solid State Commun.*, 102:631, 1997. (Cited on page 134).

V. M. Agranovich. On the influence of reabsorption on the decay of fluorescence in molecular crystals. *Opt. Spectrosk.*, 3:84, 1957. (Cited on page 143).

V. M. Agranovich and G. F. Bassani, editors. *Electronic excitations in organic based nanostructures*, volume 31 of *Thin Films and Nanostructures*. Academic Press, Amsterdam, 2003. (Cited on page 134).

V. M. Agranovich and O. A. Dubovskii. Effect of retarded interaction on the exciton spectrum in one-dimensional and two-dimensional crystals. *JETP Lett.*, 3:233, 1966. (Cited on page 134).

V. M. Agranovich, M. Litinskaia, and D. G. Lidzey. Cavity polaritons in microcavities containing disordered organic semiconductors. *Phys. Rev. B*, 67:85311, 2003. (Cited on page 334).

D. Aharonov, W. van Dam, J. Kempe, Z. Landau, S. Lloyd, and O. Regev. Adiabatic quantum computation is equivalent to standard quantum computation. *Siam J. Comput.*, 37:166, 2007. (Cited on page 513).

H. Ajiki, T. Tsuji, K. Kawano, and K. Cho. Optical spectra and exciton-light coupled modes of a spherical semiconductor nanocrystal. *Phys. Rev. B*, 66:245322, 2002. (Cited on page 172).

N. Akhmediev and A. Ankiewicz-Kik. *Dissipative solitons*. Springer, 2005. (Cited on page 460).

M. Albiez, R. Gati, J. Fölling, S. Hunsmann, M. Cristiani, and M. K. Oberthaler. Direct observation of tunneling and nonlinear self-trapping in a single bosonic josephson junction. *Phys. Rev. Lett.*, 95:010402, 2005. (Cited on page 193).

I. L. Aleiner, B. L. Altshuler, and Y. G. Rubo. Radiative coupling and weak lasing of exciton-polariton condensates. *Phys. Rev. B*, 85:121031(R), 2012. (Cited on pages 450, 452, and 525).

Z. I. Alferov. Nobel lecture: the double heterostructure concept and its applications in physics, electronics, and technology. *Rev. Mod. Phys.*, 73:767, 2001. (Cited on page 263).

R. Alicki. Master equations for a damped nonlinear oscillator and the validity of the Markovian approximation. *Phys. Rev. A*, 40:4077, 1989. (Cited on page 214).

D. Allender, J. Bray, and J. Bardeen. Model for an exciton mechanism of superconductivity. *Phys. Rev. B*, 7:1020, 1973. (Cited on pages 377 and 378).

A. E. Almand-Hunter, H. Li, S. T. Cundiff, M. Mootz, M. Kira, and S. W. Koch. Quantum droplets of electrons and holes. *Nature*, 506:471, 2014. (Cited on page 503).

E. Altman, L. M. Sieberer, L. Chen, S. Diehl, and J. Toner. Two-dimensional superfluidity of exciton polaritons requires strong anisotropy. *Phys. Rev. X*, 5:011017, 2015. (Cited on page 368).

U. Alvarez-Rodriguez, M. Sanz, L. Lamata, and E. Solano. The forbidden quantum adder. *Sci. Rep.*, 5:11983, 2015. (Cited on page 511).

A. Amo and J. Bloch. Exciton-polaritons in lattices: A non-linear photonic simulator. *C. R. Physique*, 2016. (Cited on page 473).

A. Amo, J. Lefrère, S. Pigeon, C. Adrados, C. Ciuti, I. Carusotto, R. Houdré, E. Giacobino, and A. Bramati. Superfluidity of polaritons in semiconductor microcavities. *Nat. Phys.*, 5:805, 2009a. (Cited on pages 313, 375, and 424).

A. Amo, D. Sanvitto, F. P. Laussy, D. Ballarini, E. del Valle, M. D. Martin, A. Lemaître, J. Bloch, D. N. Krizhanovskii, M. S. Skolnick, C. Tejedor, and L. Viña. Collective fluid dynamics of a polariton condensate in a semiconductor microcavity. *Nature*, 457:291, 2009b. (Cited on pages 312, 313, 314, 375, and 424).

A. Amo, T. C. H. Liew, C. Adrados, R. Houdré, E. Giacobino, A. V. Kavokin, and A. Bramati. Exciton-polariton spin switches. *Nat. Photon.*, 4:361, 2010a. (Cited on pages 414, 424, 539, and 545).

A. Amo, S. Pigeon, C. Adrados, R. Houdré, E. Giacobino, C. Ciuti, and A. Bramati. Light engineering of the polariton landscape in semiconductor microcavities. *Phys. Rev. B*, 82:081301(R), 2010b. (Cited on page 516).

A. Amo, S. Pigeon, D. Sanvitto, V. G. Sala, R. Hivet, I. Carusotto, F. Pisanello, G. Leménager, R. Houdré, E. Giacobino, C. Ciuti, and A. Bramati. Polariton superfluids reveal quantum hydrodynamic solitons. *Science*, 332:1167, 2011. (Cited on pages 317, 424, 431, 433, and 445).

A. Amo, J. Bloch, A. Bramati, I. Carusotto, C. Ciuti, B. Deveaud, E. Giacobino, G. Grosso, A. Kamchatnov, G. Malpuech, N. Pavloff, S. Pigeon, D. Sanvitto, and D. Solnyshkov. Comment on "linear wave dynamics explains observations attributed to dark solitons in a polariton quantum fluid". *Phys. Rev. Lett.*, 115:089401, 2015. (Cited on page 434).

M. Amthor, S. Weißenseel, J. Fischer, M. Kamp, C. Schneider, and S. Höfling. Electro-optical switching between polariton and cavity lasing in an InGaAs quantum well microcavity. *Opt. Express*, 22:31146, 2014. (Cited on pages 525 and 533).

M. Amthor, T. C. H. Liew, C. Metzger, S. Brodbeck, L. Worschech, M. Kamp, I. A. Shelykh, A. V. Kavokin, C. Schneider, and S. Höfling. Optical bistability in electrically driven polariton condensates. *Phys. Rev. B*, 91:081404(R), 2015. (Cited on page 539).

M. H. Anderson, J. R. Ensher, M. R. Matthews, C. E. Wieman, and E. A. Cornell. Observation of Bose-Einstein condensation in a dilute atomic vapor. *Science*, 269:198, 1995. (Cited on pages 324 and 431).

P. W. Anderson. Considerations on the flow of superfluid helium. *Rev. Mod. Phys.*, 38:298, 1966. (Cited on pages 194 and 324).

L. C. Andreani, F. Tassone, and F. Bassani. Radiative lifetime of free excitons in quantum wells. *Solid State Commun.*, 77:641, 1991. (Cited on page 148).

L. C. Andreani, G. Panzarini, A. V. Kavokin, and M. R. Vladimirova. Effect of inhomogeneous broadening on optical properties of excitons in quantum wells. *Phys. Rev. B*, 57:4670, 1998. (Cited on page 169).

D. G. Angelakis, M. Huo, E. Kyoseva, and L. Chuan Kwek. Luttinger liquid of photons and spin-charge separation in hollow-core fibers. *Phys. Rev. Lett.*, 106:153601, 2011. (Cited on page 516).

C. Antón, T. C. H. Liew, G. Tosi, M. D. Martín, T. Gao, Z. Hatzopoulos, P. S. Eldridge, P. G. Savvidis, and L. Viña. Dynamics of a polariton condensate transistor switch. *Appl. Phys. Lett.*, 101:261116, 2012. (Cited on page 542).

C. Antón, T. C. H. Liew, D. Sarkar, M. D. Martín, Z. Hatzopoulos, P. S. Eldridge, P. G. Savvidis, and L. Viña. Operation speed of polariton condensate switches gated by excitons. *Phys. Rev. B*, 89:235312, 2014. (Cited on pages 194 and 545).

L. Apker and E. Taft. Photoelectric emission from F-centers in KI. *Phys. Rev.*, 79:964, 1950. (Cited on page 133).

V. Ardizzone, M. Abbarchi, A. Lemaître, I. Sagnes, P. Senellart, J. Bloch, C. Delalande, J. Tignon, and P. Roussignol. Bunching visibility of optical parametric emission in a semiconductor microcavity. *Phys. Rev. B*, 86, 2012. (Cited on page 484).

D. K. Armani, T. J. Kippenberg, S. M. Spillane, and K. J. Vahala. Ultra-high-Q toroid microcavity on a chip. *Nature*, 421:925, 2003. (Cited on page 74).

H. F. Arnoldus and G. Nienhuis. Photon correlations between the lines in the spectrum of resonance fluorescence. *J. Phys. B.: At. Mol. Phys.*, 17:963, 1984. (Cited on page 491).

A. Askitopoulos, H. Ohadi, A. V. Kavokin, Z. Hatzopoulos, P. G. Savvidis, and P. G. Lagoudakis. Polariton condensation in an optically induced two-dimensional potential. *Phys. Rev. B*, 88:041308(R), 2013. (Cited on page 537).

A. Aspect, G. Roger, S. Reynaud, J. Dalibard, and C. Cohen-Tannoudji. Time correlations between the two sidebands of the resonance fluorescence triplet. *Phys. Rev. Lett.*, 45:617, 1980. (Cited on page 491).

M. Aßmann and M. Bayer. Nonlinearity sensing via photon-statistics excitation spectroscopy. *Phys. Rev. A*, 84:053806, 2011. (Cited on page 503).

M. Aßmann and M. Bayer. Stochastic pumping of a polariton fluid. *Phys. Rev. A*, 91:053835, 2015. (Cited on page 504).

M. Aßmann, J.-S. Tempel, F. Veita, M. Bayer, A. Rahimi-Iman, A. Löffler, S. Höfling, S. Reitzenstein, L. Worschech, and A. Forchel. From polariton condensates to highly photonic quantum degenerate states of bosonic matter. *Proc. Natl. Acad. Sci.*, 108:1804, 2011. (Cited on pages 480 and 483).

S. Azzini, D. Gerace, M. Galli, I. Sagnes, R. Braive, A. Lemaître, J. Bloch, and D. Bajoni. Ultra-low threshold polariton lasing in photonic crystal cavities. *Appl. Phys. Lett.*, 99:111106, 2011. (Cited on page 523).

A. Baas, J. Ph. Karr, H. Eleuch, and E. Giacobino. Optical bistability in semiconductor microcavities. *Phys. Rev. A*, 69:023809, 2004. (Cited on pages 304 and 539).

A. Baas, K. G. Lagoudakis, M. Richard, R. André, Le Si Dang, and B. Deveaud-Plédran. Synchronized and desynchronized phases of exciton-polariton condensates in the presence of disorder. *Phys. Rev. Lett.*, 100:170401, 2008. (Cited on page 536).

A. Badolato, K. Hennessy, M. Atature, J. Dreyser, E. Hu, P. M. Petroff, and A. İmamoğlu. Deterministic coupling of single quantum dots to single nanocavity modes. *Science*, 308:1158, 2005. (Cited on page 255).

D. Bajoni, M. Perrin, P. Senellart, A. Lemaître, B. Sermage, and J. Bloch. Dynamics of microcavity polaritons in the presence of an electron gas. *Phys. Rev. B*, 73:205344, 2006. (Cited on page 338).

D. Bajoni, P. Senellart, A. Lemaître, and J. Bloch. Photon lasing in GaAs microcavity: Similarities with a polariton condensate. *Phys. Rev. B*, 76:201305(R), 2007. (Cited on page 374).

D. Bajoni, E. Semenova, A. Lemaître, S. Bouchoule, E. Wertz, P. Senellart, and J. Bloch. Polariton light-emitting diode in a GaAs-based microcavity. *Phys. Rev. B*, 77:113303, 2008a. (Cited on pages 374 and 524).

D. Bajoni, P. Senellart, E. Wertz, I. Sagnes, A. Miard, A. Lemaître, and J. Bloch. Polariton laser using single micropillar GaAs-GaAlAs semiconductor cavities. *Phys. Rev. Lett.*, 100:047401, 2008b. (Cited on page 523).

D. Bajoni, E. Wertz, P. Senellart, A. Miard, E. Semenova, A. Lemaître, I. Sagnes, S. Bouchoule, and J. Bloch. Excitonic polaritons in semiconductor micropillars. *Acta Physica Polonica A*, 144:933, 2008c. (Cited on page 537).

R. Balili, V. Hartwell, D. Snoke, L. Pfeiffer, and K. West. Bose–Einstein condensation of microcavity polaritons in a trap. *Science*, 316:1007, 2007. (Cited on pages 372 and 521).

R. B. Balili, D. W. Snoke, L. Pfeiffer, and K. West. Actively tuned and spatially trapped polaritons. *Appl. Phys. Lett.*, 88:031110, 2006. (Cited on page 372).

D. Ballarini, M. De Giorgi, E. Cancellieri, R. Houdré, E. Giacobino, R. Cingolani, A. Bramati, G. Gigli, and D. Sanvitto. All-optical polariton transistor. *Nat. Comm.*, 4:1778, 2013. (Cited on pages 541, 542, and 543).

M. Bamba, S. Pigeon, and C. Ciuti. Quantum squeezing generation versus photon localization in a disordered planar microcavity. *Phys. Rev. Lett.*, 104:213604, 2010. (Cited on page 481).

M. Bamba, A. İmamoğlu, I. Carusotto, and C. Ciuti. Origin of strong photon antibunching in weakly nonlinear photonic molecules. *Phys. Rev. A*, 83:021802(R), 2011. (Cited on pages 488 and 489).

M. A. Bandres. Accelerating beams. *Opt. Lett.*, 34:3791, 2009. (Cited on page 307).

J. Bardeen. Tunnelling from a many-particle point of view. *Phys. Rev. Lett.*, 6:57, 1961. (Cited on page 191).

J. Bardeen, L. N. Cooper, and J. R. Schrieffer. Theory of superconductivity. *Phys. Rev.*, 108:1175, 1957. (Cited on page 377).

C.-E. Bardyn, T. Karzig, G. Refael, and T. C. H. Liew. Topological polaritons and excitons in garden-variety systems. *Phys. Rev. B*, 91:161413(R), 2015. (Cited on page 467).

C. F. Barenghi, L. Skrbek, and K. R. Sreenivasan. Introduction to quantum turbulence. *Proc. Natl. Acad. Sci.*, 111:4647, 2014. (Cited on page 425).

S. Barland, J. R. Tredicce, M. Brambilla, L. A. Lugiato, S. Balle, M. Giudici, T. Maggipinto, L. Spinelli, G. Tissoni, T. Knödl, M. Miller, and R. Jäger. Cavity solitons as pixels in semiconductor microcavities. *Nature*, 419:699, 2002. (Cited on page 460).

N. Basov. Semiconductor lasers. *Nobel Lecture*, 1964. (Cited on page 256).

G. Bastard. *Wave mechanics applied to semiconductor heterostructures*. Les éditions de physique, Paris, 1988. (Cited on page 137).

J. J. Baumberg, A. P. Heberle, A. V. Kavokin, M. R. Vladimirova, and K. Köhler. Polariton motional narrowing in semiconductor multiple quantum wells. *Phys. Rev. Lett.*, 80:3567, 1998. (Cited on pages 281 and 286).

J. J. Baumberg, P. G. Savvidis, R. M. Stevenson, A. I. Tartakovskii, M. S. Skolnick, D. M. Whittaker, and J. S. Roberts. Parametric oscillation in a vertical microcavity: A polariton condensate or micro-optical parametric oscillation. *Phys. Rev. B*, 62:R16247, 2000. (Cited on page 292).

J. J. Baumberg, A. V. Kavokin, S. Christopoulos, A. J. D. Grundy, R. Butté, G. Christmann, D. D. Solnyshkov, G. Malpuech, G. Baldassarri Höger von Högersthal, E. Feltin, J.-F. Carlin, and N. Grandjean. Spontaneous polarization buildup in a room-temperature polariton laser. *Phys. Rev. Lett.*, 101:136409, 2008. (Cited on pages 21, 326, 364, and 374).

J. Baumgartl, M. Mazilu, and K. Dholakia. Optically mediated particle clearing using Airy wavepackets. *Nat. Photon.*, 2:675, 2008. (Cited on page 307).

M. Bayer, T. L. Reinecke, F. Weidner, A. Larionov, A. McDonald, and A. Forchel. Inhibition and enhancement of the spontaneous emission of quantum dots in structured microresonators. *Phys. Rev. Lett.*, 86:3168, 2001. (Cited on pages 250 and 255).

G. Baym. The physics of Hanbury Brown–Twiss intensity interferometry: from stars to nuclear collisions. *Acta Physica Polonica B*, 29:1839, 1998. (Cited on page 43).

A. M. Beckley, T. G. Brow, and M. A. Alonso. Full poincaré beams. *Opt. Express*, 10:10777, 2010. (Cited on page 402).

G. Bel and F. L. H. Brown. Theory for wavelength-resolved photon emission statistics in single-molecule fluorescence spectroscopy. *Phys. Rev. Lett.*, 102:018303, 2009. (Cited on page 492).

J. S. Bell. On the problem of hidden variables in quantum mechanics. *Rev. Mod. Phys.*, 38:447, 1966. (Cited on page 305).

V. V. Belykh, N. N. Sibeldin, V. D. Kulakovskii, M. M. Glazov, M. A. Semina, C. Schneider, S. Höfling, M. Kamp, and A. Forchel. Coherence expansion and polariton condensate formation in a semiconductor microcavity. *Phys. Rev. Lett.*, 110:137402, 2013. (Cited on page 536).

C. H. Bennett and G. Brassard. Quantum cryptography: Public key distribution and coin tossing. *Proc. IEEE*, page 175, 1984. (Cited on page 511).

M. Bennett, M. F. Schatz, H. Rockwood, and K. Wiesenfeld. Huygens's clocks. *Proc. Math, Phys. and Eng.*, 458:563, 2002. (Cited on page 525).

V.L. Berezinskii. Violation of long range order in one-dimensional and two-dimensional systems with a continuous symmetry group. *Sov. Phys. JETP*, 32:493, 1971. (Cited on page 430).

J.D. Berger, O. Lyngnes, H.M. Gibbs, G. Khitrova, T.R. Nelson, E.K. Lindmark, A.V. Kavokin, M.A. Kaliteevski, and V.V. Zapasskii. Magnetic-field enhancement of the exciton-polariton splitting in a semiconductor quantum-well microcavity: The strong coupling threshold. *Phys. Rev. B*, 54:1975, 1996. (Cited on pages 388 and 389).

O.L. Berman, Y.E. Lozovik, D.W. Snoke, and R.D Coalson. Collective properties of indirect excitons in coupled quantum wells in a random field. *Phys. Rev. B*, 70:235310, 2004. (Cited on page 334).

M. Berry. Quantum physics on the edge of chaos. *New Scientist*, 116 (1587):44, 1987. (Cited on page 306).

M. V. Berry and N. L. Balazs. Nonspreading wave packets. *Am. J. Phys.*, 47:264, 1979. (Cited on pages 306 and 307).

P. Bertet, S. Osnaghi, P. Milman, A. Auffèves, P. Maioli, M. Brune, J. M. Raimond, and S. Haroche. Generating and probing a two-photon Fock state with a single atom in a cavity. *Phys. Rev. Lett.*, 88:143601, 2002. (Cited on page 105).

B. Besga, C. Vaneph, J. Reichel, J. Estève, A. Reinhard, J. Miguel-Sánchez, A. İmamoğlu, and T. Volz. Polariton boxes in a tunable fiber cavity. *Phys. Rev. Appl.*, 3:014008, 2015. (Cited on page 488).

H. Bethe. On the theory of metals. I. Eigenvalues and eigenfunctions of the linear chain of atoms. *Zeitschrift für Physik*, 71:205, 1931. (Cited on page 502).

P. Bhattacharya, B. Xiao, A. Das, S. Bhowmick, and J. Heo. Solid state electrically injected exciton-polariton laser. *Phys. Rev. Lett.*, 110:206403, 2013. (Cited on page 524).

P. Bhattacharya, T. Frost, S. Deshpande, M. Z. Baten, A. Hazari, and A. Das. Room temperature electrically injected polariton laser. *Phys. Rev. Lett.*, 112:236802, 2014. (Cited on pages 525 and 545).

K.M. Birnbaum, A. Boca, R. Miller, A.D. Boozer, T.E. Northup, and H.J. Kimble. Photon blockade in an optical cavity with one trapped atom. *Nature*, 436:87, 2005. (Cited on pages 485 and 486).

L. S. Bishop, J. M. Chow, J. Koch, A. A. Houck, M. H. Devoret, E. Thuneberg, S. M. Girvin, and R. J. Schoelkopf. Nonlinear response of the vacuum Rabi resonance. *Nat. Phys.*, 5:105, 2009. (Cited on page 222).

G. Björk and Y. Yamamoto. Analysis of semiconductor microcavity lasers using rate equations. *IEEE Quantum Electron.*, 27:2386, 1991. (Cited on pages 265 and 266).

G. Björk, H. Heitmann, and Y. Yamamoto. Spontaneous-emission coupling factor and mode characteristics of planar dielectric microcavity lasers. *Phys. Rev. A*, 47:4451, 1993. (Cited on page 266).

J. M. Blatt, K. W. Böer, and W. Brandt. Bose–Einstein condensation of excitons. *Phys. Rev.*, 126:1691, 1962. (Cited on pages 324 and 520).

I. Bloch, T. W. Hänsch, and T. Esslinger. Atom laser with a cw output coupler. *Phys. Rev. Lett.*, 82:3008, 1999. (Cited on page 520).

I. Bloch, J. Dalibard, and S. Nascimbène. Quantum simulations with ultracold quantum gases. *Nat. Phys.*, 8:267, 2012. (Cited on page 514).

A. Boca, R. Miller, K. M. Birnbaum, A. D. Boozer, J. McKeever, and H. J. Kimble. Observation of the vacuum Rabi spectrum for one trapped atom. *Phys. Rev. Lett.*, 93:233603, 2004. (Cited on page 210).

N. N. Bogoliubov. Theory of the weakly interacting Bose gas. *J. Phys. (Moscow)*, 11:23, 1947. (Cited on page 298).

N. N. Bogoliubov. *Lectures on quantum statistics, Vol. 1: Quantum statistics*. Gordon and Breach Science Publisher, New York, London, Paris, 1970. (Cited on page 298).

D. Bohm. A suggested interpretation of the quantum theory in terms of "hidden" variables. I. *Phys. Rev.*, 85:166, 1952a. (Cited on page 305).

D. Bohm. A suggested interpretation of the quantum theory in terms of "hidden" variables. II. *Phys. Rev.*, 85:180, 1952b. (Cited on page 305).

D. Bohm and Y. Aharonov. Discussion of experimental proof for the paradox of Einstein, Rosen, and Podolsky. *Phys. Rev.*, 108:1070, 1957. (Cited on page 88).

J. G. Bohnet, Z. Chen, J. M. Weiner, D. Meiser, M. J. Holland, and J. K. Thompson. A steady-state superradiant laser with less than one intracavity photon. *Nature*, 484:78, 2012. (Cited on page 532).

S. Boixo, T. F. Rønnow, S. V. Isakov, Z. Wang, D. Wecker, D. A. Lidar, J. M. Martinis, and M. Troyer. Evidence for quantum annealing with more than one hundred qubits. *Nat. Phys.*, 10:218, 2014. (Cited on page 514).

S. Boixo, V. N. Smelyanskiy, A. Shabani, S. V. Isakov, M. Dykman, V. S. Denchev, M. H. Amin, A. Yu Smirnov, M. Mohseni, and H. Neven. Computational multiqubit tunnelling in programmable quantum annealers. *Nat. Comm.*, 7:10327, 2016. (Cited on page 514).

M. O. Borgh, J. Keeling, and N. G. Berloff. Spatial pattern formation and polarization dynamics of a nonequilibrium spinor polariton condensate. *Phys. Rev. B*, 81:235302, 2010. (Cited on page 536).

M. Born. Quantenmechanik der stoßvorgänge. *Zeitschrift für Physik*, 38:803, 1926. (Cited on page 305).

M. Born and E. Wolf. *Principles of Optics*. Pergamon Press, 1970. (Cited on pages 38, 51, and 77).

S. N. Bose. Plancks Gesetz und Lichtquantenhypothese. *Z. Phys.*, 26: 178, 1924. (Cited on page 321).

T. Boulier, M. Bamba, A. Amo, C. Adrados, A. Lemaître, E. Galopin, I. Sagnes, J. Bloch, C. Ciuti, E. Giacobino, and A. Bramati. Polariton-generated intensity squeezing in semiconductor micropillars. *Nat. Comm.*, 5:3260, 2014. (Cited on page 481).

T. Boulier, H. Terças, D. D. Solnyshkov, Q. Glorieux, E. Giacobino, G. Malpuech, and A. Bramati. Vortex chain in a resonantly pumped polariton superfluid. *Sci. Rep.*, 5:9230, 2015. (Cited on page 453).

G. Breit and E. Teller. Metastability of hydrogen and helium levels. *Ap. J.*, 91:215, 1940. (Cited on page 496).

G. Breitenbach, S. Schiller, and J. Mlynek. Measurement of the quantum states of squeezed light. *Nature*, 387:471, 1997. (Cited on page 480).

T. Brixner and G. Gerber. Femtosecond polarization pulse shaping. *Opt. Lett.*, 26:557, 2001. (Cited on page 402).

S. Brodbeck, H. Suchomel, M. Amthor, A. Wolf, M. Kamp, C. Schneider, and S. Höfling. Impact of lateral carrier confinement on electro-optical tuning properties of polariton condensates. *Appl. Phys. Lett.*, 107:041108, 2015. (Cited on page 525).

M. Brune, F. Schmidt-Kaler, A. Maali, J. Dreyer, E. Hagley, J. M. Raimond, and S. Haroche. Quantum Rabi oscillation: A direct test of field quantization in a cavity. *Phys. Rev. Lett.*, 76:1800, 1996. (Cited on page 199).

A. Brunetti, M. Vladimirova, D. Scalbert, M. Nawrocki, A. V. Kavokin, I. A. Shelykh, and J. Bloch. Observation of spin beats at the Rabi frequency in microcavities. *Phys. Rev. B*, 74:241101(R), 2006. (Cited on pages 391, 402, and 406).

L. V. Butov, A. L. Ivanov, A. İmamoğlu, P. B. Littlewood, A. A. Shashkin, V. T. Dolgopolov, K. L. Campman, and A. C. Gossard. Stimulated scattering of indirect excitons in coupled quantum wells: Signature of a degenerate Bose-gas of excitons. *Phys. Rev. Lett.*, 86:5608, 2001. (Cited on pages 287 and 325).

L. V. Butov, C. W. Lai, A. L. Ivanov, A. C. Gossard, and D. S. Chemla. Towards Bose–Einstein condensation of excitons in potential traps. *Nature*, 417:47, 2002. (Cited on page 325).

R. Butté, G. Delalleau, A. I. Tartakovskii, M. S. Skolnick, V. N. Astratov, J. J. Baumberg, G. Malpuech, A. Di Carlo, A. V. Kavokin, and J. S. Roberts. Transition from strong to weak coupling and the onset of lasing in semiconductor microcavities. *Phys. Rev. B*, 65:205310, 2002. (Cited on page 335).

R. Butté, S. Skolnick, D. M. Whittaker, D. Bajoni, and J. S. Roberts. Dependence of stimulated scattering in semiconductor microcavities on pump power, angle, and energy. *Phys. Rev. B*, 68:115325, 2003. (Cited on page 295).

Yu. A. Bychkov and E. I. Rashba. Properties of a 2D electron gas with lifted spectral degeneracy. *JETP Lett.*, 39:78, 1984. (Cited on page 399).

T. Byrnes, P. Recher, and Y. Yamamoto. Mott transitions of exciton polaritons and indirect excitons in a periodic potential. *Phys. Rev. B*, 81:205312, 2010. (Cited on page 514).

T. Byrnes, K. Yan, and Y. Yamamoto. Accelerated optimization problem search using Bose–Einstein condensation. *New J. Phys.*, 13: 113025, 2011. (Cited on page 515).

T. Byrnes, S. Koyama, K. Yan K, and Y. Yamamoto. Neural networks using two-component Bose–Einstein condensates. *Sci. Rep.*, 3: 2531, 2013. (Cited on page 515).

E. Cancellieri, F. M. Marchetti, M. H. Szymańska, and C. Tejedor. Multistability of a two-component exciton-polariton fluid. *Phys. Rev. B*, 83:214507, 2011. (Cited on page 539).

H. J. Carmichael. Quantum trajectory theory for cascaded open systems. *Phys. Rev. Lett.*, 70:2273, 1993. (Cited on page 503).

H. J. Carmichael. *Statistical methods in quantum optics 1*. Springer, 2 edition, 2002. (Cited on page 344).

H. J. Carmichael, R. J. Brecha, and P. R. Rice. Quantum interference and collapse of the wavefunction in cavity QED. *Opt. Commun.*, 82:73, 1991. (Cited on pages 488 and 489).

I. Carusotto. Nonlinear atomic Fabry–Perot interferometer: From the mean-field theory to the atom blockade effect. *Phys. Rev. A*, 63: 023610, 2001. (Cited on page 487).

I. Carusotto and C. Ciuti. Probing microcavity polariton superfluidity through resonant Rayleigh scattering. *Phys. Rev. Lett.*, 93:166401, 2004. (Cited on pages 313 and 315).

I. Carusotto and C. Ciuti. Quantum fluids of light. *Rev. Mod. Phys.*, 85:299, 2013. (Cited on pages 315, 424, and 473).

Y. Castin and J. Dalibard. Relative phase of two Bose–Einstein condensates. *Phys. Rev. A*, 55:4330, 1997. (Cited on page 194).

I. M. Catalano, A. Cingolani, R. Cingolani, M. Lepore, and K. Ploog. Two-photon spectroscopy in GaAs/Al$_x$Ga$_{1-x}$As multiple quantum wells. *Phys. Rev. B*, 40:1312, 1989. (Cited on page 527).

R. Centeno Neelen, D. M. Boersma, M. P. van Exter, G. Nienhuis, and J. P. Woerdman. Spectral filtering within the Schawlow–Townes linewidth as a diagnostic tool for studying laser phase noise. *Opt. Commun.*, 100:289, 1993. (Cited on page 492).

E. A. Cerda-Méndez, D. Sarkar, D. N. Krizhanovskii, S. S. Gavrilov, K. Biermann, M. S. Skolnick, and P. V. Santos. Exciton-polariton gap solitons in two-dimensional lattices. *Phys. Rev. Lett.*, 111: 146401, 2013. (Cited on pages 453 and 460).

R. Cerna, Y. Léger, T.K. Paraïso, M. Wouters, F. Morier-Genoud, M.T. Portella-Oberli, and B. Deveaud. Ultrafast tristable spin memory of a coherent polariton gas. *Nat. Commn.*, 4:2008, 2013. (Cited on pages 539 and 545).

C.J. Chang-Hasnain. Tunable VCSEL. *IEEE Quantum Electron.*, 6: 978, 2000. (Cited on page 14).

W. Chen and D. L. Mills. Gap solitons and the nonlinear optical response of superlattices. *Phys. Rev. Lett.*, 58:160, 1987. (Cited on pages 425 and 460).

W.-J. Chen, S.-J. Jiang, X.-D. Chen, B. Zhu, L. Zhou, J.-W. Dong, and C. T. Chan. Experimental realization of photonic topological insulator in a uniaxial metacrystal waveguide. *Nat. Comm.*, 5:5782, 2014. (Cited on page 467).

I. Yu. Chestnov, S. S. Demirchyan, A. P. Alodjants, Yu. G. Rubo, and A. V. Kavokin. Permanent Rabi oscillations in coupled exciton-photon systems with PT-symmetry. *Sci. Rep.*, 6:19551, 2016. (Cited on page 480).

W. C. Chew. *Waves and fields in inhomogeneous media*. IEEE Press, New York, 1995. (Cited on pages 74 and 77).

A. Chiocchetta and I. Carusotto. Non-equilibrium quasi-condensates in reduced dimensions. *Europhys. Lett.*, 102:67007, 2013. (Cited on page 368).

S. Choi, S. A. Morgan, and K. Burnett. Phenomenological damping in trapped atomic Bose–Einstein condensates. *Phys. Rev. A*, 57: 4057, 1998. (Cited on page 368).

G. Christmann, R. Butté, E. Feltin, J.-F Carlin, and N. Grandjean. Room temperature polariton lasing in a GaN/AlGaN multiple quantum well microcavity. *Appl. Phys. Lett.*, 93:051102, 2008. (Cited on pages 373 and 374).

G. Christmann, C. Coulson, J. J. Baumberg, N. T. Pelekanos, Z. Hatzopoulos, S. I. Tsintzos, and P. G. Savvidis. Control of polariton scattering in resonant-tunneling double-quantum-well semiconductor microcavities. *Phys. Rev. B*, 82:113308, 2010. (Cited on page 533).

G. Christmann, G. Tosi, N. G. Berloff, P. Tsotsis, P. S. Eldridge, Z. Hatzopoulos, P. G. Savvidis, and J. J. Baumberg. Polariton ring condensates and sunflower ripples in an expanding quantum liquid. *Phys. Rev. B*, 85:235303, 2012. (Cited on pages 507 and 537).

S. Christopoulos, G. Baldassarri Höger von Högersthal, A. J. D. Grundy, P. G. Lagoudakis, A. V. Kavokin, J. J. Baumberg, G. Christmann, R. Butté, E. Feltin, J.-F. Carlin, and N. Grandjean. Room-temperature polariton lasing in semiconductor microcavities. *Phys. Rev. Lett.*, 98:126405, 2007. (Cited on pages 21, 374, and 523).

P. Cilibrizzi, H. Ohadi, T. Ostatnicky, A. Askitopoulos, W. Langbein, and P. Lagoudakis. Linear wave dynamics explains observations attributed to dark solitons in a polariton quantum fluid. *Phys. Rev. Lett.*, 113:103901, 2014. (Cited on page 434).

P. Cilibrizzi, H. Ohadi, T. Ostatnicky, A. Askitopoulos, W. Langbein, and P. Lagoudakis. Cilibrizzi *et al.* reply. *Phys. Rev. Lett.*, 115: 089402, 2015. (Cited on page 434).

C. Ciuti. Branch-entangled polariton pairs in planar microcavities and photonic wires. *Phys. Rev. B*, 69:245304, 2004. (Cited on pages 294 and 483).

C. Ciuti, V. Savona, C. Piermarocchi, A. Quattropani, and P. Schwendimann. Role of the exchange of carriers in elastic exciton-exciton scattering in quantum wells. *Phys. Rev. B*, 58: 7926, 1998. (Cited on page 552).

C. Ciuti, P. Schwendimann, B. Deveaud, and A. Quattropani. Theory of the angle-resonant polariton amplifier. *Phys. Rev. B*, 62:R4825, 2000. (Cited on pages 297, 298, 299, 303, and 483).

C. Ciuti, P. Schwendimann, and A. Quattropani. Parametric luminescence of microcavity polaritons. *Phys. Rev. B*, 63:041303R, 2001. (Cited on pages 297 and 483).

C. Ciuti, G. Bastard, and I. Carusotto. Quantum vacuum properties of the intersubband cavity polariton field. *Phys. Rev. B*, 72:115303, 2005. (Cited on page 283).

R. Cleve, A. Ekert, C. Macchiavello, and M. Mosca. Quantum algorithms revisited. *Proc. Roy. Soc. A*, 454:339, 1998. (Cited on page 509).

M. H. Cohen, L. M. Falicov, and J. C. Phillips. Superconductive tunneling. *Phys. Rev. Lett.*, 8:316, 1962. (Cited on page 191).

C. Cohen-Tannoudji and S. Reynaud. Atoms in strong light-fields: Photon antibunching in single atom fluorescence. *Phil. Trans. R. Soc. Lond. A*, 293:223, 1979. (Cited on page 491).

C. Cohen-Tannoudji, J Dupont-Roc, and G. Grynberg. *Photons et atomes*. EDP Sciences, 2001. (Cited on page 122).

D. Colas and F.P. Laussy. Self-interfering wave packets. *Phys. Rev. Lett.*, 116:026401, 2016. (Cited on pages 309, 310, and 311).

D. Colas, L. Dominici, S. Donati, A. A Pervishko, T.C.H. Liew, I. A. Shelykh, D. Ballarini, M. de Giorgi, A. Bramati, G. Gigli, E. del Valle, F. P Laussy, A. V. Kavokin, and D. Sanvitto. Polarization shaping of Poincaré beams by polariton oscillations. *Light: Sci. & App.*, 4:e350, 2015. (Cited on pages 403, 404, and 480).

L. A. Coldren and S. W. Corzine. *Diode lasers and photonic integrated circuits.* Wiley, 1995. (Cited on page 520).

S. Coleman. There are no Goldstone bosons in two dimensions. *Comm. math. Phys*, 31:259, 1973. (Cited on page 326).

L. Collot, V. Lefevre-Seguin, M. Brune, J. M. Raimond, and S. Haroche. Very high-Q whispering-gallery mode resonances observed on fused silica microspheres. *Europhys. Lett.*, 23:327, 1993. (Cited on page 76).

M. Combescot and S.-Y. Shiau. *Excitons and Cooper Pairs: Two Composite Bosons in Many-Body Physics.* Oxford Graduate Texts, 2015. (Cited on page 379).

M. Combescot, O. Betbeder-Matibet, and R. Combescot. Exciton-exciton scattering: Composite boson versus elementary boson. *Phys. Rev. B*, 75:174305, 2007a. (Cited on page 552).

M. Combescot, O. Betbeder-Matibet, and R. Combescot. Bose–Einstein condensation in semiconductors: The key role of dark excitons. *Phys. Rev. Lett.*, 99:176403, 2007b. (Cited on page 384).

C. Comte and P. Nozières. Exciton Bose condensation – the ground state of an electron hole gas: Mean field description of a simplified model. *J. Phys.*, 43:1069, 1982. (Cited on page 325).

L. N. Cooper. Bound electron pairs in a degenerate Fermi gas. *Phys. Rev.*, 104:1189, 1956. (Cited on page 377).

Y. Couder, E. Fort, C.-H. Gautier, and A. Boudaoud. From bouncing to floating: Noncoalescence of drops on a fluid bath. *Phys. Rev. Lett.*, 94:177801, 2005. (Cited on page 305).

Y. Couder, A. Boudaoud, S. Protière, and E. Fort. Walking droplets, a form of wave-particle duality at macroscopic scale? *Europhys. News*, 41:14, 2010. (Cited on page 306).

J. D. Cresser. Intensity correlations of frequency-filtered light fields. *J. Phys. B.: At. Mol. Phys.*, 20:4915, 1987. (Cited on page 491).

P. Cristofolini, G. Christmann, S. I. Tsintzos, G. Deligeorgis, G. Konstantinidis, Z. Hatzopoulos, P. G. Savvidis, and J. J. Baumberg. Coupling quantum tunneling with cavity photons. *Science*, 336:704, 2012. (Cited on pages 517 and 530).

P. Cristofolini, A. Dreismann, G. Christmann, G. Franchetti, N. G. Berloff, P. Tsotsis, Z. Hatzopoulos, P. G. Savvidis, and J. J. Baumberg. Optical superfluid phase transitions and trapping of polariton condensates. *Phys. Rev. Lett.*, 110:186403, 2013. (Cited on page 537).

G. Cui, J. M. Hannigan, R. Loeckenhoff, F. M. Matinaga, M. G. Raymer, S. Bhongale, M. Holland, S. Mosor, S. Chatterjee, H. M. Gibbs, and G. Khitrova. A hemispherical, high-solid-angle optical micro-cavity for cavity-QED studies. *Opt. Express*, 14:2289, 2006. (Cited on page 11).

F. W. Cummings. Reminiscing about thesis work with E. T. Jaynes at Stanford in the 1950s. *J. Phys. B.: At. Mol. Phys.*, 46:220202, 1950. (Cited on page 196).

O. El Daïf, A. Baas, T. Guillet, J.-P. Brantut, R. Idrissi Kaitouni, J. L. Staehli, F. Morier-Genoud, and B. Deveaud. Polariton quantum boxes in semiconductor microcavities. *Appl. Phys. Lett.*, 88:061105, 2006. (Cited on page 515).

J. Dalibard and S. Reynaud. Correlation signals in resonance fluorescence : interpretation via photon scattering amplitudes. *J. Phys. France*, 44:1337, 1983. (Cited on page 491).

R. Dall, M. D. Fraser, A. S. Desyatnikov, G. Li, S. Brodbeck, M. Kamp, C. Schneider, S. Höfling, and E. A. Ostrovskaya. Creation of orbital angular momentum states with chiral polaritonic lenses. *Phys. Rev. Lett.*, 113:200404, 2014. (Cited on page 537).

C. G. Darwin. Free motion in the wave mechanics. *Proc. Roy. Soc*, A117:258, 1928. (Cited on page 305).

G. Dasbach, A. A. Dremin, M. Bayer, V. D. Kulakovskii, N. A. Gippius, and A. Forchel. Oscillations in the differential transmission of a semiconductor microcavity with reduced symmetry. *Phys. Rev. B*, 65:245316, 2003. (Cited on page 289).

K. S. Daskalakis, P. S. Eldridge, G. Christmann, E. Trichas, R. Murray, E. Iliopoulos, E. Monroy, N. T. Pelekanos, J. J. Baumberg, and P. G. Savvidis. All-dielectric GaN microcavity: Strong coupling

and lasing at room temperature. *Appl. Phys. Lett.*, 102:101113, 2013. (Cited on page 523).

K. S. Daskalakis, S. A. Maier, R. Murray, and S. Kéna-Cohen. Nonlinear interactions in an organic polariton condensate. *Nat. Mater.*, 13:271, 2014. (Cited on page 523).

P. C. W. Davies. The implications of a cosmological information bound for complexity, quantum information and the nature of physical law. *Fluct. Noise Lett.*, 7:C37, 2007. (Cited on page 511).

L. de Broglie. La mécanique ondulatoire et la structure atomique de la matière et du rayonnement. *J. Phys. Radium*, 8:225, 1927. (Cited on page 305).

F. de Martini, G. Innocenti, G. R. Jacobovitz, and P. Mataloni. Anomalous spontaneous emission time in a microscopic optical cavity. *Phys. Rev. Lett.*, 59:2955, 1987. (Cited on page 255).

E. del Valle. Strong and weak coupling of two coupled qubits. *Phys. Rev. A*, 81:053811, 2010a. (Cited on page 226).

E. del Valle. *Microcavity Quantum Electrodynamics.* VDM Verlag, 2010b. (Cited on pages 112 and 226).

E. del Valle. Distilling one, two and entangled pairs of photons from a quantum dot with cavity QED effects and spectral filtering. *New J. Phys.*, 15:025019, 2013. (Cited on page 496).

E. del Valle and F. P. Laussy. Effective cavity pumping from weakly coupled quantum dots. *Superlatt. Microstruct.*, 49:241, 2010. (Cited on pages 212 and 228).

E. del Valle and F. P. Laussy. Regimes of strong light-matter coupling under incoherent excitation. *Phys. Rev. A*, 84:043816, 2011. (Cited on page 223).

E. del Valle, F. P. Laussy, and C. Tejedor. Luminescence spectra of quantum dots in microcavities. II. Fermions. *Phys. Rev. B*, 79:235326, 2009a. (Cited on pages 201, 226, 227, 358, 364, and 365).

E. del Valle, D. Sanvitto, A. Amo, F. P. Laussy, R. André, C. Tejedor, and L. Viña. Dynamics of the formation and decay of coherence in a polariton condensate. *Phys. Rev. Lett.*, 103:096404, 2009b. (Cited on page 358).

E. del Valle, A. González-Tudela, E. Cancellieri, F. P. Laussy, and C. Tejedor. Generation of a two-photon state from a quantum dot in a microcavity. *New J. Phys.*, 13:113014, 2011. (Cited on page 500).

E. del Valle, A. González-Tudela, F. P. Laussy, C. Tejedor, and M. J. Hartmann. Theory of frequency-filtered and time-resolved n-photon correlations. *Phys. Rev. Lett.*, 109:183601, 2012. (Cited on pages 492, 493, and 494).

C. Dembowski, H.-D. Gräf, H. L. Harney, A. Heine, W. D. Heiss, H. Rehfeld, and A. Richter. Experimental observation of the topological structure of exceptional points. *Phys. Rev. Lett.*, 86:787, 2001. (Cited on page 376).

S. S. Demirchyan, I. Yu. Chestnov, A. P. Alodjants, M. M. Glazov, and A. V. Kavokin. Qubits based on polariton rabi oscillators. *Phys. Rev. Lett.*, 112:196403, 2014. (Cited on page 480).

H. Deng, G. Weihs, C. Santori, J. Bloch, and Y. Yamamoto. Condensation of semiconductor microcavity exciton polaritons. *Science*, 298:199, 2002. (Cited on pages 481, 482, and 521).

H. Deng, D. Press, S. Götzinger, G. S. Solomon, R. Hey, K. H. Ploog, and Y. Yamamoto. Quantum degenerate exciton-polaritons in thermal equilibrium. *Phys. Rev. Lett.*, 97:146402, 2006. (Cited on page 372).

D. Deutsch. Quantum theory, the Church-Turing principle and the universal quantum computer. *Proc. R. Soc. Lond. A*, 400:97, 1985. (Cited on page 508).

D. Deutsch. *The Fabric of Reality.* Viking Adult, 1997. (Cited on page 508).

D. Deutsch and R. Jozsa. Rapid solutions of problems by quantum computation. *Proc. R. Soc. Lond. A*, 439:553, 1992. (Cited on page 509).

B. Deveaud, F. Clérot, N. Roy, K. Satzke, B. Sermage, and D. S. Katzern. Enhanced radiative recombination of free excitons in GaAs quantum wells. *Phys. Rev. Lett.*, 67:2355, 1991. (Cited on pages 152 and 153).

A. Di Carlo, S. Pescetelli, A. Kavokin, M. Vladimirova, and P. Lugli. Off-resonance γ-X mixing in semiconductor quantum wires. *Phys. Rev. B*, 57:9770, 1998. (Cited on page 142).

R. H. Dicke. Coherence in spontaneous radiation processes. *Phys. Rev.*, 93:99, 1954. (Cited on pages 202 and 532).

C. Diederichs and J. Tignon. Design for a triply resonant vertical-emitting micro-optical parametric oscillator. *Appl. Phys. Lett.*, 87:251107, 2005. (Cited on page 294).

C. Diederichs, J. Tignon, G. Dasbach, C. Ciuti, A. Lemaître, J. Bloch, Ph. Roussignol, and C. Delalande. Parametric oscillation in vertical triple microcavities. *Nature*, 440:904, 2006. (Cited on page 484).

P. A. M. Dirac. The quantum theory of the emission and absorption of radiation. *Proc. Roy. Soc. A*, 114:243, 1927. (Cited on page 104).

P. A. M. Dirac. *The principles of quantum mechanics*. Oxford Science Publications, 1930. (Cited on page 80).

P. A. M. Dirac. Quantised singularities in the electromagnetic field. *Proc. Roy. Soc. A*, 133:60, 1931. (Cited on page 441).

T. D. Doan, H. Thien Cao, D. B. Tran Thoai, and H. Haug. Coherence of condensed microcavity polaritons calculated within Boltzmann-master equations. *Phys. Rev. B*, 78:205306, 2008. (Cited on page 481).

L. Dominici, D. Colas, S. Donati, J. P. Restrepo Cuartas, M. De Giorgi, D. Ballarini, G. Guirales, J. C. López Carreño, A. Bramati, G. Gigli, E. del Valle, F. P. Laussy, and D. Sanvitto. Ultrafast control and Rabi oscillations of polaritons. *Phys. Rev. Lett.*, 113:226401, 2014. (Cited on pages 193, 478, and 480).

L. Dominici, M. Petrov, M. Matuszewski, D. Ballarini, M. De Giorgi, D. Colas, E. Cancellieri, B. Silva Fernández, A. Bramati, G. Gigli, A. Kavokin, F. Laussy, and D. Sanvitto. Real-space collapse of a polariton condensate. *Nat. Comm.*, 6:8993, 2015. (Cited on pages 317 and 318).

A. Dousse, L. Lanco, J. Suffczyński, E. Semenova, A. Miard, A. Lemaître, I. Sagnes, C. Roblin, J. Bloch, and P. Senellart. Controlled light–matter coupling for a single quantum dot embedded in a pillar microcavity using far-field optical lithography. *Phys. Rev. Lett.*, 101:267404, 2008. (Cited on page 255).

D. Dragoman and M. Dragoman. Terahertz fields and applications. *Progress in Quantum Eletronics*, 28:1, 2004. (Cited on page 526).

P. D. Drummond and D. F. Walls. Quantum theory of optical bistability. I. Nonlinear polarisalibity model. *J. Phys. A.: Math. Gen.*, 13:725, 1980. (Cited on page 487).

Q. Duan, D. Xu, W. Liu, J. Lu, L. Zhang, J. Wang, Y. Wang, J. Gu, T. Hu1, W. Xie, X. Shen, and Z. Chen. Polariton lasing of quasi-whispering gallery modes in a ZnO microwire. *Appl. Phys. Lett.*, 103:022103, 2013. (Cited on page 523).

M. I. Dyakonov. Quantum computing: A view from the enemy camp. *Opt. Spectrosc.*, 95:261, 2003. (Cited on page 511).

M. I. Dyakonov. *Future Trends In Microelectronics: The Nano, the Giga, and the Ultra*, chapter Spintronics. IEEE, 2004. (Cited on page 421).

M. I. Dyakonov. *Future Trends in Microelectronics. Frontiers and Innovations.*, chapter State of the art and prospects for quantum computing, page 266. Wiley, 2013. (Cited on pages 511 and 512).

M.I. Dyakonov and V.I. Perel. Current-induced spin orientation of electrons in semiconductors. *Phys. Lett. A*, 35A:459, 1971. (Cited on pages 385 and 399).

P. R. Eastham and P. B. Littlewood. Bose condensation of cavity polaritons beyond the linear regime: The thermal equilibrium of a model microcavity. *Phys. Rev. B*, 64:235101, 2001. (Cited on page 332).

J.H. Eberly and K. Wódkiewicz. The time-dependent physical spectrum of light. *J. Opt. Soc. Am.*, 67:1252, 1977. (Cited on pages 112, 490, and 491).

B. J. Eggleton, R. E. Slusher, C. Martijn de Sterke, P. A. Krug, and J. E. Sipe. Bragg grating solitons. *Phys. Rev. Lett.*, 76:1627, 1996. (Cited on pages 425 and 460).

O. A. Egorov, D. V. Skryabin, A. V. Yulin, and F. Lederer. Bright cavity polariton solitons. *Phys. Rev. Lett.*, 102:153904, 2009. (Cited on pages 375 and 460).

B. Eiermann, Th. Anker, M. Albiez, M. Taglieber, P. Treutlein, K.-P. Marzlin, and M. K. Oberthaler. Bright Bose–Einstein gap solitons of atoms with repulsive interaction. *Phys. Rev. Lett.*, 92:230401, 2004. (Cited on page 460).

A. Einstein. Strahlungs-emission und absorption nach der quantentheorie. *Verh. Deutsch. Phys. Gesell.*, 18:318, 1916. (Cited on page 520).

A. Einstein. Quantentheorie des einatomigen idealen Gases. *Sitzungsberichte der Preussischen Akademie der Wissenschaften*, 1:3, 1925. (Cited on page 321).

A. K. Ekert. Quantum cryptography based on Bell's theorem. *Phys. Rev. Lett.*, 67:661, 1991. (Cited on page 511).

G. A. El, A. Gammal, and A. M. Kamchatnov. Oblique dark solitons in supersonic flow of a Bose–Einstein condensate. *Phys. Rev. Lett.*, 97:180405, 2006. (Cited on pages 426 and 432).

C. Ell, J. Prineas, Jr. T. R. Nelson, S. Park, H. M. Gibbs, G. Khitrova, S. W. Koch, and R. Houdré. Influence of structural disorder and light coupling on the excitonic response of semiconductor microcavities. *Phys. Rev. Lett.*, 21:4795, 1998. (Cited on page 167).

R. J. Elliot. Spin-orbit coupling in band theory-character tables for some "double" space groups. *Phys. Rev.*, 96:280, 1954. (Cited on page 385).

T. Espinosa-Ortega and T. C. H. Liew. Perceptrons with Hebbian learning based on wave ensembles in spatially patterned potentials. *Phys. Rev. Lett.*, 114:118101, 2015. (Cited on page 515).

H. Everett III. "relative state" formulation of quantum mechanics. *Rev. Mod. Phys.*, 29:454, 1957. (Cited on page 508).

J. Faist, F. Capasso, D. L. Sivco, C. Sirtori, A. L. Hutchinson, and A. Y. Cho. Quantum cascade laser. *Science*, 264:553, 1994. (Cited on pages 521 and 527).

M. Faraday. On the magnetization of light, and the illumination of magnetic lines of force. *J. Franklin Inst.*, 41:139, 1846. (Cited on page 46).

A. Faraon, I. Fushman, D. Englund, N. Stoltz, P. Petroff, and J. Vučković. Coherent generation of non-classical light on a chip via photon-induced tunnelling and blockade. *Nat. Phys.*, 4:859, 2008. (Cited on pages 222 and 485).

J. Federici and L. Moeller. Review of terahertz and subteraherz wireless communications. *J. Appl. Phys.*, 107:111101, 2010. (Cited on page 527).

E. Fermi. Quantum theory of radiation. *Rev. Mod. Phys.*, 4:87, 1932. (Cited on page 104).

S. Fernández, F. B. Naranjo, F. Calle, M. A. Sánchez-García, E. Calleja, P. Vennegues, A. Trampert, and K. H. Ploog. MBE-grown high-quality (Al,Ga)N/GaN distributed Bragg reflectors for resonant cavity LEDs. *Semicond. Sci. Technol.*, 16:913, 2001. (Cited on page 56).

S. Ferretti, V. Savona, and D. Gerace. Optimal antibunching in passive photonic devices based on coupled nonlinear resonators. *New J. Phys.*, 15:025012, 2013. (Cited on page 490).

L. Ferrier, E. Wertz, R. Johne, D. D. Solnyshkov, P. Senellart, I. Sagnes, A. Lemaître, G. Malpuech, and J. Bloch. Interactions in confined polariton condensates. *Phys. Rev. Lett.*, 106:126401, 2011. (Cited on page 507).

A. L. Fetter and J. D. Walecka. *Quantum theory of many-particle systems*. Dover, 2003. (Cited on page 122).

R. P. Feynman. Simulating physics with computers. *Int. J. of Th. Phys.*, 21:467, 1982. (Cited on pages 514 and 516).

J. M. Fink, M. Göppl, M. Baur, R. Bianchetti, P. J. Leek, A. Blais, and A. Wallraff. Climbing the Jaynes–Cummings ladder and observing its $\sqrt{n}$ nonlinearity in a cavity QED system. *Nature*, 454:315, 2008. (Cited on pages 222 and 485).

J. Fischer, I. G. Savenko, M. D. Fraser, S. Holzinger, S. Brodbeck, M. Kamp, I. A. Shelykh, C. Schneider, and S. Höfling. Spatial coherence properties of one dimensional exciton-polariton condensates. *Phys. Rev. Lett.*, 113:203902, 2014. (Cited on page 537).

D. S. Fisher and P. C. Hohenberg. Dilute Bose gas in two dimensions. *Phys. Rev. B*, 37:4936, 1988. (Cited on pages 326 and 331).

M.P.A. Fisher, P.B. Weichman, G. Grinstein, and D.S. Fisher. Boson localization and the superfluid-insulator transition. *Phys. Rev. B*, 40:546, 1989. (Cited on page 334).

T. A. Fisher, A. M. Afshar, D. M. Whittaker, J. S. Roberts, G. Hill, and M. A. Pate. Electric-field and temperature tuning of exciton-photon coupling in quantum microcavity structures. *Phys. Rev. B*, 51:2600, 1995. (Cited on pages 281, 282, and 283).

T. A. Fisher, A. M. Afshar, M. S. Skolnick, D. M. Whittaker, and J. S. Roberts. Vacuum Rabi coupling enhancement and Zeeman splitting in semiconductor quantum microcavity structures in a high magnetic field. *Phys. Rev. B*, 53:R10469, 1996. (Cited on page 282).

H. Flayac. *New trends in the physics of spinor exciton-polariton condensates : topological defects and low dimensional structures.* PhD thesis, Université Blaise Pascal — Clermont–Ferrand II, 2012. (Cited on page 473).

H. Flayac, D. D. Solnyshkov, and G. Malpuech. Oblique half-solitons and their generation in exciton-polariton condensates. *Phys. Rev. B*, 83:193305, 2011. (Cited on pages 425, 440, and 445).

H. Flayac, D. D. Solnyshkov, and G. Malpuech. Separation and acceleration of magnetic monopole analogs in semiconductor microcavities. *New J. Phys.*, 14:085018, 2012. (Cited on page 440).

H. Flayac, D. D. Solnyshkov, I. A. Shelykh, and G. Malpuech. Transmutation of skyrmions to half-solitons driven by the nonlinear optical spin Hall effect. *Phys. Rev. Lett.*, 110:016404, 2013. (Cited on pages 434 and 465).

H. Flayac, D. Gerace, and V. Savona. An all-silicon single-photon source by unconventional photon blockade. *Sci. Rep.*, 5:11223, 2015. (Cited on page 490).

G. T. Foster, S. L. Mielke, and L. A. Orozco. Intensity correlations in cavity QED. *Phys. Rev. A*, 61:053821, 2000. (Cited on pages 488 and 489).

A. G. Fowler, M. Mariantoni, J. M. Martinis, and A. N. Cleland. Surface codes: Towards practical large-scale quantum computation. *Phys. Rev. A*, 86:032324, 2012. (Cited on page 518).

A. M. Fox, J. J. Baumberg, M. Dabbicco, B. Huttner, and J. F. Ryan. Squeezed light generation in semiconductors. *Phys. Rev. Lett.*, 74:1728, 1995. (Cited on page 480).

N. C. Frateschi, A. P. Kanjamala, A. F. J. Levi, and T. Tanbun-Ek. Polarization of lasing emission in microdisk laser diodes. *Appl. Phys. Lett.*, 66:1859, 1995. (Cited on page 16).

T. Freixanet, B.Sermage, J. Bloch, J. Y. Marzin, and R. Planel. Annular resonant Rayleigh scattering in the picosecond dynamics of cavity polaritons. *Phys. Rev. B*, 60:R8509, 1999. (Cited on page 553).

J. Frenkel. On the transformation of light into heat in solids. I. *Phys. Rev.*, 37:17, 1931. (Cited on page 133).

Ya. I. Frenkel. On the solid body model of heavy nuclei. *Phys. Z. Soviet Union*, 9:158, 1936. (Cited on page 133).

T. Frisch, Y. Pomeau, and S. Rica. Transition to dissipation in a model of superflow. *Phys. Rev. Lett.*, 69:1644, 1992. (Cited on page 433).

H. Frohlich. Theory of electrical breakdown in ionic crystals. *Proc. Roy. Soc. A*, 160:230, 1937. (Cited on page 548).

M. Fujita and T. Baba. Microgear laser. *Appl. Phys. Lett.*, 80:2051, 2002. (Cited on page 17).

T. A. Fulton and G. J. Dolan. Observation of single-electron charging effects in small tunnel junctions. *Phys. Rev. Lett.*, 59:109, 1987. (Cited on page 484).

A. Gahl, S. Balle, and M. San Miguel. Polarization dynamics of optically pumped VCSELs. *IEEE Quantum Electron.*, 35:342, 1999. (Cited on page 265).

M. Galbiati, L. Ferrier, D. D. Solnyshkov, D. Tanese, E. Wertz, A. Amo, M. Abbarchi, P. Senellart, I. Sagnes, A. Lemaître, E. Galopin, G. Malpuech, and J. Bloch. Polariton condensation in photonic molecules. *Phys. Rev. Lett.*, 108:126403, 2012. (Cited on pages 454, 455, and 537).

P. Gallo, M. Felici, B. Dwir, K. A. Atlasov, K. F. Karlsson, A. Rudra, A. Mohan, G. Biasiol, L. Sorba, and E. Kapon. Integration of site-controlled pyramidal quantum dots and photonic crystal membrane cavities. *Appl. Phys. Lett.*, 92:263101, 2008. (Cited on page 255).

T. Gao, P. S. Eldridge, T. C. H. Liew, S. I. Tsintzos, G. Stavrinidis, G. Deligeorgis, Z. Hatzopoulos, and P. G. Savvidis. Polariton condensate transistor switch. *Phys. Rev. B*, 85:235102, 2012. (Cited on page 540).

T. Gao, C. Antón, T. C. H. Liew, M. D. Martín, Z. Hatzopoulos, L. Viña, P. S. Eldridge, and P. G. Savvidis. Spin selective filtering of polariton condensate flow. *Appl. Phys. Lett.*, 107:011106, 2015a. (Cited on pages 541 and 542).

T. Gao, E. Estrecho, K. Y. Bliokh, T. C. H. Liew, M. D. Fraser, S. Brodbeck, M. Kamp, C. Schneider, S. Höfling, Y. Yamamoto, F. Nori, Y. S. Kivshar, A. G. Truscott, R. G. Dall, and E. A. Ostrovskaya. Observation of non-hermitian degeneracies in a chaotic exciton-polariton billiard. *Nature*, 526:554, 2015b. (Cited on pages 376 and 377).

B. T. Gard, K. R. Motes, J. P. Olson, P. P. Rohde, and J. P. Dowling. *An introduction to boson-sampling*, chapter 8. World Scientific Publishing Co, 2015. (Cited on page 512).

C. W. Gardiner. Driving a quantum system with the output field from another driven quantum system. *Phys. Rev. Lett.*, 70:2269, 1993. (Cited on page 503).

C. W. Gardiner and P. Zoller. Quantum kinetic theory: A quantum kinetic master equation for condensation of a weakly interacting Bose gas without a trapping potential. *Phys. Rev. A*, 55:2902, 1997. (Cited on page 351).

G. W. Gardiner and P. Zoller. *Quantum Noise*. Springer-Verlag, Berlin, 2nd edition, 2000. (Cited on pages 213 and 504).

S. Garnerone, P. Zanardi, and D. A. Lidar. Adiabatic quantum algorithm for search engine ranking. *Phys. Rev. Lett.*, 108:230506, 2012. (Cited on page 510).

R. Gati and M. K. Oberthaler. A bosonic Josephson junction. *J. Phys. B.: At. Mol. Phys.*, 40:R61, 2007. (Cited on page 192).

S. S. Gavrilov, A. S. Brichkin, S. I. Novikov, S. Höfling, C. Schneider, M. Kamp, A. Forchel, and V. D. Kulakovskii. Nonlinear route to intrinsic Josephson oscillations in spinor cavity-polariton condensates. *Phys. Rev. B*, 90:235309, 2014. (Cited on page 194).

L. Ge, A. Nersisyan, B. Oztop, and H. E. Tureci. Pattern formation and strong nonlinear interactions in exciton-polariton condensates. *arXiv:1311.4847*, 2013. (Cited on page 536).

L. J. Geerligs, V. F. Anderegg, P. A. M. Holweg, J. E. Mooij, H. Pothier, D. Esteve, C. Urbina, and M. H. Devoret. Frequency-locked turnstile device for single electrons. *Phys. Rev. Lett.*, 64:2691, 1990. (Cited on page 485).

I. M. Georgescu, S. Ashhab, and Franco Nori. Quantum simulation. *Rev. Mod. Phys.*, 86:153, 2014. (Cited on page 514).

D. Gerace and V. Savona. Unconventional photon blockade in doubly resonant microcavities with second-order nonlinearity. *Phys. Rev. A*, 89:031803(R), 2014. (Cited on page 490).

D. Gerace, M. Galli, D. Bajoni, G. Guizzetti, L. C. Andreani, F. Riboli, M. Melchiorri, N. Daldosso, L. Pavesi, G. Pucker, S. Cabrini, L. Businaro, and E. Di Fabrizio. Wide-band transmittance of one-dimensional photonic crystals carved in Si_3N_4/SiO_2 channel waveguides. *Appl. Phys. Lett.*, 87:211116, 2005. (Cited on page 56).

J. M. Gérard, D. Barrier, J. Y. Marzin, R. Kuszelewicz, L. Manin, E. Costard, V. Thierry-Mieg, and T. Rivera. Quantum boxes as active probes for photonic microstructures: The pillar microcavity case. *Appl. Phys. Lett.*, 69:449, 1996. (Cited on page 13).

J.-M. Gérard, B. Sermage, B. Gayral, B. Legrand, E. Costard, and V. Thierry-Mieg. Enhanced spontaneous emission by quantum boxes in a monolithic optical microcavity. *Phys. Rev. Lett.*, 81:1110, 1998. (Cited on pages 244 and 250).

V. A. Gergel', R. F. Kazarinov, and R. A. Suris. Optical properties of an exciton condensate in a semiconductor. *Sov. Phys. JETP*, 26:354, 1968a. (Cited on page 520).

V. A. Gergel', R. F. Kazarinov, and R. A. Suris. Superfluidity of excitons in semiconductors. *Sov. Phys. JETP*, 27:159, 1968b. (Cited on page 536).

H. M. Gibbs. *Optical Bistability: Controlling Light with Light*. Academic Press, Orlando, Fla., USA, 1985. (Cited on page 274).

B. Gil and A. V. Kavokin. Giant exciton-light coupling in ZnO quantum dots. *Appl. Phys. Lett.*, 81:748, 2002. (Cited on page 155).

V. L. Ginzburg. The problem of high-temperature superconductivity. *Annu. Rev. Mater. Sci.*, 2:663, 1972. (Cited on page 378).

V. L. Ginzburg. *The Physics of a Lifetime: Reflections on the Problems and Personalities of 20th Century Physics.* Springer, 2001. (Cited on page 67).

M. De Giorgi, D. Ballarini, E. Cancellieri, F. M. Marchetti, M. H. Szymańska, C. Tejedor, R. Cingolani, E. Giacobino, A. Bramati, G. Gigli, and D. Sanvitto. Control and ultrafast dynamics of a two-fluid polariton switch. *Phys. Rev. Lett.*, 109:266407, 2012. (Cited on pages 539, 543, and 545).

N. A. Gippius, S. G. Tikhodeev, V. D. Kulakovskii, D. N. Krizhanovskii, and A. I. Tartakovskii. Nonlinear dynamics of polariton scattering in semiconductor microcavity: Bistability vs. stimulated scattering. *Europhys. Lett.*, 67:997, 2004. (Cited on page 304).

N. A. Gippius, I. A. Shelykh, D. D. Solnyshkov, S. S. Gavrilov, Yuri G. Rubo, A. V. Kavokin, S. G. Tikhodeev, and G. Malpuech. Polarization multistability of cavity polaritons. *Phys. Rev. Lett.*, 98: 236401, 2007. (Cited on page 539).

Yu. G. Gladush, A. M. Kamchatnov, Z. Shi, P. G. Kevrekidis, D. J. Frantzeskakis, and B. A. Malomed. Wave patterns generated by a supersonic moving body in a binary Bose–Einstein condensate. *Phys. Rev. A*, 79:033623, 2009. (Cited on pages 434 and 445).

R. J. Glauber. Photon correlations. *Phys. Rev. Lett.*, 10:84, 1963. (Cited on pages 106, 113, and 114).

J. Goldstone. Spontaneous symmetry breaking. *Nuovo Cimento*, 19: 154, 1961. (Cited on page 324).

J. Goldstone, A. Salam, and S. Weinberg. Broken symetries. *Phys. Rev.*, 127:965, 1962. (Cited on page 324).

L. E. Golub, E. L. Ivchenko, and R. Y. Rasulov. Intersubband absorption of light in a semiconductor quantum-well with a complex band-structure. *Semiconductors*, 29:566, 1995. (Cited on page 533).

A. González-Tudela, E. del Valle, E. Cancellieri, C. Tejedor, D. Sanvitto, and F. P. Laussy. Effect of pure dephasing on the Jaynes-Cummings nonlinearities. *Opt. Express*, 18:7002, 2010a. (Cited on page 229).

A. González-Tudela, E. del Valle, C. Tejedor, and F.P. Laussy. Anticrossing in the PL spectrum of light–matter coupling under incoherent continuous pumping. *Superlatt. Microstruct.*, 47:16, 2010b. (Cited on page 221).

A. González-Tudela, F. P. Laussy, C. Tejedor, M. J Hartmann, and E. del Valle. Two-photon spectra of quantum emitters. *New J. Phys.*, 15:033036, 2013. (Cited on pages 495, 496, and 497).

A. González-Tudela, E. del Valle, and F. P. Laussy. Optimization of photon correlations by frequency filtering. *Phys. Rev. A*, 91: 043807, 2015. (Cited on pages 496 and 497).

M. Göppert-Mayer. Über Elementarakte mit zwei Quantensprüngen. *Annalen der Physik*, 401:273, 1931. (Cited on page 496).

L.P. Gor'kov and I.E. Dzialoshinskii. On the theory of Mott excitons in high magnetic fields. *J. Exp. Th. Phys.*, 53:717, 1967. (Cited on page 557).

P. Goy, J. M. Raimond, M. Gross, and S. Haroche. Observation of cavity-enhanced single-atom spontaneous emission. *Phys. Rev. Lett.*, 50:1903, 1983. (Cited on page 244).

A. Gozar, G. Logvenov, L. Fitting Kourkoutis, A. T. Bollinger, L. A. Giannuzzi, D. A. Muller, and I. Bozovic. High-temperature interface superconductivity between metallic and insulating copper oxides. *Nature*, 455:782, 2008. (Cited on page 378).

H. Grabert and M. H. Devoret, editors. *Single Charge Tunneling*, volume 294 of *NATO ASI Series*. Springer US, 1992. (Cited on page 485).

A. D. Greentree, C. Tahan, J. H. Cole, and L. C. L. Hollenberg. Quantum phase transitions of light. *Nat. Phys.*, 2:856, 2006. (Cited on page 516).

M. Greiner, O. Mandel, T. Esslinger, T. W. Hänsch, and I. Bloch. Quantum phase transition from a superfluid to a Mott insulator in a gas of ultracold atoms. *Nature*, 415:39, 2002. (Cited on page 514).

A. Griffin. Conserving and gapless approximations for an inhomogeneous Bose gas at finite temperatures. *Phys. Rev. B*, 53:9341, 1996. (Cited on page 368).

A. Grifin, D. W. Snoke, and S. Stringari, editors. *Bose–Einstein Condensation.* Cambridge University Press, Cambridge, 1996. (Cited on page 379).

J. Gripp, S. L. Mielke, and L. A. Orozco. Evolution of the vacuum Rabi peaks in a detuned atom-cavity system. *Phys. Rev. A*, 56: 3262, 1997. (Cited on pages 274 and 275).

E. F. Gross, B. P. Zakharchenia, and N. M. Reinov. Linear and quadratic Zeeman effect and the diamagnetism of excitons in CuO_2 crystals. *Dokl. Acad. Sci. USSR*, 111:564, 1956. (Cited on page 137).

L. K Grover. A fast quantum mechanical algorithm for database search. *Proc. ACM Symp.*, 28:112, 1996. (Cited on page 510).

G. Günter, A. A. Anappara, J. Hees, A. Sell, G. Biasiol, L. Sorba, S. De Liberato, C. Ciuti, A. Tredicucci, A. Leitenstorfer, and R. Huber. Sub-cycle switch-on of ultrastrong light–matter interaction. *Nature*, 458:178, 2009. (Cited on pages 283 and 284).

M. Gurioli, P. Borri, M. Colocci, M. Gulia, F. Rossi, E. Molinari, P. E. Selbmann, and P. Lugli. Exciton formation and relaxation in GaAs epilayers. *Phys. Rev. B*, 58:R13403, 1998. (Cited on page 335).

T. Gutbrod, M. Bayer, A. Forchel, J. P. Reithmaier, T. L. Reinecke, S. Rudin, and P. A. Knipp. Weak and strong coupling of photons and excitons in photonic dots. *Phys. Rev. B*, 57:9950, 1998. (Cited on page 537).

T. Gutbrod, M. Bayer, A. Forchel, P. A. Knipp, T. L. Reinecke, A. Tartakovskii, V. D. Kulakovskii, N. A. Gippius, and S. G. Tikhodeev. Angle dependence of the spontaneous emission from confined optical modes in photonic dots. *Phys. Rev. B*, 59:2223, 1999. (Cited on page 452).

X. Hachair, S. Barland, L. Furfaro, M. Giudici, S. Balle, J. R. Tredicce, M. Brambilla, T. Maggipinto, I. M. Perrini, G. Tissoni, and L. Lugiato. Cavity solitons in broad-area vertical-cavity surface-emitting lasers below threshold. *Phys. Rev. A*, 69:43817, 2004. (Cited on page 275).

Z. Hadzibabic, P. Krüger, M. Cheneau, B. Battelier, and J. Dalibard. Berezinskii-Kosterlitz-Thouless crossover in a trapped atomic gas. *Nature*, 441:1118, 2006. (Cited on page 431).

M. Hafezi, E. A. Demler, M. D. Lukin, and J. M. Taylor. Robust optical delay lines with topological protection. *Nat. Phys.*, 7:907, 2011. (Cited on page 467).

F. D. M. Haldane and S. Raghu. Possible realization of directional optical waveguides in photonic crystals with broken time-reversal symmetry. *Phys. Rev. Lett.*, 100:013904, 2008. (Cited on pages 465 and 467).

F.D.M. Haldane. Model for a quantum Hall effect without Landau levels: Condensed-matter realization of the "parity anomaly". *Phys. Rev. Lett.*, 61:2015, 1988. (Cited on page 465).

R. N. Hall, G. E. Fenner, J. D. Kingsley, T. J. Soltys, and R. O. Carlson. Coherent light emission from GaAs junctions. *Phys. Rev. Lett.*, 9:366, 1962. (Cited on page 261).

R. Hanbury Brown and R. Twiss. Interferometry of the intensity fluctuations in light. III. Applications to astronomy. *Proc. Roy. Soc*, 248:199, 1958. (Cited on page 44).

R. Hanbury Brown and R. Q. Twiss. A test of a new type of stellar interferometer on Sirius. *Nature*, 178:1046, 1956a. (Cited on page 43).

R. Hanbury Brown and R. Q. Twiss. Correlation between photons in two coherent beams of light. *Nature*, 177:27, 1956b. (Cited on page 43).

R. Hanbury Brown and R. Q. Twiss. The question of correlation between photons in coherent light rays. *Nature*, 178:1447, 1956c. (Cited on page 43).

R. Hanbury Brown and R.Q. Twiss. A new type of interferometer for use in radio astronomy. *Phil. Mag.*, 45:663, 1954. (Cited on page 43).

R. Hanbury Brown, R. C. Jennison, and M. K. Das Gupta. Apparent angular sizes of discrete radio sources: Observations at Jodrell Bank, Manchester. *Nature*, 170:1061, 1952. (Cited on page 43).

W. Hanle. Über magnetische Beeinflussung der Polarisation der Resonanzfluoreszenz. *Z. Phys.*, 30:93, 1924. (Cited on page 421).

R. Harel, E. Cohen, E. Linder, Arza Ron, and L. N. Pfeiffer. Absolute transmission, reflection, and absorption studies in GaAs/AlAs quantum wells containing a photoexcited electron gas. *Phys. Rev. B*, 53:7868, 1996. (Cited on page 337).

F. Hargart, M. Müller, K. Roy-Choudhury, S. L. Portalupi, C. Schneider, S. Höfling, M. Kamp, S. Hughes, and P. Michler. Cavity-enhanced simultaneous dressing of quantum dot exciton and biexciton states. *Phys. Rev. B*, 93:115308, 2016. (Cited on page 501).

S. Haroche. Nobel lecture: Controlling photons in a box and exploring the quantum to classical boundary. *Rev. Mod. Phys.*, 85:1083, 2013. (Cited on page 476).

A. W. Harrow, A. Hassidim, and S. Lloyd. Quantum algorithm for linear systems of equations. *Phys. Rev. Lett.*, 103:150502, 2009. (Cited on page 510).

M. J. Hartmann, F. G. S. L. Brandão, and M. B. Plenio. Strongly interacting polaritons in coupled arrays of cavities. *Nat. Phys.*, 2:849, 2006. (Cited on page 516).

M. J. Hartmann, F. G. S. L. Brandão, and M. B. Plenio. Quantum many-body phenomena in coupled cavity arrays. *Laser Photon. Rev.*, 2:527, 2008. (Cited on page 516).

Z. Hasan and C. L. Kane. Topological insulators. *Rev. Mod. Phys.*, 82:3045, 2010. (Cited on pages 425, 465, 468, 469, and 473).

H. Haug and E. Hanamura. Derivation of the two-fluid model for Bose-condensed excitons. *Phys. Rev. B*, 11:3317, 1975. (Cited on page 325).

H. Haug and S. W. Koch. *Quantum theory of the optical and electronic properties of semiconductors*. World Scientific, 1990. (Cited on page 144).

N. Hauke, T. Zabel, K. Müller, M. Kaniber, A. Laucht, D. Bougeard, G. Abstreiter, J. J. Finley, and Y. Arakawa. Enhanced photoluminescence emission from two-dimensional silicon photonic crystal nanocavities. *New J. Phys.*, 12:053005, 2010. (Cited on pages 19 and 20).

A. Hayat, P. Ginzburg, and M. Orenstein. Observation of two-photon emission from semiconductors. *Nat. Photon.*, 2:256, 2008. (Cited on page 496).

D. J. Heinzen, J. J. Childs, J. E. Thomas, and M. S. Feld. Enhanced and inhibited visible spontaneous emission by atoms in a confocal resonator. *Phys. Rev. Lett.*, 58:1320, 1987. (Cited on page 254).

W. Heisenberg. Über quantentheoretische Umdeutung kinematischer und mechanischer Beziehungen. *Zeitschrift für Physik*, 33:879, 1925. (Cited on page 305).

W D Heiss. The physics of exceptional points. *J. Phys. A.: Math. Gen.*, 45:444016, 2012. (Cited on page 376).

K. Hennessy, A. Badolato, P.M. Petroff, and E. Hu. Positioning photonic crystal cavities to single InAs quantum dots. *Photonics and Nanostructures*, 2:65, 2004. (Cited on pages 255 and 256).

K. Hennessy, A. Badolato, M. Winger, D. Gerace, M. Atature, S. Gulde, S. Fält, E. L. Hu, and A. İmamoğlu. Quantum nature of a strongly coupled single quantum dot–cavity system. *Nature*, 445:896, 2007. (Cited on pages 211, 228, 229, and 491).

M. Hennrich, A. Kuhn, and G. Rempe. Transition from antibunching to bunching in cavity QED. *Phys. Rev. Lett.*, 94:053604, 2005. (Cited on page 116).

T. J. Herzog, P. G. Kwiat, H. Weinfurter, and A. Zeilinger. Complementarity and the quantum eraser. *Phys. Rev. Lett.*, 75:3034, 1995. (Cited on page 484).

R. Hivet, H. Flayac, D. D. Solnyshkov, D. Tanese, T. Boulier, D. Andreoli, E. Giacobino, J. Bloch, A. Bramati, G. Malpuech, and A. Amo. Half-solitons in a polariton quantum fluid behave like magnetic monopoles. *Nat. Phys.*, 8:724, 2012. (Cited on pages 425, 440, and 445).

D. R. Hofstadter. A nose for depth: Gregory Wannier's style in physics. *Phys. Rep.*, 110:273, 1984. (Cited on page 135).

P.C. Hohenberg. Existence of long-range order in one and two dimensions. *Phys. Rev.*, 158:383, 1967. (Cited on page 326).

U. Hohenester. Cavity quantum electrodynamics with semiconductor quantum dots: Role of phonon-assisted cavity feeding. *Phys. Rev. B*, 81:155303, 2010. (Cited on page 211).

U. Hohenester, A. Laucht, M. Kaniber, N. Hauke, A. Neumann, A. Mohtashami, M. Seliger, M. Bichler, and J. J. Finley. Phonon-assisted transitions from quantum dot excitons to cavity photons. *Phys. Rev. B*, 80:201311(R), 2009. (Cited on page 211).

N. Holonyak and S. F. Bevacqua. Coherent (visible) light emission from $Ga(As_{1-x}P_x)$ junctions. *Appl. Phys. Lett.*, 1:80, 1962. (Cited on pages 261 and 262).

S. Hoogland, J. J. Baumberg, S. Coyle, J. Baggett, M. J. Coles, and H. J. Coles. Self-organized patterns and spatial solitons in liquid-crystal microcavities. *Phys. Rev. A*, 66:055801, 2002. (Cited on page 275).

J. J. Hopfield. Theory of the contribution of excitons to the complex dielectric constant of crystals. *Phys. Rev.*, 112:1555, 1958. (Cited on pages 143 and 144).

T. Horikiri, P. Schwendimann, A. Quattropani, S. Höfling, A. Forchel, and Y. Yamamoto. Higher order coherence of exciton-polariton condensates. *Phys. Rev. B*, 81:033307, 2010. (Cited on page 482).

R. Houdré, C. Weisbuch, R. P. Stanley, U. Oesterle, P. Pellandin, and M. Ilegems. Measurement of cavity-polariton dispersion curve from angle-resolved photoluminescence experiments. *Phys. Rev. Lett.*, 73:2043, 1994. (Cited on page 335).

R. Houdré, J. L. Gibernon, P. Pellandini, R. P. Stanley, U. Oesterle, C. Weisbuch, J. O'Gorman, B. Roycroft, and M. Ilegems. Saturation of the strong-coupling regime in a semiconductor microcavity: Free-carrier bleaching of cavity polaritons. *Phys. Rev. B*, 52:7810, 1995. (Cited on page 507).

D. Hsieh, D. Qian, L. Wray, Y. Xia, Y. S. Hor, R. J. Cava, and M. Z. Hasan. A topological Dirac insulator in a quantum spin Hall phase. *Nature*, 452:970, 2008. (Cited on page 467).

M. Hübner, J. Kuhl, T. Stroucken, A. Knorr, S.W. S.W. Koch, R. Hey, and K. Ploog. Collective effects of excitons in multiple-quantum-well Bragg and anti-Bragg structures. *Phys. Rev. Lett.*, 76:4199, 1996. (Cited on page 280).

S. Hughes and P. Yao. Theory of quantum light emission from a strongly-coupled single quantum dot photonic-crystal cavity system. *Opt. Express*, 17:3322, 2009. (Cited on page 229).

S. Huppert, O. Lafont, E. Baudin, J. Tignon, and R. Ferreira. Terahertz emission from multiple-microcavity exciton-polariton lasers. *Phys. Rev. B*, 90:241302(R), 2014. (Cited on page 527).

D. A. W. Hutchinson, E. Zaremba, and A. Griffin. Finite temperature excitations of a trapped Bose gas. *Phys. Rev. Lett.*, 78, 1997. (Cited on page 368).

C. Huygens. Letter to de Sluse. *Œuvres Completes de Christian Huygens*, 1665. (Cited on page 525).

A. İmamoğlu and R. J. Ram. Quantum dynamics of exciton lasers. *Physics Letter A*, 214:193, 1996. (Cited on page 320).

A. İmamoğlu and Y. Yamamoto. Turnstile device for heralded single photons: Coulomb blockade of electron and hole tunneling in quantum confined p-i-n heterojunctions. *Phys. Rev. Lett.*, 72:210, 1994. (Cited on page 485).

A. İmamoğlu, R. J. Ram, S. Pau, and Y. Yamamoto. Nonequilibrium condensates and lasers without inversion: Exciton-polariton lasers. *Phys. Rev. A*, 53:4250, 1996. (Cited on pages 520 and 544).

A. İmamoğlu, H. Schmidt, G. Woods, and M. Deutsch. Strongly interacting photons in a nonlinear cavity. *Phys. Rev. Lett.*, 79:1467, 1997. (Cited on page 485).

M. Inguscio, S. Stringari, and C. E. Wieman, editors. *Bose–Einstein condensation in atomic gases*. IOS Press, Amsterdam, 1999. (Cited on page 379).

A. L. Ivanov, P. Borri, W. Langbein, and U. Woggon. Radiative corrections to the excitonic molecule state in GaAs microcavities. *Phys. Rev. B*, 69:075312, 2004. (Cited on page 407).

E. L. Ivchenko. Exchange interaction and scattering of light with reversal of the hole angular momentum at an acceptor in quantum-well structures. *Sov. Phys. Solid State*, 34:254, 1992. (Cited on page 148).

E. L. Ivchenko. *Optical spectroscopy of semiconductor nanostructures*. Alpha Science, 2005. (Cited on page 137).

E. L. Ivchenko and G. E. Pikus. *Superlattices and other heterostructures. Symmetry and optical phenomena*. Springer, Berlin, 1997. (Cited on page 137).

E. L. Ivchenko, A. I. Nesvizhskii, and S. Jorda. Resonant Bragg reflection from quantum-well structures. *Superlatt. Microstruct.*, 16:17, 1994. (Cited on page 280).

J. D. Jackson. *Classical electrodynamics*. Wiley, 2 edition, 1975. (Cited on page 77).

T. Jacqmin, I. Carusotto, I. Sagnes, M. Abbarchi, D. D. Solnyshkov, G. Malpuech, E. Galopin, A. Lemaître, J. Bloch, and A. Amo. Direct observation of Dirac cones and a flatband in a honeycomb lattice for polaritons. *Phys. Rev. Lett.*, 112:116402, 2014. (Cited on pages 424, 453, 460, 461, and 537).

F. Jahnke, M. Kira, S. W. Koch, G. Khitrova, E. K. Lindmark, T. R. Nelson, Jr. D. V. Wick, J. D. Berger, O. Lyngnes, H. M. Gibbs, and K. Tai. Excitonic nonlinearities of semiconductor microcavities in the nonperturbative regime. *Phys. Rev. Lett.*, 77:5257, 1996. (Cited on page 285).

D. Jaksch, C. W. Gardiner, and P. Zoller. Quantum kinetic theory. II. Simulation of the quantum Boltzmann master equation. *Phys. Rev. A*, 56:575, 1997. (Cited on page 361).

J. Javanainen. Oscillatory exchange of atoms between traps containing Bose condensates. *Phys. Rev. Lett.*, 57:3164, 1986. (Cited on page 194).

J. Javanainen and R. Rajapakse. Bayesian inference to characterize Josephson oscillations in a double-well trap. *Phys. Rev. A*, 92: 023613, 2015. (Cited on page 194).

E.T. Jaynes. Electrodynamics today. In L. Mandel and E. Wolf, editors, *Proceedings of the Second Rochester Conference on Coherence and Quantum Optics*, page 21. Plenum, New York, 1977. (Cited on page 476).

E.T. Jaynes and F.W. Cummings. Comparison of quantum and semiclassical radiation theory with application to the beam maser. *Proc. IEEE*, 51:89, 1963. (Cited on pages 196, 476, and 477).

D. Jerison, I. M. Singer, and D. W. Stroock, editors. *The Legacy of Norbert Wiener: A Centennial Symposium*. American Mathematical Society, 1997. (Cited on page 41).

W. Jhe, E. A. Hinds, D. Meschede, L. Moi, and S. Haroche. Suppression of spontaneous decay at optical frequencies: Test of vacuum-field anisotropy in confined space. *Phys. Rev. Lett.*, 58: 666, 1987. (Cited on page 254).

J. D. Joannopoulos, R. D. Meade, and J. N. Winn. *Photonic Crystals: Molding the Flow of Light*. Princeton University Press, 1995. (Cited on pages 70 and 77).

M. W. Johnson, M. H. S. Amin, S. Gildert, T. Lanting, F. Hamze, N. Dickson, R. Harris, A. J. Berkley, J. Johansson, P. Bunyk, E. M. Chapple, C. Enderud, J. P. Hilton, K. Karimi, E. Ladizinsky, N. Ladizinsky, T. Oh, I. Perminov, C. Rich, M. C. Thom, E. Tolkacheva, C. J. S. Truncik, S. Uchaikin, J. Wang, B. Wilson, and G. Rose. Quantum annealing with manufactured spins. *Nature*, 473:194, 2011. (Cited on page 514).

M. B. Johnston, D. M. Whittaker, A. Corchia, A. G. Davies, and E. H. Linfield. Simulation of terahertz generation at semiconductor surfaces. *Phys. Rev. B*, 65:165301, 2002. (Cited on page 533).

R. Clark Jones. A new calculus for the treatment of optical systems I. Description and discussion of the calculus. *J. Opt. Soc. Am.*, 31: 488, 1941. (Cited on page 30).

K. Joosten and G. Nienhuis. Influence of spectral filtering on the quantum nature of light. *J. Phys. B.: At. Mol. Phys.*, 2:158, 2000. (Cited on page 492).

B.D. Josephson. Possible new effects in superconductive tunnelling. *Phys. Lett.*, 1:251, 1962. (Cited on pages 191 and 194).

T. Kadowaki and H. Nishimori. Quantum annealing in the transverse Ising model. *Phys. Rev. E*, 58:5355, 1998. (Cited on page 513).

P. Kaer, T. R. Nielsen, P. Lodahl, A.-P. Jauho, and J. Mørk. Non-Markovian model of photon-assisted dephasing by electron-phonon interactions in a coupled quantum-dot–cavity system. *Phys. Rev. Lett.*, 104:157401, 2010. (Cited on page 212).

R. A. Kaindl, D. Hägele, M. A. Carnahan, and D. S. Chemla. Transient terahertz spectroscopy of excitons and unbound carriers in quasi-two-dimensional electron-hole gases. *Phys. Rev. B*, 79:045320, 2009. (Cited on page 527).

R. Idrissi Kaitouni, O. El Daïf, A. Baas, M. Richard, T. Paraiso, P. Lugan, T. Guillet, F. Morier-Genoud, J. D. Ganière, J. L. Staehli, V. Savona, and B. Deveaud. Engineering the spatial confinement of exciton polaritons in semiconductors. *Phys. Rev. B*, 74:155311, 2006. (Cited on page 515).

G. Kalai. How quantum computers fail: Quantum codes, correlations in physical systems, and noise accumulation. *arXiv:1106.0485*, 2011. (Cited on page 511).

M. Kaliteevski, I. Iorsh, S. Brand, R. A. Abram, J. M. Chamberlain, A. V. Kavokin, and I. A. Shelykh. Tamm plasmon-polaritons: Possible electromagnetic states at the interface of a metal and a dielectric Bragg mirror. *Phys. Rev. B*, 76:165415, 2007. (Cited on page 69).

M. A. Kaliteevski, S. Brand, R. A. Abram, V. V. Nikolaev, M. V. Maximov, C. M. Sotomayor Torres, and A. V. Kavokin. Electromagnetic theory of the coupling of zero-dimensional exciton and photon states: A quantum dot in a spherical microcavity. *Phys. Rev. B*, 64, 2001. (Cited on pages 171 and 172).

M. A. Kaliteevski, K. A. Ivanov, G. Pozina, and A. J. Gallant. Single and double bosonic stimulation of THz emission in polaritonic systems. *Sci. Rep.*, 4:5444, 2014. (Cited on page 533).

A. M. Kamchatnov and N. Pavloff. Interference effects in the two-dimensional scattering of microcavity polaritons by an obstacle: phase dislocations and resonances. *Eur. Phys. J. D*, 69:32, 2015. (Cited on page 434).

A. M. Kamchatnov and L. P. Pitaevskii. Stabilization of solitons generated by a supersonic flow of Bose–Einstein condensate past an obstacle. *Phys. Rev. Lett.*, 100:160402, 2008. (Cited on page 432).

C. L. Kane and E. J. Mele. Quantum spin Hall effect in graphene. *Phys. Rev. Lett.*, 95:226801, 2005a. (Cited on pages 464 and 467).

C. L. Kane and E. J. Mele. Z_2 topological order and the quantum spin hall effect. *Phys. Rev. Lett.*, 95:146802, 2005b. (Cited on pages 464 and 467).

M. Kaniber, A. Laucht, A. Neumann, J. M. Villas-Bôas, M. Bichler, M.-C. Amann, and J. J. Finley. Investigation of the nonresonant dot-cavity coupling in two-dimensional photonic crystal nanocavities. *Phys. Rev. B*, 77:161303(R), 2008. (Cited on pages 211 and 491).

D. V. Karpov, I. G. Savenko, H. Flayac, and N. N. Rosanov. Dissipative soliton protocols in semiconductor microcavities at finite temperatures. *Phys. Rev. B*, 92:075305, 2015. (Cited on page 536).

J. Ph. Karr, A. Baas, and E. Giacobino. Twin polaritons in semiconductor microcavities. *Phys. Rev. A*, 69:063807, 2004a. (Cited on page 483).

J. Ph. Karr, A. Baas, R. Houdré, and E. Giacobino. Squeezing in semiconductor microcavities in the strong-coupling regime. *Phys. Rev. A*, 69:R031802, 2004b. (Cited on page 481).

Y. V. Kartashov, V. V. Konotop, and L. Torner. Compactons and bistability in exciton-polariton condensates. *Phys. Rev. B*, 86:205313, 2012. (Cited on page 507).

T. Karzig, C.-E. Bardyn, N. H. Lindner, and G. Refael. Topological polaritons. *Phys. Rev. X*, 5:031001, 2015. (Cited on pages 465 and 467).

K. Kasamatsu, M. Tsubota, and M. Ueda. Spin textures in rotating two-component Bose–Einstein condensates. *Phys. Rev. A*, 71: 043611, 2005. (Cited on pages 434 and 439).

J. Kasprzak, M. Richard, S. Kundermann, A. Baas, P. Jeambrun, J. M. J. Keeling, F. M. Marchetti, M. H. Szymańska, R. André, J. L. Staehli, V. Savona, P. B. Littlewood, B. Deveaud, and Le Si Dang. Bose–Einstein condensation of exciton polaritons. *Nature*, 443:409, 2006. (Cited on pages 326, 372, 373, 412, 424, 481, and 521).

J. Kasprzak, M. Richard, A. Baas, B. Deveaud, R. André, J.-Ph. Poizat, and Le Si Dang. Second-order time correlations within a polariton Bose–Einstein condensate in a CdTe microcavity. *Phys. Rev. Lett.*, 100:067402, 2008a. (Cited on pages 481 and 482).

J. Kasprzak, D. D. Solnyshkov, R. André, Le Si Dang, and G. Malpuech. Formation of an exciton polariton condensate: Thermodynamic versus kinetic regimes. *Phys. Rev. Lett.*, 101: 146404, 2008b. (Cited on pages 335, 342, and 343).

J. Kasprzak, S. Reitzenstein, E. A. Muljarov, C. Kistner, C. Schneider, M. Strauss, S. Höfling, A. Forchel, and W. Langbein. Up on the Jaynes-Cummings ladder of a quantum-dot/microcavity system. *Nat. Mater.*, 9:304, 2010. (Cited on pages 222 and 228).

T. Kato. *Perturbation Theory of Linear Operators*. Berlin: Springer, 1980. (Cited on page 376).

Y. K. Kato, R. C. Myers, A. C. Gossard, and D. D. Awschalom. Observation of the spin Hall effect in semiconductors. *Science*, 306: 1910, 2004. (Cited on page 399).

A. Kavokin and B. Gil. GaN microcavities: Giant Rabi splitting and optical anisotropy. *Appl. Phys. Lett.*, 72:2880, 1998. (Cited on page 164).

A. Kavokin and G. Malpuech. *Cavity polaritons*, volume 32 of *Thin films and nanostructures*. Elsevier, 2003. (Cited on pages 60 and 77).

A. Kavokin, P. G. Lagoudakis, G. Malpuech, and J. J. Baumberg. Polarization rotation in parametric scattering of polaritons in semiconductor microcavities. *Phys. Rev. B*, 67:195321, 2003a. (Cited on pages 386 and 394).

A. Kavokin, G. Malpuech, and F. P. Laussy. Polariton laser and polariton superfluidity in microcavities. *Phys. Lett. A*, 306:187, 2003b. (Cited on pages 332 and 336).

A. Kavokin, G. Malpuech, and M. Glazov. Optical spin Hall effect. *Phys. Rev. Lett.*, 95:136601, 2005a. (Cited on pages 401, 424, and 465).

A. Kavokin, D. Solnyshkov, and G. Malpuech. Quatron-polaritons: charged quasi-particles having the bosonic statistics. *J. Phys.: Condens. Matter*, 19:295212, 2007. (Cited on page 379).

A. V. Kavokin. Motional narrowing of inhomogeneously broadened excitons in a semiconductor microcavity: Semiclassical treatment. *Phys. Rev. B*, 57:3757, 1997. (Cited on pages 167 and 391).

A. V. Kavokin and M. A. Kaliteevski. Excitonic light reflection and absorption in semiconductor microcavities at oblique incidence. *Solid State Commun.*, 95:859, 1995. (Cited on page 163).

A. V. Kavokin, G. Malpuech, and W. Langbein. Theory of propagation and scattering of exciton-polaritons in quantum wells. *Solid State Commun.*, 120:259, 2001. (Cited on page 333).

A. V. Kavokin, I. A. Shelykh, and G. Malpuech. Lossless interface modes at the boundary between two periodic dielectric structures. *Phys. Rev. B*, 72:233102, 2005b. (Cited on page 69).

A. V. Kavokin, I. A. Shelykh, T. Taylor, and M. M. Glazov. Vertical cavity surface emitting terahertz laser. *Phys. Rev. Lett.*, 108:197401, 2012. (Cited on page 527).

A. V. Kavokin, A. S. Sheremet, I. A. Shelykh, P. G. Lagoudakis, and Y. G. Rubo. Exciton-photon correlations in bosonic condensates of exciton-polaritons. *Sci. Rep.*, 5:12020, 2015. (Cited on page 483).

K. V. Kavokin, I. A. Shelykh, A. V. Kavokin, G. Malpuech, and P. Bigenwald. Quantum theory of spin dynamics of exciton-polaritons in microcavities. *Phys. Rev. Lett.*, 92:017401, 2004. (Cited on page 399).

K. V. Kavokin, M. A. Kaliteevski, R. A. Abram, A. V. Kavokin, S. Sharkova, and I. A. Shelykh. Stimulated emission of terahertz radiation by exciton-polariton lasers. *Appl. Phys. Lett.*, 97:201111, 2010. (Cited on page 527).

K.V. Kavokin, P. Reṇucci, T. Amand, X. Marie, P. Senellart, J. Bloch, and B. Sermage. Linear polarisation inversion: A signature of Coulomb scattering of cavity polaritons with opposite spins. *Phys. Stat. Sol. C*, 2:763, 2005c. (Cited on pages 399 and 407).

Y. Kawaguchia and M. Ueda. Spinor Bose–Einstein condensates. *Phys. Rep.*, 520:253, 2012. (Cited on page 434).

T. Kazimierczuk, J. Schmutzler, M. Aßmann, C. Schneider, M. Kamp, S. Höfling, and M. Bayer. Photon-statistics excitation spectroscopy of a quantum-dot micropillar laser. *Phys. Rev. Lett.*, 115:027401, 2015. (Cited on page 503).

J. Keeling. Response functions and superfluid density in a weakly interacting Bose gas with nonquadratic dispersion. *Phys. Rev. B*, 74:155325, 2006. (Cited on page 331).

J. Keeling and N. G. Berloff. Spontaneous rotating vortex lattices in a pumped decaying condensate. *Phys. Rev. Lett.*, 100:250401, 2008. (Cited on pages 368, 451, 536, and 537).

J. Keeling, P. R. Eastham, M. H. Szymańska, and P. B. Littlewood. Polariton condensation with localized excitons and propagating photons. *Phys. Rev. Lett.*, 93:226403, 2004. (Cited on page 333).

J. Keeling, P. R. Eastham, M. H. Szymańska, and P. B. Littlewood. BCS–BEC crossover in a system of microcavity polaritons. *Phys. Rev. B*, 72:115320, 2005. (Cited on page 332).

L. V. Keldysh and A. N. Kozlov. Collective properties of large-radius excitons. *JETP Lett.*, 5:190, 1967. (Cited on page 520).

L. V. Keldysh and A. N. Kozlov. Collective properties of excitons in semiconductors. *J. Exp. Th. Phys.*, 27:521, 1968. (Cited on page 325).

L. V. Keldysh, V. D. Kulakovskii, S. Reitzenstein, M. N. Makhonin, and A. Forchel. Interference effects in the emission spectra of QD's in high quality cavities. *Pis'ma ZhETF*, 84:584, 2006. (Cited on page 220).

S. Kéna-Cohen and S. R. Forrest. Room-temperature polariton lasing in an organic single-crystal microcavity. *Nat. Photon.*, 4:371, 2010. (Cited on page 523).

J. Kerr. On rotation of the plane of polarization by reflection from the pole of a magnet. *Phil. Mag.*, 3:321, 1877. (Cited on page 46).

P. G. Kevrekidis, H. E. Nistazakis, D. J. Frantzeskakis, B. A. Malomed, and R. Carretero-González. Families of matter-waves in two-component Bose–Einstein condensates. *Eur. Phys. J. D*, 28:181, 2004. (Cited on page 434).

P. G. Kevrekidis, D. J. Frantzeskakis, and R. Carretero-González. Emergent nonlinear phenomena in Bose–Einstein condensates. *Springer*, 2008. (Cited on pages 425 and 473).

K. M. Khalatnikov. *An Introduction to the theory of superfluidity*. Benjamin, New York, 1965. (Cited on page 379).

A. A. Khalifa, A. P. D. Love, D. N. Krizhanovskii, M. S. Skolnick, and J. S. Roberts. Electroluminescence emission from polariton states in GaAs-based semiconductor microcavities. *Appl. Phys. Lett.*, 2008. (Cited on pages 374 and 524).

A. B. Khanikaev, S. Hossein Mousavi, W.-K. Tse, M. Kargarian, A. H. MacDonald, and G. Shvets. Photonic topological insulators. *Nat. Mater.*, 12:233, 2013. (Cited on page 467).

U. Al Khawaja and H. Stoof. Skyrmions in a ferromagnetic Bose–Einstein condensate. *Nature*, 411:918, 2001. (Cited on pages 434 and 439).

A. Khintchine. Korrelationstheorie der stationären stochastischen Prozesse. *Math. Ann.*, 109:604, 1934. (Cited on page 40).

G. Khitrova, H. M. Gibbs, F. Jahnke, M. Kira, and S. W. Koch. Nonlinear optics of normal-mode-coupling semiconductor microcavities. *Rev. Mod. Phys*, 71:1591, 1999. (Cited on page 286).

C. Khripkov, C. Piermarocchi, and A. Vardi. Dynamics of microcavity exciton polaritons in a Josephson double dimer. *Phys. Rev. B*, 88:235305, 2013. (Cited on page 194).

T. W. B. Kibble. Topology of cosmic domains and strings. *J. Phys. A.: Math. Gen.*, 9:1387, 1976. (Cited on pages 482 and 537).

N. Y. Kim and Y. Yamamoto. Exciton-polariton quantum simulators. *arXiv:1510.08203*, 2015. (Cited on page 518).

N. Y. Kim, K. Kusudo, C. Wu, N. Masumoto, A. Löffler, S. Höfling, N. Kumada, L. Worschech, A. Forchel, and Y. Yamamoto. Dynamical d-wave condensation of exciton-polaritons in a two-dimensional square-lattice potential. *Nat. Phys.*, 7:681, 2011. (Cited on pages 453 and 537).

N. Y. Kim, K. Kusudo, A. Löffler, S. Höfling, A. Forchel, and Y. Yamamoto. Exciton-polariton condensates near the Dirac point in a triangular lattice. *New J. Phys.*, 15:035032, 2013. (Cited on page 537).

H. J. Kimble, M. Dagenais, and L. Mandel. Photon antibunching in resonance fluorescence. *Phys. Rev. Lett.*, 39:691, 1977. (Cited on page 116).

M. Kira and S. W. Koch. Quantum-optical spectroscopy of semiconductors. *Phys. Rev. A*, 73:013813, 2006. (Cited on page 286).

S. Kirkpatrick, C. D. Gelatt Jr., and M. P. Vecchi. Optimization by simulated annealing. *Science*, 220:671, 1983. (Cited on page 513).

J. R. Kirtley, C. C. Tsuei, M. Rupp, J. Z. Sun, L. See Yu-Jahnes, A. Gupta, M. B. Ketchen, T. J. Watson, Y. Heights, K. A. Moler, and M. Bhushan. Direct imaging of integer and half-integer Josephson vortices in high-T_C grain boundaries. *Phys. Rev. Lett.*, 76:1336, 1996. (Cited on page 434).

V. A. Kiselev, B. S. Razbirin, and I. N. Ural'tsev. Interference states of optical excitons. observation of additional waves. *JETP Lett.*, 18:504, 1973. (Cited on page 145).

A. Kitaev. Fault-tolerant quantum computation by anyons. *Annals of Physics*, 303:2, 2003. (Cited on page 518).

Y. Kivshar and G. Agrawal. *Optical solitons: from fibers to photonic crystals*. Academic Press, 2001. (Cited on page 473).

S. Klembt, E. Durupt, S. Datta, T. Klein, A. Baas, Y. Léger, C. Kruse, D. Hommel, A. Minguzzi, and M. Richard. Exciton-polariton gas as a nonequilibrium coolant. *Phys. Rev. Lett.*, 114:186403, 2015. (Cited on page 318).

D. Kleppner. Inhibited spontaneous emission. *Phys. Rev. Lett.*, 47: 233, 1981. (Cited on page 244).

E. Knill, R. Laflamme, and G. J. Milburn. A scheme for efficient quantum computation with linear optics. *Nature*, 409:46, 2001. (Cited on page 510).

L. Knöll and G. Weber. Theory of n-fold time-resolved correlation spectroscopy and its application to resonance fluorescence radiation. *J. Phys. B.: At. Mol. Phys.*, 19:2817, 1986. (Cited on pages 491 and 492).

L. Knöll, G. Weber, and T. Schafer. Theory of time-resolved correlation spectroscopy and its application to resonance fluorescence radiation. *J. Phys. B.: At. Mol. Phys.*, 17:4861, 1984. (Cited on page 491).

L. Knöll, W. Vogel, and D. G. Welsch. Quantum noise in spectral filtering of light. *J. Opt. Soc. Am. B*, 3:1315, 1986. (Cited on page 491).

L. Knöll, W. Vogel, and D.-G. Welsch. Spectral properties of light in quantum optics. *Phys. Rev. A*, 42:503, 1990. (Cited on page 492).

O. Knudsen. The Faraday effect and physical theory, 1845-1873. *Arch. Hist. Exact Sci.*, 15:235, 1976. (Cited on page 47).

Z. G. Koinov. Bose–Einstein condensation of excitons in a single quantum well. *Phys. Rev. B*, 61:8411, 2000. (Cited on pages 326 and 330).

M. König, S. Wiedmann, C. Brüne, A. Roth, H. Buhmann, L. W. Molenkamp, X.-L. Qi, and S.-C. Zhang. Quantum spin Hall insulator state in HgTe quantum wells. *Science*, 318:766, 2007. (Cited on page 467).

V. V. Konotop and M. Salerno. Small-amplitude excitations in a deformable discrete nonlinear Schrödinger equation. *Phys. Rev. E*, 55:4706, 1997. (Cited on page 426).

V. E. Korepin and N. M. Bololiubov. *Quantum Inverse Scattering Method and Correlation Functions*. Cambridge University Press, Cambridge, 1997. (Cited on page 502).

D. J. Korteweg and G. de Vries. On the change of form of long waves advancing in a rectangular canal, and on a new type of long stationary waves. *Phil. Mag.*, 39:422, 1895. (Cited on page 306).

J. M. Kosterlitz and D. J. Thouless. Ordering, metastability and phase transitions in two-dimensional systems. *J. Phys. C.: Solid State Phys.*, 6:1181, 1973. (Cited on pages 326 and 430).

D. N. Krizhanovskii, D. Sanvitto, I. A. Shelykh, M. M. Glazov, G. Malpuech, D. D. Solnyshkov, A. Kavokin, S. Ceccarelli, M. S. Skolnick, and J. S. Roberts. Rotation of the plane of polarization of light in a semiconductor microcavity. *Phys. Rev. B*, 73:073303, 2006. (Cited on pages 375 and 406).

D. N. Krizhanovskii, K. G. Lagoudakis, M. Wouters, B. Pietka, R. A. Bradley, K. Guda, D. M. Whittaker, M. S. Skolnick, B. Deveaud-Plédran, M. Richard, R. André, and Le Si Dang. Coexisting nonequilibrium condensates with long-range spatial coherence in semiconductor microcavities. *Phys. Rev. B*, 80:045317, 2009. (Cited on page 526).

D. N. Krizhanovskii, D. M. Whittaker, R. A. Bradley, K. Guda, D. Sarkar, D. Sanvitto, L. Viña, E. Cerda, P. Santos, K. Biermann, R. Hey, and M. S. Skolnick. Effect of interactions on vortices in a nonequilibrium polariton condensate. *Phys. Rev. Lett.*, 104: 126402, 2010. (Cited on page 375).

H. Kroemer. A proposed class of hetero-junction injection lasers. *Proc. IEEE*, 51:1782, 1963. (Cited on page 261).

A. Kronenberger and P. Pringsheim. über das Absorptionsspektrum des festen Benzols bei $-180°$. *Z. Phys. B*, 40:75, 1926. (Cited on page 133).

S. Kundermann, M. Saba, C. Ciuti, T. Guillet, U. Oesterle, J. L. Staehli, and B. Deveaud. Coherent control of polariton parametric scattering in semiconductor microcavities. *Phys. Rev. Lett.*, 91: 107402, 2003. (Cited on page 288).

K. Kuruma, Y. Ota, M. Kakuda, D. Takamiya, S. Iwamoto, and Y. Arakawa. Position dependent optical coupling between single quantum dots and photonic crystal nanocavities. *Appl. Phys. Lett.*, 109:071110, 2016. (Cited on page 255).

K. Kusudo, N. Y. Kim, A. Löffler, S. Höfling, A. Forchel, and Y. Yamamoto. Stochastic formation of polariton condensates in two degenerate orbital states. *Phys. Rev. B*, 87:214503, 2013. (Cited on page 539).

P. G. Kwiat, K. Mattle, H. Weinfurter, A. Zeilinger, A. V. Sergienko, and Y. Shih. New high-intensity source of polarization-entangled photon pairs. *Phys. Rev. Lett.*, 75:4337, 1995. (Cited on page 483).

O. Kyriienko and T. C. H. Liew. Exciton-polariton quantum gates based on continuous variables. *Phys. Rev. A*, 93:035301, 2016. (Cited on pages 516, 517, and 518).

O. Kyriienko, A. V. Kavokin, and I. A. Shelykh. Superradiant terahertz emission by dipolaritons. *Phys. Rev. Lett.*, 111:176401, 2013. (Cited on pages 529 and 531).

O. Kyriienko, E. A. Ostrovskaya, O. A. Egorov, I. A. Shelykh, and T. C. H. Liew. Bistability in microcavities with incoherent optical or electrical excitation. *Phys. Rev. B*, 90:125407, 2014. (Cited on pages 490 and 539).

K. G. Lagoudakis, M. Wouters, M. Richard, A. Baas, I. Carusotto, R. André, Le Si Dang, and B. Deveaud-Plédran. Quantized vortices in an exciton-polariton condensate. *Nat. Phys.*, 4:706, 2008. (Cited on pages 375, 424, 431, 432, 434, and 536).

K. G. Lagoudakis, T. Ostatnický, A. V. Kavokin, Y. G. Rubo, R. André, and B. Deveaud-Plédran. Observation of half-quantum vortices in an exciton-polariton condensate. *Science*, 326:974, 2009. (Cited on pages 375, 425, 435, and 537).

K. G. Lagoudakis, B. Pietka, M. Wouters, R. André, and B. Deveaud-Plédran. Coherent oscillations in an exciton-polariton Josephson junction. *Phys. Rev. Lett.*, 105:120403, 2010. (Cited on page 194).

K. G. Lagoudakis, F. Manni, B. Pietka, M. Wouters, T. C. H. Liew, V. Savona, A. V. Kavokin, R. André, and B. Deveaud-Plédran. Probing the dynamics of spontaneous quantum vortices in polariton superfluids. *Phys. Rev. Lett.*, 106:115301, 2011. (Cited on pages 537 and 538).

P. G. Lagoudakis, P. G. Savvidis, J. J. Baumberg, D. M. Whittaker, P. R. Eastham, M. S. Skolnick, and J. S. Roberts. Stimulated spin dynamics of polaritons in semiconductor microcavities. *Phys. Rev. B*, 65:161310R, 2002. (Cited on page 394).

C. W. Lai, N. Y. Kim, S. Utsunomiya, G. Roumpos, H. Deng, M. D. Fraser, T. Byrnes, P. Recher, N. Kumada, T. Fujisawa, and Y. Yamamoto. Coherent zero-state and π-state in an exciton-polariton condensate array. *Nature*, 450:529, 2007. (Cited on pages 515, 521, and 537).

W. E. Lamb Jr. Theory of an optical maser. *Phys. Rev.*, 134:A1429, 1964. (Cited on page 269).

L. D. Landau. Theory of the superfluidity of helium II. *Phys. Rev.*, 60: 356, 1941. (Cited on pages 327 and 368).

L. D. Landau. The theory of superfluidity of helium II. *J. Phys. (USSR)*, 11:91, 1947. (Cited on page 327).

L. D. Landau and E. M. Lifshitz. *The Classical Theory of Fields*. Butterworth-Heinemann, 1980. (Cited on page 531).

W. Langbein. Spontaneous parametric scattering of microcavity polaritons in momentum space. *Phys. Rev. B*, 70:205301, 2004. (Cited on page 484).

W. Langbein. Polariton correlation in microcavities produced by parametric scattering. *Phys. Stat. Sol. B*, 242:2260, 2005. (Cited on page 484).

W. Langbein and J. M. Hvam. Elastic scattering dynamics of cavity polaritons: Evidence for time-energy uncertainty and polariton localization. *Phys. Rev. Lett.*, 88:47401, 2002. (Cited on page 553).

W. Langbein, I. Shelykh, D. Solnyshkov, G. Malpuech, Yu. Rubo, and A. Kavokin. Polarization beats in ballistic propagation of exciton-polaritons in microcavities. *Phys. Rev. B*, 75:075323, 2007. (Cited on page 395).

P.-É. Larré, N. Pavloff, and A. M. Kamchatnov. Wave pattern induced by a localized obstacle in the flow of a one-dimensional polariton condensate. *Phys. Rev. B*, 86:165304, 2012. (Cited on page 368).

N. Laskin. Fractional quantum mechanics and Lévy path integrals. *Phys. Lett. A*, 268:298, 2000. (Cited on page 309).

A. Laucht, N. Hauke, J. M. Villas-Bôas, F. Hofbauer, G. Böhm, M. Kaniber, and J. J. Finley. Dephasing of exciton polaritons in photoexcited InGaAs quantum dots in GaAs nanocavities. *Phys. Rev. Lett.*, 103:087405, 2009a. (Cited on pages 221 and 228).

A. Laucht, F. Hofbauer, N. Hauke, J. Angele, S. Stobbe, M. Kaniber, G. Böhm, P. Lodahl, M.-C. Amann, and J. J. Finley. Electrical control of spontaneous emission and strong coupling for a single quantum dot. *New J. Phys.*, 11:023034, 2009b. (Cited on pages 210 and 211).

A. Laucht, M. Kaniber, A. Mohtashami, N. Hauke, M. Bichler, and J. J. Finley. Temporal monitoring of nonresonant feeding of semiconductor nanocavity modes by quantum dot multiexciton transitions. *Phys. Rev. B*, 81:241302(R), 2010. (Cited on page 211).

F. P. Laussy, G. Malpuech, and A. Kavokin. Spontaneous coherence buildup in a polariton laser. *Phys. Stat. Sol. C*, 1:1339, 2004a. (Cited on pages 353 and 481).

F. P. Laussy, G. Malpuech, A. Kavokin, and P. Bigenwald. Spontaneous coherence buildup in a polariton laser. *Phys. Rev. Lett.*, 93:016402, 2004b. (Cited on pages 363 and 364).

F. P. Laussy, G. Malpuech, A. V. Kavokin, and P. Bigenwald. Coherence dynamics in microcavities and polariton lasers. *J. Phys.: Condens. Matter*, 16:S3665, 2004c. (Cited on page 347).

F. P. Laussy, M. M. Glazov, A. V. Kavokin, and G. Malpuech. Multiplets in the optical emission spectra of Dicke states of quantum dots excitons coupled to microcavity photons. *Phys. Stat. Sol. C*, 2:3819, 2005. (Cited on page 202).

F. P. Laussy, M. M. Glazov, A. Kavokin, D. M. Whittaker, and G. Malpuech. Statistics of excitons in quantum dots and their effect on the optical emission spectra of microcavities. *Phys. Rev. B*, 73:115343, 2006a. (Cited on pages 236 and 238).

F. P. Laussy, I. A. Shelykh, G. Malpuech, and A. Kavokin. Effects of Bose–Einstein condensation of exciton polaritons in microcavities on the polarization of emitted light. *Phys. Rev. B*, 73:035315, 2006b. (Cited on pages 349, 364, 365, and 366).

F. P. Laussy, E. del Valle, and C. Tejedor. Strong coupling of quantum dots in microcavities. *Phys. Rev. Lett.*, 101:083601, 2008. (Cited on page 221).

F. P. Laussy, E. del Valle, and C. Tejedor. Luminescence spectra of quantum dots in microcavities. I. Bosons. *Phys. Rev. B*, 79:235325, 2009. (Cited on pages 215 and 218).

F. P. Laussy, A. V. Kavokin, and I. A. Shelykh. Exciton-polariton mediated superconductivity. *Phys. Rev. Lett.*, 104:106402, 2010. (Cited on pages 377 and 379).

F. P. Laussy, E. del Valle, M. Schrapp, A. Laucht, and J. J. Finley. Climbing the Jaynes–Cummings ladder by photon counting. *J. Nanophoton.*, 6:061803, 2012. (Cited on pages 486 and 498).

F.P. Laussy. *Exciton-polaritons in microcavities*, volume 172, chapter 1. Quantum Dynamics of Polariton Condensates, pages 1–42. Springer Berlin Heidelberg, 2012. ISBN 978-3-642-24186-4. (Cited on pages 308, 309, 356, and 358).

Le Si Dang, D. Heger, R. André, F. Boeuf, and R. Romestain. Stimulation of polariton photoluminescence in semiconductor microcavity. *Phys. Rev. Lett.*, 81:3920, 1998. (Cited on pages 284, 331, 374, and 521).

A. J. Leggett. Bose-Einstein condensation in the alkali gases: Some fundamental concepts. *Rev. Mod. Phys.*, 73:307, 1975. (Cited on page 192).

A. J. Leggett. BEC: The alkali gases from the perspective of research on liquid helium. *AIP Conf. Proc.*, 477:154, 1999. (Cited on page 193).

A. J. Leggett. *Quantum Liquids*. Oxford University Press, 2006. (Cited on page 472).

A. J. Leggett. Quantum liquids. *Science*, 319:1203, 2008. (Cited on page 424).

A. J. Leggett and F. Sols. On the concept of spontaneously broken gauge symmetry in condensed matter physics. *Found. Phys.*, 21:353, 1991. (Cited on page 194).

M.-A. Lemonde, N. Didier, and A. A. Clerk. Antibunching and unconventional photon blockade with gaussian squeezed states. *Phys. Rev. A*, 90:063824, 2014. (Cited on page 488).

L. A. Levin. The tale of one-way functions. *Prob. Inf. Trans.*, 39:92, 2003. (Cited on page 511).

J. Levrat, R. Butté, E. Feltin, J.-F. Carlin, N. Grandjean, D. Solnyshkov, and G. Malpuech. Condensation phase diagram of cavity polaritons in GaN-based microcavities: Experiment and theory. *Phys. Rev. B*, 81:125305, 2010. (Cited on pages 335 and 342).

C. Leyder, M. Romanelli, J. Ph. Karr, E. Giacobino, T. C. H. Liew, M. M. Glazov, A. V. Kavokin, G. Malpuech, and A. Bramati. Observation of the optical spin Hall effect. *Nat. Phys.*, 3:628, 2007. (Cited on pages 399, 424, and 465).

G. Lheureux, S. Azzini, C. Symonds, P. Senellart, A. Lemaître, C. Sauvan, J.-P. Hugonin, J.-J. Greffet, and J. Bellessa. Polarization-controlled confined Tamm plasmon lasers. *ACS Photonics*, 2:842, 2015. (Cited on page 69).

F. Li, L. Orosz, O. Kamoun, S. Bouchoule, C. Brimont, P. Disseix, T. Guillet, X. Lafosse, M. Leroux, J. Leymarie, M. Mexis, M. Mihailovic, G. Patriarche, F. Réveret, D. Solnyshkov, J. Zuniga-Perez, and G. Malpuech. From excitonic to photonic polariton condensate in a ZnO-based microcavity. *Phys. Rev. Lett.*, 110:196406, 2013. (Cited on pages 335, 342, and 523).

G. Li, M. D. Fraser, A. Yakimenko, and E. A. Ostrovskaya. Stability of persistent currents in open dissipative quantum fluids. *Phys. Rev. B*, 91:184518, 2015a. (Cited on page 539).

G. Li, T. C. H. Liew, O. A. Egorov, and E. A. Ostrovskaya. Incoherent excitation and switching of spin states in exciton-polariton condensates. *Phys. Rev. B*, 92:064304, 2015b. (Cited on page 539).

S. De Liberato, C. Ciuti, and C. C. Phillips. Terahertz lasing from intersubband polariton-polariton scattering in asymmetric quantum wells. *Phys. Rev. B*, 87:241304(R), 2013. (Cited on page 527).

T. C. H. Liew and V. Savona. Single photons from coupled quantum modes. *Phys. Rev. Lett.*, 104:183601, 2010. (Cited on pages 488 and 489).

T. C. H. Liew and V. Savona. Multipartite polariton entanglement in semiconductor microcavities. *Phys. Rev. B*, 84:032301, 2011. (Cited on page 518).

T. C. H. Liew, A. V. Kavokin, and I. A. Shelykh. Optical circuits based on polariton neurons in semiconductor microcavities. *Phys. Rev. Lett.*, 101:016402, 2008. (Cited on page 544).

T. C. H. Liew, M. M. Glazov, K. V. Kavokin, I. A. Shelykh, M. A. Kaliteevski, and A. V. Kavokin. Proposal for a bosonic cascade laser. *Phys. Rev. Lett.*, 110:047402, 2013. (Cited on pages 532 and 534).

T. C. H. Liew, O. A. Egorov, M. Matuszewski, O. Kyriienko, X. Ma, and E. A. Ostrovskaya. Instability-induced formation and nonequilibrium dynamics of phase defects in polariton condensates. *Phys. Rev. B*, 91:085413, 2015. (Cited on pages 536 and 539).

T. C. H. Liew, Y. G. Rubo, A. S. Sheremet, S. De Liberato, I. A. Shelykh, F. P. Laussy, and A. V. Kavokin. Quantum statistics of bosonic cascades. *New J. Phys.*, 18:023041, 2016. (Cited on pages 532 and 533).

L. Lifshitz and L. Pitaevskii. *Statistical Physics II*. Pergamon Press, 1980. (Cited on page 416).

N. H. Lindner, G. Refael, and V. Galitski. Floquet topological insulator in semiconductor quantum wells. *Nat. Phys.*, 7:490, 2011. (Cited on page 467).

W. A. Little. Possibility of synthesizing an organic superconductor. *Phys. Rev.*, 134:A1416, 1964. (Cited on page 377).

S. Lloyd. Universal quantum simulators. *Science*, 273:1073, 1996. (Cited on page 514).

S. Lloyd and S. L. Braunstein. Quantum computation over continuous variables. *Phys. Rev. Lett.*, 82:1784, 1999. (Cited on page 516).

F. London. The λ-phenomenon of liquid helium and the Bose–Einstein degeneracy. *Nature*, 141:643, 1938. (Cited on page 323).

S. L. Lopatnikov and A. H.-D. Cheng. If you ask a physicist from any country: A tribute to Yacov Il'ich Frenkel. *J. Eng. Mech.*, 131:875, 2005. (Cited on page 135).

J. C. López Carreño and F. P. Laussy. Excitation with quantum light. I. Exciting a harmonic oscillator. *Phys. Rev. A*, 94:063825, 2016. (Cited on pages 505, 506, and 507).

J. C. López Carreño, C. Sánchez Muñoz, D. Sanvitto, E. del Valle, and F. P. Laussy. Exciting polaritons with quantum light. *Phys. Rev. Lett.*, 115:196402, 2015. (Cited on page 507).

J. C. López Carreño, C. Sánchez Muñoz, E. del Valle, and F. P. Laussy. Excitation with quantum light. II. Exciting a two-level system. *Phys. Rev. A*, 94:063826, 2016. (Cited on pages 505 and 507).

R. Loudon. *The quantum theory of light*. Oxford Science Publications, 3 edition, 2000. (Cited on pages 122 and 299).

W. H. Louisell, A. Yariv, and A. E. Siegman. Quantum fluctuations and noise in parametric processes. I. *Phys. Rev.*, 124:1646, 1961. (Cited on pages 297 and 299).

A. P. D. Love, D. N. Krizhanovskii, D. M. Whittaker, R. Bouchekioua, D. Sanvitto, S. Al Rizeiqi, R. Bradley, M. S. Skolnick, P. R. Eastham, R. André, and Le Si Dang. Intrinsic decoherence mechanisms in the microcavity polariton condensate. *Phys. Rev. Lett.*, 101:067404, 2008. (Cited on pages 482 and 536).

Yu. E. Lozovik and I. V. Ovchinnikov. Many-photon coherence of Bose-condensed excitons: Luminescence and related nonlinear optical phenomena. *Phys. Rev. B*, 66:075124, 2002. (Cited on page 325).

Yu. E. Lozovik and V.I. Yudson. On the possibility of superfluidity of spatially separated electrons and holes due to their coupling: new mechanism of superfluidity. *JETP Lett.*, 22:26, 1975. (Cited on page 325).

Yu. E. Lozovik and V.I. Yudson. New mechanism of superconductivity: coupling between spatially separated electrons and holes. *J. Exp. Th. Phys.*, 44:389, 1976a. (Cited on page 325).

Yu. E. Lozovik and V.I. Yudson. Superconductivity at dielectric pairing of spatially separated quasiparticles. *Solid State Commun.*, 19: 391, 1976b. (Cited on page 325).

Yu. E. Lozovik, O. L. Bergmann, and A. A. Panfinov. The excitation spectra and superfluidity of the electron-hole system in coupled quantum wells. *Phys. Stat. Sol. B*, 209:287, 1998. (Cited on page 326).

L. Lu, J. D. Joannopoulos, and M. Soljaçić. Topological photonics. *Nat. Photon.*, 8:821, 2014. (Cited on pages 467 and 473).

Q.-Y. Lu, X.-H. Chen, W.-H. Guo, L.-J. Yu, Y.-Z. Huang, J. Wang, and Y. Luo. Mode characteristics of semiconductor equilateral triangle microcavities with side length of 5–20 μm. *IEEE Phot. Tech. Lett.*, 16:359, 2004. (Cited on page 16).

J. M. Luttinger and W. Kohn. Motion of electrons and holes in perturbed periodic fields. *Phys. Rev.*, 97:869, 1955. (Cited on page 131).

Jr. M. M. de Lima, M. van der Poel, P. V. Santos, and J. M. Hvam. Phonon-induced polariton superlattices. *Phys. Rev. Lett.*, 97: 045501, 2006. (Cited on page 515).

X. Ma, I. Yu. Chestnov, M. V. Charukhchyan, A. P. Alodjants, and O. A. Egorov. Oscillatory dynamics of nonequilibrium dissipative exciton-polariton condensates in weak-contrast lattices. *Phys. Rev. B*, 91:214301, 2015. (Cited on page 194).

M. Z. Maialle, E. A. de Andrada e Silva, and L. J. Sham. Exciton spin dynamics in quantum wells. *Phys. Rev. B*, 47:15776, 1993. (Cited on page 387).

A. Majumdar, M. Bajcsy, A. Rundquist, and J. Vučković. Loss-enabled sub-poissonian light generation in a bimodal nanocavity. *Phys. Rev. Lett.*, 108:183601, 2012. (Cited on page 490).

G. Malpuech and A. Kavokin. Picosecond beats in coherent optical spectra of semiconductor heterostructures: photonic bloch and exciton-polariton oscillations. *Semicond. Sci. Technol.*, 16:R1, 2001. (Cited on page 333).

G. Malpuech, A. Di Carlo, A. Kavokin, J. J. Baumberg, M. Zamfirescu, and P. Lugli. Room-temperature polariton lasers based on GaN microcavities. *Appl. Phys. Lett.*, 81:412, 2002a. (Cited on pages 331 and 334).

G. Malpuech, A. Kavokin, A. Di Carlo, and J. J. Baumberg. Polariton lasing by exciton-electron scattering in semiconductor microcavities. *Phys. Rev. B*, 65:153310, 2002b. (Cited on pages 334, 336, and 338).

G. Malpuech, D. D. Solnyshkov, H. Ouerdane, M. M. Glazov, and I. Shelykh. Bose glass and superfluid phases of cavity polaritons. *Phys. Rev. Lett.*, 98:206402, 2007. (Cited on page 334).

G. B. Malykin. Use of the Poincaré sphere in polarization optics and classical and quantum mechanics. review. *Radiophys. Quantum Electron.*, 40:175, 1997. (Cited on page 33).

L. Mandel and E. Wolf. Correlation in the fluctuating outputs from two square-law detectors illuminated by light of any state of coherence and polarization. *Phys. Rev.*, 124:1696, 1961. (Cited on pages 106 and 299).

L. Mandel and E. Wolf. Photon correlations. *Phys. Rev. Lett.*, 10:176, 1963. (Cited on page 113).

L. Mandel and E. Wolf. *Optical coherence and quantum optics*. Cambridge University Press, Cambridge, 1995. (Cited on pages 38, 122, 270, and 347).

L Mandel, E C G Sudarshan, and E Wolf. Theory of photoelectric detection of light fluctuations. *Proc. Phys. Soc.*, 84:435, 1964. (Cited on pages 45 and 114).

F. Manni, K. G. Lagoudakis, T. C. H. Liew, R. André, , and B. Deveaud-Plédran. Spontaneous pattern formation in a polariton condensate. *Phys. Rev. Lett.*, 107:106401, 2011. (Cited on page 537).

F. Manni, K. G. Lagoudakis, T. C. H Liew, R. André, V. Savona, and B. Deveaud. Dissociation dynamics of singly charged vortices into half-quantum vortex pairs. *Nat. Comm.*, 3:1309, 2012. (Cited on pages 537 and 539).

F. Manni, T. C. H. Liew, K. G. Lagoudakis, C. Ouellet-Plamondon, R. André, V. Savona, and B. Deveaud. Spontaneous self-ordered states of vortex-antivortex pairs in a polariton condensate. *Phys. Rev. B*, 88:201303(R), 2013. (Cited on page 537).

M. Maragkou, A. J. D. Grundy, T. Ostatnický, and P. G. Lagoudakis. Longitudinal optical phonon assisted polariton laser. *Appl. Phys. Lett.*, 97:111110, 2010. (Cited on page 521).

F. M. Marchetti, B. D. Simons, and P. B. Littlewood. Condensation of cavity polaritons in a disordered environment. *Phys. Rev. B*, 70: 155327, 2004. (Cited on page 332).

F. M. Marchetti, J. Keeling, M. H. Szymańska, and P. B. Littlewood. Thermodynamics and excitations of condensed polaritons in disordered microcavities. *Phys. Rev. Lett.*, 96:066405, 2006. (Cited on page 333).

F. M. Marchetti, M. H. Szymańska, C. Tejedor, and D. M. Whittaker. Spontaneous and triggered vortices in polariton optical-parametric-oscillator superfluids. *Phys. Rev. Lett.*, 105:063902, 2010. (Cited on page 431).

G. I. Márk. Analysis of the spreading Gaussian wavepacket. *Eur. Phys. J. B*, 18:247, 1997. (Cited on page 305).

M. D. Martin, G. Aichmayr, L. Viña, and R. André. Polarization control of the nonlinear emission of semiconductor microcavities. *Phys. Rev. Lett.*, 89:077402, 2002. (Cited on pages 391, 392, and 398).

M. R. Matthews, B. P. Anderson, P. C. Haljan, D. S. Hall, C. E. Wieman, and E. A. Cornell. Vortices in a Bose–Einstein condensate. *Phys. Rev. Lett.*, 83:2498, 1999. (Cited on page 431).

M. Matuszewski and E. Witkowska. Universality in nonequilibrium condensation of exciton-polaritons. *Phys. Rev. B*, 89:155318, 2014. (Cited on page 537).

J. C. Maxwell. A dynamical theory of the electromagnetic field. *Philos. Trans. Roy. Soc. London*, 155:459, 1865. (Cited on page 26).

E. McCann and M. Koshino. The electronic properties of bilayer graphene. *Rep. Prog. Phys.*, 76:056503, 2013. (Cited on pages 468 and 472).

J. McKeever, A. Boca, A. D. Boozer, J. R. Buck, and H. J. Kimble. Experimental realization of a one-atom laser in the regime of strong coupling. *Nature*, 425:268, 2003. (Cited on page 209).

B. Meier and B. P. Zakharchenia, editors. *Optical orientation*. Elsevier Science, North Holland, Amsterdam, 1984. (Cited on page 384).

N. C. Menicucci, P. van Loock, M. Gu, C. Weedbrook, T. C. Ralph, and M. A. Nielsen. Universal quantum computation with continuous-variable cluster states. *Phys. Rev. Lett.*, 97:110501, 2006. (Cited on page 518).

N.D Mermin and H. Wagner. Absence of ferromagnetism or antiferromagnetism in one or two-dimensional isotropic Heisenberg models. *Phys. Rev. Lett.*, 22:1133, 1966. (Cited on page 326).

E. Merzbacher. *Quantum mechanics*. John Wiley & Sons, Inc., 3 edition, 1998. (Cited on page 122).

A. M. L. Messiah and O. W. Greenberg. Symmetrization postulate and its experimental foundation. *Phys. Rev.*, 136:B248, 1964. (Cited on page 88).

G. Messin, J. Ph. Karr, A. Baas, G. Khitrova, R. Houdré, R. P. Stanley, U. Oesterle, and E. Giacobino. Parametric polariton amplification in semiconductor microcavities. *Phys. Rev. Lett.*, 87:127403, 2001. (Cited on page 293).

B. R. Mollow. Power spectrum of light scattered by two-level systems. *Phys. Rev.*, 188:1969, 1969. (Cited on pages 199, 228, and 491).

K. Molmer. Optical coherence: A convenient fiction. *Phys. Rev. A*, 55: 3195, 1997. (Cited on page 194).

A. Montanaro. Quantum algorithms: an overview. *npj Quantum Information*, 2:15023, 2016. (Cited on page 518).

S.A. Moskalenko. Reversible optico-hydrodynamic phenomena in a non ideal exciton gas. *Sov. Phys. Solid State*, 4:199, 1962. (Cited on pages 324 and 520).

S.A. Moskalenko and D.W. Snoke. *Bose–Einstein condensation of excitons and biexcitons.* Cambridge University Press, Cambridge, 2000. (Cited on page 379).

N. F. Mott. Conduction in polar crystals. II. The conduction band and ultra-violet absorption of alkali-halide crystals. *Trans. Faraday Soc.*, 34:500, 1938. (Cited on page 133).

S. Mukamel and K. E. Dorfman. Nonlinear fluctuations and dissipation in matter revealed by quantum light. *Phys. Rev. A*, 91:053844, 2015. (Cited on page 503).

K. Müller, A. Rundquist, K. A. Fischer, T. Sarmiento, K. G. Lagoudakis, Y. A. Kelaita, C. Sánchez Muñoz, E. del Valle, F. P. Laussy, and J. Vučković. Coherent generation of nonclassical light on chip via detuned photon blockade. *Phys. Rev. Lett.*, 114: 233601, 2015. (Cited on pages 487 and 498).

M. Muller, J. Bleuse, R. Andre, and H. Ulmer-Tuffigo. Observation of bottleneck effects on the photoluminescence from polaritons in II-VI microcavities. *Physica B*, 272:467, 1999. (Cited on pages 284 and 285).

S. Münch, S. Reitzenstein, P. Franeck, A. Löffler, T. Heindel, S. Höfling, L. Worschech, and A. Forchel. The role of optical excitation power on the emission spectra of a strongly coupled quantum dot-micropillar system. *Opt. Express*, 17:12821, 2009. (Cited on page 221).

N. Na and Y. Yamamoto. Massive parallel generation of indistinguishable single photons via the polaritonic superfluid to Mott-insulator quantum phase transition. *New J. Phys.*, 12:123001, 2010. (Cited on page 514).

A. V. Nalitov, G. Malpuech, H. Terças, and D. D. Solnyshkov. Spin-orbit coupling and the optical spin Hall effect in photonic graphene. *Phys. Rev. Lett.*, 114:026803, 2015a. (Cited on page 468).

A. V. Nalitov, D. D. Solnyshkov, and G. Malpuech. Polariton $\mathbf{Z}$ topological insulator. *Phys. Rev. Lett.*, 114:116401, 2015b. (Cited on pages 453, 465, 466, 467, 469, and 471).

G. Nardin, G. Grosso, Y. Léger, B. Piętka, F. Morier-Genoud, and B. Deveaud-Plédran. Hydrodynamic nucleation of quantized vortex pairs in a polariton quantum fluid. *Nat. Phys.*, 7:635, 2011. (Cited on pages 424, 431, and 433).

M. Nathan, W. P. Dumke, G. Burns, Jr. F. H. Dill, and G. Lasher. Stimulated emission of radiation from GaAs p-n junctions. *Appl. Phys. Lett.*, 1:62, 1962. (Cited on page 261).

J. W. Negele and H. Orland. *Quantum many-particle systems.* Perseus, 1998. (Cited on page 122).

D. R. Nelson and J. M. Kosterlitz. Universal jump in the superfluid density of two dimensional superfluid. *Phys. Rev. Lett.*, 39, 1977. (Cited on page 330).

M. A. Nielsen. Cluster-state quantum computation. *Reports on Mathematical Physics*, 57:147, 2006. (Cited on page 518).

G. Nienhuis. Spectral correlations in resonance fluorescence. *Phys. Rev. A*, 47:510, 1993. (Cited on pages 492, 493, and 494).

P. B. R. Nisbet-Jones, J. Dilley, D. Ljunggren, and A. Kuhn. Highly efficient source for indistinguishable single photons of controlled shape. *New J. Phys.*, 13:103036, 2011. (Cited on page 402).

J. U. Nöckel. Mikrolaser als photonen-billard: wie chaos ans licht kommt. *Physikalische Blätter*, 54:927, 1998. (Cited on page 17).

S. Noda, K. Tomoda, N. Yamamoto, and A. Chutinan. Full three-dimensional photonic bandgap crystals at near-infrared wavelengths. *Science*, 289:604, 2000. (Cited on page 71).

M. Nomura, N. Kumagai, S. Iwamoto, Y. Ota, and Y. Arakawa. Laser oscillation in a strongly coupled single-quantum-dot–nanocavity system. *Nat. Phys.*, 6:279, 2010. (Cited on pages 229 and 230).

K. S. Novoselov, Z. Jiang, Y. Zhang, S. V. Morozov, H. L. Stormer, U. Zeitler, J. C. Maan, G. S. Boebinger, P. Kim, and A. K. Geim. Room-temperature quantum Hall effect in graphene. *Science*, 315:1379, 2007. (Cited on page 465).

H. Ohadi, A. Dreismann, Y. G. Rubo, F. Pinsker, Y. del Valle-Inclan Redondo, S. I. Tsintzos, Z. Hatzopoulos, P. G. Savvidis, and J. J. Baumberg. Spontaneous spin bifurcations and ferromagnetic phase transitions in a spinor exciton-polariton condensate. *Phys. Rev. X*, 5:031002, 2015. (Cited on pages 421, 449, 450, 451, 539, and 540).

P. Öhberg and L. Santos. Dark solitons in a two-component Bose–Einstein condensate. *Phys. Rev. Lett.*, 86:2918, 2001. (Cited on page 434).

L. Onsager. Statistical hydrodynamics. *Nuovo Cimento*, 6:279, 1949. (Cited on page 428).

E. A. Ostrovskaya, J. Abdullaev, A. S. Desyatnikov, . D. Fraser, and Y. S. Kivshar. Dissipative solitons and vortices in polariton Bose–Einstein condensates. *Phys. Rev. A*, 86:013636, 2012. (Cited on pages 536 and 537).

Y. Ota, N. Kumagai, S. Ohkouchi, M. Shirane, M. Nomura, S. Ishida, S. Iwamoto, S. Yorozu, and Y. Arakawa. Investigation of the spectral triplet in strongly coupled quantum dot–nanocavity system. *Appl. Phys. Express*, 2:122301, 2009. (Cited on pages 228 and 229).

T. Ozawa and I. Carusotto. Anomalous and quantum Hall effects in lossy photonic lattices. *Phys. Rev. Lett.*, 112:133902, 2014. (Cited on page 467).

G. Panzarini and L. C. Andreani. Bulk polariton beatings and two-dimensional radiative decay: Analysis of time-resolved transmission through a dispersive film. *Solid State Commun.*, 102:505, 1997. (Cited on page 149).

G. Panzarini, L. C. Andreani, A. Armitage, D. Baxter, M. S. Skolnick, V. N. Astratov, J. S. Roberts, A. V. Kavokin, M. R. Vladimirova, and M. A. Kaliteevski. Exciton-light coupling in single and coupled semiconductor microcavities: Polariton dispersion and polarization splitting. *Phys. Rev. B*, 59:5082, 1999a. (Cited on page 165).

G. Panzarini, L. C. Andreani, A. Armitage, D. Baxter, M. S. Skolnick, V. N. Astratov, J. S. Roberts, A. V. Kavokin, M. R. Vladimirova, and M. A. Kaliteevski. Cavity-polariton dispersion and polarization splitting in single and coupled semiconductor microcavities. *Phys. Solid State*, 41:1223, 1999b. (Cited on pages 66 and 166).

T. K. Paraïso, M. Wouters, Y. Léger, F. Mourier-Genoud, and B. Deveaud-Plédran. Multistability of a coherent spin ensemble in a semiconductor microcavity. *Nat. Mater.*, 10:80, 2010. (Cited on page 539).

G. Parascandolo and V. Savona. Long-range radiative interaction between semiconductor quantum dots. *Phys. Rev. B*, 71:045335, 2005. (Cited on page 154).

M. Patrini, M. Galli, F. Marabelli, M. Agio, L. C. Andreani, D. Peyrade, and Y. Chen. Photonic bands in patterned silicon-on-insulator waveguides. *IEEE Quantum Electron.*, 38:885, 2002. (Cited on page 72).

W. Pauli. The connection between spin and statistics. *Phys. Rev.*, 58: 716, 1940. (Cited on page 89).

G. Pavlovic, G. Malpuech, and I. A. Shelykh. Pseudospin dynamics in multimode polaritonic Josephson junctions. *Phys. Rev. B*, 87: 125307, 2013. (Cited on page 194).

M. Peiris, B. Petrak, K. Konthasinghe, Y. Yu, Z. C. Niu, and A. Muller. Two-color photon correlations of the light scattered by a quantum dot. *Phys. Rev. B*, 91:195125, 2015. (Cited on pages 495 and 496).

S. I. Pekar. Theory of electromagnetic waves in a crystal in which excitons arise. *J. Exp. Th. Phys.*, 33:1022, 1957. (Cited on pages 143 and 145).

X. Peng, Z. Liao, N. Xu, G. Qin, X. Zhou, D. Suter, and J. Du. Quantum adiabatic algorithm for factorization and its experimental implementation. *Phys. Rev. Lett.*, 101:220405, 2008. (Cited on page 513).

O. Penrose and L. Onsager. Bose–Einstein condensation and liquid helium. *Phys. Rev.*, 104:576, 1954. (Cited on page 324).

M. Perrin, P. Senellart, A. Lemaître, and J. Bloch. Polariton relaxation in semiconductor microcavities: Efficiency of electron-polariton scattering. *Phys. Rev. B*, 72:075340, 2005. (Cited on page 338).

A. A. Pervishko, T. C. H. Liew, A. V. Kavokin, and I. A. Shelykh. Bistability in bosonic terahertz lasers. *J. Phys.: Condens. Matter*, 26:085303, 2014. (Cited on page 528).

E. Peter, P. Senellart, D. Martrou, A. Lemaître, J. Hours, J. M. Gérard, and J. Bloch. Exciton-photon strong-coupling regime for a single quantum dot embedded in a microcavity. *Phys. Rev. Lett.*, 95: 067401, 2005. (Cited on pages 171 and 210).

C. J. Pethick and H. Smith. *Bose–Einstein condensation in dilute gases*. Cambridge University Press, 2001. (Cited on page 379).

S. Pigeon, I. Carusotto, and C. Ciuti. Hydrodynamic nucleation of vortices and solitons in a resonantly excited polariton superfluid. *Phys. Rev. B*, 83:144513, 2011. (Cited on page 433).

G.E. Pikus and G.L. Bir. Exchange interaction in excitons in semiconductors. *J. Exp. Th. Phys.*, 33:108, 1971. (Cited on pages 385 and 398).

F. Pinsker and H. Flayac. On-demand dark soliton train manipulation in a spinor polariton condensate. *Phys. Rev. Lett.*, 112:140405, 2014. (Cited on page 536).

F. Pinsker, W. Bao, Y. Zhang, H. Ohadi, A. Dreismann, and J. J. Baumberg. Fractional quantum mechanics in polariton condensates with velocity-dependent mass. *Phys. Rev. B*, 92:195310, 2015. (Cited on page 309).

L. Pitaevskii and S. Stringari. *Bose–Einstein Condensation*. Oxford University Press, 2003. (Cited on pages 379, 428, and 472).

L.P. Pitaevskii. Phenomenological theory of superfluidity near the λ point. *Sov. Phys. JETP*, 35:282, 1959. (Cited on page 368).

J. D. Plumhof, T. Stöferle, L. Mai, U. Scherf, and R. F. Mahrt. Room-temperature Bose–Einstein condensation of cavity exciton-polaritons in a polymer. *Nat. Mater.*, 13:247, 2014. (Cited on page 523).

D. Porras and C. Tejedor. Linewidth of a polariton laser: Theoretical analysis of self-interaction effects. *Phys. Rev. B*, 67:161310(R), 2003. (Cited on pages 365 and 366).

D. Porras, C. Ciuti, J. J. Baumberg, and C. Tejedor. Polariton dynamics and Bose–Einstein condensation in semiconductor microcavities. *Phys. Rev. B*, 66:085304, 2002. (Cited on pages 334 and 521).

S. Portolan, L. Einkemmer, Z. Vörš, G. Weihs, and P. Rabl. Generation of hyper-entangled photon pairs in coupled microcavities. *New J. Phys.*, 16:063030, 2014. (Cited on page 484).

G. V. Prakash, L. Besombes, T. Kelf, J. J. Baumberg, P. N. Bartlett, and M. E. Abdelsalam. Tunable resonant optical microcavities by self-assembled templating. *Opt. Lett.*, 29:1500, 2004. (Cited on pages 11 and 12).

D. Press, S. Götzinger, S. Reitzenstein, C. Hofmann, A. Löffler, M. Kamp, A. Forchel, and Y. Yamamoto. Photon antibunching from a single quantum dot-microcavity system in the strong coupling regime. *Phys. Rev. Lett.*, 98:117402, 2007. (Cited on page 491).

J. P. Prineas, J. Y. Zhou, J. Kuhl, H. M. Gibbs, G. Khitrova, S. W. Koch, and A. Knorr. Ultrafast ac Stark effect switching of the active photonic band gap from Bragg-periodic semiconductor quantum wells. *Appl. Phys. Lett.*, 81:4332, 2002. (Cited on page 280).

A. M. Prokhorov. Quantum electronics. *Nobel Lecture*, 1964. (Cited on page 256).

E. M. Purcell. Spontaneous emission probabilities at radio frequencies. *Phys. Rev.*, 69:681, 1946. (Cited on page 244).

E. M. Purcell. The question of correlation between photons in coherent light rays. *Nature*, 178:1449, 1956. (Cited on page 113).

E. M. Purcell, H. C. Torrey, and R. V. Pound. Resonance absorption by nuclear magnetic moments in a solid. *Phys. Rev.*, 69:37, 1946. (Cited on page 249).

A. Qarry, G. Ramon, R. Rapaport, E. Cohen, Arza Ron, A. Mann, E. Linder, and L. N. Pfeiffer. Nonlinear emission due to electron-polariton scattering in a semiconductor microcavity. *Phys. Rev. B*, 67:115320, 2003. (Cited on page 338).

X.-L. Qi and S.-C. Zhang. Topological insulators and superconductors. *Rev. Mod. Phys.*, 83:1057, 2011. (Cited on pages 425 and 473).

W. Que. Excitons in quantum dots with parabolic confinement. *Phys. Rev. B*, 45:11036, 1992. (Cited on page 235).

T. M. Quist, R. H. Rediker, R. J. Keyes, W. E. Krag, B. Lax, A. L. McWhorter, and H. J. Zeigler. Semiconductor MASER of GaAs. *Appl. Phys. Lett.*, 1:91, 1962. (Cited on page 261).

D. Racine and P. R. Eastham. Quantum theory of multimode polariton condensation. *Phys. Rev. B*, 90:085308, 2014. (Cited on page 194).

S. Raghavan, A. Smerzi, S. Fantoni, and S. R. Shenoy. Coherent oscillations between two weakly coupled Bose–Einstein condensates: Josephson effects, π oscillations, and macroscopic quantum self-trapping. *Phys. Rev. A*, 59:620, 1999. (Cited on page 193).

A. Rahmani and F.P. Laussy. Polaritonic Rabi and Josephson oscillations. *Sci. Rep.*, 6:28930, 2016. (Cited on pages 192, 194, and 195).

M. G. Raizen, R. J. Thompson, R. J. Brecha, H. J. Kimble, and H. J. Carmichael. Normal-mode splitting and linewidth averaging for two-state atoms in an optical cavity. *Phys. Rev. Lett.*, 63:240, 1989. (Cited on page 209).

P. Rakyta, A. Kormányos, and J. Cserti. Trigonal warping and anisotropic band splitting in monolayer graphene due to Rashba spin-orbit coupling. *Phys. Rev. B*, 82:113405, 2010. (Cited on pages 468 and 472).

G. Ramon, A. Mann, and E. Cohen. Theory of neutral and charged exciton scattering with electrons in semiconductor quantum wells. *Phys. Rev. B*, 67:045323, 2003. (Cited on page 550).

R. Rapaport, R. Harel, E. Cohen, A. Ron, and E. Linder. Negatively charged quantum well polaritons in a GaAs/AlAs microcavity: An analog of atoms in a cavity. *Phys. Rev. Lett.*, 84:1607, 2000. (Cited on page 337).

E. I. Rashba and G. E. Gurgenishvili. Edge absorption theory in semiconductors. *Sov. Phys. Solid State*, 4:759, 1962. (Cited on page 155).

E. I. Rashba and M. E. Portnoi. Anyon excitons. *Phys. Rev. B*, 70: 3315, 1993. (Cited on page 134).

K. Rayanov, B. L. Altshuler, Y. G. Rubo, and S. Flach. Frequency combs with weakly lasing exciton-polariton condensates. *Phys. Rev. Lett.*, 114:193901, 2015. (Cited on page 194).

D. Read, T. C. H. Liew, Yu. G. Rubo, and A. V. Kavokin. Stochastic polarization formation in exciton-polariton Bose–Einstein condensates. *Phys. Rev. B*, 80:195309, 2009. (Cited on page 364).

M. C. Rechtsman, J. M. Zeuner, Y. Plotnik, Y. Lumer, D. Podolsky, F. Dreisow, S. Nolte, M. Segev, and A. Szameit. Photonic floquet topological insulators. *Nature*, 496:196, 2013. (Cited on page 467).

A. Reinhard, T. Volz, M. Winger, A. Badolato, K. J. Hennessy, E. L. Hu, and A. İmamoğlu. Strongly correlated photons on a chip. *Nat. Photon.*, 6:93, 2012. (Cited on page 485).

J. P. Reithmaier, G. Sek, A. Löffler, C. Hofmann, S. Kuhn, S. Reitzenstein, L. V. Keldysh, V. D. Kulakovskii, T. L. Reinecker, and A. Forchel. Strong coupling in a single quantum dot–semiconductor microcavity system. *Nature*, 432:197, 2004. (Cited on pages 171, 210, and 221).

G. Rempe, R. J. Thompson, R. J. Brecha, W. D. Lee, , and H. J. Kimble. Optical bistability and photon statistics in cavity quantum electrodynamics. *Phys. Rev. Lett.*, 67(1727), 1991. (Cited on page 488).

G. Rempe, R. J. Thompson, H. J. Kimble, and R. Lalezari. Measurement of ultralow losses in an optical interferometer. *Opt. Lett.*, 17: 363, 1992. (Cited on page 11).

P. Renucci, T. Amand, X. Marie, P. Senellart, J. Bloch, B. Sermage, and K. V. Kavokin. Microcavity polariton spin quantum beats without a magnetic field: A manifestation of Coulomb exchange in dense and polarized polariton systems. *Phys. Rev. B*, 72: 075317, 2005. (Cited on page 395).

S. Reynaud. La fluorescence de résonance: étude par la méthode de l'atome habillé. *Annales de Physique*, 8:315, 1983. (Cited on page 491).

M. Richard, J. Kasprzak, R. André, R. Romestain, Le Si Dang, G. Malpuech, and A. Kavokin. Experimental evidence for nonequilibrium Bose condensation of exciton polaritons. *Phys. Rev. B*, 72:201301R, 2005a. (Cited on page 413).

M. Richard, J. Kasprzak, R. Romestain, R. André, and Le Si Dang. Spontaneous coherent phase transition of polaritons in CdTe microcavities. *Phys. Rev. Lett.*, 94:187401, 2005b. (Cited on page 334).

S. Ritter, P. Gartner, C. Gies, and F. Jahnke. Emission properties and photon statistics of a single quantum dot laser. *Opt. Express*, 18: 9909, 2010. (Cited on page 211).

A. S. Rodrigues, P. G. Kevrekidis, R. Carretero-González, D. J. Frantzeskakis J. Cuevas-Maraver, and F. Palmero. From nodeless clouds and vortices to gray ring solitons and symmetry-broken states in two-dimensional polariton condensates. *J. Phys.: Condens. Matter*, 26:155801, 2014. (Cited on page 536).

M. Romanelli, C. Leyder, J.-P. Karr, E. Giacobino, and A. Bramati. Generation of correlated polariton modes in a semiconductor microcavity. *Phys. Stat. Sol. C*, 2:3924, 2005. (Cited on page 289).

M. Romanelli, C. Leyder, J. Ph. Karr, E. Giacobino, and A. Bramati. Four wave mixing oscillation in a semiconductor microcavity: Generation of two correlated polariton populations. *Phys. Rev. Lett.*, 98:106401, 2007. (Cited on page 484).

M. Romanelli, J. Ph. Karr, C. Leyder, E. Giacobino, and A. Bramati. Two-mode squeezing in polariton four-wave mixing. *Phys. Rev. B*, 82:155313, 2010. (Cited on page 481).

G. Roumpos, W. H. Nitsche, S. Höfling, A. Forchel, and Y. Yamamoto. Gain-induced trapping of microcavity exciton polariton condensates. *Phys. Rev. Lett.*, 104:126403, 2010. (Cited on page 537).

G. Roumpos, M. D. Fraser, A. Löffler, S. Höfling, A. Forchel, and Y. Yamamoto. Single vortex-antivortex pair in an exciton-polariton condensate. *Nat. Phys.*, 7:129, 2011. (Cited on pages 431 and 537).

Yu. G. Rubo. Half vortices in exciton polariton condensates. *Phys. Rev. Lett.*, 99:106401, 2007. (Cited on pages 425, 434, 436, 437, and 440).

Yu. G. Rubo, F. P. Laussy, G. Malpuech, A. Kavokin, and P. Bigenwald. Dynamical theory of polariton amplifiers. *Phys. Rev. Lett.*, 91:156403, 2003. (Cited on pages 347 and 349).

Yu. G. Rubo, A.V. Kavokin, and I.A. Shelykh. Suppression of superfluidity of exciton-polaritons by magnetic field. *Phys. Lett. A*, 358: 227, 2006. (Cited on pages 417, 418, and 436).

V. I. Rupasov and V. I. Yudson. Exact Dicke superradiance theory: Bethe wavefunctions in the discrete atom model. *Zh. Eksp. Teor. Fiz*, 86:819, 1984a. (Cited on page 502).

V. I. Rupasov and V. I. Yudson. Rigorous theory of cooperative spontaneous emission of radiation from a lumped system of two-level atoms: Bethe ansatz method. *Zh. Eksp. Teor. Fiz*, 87:1617, 1984b. (Cited on page 502).

J. Scott Russell. Report on waves. *Fourteenth meeting of the British Association for the Advancement of Science*, 1844. (Cited on page 306).

M. Saba, C. Ciuti, J. Bloch, V. Thierry-Mieg, R. Andre, Le Si Dang, S. Kundermann, A. Mura, G. Bongiovanni, J. L. Staehli, and B. Deveaud. High temperature ultrafast polariton amplification in semiconductor microcavities. *Nature*, 414:731, 2003. (Cited on pages 288 and 289).

G. Sallen, A. Tribu, T. Aichele, R. André, L. Besombes, C. Bougerol, M. Richard, S. Tatarenko, K. Kheng, and J.-Ph. Poizat. Subnanosecond spectral diffusion measurement using photon correlation. *Nat. Photon.*, 4:696, 2010. (Cited on page 491).

M. M. Salomaa and G. E. Volovik. Half-solitons in superfluid ^{3}he-a: Novel $\pi/2$-quanta of phase slippage. *J. Low Temp. Phys.*, 74: 319, 1989. (Cited on pages 434 and 440).

C. Sánchez Muñoz, E. del Valle, C. Tejedor, and F.P. Laussy. Violation of classical inequalities by photon frequency filtering. *Phys. Rev. A*, 90:052111, 2014a. (Cited on pages 494 and 495).

C. Sánchez Muñoz, E. del Valle, A. González Tudela, K. Müller, S. Lichtmannecker, M. Kaniber, C. Tejedor, J.J. Finley, and F.P. Laussy. Emitters of N-photon bundles. *Nat. Photon.*, 8:550, 2014b. (Cited on pages 497, 498, 499, 500, 501, and 502).

C. Sánchez Muñoz, F. P. Laussy, C. Tejedor, and E del Valle. Enhanced two-photon emission from a dressed biexciton. *New J. Phys.*, 17: 123021, 2015. (Cited on page 501).

D. Sanvitto and V. Timofeev, editors. *Exciton Polaritons in Microcavities: New Frontiers*, volume 172 of *Springer Series on Solid-State Sciences*. Springer-Verlag,, Berlin, Heidelberg, 2012. (Cited on page 545).

D. Sanvitto, A. Daraei, A. Tahraoui, M. Hopkinson, P. W. Fry, D. M. Whittaker, and M. S. Skolnick. Observation of ultrahigh quality factor in a semiconductor microcavity. *Appl. Phys. Lett.*, 86: 191109, 2005. (Cited on page 294).

D. Sanvitto, D. N. Krizhanovskii, D. M. Whittaker, S. Ceccarelli, M. S. Skolnick, and J. S. Roberts. Spatial structure and stability of the macroscopically occupied polariton state in the microcavity optical parametric oscillator. *Phys. Rev. B*, 73:241308(R), 2006. (Cited on page 294).

D. Sanvitto, F. M. Marchetti, M. H. Szymańska, G. Tosi, M. Baudisch, F. P. Laussy, D. N. Krizhanovskii, M. S. Skolnick, L. Marrucci, A. Lemaître, J. Bloch, C. Tejedor, and L. Viña. Persistent currents and quantized vortices in a polariton superfluid. *Nat. Phys.*, 6:527, 2010. (Cited on page 375).

D. Sanvitto, S. Pigeon, A. Amo, D. Ballarini, M. De Giorgi, I. Carusotto, R. Hivet, F. Pisanello, V. G. Sala, P. S. S. Guimaraes, R. Houdré, E. Giacobino, C. Ciuti, A. Bramati, and G. Gigli. All-optical control of the quantum flow of a polariton condensate. *Nat. Photon.*, 5:610, 2011. (Cited on pages 431 and 433).

D. Sarchi, I. Carusotto, M. Wouters, and V. Savona. Coherent dynamics and parametric instabilities of microcavity polaritons in double-well systems. *Phys. Rev. B*, 77:125324, 2008a. (Cited on page 194).

D. Sarchi, P. Schwendimann, and A. Quattropani. Effects of noise in different approaches for the statistics of polariton condensates. *Phys. Rev. B*, 78:073404, 2008b. (Cited on pages 481 and 483).

M. E. Sasin, R. P. Seisyan, M. A. Kaliteevski, S. Brand, R. A. Abram, J. M. Chamberlain, A. Yu. Egorov, A. P. Vasil'ev, V. S. Mikhrin, and A. V. Kavokin. Tamm-plasmon polaritons: Slow and spatially compact light. *Appl. Phys. Lett.*, 92:251112, 2008. (Cited on page 69).

S. Savasta, O. Di Stefano, V. Savona, and W. Langbein. Quantum complementarity of microcavity polaritons. *Phys. Rev. Lett.*, 94: 246401, 2005. (Cited on pages 294 and 484).

I. G. Savenko, I. A. Shelykh, and M. A. Kaliteevski. Nonlinear terahertz emission in semiconductor microcavities. *Phys. Rev. Lett.*, 107:027401, 2011. (Cited on page 527).

V. Savona, L. C. Andreani, P. Schwendimann, and A. Quattropani. Quantum well excitons in semiconductor microcavities: Unified treatment of weak and strong coupling regimes. *Solid State Commun.*, 93:733, 1995. (Cited on page 163).

V. Savona, C. Piermarocchi, A. Quattropani, P. Schwendimann, and F. Tassone. Optical properties of microcavity polaritons. *Phase Trans.*, 68:169, 1998. (Cited on pages 240 and 241).

P. G. Savvidis, J. J. Baumberg, R. M. Stevenson, M. S. Skolnick, D. M. Whittaker, and J. S. Roberts. Angle-resonant stimulated polariton amplifier. *Phys. Rev. Lett.*, 84:1547, 2000. (Cited on pages 9, 286, 287, and 483).

P. G. Savvidis, C. Ciuti, J. J. Baumberg, D. M. Whittaker, M. S. Skolnick, and J. S. Roberts. Off-branch polaritons and multiple scattering in semiconductor microcavities. *Phys. Rev. B*, 64: 075311, 2001. (Cited on page 289).

H. Schmidt and A. İmamoğlu. Giant Kerr nonlinearities obtained by electromagnetically induced transparency. *Opt. Lett.*, 21:1936, 1996. (Cited on page 485).

S. Schmitt-Rink, D. S. Chemla, and D. A. B. Miller. Theory of transient excitonic optical nonlinearities in semiconductor quantum-well structures. *Phys. Rev. B*, 32:6601, 1985. (Cited on page 507).

C. Schneider, M. Strauß, T. Sünner, A. Huggenberger, D. Wiener, S. Reitzenstein, M. Kamp, S. Höfling, and A. Forchel. Lithographic alignment to site-controlled quantum dots for device integration. *Appl. Phys. Lett.*, 92:183101, 2008. (Cited on page 255).

C. Schneider, A. Rahimi-Iman, N. Y. Kim, J. Fischer, I. G. Savenko, M. Amthor, M. Lermer, A. Wolf, L. Worschech, V. D. Kulakovskii, I. A. Shelykh, M. Kamp, S. Reitzenstein, A. Forchel, Y. Yamamoto, and S. Höfling. An electrically pumped polariton laser. *Nature*, 497:348, 2013. (Cited on pages 524 and 544).

C. A. Schrama, G. Nienhuis, H. A. Dijkerman, C. Steijsiger, and H. G. M. Heideman. Destructive interference between opposite time orders of photon emission. *Phys. Rev. Lett.*, 67, 1991. (Cited on page 491).

C. A. Schrama, G. Nienhuis, H. A. Dijkerman, C. Steijsiger, and H. G. M. Heideman. Intensity correlations between the components of the resonance fluorescence triplet. *Phys. Rev. A*, 45:8045, 1992. (Cited on page 494).

E. Schrödinger. Der stetige übergang von der Mikro- zur Makromechanik. *Naturwissenschaften*, 14:664, 1926. (Cited on page 305).

B. Schumacher. Quantum coding. *Phys. Rev. A*, 51:2738, 1995. (Cited on page xxviii).

I. Schuster, A. Kubanek, A. Fuhrmanek, T. Puppe, P. W. H. Pinkse, K. Murr, and G. Rempe. Nonlinear spectroscopy of photons bound to one atom. *Nat. Phys.*, 4:382, 2008. (Cited on page 222).

P. Schwendimann and A. Quattropani. Amplification and quantum statistics of microcavity polaritons under nonresonant excitation. *Phys. Rev. B*, 74:045324, 2006. (Cited on page 481).

P. Schwendimann and A. Quattropani. Statistics of the polariton condensate. *Phys. Rev. B*, 77:085317, 2008. (Cited on page 481).

P. Schwendimann, C. Ciuti, and A. Quattropani. Statistics of polaritons in the nonlinear regime. *Phys. Rev. B*, 68:165324, 2003. (Cited on page 483).

M. O. Scully. Condensation of N bosons and the laser phase transition analogy. *Phys. Rev. Lett.*, 82:3927, 1999. (Cited on page 351).

M. O. Scully and W. E. Lamb Jr. Quantum theory of an optical maser. I. general theory. *Phys. Rev.*, 159:208, 1967. (Cited on page 269).

M. O. Scully and M. S. Zubairy. *Quantum optics*. Cambridge University Press, 2002. (Cited on pages 122 and 528).

R. P. Seisyan. Diamagnetic excitons in semiconductors (review). *Phys. Solid State*, 58:859, 2016. (Cited on page 558).

P. E. Selbmann, M. Gulia, F. Rossi, E. Molinari, and P. Lugli. Coupled free-carrier and exciton relaxation in optically excited semiconductors. *Phys. Rev. B*, 54:4660, 1996. (Cited on page 335).

P. Senellart and J. Bloch. Nonlinear emission of microcavity polaritons in the low density regime. *Phys. Rev. Lett.*, 82:1233, 1999. (Cited on pages 284, 331, and 374).

P. Senellart, J. Bloch, B. Sermage, and J. Y. Marzin. Microcavity polariton depopulation as evidence for stimulated scattering. *Phys. Rev. B*, 62:R16263, 2000. (Cited on page 331).

J. Shan and T. F. Heinz. *Ultrafast Dynamical Processes in Semiconductors*, chapter Terahertz Radiation from Semiconductors. Springer Berlin Heidelberg, 2004. (Cited on page 531).

I. Schneider, K. V. Kavokin, A. V. Kavokin, G. Malpuech, P. Bigenwald, H. Deng, G. Weihs, and Y. Yamamoto. Semiconductor microcavity as a spin-dependent optoelectronic device. *Phys. Rev. B*, 70:035320, 2004. (Cited on pages 392 and 393).

I. A. Shelykh, M. M. Glazov, D. D. Solnyshkov, N. G. Galkin, A. V. Kavokin, and G. Malpuech. Spin dynamics of polariton parametric amplifiers. *Phys. Stat. Sol. C*, 2:768, 2005. (Cited on page 408).

I. A. Shelykh, Yuri G. Rubo, G. Malpuech, D. D. Solnyshkov, and A. Kavokin. Polarization and propagation of polariton condensates. *Phys. Rev. Lett.*, 97:066402, 2006. (Cited on page 364).

I. A. Shelykh, D. D. Solnyshkov, G. Pavlovic, and G. Malpuech. Josephson effects in condensates of excitons and exciton polaritons. *Phys. Rev. B*, 78:041302(R), 2008. (Cited on page 194).

I. A. Shelykh, A. V. Kavokin, Yu. G. Rubo, T. C. H. Liew, and G. Malpuech. Polariton polarization-sensitive phenomena in planar semiconductor microcavities. *Semicond. Sci. Technol.*, 25: 013001, 2010. (Cited on page 305).

Y. R. Shen. Quantum statistics of nonlinear optics. *Phys. Rev.*, 155: 921, 1967. (Cited on page 344).

P. W. Shor. Polynomial-time algorithms for prime factorization and discrete logarithms on a quantum computer. *Siam J. Comput.*, 26: 1484, 1997. (Cited on page 509).

M. Sich, D.N. Krizhanovskii, M.S. Skolnick, A.V. Gorbach, R. Hartley, D. V. Skryabin, E. A. Cerda-Méndez, K. Biermann, R. Hey, and P.V. Santos. Observation of bright polariton solitons in a semiconductor microcavity. *Nat. Photon.*, 6:50, 2012. (Cited on page 460).

M. Sich, F. Fras, J. K. Chana, M. S. Skolnick, D. N. Krizhanovskii, A. V. Gorbach, R. Hartley, D. V. Skryabin, S. S. Gavrilov, E. A. Cerda-Méndez, K. Biermann, R. Hey, , and P. V. Santos. Effects of spin-dependent interactions on polarization of bright polariton solitons. *Phys. Rev. Lett.*, 112:046403, 2014. (Cited on page 460).

H. Sigurdsson, O. A. Egorov, X. Ma, I. A. Shelykh, and T. C. H. Liew. Information processing with topologically protected vortex memories in exciton-polariton condensates. *Phys. Rev. B*, 90:014504, 2014. (Cited on page 539).

E. A. Silinish. *Organic molecular crystals: Their electronic states*. Springer-Verlag, Berlin, 1980. (Cited on page 134).

B. Silva, C. Sánchez Muñoz, D. Ballarini, A. González-Tudela, M. de Giorgi, G. Gigli, K. West, L. Pfeiffer, E. del Valle, D. Sanvitto, and F. P. Laussy. The colored Hanbury Brown–Twiss effect. *Sci. Rep.*, 6:37980, 2016. (Cited on page 497).

B. Simon. Holonomy, the quantum adiabatic theorem, and Berry's phase. *Phys. Rev. Lett.*, 51:2167, 1983. (Cited on page 465).

G. A. Siviloglou and D. N. Christodoulides. Accelerating finite energy Airy beams. *Opt. Lett.*, 32:979, 2007. (Cited on page 307).

R. E. Slusher, L. W. Hollberg, B. Yurke, J. C. Mertz, and J. F. Valley. Observation of squeezed states generated by four-wave mixing in an optical cavity. *Phys. Rev. Lett.*, 55:2409, 1985. (Cited on page 480).

L. A. Smirnov, D. A. Smirnova, E. A. Ostrovskaya, and Y. S. Kivshar. Dynamics and stability of dark solitons in exciton-polariton condensates. *Phys. Rev. B*, 89:235310, 2014. (Cited on page 536).

J. A. Smolin, G. Smith, and A. Vargo. Oversimplifying quantum factoring. *Nature*, 499:163, 2013. (Cited on page 510).

D. Snoke. Spontaneous Bose coherence of excitons and polaritons. *Science*, 298:5597, 2002. (Cited on page 302).

D. Snoke, S. Denev, Y. Liu, L. Pfeiffer, and K. West. Long-range transport in excitonic dark states in coupled quantum wells. *Nature*, 418:754, 2002. (Cited on page 325).

D. D. Solnyshkov, H. Flayac, and G. Malpuech. Stable magnetic monopoles in spinor polariton condensates. *Phys. Rev. B*, 85: 073105, 2012. (Cited on pages 425, 440, and 443).

D. D. Solnyshkov, H. Terças, K. Dini, and G. Malpuech. Hybrid Boltzmann–Gross-Pitaevskii theory of Bose–Einstein condensation and superfluidity in open driven-dissipative systems. *Phys. Rev. A*, 89:033626, 2014. (Cited on pages 369, 370, 371, and 372).

R. D. Somma, D. Nagaj, and M. Kieferová. Quantum speedup by quantum annealing. *Phys. Rev. Lett.*, 109:050501, 2012. (Cited on page 513).

J.-H. Song, Y. He, A. V. Nurmikko, J. Tischler, and V. Bulovic. Exciton-polariton dynamics in a transparent organic semiconductor microcavity. *Phys. Rev. B*, 69:235330, 2004. (Cited on page 63).

L. Spitzer Jr. and J. L. Greenstein. Continuous emission from planetary nebulae. *Ap. J.*, 114:407, 1951. (Cited on page 496).

R. J. C. Spreeuw. A classical analogy of entanglement. *Found. Phys.*, 28:361, 1998. (Cited on page 477).

R. M. Stevenson, V. N. Astratov, M. S. Skolnick, D. M. Whittaker, M. Emam-Ismail, A. I. Tartakovskii, P. G. Savvidis, J. J. Baumberg, and J. S. Roberts. Continuous wave observation of massive polariton redistribution by stimulated scattering in semiconductor microcavities. *Phys. Rev. Lett.*, 85:3680, 2000. (Cited on pages 292 and 293).

G. G. Stokes. On the composition and resolution of streams of polarized light from different sources. *Trans. Camb. Phil. Soc*, 9:399, 1852. (Cited on page 31).

N. S. Stoyanov, D. W. Ward, T. Feurer, and K. A. Nelson. Terahertz polariton propagation in patterned materials. *Nat. Mater.*, 1:95, 2002. (Cited on page 143).

D. V. Strekalov. Cavity quantum electrodynamics: A bundle of photons, please. *Nat. Photon.*, 8:500, 2014. (Cited on page 498).

C. Sturm, D. Tanese, H.S. Nguyen, H. Flayac, E. Galopin, A. Lemaître, I. Sagnes, D. Solnyshkov, A. Amo, G. Malpuech, and J. Bloch. All-optical phase modulation in a cavity-polariton Mach–Zehnder interferometer. *Nat. Comm.*, 5:3278, 2014. (Cited on page 540).

C. Sturm, D. Solnyshkov, O. Krebs, A. Lemaître, I. Sagnes, E. Galopin, A. Amo, G. Malpuech, and J. Bloch. Nonequilibrium polariton condensate in a magnetic field. *Phys. Rev. B*, 91:155130, 2015. (Cited on pages 454 and 455).

E. C. G. Sudarshan. Equivalence of semiclassical and quantum mechanical descriptions of statistical light beams. *Phys. Rev. Lett.*, 10:277, 1963. (Cited on page 107).

J. Suffczyński, A. Dousse, K. Gauthron, A. Lemaître, I. Sagnes, L. Lanco, J. Bloch, P. Voisin, and P. Senellart. Origin of the optical emission within the cavity mode of coupled quantum dot-cavity systems. *Phys. Rev. Lett.*, 103:027401, 2009. (Cited on page 211).

G. 't Hooft. Quantum gravity as a dissipative deterministic system. *Classical and Quantum Gravity*, 16:3263, 1999. (Cited on page 511).

D. Tanese, H. Flayac, D. Solnyshkov, A. Amo, A. Lemaître, E. Galopin, R. Braive, P. Senellart, I. Sagnes, G. Malpuech, and J. Bloch. Polariton condensation in solitonic gap states in a one-dimensional periodic potential. *Nat. Comm.*, 4:1749, 2013. (Cited on pages 460 and 537).

D. Tanese, E. Gurevich, F. Baboux, T. Jacqmin, A. Lemaître, E. Galopin, I. Sagnes, A. Amo, J. Bloch, and E. Akkermans. Fractal energy spectrum of a polariton gas in a Fibonacci quasiperiodic potential. *Phys. Rev. Lett.*, 112:146404, 2014. (Cited on page 457).

A. I. Tartakovskii, V. D. Kulakovskii, A. Forchel, and J. P. Reithmaier. Exciton-photon coupling in photonic wires. *Phys. Rev. B*, 57:R6807(R), 1998. (Cited on page 452).

A. I. Tartakovskii, D. N. Krizhanovskii, G. Malpuech, M. Emam-Ismail, A. V. Chernenko, A. V. Kavokin, V. D. Kulakovskii, M. S. Skolnick, and J. S. Roberts. Giant enhancement of polariton relaxation in semiconductor microcavities by polariton-free carrier interaction: Experimental evidence and theory. *Phys. Rev. B*, 67:16, 2003. (Cited on page 338).

F. Tassone and Y. Yamamoto. Exciton-exciton scattering dynamics in a semiconductor microcavity and stimulated scattering into polaritons. *Phys. Rev. B*, 59:10830, 1999. (Cited on pages 334 and 552).

F. Tassone and Y. Yamamoto. Lasing and squeezing of composite bosons in a semiconductor microcavity. *Phys. Rev. A*, 62:063809, 2000. (Cited on page 481).

F. Tassone, F. Bassani, and L.C. Andreani. Quantum-well reflectivity and exciton-polariton dispersion. *Phys. Rev. B*, 45:6023, 1992. (Cited on page 398).

F. Tassone, C. Piermarocchi, V. Savona, A. Quattropani, and P. Schwendimann. Bottleneck effects in the relaxation and photoluminescence of microcavity polaritons. *Phys. Rev. B*, 56:7554, 1997. (Cited on pages 337 and 521).

M. Tavis and F. W. Cummings. Exact solution for an n-molecule-radiation-field hamiltonian. *Phys. Rev.*, 170:379, 1968. (Cited on page 196).

H. Terças, D. D. Solnyshkov, and G. Malpuech. Topological Wigner crystal of half-solitons in a spinor Bose–Einstein condensate. *Phys. Rev. Lett.*, 110:035303, 2013. (Cited on page 445).

H. Terças, D. D. Solnyshkov, and G. Malpuech. High-speed DC transport of emergent monopoles in spinor photonic fluids. *Phys. Rev. Lett.*, 113:036403, 2014. (Cited on page 445).

H. Terças, H. Flayac, D. D. Solnyshkov, and G. Malpuech. Non-Abelian gauge fields in photonic cavities and photonic superfluids. *Phys. Rev. Lett.*, 112:066402, 2014. (Cited on page 465).

S. M. Thon, M. T. Rakher, H. Kim, J. Gudat, W. T. M. Irvine, P. M. Petroff, and D. Bouwmeester. Strong coupling through optical positioning of a quantum dot in a photonic crystal cavity. *Appl. Phys. Lett.*, 94:111115, 2009. (Cited on page 255).

S. T. Thornton and J. B. Marion. *Classical Dynamics of Particles and Systems*. Brooks/Cole, 5th edition, 2003. (Cited on page 84).

D. J. Thouless, M. Kohmoto, M. P. Nightingale, and M. den Nijs. Quantized Hall conductance in a two-dimensional periodic potential. *Phys. Rev. Lett.*, 49:405, 1982. (Cited on page 465).

J. Tignon, P. Voisin, C. Delalande, M. Voos, R. Houdré, U. Oesterle, and R. P. Stanley. From Fermi's golden rule to the vacuum Rabi splitting: Magnetopolaritons in a semiconductor optical microcavity. *Phys. Rev. Lett.*, 74:3967, 1995. (Cited on pages 282 and 387).

S. G. Tikhodeev. Comment on "critical velocities in exciton superfluidity". *Phys. Rev. Lett.*, 84:352, 2000. (Cited on page 325).

J. R. Tischler, M. Scott Bradley, V. Bulović, J. H. Song, and A. Nurmikko. Strong coupling in a microcavity LED. *Phys. Rev. Lett.*, 95:036401, 2005. (Cited on page 524).

J. L. Tomaino, A. D. Jameson, Yun-Shik Lee, G. Khitrova, H. M. Gibbs, A. C. Klettke, M. Kira, and S. W. Koch. Terahertz excitation of a coherent λ-type three-level system of exciton-polariton modes in a quantum-well microcavity. *Phys. Rev. Lett.*, 108:267402, 2012. (Cited on page 527).

P. Török, P. D. Higdon, R. Jušmagekaitis, and T. Wilson. Optimising the image contrast of conventional and confocal optical microscopes imaging finite sized spherical gold scatterers. *Opt. Commun.*, 155:335, 1998. (Cited on page 60).

G. Tosi, F. M. Marchetti, D. Sanvitto, C. Antón, M. H. Szymańska, A. Berceanu, C. Tejedor, L. Marrucci, A. Lemaître, J. Bloch, and L. Viña. Onset and dynamics of vortex-antivortex pairs in polariton optical parametric oscillator superfluids. *Phys. Rev. Lett.*, 107:036401, 2011. (Cited on page 431).

G. Tosi, G. Christmann, N. G. Berloff, P. Tsotsis, T. Gao, Z. Hatzopoulos, P. G. Savvidis, and J. J. Baumberg. Sculpting oscillators with light within a nonlinear quantum fluid. *Nat. Phys.*, 8:190, 2012. (Cited on pages 424, 453, 516, and 534).

C. H. Townes. Production of coherent radiation by atoms—and molecules. *Nobel Lecture*, 1964. (Cited on page 256).

A. Tredicucci, Yong Chen, Vittorio Pellegrini, Marco Börger, Lucia Sorba, Fabio Beltram, and Franco Bassani. Controlled exciton-photon interaction in semiconductor bulk microcavities. *Phys. Rev. Lett.*, 75:3906, 1995. (Cited on pages 280 and 281).

A. Tredicucci, Y. Chen, V. Pellegrini, M. Börger, and F. Bassani. Optical bistability of semiconductor microcavities in the strong-coupling regime. *Phys. Rev. A*, 54:3493, 1996. (Cited on page 304).

S. I. Tsintzos, N. T. Pelekanos, G. Konstantinidis, Z. Hatzopoulos, and P. G. Savvidis. A GaAs polariton light-emitting diode operating near room temperature. *Nature*, 453:372, 2008. (Cited on pages 374 and 524).

P. Tsotsis, S. I. Tsintzos, G. Christmann, P. G. Lagoudakis, O. Kyriienko, I A. Shelykh, J. J. Baumberg, A. V. Kavokin, Z. Hatzopoulos, P. S. Eldridge, and P. G. Savvidis. Tuning the energy of a polariton condensate via bias-controlled Rabi splitting. *Phys. Rev. Appl.*, 2:014002, 2014. (Cited on page 525).

G. E. Uhlenbeck and L. Gropper. The equation of state of a non-ideal Einstein–Bose or Fermi–Dirac gas. *Phys. Rev.*, 41:79, 1932. (Cited on page 338).

A. Ulhaq, S. Weiler, S. M. Ulrich, R. Roßbach, M. Jetter, and P. Michler. Cascaded single-photon emission from the Mollow triplet sidebands of a quantum dot. *Nat. Photon.*, 6:238, 2012. (Cited on page 491).

R. O. Umucalılar and I. Carusotto. Fractional quantum Hall states of photons in an array of dissipative coupled cavities. *Phys. Rev. Lett.*, 108:206809, 2012. (Cited on page 467).

D. C. Unitt, A. J. Bennett, P. Atkinson, D. A. Ritchie, and A. J. Shields. Polarization control of quantum dot single-photon sources via a dipole-dependent Purcell effect. *Phys. Rev. B*, 72:033318, 2005. (Cited on page 250).

W. G. Unruh. Experimental black-hole evaporation? *Phys. Rev. Lett.*, 46:1351, 1981. (Cited on page 424).

S. Utsunomiya, L. Tian, G. Roumpos, C. W. Lai, N. Kumada, T. Fujisawa, M. Kuwata-Gonokami, A. Löffler, S. Höfling, A. Forchel, and Y. Yamamoto. Observation of Bogoliubov excitations in exciton-polariton condensates. *Nat. Phys.*, 4:700, 2008. (Cited on page 375).

K. J. Vahala. Optical microcavities. *Nature*, 424:839, 2003. (Cited on page 18).

L. M. K. Vandersypen, M. Steffen, G. Breyta, C. S. Yannoni, M. H. Sherwood, and I. L. Chuang. Experimental realization of Shor's quantum factoring algorithm using nuclear magnetic resonance. *Nature*, 414:883, 2001. (Cited on page 509).

V. Vedral, A. Barenco, and A. Ekert. Quantum networks for elementary arithmetic operations. *Phys. Rev. A*, 54:147, 1996. (Cited on page 511).

A. Verger, C. Ciuti, and I. Carusotto. Polariton quantum blockade in a photonic dot. *Phys. Rev. B*, 73:193306, 2006. (Cited on pages 487 and 488).

G. F. Viamontes, I. L. Markov, and J. P. Hayes. Is quantum search practical? *Comput. Sci. Eng.*, 7:62, 2005. (Cited on page 510).

D. V. Vishnevsky, H. Flayac, A. V. Nalitov, D. D. Solnyshkov, N. A. Gippius, and G. Malpuech. Skyrmion formation and optical spin-Hall effect in an expanding coherent cloud of indirect excitons. *Phys. Rev. Lett.*, 110:246404, 2013. (Cited on page 465).

M. Vladimirova, S. Cronenberger, D. Scalbert, K. V. Kavokin, A. Miard, A. Lemaître, J. Bloch, D. Solnyshkov, G. Malpuech, and A. V. Kavokin. Polariton-polariton interaction constants in microcavities. *Phys. Rev. B*, 82:075301, 2010. (Cited on pages 411 and 412).

M. R. Vladimirova, A. V. Kavokin, and M. A. Kaliteevski. Dispersion of bulk exciton polaritons in a semiconductor microcavity. *Phys. Rev. B*, 54:14566, 1996. (Cited on pages 146, 159, and 160).

W. Vogel and D.-G. Welsch. *Quantum Optics*. Wiley-VCH, 3, 2006. (Cited on pages 490 and 492).

N. Voloch-Bloch, Y. Lereah, Y. Lilach, A. Gover, and A. Arie. Generation of electron Airy beams. *Nature*, 494:331, 2013. (Cited on page 307).

G. E. Volovik. *The Universe in a Helium Droplet*. Oxford University Press, 2003. (Cited on pages 424, 443, and 473).

G.E. Volovik and V.P. Mineev. Line and point singularities in superfluid He^3. *JETP Lett.*, 24:605, 1976. (Cited on page 434).

K. von Klitzing, G. Dorda, and M. Pepper. New method for high-accuracy determination of the fine-structure constant based on quantized hall resistance. *Phys. Rev. Lett.*, 45:494, 1980. (Cited on page 465).

J. von Neumann. *Mathematical foundations of quantum mechanics*. Princeton University Press, 1932. (Cited on page 80).

N. S. Voronova, A. A. Elistratov, and Y. E. Lozovik. Detuning-controlled internal oscillations in an exciton-polariton condensate. *Phys. Rev. Lett.*, 115:186402, 2015. (Cited on pages 193 and 194).

I. Vurgaftman, J. R. Meyer, and L. R. Ram-Mohan. Band parameters for III–V compound semiconductors and their alloys. *J. Appl. Phys.*, 89:5815, 2001. (Cited on page 131).

P. M. Walker, L. Tinkler, D. V. Skryabin, A. Yulin, B. Royall, I. Farrer, D. A. Ritchie, M. S. Skolnick, and D. N. Krizhanovskii. Ultra-low-power hybrid light-matter solitons. *Nat. Comm.*, 6:8317, 2015. (Cited on page 460).

P. R. Wallace. The band theory of graphite. *Phys. Rev.*, 71:622, 1947. (Cited on page 464).

D. F. Walls. Squeezed states of light. *Nature*, 306:141, 1983. (Cited on page 503).

C. Walther, G. Scalari, M. I. Amanti, M. Beck, and J. Faist. Microcavity laser oscillating in a circuit-based resonator. *Science*, 327:1495, 2010. (Cited on page 527).

X. Wang. Continuous-variable and hybrid quantum gates. *J. Phys. A.: Math. Gen.*, 34:9577, 2001. (Cited on page 516).

Z. Wang, Y. D. Chong, J. D. Joannopoulos, and M. Soljačić. Reflection-free one-way edge modes in a gyromagnetic photonic crystal. *Phys. Rev. Lett.*, 100:013905, 2008. (Cited on page 467).

Z. Wang, Y. Chong, J. D. Joannopoulos, and M. Soljačić. Observation of unidirectional backscattering-immune topological electromagnetic states. *Nature*, 461:772, 2009. (Cited on page 465).

G. H. Wannier. The structure of electronic excitation levels in insulating crystals. *Phys. Rev.*, 52:191, 1937. (Cited on page 133).

G. Weihs, T. Jennewein, C. Simon, H. Weinfurter, and A. Zeilinger. Violation of Bell's inequality under strict Einstein locality conditions. *Phys. Rev. Lett.*, 81:5039, 1998. (Cited on page 88).

P. Weinberger. John Kerr and his effects found in 1877 and 1878. *Phil. Mag. Lett.*, 88:897, 2008. (Cited on page 46).

C. Weisbuch and H. Benisty. Microcavities in Ecole Polytechnique Fédérale de Lausanne, Ecole Polytechnique (France) and elsewhere: past, present and future. *Phys. Stat. Sol. B*, 242:2345, 2005. (Cited on page 279).

C. Weisbuch, M. Nishioka, A. Ishikawa, and Y. Arakawa. Observation of the coupled exciton-photon mode splitting in a semiconductor quantum microcavity. *Phys. Rev. Lett.*, 69:3314, 1992. (Cited on pages 209, 210, 278, 476, 483, and 521).

E. Wertz, L. Ferrier, D. D. Solnyshkov, P. Senellart, D. Bajoni, A. Miard, A. Lemaître, G. Malpuech, and J. Bloch. Spontaneous formation of a polariton condensate in a planar gaas microcavity. *Appl. Phys. Lett.*, 95:051108, 2009. (Cited on pages 335 and 342).

E. Wertz, L. Ferrier, D. D. Solnyshkov, R. Johne, D. Sanvitto, A. Lemaître, I. Sagnes, R. Grousson, A. V. Kavokin, P. Senellart, and G. Malpuech an J. Bloch. Spontaneous formation and optical manipulation of extended polariton condensates. *Nat. Phys.*, 6: 860, 2010. (Cited on pages 424, 453, and 537).

E. Wertz, A. Amo, D. D. Solnyshkov, L. Ferrier, T. C. H. Liew, D. Sanvitto, P. Senellart, I. Sagnes, A. Lemaître, A. V. Kavokin, G. Malpuech, and J. Bloch. Propagation and amplification dynamics of 1D polariton condensates. *Phys. Rev. Lett.*, 109:216404, 2012. (Cited on page 534).

D. M. Whittaker. Classical treatment of parametric processes in a strong-coupling planar microcavity. *Phys. Rev. B*, 63:193305, 2001. (Cited on page 299).

D. M. Whittaker. Effects of polariton-energy renormalization in the microcavity optical parametric oscillator. *Phys. Rev. B*, 71:115301, 2005. (Cited on page 304).

D. M. Whittaker, P. Kinsler, T. A. Fisher, M. S. Skolnick, A. Armitage, A. M. Afshar, M. D. Sturge, and J. S. Roberts. Motional narrowing in semiconductor microcavities. *Phys. Rev. Lett.*, 77:4792, 1996. (Cited on pages 166 and 167).

F. Widmann, B. Daudin, G. Feuillet, Y. Samson, M. Arley, and J. L. Rouviere. Evidence of 2D–3D transition during the first stages of GaN growth on AlN. *MRS Internet J. of Nitr. Semic. Res.*, 2:20, 1997. (Cited on pages 142 and 235).

N. Wiebe, D. Braun, and S. Lloyd. Quantum algorithm for data fitting. *Phys. Rev. Lett.*, 109:050505, 2012. (Cited on page 510).

N. Wiener. Generalized harmonic analysis. *Acta Mathematica*, 55: 117, 1930. (Cited on page 40).

J. Wiersig, C. Gies, F. Jahnke, M. Aßmann, T. Berstermann, M. Bayer, C. Kistner, S. Reitzenstein, C. Schneider, S. Höfling, A. Forchel, C. Kruse, J. Kalden, and S. Homme. Direct observation of correlations between individual photon emission events of a microcavity laser. *Nature*, 460:245, 2009. (Cited on pages 117 and 492).

S. Wiesner. Conjugate coding. *ACM SIGACT*, 15:78, 1983. (Cited on page 512).

M. Winger, T. Volz, G. Tarel, S. Portolan, A. Badolato, K. J. Hennessy, E. L. Hu, A. Beveratos, J. Finley, V. Savona, and A. İmamoğlu. Explanation of photon correlations in the far-off-resonance optical emission from a quantum-dot-cavity system. *Phys. Rev. Lett.*, 103:207403, 2009. (Cited on page 211).

K. Winkler, J. Fischer, A. Schade, M. Amthor, R. Dall, J. Geßler, M. Emmerling, E. A. Ostrovskaya, M. Kamp, C. Schneider, and S. Höfling. A polariton condensate in a photonic crystal potential landscape. *New J. Phys.*, 17:023001, 2015. (Cited on page 537).

K. Winkler, O. A. Egorov, I. G. Savenko, X. Ma, E. Estrecho, T. Gao, S. Müller, M. Kamp, T. C. H. Liew, E. A. Ostrovskaya, S. Höfling, and C. Schneider. Collective state transitions of exciton-polaritons loaded into a periodic potential. *Phys. Rev. B*, 93:121303(R), 2016. (Cited on page 537).

S. A. Wolf, D. D. Awschalom, R. A. Buhrman, J. M. Daughton, S. von Molnár, M. L. Roukes, A. Y. Chtchelkanova, and D. M. Treger. Spintronics: A spin-based electronics vision for the future. *Science*, 294:1488, 2001. (Cited on page 421).

R.W. Wood and A. Ellett. Polarized resonance radiation in weak magnetic fields. *Phys. Rev.*, 24:243, 1924. (Cited on page 387).

M. Wouters. Synchronized and desynchronized phases of coupled nonequilibrium exciton-polariton condensates. *Phys. Rev. B*, 77: 121302(R), 2008. (Cited on page 194).

M. Wouters. Energy relaxation in the mean-field description of polariton condensates. *New J. Phys.*, 14:075020, 2012. (Cited on page 541).

M. Wouters and I. Carusotto. Excitations in a nonequilibrium Bose–Einstein condensate of exciton polaritons. *Phys. Rev. Lett.*, 99: 140402, 2007. (Cited on pages 368 and 536).

M. Wouters and I. Carusotto. Superfluidity and critical velocities in nonequilibrium Bose–Einstein condensates. *Phys. Rev. Lett.*, 105: 020602, 2010. (Cited on pages 368 and 375).

M. Wouters and V. Savona. Superfluidity of a nonequilibrium Bose–Einstein condensate of polaritons. *Phys. Rev. B*, 81:054508, 2010. (Cited on page 375).

M. Wouters, T. C. H. Liew, and V. Savona. Energy relaxation in one-dimensional polariton condensates. *Phys. Rev. B*, 82:245315, 2010. (Cited on page 368).

W. Xie, H. Dong, S. Zhang, L. Sun, W. Zhou, Y. Ling, J. Lu, X. Shen, and Z. Chen. Room-temperature polariton parametric scattering driven by a one-dimensional polariton condensate. *Phys. Rev. Lett.*, 108:166401, 2012. (Cited on page 523).

Q. Xu, B. Schmidt, S. Pradhan, and M. Lipson. Micrometre-scale silicon electro-optic modulator. *Nature*, 435:325, 2005. (Cited on page 17).

X.-W. Xu and Y.-J Li. Antibunching photons in a cavity coupled to an optomechanical system. *J. Phys. B.: At. Mol. Phys.*, 46:035502, 2013. (Cited on page 490).

Y. Xue and M. Matuszewski. Creation and abrupt decay of a quasistationary dark soliton in a polariton condensate. *Phys. Rev. Lett.*, 112:216401, 2014. (Cited on page 536).

E. Yablonovitch. Inhibited spontaneous emission in solid-state physics and electronics. *Phys. Rev. Lett.*, 58:2059, 1987. (Cited on page 69).

E. Yablonovitch. Photonic crystals: Semiconductors of light. *Sci. Am.*, 285, 2001. (Cited on page 76).

M. Yamaguchi, T. Asano, K. Kojima, and S. Noda. Quantum electrodynamics of a nanocavity coupled with exciton complexes in a quantum dot. *Phys. Rev. B*, 80:155326, 2009. (Cited on pages 211 and 229).

C. C. Yan. Soliton like solutions of the Schrödinger equation for simple harmonic oscillator. *Am. J. Phys.*, 62:147, 1994. (Cited on page 306).

B. Yang, J.-B. Trebbia, R. Baby, Ph. Tamarat, and B. Lounis. Optical nanoscopy with excited state saturation at liquid helium temperatures. *Nat. Photon.*, 9:658, 2015. (Cited on page 255).

C. N. Yang. Concept of off-diagonal long-range order and the quantum phases of liquid He and of superconductors. *Rev. Mod. Phys.*, 34:694, 1962. (Cited on page 324).

L. Yang, S. G. Carter, A. S. Bracker, M. K. Yakes, M. Kim, C. Soo Kim, P. M. Vora, and D. Gammon. Optical spectroscopy of site-controlled quantum dots in a Schottky diode. *Appl. Phys. Lett.*, 108:233102, 2016. (Cited on page 255).

X. L. Yang, S. H. Guo, F. T. Chan, K. W. Wong, and W. Y. Ching. Analytic solution of a two-dimensional hydrogen atom. I. Nonrelativistic theory. *Phys. Rev. A*, 43:1186, 1991. (Cited on page 528).

A. Yariv and P. Yeh. *Optical Waves in Crystals: Propagation and Control of Laser Radiation.* Wiley-Interscience, reprint edition, 2002. (Cited on pages 70 and 77).

T. Yoshie, A. Scherer, J. Heindrickson, G. Khitrova, H. M. Gibbs, G. Rupper, C. Ell, O. B. Shchekin, and D. G. Deppe. Vacuum Rabi splitting with a single quantum dot in a photonic crystal nanocavity. *Nature*, 432:200, 2004. (Cited on pages 171 and 210).

V. I. Yudson. Dynamics of integrable quantum systems. *Zh. Eksp. Teor. Fiz*, 88:1757, 1985. (Cited on page 502).

V. I. Yukalov and E. P. Yukalova. Dynamics of quantum dot superradiance. *Phys. Rev. B*, 81:075308, 2010. (Cited on page 532).

V. A. Zagrebnov and J. B. Bru. The Bogoliubov model of weakly imperfect Bose gas. *Phys. Rep.*, 250, 2001. (Cited on page 379).

B. P. Zakharchenya. Discovery of excitons. *Phys.-Usp.*, 37:324, 1994. (Cited on page 133).

M. Zamfirescu, A. Kavokin, B. Gil, G. Malpuech, and M. Kaliteevski. ZnO as a material mostly adapted for the realization of room-temperature polariton lasers. *Phys. Rev. B*, 65:R161205, 2002. (Cited on page 331).

E. Zaremba, A. Griffin, and T. Nikuni. Two-fluid hydrodynamics for a trapped weakly interacting Bose gas. *Phys. Rev. A*, 57:4695, 1998. (Cited on page 368).

B. Ya Zel'dovich, A. M. Perelomov, and V. S. Popov. Relaxation of a quantum oscillator. *J. Exp. Th. Phys.*, 55:589, 1968. (Cited on page 344).

B. Zhang, Z. Wang, S. Brodbeck, C. Schneider, M. Kamp, S. Höfling, and H. Deng. Zero-dimensional polariton laser in a subwavelength grating-based vertical microcavity. *Light: Sci. & App.*, 3: e135, 2014. (Cited on page 523).

C. Zhang and W. Zhang. Exciton-polariton Josephson interferometer in a semiconductor microcavity. *Europhys. Lett.*, 108:27002, 2014. (Cited on page 194).

J. Zhang and S. L. Braunstein. Continuous-variable Gaussian analog of cluster states. *Phys. Rev. A*, 73:032318, 2006. (Cited on page 518).

L. Zhang, W. Xie, J. Wang, A. Poddubny, J. Lu, Y. Wang, J. Gu, W. Liu, D. Xu, X. Shen, Y. Rubo, B. Altshuler, A. V. Kavokin, and Z. Chen. Weak lasing in one-dimensional polariton superlattices. *Proc. Natl. Acad. Sci.*, 112:E1516, 2015. (Cited on pages 526 and 537).

Y. Zhang, L. Shi, G. Jin, and B. Zou. Magnetic-field modulated exciton-exciton interaction in semiconductor microcavities. *J. Appl. Phys.*, 107:053527, 2010. (Cited on page 533).

Y. H. Zhou, H. Z. Shen, and X. X. Yi. Unconventional photon blockade with second-order nonlinearity. *Phys. Rev. A*, 92:023838, 2015. (Cited on page 490).

Y. Zhu, D. J. Gauthier, S. E. Morin, Q. Wu, H. J. Carmichael, and T. W. Mossberg. Vacuum Rabi splitting as a feature of linear-dispersion theory: Analysis and experimental observations. *Phys. Rev. Lett.*, 64:2499, 1990. (Cited on page 476).

Z. Zou, D. L. Huffaker, and D. G. Deppe. Ultralow-threshold cryogenic vertical-cavity surface-emitting laser. *IEEE Phot. Tech. Lett.*, 12, 2000. (Cited on page 265).

W. H. Zurek. Cosmological experiments in superfluid helium? *Nature*, 317:505, 1985. (Cited on pages 482 and 537).

W. H. Zurek. Cosmological experiments in condensed matter systems. *Phys. Rep.*, 276:177, 1996. (Cited on page 424).

INDEX

583

AUTHORS

Prof. Alexey Kavokin

Chair of Nanophysics and Photonics at the University of Southampton. Head of the Spin Optics laboratory at the State University of St-Petersburg and the Quantum Polaritonics group at the Russian Quantum Center. One of the world-leading theoreticians specialising in the optics of semiconductors and spintronics. Ioffe institute prizes for the Best scientific work of the year in 1995 and 1999. Founder of several series of international conferences and schools. Scientific Director of the Mediterranean Institute of Fundamental Physics (Rome). Previously worked in Russia, France, Italy. Married with four children, his hobbies include writing (published fairy tale "Saladin the Cat", in Russian and English), chess, drawing.

Prof. Jeremy Baumberg

Professor of Nanophotonics in Cambridge, UK, director of the Nanophotonics Centre and fellow of Jesus College. Innovator in NanoPhotonics and strategic advisor to the UK Research Councils. 2004 Royal Society Mullard Prize, 2004 Mott Lectureship of the Institute of Physics and the Charles Vernon Boys Medal in 2000. Previously worked for Hitachi and IBM, spun-out his research into the Mesophotonics Ltd company and, for the past 10 years, directed the NanoTechnology group at the University of Southampton. Wide range of research interests including ultra-fast coherent control, magnetic semiconductors, ultrafast phonon propagation, photonic crystals, single quantum dots, microcavities, and self-assembled photonic and plasmonic nanostructures. Married with two children, his hobbies include playing the piano, tennis and making kinetic sculptures.

Dr. Guillaume Malpuech

Researcher at the Centre National de la Recherche Scientifique (CNRS) since 2002. Postdoctoral researcher in Rome and Southampton in 2001 and 2002. Presently head of the Quantum Optolectronics and Nanophotonics group at the Institut Pascal, at the Université Clermont Auvergne (Clermont–Ferrand, France). Authors of over 200 research papers on optical properties of semiconductors and of the monograph "cavity polaritons" (with Alexey Kavokin). Among other contributions, he developed the theory of Bose–Einstein condensation of cavity polaritons, and proposed the concept of spin-optronic devices. His hobbies include skiing, alpinism and hiking.

Dr. Fabrice P. Laussy

Principal Lecturer and Director of Studies for Physics at the University of Wolverhampton, UK, and Senior Staff at the Russian Quantum Center, Moscow, Russia, working on microcavity quantum electrodynamics. Previously Ramón y Cajal group leader at the Universidad Autónoma de Madrid, Spain, Marie Curie fellow in the Walter Schottky Institut of the Technische Universität München, Germany and with post-doctoral experience in Sheffield, Madrid and Southampton. PhD of Université Blaise Pascal, Clermont-Ferrand, 2005. Made significant contributions to the quantum theory of polariton lasers, Bose condensation and superfluidity of exciton-polaritons. Designed a new source of quantum light: the "bundler", that replaces the fundamental brick of light, the photon, by a group (or bundle) of a tunable integer number of them. His hobbies include chess, going everywhere and collecting collections.

591

The Authors (2006)
Alexey KAVOKIN, Jeremy BAUMBERG, Guillaume MALPUECH and Fabrice LAUSSY